NONPOLAR COVALENT RADII

Atom	radius (pm)
H	32
Li	134
Be	91
B	82
C	77
N	74
O	70
F	68
Na	154
Mg	138
Al	126
Si	117
P	110
S	104
Cl	99
Ge	122
As	119
Se	116
Br	114
Sn	140
Sb	138
Te	135
I	133
Tl	148
Pb	147
Bi	146

an introduction to INORGANIC CHEMISTRY

KEITH F. PURCELL

Kansas State University
Manhattan, Kansas

JOHN C. KOTZ

State University of New York
College at Oneonta

 **SAUNDERS GOLDEN SUNBURST SERIES**

1980   SAUNDERS COLLEGE   Philadelphia

Saunders College
West Washington Square
Philadelphia, PA 19105

Cover illustration by Joanne Shultis

An Introduction to Inorganic Chemistry   ISBN 0-03-056-768-8

0123   074   9 8 7 6 5 4 3 2

This book is dedicated to the many students and colleagues with whom we have worked over the years, but most especially it is dedicated to those who have had the greatest impact upon our careers:

Ken	*Shirley*
Marge	*Syd*
Susan	*Katie*
Russ	*Lauby*
	Mike
	Riley

PREFACE

TO THE STUDENT:

Inorganic chemistry is an exciting field! We wish to share with you the fact that inorganic chemistry has reached a point where an attack can begin on problems of considerable scientific and social significance. For example, watch for significant developments in the communications, electronics, and energy industries in the near future. These advances will derive from current inorganic research on metal catalysts for production of hydrogen and organic fuels, on solid-state, one-dimensional conductors and solid-state batteries, and on metal catalysts for solar energy conversion. These problems, and many others like them, require a broad knowledge of inorganic chemistry.

Inorganic chemistry—contrary to the belief of some students after taking general chemistry—is more than sulfuric acid and sodium chloride. It is understanding molecular topology and its relation to the reactivities of non-metal and metal compounds and the mechanisms of their reactions. Our goal in this text is to provide a firm foundation in these principles and to lead you to new insight into the chemistry of the elements other than carbon. We have attempted to write a book that is, above all, correct and clear, but one that is also insightful and illustrative of the discoveries that lie just ahead. An important goal is to help you share the enthusiasm of the international community of scholars in our field. We hope that you come away from your course in inorganic chemistry with this feeling, confident that the principles discussed will be of value to you throughout your career in chemistry.

TO THE INSTRUCTOR:

"Introduction to Inorganic Chemistry" is intended for use by college junior and senior chemistry majors; we assume that the student possesses the background of a first year, college level course and some knowledge of organic chemistry.

The present book is clearly an outgrowth of our previous book, "Inorganic Chemistry" (W.B. Saunders Co., 1977). Our goal in both texts is to promote the unification of descriptive chemistry with structure, bonding, and reaction mechanism concepts. In both instances, we attempted to write books that are correct and clear, but which also illustrate the excitement of modern inorganic chemistry and our pleasure in being involved with this field and with the world-wide community of inorganic chemists.

How do the two texts differ? First of all, the 40% reduction in size comes at the expense of the higher level theoretical and descriptive material. Comments on special items of interest follow:

The atomic structure chapter has been written at a lower level. Our students report that this material clarifies many of their misconceptions about atomic orbitals and gives them new insights into their importance in chemistry.

In the second chapter, the symmetry properties of molecules are described, but group theory has been omitted; solid-state topics and contemporary applications appear in this chapter.

The third chapter (hybrid orbital concepts and spectroscopy applications) has been shortened and rewritten to some extent, as has the molecular orbital chapter. We have removed from the latter the discussions of polyhedral molecules, forecasting fewer applications of orbital concepts to descriptive chemistry in the remainder of the text.

Chapters 5 to 17 have been shortened by eliminating the more difficult descriptive and interpretive sections.

Instructors seeking to give coverage of, but not emphasis to, descriptive non-metal chemistry can utilize the first part of Chapter 5 and one of the three succeeding chapters; the selection of the latter should derive from one's own convictions regarding the importance of thermochemistry, kinetics, or rational synthesis.

Catalysis has become such an important topic that it has been made into a separate chapter (Chapter 18). This chapter could be used without first covering the previous chapters on organometallic chemistry if the students are first introduced to the 16-18 electron rule (Chapter 15) and then to some common classes of organometallic compounds (e.g., metal carbonyls and metal-olefin complexes) as well as to some standard reaction types (oxidative addition and reductive elimination, carbonylation).

The basic organization of "Introduction to Inorganic Chemistry" is:

Part 1: The Tools of Chemical Interpretation

It is in this section that **atomic orbitals** and **electron configurations** of the elements are described. The student is led to discover the factors that determine the **structures** of simple **non-metal compounds** and to describe these structures in terms of their **symmetry properties**. The basic ideas of **solid-state** structures are outlined, and the student is exposed to covalent bonding models (**hybrid orbital** and **molecular orbital**) and examples of their application.

Part 2: Descriptive Non-Metal Chemistry with Interpretations

It is here that we illustrate the power of the concepts of **Lewis acid-base** interactions, **structure**, **bonding**, and **kinetics** to unify **non-metal chemistry**. One entire chapter is devoted to the techniques and systematics of the **synthesis** of non-metal compounds.

Part 3: Descriptive Transition Metal Chemistry with Interpretations

The final portion of the text is devoted to metal chemistry, again with stress on **bonding**, **structure**, and **reaction mechanisms**. Of particular interest today are the topics of **organometallic chemistry**, **catalysis**, and **bio-inorganic chemistry**. The student who is bound for either industry or graduate school should have exposure to at least the first two areas. Some aspects of catalysis are covered in a chapter devoted entirely to that subject.

Since "Inorganic Chemistry" was published in 1977, the adoption of the SI system of units has been rapid. Therefore, this text uses joules, meters, and so on.

And finally, we note that, for those who seek more detailed and extensive coverage, our previous book will continue to be available. The current "Introduction to Inorganic Chemistry" is not intended to supplant the former book.

ACKNOWLEDGMENTS

First, we wish to thank all of our fellow inorganic chemists who have passed on their experiences with our previous book. The compliments were appreciated; more importantly, the criticisms were constructive and invaluable in producing this new book.

We again thank the reviewers of the previous book (Mike Bellema, Bob Fay, Ron Gillespie, Bill Hatfield, Galen Stucky, and Jerry Zuckerman). They contributed to the success of the first book and, as "Introduction to Inorganic Chemistry" is an outgrowth of that first effort, these reviewers have contributed to the present book as well. The same is true of our students and colleagues (Bill Fateley, Bruce Knauer, and Tay Tahk especially) who supported these efforts in a variety of ways. We wish to thank Alberto Romão Dias, Carlos Romão, and the students of the Instituto Superior Tecnico in Lisbon, Portugal for their generous hospitality during a stay in which portions of the current manuscript were completed.

We also acknowledge a very pleasant, continuing association with the W. B. Saunders Company. These projects could not have been completed without the encouragement and assistance of John Vondeling, John's assistants, Kay Dowgun and Jeannie Shoch, and Jay Freedman, the best copy editor in the business.

Finally, our greatest debt of gratitude is to our families (Susan, Kristan, and Karen; Katie, David, and Peter) for being supportive in times of stress and for helping with the duller tasks of manuscript preparation.

KEITH F. PURCELL
JOHN C. KOTZ

CONTENTS

3

4

PART 2

DESCRIPTIVE NON-METAL CHEMISTRY WITH INTERPRETATIONS

5

6

7

8

PART 3

DESCRIPTIVE TRANSITION METAL CHEMISTRY WITH INTERPRETATIONS

9

FUNDAMENTAL CONCEPTS FOR TRANSITION METAL COMPLEXES

10

COORDINATION CHEMISTRY: STRUCTURAL ASPECTS

11

12

13

PROLOGUE
THE COMING OF AGE:
PERSPECTIVE

What is modern inorganic chemistry and what direction is it likely to take in the near future? What does one teach undergraduate and graduate students in general, and prospective inorganic chemists in particular, that they gain perspective on the past and are prepared for the future? It is these questions that inorganic chemists ask one another at American Chemical Society meetings, at Gordon Research Conferences, and at other "gatherings of the clan." Not surprisingly, there are nearly as many answers as there are inorganic chemists, and the answers change with time.

From an historical view inorganic chemistry is synonymous with "general" chemistry; in addition to his area of specialization the "inorganic" chemist was expected to be conversant, if not knowledgeable, with the chemistries of all the elements. At various times the inorganic chemist has been closely allied with analytical chemistry (both qualitative and quantitative chemical methods), with physical chemistry, and even with organic chemistry. These strong ties are still existent today, but the rapid development of our field since the late fifties has created great depth in research specialties so that the "generalist" posture has waned slightly. Concomitant with this evolution, the field gained formal recognition in this country with the establishment of the journal *Inorganic Chemistry* by the American Chemical Society in 1962; since then the "Division of Inorganic Chemistry of the American Chemical Society" has produced offspring in the form of the "Organometallic" and "Solid State" subdivisions.

To illustrate what modern inorganic chemistry has become, we surveyed the papers published on topics in the field in *Accounts of Chemical Research* since the inception of the journal in 1968. Such a survey at least partially measures the current status of inorganic chemistry and the probable direction of future work because *Accounts* . . . "publishes concise, critical reviews of research areas currently under active investigation . . . written by scientists personally contributing to the area reviewed." Although our division of papers was perhaps arbitrary at times, the results summarized in Table 1 are quite enlightening. Metals make up about 80% of the Periodic Table, and it is interesting that the fraction of papers on metal chemistry is approximately the same, although all but seven of the papers are concerned with transition metal chemistry. Nonetheless, the results of this crude survey give a fairly clear indication of the current emphasis—the bulk of the papers is on organometallic chemistry, with a bias toward catalysis, while a large number are concerned with coordination chemistry and the biochemical role of metals.

The chemistry of organometallic compounds, especially of the transition metals, has become increasingly important in the past twenty years, as attested to by the fact that the Nobel Prizes in Chemistry in 1963 and 1973 were awarded for work falling in this general area.[1,2] Karl Ziegler of Germany and Giulio Natta of Italy shared the Prize in 1963 for their finding that the stereospecific polymerization of olefins was catalyzed by alkyl aluminum–transition metal halide mixtures.[3-5]

TABLE 1

REVIEWS ON INORGANIC CHEMISTRY IN
ACCOUNTS OF CHEMICAL RESEARCH,
1968–1978

General Area and Sub-area	Number of Reviews
Metal chemistry	Total = 104
Transition metals	Total = 97
Organometallic chemistry (incl. catalysis)	47
Coordination chemistry	13
Bioinorganic chemistry	19
Theoretical and physical papers	18
Non-transition metals: Organometallic chemistry	7
Non-metal chemistry	Total = 27
Boron chemistry (5 with significant metal chemistry)	10
Silicon chemistry	5
Theoretical papers	4
Other (chiefly halogen compounds)	8
TOTAL REVIEWS	131

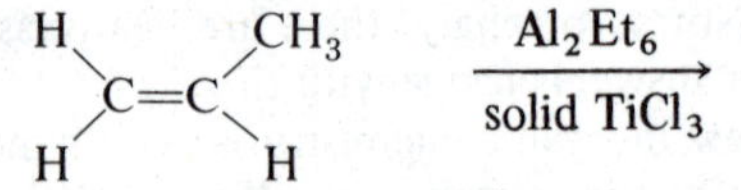

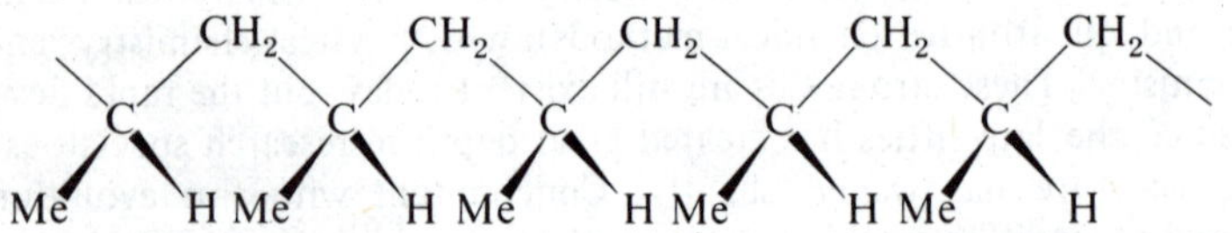

(polypropylene)

This example serves to illustrate much of the fascination of inorganic chemistry. For example, why is the alkylating agent, triethylaluminum (**1**), a dimeric species under normal conditions, whereas triethylborane, $B(CH_2CH_3)_3$, is monomeric? How is one AlR_3 unit bound to the other? In what way is an olefin bound to a metal, and why do metals form kinetically stable olefin complexes only when the metal is in a low oxidation state?

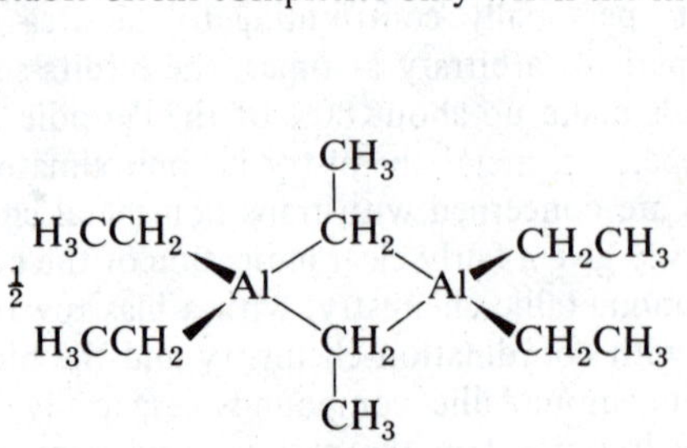

1

Professor Wilkinson at Imperial College, London, and Professor Fischer at Munich earned the Nobel Prize in 1973 for the impetus they provided to the study of metallocenes, or "metal sandwiches" as they are often called. Ferrocene (**2**) in particular has been widely studied, as it is readily synthesized or can be purchased inexpensively. The compound is electron-rich and has an extensive chemistry based on electrophilic substitution reactions.[8] As prototypes of many other similar complexes, **2** and **3** have been thoroughly studied to uncover the reasons for the tremendous stability achieved when a low-valence metal interacts with the π electrons of an unsaturated or aromatic molecule.

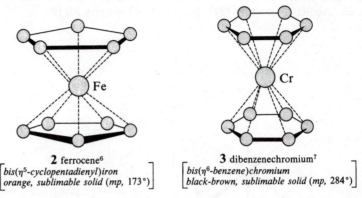

2 ferrocene[6]
[bis(η^5-cyclopentadienyl)iron]
[orange, sublimable solid (mp, 173°)]

3 dibenzenechromium[7]
[bis(η^6-benzene)chromium]
[black-brown, sublimable solid (mp, 284°)]

Perhaps the most active area of inorganic chemistry in the 1950's and 1960's was coordination chemistry: the study of complexes of higher valence metals. The most important development during that period was in our understanding of the electronic spectra of these complexes, and many excellent books have been published on "ligand field theory" as a result.[9-11] Another area of special activity in coordination chemistry has been the study of the stereochemistry of chiral complexes (**4**),[12] or those with high coordination numbers (**5**)[13] or with unusual coordination geometries (**6**).[14]

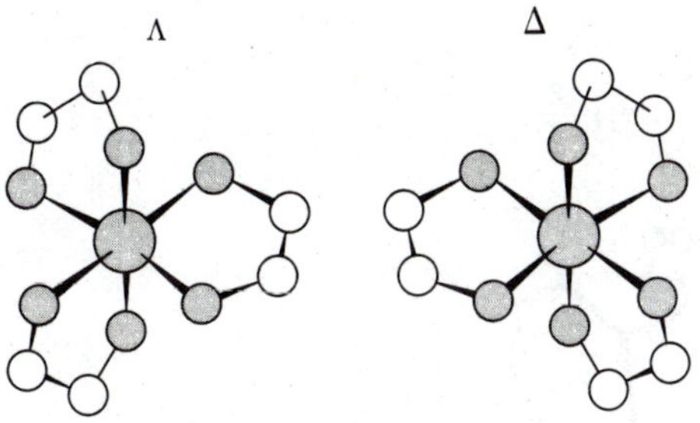

Λ   Δ

4

$M\{C_2H_4(NH_2)_2\}_3^{n+}$

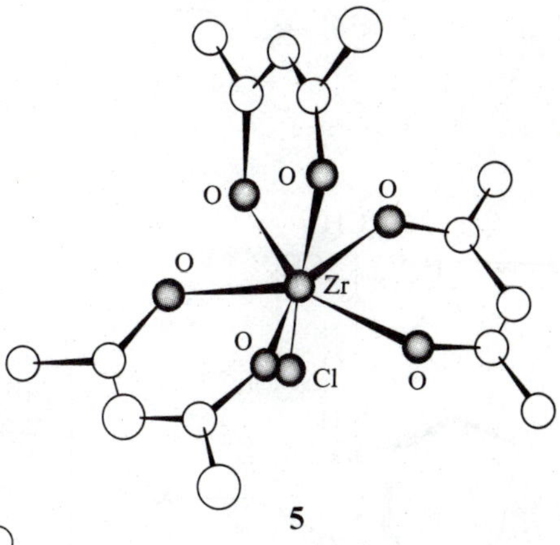

5

$Zr(acac)_3Cl$[15]

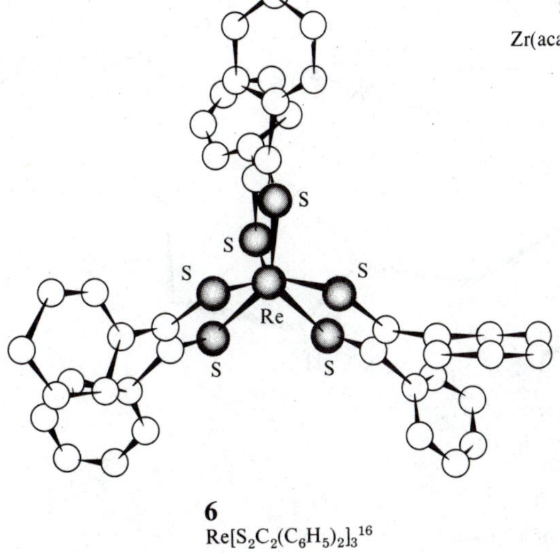

6

$Re[S_2C_2(C_6H_5)_2]_3$[16]

Complexes of unsaturated dithiolates **(6)** were also widely studied in the 1960's, not only because such ligands led to unusual geometries, but also because their complexes could often exist in several stable oxidation states.[17] More recently, the discovery of metal complexes of molecular nitrogen **(7)** has had considerable impact on inorganic chemistry[18] and the search for the mysteries of natural nitrogen fixation. New light has been shed on metal-O_2 complexes **(8)**[19] and on metalloporphyrins **(9)**.[20]

7 $[Os(NH_3)_5N_2]Cl_2$[21]

8 An O_2 adduct of a "picket fence porphyrin"[22]

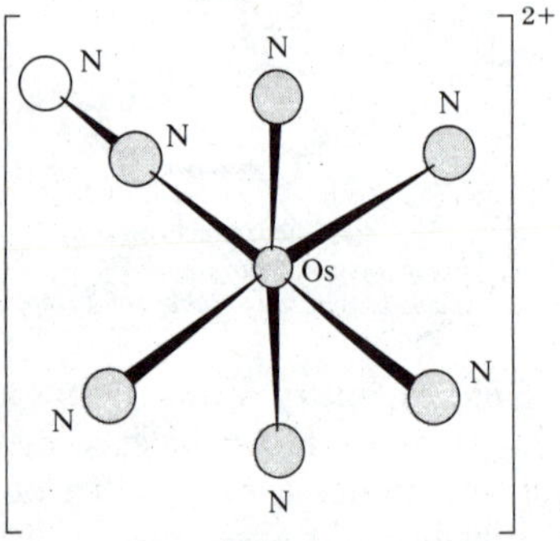

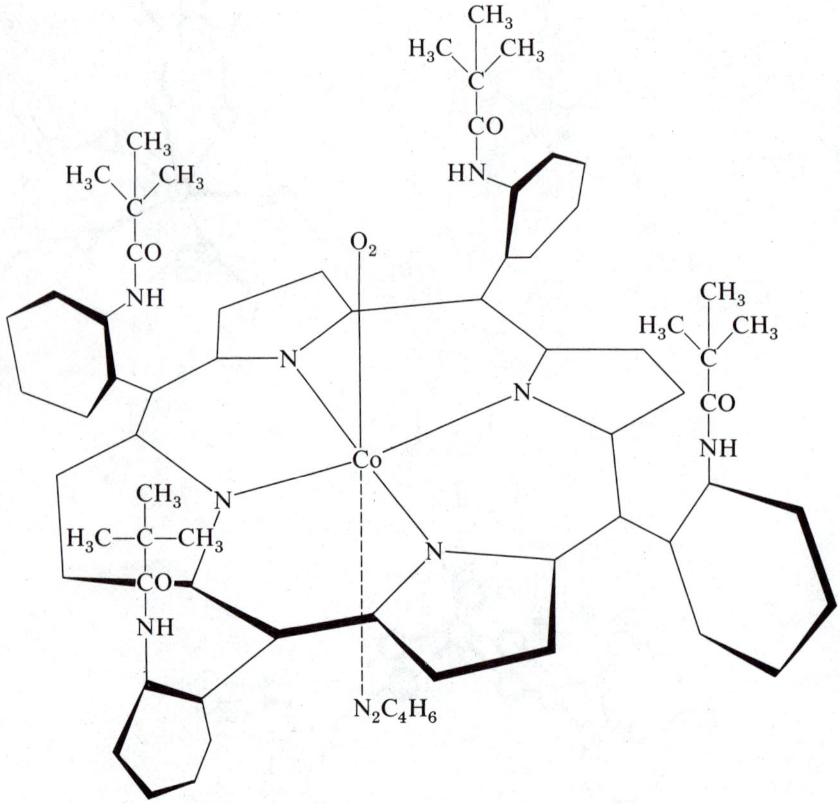

9 Zinc tetraphenylporphine dihydrate[23]
$Zn[(C_6H_5)_4C_{20}H_7N_4] \cdot 2H_2O$

By now our knowledge of reaction mechanisms,[24] metal–carbon bond chemistry, coordination complex stereochemistry, metal–sulfur complexes, and complexes with O_2 and N_2 has increased to the point where a systematic attack on the vast area of bioinorganic chemistry seems possible and potentially fruitful.[25,26] Indeed, examination of model complexes **(10)**[27] has led to a better understanding of the Co–C bond in vitamin B_{12} **(11)** and the B_{12} coenzymes,[28] and models have recently been generated that duplicate some of the properties of O_2-hemoproteins.[29,30] Further, models of the iron-sulfur proteins (e.g., **12**) have been discovered,[31] and simple platinum complexes such as cis-$Pt(NH_3)_2Cl_2$ **(13)** have been found to be active against certain tumors.[32]

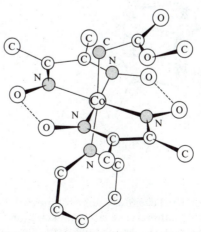

10 O-methyl-(Co-C)-carboxymethyl-[bis(dimethylglyoximato)pyridine] cobalt

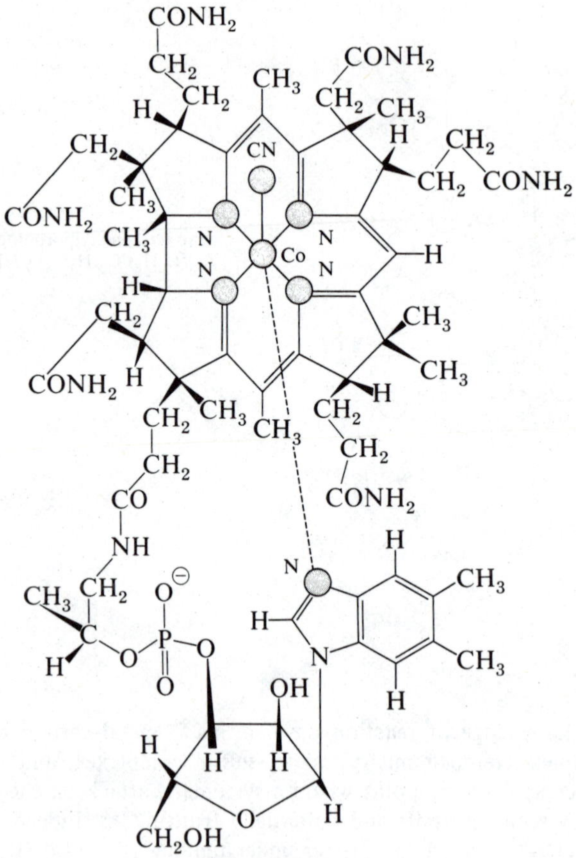

11   Vitamin B$_{12}$

12   [Fe$_4$S$_4$(SC$_6$H$_5$)$_4$]$^{2-}$ [33]

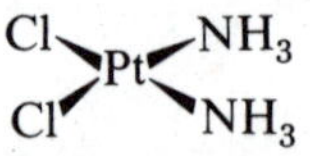

13   *cis*-PtCl$_2$(NH$_3$)$_2$

The recent emphasis on metal chemistry does not mean that non-metal chemistry has been neglected. When your authors were at your stage of their chemical education (~ 1960), we were taught that the compounds HOF (hypofluorous acid, **14**), BrO$_4^-$ (perbromate ion, **15**), and XeF$_n$ (the xenon fluorides, **16**) were not likely to exist or would have only the most fleeting existence at room temperature, and would be of little interest to the chemist. The intervening 20 years have seen the syntheses of all these compounds![34,35,36] These impressive achievements accentuate the need by all chemists for a good understanding of theoretical, thermodynamic, and mechanistic concepts in application to chemical synthesis.

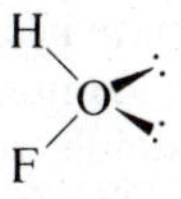

14 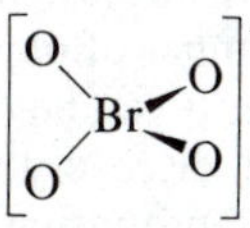   **15**

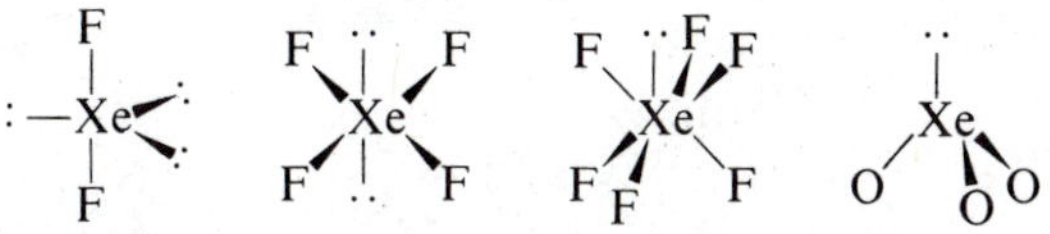

16

The need of industries for new stable polymers, particularly those that could withstand extremes in thermal and physical stress, has spurred great interest in inorganic polymers with skeletons consisting of boron, aluminum, silicon, and phosphorus atoms **(17–21)**.[37]

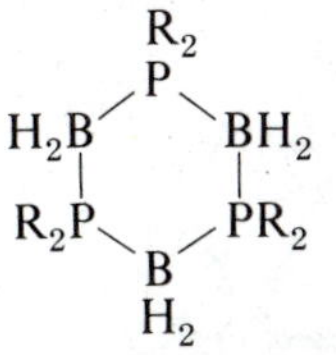

17 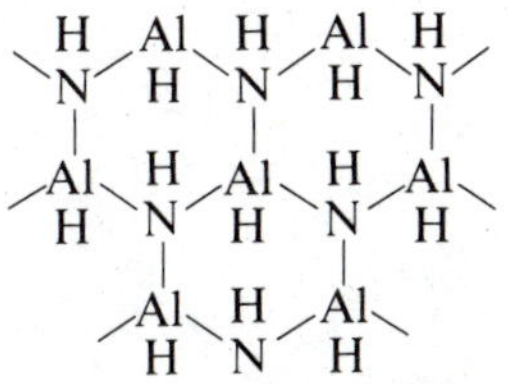   **18**

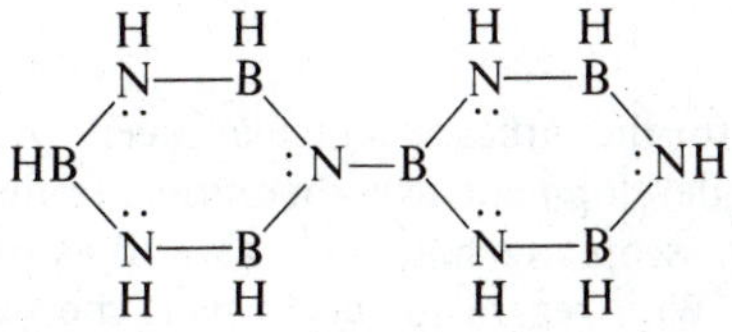

19

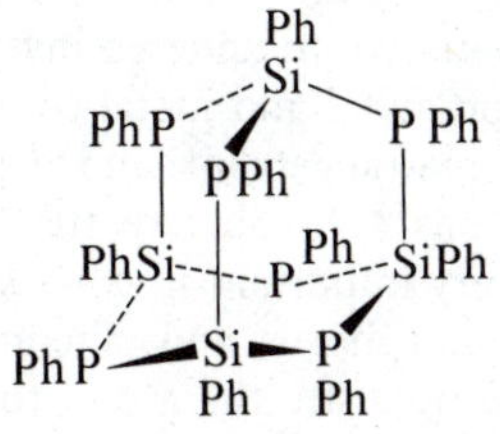

20 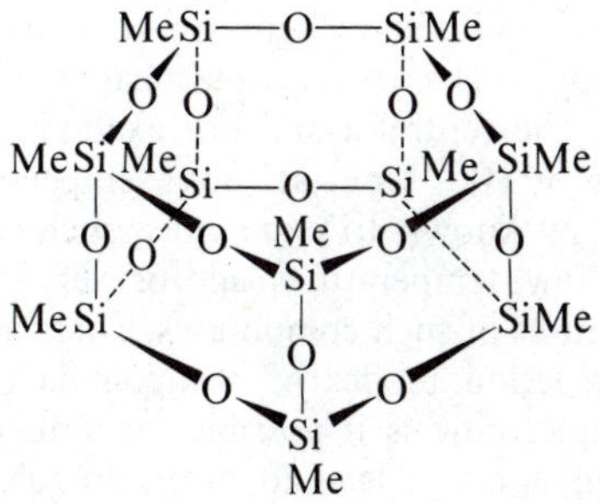   **21**

The agricultural problems of the world have spurred the development of many phosphorus, sulfur, and halogen compounds useful as toxins for control of insects, fungi, and weeds. The space program in particular, and industrial and military concerns in general, have generated continued interest in the development of techniques for synthesizing and handling high energy oxidizers (22–24),[38] while the pharmaceutical promise of fluorine-containing compounds has sustained interest in stable organo compounds containing fluorine. Although a considerable number of papers have been published concerning the boron hydrides, additional developments can be expected in this area as new order was recently imposed on the field by the discovery of simple rules for interrelating structure and bonding.[39,40] Indeed, a most promising area for future development in boron hydride chemistry is in the metal complexes of the boranes and carboranes (25, 26).[41]

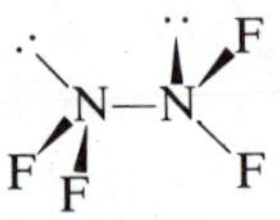

22 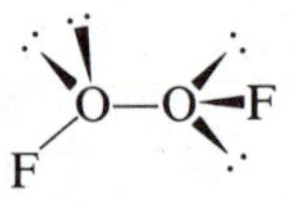   23 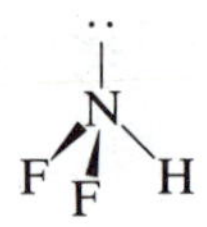   24

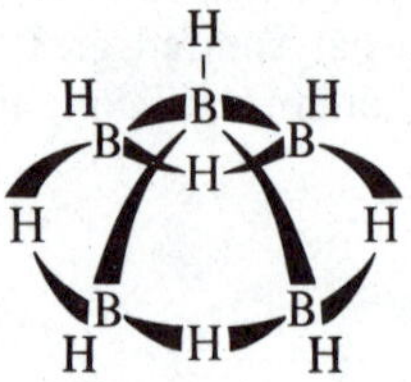

B_5H_9
25

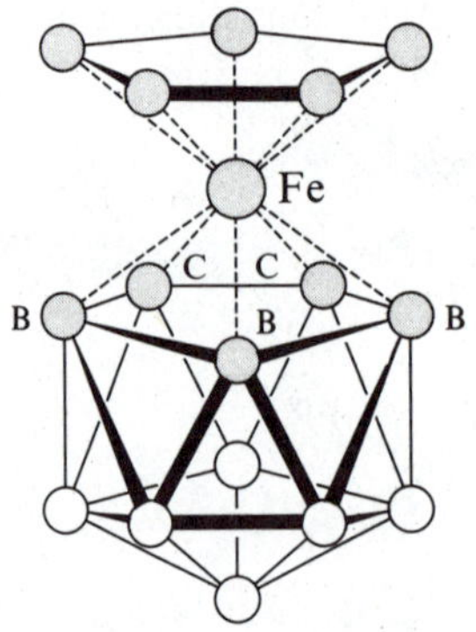

$[\pi\text{-}C_5H_5]Fe[\pi\text{-}(3)\text{-}1,2\text{-}B_9C_2H_{11}]$
26

Some of the reasons for the rather spectacular recent progress in inorganic chemistry have been the continuing development of a theoretical framework and the application of crystallographic and spectroscopic techniques[42] as well as other instrumental techniques (e.g., electrochemistry[43]). With regard to the former, the recognition of symmetry constraints on chemical problems has had a great impact on inorganic chemistry[44] and has encouraged the wider application of molecular orbital concepts to inorganic structures and reactions. Among the more important advances in the area of structures of molecules are models[45] for predicting and correlating the molecular topologies of non-metal compounds and the recognition that many of these molecules are not "rigid" but rather "floppy." The more sophisticated expression is "pseudorotation" of terminal groups about the central atom. For example, NH_3 has long been known to undergo inversion at a rate of 10^{10} sec^{-1} at room temperature. Many other non-metal and metal compounds are now known to experience such rapid intramolecular rearrangement, and it requires very low temperatures and/or very fast detection methods[46] to observe the "static" structures of such compounds. These advances have been very important in understanding the reaction modes of compounds of four-coordinate phosphorus and silicon.[47] For example, how is it possible for a nucleophile to attack Si opposite the group to be displaced and *not* lead to inversion (change in molecular chirality) of the silicon stereochemistry?

$$\underset{H}{\overset{\cdot\cdot}{N}}\!\!-\!H \rightleftharpoons \left[H\!-\!N\!\!\overset{H}{\underset{H}{\big|}}\right]^{\ddagger} \rightleftharpoons \overset{H\ H}{\underset{\cdot\cdot}{N}}\!\!H$$

$$F\!-\!\overset{F}{\underset{F}{P}}\!\!\overset{F}{\underset{}{F}} \rightleftharpoons \left[F\!-\!\overset{F\ F}{\underset{F}{P}}\right]^{\ddagger} \rightleftharpoons F\!-\!\overset{F}{\underset{F}{P}}\!\!\overset{F}{\underset{}{F}}$$

$$Y\!: +\ \underset{B}{\overset{A}{C}}\!Si\!-\!X \rightarrow Y\!-\!\underset{B}{\overset{A}{C}}\!Si\!-\!X \rightarrow \underset{B}{\overset{A}{C}}\!Si\!-\!Y\ +\ :X$$

The scope of inorganic chemistry is immense, and the texture of the field is changing rapidly. In the chapters that follow we cannot hope to be comprehensive. Rather, we have attempted first to present you with an invaluable theoretical framework upon which modern concepts of structure, bonding, and reactivity in inorganic chemistry are built. In later chapters, you will sample—in as un-superficial a manner as possible—both "established" areas and those of greatest current interest, areas sharing in common the promise to dominate the field in the coming years.

REFERENCES

[1] C. G. Overberger, *Science*, **142**, 938 (1963).

[2] D. Seyferth and A. Davison, *Science*, **182**, 669 (1973).

[3] G. Natta and I. Pasquon, *Advan. Catalysis*, **11**, 1 (1959).

[4] P. Cossee, *in* "The Stereochemistry of Macromolecules," A. D. Ketley (ed.), Vol. 1, Marcel Dekker, New York, 1967, p. 145.

[5] T. Mole and E. A. Jeffrey, "Organoaluminum Compounds," Elsevier, New York, 1972.

[6] G. Wilkinson, *Org. Syn.*, **36**, 31 (1956).

[7] E. O. Fischer, *Inorg. Syn.*, **6**, 132 (1960).

[8] M. Rosenblum, "Chemistry of the Iron Group Metallocenes," Interscience Publishers, New York, 1965.

[9] C. J. Ballhausen, "Introduction to Ligand Field Theory," McGraw–Hill, New York, 1962.

[10] B. N. Figgis, "Introduction to Ligand Fields," Interscience, New York, 1966.

[11] A. B. P. Lever, "Inorganic Electronic Spectroscopy," Elsevier, New York, 1968.

[12] C. J. Hawkins, "Absolute Configuration of Metal Complexes," Interscience, New York, 1971.

[13] S. J. Lippard, "Eight Coordination Chemistry," *Prog. Inorg. Chem.*, **8**, 109 (1967); E. L. Muetterties and C. M. Wright, *Quart. Revs.*, **21**, 109 (1967).

[14] E. Larsen, G. N. LaMar, B. E. Wagner, J. E. Parks, and R. H. Holm, *Inorg. Chem.*, **11**, 2652 (1972).

[15] R. B. VonDreele, J. J. Stezowski, and R. C. Fay, *J. Amer. Chem. Soc.*, **93**, 2887 (1971).

[16] R. Eisenberg and J. A. Ibers, *Inorg. Chem.*, **5**, 411 (1966).

[17] J. A. McCleverty, "Metal 1,2-Dithiolene and Related Complexes," *Prog. Inorg. Chem.*, **10**, 49–221 (1969); G. N. Schrauzer, *Acc. Chem. Res.*, **2**, 72 (1969).

[18] A. D. Allen and F. Bottomley, *Acc. Chem. Res.*, **1**, 360 (1968); see also E. E. van Tamelen, J. A. Gladysz, and C. R. Brulet, *J. Amer. Chem. Soc.*, **96**, 3020 (1974) for a recent publication dealing with the fixation of N_2 by an inorganic "model" system.

[19] J. P. Collman, R. R. Gagne, J. Kouba, and H. Ljusberg-Wahren, *J. Amer. Chem. Soc.*, **96**, 6800 (1974); J. Valentine, *Chem. Rev.*, **73**, 235 (1973).

[20] E. B. Fleischer, *Acc. Chem. Res.*, **3**, 105 (1970).

[21] J. E. Fergusson, J. L. Love, and W. T. Robinson, *Inorg. Chem.*, **11**, 1662 (1972).

[22] J. P. Collman, *et al.*, *J. Amer. Chem. Soc.*, **97**, 1427 (1975).

[23] E. B. Fleischer, C. K. Miller, and L. E. Webb, *J. Amer. Chem. Soc.*, **86**, 2342 (1964).

[24] F. Basolo and R. Pearson, "Mechanisms of Inorganic Reactions," Wiley, New York, 1967.

[25] M. N. Hughes, "The Inorganic Chemistry of Biological Processes," John Wiley and Sons, New York, 1972.

[26] G. L. Eichhorn (ed.), "Inorganic Biochemistry," Elsevier, New York, 1973.

[27] G. N. Schrauzer, *Acc. Chem. Res.*, **1**, 97 (1968).

[28] J. M. Pratt, "Inorganic Chemistry of Vitamin B_{12}," Academic Press, New York, 1972.

[29] See ref. 22.

[30] *Newsweek*, March 24, 1975, page 91; J. Almog, J. E. Baldwin, and J. Huff, *J. Amer. Chem. Soc.*, 97, 227 (1975).

[31] S. J. Lippard, *Acc. Chem. Res.*, 6, 282 (1973); J. J. Mayerle, S. E. Denmark, B. V. DePamphilis, J. A. Ibers, and R. H. Holm, *J. Amer. Chem. Soc.*, 97, 1032 (1975).

[32] A. J. Thompson, R. J. P. Williams, and S. Reslova, *Structure and Bonding*, 11, 1 (1972).

[33] L. Que, Jr., M. A. Bobrik, J. A. Ibers, and R. H. Holm, *J. Amer. Chem. Soc.*, 96, 4168 (1974).

[34] M. H. Studier and E. H. Appleman, *J. Amer. Chem. Soc.*, 93, 2349 (1971).

[35] G. K. Johnson, *et al.*, *Inorg. Chem.*, 9, 119 (1970); J. R. Brand and S. A. Bunck, *J. Amer. Chem. Soc.*, 91, 6500 (1969).

[36] H. H. Hyman (ed.), "Noble Gas Compounds," University of Chicago Press, Chicago (1963) lets you feel the pulse of these history making events. For synthesis methods see E. H. Appleman and J. G. Malm, *Prep. Inorg. Reactions*, 2, 341 (1965); J. H. Holloway, "Noble-Gas Chemistry," Methuen, London (1968).

[37] D. A. Armitage, "Inorganic Rings and Cages," Edward Arnold, London (1972).

[38] E. W. Lawless and I. C. Smith, "Inorganic High Energy Oxidizers," Edward Arnold, London (1968).

[39] K. Wade, *Adv. Inorg. Chem. Radiochem.*, 18, 1 (1976).

[40] R. W. Rudolph and W. R. Pretzer, *Inorg. Chem.*, 11, 1974 (1972).

[41] G. B. Dunks and M. F. Hawthorne, *Acc. Chem. Res.*, 6, 124 (1973).

[42] R. S. Drago, "Physical Methods in Chemistry," W. B. Saunders Company, Philadelphia, 1977.

[43] R. E. Dessy and L. A. Bares, *Acc. Chem. Res.*, 5, 415 (1972); J. B. Headridge, "Electrochemical Techniques for Inorganic Chemists," Academic Press, New York, 1969.

[44] F. A. Cotton, "Chemical Applications of Group Theory," 2nd edition, John Wiley and Sons, New York, 1971.

[45] R. J. Gillespie, "Molecular Geometry," Van Nostrand Reinhold Co., London (1972); *J. Chem. Educ.*, 51, 367 (1974); 47, 18 (1970); L. S. Bartell, *J. Chem. Educ.*, 45, 457 (1968); *Inorg. Chem.*, 5, 1635 (1966); R. M. Gavin, *J. Chem. Educ.*, 46, 413 (1969).

[46] See ref. 42.

[47] F. H. Westheimer, *Accts. Chem. Res.*, 1, 70 (1968); K. Mislow, *ibid.*, 3, 321 (1970); *J. Amer. Chem. Soc.*, 91, 7031 (1969).

Part 1

THE TOOLS OF CHEMICAL INTERPRETATION

1. USEFUL ATOMIC CONCEPTS

The need for the inorganic chemist to have a rudimentary understanding of the presently accepted theory of atomic structure cannot be disputed. In fact, many of the results of this theory have greatly spurred the development of synthetic and physical inorganic chemistry. The point of this chapter is to present the principal ideas and identify their utility to you in interpreting chemistry. The proof of any theory is its success in helping the understanding of experimental observations and, in this capacity, the wave model of the atom has been eminently successful.

PROBABILITY DENSITY FUNCTIONS

The Electron as a Matter Wave

The Bohr model[1] of the atom was the first successful model of the hydrogen atom, in that it quantitatively accounted for the emission spectrum and ionization potential of atomic hydrogen. To achieve this quantitative agreement, Niels Bohr followed the lead of Planck in adopting the **quantum** concept and applied it to the hydrogen atom. Bohr had to make assumptions that ran counter to the accepted dogma of classical mechanics. Nevertheless, his postulates yielded some results that agreed quantitatively with experiment and were not to be taken lightly.

In a tremendously significant theoretical advance, de Broglie[2] proposed his relation for the connection between the momentum of a classical particle (momentum $\equiv p = mv$) and the length of the wave that describes its motion, λ:

$$\lambda = h/p$$

Davisson and Germer[3] were able to verify de Broglie's relation experimentally by showing that electrons are diffracted with a wavelength inversely related to their momentum. As a result of these developments, the previously derived theory of wave motion for light was then borrowed to describe electrons and their motion in the form of Schrödinger's wave model[4] of the hydrogen atom. These events marked the birth of the concept of a **matter wave**.

Our goal in this chapter is to learn how the chemist interprets the Schrödinger electron waves for the hydrogen atom. These concepts can then be elaborated upon for a qualitative "feel" for the physical and chemical properties of all atoms. To summarize the properties of **electron matter waves**:

1. In wave theories, where ψ is the wave function of space and time variables, the important wave property of intensity or magnitude is given by ψ^2. The classical analog of wave intensity is particle density. For example, light intensity and photon density (photons/volume) are equivalent concepts; the conceptual dualism arises because some experiments with light are best understood in terms of light as waves, while others make more sense if light is thought of as particulate.

2. Any theory of the motion of the electron must provide us with information about its distribution in space and its energy. In classical theory, energy, position, and time are uniquely related; in small particle experiments this is not true. **Heisenberg's Uncertainty Principle** for small particles is the statement that an experiment designed to measure the electron's energy will be vague about the position of the electron at the time of the measurement. The matter-wave theory is fully compatible with this requirement: solution of the equations of motion of an electron in any system provides an *exact* value for its energy, E, and a *function*, ψ, describing its position in space. As noted in 1, ψ^2 should be treated as a density function, but it must also be considered as a probability function. Both attributes are embodied in the description of ψ^2 as a **probability density function.**

3. For a single electron, the sum probability at all points in space must equal unity. Mathematically,

$$\int\limits^{\text{all space}} \psi^2(x,y,z)\,d\tau = 1$$

This is the **normalization condition** whereby, for consistency with the notion that the electron must be somewhere in space, the *normalization is to unity.*

4. Schrödinger's matter-wave analog of the classical wave equation leads to the electron space wavefunction in terms of polar rather than cartesian coordinates (Figure 1-1). The use of polar coordinates is for convenience in solving the matter wave equation. The result Schrödinger obtained has the form

$$\psi(r,\theta,\phi) = R_{n,\ell}(r)Y_\ell^m(\theta,\phi)$$

5. The function $R(r)$ is called the **radial function**; its contribution, $R^2(r)$, to the probability density function ψ^2 tells us how the electron probability rises and falls with distance from the nucleus. The primary and secondary quantum numbers n and ℓ appear in this function.

6. The probability variation for the electron about the nucleus (irrespective of distance away) is given by $Y^2(\theta, \phi)$. The functions $Y_\ell^m(\theta, \phi)$ are called the **spherical harmonics**, and they depend on the secondary and magnetic quantum numbers ℓ and m. $Y^2(\theta, \phi)$ tells how the probability density varies as one circumnavigates the nucleus at any (and all) distances from the nucleus.

With this brief summary behind us, we may proceed with the important business of analyzing the total density function $R^2(r)Y^2(\theta,\phi)$.

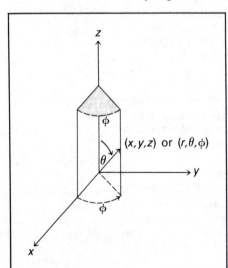

Figure 1-1. Spherical polar coordinates. θ and ϕ are used to locate a direction in space from the nucleus at the origin (as latitude and longitude are used by navigators), and r specifies "how far out" from the nucleus. If ϕ is allowed values from 0 to 2π, then ϕ need only range from 0 to π to describe any direction in space; r takes values from 0 to ∞.

As suggested in the preceding outline, by calculating the value of ψ^2 at some point (r, θ, ϕ) in space we determine the **probability density** for the electron at that point.

$$\psi^2(r, \theta, \phi) = \text{probability density for the point } (r, \theta, \phi): \text{units} = (\text{vol})^{-1}$$

The unit is 1/volume, as you will see in the next section, and this is why we think of ψ^2 as a probability *density* function. Multiplication of this density function by some small volume of space yields the **occupation number** for the electron in that small volume element, $d\tau$:

$$\psi^2(r, \theta, \phi)\, d\tau = \text{occupation number for the infinitesimal volume } d\tau$$
$$\text{about the point } (r, \theta, \phi): \text{units} = \text{none}$$

This function is dimensionless. Finally, by integrating the occupation number for each infinitesimal volume element in all of the space over all space, you will obtain the total occupation of the electron in all space, which physically must equal unity.

$$\int \psi^2(r, \theta, \phi)\, d\tau = 1$$

The volume element $d\tau$ represents $(dx\, dy\, dz)$ if the coordinates are the conventional cartesian coordinates. On conversion to spherical coordinates, $(dx\, dy\, dz)$ becomes $(r^2 \sin\theta\, dr\, d\theta\, d\phi)$. Because of the separability of the wavefunction into $R(r)Y(\theta, \phi)$, the density integral takes the form

$$\int_0^\infty R^2(r)r^2\, dr \int_0^{2\pi} \int_0^\pi Y^2(\theta, \phi)\, \sin\theta\, d\theta\, d\phi = 1$$

Thus, we think of the total probability density function as being a product of a **radial density function**, $R^2(r)$, and an **angular density function**, $Y^2(\theta, \phi)$. For conveniene, each of these is separately normalized.

$$\int_0^\infty R^2(r)r^2\, dr = 1 \tag{1-1}$$

$$\int_0^{2\pi} \int_0^\pi Y^2(\theta, \phi)\, \sin\theta\, d\theta\, d\phi = 1 \tag{1-2}$$

In the next sections we will examine these density functions in some detail for certain values of n, ℓ, and m. To do this will give you some feeling for the *shapes and symmetries of atomic orbitals* through analysis of the *angular density functions*. The *radial density functions* are important in developing a link between theory and experiment on the subjects of *atom sizes*, *ionization potentials*, and *electronegativity*.

Surface Density Functions and Orbital Energies

Many properties of the atom derive from the way in which the electron is distributed as a function of distance from the nucleus, taking all directions into account. To find a probability function that gives this information, all we need do is start with the probability density function ψ^2 and sum (integrate) over all directions (θ, ϕ).

$$\int_0^{2\pi} \int_0^\pi \psi^2\, d\tau = R^2(r)r^2\, dr \int_0^{2\pi} \int_0^\pi Y^2(\theta, \phi) \sin\theta\, d\theta\, d\phi$$

$$= R^2(r)r^2\, dr = S(r)\, dr$$

This is the *electron occupation function for infinitesimally thin spherical shells of radius r and thickness dr.* The function $R^2(r)r^2$ represents a *spherical* surface density function, with units = $(\text{distance})^{-1}$, for a sphere of radius r and center at the nucleus. *We will call this the* **surface density function, S(r).**

Table 1-1 lists the radial density functions for (n, ℓ) pairs through $n = 4$. When distances (r) are measured in pm, a_0 has the value 52.9 pm; otherwise, r can be measured in atomic units, with $a_0 = 1$. Figure 1-2 shows plots of $R_{1,0}^2(r)$ and $S_{1,0}(r)$. An important, unique characteristic of all "s" radial functions is that $R_{n,0}^2(r = 0) \neq 0$; *only "s" electron waves have amplitude (probability) at the nucleus.* On the other hand, all $S_{n,\ell}(r = 0) = 0$ because $R^2(r) \cdot r^2 = 0$ when $r = 0$, regardless of the value of $R^2(r = 0)$.

You should be careful to notice how the functions $R^2(r)$ and $S(r)$ differ. The $R^2(r)$ function is an exponential function decreasing from a maximum value at $r = 0$ to zero at $r = \infty$. As far as density is concerned, the electron with quantum numbers $n = 1$, $\ell = 0$, and $m = 0$ is most highly concentrated at the nucleus. If you ask the question, "What is

TABLE 1-1

RADIAL DENSITY FUNCTIONS $R_{n,\ell}^2(r)$ FOR HYDROGEN.

$a_0 \equiv$ atomic distance unit = 52.92 pm; $\rho = r/a_0$.

n, ℓ	Orbital Symbol	$R_{n,\ell}^2(r)$	NOTE: R^2 has units $(\text{vol})^{-1}$
1, 0	1s	$N^2 [e^{-2/\rho}]$	where $N = 2a_0^{-3/2}$
2, 0	2s	$N^2 (2 - \rho)^2 [e^{-\rho}]$	$N = \dfrac{1}{2\sqrt{2}} a_0^{-3/2}$
2, 1	2p	$N^2 \rho [e^{-\rho}]$	$N = \dfrac{1}{2\sqrt{6}} a_0^{-3/2}$
3, 0	3s	$N^2 (27 - 18\rho + 2\rho^2)^2 [e^{-2\rho/3}]$	$N = \dfrac{2}{81\sqrt{3}} a_0^{-3/2}$
3, 1	3p	$N^2 (6\rho - \rho^2)^2 [e^{-2\rho/3}]$	$N = \dfrac{4}{81\sqrt{6}} a_0^{-3/2}$
3, 2	3d	$N^2 \rho^4 [e^{-2\rho/3}]$	$N = \dfrac{4}{81\sqrt{30}} a_0^{-3/2}$
4, 0	4s	$N^2 (192 - 144\rho + 24\rho^2 - \rho^3)^2 [e^{-\rho/2}]$	$N = \dfrac{1}{768} a_0^{-3/2}$
4, 1	4p	$N^2 (80 - 20\rho + \rho^2)^2 \rho^2 [e^{-\rho/2}]$	$N = \dfrac{3}{768\sqrt{15}} a_0^{-3/2}$
4, 2	4d	$N^2 (12 - \rho)^2 \rho^4 [e^{-\rho/2}]$	$N = \dfrac{1}{768\sqrt{5}} a_0^{-3/2}$
4, 3	4f	$N^2 \rho^6 [e^{-\rho/2}]$	$N = \dfrac{1}{768\sqrt{35}} a_0^{-3/2}$

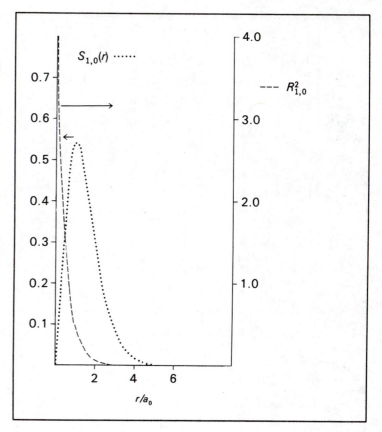

Figure 1-2. The $1s$ radial density function, in units of $(1/a_0^3)$, and surface density function, in units of $(1/a_0)$.

the radius of the spherical surface over which the electron is most likely to be found?" the answer comes from the maximum in the $S(r)$ curve and is 1 a_0, or 52.92 pm. Don't miss the further point that *the area under the $S(r)$ curve is unity*, since $\int_0^\infty S(r)\,dr = 1$ [from (1-1)].

Figure 1-3 shows the $1s$, $2s$, and $2p$ surface density functions for the hydrogen atom. Several important points emerge from these graphs. First, an $n = 2$ electron is *radially more diffuse* than is a $1s$ (its most probable position is further from the nucleus). This reflects the fact that a $2s$ or $2p$ electron is less tightly held to the nucleus than is a $1s$ electron. Secondly, the $2s$ probability goes to zero at $r = 2a_0$, with a second, weak maximum probability at $0.7a_0$.† The zero probability at $r = 2a_0$ corresponds to a wave **node** for $R_{2,0}(r)$; nodes are a familiar property of waves.

---

†When we think of the electron in terms of its particle (classical) properties, it is difficult to imagine surfaces of zero probability (nodal surfaces). The fallacy of this artificial dilemma lies in the duality concept—if we must describe the electron in terms of its wave property, we must automatically abandon the classical particle concept. When conceptualizing experiments requiring a classical view, we must abandon the wave concept. The classical view of the electron as a particle is inappropriate for the Schrödinger atom and for the diffraction part of the Davisson-Germer experiment. An interesting point, of little *chemical* consequence, is made by A. Szabo in *J. Chem. Educ.*, **46**, 678 (1969). Here it is pointed out that the Schrödinger treatment ignores relativity (incurring a negligible error for chemical purposes); were relativity to be incorporated, the nodal surfaces would become surfaces of low, but not zero, probability. In this case, the "dilemma" of mixing classical and wave concepts doesn't even arise, though it still is improper to invoke both simultaneously.

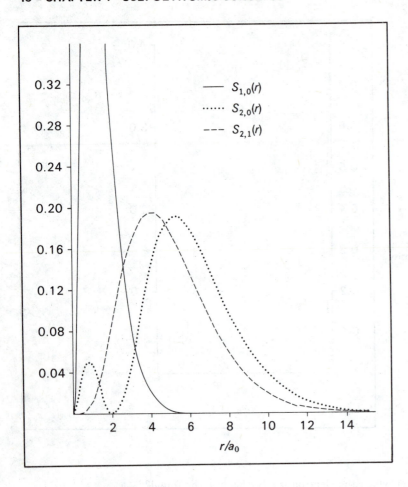

Figure 1-3. Comparison of the 1s, 2s, and 2p surface density functions, units = a_0^{-1}.

Of considerable significance, as you shall see later, is the greater concentration of the 2s function than of the 2p function in the range $r = 0$ to $1.3\, a_0$, in contrast to the smaller values over the range $r = 1.3$ to $4.8\, a_0$. *With reference to Figure 1-3, the 2s and 2p both penetrate the region of 1s distribution, but the 2p does so to a much greater extent than the 2s* (56% of the 2p and 35% of the 2s densities occur within the sphere of radius $5a_0$, which contains essentially all of the 1s density). This **penetration effect** will be of great importance later when we discuss the order of filling of atomic orbitals in an attempt to account for the electronic structures of atoms.

In anticipation of that discussion, let us look into the physical significance of the minor hump at small r for the 2s orbital. From the matter-wave theory, the electron energy is simply half of the potential energy:

$$E = -\tfrac{1}{2}\frac{Ze^2}{r} \qquad (-Ze^2/r \text{ is the Coulomb potential energy})$$

where Z = the atomic number. Since an electron is distributed over r between 0 and ∞, there is no single value of r to use to find E. We must find the average value of E, given the symbol $\langle E \rangle$. When the electron is at a specific distance r from the nucleus, it has an energy computed by using that value of r in the expression $-Ze^2/2r$. The average value of any quantity (exam scores or whatever) is calculated by weighting each of its specific values by their probability of occurrence, summing these weighted, specific values, and dividing by the sum of their probabilities:

$$\langle x \rangle = \frac{\sum\limits_{x} \left(\begin{array}{c} \text{probability} \\ \text{of the value } x \end{array}\right) \cdot x}{\sum\limits_{x} \left(\begin{array}{c} \text{probability} \\ \text{of the value } x \end{array}\right)} = \frac{\sum\limits_{x} P(x) \cdot x}{\sum\limits_{x} P(x)}$$

If x and its probability function, $P(x)$, are continuous,

$$\langle x \rangle = \frac{\int P(x) \cdot x \, dx}{\int P(x) \, dx}$$

Applying this idea to the average electron energy,

$$\langle E \rangle = \frac{\int\limits_{0}^{\infty} S(r) \cdot \left(-\dfrac{Ze^2}{2r}\right) dr}{\int\limits_{0}^{\infty} S(r) \, dr} = -\frac{Ze^2}{2} \int\limits_{0}^{\infty} S(r) \cdot \frac{1}{r} \, dr$$

In Figure 1–4 you will find plots of the probability-weighted electron-nuclear Coulomb potential energy, $-Ze^2 \cdot S(r) \cdot \dfrac{1}{r}$, for the $2s$ and $2p$ electrons of the hydrogen atom. The striking feature of these graphs is the reversal in importance of the $2s$ humps! At small r,

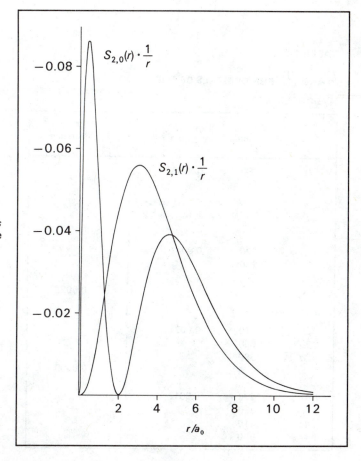

Figure 1–4. Potential energy, $V(r)$, plots for $2s$ and $2p$ wave forms. The ordinate units are (e^2/a_0^2).

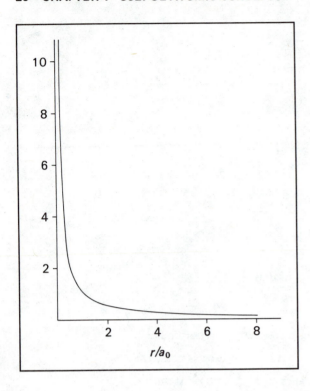

Figure 1-5. Plot showing that $\dfrac{1}{r}$ has greatest magnitude in the region $0 < r < 2$. Ordinate units $= 1/a_0$.

TABLE 1-2

$\langle r \rangle$ AND $\left\langle \dfrac{1}{r} \right\rangle$ FOR ORBITALS OF THE HYDROGEN ATOM

	$\langle r \rangle$, pm	$\left\langle \dfrac{1}{r} \right\rangle$, pm^{-1}	Difference
1s	80	0.0189	
			0.0142
2s	320	0.0047	
2p	260		
			0.0026
3s	710	0.0021	
3p	660		
3d	560		
			0.0009
4s	1270	0.0012	
4p	1220		
4d	1110		
4f	950		

where $S_{2s}(r)$ is small but $1/r$ is very large (see Figure 1-5), $S_{2s}(r) \cdot \dfrac{1}{r}$ is large; at large r,

where $S_{2s}(r)$ is large but $1/r$ is small, $S_{2s}(r) \cdot \dfrac{1}{r}$ is smaller. The areas under the 2s and 2p

curves give their potential energies—it turns out that their areas are the same and $\langle E \rangle_{2s} = \langle E \rangle_{2p}$. Were it not for the inner probability band, the 2s electron would be less tightly held to the nucleus than the 2p.

For future reference, Table 1–2 lists values of $\langle r \rangle$ and of $\langle 1/r \rangle$ for the orbitals of the hydrogen atom. Note that $\langle r \rangle$ decreases as ℓ increases for any particular principal quantum number n ($ns > np > nd > nf$), that $\langle r \rangle$ increases as n increases for a given ℓ quantum number ($1s < 2s < 3s \ldots$), and that $\langle 1/r \rangle$ (energy) gets smaller (approaches zero) as n increases. These are the trends in atom size and ionization potential you learned in freshman chemistry.

1. Construct plots of $S(r) \cdot 1/r$ for the hydrogen atom 3s, 3p, and 3d functions. Use these plots to explain why $\langle r^{-1} \rangle$ is the same for each.

STUDY QUESTION

Angular Functions and Orbital Shapes[5]

Before inspecting the angular probability functions, it is important to discuss the symmetries of the orbital functions themselves, *a class of functions widely known as the* **spherical harmonics**, *symbolized* Y_ℓ^m, where ℓ and m are quantum numbers. We could begin the inspection by retaining the angles θ and ϕ or by converting to the cartesian coordinates x, y, and z. As you will soon see, the cartesian forms of the spherical harmonics are more easily interpreted, especially since the cartesian coordinates are more familiar to you. The real spherical harmonics shown in Table 1–3 result.

TABLE 1–3

THE REAL SPHERICAL HARMONICS

	Symbol	Real
$Y_{0,0}$	s	$1/\sqrt{4\pi}$
$Y_{1,z}$	p_z	$\sqrt{\dfrac{3}{4\pi}}\,[z/r]$
$Y_{1,x}$	p_x	$\sqrt{\dfrac{3}{4\pi}}\,[x/r]$
$Y_{1,y}$	p_y	$\sqrt{\dfrac{3}{4\pi}}\,[y/r]$
Y_{2,z^2}	d_{z^2}	$\sqrt{\dfrac{5}{16\pi}}\,[(2z^2 - (x^2 + y^2))/r^2]$
$Y_{2,xz}$	d_{xz}	$\sqrt{\dfrac{60}{16\pi}}\,[xz/r^2]$
$Y_{2,yz}$	d_{yz}	$\sqrt{\dfrac{60}{16\pi}}\,[yz/r^2]$
Y_{2,x^2-y^2}	$d_{x^2-y^2}$	$\sqrt{\dfrac{15}{16\pi}}\,[(x^2 - y^2)/r^2]$
$Y_{2,xy}$	d_{xy}	$\sqrt{\dfrac{60}{16\pi}}\,[xy/r^2]$

The cartesian forms of the spherical harmonics make very easy your visualization of the angular functions. *For each harmonic, those coordinates x, y, and z that cause the harmonic to have zero value define nodal surfaces.* For the p_x harmonic, the wave value

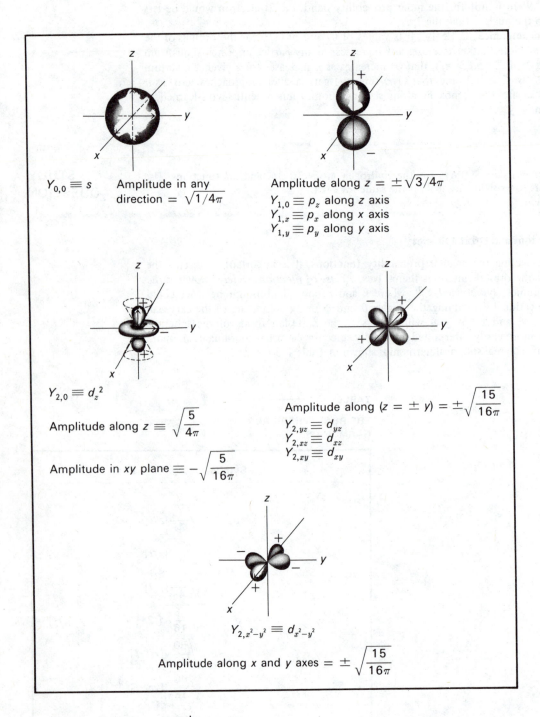

$Y_{0,0} \equiv s$ Amplitude in any direction $= \sqrt{1/4\pi}$

Amplitude along $z = \pm\sqrt{3/4\pi}$

$Y_{1,0} \equiv p_z$ along z axis
$Y_{1,x} \equiv p_x$ along x axis
$Y_{1,y} \equiv p_y$ along y axis

$Y_{2,0} \equiv d_{z^2}$

Amplitude along $z \equiv \sqrt{\dfrac{5}{4\pi}}$

Amplitude in xy plane $\equiv -\sqrt{\dfrac{5}{16\pi}}$

Amplitude along $(z = \pm y) = \pm\sqrt{\dfrac{15}{16\pi}}$

$Y_{2,yz} \equiv d_{yz}$
$Y_{2,xz} \equiv d_{xz}$
$Y_{2,xy} \equiv d_{xy}$

$Y_{2,x^2-y^2} \equiv d_{x^2-y^2}$

Amplitude along x and y axes $= \pm\sqrt{\dfrac{15}{16\pi}}$

Figure 1-6. Sketches of the Y_ℓ^{real} in Table 1-3.

is zero whenever the x coordinate is zero (regardless of the values of y and z). Consequently, the yz plane is a nodal plane; the function is positive in the $+x$ direction and negative in the $-x$ direction. The d_{xy} harmonic has two nodal surfaces (as do all the d harmonics); the xz and yz planes define surfaces over which d_{xy} has zero value. When x and y are both positive or both negative, the function has positive phase, and when x and y are of opposite sign, the harmonic lobes are negatively phased. Figure 1-6 summarizes these features for all the s, p, and d (real) harmonics. You should compare the real functions in Table 1-3 with their sketches in Figure 1-6 to verify the location of the nodal surfaces and the signs of the orbital lobes. The harmonic plots of Figure 1-6 are simply surfaces in which each point on the surface defines a direction in space from the origin (nucleus) and the distance of the surface from the origin is the magnitude of Y in that direction (these are polar plots).

To obtain the spherical harmonic probability functions for the Y's of Table 1-3 we need simply square the Y_{ϱ}. For our purposes, there is little difference between the form of the real harmonic and that of its probability function. That is, the nodal surfaces do not change while the probability functions become positive everywhere (there are no negative lobes).

STUDY QUESTION

2. Which has the greater probability:

   a) along the x axis: an s orbital or a p_y orbital?
   b) along the line $x = z$, $y = 0$: a p_z orbital or a d_{yz} orbital?

Total Density Functions

As a conclusion to this section on probability functions, we wish to put the radial and angular probability functions together to arrive at a *total* density distribution function for the electron. The graphical presentation of the total density function presents a special problem; one must give the density as a function of three space variables, and such a situation requires a four-dimensional graph. Many techniques have been used, but the most precise technique—and the one that most clearly emphasizes the orbital nodal characteristics—is the construction of contour maps. For any arbitrary plane through the atom, one gets the corresponding cross section of the three-dimensional density function. *All points in this plane for which the electron density is constant are connected by a smooth curve or curves.* These curves are called contour lines, and the entire figure is referred to as a contour map. Such maps are illustrated in Figure 1-7 for $2p_z$, $3p_z$ and some $3d$ orbitals. The point near the center of each closed set of contour curves locates the point of greatest probability. A clear advantage of such contour cross sections is that the location of radial nodal surfaces is very simple. The dashed circle in the map for the $3p_z$ function is the nodal line that arises from the node in the $3p$ radial function. *In three dimensions, all radial nodal surfaces are spheres.* The $3p_z$ function also has an angular nodal plane, the xy plane.†

---

†It is our experience that students do best by first visualizing the nodal surfaces in three dimensions. Then introduce the intersecting contour plane and deduce the forms of the lines of intersection. These can be sketched as dashed line curves.

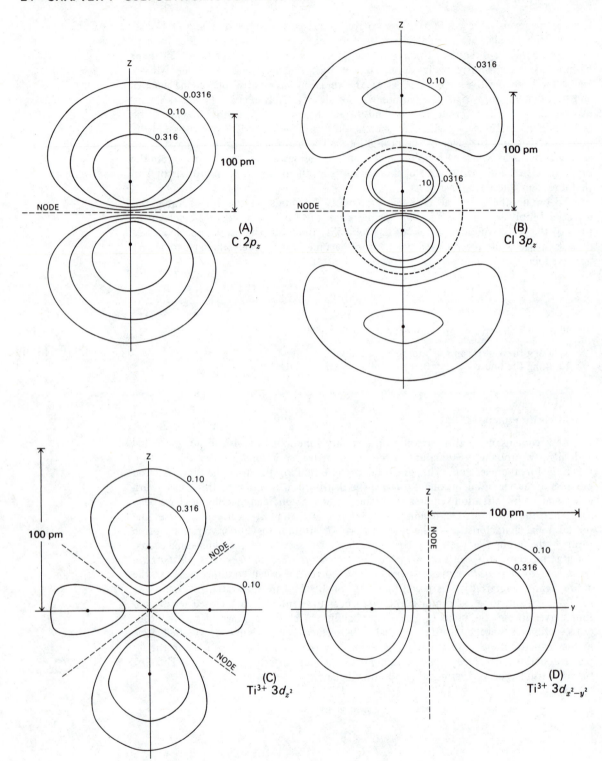

Figure 1-7. Contour maps for various a.o.'s. For (a) through (d) the cross-sectional plane could be any plane containing the z axis. For (e) the cross-sectional plane is the *xy* plane. (From E. A. Ogryzlo and G. B. Porter, *J. Chem. Educ.,* **40**, 258 (1963).

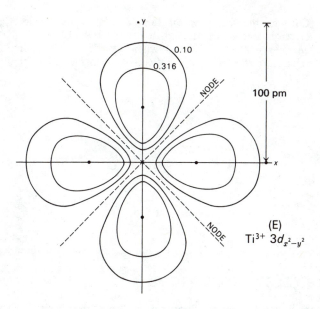

(E)

Ti^{3+} $3d_{x^2-y^2}$

Figure 1-7. *Continued.*

We close this analysis of electron matter waves for the atom by noting the relationship between the quantum members n, ℓ, m and the wave nodal types and electron energy.

$n - \ell - 1$: derives from the radial function and equals the number of spherical nodal surfaces (each *radial* nodal point distant from the nucleus becomes a spherical surface in three dimensions).

ℓ: derives from the angular functions and equals the number of nodal surfaces through the nucleus, for real orbitals.

$n - 1$ = the total number of nodal surfaces.

A little reflection also reveals that the more nodal surfaces there are in the electron's wave, the higher is the energy of the electron. In fact, the energy order $1s < 2s = 2p < 3s = 3p = 3d \ldots$ correlates exactly with increasing number of nodes: $0, 1, 2, \ldots$ A more explicit physical consequence of n derives from the energy expression

$$E = -\frac{1}{n^2}(e^2/2a_0)Z^2$$

STUDY QUESTIONS

3. Calculate the probability density for a H $2p_z$ electron at the point $r = a_0$, $\theta = 60°$, $\phi = 45°$. What are the units of the probability density? Does the probability density change with the angle ϕ?

4. Sketch a contour diagram for $4d_{x^2-y^2}$ in the xy plane. Show nodal lines as dashed lines.

5. Carefully sketch the xy plane contour maps for the $5d_{z^2}$ and $5d_{xy}$ orbitals. Indicate nodal lines as dashed lines and identify their source (radial or angular).

6. Consider the following diagram of an atom in the electric field of a "spy satellite" electron located on the positive z axis. Because of e^-, e^- repulsions, the d orbital energies will not be equal when the "spy" is present. Sketch the $3d$ orbitals in relation to this "spy" and indicate the order of their energies.

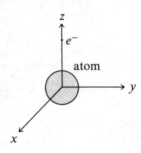

POLYELECTRONIC ATOMS

Atom Electron Configurations and the Long Form Periodic Table

For most of our purposes in inorganic chemistry, often all that is needed is a basic understanding of the qualitative aspects of the electronic structures of many-electron atoms. In this section we will use our understanding of the hydrogen wave functions to develop an appreciation for the electronic structures of the elements and their energy states.

An important consequence of electron-electron repulsions is that all electrons are less tightly held to the nucleus than would be the case were there no coulomb repulsions between them. In effect, the nuclear charge as experienced by each electron is less than Z. The reduction is different for different kinds of electrons (for example, $2s$ and $2p$ electron waves are no longer equi-energetic). A useful approximation to the repulsion problem is that each electron screens all others (to different extents) from the positive nuclear charge.

The variation in ability of inner electrons to screen nuclear charge from the outer electrons lies at the heart of an understanding of the electron configurations and their relation to the periodic table. In building up the periodic table by starting with the hydrogen atom and successively adding a proton to the nucleus and an electron to the extranuclear shell (this process is called the *Aufbau principle*), we must allow for the screening effects. In other words, we must account for the fact that orbitals of the same n quantum number but different ℓ quantum number may have different energies in a given atom. The only *ad hoc* assumption we need introduce at this point is that of the Pauli exclusion principle, which permits only two electrons to occupy the same spatial orbital or to have the same spatial wave function. From various experiments, we may deduce the following *empirical rule concerning the order of filling of the hydrogen atom orbitals:*

(a) *Orbital energies increase as $(n + \ell)$ increases.*

(b) *For two orbitals of the same $(n + \ell)$, the one with the smaller n lies lower in energy.*

Thus, the order of filling becomes

$$1s, 2s, \underbrace{2p, 3s}, \underbrace{3p, 4s}, \underbrace{3d, 4p, 5s}, \ldots$$
$$n + \ell = \quad 1 \quad 2 \qquad 3 \qquad 4 \qquad 5$$

The radial wave functions, which depend on n and ℓ, may be used to formulate a basis for this rule. In what follows, your attention will be directed to the penetration of inner electron density by outer radial functions, thereby defining an **effective nuclear** charge (Z^*) for the outer electron. A qualitative estimate of the relative energies of outer electrons can be obtained from estimates of their potential energies as given (classically) by

$$V = -e^2 \, \frac{Z^*}{r}$$

According to the wave formulation of the problem, we need to consider the average potential energy given by

$$\langle V \rangle = -e^2 \left\langle \frac{Z^*}{r} \right\rangle = -e^2 \int_0^\infty \frac{Z^*}{r} \cdot S(r) \, dr$$

A graphical evaluation of the potential energy then requires a determination of the area under a plot of $Z^*/r \cdot S(r)$ as a function of r. To compare the potential energies of an electron in one outer orbital to those of another, we need to compare the areas under the appropriate potential energy curves. Note that each point on such a curve gives the potential energy of the electron for the given distance from the nucleus.

According to the wave model expression for the point-by-point potential energy of an electron, we need to evaluate the potential energy of the electron at each point ($= -e^2 Z^*/r$) and weight this by the probability that the electron will be at that point. In a polyelectronic atom, the nuclear positive charge experienced by an electron will be Z when $r = 0$ and will decrease toward unity as r approaches infinity. The decrease in Z^*

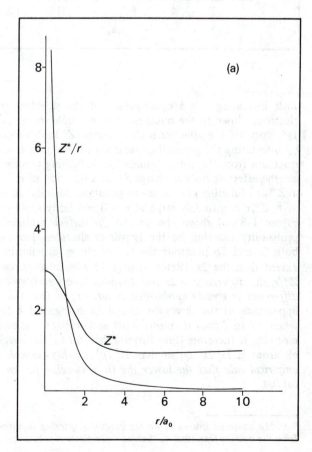

Figure 1-8. (a) Z^* and Z^*/r versus r for the third electron of Li. Ordinate units = $1/a_0$.

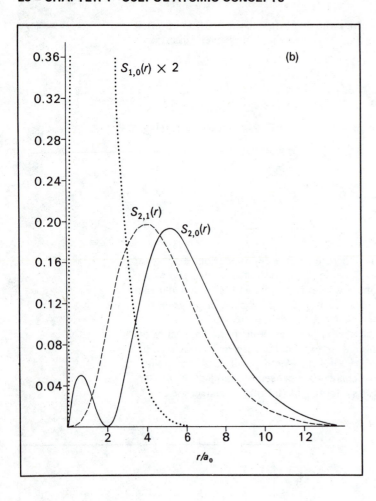

Figure 1-8. *Continued.*

(b) Surface density plots for *2s* and *2p* functions in relation to the same plot for two *1s* electrons.

with increasing *r* is a consequence of the shielding of nuclear charge by the other electrons closer to the nucleus.† For example, consider the Li^+ ion with configuration $1s^2$; calculate the effective nuclear charge (Z^*) as a function of distance from the nucleus by integrating the probability function for the two *1s* electrons from zero to *r*, and subtract this from the atomic number *Z*. Repeating this for every *r* value from *r* = 0 on, you get the effective nuclear charge Z^* as a function of *r*. Figure 1-8 (a) shows the variation in Z^* as a function of *r* for such a situation. Also shown in this figure is Z^*/r as a function of *r*. Z^*/r is infinitely large at *r* = 0 and fairly rapidly decreases to a very small value. Figure 1-8 (b) shows the *2s* and *2p* surface probability functions in relation to the probability function for the *1s* pair of electrons, from which it is readily apparent that both *2s* and *2p* penetrate the *1s* pair charge distribution, the *2p* to a seemingly greater extent than the *2s*. (Refer to page 18.) *However, because of the form of the function Z^*/r, the difference in 2s and 2p radial functions is greatly magnified at small r while the difference is greatly diminished at large r*, so that the function $Z^*/r \cdot S(r)$ takes on the appearance of the curves for *2s* and *2p* in Figure 1-8 (c). *A 2s electron penetrates best where (0 to $1.5a_0$) it counts most and realizes a significantly lower energy than a 2p* electron. It therefore is no surprise that, for Li, the lowest energy configuration for three electrons is $1s^2 2s^1$ rather than $1s^2 2p^1$. *In this example we find a physical basis for the empirical rule that the lower the (n + ℓ) value for an orbital, the more stable is that orbital.*

---

†In adopting this view, we are implicitly ignoring repulsion from electron density further out from the nucleus than (that is, "behind") the electron whose energy we are calculating.

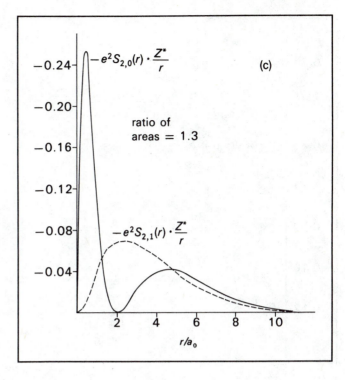

Figure 1-8. *Continued.*

(c) The 2s and 2p potential energy curves for $Z = 3$ with a pair of 1s electrons present. Units $= (e/a_0)^2$.

The second part of the rule deals with the eventuality that $(n + \ell)$ is the same for two orbitals. Such a situation is first encountered with 2p and 3s. That is, the proper configuration for boron is $1s^2 2s^2 2p^1$, not $1s^2 2s^2 3s^1$. An analysis of the effective potential energies of 2p and 3s waves in the presence of two 1s and two 2s electrons shows that the 2p is lower in energy because it has larger amplitude in the regions where Z^* is still large; see Figure 1-9.

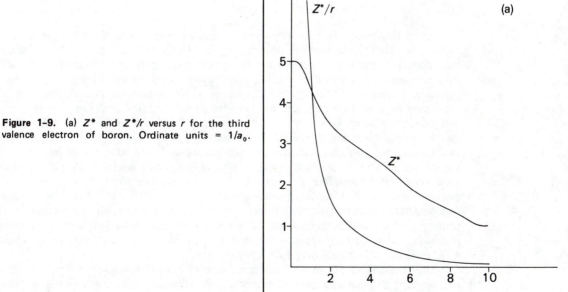

Figure 1-9. (a) Z^* and Z^*/r versus r for the third valence electron of boron. Ordinate units $= 1/a_0$.

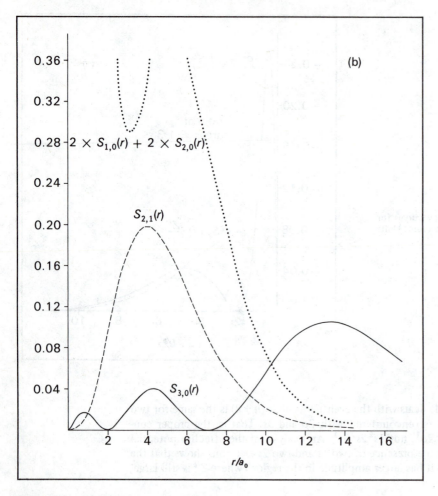

Figure 1-9. *Continued.*

(b) Surface density plots for 2*p* and 3*s* functions in relation to the same plot for two 1*s* and 2*s* electrons.

Up to this point, we have pretty much glossed over a very important property of an electron. That is, the electron possesses a spin angular momentum; (relativistic quantum) theory, backed by experiment, tells us that this angular momentum is quantized, with the quantum number m_s taking only two possible values, +1/2 or −1/2. Thus, a general statement of the Pauli exclusion principle is that no two electrons in an atom may have all four quantum numbers the same. Two electrons can have the three space quantum numbers n, ℓ, and m the same, but they must differ in the value of the fourth or spin quantum number. This leads directly, as we apply the Aufbau principle, to the statement that only two electrons are permitted in the same orbital. The physical phenomenon on which the Pauli principle is based is that *two particles with the same spin do a better job of avoiding each other in space than do two electrons of opposite spin quantum number.* In fact, this phenomenon is a spin-quantum effect, has no classical counterpart, and is observed for any particles that possess spin angular momentum. Even neutrons, which are uncharged and so do not experience a Coulomb or electrostatic repulsion, still avoid each other better in their motions if they have the same spin (both + 1/2 or both −1/2) than if they have opposite spins.

The chemical significance of this **spin correlation of electron motions** is that, since electrons are charged and repel one another, *the repulsion is less, by an amount called the "exchange energy," for electrons of like spin than for electrons of different spin.* We have an opportunity to see this effect in operation on passing from boron to carbon,

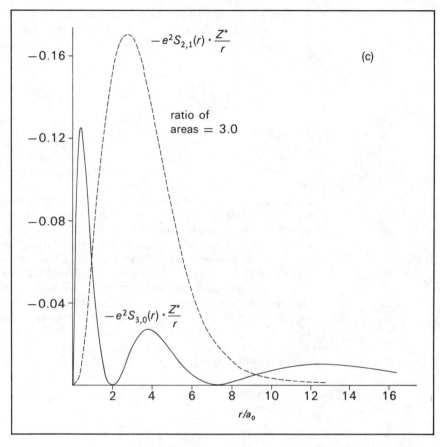

Figure 1–9. *Continued.*
(c) The 2p and 3s potential energy curves for $Z = 5$ with 1s and 2s electron pairs present.

which has two electrons in the 2p orbitals. **Charge correlation energy** (e^2/r) *dictates that a lower energy situation exists if the two p electrons have different angular functions.* **Spin correlation** *further affords a lower (more negative) energy for like spins.*[6] In the conventional box diagram notation, the lowest energy configuration of the carbon atom will be $1s^2 2s^2 2p^2$, with the requirement that the two p electrons reside in different orbitals with the same m_s value.

$$\boxed{\uparrow\downarrow}\quad\boxed{\uparrow\downarrow}\quad\boxed{\uparrow}\;\boxed{\uparrow}\;\boxed{}$$
$$1s\qquad 2s\qquad\;\; 2p$$

Similarly, in the lowest energy configuration for nitrogen, all three p electrons have the same spin; therefore, the configuration is one of filled 1s and 2s orbitals and a half-filled 2p shell.

$$\boxed{\uparrow\downarrow}\quad\boxed{\uparrow\downarrow}\quad\boxed{\uparrow}\;\boxed{\uparrow}\;\boxed{\uparrow}$$
$$1s\qquad 2s\qquad\;\; 2p$$

Spin correlation and its "exchange energy" are the physical basis for the often cited stability of the half-filled shell. It is simply a matter of choosing the electron arrangement that leads to the lowest electron-electron repulsions. Obviously, placing two electrons in the same orbital will lead to higher repulsions than if they move in different spatial orbitals (the electrons have the same charge). In addition to the lower energy made possible by allowing separate orbital motions for the two charged particles, the spin correlation allows for a further reduction in repulsion if both have the same spin. To summarize, we would rank the stability of p^2 electron arrangements for carbon as follows:

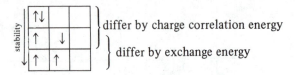

Very often, then, the electron configuration and energy of an atom are dictated by fine details of electron repulsions that are not apparent in the averaged repulsion (screening) concept of orbital energies. A good illustration can be found in the trends in ionization potential and electron affinity[7] (see p. 94) for the main group elements from groups III through VIII. (*Electron affinity, E, is defined as the energy for the process atom⁻ → atom + e⁻, just as ionization potential, I, is defined for atom → atom⁺ + e⁻.*) The valence configurations of these elements are $ns^2 np^x$, where x progresses from 1 through 6. The trend of increasing I and E across this series is broken at the configuration $s^2 p^4$. That is, the configurations with $x = 1, 2, 3$ follow a smoothly increasing trend in I, as do the configurations with $x = 4, 5, 6$. The latter series is simply shifted to lower I from the values that would be extrapolated from $x = 1, 2, 3$. The origin of the dislocation at $s^2 p^4$ is easily traced to a discontinuity of electron repulsions, as the nuclear charge smoothly increases by 1 across the entire series. The following box diagrams should clarify the anomaly.

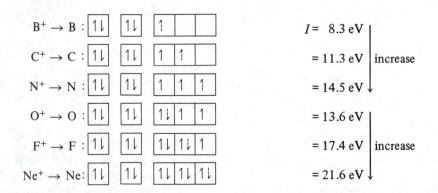

The trend in I from $x = 1, 2, 3$ is determined by a balance between regularly increasing nuclear charge and regularly increasing electron repulsion for the added electron; in combination, these factors establish a trend of smoothly increasing effective nuclear charge and I. In each case, the added electron has the same spin as those already present in the p sub-shell. On adding a p electron to the $s^2 p^3$ configuration to produce the $s^2 p^4$ configuration, the effective nuclear charge for the added electron will appear to drop because this electron must be added with spin opposite to those already present and must be placed in an orbital already occupied by an opposite-spin electron. This electron does not benefit from the "exchange energy" stabilization possible for the preceding three members of the series, and further suffers from larger coulombic repulsion. This configuration establishes a new "origin" for a smooth trend, shifted from the original, in effective nuclear charge and I for the remaining series members.

Another illustration of the breakdown of the screening concept arises for the transition elements.[8] The orbital energies estimated from shielding or penetration effects adequately account for the fact that the $4s$ orbital is filled prior to the $3d$ on passing from Ar to K to Ca. However, proceeding across the $3d$ series of metals (Sc through Zn), there are so-called anomalies at Cr and Cu, which have [Ar] $4s^1 3d^5$ and [Ar] $4s^1 3d^{10}$ configurations. Reference to Figure 1–10 shows just how complex the situation becomes;

of the ten transition metal groups, five show irregularities in electron configuration. Thus, while "the stability of half-filled shells" seems a good phrase to invoke for Cr, the "principle" fails for W. It is certainly better to realize that the electron configurations of these elements depend on fine points of electron interactions (the balance between d/d, d/s, and s/s repulsions) and accept them for what they are than to propose a rule that must be violated 50% of the time.

Sc d^1s^2	Ti d^2s^2	V d^3s^2	Cr d^5s^1	Mn d^5s^2	Fe d^6s^2	Co d^7s^2	Ni d^8s^2	Cu $d^{10}s^1$	Zn $d^{10}s^2$
Y d^1s^2	Zr d^2s^2	Nb d^4s^1	Mo d^5s^1	Tc d^5s^2	Ru d^7s^1	Rh d^8s^1	Pd d^{10}	Ag $d^{10}s^1$	Cd $d^{10}s^2$
La d^1s^2	Hf d^2s^2	Ta d^3s^2	W d^4s^2	Re d^5s^2	Os d^6s^2	Ir d^7s^2	Pt d^9s^1	Au $d^{10}s^1$	Hg $d^{10}s^2$

Figure 1-10. The nd and $(n + 1)s$ electron configurations for transition elements.

While the electron configurations of the transition elements in the zero oxidation states are difficult to interpret, the configurations of the divalent and trivalent ions are much simpler. *All di- and trivalent ions of the transition elements have d^n configurations. For the ions*, the energies of the d orbitals are considerably lower than the energies of the next higher s orbital. There is a simple interpretation possible here. After ionization of two or three electrons, we reexamine the relative stabilities of the d and s orbitals in the ion. The ionization process that forms a 2+ cation produces an ion that could also have arisen by the addition of two protons to the nucleus of an atom of the preceding element with the same number of d and s electrons. *Regardless of how the cation is formed, it has but a single most stable electron configuration.* The effective nuclear charge experienced

$$V(d^3s^2) \xrightarrow{-2e^-} V^{2+}(d^3) \xleftarrow{+2p} Sc(d^1s^2)$$

by the $3d$ electrons is greatly enhanced over that of any $4s$ electrons as a direct consequence of the greater penetration of the $3s^2 3p^6$ core by a $3d$ electron than by a $4s$ electron. Consequently, the $3d$ orbitals are expected to drop significantly below the $4s$ (the orbital energy concept is reinstated), and any $4s$ electrons will move to the $3d$ orbitals. Thus, ionization of two or more electrons from an atom of a transition element will make it appear as though the s electrons have been preferentially removed.

Summarizing the relation between the positions of the elements in the periodic table and their electron configurations, we can make the following generalization. As far as the main-group elements are concerned, all elements in a group have the same valence orbital configuration. This is not always true for the transition elements. Ionization of the representative elements results in ions that appear to have simply lost the least stable electrons in the valence shell. For the transition ions, the electron configurations are those to be expected because the d orbitals are clearly more stable than the s orbitals of one greater quantum number, n. Finally, you should be aware that elements, like potassium and copper, that have the same number of unfilled valence orbitals of the same type would not necessarily be expected to show strong similarities in their chemistry. Using the potassium-copper pair as an example, both have $4s^1$ valence configurations, but this electron in potassium is outside a $1s^2 2s^2 2p^6 3s^2 3p^6$ core that the $4s$ electron does not penetrate as well as the $4s$ electron of copper penetrates the $1s^2 2s^2 2p^6 3s^2 3p^6 3d^{10}$ core of that atom. The 10 extra d electrons of copper do not completely screen the additional 10 units of nuclear charge. Clearly, the loss of an electron (the $4s$) from potassium should require much less energy than from copper. Thus, the correlation between valence configuration and group in the periodic table is not perfect (hence the use of A and B groups in the long form of the periodic table), but at least it is understandable in terms of the hydrogen radial functions.

One more consequence of orbital penetration is orbital contraction. Understandably, the greater the orbital penetration of the core electrons, the greater will be the contraction of the radial function—that is, the maximum in the probability curve will occur at a smaller distance from the nucleus, and the wave amplitude (electron probability) will increase at small r and decrease at large r. For example, the radial functions for the hydrogen $2s$ and $2p$ orbitals will be altered in the lithium atom so that the penetration differences discussed above will be reinforced, and it is found that the $2s$ function is not as radially diffuse as the $2p$. This phenomenon is of some importance in interpretations of atom hybridization, to be discussed in Chapter 3.

STUDY QUESTIONS

7. In the discussion of why the $2s$ orbital of Li^+ fills with an electron before the $2p$ orbitals, we encountered the idea that the $2s$ penetration of the $1s^2$ cloud is greater than that of the $2p$. Would the nuclear charge of Li, being greater than that of H, cause a change in the Z^* versus r plot of Figure 1–9 (for which the *hydrogen* $1s$ radial function was used to calculate the $1s^2$ core distribution)? Would this further enhance the difference in $2s$ and $2p$ probabilities at low r? Which orbital, $2s$ or $2p$, would be more greatly contracted toward the nucleus? Can we still argue that $1s^2 2s^1$ is a lower-energy electron configuration than $1s^2 2p^1$?

8. What are the *spdf* configurations for Tl^+, Ca, Sc, O^+, Cr, Ni, Cu^+, Mn^{2+}, and Fe^{3+}?

9. What is the valence shell configuration for Tl? Pt^{2+}? Group III main-group elements? Group V main-group elements?

10. Why is it more difficult to remove the valence electron from K than to remove the valence electron from Na after the latter has been excited to the $4s$ level?

11. Why has Zn a smaller atomic volume than Ca?

12. Which atom in each pair has the higher experimental ionization potential (see p. 94) and why?

   S or P   Al or Mg
   K or Cu   Cs or At
   Ca or Rb

CHECKPOINTS

1. Atomic orbitals are matter waves that consist of
   a. a radial part
   b. an angular part
2. The radial wave determines
   a. the average distance of the electron from the nucleus (atom size)
   b. the effective positive nuclear charge felt by the electron
3. s electrons experience greater effective nuclear charge than do $p, d, \ldots$ electrons of the same n quantum number.
4. Electrons prefer to have different angular distributions because of
   a. charge correlation (Coulomb's law)
   b. spin correlation (like-spin particles avoid each other better than opposite-spin particles)
5. The electron-nuclear attraction is partially offset by electron-electron repulsion. This is handled qualitatively by invoking the idea of an effective nuclear charge,

less than the true atomic number. The effective nuclear charge for an electron depends on its n and ℓ quantum numbers.

6. The organization of the long form Periodic Table correlates directly with the wave model of the atom electron motions, under the premise that electrons move so as to maximize the effective nuclear charge each experiences. This results in the order of filling: $1s, 2s, 2p, 3s, 3p, 4s, 3d, 4p, \ldots$

Appendix to Chapter 1

Hydrogen Type Wavefunctions ($\rho = Zr/a_0$, $\xi = Z/a_0$)

$1s$	$(\xi^{3/2}/\sqrt{\pi})\,e^{-\rho}$	
$2s$	$(2 - \rho)$	
$2p_x$	ξx	
$2p_y$	ξy	$\times (\xi^{3/2}/\sqrt{32\pi})\,e^{-\rho/2}$
$2p_z$	ξz	
$3s$	$(27 - 18\rho + 2\rho^2)/\sqrt{3}$	
$3p_x$	$\xi\sqrt{2}(6 - \rho)x$	
$3p_y$	$\xi\sqrt{2}(6 - \rho)y$	
$3p_z$	$\xi\sqrt{2}(6 - \rho)z$	
$3d_{z^2}$	$\xi^2(3z^2 - r^2)/\sqrt{6}$	$\times (\xi^{3/2}/81\sqrt{\pi})\,e^{-\rho/3}$
$3d_{xz}$	$\xi^2\sqrt{2}(xz)$	
$3d_{yz}$	$\xi^2\sqrt{2}(yz)$	
$3d_{xy}$	$\xi^2\sqrt{2}(xy)$	
$3d_{x^2-y^2}$	$\xi^2(x^2 - y^2)/\sqrt{2}$	
$4s$	$(1 - 3\rho/4 + \rho^2/8 - \rho^3/192)/8$	
$4p_x$	ξx	
$4p_y$	ξy	$\times (\sqrt{5}/32)(1 - \rho/4 + \rho^2/80)$
$4p_z$	ξz	
$4d_{z^2}$	$\xi^2(3z^2 - r^2)$	
$4d_{xz}$	$\xi^2\sqrt{12}(xz)$	
$4d_{yz}$	$\xi^2\sqrt{12}(yz)$	$\times (1 - \rho/12)/256$
$4d_{xy}$	$\xi^2\sqrt{12}(xy)$	
$4d_{x^2-y^2}$	$\xi^2\sqrt{3}(x^2 - y^2)$	
$4f_{z^3}$	$(1/\sqrt{5})(5z^2 - 3r^2)z$	
$4f_{xz^2}$	$(3/\sqrt{30})(5z^2 - r^2)x$	
$4f_{yz^2}$	$(3/\sqrt{30})(5z^2 - r^2)y$	
$4f_{xyz}$	$(6/\sqrt{3})xyz$	$\times (\xi^3/3072)$
$4f_{z(x^2-y^2)}$	$\sqrt{3}(x^2 - y^2)z$	
$4f_{x(x^2-3y^2)}$	$(1/\sqrt{2})(x^2 - 3y^2)x$	
$4f_{y(3x^2-y^2)}$	$(1/\sqrt{2})(3x^2 - y^2)y$	

The $4d$, $4p$, $4s$ and $4f$ groups are multiplied by $\times (\xi^{3/2}/\sqrt{\pi})\,e^{-\rho/4}$.

REFERENCES

[1] N. Bohr, *Phil. Mag.*, **26**, 1, 476, 875 (1913).
[2] L. de Broglie, *Ann. de Phys.*, **3**, 22 (1925). de Broglie argued that, for a photon, $E = mc^2$ and $E = h\nu$ implies $mc = h\nu/c = h/\lambda$, where mc = the momentum of the photon. He felt that a similar relationship was appropriate for matter.
[3] C. J. Davisson and L. H. Germer, *Nature*, **119**, 558 (1927).
[4] E. Schrödinger, *Ann. Phys.*, **81**, 109 (1926).
[5] A more complete discussion of this is given in K. F. Purcell and J. C. Kotz, "Inorganic Chemistry," p. 25, W. B. Saunders Co., Philadelphia (1977).
[6] While it is true that "like spins" afford lower energy than "opposite spins," the origin for this energy difference is being closely examined. R. L. Snow and J. L. Bills, *J. Chem. Educ.*, **51**, 585 (1974) have argued that, in some atoms, the "like spin" case may have a lower energy because of a lower *kinetic energy*, not lower electron repulsion energy.
[7] E. C. M. Chen and W. E. Wentworth, *J. Chem. Educ.*, **52**, 486 (1975).
[8] An excellent discussion of the transition metal electron configurations is presented by R. M. Hochstrasser, *J. Chem. Educ.*, **42**, 154 (1965).

2. BASIC CONCEPTS OF MOLECULAR TOPOLOGIES

It is the goal of this and the following two chapters to develop the commonly used models for the electronic and geometric structures of molecules. We begin here with a review of basic electron accounting procedures familiar to you from a general chemistry course, and proceed to a model for predicting three-dimensional molecular structures and to the use of symmetry to classify molecular shapes. The succeeding two chapters will explore the application of the wave model to these structures.

SHARED AND LONE PAIRS AND LEWIS STRUCTURES

One of the very first concepts one encounters in describing the electronic structures of molecules is that of the shared pair of electrons defining a bond between two atoms. This "perfect pairing of electrons" concept of G. N. Lewis[1] is where we start in developing the hierarchy of models for molecular structure and models for chemical bonding.

Drawing Lewis structures for electron pairs in molecules is simply a matter of deciding how to partition the valence electrons of a molecule into pairs—either bond pairs to be shared by adjacent atoms or lone pairs to be localized about one atom. To make these decisions, the following guidelines are useful:

A. Given the molecular formula, what is the atom sequencing?
  1. Draw analogies with known, related compounds.
  2. Atoms with higher electronegativity (high Z^*) are terminal, as is hydrogen.
B. Given the atom sequencing, how does one assign bond and lone pairs?
  1. All adjacent atoms are connected by at least one bond pair (the sigma, σ, pair).
  2. Elements from the second period tend to have only 4 pairs (lone and/or bond) in their valence orbitals.
  3. Elements from the third and higher periods may exhibit expanded valence shell pairs of 5 or 6.
  4. If alternative structures are possible, compute the atom formal charges; zero formal charges are preferred and atoms of higher Z^* should bear negative, not positive, formal charge.
  5. For main-group elements, the number of valence electrons is given by the Group number. *The number of molecular valence electrons = the sum of the atom Group numbers, corrected for any ionic charge.*

Let us illustrate the application of these guidelines with some examples. A tetraatomic molecule like S_2Cl_2 could have one of two topographies

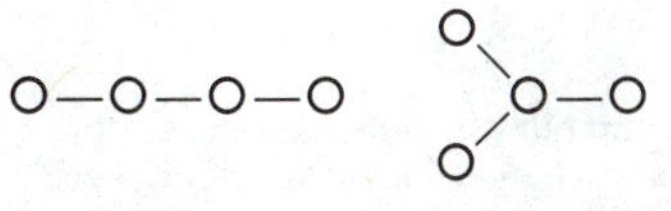

(chain)   (branched)

With reference to guide A-2, we expect sulfur to be central and chlorine to be terminal, where possible. Invoking guide B-1, we have

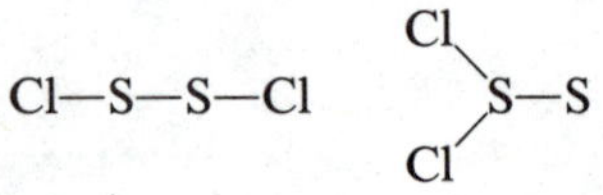

$$Cl-S-S-Cl \qquad \begin{array}{c} Cl \\ \diagdown \\ \diagup \\ Cl \end{array} S-S$$

Better known analogs for the first two topographies would be H_2O_2 and R_2SO. Following the octet rule for assigning the remaining 10 electron pairs (= 13 – 3 = $V - n_\sigma$), we determine the Lewis structures:

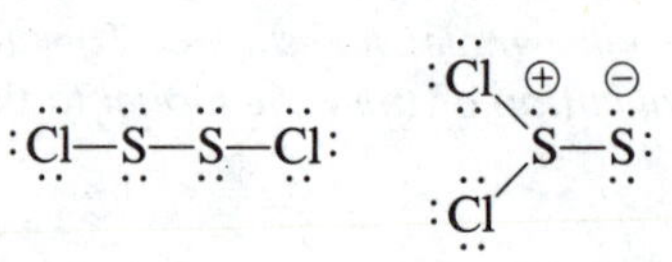

$$:\ddot{Cl}-\ddot{S}-\ddot{S}-\ddot{Cl}: \qquad \begin{array}{c} :\ddot{Cl} \quad \oplus \quad \ominus \\ \diagdown \\ \diagup \quad S-\ddot{S}: \\ :\ddot{Cl} \end{array}$$

$n_\sigma = 3$	$n_\sigma = 3$	($n_\sigma \equiv$ number of σ electron *pairs*)
$n_\pi = 0$	$n_\pi = 0$	($n_\pi \equiv$ number of π electron *pairs*)
$n_\varrho = 10$	$n_\varrho = 10$	($n_\varrho \equiv$ number of lone electron *pairs*)
$V = 13$	$V = 13$	($V \equiv$ number of valence electron *pairs*)

Of the two structures, only the first is found in the laboratory.

The sulfoxide-like structure shows atom formal charges. *The **formal charge** of an atom is defined as the sum of atomic valence core charge* (always positive and equal to the ion charge developed upon removal of all valence electrons†) *and valence electron charge* (always negative and equal to the number of lone electrons plus half the number of shared electrons) about the atom.§ For atom a:

$$(\text{F.C.})_a = Z_a^{core} - (2n_l + n_\sigma + n_\pi)_a$$

As another example, consider the compound of molecular formula NOH_3. The hydrogen atoms may be assumed to be terminal atoms, so we have the two possible topographies:

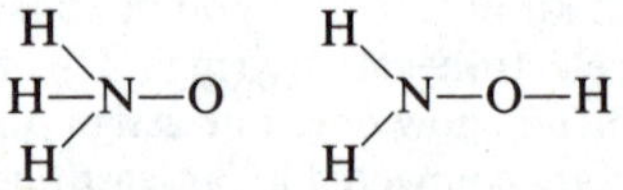

$$\begin{array}{c} H \\ \diagdown \\ H-N-O \\ \diagup \\ H \end{array} \qquad \begin{array}{c} H \\ \diagdown \\ N-O-H \\ \diagup \\ H \end{array}$$

(You may wish to consider the topography H_2ONH on your own.) Following the octet rule, we get

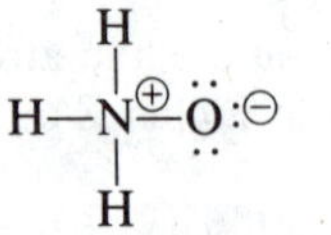

$$\begin{array}{c} H \\ | \\ H-N\overset{\oplus}{}\ddot{O}:^{\ominus} \\ | \\ H \end{array} \qquad \begin{array}{l} n_\sigma = 4 \\ n_\pi = 0 \\ n_l = 3 \end{array} \qquad H-\ddot{N}-\ddot{O}-H \\ \qquad\qquad\qquad\qquad | \\ \qquad\qquad\qquad\qquad H$$

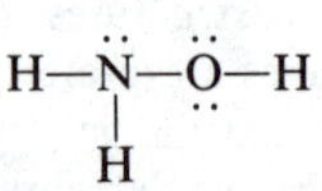

Nature prefers the second of these (hydroxyl amine).

Finally, we examine thiosulfate, $S_2O_3^{2-}$. There are many topographical possibilities for a five-atom system. Perhaps the most obvious choice is the one with a four-branch

---

†Simply identifying the group number tells the valence core charge. Note that summing the atom formal charges gives the molecular charge. This is a useful check to avoid "lost" or excess electrons.

§Note that "formal charge" is not the same as "oxidation number," which is defined by a different electron book-keeping scheme.

point (the tetrahedron). Analogy with SO_4^{2-} (hence the name thiosulfate) strongly favors this choice (note that all atoms come from Group VI). Following the octet rule, we develop

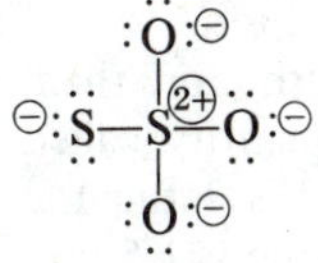

Further examples are given in Table 2-1. Perhaps needless to say, the "derivation" of possible molecular topographies by this procedure is cumbersome for large molecules; after a little experience one learns to rely heavily on analogy with simpler topographies and to build up complex ones from these.

TABLE 2-1

FORMAL CHARGES

Species	Structure		Formal Charge
SF_4 (Sulfur tetrafluoride)	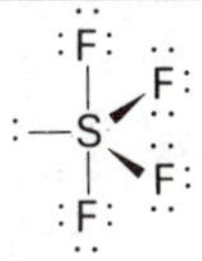	S F	0 0
XeF_2 (Xenon difluoride)	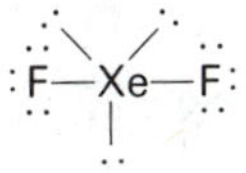	Xe F	0 0
NO (Nitrogen oxide)	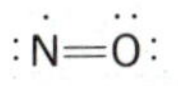 $:N{=}O:$	N O	0 0
H_3NO (Hydroxyl amine)	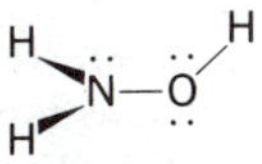	N O	0 0
$POCl_3$ (Phosphoryl chloride)	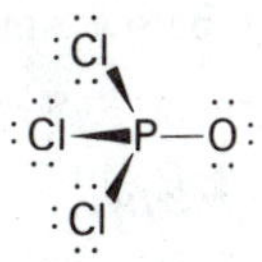	P Cl O	+1 0 −1
NO_2^+ (Nitronium ion)	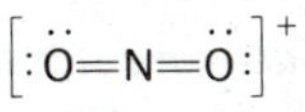 $\left[:O{=}N{=}O:\right]^+$	N O	+1 0
$S_2O_8^{2-}$ (Peroxydisulfate ion)	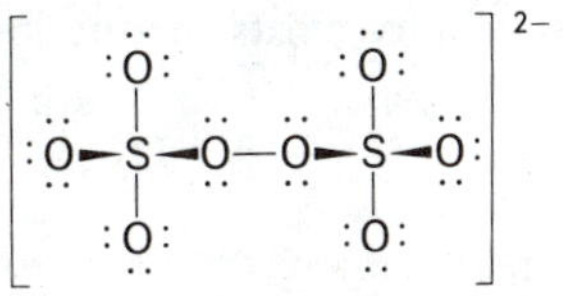	S $O_c{}^*$ $O_t{}^*$	+2 0 −1

*The subscripts c and t stand for central and terminal.

In drawing electron dot structures for molecules with $n_\pi > 0$, one frequently encounters the possibility of writing more than one such structure. Often the possibilities are equivalent, as in the case of NO_2^-. (Note that the net oxygen formal charges = −0.5.)

In connection with such structures with more than one shared pair per pair of atoms, the chemist finds it useful to define a quantity called the **bond order** for two atoms. The bond order for two connected atoms is defined as *the average number of shared pairs.* In the NO_2^- example, each NO bond order is 1.5.

Another example arises for the boron trihalides; in BF_3, for example, the fluorine atoms are equivalent. This is an interesting case; satisfying the octet rule for boron requires net F formal charges between 0 and +0.33. Three equivalent structures need to be given, and each BF bond order is between 1.0 and 1.33.

A more complex example of resonance structures is found in $S_2O_3^{2-}$. The thiosulfate anion with the "all octet" resonance structure yields a rather excessive formal charge for sulfur of +2. Many more structures may be written to reduce the sulfur charge by allowing some *bonding character for lone pairs;* this is done by designating multiple bonds between the central sulfur and the terminal sulfur and oxygen atoms.

From the last example given, it can be appreciated that some guidelines must be followed to estimate the relative importance of the structures and thus limit the often unbelievably large number of structures that can be written for some molecules. Four such guidelines are listed below and illustrated with one final example, that of thio-cyanate ion, SCN^-.

1. *Do not change the relative positions of the nuclei in writing resonance structures.* To do so involves writing structures for different isomers of the same molecular formula (refer to H_3NO, H_2NOH; ClS_2Cl, Cl_2SS).
2. *All structures must have the same number of unpaired electrons.* That is, do not draw one structure with no unpaired electrons and another with two unpaired electrons. Only one possibility exists for the lowest energy electron configuration of the molecule.

3. Generally, *adhere to the octet or inert gas rule*, but remember that elements from the third or later periods may have expanded octets.
4. Generally, *zero formal charge at each atom is preferred.* In deference to the other rules above, and certainly for ions, it may be reasonable to write structures with non-zero atom formal charges. *Structures that locate negative formal charge at the more electro-negative atoms are to be preferred.* In such cases, *avoid a large separation of opposite charge and close proximity of like charges.*

Now let's consider several structures that one might draw for SCN^-, and estimate with the aid of the preceding guides the importance of each. We have (I), (II), and (III).

(I) $\overset{\ominus}{:}\ddot{S}{-}C{\equiv}N:$ (II) $:\ddot{S}{=}C{=}\ddot{N}:^{\ominus}$ (III) $\overset{\oplus}{:}S{\equiv}C{-}\ddot{N}:\textcircled{2-}$

Structures (I) and (II) are sound choices, as they adhere to all the rules. (II) is preferred to (I) by item 4. That at least one atom must carry a formal negative charge is dictated by the fact that the species is an anion. *A useful check of any structure drawn is that the sum of the atom formal charges must equal the overall charge on the molecule.* This is simply a consequence of the fact that the algebraic sum of core charges and electron charges must equal the molecule charge. (III) is judged by item 4 not to be very important.

With structures (I) and (II) as the only important ones, and with (II) somewhat more important than (I), we may estimate the SC bond order to be between $1\frac{1}{2}$ and 2. The sulfur formal charge is taken to be less negative than -0.5, while the nitrogen atom formal charge is estimated to be a little more negative than -0.5.

Before continuing to the next section, it is well to point out here that formal charges are often somewhat unrealistic and that use of them in chemical arguments must be made judiciously. The difficulty lies in the convention of considering each shared pair to be *equally* shared by the two atoms in question. For example, reconsidering the molecule BF_3, we estimate from the three double-bond and one single-bond resonance structures that the boron atom carries a negative formal charge (between zero and -1), while each fluorine bears a positive formal charge (between zero and $+0.33$). The assumption that the BF σ pair is equally shared is not a fair representation of the electron distribution in the BF_3 molecule. It is for this reason that the phrase "formal charge" has been used, and such charges should rarely be taken to give an accurate representation of atom charges.

1. Two of the three possible isomers of the anion NCO^- are known (cyanate and fulminate). Develop Lewis structures for these, and compare atom formal charges and bond orders.

2. Draw a Lewis dot structure and give the atom formal charges for $XeOF_4$. Repeat for XeO_3 and ICl_2^+.

3. Draw the important resonance structures for

   (a) NNO (nitrous oxide)
   (b) allyl radical (H_2CCHCH_2)
   (c) F_2SO (thionyl fluoride)
   (d) N,N-dimethylacetamide

   (e) HOCN and OCN^-; how do you expect the OC and CN bond strengths to differ in these two molecules?

STUDY QUESTIONS

ELECTRON PAIR REPULSION MODEL

The next step in the construction of a working model of molecular structure concerns the three-dimensional aspects of structure. Here we will use the qualitative but highly useful idea that geometries of molecules can be predicted from consideration of repulsions between pairs of electrons about an atom.[2] Simply put, *we assume (i) that each valence electron pair of an atom is stereochemically significant, and (ii) that repulsions between these pairs determine molecular shape.* The orientations of bond pair electrons will serve to define the spatial arrangement of nuclei in a molecule.

The angular distribution of valence electron pair clouds about a nucleus follows the dictates of some force law which must reflect the repulsive forces between electrons.[3] The exact form of the force law ($F = e^2/r^2$ is appropriate for *point* charges) seems to be of little importance for two to six such electron pairs; for seven pairs, however, a unique result is not possible. Table 2-2 summarizes the results of this model. The first six basic structural types are common; you should become facile in recognizing into which basic structural pattern a molecule will fall, once you have written an electron dot structure for the molecule.

TABLE 2-2

IDEALIZED ORIENTATIONS OF e⁻ PAIRS ABOUT AN ATOM

No. of e⁻ pairs	Idealized Structure of e⁻ pairs		Angles
2	linear	:—E—:	180°
3	trigonal plane	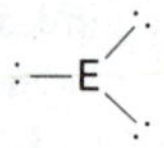 :—E⟨	120°
4	tetrahedron	 ⟩E⟨	109.5°
5	trigonal bipyramid	 :—E⟨	90°, 120°
6	octahedron	 ⟩E⟨	90°
7	pentagonal bipyramid	 ⟩E═:	90°, 72°
	capped octahedron "1:3:3"		
	capped trigonal prism "1:4:2"	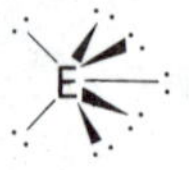	

Considerable refinement of these basic structural units can be made by distinguishing between lone pairs of electrons and bond pairs of electrons. With good justification, *we assume that the spatial requirements of lone pairs are greater than those of bond pairs.* Such a premise is not unreasonable when it is realized that lone pairs move under the effective nuclear charge of only one atom, while bond pair electrons experience the effective nuclear charges of two atoms. We expect bond pairs to be less angularly diffuse and more radially distended from the nucleus than lone pairs about an atom.

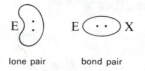

lone pair   bond pair

An equivalent view of this difference between lone and bond pairs is that the amount of electron density near the valence core is different. Lone pairs are assumed to be localized at the atom in question and thus create two negative charges near the valence core. Bond pairs are shared, with the result that the net electron density near a core is less than two electrons. Consequently, *we expect the order of repulsion between pairs to decrease as*

$$lp/lp \gg lp/bp > bp/bp$$

| lp/lp | lp/bp | bp/bp |

A final aspect of the model concerns the interpair angles in Table 2-2. The repulsive forces should markedly decrease with increasing angle. Thus, while repulsions may be significant at all angles between 90° and 180°, they are much more severe at 90° than at 120° and are lowest at 180°. As you will see in the following examples, repulsions at 90° provide a key to molecular geometry. All the foregoing conceptualizations and assumptions define what is referred to as the *valence shell electron pair repulsion model* (**VSEPR model**).

A consequence of these hypotheses, considering the stereochemistry of four pairs for example, is that four lone pairs or four bond pairs will be distributed with interpair angles of 109° 28′. If one of the pairs is a lone pair while the other three are bond pairs, the forces acting on the bond pairs will not be balanced in the regular tetrahedron configuration; the bond pairs will be forced together slightly until the forces on all pairs do balance. In this case we expect the *interbond pair angles* to be less than 109.5°.

To give two examples of the utility of the electron repulsion concept in distinguishing structural isomers, we will consider ICl_4^- and ClF_3. In drawing an electron dot structure for ICl_4^-, one quickly sees that there are six electron pairs about the iodine atom.

Consequently, the basic stereochemical arrangement for the six pairs is the octahedron. Interestingly, two possibilities arise when a stereochemical view of the ion is sketched. One possibility is that all four chlorines are coplanar with the iodine (see sketch A below). The other possibility has one chlorine perpendicular to the plane defined by the iodine and the remaining three chlorine atoms. The observed structure is the square planar arrangement of atoms; this is nicely rationalized by the electron repulsion model, which predicts a higher molecular energy for the juxtaposition of two lone pairs at 90° to each other (as in sketch B) than for their separation by 180°.

In making the valence shell electron pair repulsion (VSEPR) analysis of the two structures, we list below each structure the number of pair interactions of each type (all at 90° in the ICl_4^- example). On comparison of (A) with (B), we see that two lp/bp repulsions in (A) are replaced by one lp/lp and one bp/bp interaction. The lp/lp repulsion at 90° is expected to be quite large and disfavors (B).

lp/lp = 0	→ 1 = lp/lp
lp/bp = 8	6 = lp/bp
bp/bp = 4	→ 5 = bp/bp

The second example, that of ClF_3, leads to the prediction of the basic trigonal bipyramidal orientation of electron pairs of chlorine. Three isomeric structures are encountered: (A), the two lone pairs equatorial; (B), one lone pair equatorial and the other axial; and (C), both lone pairs axial. The observed structure is that of a T-shaped molecule, i.e., the first possibility. The summary of interpair repulsion types below each sketch should make it clear that the order of stability is (A) > (B) or (C). In making the VSEPR analysis of ClF_3, we focus our attention on pair interactions at 90°, recognizing that the interactions at 120° will be much less important. The (A) → (B) comparison reveals conversion of a lp/bp repulsion into a lp/lp repulsion. For (A) → (C) we find conversion of two bp/bp interactions into two of the lp/bp type. Hence, (A) is preferred.

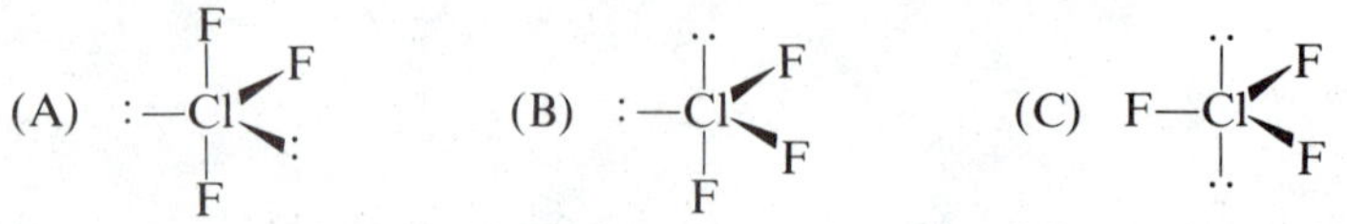

(A)

lp/lp = 0
lp/bp = 4
bp/bp = 2

(B)

lp/lp = 1
lp/bp = 3
bp/bp = 2

(C)

lp/lp = 0
lp/bp = 6
bp/bp = 0

As a finer point of the geometry of (A), we expect the axial Cl—F bonds to be somewhat longer than the equatorial bond because the axial pairs are acted upon by two lp and one bp (at 90°), whereas the equatorial pair is acted upon by only two bp (at 90°). Similarly, the FClF angles should be less than 90°. Both expectations are borne out by the experimental structure (Cl—F_{eq} = 160 pm, Cl—F_{ax} = 170 pm; FClF = 87.5°).

Be sure to note that in the preceding examples we have carefully distinguished between *molecular structure* and *electron pair orientations*. The former describes the relative nuclear positions and is subject to experimental determination; electron pair orientations cannot be experimentally determined.

Thus far we have dealt only with examples for which we conceive of a single bond pair. The occurrence of double and triple pairs in a given internuclear region certainly enhances repulsions between that region's bond pair(s) and other pairs. Note that second and third pairs do not define new stereochemical directions, for their orientation mimics that of the first pair. An interesting example of this is found in the series NO_2^+, NO_2, and NO_2^-, which differ progressively by one electron and by a large reduction in NO bond order from NO_2^+ to NO_2. The most important resonance structures, combined with the electron repulsion model, lead us to expect—in agreement with experiment—a progressively decreasing ONO angle over the series.

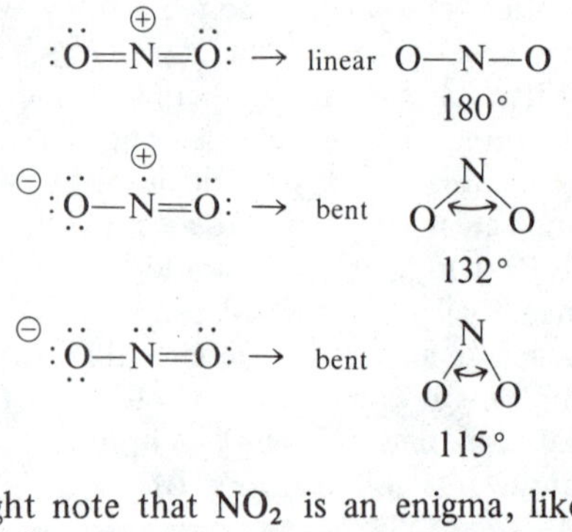

$$:\overset{..}{O}=\overset{\oplus}{N}=\overset{..}{O}: \rightarrow \text{linear } O—N—O$$
180°

$$\overset{\ominus}{\underset{}{:}}\overset{..}{O}—\overset{\oplus}{\overset{.}{N}}=\overset{..}{O}: \rightarrow \text{bent}$$
132°

$$\overset{\ominus}{:}\overset{..}{O}—\overset{..}{N}=\overset{..}{O}: \rightarrow \text{bent}$$
115°

As an aside, you might note that NO_2 is an enigma, like NO, when one tries to satisfy the octet rule for all atoms. To avoid exceeding the octet for oxygen the odd electron is assigned to nitrogen.

A more subtle effect of π bonding arises for atoms which possess a lone pair of electrons in one resonance form, and for which this lone pair becomes a π bond pair in another, important structure. The near planarity of formamide is an excellent example:

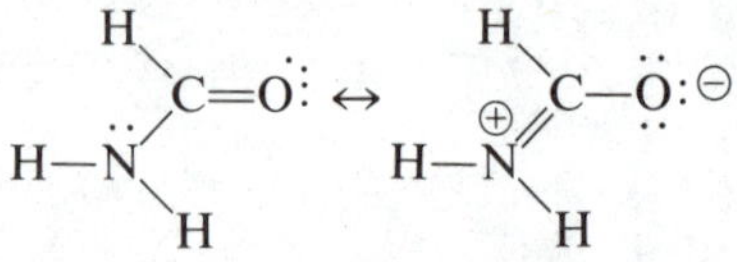

The N—C resonance form suggests a pyramidal structure for ($H_2\overset{..}{N}C\cdots$), while the N=C form corresponds to a planar arrangement for ($H_2\overset{..}{N}=C\cdots$).

4. Sketch the possible isomers of SF_4, predict the stable isomer, and describe the FSF angles as $>$, $=$, or $<$ the "ideal" angles.

5. Which in each pair will have the larger bond angle?

   (a) CH_4 or NH_3
   (b) NO_2^- or O_3

6. Noting that Si (as a third period element) may exhibit an expanded octet, show how $N{=}Si$ resonance structures arise and how they may be used to explain the planarity of the NSi_3 skeleton of $N(SiH_3)_3$ (trisilylamine).

7. Describe the bond angles of $XeOF_4$ in relation to the "ideal" angles. Repeat for $SnCl_3^-$.

SYMMETRY CONCEPTS[4]

Point Groups

Now that we have discussed a working hierarchy of models involved in reaching a prediction of a molecule's structure, it is convenient to introduce a formal system for

Figure 2-1. Two examples (*Circle Limit IV* and *Whirlpools*) of symmetry in art from M. C. Escher, famous for his graphic (and sometimes paradoxical) designs. (With permission of the Escher Foundation—Haags Gemeentemuseum—The Hague.)

Illustration continued on p. 46

Figure 2-1. *Continued.*

classifying molecules according to their structures. To be specific, we will characterize a molecular structure in terms of the elements of symmetry that it possesses and introduce a system of special symbols, called **point group symbols**, that convey all the symmetry elements possessed by a molecule. *Molecular symmetry elements are planes, axes, and points that relate equivalent nuclear positions.* Every molecule belongs to a **point group**, which is defined by the numbers and kinds of symmetry elements (called a group of elements) having the property of a common intersection, usually a point (hence the name point group).

Fortunately, there are only two basic symmetry elements used to classify the structure of a molecule—the **proper rotation axis** (C_n) and the **improper rotation axis** (S_n), as defined in Table 2–3. The special cases of C_1, S_1. and S_2 are commonly identified as the **identity** (E), the **reflection** (σ), and the **inversion** (i), respectively. Notice that subscripts (to be explained later) are given for the reflection element.

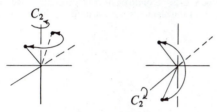
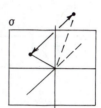

proper rotation axes

$$S_2 = \sigma^{\perp} \cdot C_2 = i$$

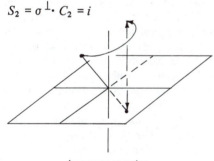
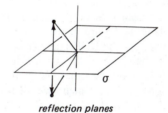

reflection planes

improper rotation

TABLE 2–3

MOLECULAR SYMMETRY ELEMENTS, OPERATIONS, AND SCHOENFLIES SYMBOLS

Element	Operation	Symbol
Axis	Rotation by 2π	E^*
	Rotation by $\dfrac{2\pi}{n}$	C_n
Axis $+$ $\perp$ plane	(Improper rotation)	S_n
	Rotation by $\dfrac{2\pi}{n}$ followed by reflection through a plane perpendicular to rotation axis	
Plane	Reflection	σ^* ($\sigma, \sigma_v, \sigma_d, \sigma_h$)
Point	Inversion	i^*

*Note that E is a special case of C_n, and that i and σ are special cases of S_n: $S_1 = \sigma$ and $S_2 = i$.

We will now proceed to examine some specific examples. The molecule PF_2H is predicted by the electron repulsion scheme to be a pyramidal molecule, as shown in

Figure 2-2. Only one symmetry element, other than the identity, can be identified here—a reflection plane containing the phosphorus and hydrogen atoms and bisecting the F—P—F angle. A molecule possessing these two elements, and only these two, is said to belong to the C_s point group. Conversely, a molecule with C_s symmetry possesses only these two symmetry elements. A molecule with no symmetry beyond the identity is classified as C_1; an example is $NDHCH_3$, shown in Figure 2-2.

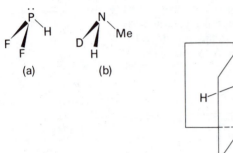

(a)   (b)

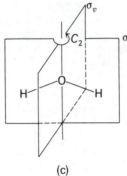

(c)

Figure 2-2. Examples of (a) C_s and (b) C_1 symmetry, and (c) the elements constituting C_{2v} symmetry.

The water molecule is the next example. As shown in Figure 2-2, H_2O possesses three distinct symmetry elements beyond the identity. First, there is the two-fold rotation axis (after rotation by 180° about this axis one could not tell that the operation had been effected); secondly, there is a reflection plane coincident with the molecular plane; and thirdly, there is a reflection plane perpendicular to the molecular plane and bisecting the HOH angle (the space to one side of a reflection plane is the mirror image of the space to the other side). As the two reflection planes are *not* equivalent (related by any of the other symmetry elements of the group), they are given separate designations. The first plane we identify as σ'_v, and the second plane is given the symbol σ_v to distinguish it from the first; the axis is called C_2. The collection of symmetry elements E, C_2, σ_v, and σ'_v is referred to as the C_{2v} point group. The Schoenflies point group symbol itself carries the complete information as to which elements are present, that is, the C_2 rotation axis and two σ_v planes. *The meaning of the subscript v (for vertical) for the reflection planes is that the planes contain (their intersection defines) the C_2 rotation axis.* Note that the "v" subscript was not used to describe the plane in the C_s point goup, which has no rotation axis. The molecule ClF_3 is another example of C_{2v} symmetry (see structure A, page 44).

The next example is BF_3. On the basis of the VSEPR model, the molecule is predicted to be precisely trigonal planar. As shown in Figure 2-3, there are several symmetry elements associated with the trigonal planar structure. First, there is the three-fold· rotation axis, C_3. Actually there are two coincident three-fold axes, one corresponding to the operation of clockwise rotation and the other to the operation of counterclockwise rotation of the molecule by 120°. These are related by each of the C_2 elements to be discussed next (any of the C_2 operations converts a clockwise C_3 arrow into a counterclockwise arrow), and are therefore not distinguished from each other but are identified as the class $2C_3$. *Elements interconverted by other operations define a* **class**, *designated by* n(*Schoenflies symbol*). Unlike the σ_v and σ'_v elements of C_{2v}, such elements share a common designation (Schoenflies symbol).

Second, there are the two-fold rotation axes defined by the BF internuclear axes. As these are related by the C_3 axes, they are not separately designated but are identified as the class $3C_2$. There are no further proper rotation axes.

Examining the predicted structure for reflection planes, we find several. Perhaps the most obvious is the molecular plane. *As this plane is perpendicular to the highest-fold rotation axis*, which is the C_3 axis, *we designate it as σ_h* (for horizontal) *to avoid confusion with reflection planes*, identified with the v subscript, *which contain the highest-*

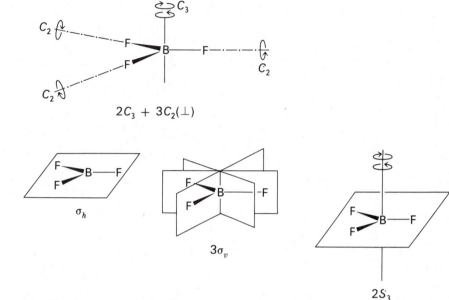

$$2C_3 + 3C_2(\perp)$$

Figure 2–3. The elements constituting D_{3h} symmetry.

fold rotation axis. In the vertical category there are three reflection planes, all related by the C_3 element, and these are identified as the class $3\sigma_v$. There is no inversion element for this molecule, but there are further improper rotation elements. The C_3 axis combined with the σ_h reflection plane serves to define S_3. As there are two of these (clockwise and counterclockwise C_3 rotation), and they are related by the C_2 axes, they are designated collectively as the class $2S_3$. The identity (E) completes the specification of symmetry elements for BF_3. This collection of elements is identified by the symbol D_{3h}. *The symbol D_n is meant to convey the presence of an n-fold rotation axis, C_n, and n 2-fold axes perpendicular to C_n.* The presence of a horizontal reflection plane is identified by adding the subscript h to the point group symbol. The combination of C_n, n 2-fold axes perpendicular to it, and σ_h guarantees the existence of n σ_v planes (see Study Questions). The presence of σ_h combined with C_n also guarantees the presence of several S_n elements. Hence, the basic information of a C_n, n 2-folds perpendicular to C_n, and the σ_h is all that is needed to characterize the collection of elements for BF_3, and the point group symbol becomes D_{3h}. Compare this with the pyramidal molecule NF_3, which lacks σ_h and n C_2 axes perpendicular to C_3. Only E, $2C_3$, and $3\sigma_v$ are left; therefore, the point group symbol is C_{3v}.

Our final detailed example is for the ICl_4^- molecule (Figure 2–4). The considerations of the VSEPR model led us to predict that this molecule is planar with equivalent I—Cl bonds. The highest-fold rotation axis is C_4. There are two of these (clockwise and counterclockwise rotations by 90°) related by any of four perpendicular rotation axes, so they are grouped together as the class $2C_4$. Coincident with the C_4 elements, there is a C_2 axis. In the plane of the molecule we find four C_2 axes, which may be grouped into two pairs (members of different pairs are not symmetry-related). To avoid confusion, one pair (those along the I—Cl axes) are designated as $2C_2'$, while the other pair (those bisecting the Cl—I—Cl axes) are represented by $2C_2''$.

For reflection planes, we find first the molecular plane, called σ_h in keeping with the convention encountered with BF_3. *Recognition of these elements is sufficient to identify the point group as D_{4h}.* The four reflection planes defined by the C_4 axis and the perpendicular C_2 axes (and required by C_4, $4 \perp C_2$, and σ_h) may be grouped into pairs corresponding to the two pairs of C_2 axes. One pair is identified as $2\sigma_v$ and the other pair is identified as $2\sigma_d$. Out of deference to the fact that the second pair of reflection planes has the special property of *bisecting the angles between equivalent C_2 axes* (the

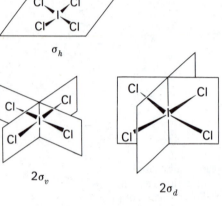

$C_2 + 2C_4$

$2C'_2 + 2C''_2$

Figure 2-4. The planes and axes of D_{4h} symmetry.

σ_h

$2\sigma_v$

$2\sigma_d$

pair $2C'_2$), they are given the special d (for diagonal) subscript rather than a prime superscript as in the H_2O example. There is a center of inversion (at the iodine atom) for this molecule (a result of $C_2 + \sigma_h$), and we identify the element i. Finally, there are two S_4 axes, defined by the two C_4 axes and σ_h, and we group them as $2S_4$. This collection, including the identity, is given the point group symbol D_{4h}. The analogy with the D_{3h} example just given is appropriate. With an even value for n of D_{nh}, there must be a C_2 coincident with C_n, and this requires the presence of i (or S_2).

To summarize our discussion, thus far we have identified the special point groups C_1 (which consists of only the identity, E) and C_s (which consists of the identity, E, and a single reflection plane σ). The two general point groups C_n and D_{nh} were discussed because they are appropriate for a great number of molecular structures. Two further special cases need to be mentioned: the tetrahedron, with point group symbol T_d, and the octahedron, with symbol O_h. A summary of the elements of these point groups is given in Table 2-4. Figures that will help in location of the symmetry elements are included for the tetrahedron and octahedron.

For practice in identifying the symmetries of commonly occurring molecular structures, Table 2-5 gives examples of molecules identified by the number of stereochemically active electron pairs and by point group. You should familiarize yourself with the content of this table, paying particular attention to the relation of the stereochemical structural types to their associated point groups.

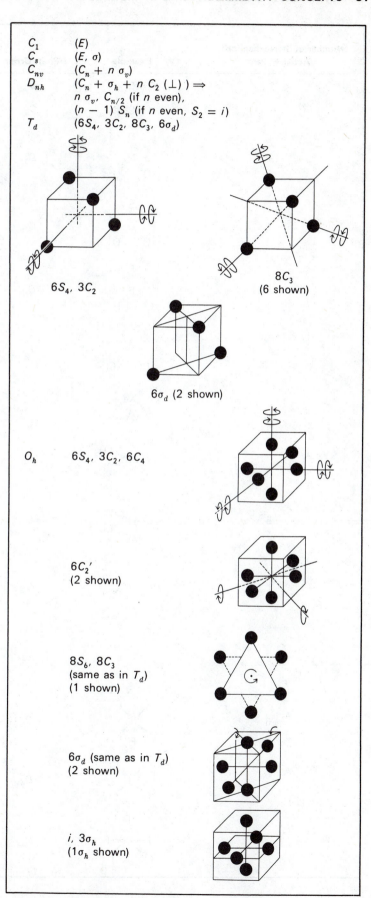

C_1 (E)
C_s (E, σ)
C_{nv} $(C_n + n\,\sigma_v)$
D_{nh} $(C_n + \sigma_h + n\,C_2\,(\perp))\Rightarrow$
           $n\,\sigma_v,\; C_{n/2}$ (if n even),
           $(n-1)\,S_n$ (if n even, $S_2 = i$)
T_d $(6S_4,\ 3C_2,\ 8C_3,\ 6\sigma_d)$

$6S_4,\ 3C_2$

$8C_3$
(6 shown)

$6\sigma_d$ (2 shown)

TABLE 2–4

**COMMON POINT GROUPS
AND THEIR ELEMENTS**

O_h $6S_4,\ 3C_2,\ 6C_4$

$6C_2'$
(2 shown)

$8S_6,\ 8C_3$
(same as in T_d)
(1 shown)

$6\sigma_d$ (same as in T_d)
(2 shown)

$i,\ 3\sigma_h$
($1\sigma_h$ shown)

Number of Stereochemically Active e^- Pairs	Example	Point Group
2	$O{=}C{=}O$	$D_{\infty h}$
	$H{-}C{\equiv}N$	$C_{\infty v}$
3	$\left[\begin{array}{c} O \\ \| \\ N \\ O \quad\quad O \end{array}\right]^{-}$	D_{3h}
	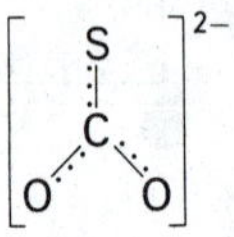 $\left[\begin{array}{c} S \\ \| \\ C \\ O \quad\quad O \end{array}\right]^{2-}$	C_{2v}
	$\begin{array}{c} F \\ \| \\ B \\ Cl \quad H \end{array}$	C_s
4	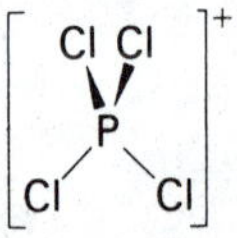 $\left[\begin{array}{c} Cl \quad Cl \\ P \\ Cl \quad Cl \end{array}\right]^{+}$	T_d
	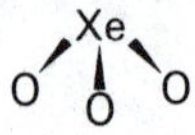 $\begin{array}{c} Xe \\ O \quad\ O \\ O \end{array}$	C_{3v}
	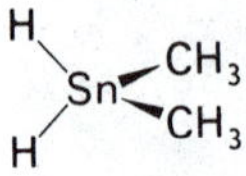 $\begin{array}{c} H \qquad CH_3 \\ Sn \\ H \qquad CH_3 \end{array}$	C_{2v}
5	$\begin{array}{c} F \\ \| \\ Cl{-}P{-}Cl \\ \| \\ F \end{array}$ (Cl)	D_{3h}
	$\begin{array}{c} F \\ \| \\ Cl{-}P{-}Cl \\ \| \\ Cl \end{array}$ (Cl)	C_{3v}
	$\begin{array}{c} F \\ \| \\ Cl{-}F \\ \| \\ F \end{array}$	C_{2v}
6	 $\begin{array}{c} F \quad F \\ S \\ F \quad F \\ F \quad F \end{array}$	O_h
	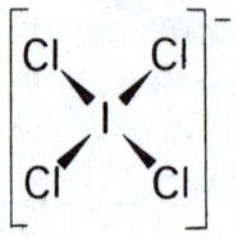 $\left[\begin{array}{c} Cl \quad Cl \\ I \\ Cl \quad Cl \end{array}\right]^{-}$	D_{4h}
	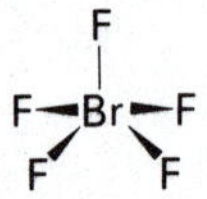 $\begin{array}{c} F \\ F{-}Br{-}F \\ F \quad F \end{array}$	C_{4v}

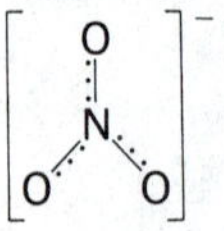

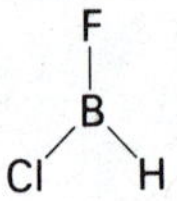

TABLE 2-5

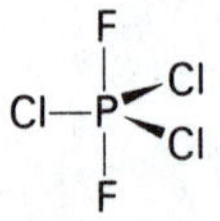

EXAMPLES OF STRUCTURE-POINT GROUP RELATIONS

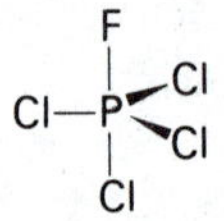

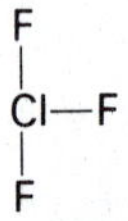

The following flow chart will help you place molecules into the point groups C_1, C_s, C_n, C_{nv}, D_n, and D_{nh}.

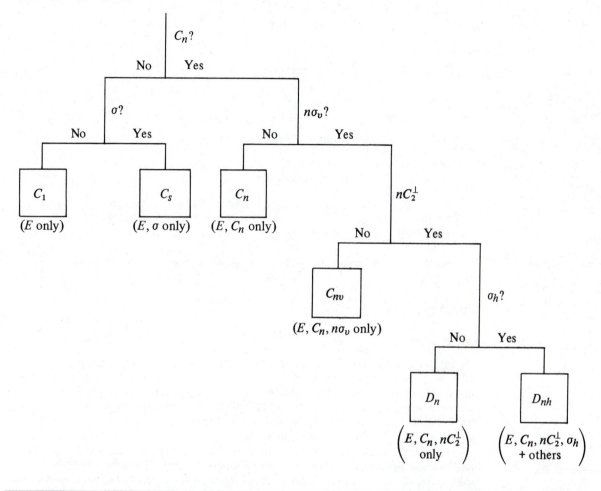

8. State the point groups of the following species.

   (a) XeF_2
   (b) HOCN
   (c) an atomic d_{z^2} orbital
   (d) $SnCl_3^-$
   (e) BrF_5
   (f) FNH_2
   (g) Cl_3PO

   (h) the stable isomer of ICl_2^-
   (i) the stable isomer of ClF_3
   (j) $XeOF_4$
   (k) SF_4
   (l) ICl_2^+
   (m) XeO_3F_2

STUDY QUESTIONS

9. How many chlorines of CCl_4 would need to be replaced with hydrogens to give molecules of C_{3v} and C_{2v} symmetries? Repeat for BF_3, to give C_{2v} symmetry. Are any of these molecules without S_n axes (remember $S_1 = \sigma$, $S_2 = i$) and therefore optically active?

10. Using the adjacent figure, demonstrate by successive application of C_3, $C_2^\perp$, and σ_h that the result is identical to that achieved by a σ_v alone (proving that C_3, $C_2^\perp$, σ_h imply the presence of σ_v). The flag rises out of the plane of the paper.

11. What point group describes the symmetry of each of the following atomic orbitals?

   (a) p_z
   (b) p_x
   (c) d_{z^2}

   (d) d_{xz}
   (e) $d_{x^2-y^2}$

12. Using the coordinate system below and centering the first orbital in each pair at point 1 and the second orbital at point 2, state whether the orbitals have a common symmetry for the operation given.

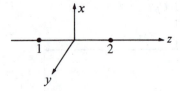

(a) s, p_z; $180°$ rotation about the z axis
(b) s, p_z; reflection in the yz plane
(c) p_x, s; reflection in the xz plane
(d) p_x, s; $180°$ rotation about the z axis

(Hint: Each function keeps the same phase or changes phase (i.e., sign) after the operation is executed. If both functions behave the same way, they have the same symmetry.)

IONIC SOLIDS

Lattice Energies

In these sections of the chapter our attention will focus on compounds that are stable as crystalline materials at ordinary conditions, with specific interest in crystalline forms of ionic compounds rather than those of uncharged compounds. The latter usually have relatively low condensation energies, which are determined by weak London and dipole-dipole forces between molecules; consequently, they have lattice energies that are not particularly decisive in the course of a chemical reaction.

As a point of contrast with the structures of covalent species, the structures of ionic materials are determined by the concept of sphere packing, rather than by stereochemically-active electron pair interactions. This is because the forces between charged atoms and molecules are largely of the non-directional, coulombic type, there being little sharing of electron pairs between such ions. The electrostatic forces acting to hold ions together in an ordered array are quite large, as you will see in the section of Chapter 6 on condensation energies for the alkali metal fluorides (p. 173); in fact, on a formula weight basis the electrostatic energy of interacting ions is comparable to the "covalent" bond energies in Table 6–3.

The electrostatic attraction between two *point charge* ions separated by a distance d is given by Coulomb's Law:†

$$E = q_+ q_- e^2 / 4\pi \epsilon_0 \, d$$

For a mole of such ion pairs, the energy is simply Avogadro's number times the energy for a single pair. The Coulomb's Law approach has to be incomplete because no allowance is made for the repulsion between the electron clouds of the ions, and this repulsion becomes increasingly important as the two ions draw closer together. (This is the same force responsible for steric effects in covalent molecules and is what limits the shortness of covalent bonds.) This repulsion energy has been treated by Born in terms of the

---

†This form is appropriate for the SI system; $\epsilon_0 = 8.854 \times 10^{-12}$ $C^2 m^{-1} J^{-1}$. See N. H. Davies, *Chem. Britain*, 7, 331 (1971). Corrections appear in *ibid.*, 8, 36 (1972).

function B/r^n, where n is characteristic of each cation/anion pair and can be determined by compressibility measurements. Incorporation of this repulsion term into the expression for the energy of attraction of two ions leads to the equation:

$$E = Ne^2 q_+ q_- / 4\pi\epsilon_0 \, d + B/d^n$$

for a mole of ion pairs. The equilibrium value of d is determined by the balance of electrostatic attraction and repulsion between the ions; minimization of E with respect to d, by setting $dE/dd = 0$, gives a relation between the coefficient B and the other parameters in the equation for E. Substitution of this expression for B back into the original equation for E yields the final form of the Born equation for the ion-pair energy:

$$E_{min} = \frac{Ne^2 q_+ q_-}{4\pi\epsilon_0 \, d_0} \left(1 - \frac{1}{n}\right)$$

By signifying the equilibrium value of d as d_0, we emphasize that this equation is valid only at the equilibrium ion-pair distance. To get a feeling for how large this energy is, you may insert the nominal values of $e = 1.60 \times 10^{-19}$ C, $q_+ = +1$, $q_- = -1$, $d_0 = 300$ pm, and $n = 9$ into the Born equation to calculate a value for E_{min} on the order of -400 kJ/mole. This is clearly of sufficient magnitude to make the formation of MX ion-pairs in the gas phase thermodynamically feasible.

To consider further the problem of formation of the ionic lattice, we need only extend the equation for ion pairs to accommodate the formation of an ordered crystal of such pairs. To do this, however, requires that we know the structure of the crystal lattice. By way of example, we will consider in some detail the formation of a compound with the **face-centered cubic** structure of the NaCl lattice depicted below. (*As an aside, you should realize that terms such as "face-centered" and "body-centered" are given with reference to ions of the same type. The NaCl structure is that of two interpenetrating face-centered lattices, one for Na^+ and the other for Cl^-; see* **2**, *p. 57.*)

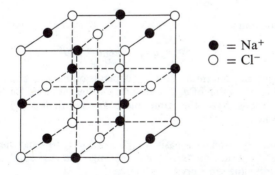

$\bullet$ = Na^+
$\bigcirc$ = Cl^-

The geometric characteristics of this lattice are the following: the lattice consists of Na^+ and Cl^- ions, each of which is surrounded by six counter-ions at a radius of d_0; furthermore, beyond this first coordination sphere there are found twelve ions of the same charge as that of that at the center and at a distance of $\sqrt{2}\,d_0$ from it; the next sphere of ions has eight counter-ions at a distance of $\sqrt{3}\,d_0$; the next sphere consists of six ions of the same type as that at the origin and at a distance of $2d_0$; and so on. Expressing each of these interactions in the same form as that used for the ion pairs, we arrive at a series of terms of the form

$$E = \frac{Ne^2 q_+ q_-}{4\pi\epsilon_0 d_0} \cdot \left(1 - \frac{1}{n}\right) \cdot \left[6 - \frac{12}{\sqrt{2}} + \frac{8}{\sqrt{3}} - \frac{6}{2} + \frac{24}{\sqrt{5}} + \cdots\right]$$

The series of fractions in this expression is convergent on the value 1.748, so that the expression for the lattice energy of a compound with the NaCl structure reduces to the simple expression

$$E = 1.748\, N \frac{e^2 q_+ q_-}{4\pi\epsilon_0 d_0} \left(1 - \frac{1}{n}\right)$$

The quantity 1.748 for the NaCl structure is, of course, dependent solely on the geometry of the crystal and is known as the **Madelung constant** for the NaCl structure. It is interesting to note that the condensation of the ion-pairs to form a three-dimensional NaCl type lattice

$$N\, NaCl_{(g)} \rightarrow (NaCl)_{N\,(c)}$$

results in a further electrostatic stabilization of the ions of ~75 per cent. Thus, more than 50 per cent of the lattice energy is accounted for simply in terms of the basic ion-pair attraction.

Other crystal geometries have derivable Madelung constants; these have been calculated for the common lattice types and are presented in Table 2-6 for future reference.

Structure	M
NaCl	1.748
CsCl	1.763
ZnS	1.638
CaF_2	2.519
TiO_2	2.408

TABLE 2-6

MADELUNG CONSTANTS FOR COMMON LATTICES

STUDY QUESTIONS

13. Using the expression

$$E = e^2 q_+ q_- / 4\pi\epsilon_0 d + B/d^n$$

for a single ion pair, determine how d_0 depends on q_+, q_-, and B. What is the effect of ion charge on the value of d_0? Assuming n to be the same for NaF and MgO (both cations and anions have the same inert gas configuration), what is the ratio $d_0(NaF)/d_0(MgO)$?

14. Using the Born equation as a basis for your answer, to what can you attribute the facts that MgO is much harder, higher melting, and less soluble in H_2O than NaF (they both exhibit the same crystal structure)?

15. What plane through a face-centered cubic unit cell contains the largest number of ion nuclei? How many nuclei are in that plane?

Structure and Stoichiometry[5]

As we begin to survey the common lattice structures of ionic solids, we need to realize that these materials have highly ordered structures in which each ion (cation or anion) appears in a regular array or lattice in three dimensions. It is on the basis of the *lattice of each ion* that we may systematize our thoughts on structures of solids. By far the simplest close packed arrangement of ions of one type is that of the cube; it is interesting, however, that two different types of cubic lattices are commonly found in nature—one of these is called **simple cubic** (sc) and the other is called **face centered cubic** (fcc), as diagrams **1** and **2** illustrate. The most common structures of salts with formula MX, MX_2, and M_2X are best visualized as interpenetrating sub-lattices of these sc and/or fcc types. (The diagrams **(1)** and **(2)** are for only one charge type of ion. For specific compounds, more complete sketches showing both cations and anions are required.)

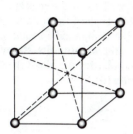

The simple cubic lattice showing the cubic "hole"

1

The face centered cubic lattice showing the complete octahedral hole and one of the tetrahedral holes

2

The solid lines connecting the ions are not meant to represent bonds in the sense of the Lewis system of structure drawing, but rather help to outline a region in the lattice known as the unit cell. A proper definition of a **unit cell** is that it be of the smallest possible dimensions, have opposite sides parallel, and exhibit the full symmetry of the cation and anion lattice networks. A unit cell completely defines the full lattice structure *by simple translation of the cell along its edges.*

One important property of the unit cell is the number of ions contained within the boundaries of the cell, and discussion of a few general rules concerning the method of counting the ions within the boundaries of the cell is in order. Note that any ion located at a corner of the unit cell contributes equally to a total of eight adjacent cells, so each corner ion is counted as one-eighth of an ion for its contribution to one cell. An ion situated in a face of the cell contributes to only two adjacent cells and so is counted as one-half of an ion for any one cell; finally, an ion located along a cell edge contributes symmetrically to four adjacent cells, and thus is counted as one-fourth of an ion for any one cell. In the examples given above, each unit cell of the sc structure contains a total of one ion of the type defining the simple cubic lattice, while the face centered lattice contains four ions of the type defining the cell.

Another geometrical property of the ions defining the unit cell is the number and symmetries of the holes in the lattice. For the *simple cubic lattice* **(1)** there is only one hole of any reasonable size (at the cell center) and we say the hole is cubic, meaning that the arrangement of ions about the hole is cubic. With one ion/unit cell and one cubic hole/unit cell, the ratio of ions to holes for the sc lattice is 1:1. Therefore, only salts of formula MX can exhibit an sc structure of anions with *all* cubic holes filled by cations. Such a salt is CsCl, and the structure **3** is referred to as the **CsCl structure.**

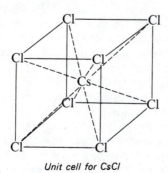

Unit cell for CsCl

3

For the *face centered lattice* (2) there are two distinct types of holes. A complete hole of octahedral symmetry is seen at the center of the cell, and 12 one-quarter holes (one at the mid-point of each edge) define three more octahedral holes/unit cell; furthermore, there are eight holes of tetrahedral symmetry, one each at the centers of the cell octants. For the fcc lattice we conclude that there are two T_d holes and one O_h hole *per ion defining the fcc structure.* Consequently, only salts of formula MX can be expected in which *all* octahedral holes are filled with cations. This structure is appropriate for common salt and is referred to as the **NaCl structure (4).** If all of the tetrahedral holes created by the fcc anion lattice are filled with cations, the compound formula must be M_2X. As will be clarified later, this structure is known as the **anti-fluorite structure (5).** Finally, if only half of the tetrahedral holes in the fcc anion lattice are filled, the formula must be MX. This structure is identified as the **ZnS structure (6).**

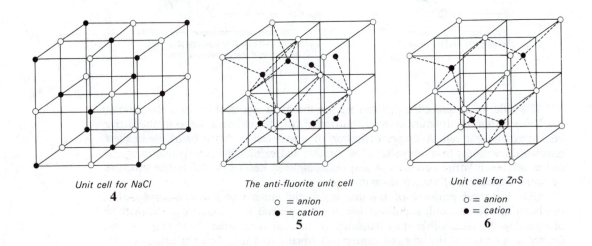

Unit cell for NaCl
4

The anti-fluorite unit cell

○ = anion
● = cation
5

Unit cell for ZnS

○ = anion
● = cation
6

A somewhat different way to express these same ideas is to identify the coordination number of an ion, were one to be located at a lattice hole: the cubic hole of the simple cubic lattice (**1** and **3**) has a coordination number of 8; each octahedral hole in the face centered cubic lattice has a C.N. of 6 (**2** and **4**), and each of the tetrahedral holes has a C.N. of 4 (**2, 5,** and **6**).

In view of the above considerations, a compound MX could be expected to exhibit the CsCl, NaCl, or ZnS structure. The simplest structure possible for an ionic compound of formula MX is that exhibited by CsCl, CsBr, and CsI (CsF has, interestingly, a different structure). This structure (**3**) is simply that of *two interpenetrating simple cubic lattices* of Cs and Cl ions. Both Cs and Cl ions sit in cubic holes formed by the sc lattice of the other, each ion has a C.N. of 8, and the unit cell contains one pair of ions. Salts of complex ions of high symmetry such as $[Be(H_2O)_4]SO_4$, $K[SbF_6]$, and $[Ni(H_2O)_6][SnCl_6]$ have been found to exhibit this type of lattice structure, as have compounds of ions of considerably less symmetry such as CsCN and CsSH, where the non-spherical anions in these examples achieve an effective spherical shape at temperatures near ambient by rapid tumbling. Notice in Table 2–6 that the Madelung constant for the CsCl structure is only slighly different from (greater than) that of the NaCl structure.

The next most common structure for MX is that formed by *two interpenetrating face centered cubic lattices,* as exemplified by NaCl (**4**). Notice in the sketch of the unit cell for this lattice that only the octahedral holes of one ion's lattice are occupied by the counter-ions. Each ion, therefore, has a C.N. of 6 and the unit cell contains four NaCl ion pairs. Examples of salts with the NaCl structure are all the alkali metal halides (with the exception of the heavier cesium halides), the alkaline earth chalcogens, the silver halides (except AgI), and more complex materials such as $[Co(NH_3)_6][TiCl_6]$ and NH_4I. Even salts such as KOH, KCN, and KSH exhibit the NaCl structure by rapid tumbling of the otherwise non-spherical anions.

Finally, there are numerous examples of MX salts with the *fcc lattice for anions* and with only *half of the tetrahedral holes* in the face centered lattice *occupied by cations* (6). The S, Se, and Te salts of Zn, Cd, and Hg form this type of lattice (known as the **zinc blende** structure after the geologist's name for ZnS). Note that this structure is appropriate to compounds of the formula MX, as the ratio of face centered ions to those in the tetrahedral holes is 1:1. Each ion in the zinc blende structure has a C.N. of 4, and each is located in a tetrahedral environment of the counter-ions. Refer to Table 2-6 and notice that the Madelung constant for zinc blende is not too different from (smaller than) those of NaCl and CsCl.

We now turn to the common structures for salts of formula MX_2 and M_2X. The simplest such structure is based on interpenetrating simple cubic and face centered cubic lattices of ions. These structures are called the **fluorite** (the common name for CaF_2) structure or the **anti-fluorite** structure, depending on the anion and cation sub-lattice structure:

sc = anion	fcc = anion
fcc = cation	sc = cation
fluorite	anti-fluorite

The unit cell for the fluorite structure may be represented in two ways, each having a particular visual advantage. Cell **7** emphasizes the sc lattice of anions (○) with cations (●) in half of the cubic holes. This structure bears a strong resemblance to the ZnS cell (6) in that the cation arrangement is the same; the difference is that ZnS has a fcc anion lattice while CaF_2 has a sc anion lattice. The alternate cell for representing the fluorite structure is **5**, also used for the anti-fluorite structure, except that for fluorite the open circles are to represent fcc cations, and the solid circles represent sc anions. The advantage to **5** is that it emphasizes the cation/anion interchange relation between fluorite and anti-fluorite. The primary differences in these two structures are that: (1) as you shall see, the lattice holes are bigger in relation to the anion size for the sc anion lattice, and (2) the stoichiometry changes from MX_2 to M_2X. The fluorite structure is common to difluorides and is found for *some* dioxides of metals, while the chalcogenides of Li, Na, and K (i.e., all M_2X) are known to have the anti-fluorite structure.

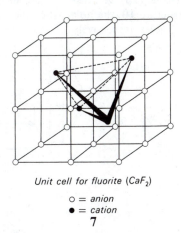

Unit cell for fluorite (CaF_2)

○ = anion
● = cation

7

At this point you should be wondering whether there are any examples of salts of fcc anions in which all O_h and T_d holes are filled with cations. Such salts would of necessity have a formula M_3X, or, with the roles of cation and anion reversed, MX_3. In fact, several compounds are known to have the **cryolite** structure ($Na_3[AlF_6]$) and the **anti-cryolite** structure (an example is $[Co(NH_3)_6]I_3$).

Finally, we should note that the most common lattice structure for MX_2 compounds that permits "octahedral" sites for the cations (fluorite provides for cubic sites) is the **rutile** structure (8). Rutile (the common name for TiO_2) presents a low symmetry sub-lattice of anions that does not fit well into the systematics of sc and fcc lattices discussed before; on the other hand, the cation sub-lattice of **8** is **body centered**.

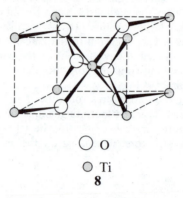

O O

○ Ti

8

Structure and the Concept of Ion Radii

In the preceding discussion of the most basic, simplest structural types it was convenient to make use of the assumption that the ions of a lattice could be thought of as hard spherical objects. If such a view is reasonable, and there will be more about this later, then we might attempt to define the holes (cubic, octahedral, and tetrahedral) in terms of the size of an ion that will fit into them. Simple geometrical considerations then allow us to determine the radius of the hole in relation to the radius of the ion defining the simple cubic and face centered cubic lattices.

For the ideal sc structure (**9**) in which the hard spheres constituting the lattice are just touching, the length of the edge of the cube is $2r_\varrho$, where r_ϱ is the radius of the lattice ion. The body diagonal of the cube is $2(r_\varrho + r_h)$, where r_h stands for the radius of the hole. The Pythagorean Theorem gives an equation that can be solved for the ratio r_h/r_ϱ, with the result that the ratio in the ideal case of just-touching lattice spheres is 0.73—the radius of the hole is about ¾ that of the ions defining it.

For the face centered lattice (**10**) we must consider both the octahedral and tetrahedral holes. Referring to structure **10**, the hole/lattice radius ratio is 0.41 and 0.22 for the octahedral and tetrahedral holes, respectively.

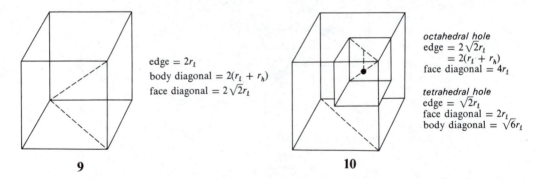

edge = $2r_l$
body diagonal = $2(r_l + r_h)$
face diagonal = $2\sqrt{2}r_l$

octahedral hole
edge = $2\sqrt{2}r_l$
  = $2(r_l + r_h)$
face diagonal = $4r_l$

tetrahedral hole
edge = $\sqrt{2}r_l$
face diagonal = $2r_l$
body diagonal = $\sqrt{6}r_l$

9

10

In summary, we expect that when the cation has a radius not much bigger than one-fourth that of the cubic lattice anion, the cation will take up a tetrahedral site in an anion face centered lattice. When the cation has a radius in the range from one-fourth to approximately one-half that of the anion, it will prefer the larger octahedral site in an anion fcc lattice. Finally, as the cation size approaches that of the cubic lattice hole, we expect the fcc anion lattice to open up to the simple cubic structure so that the cation will occupy a cubic hole.

While these observations of relative hole sizes to ion sizes for the different lattice structures are interesting, we have engaged in a rather impractical exercise if we cannot do something about defining the radii of various cations and anions. We are seeking an operational definition of ion radius from which we need only semi-quantitative information.

Now, there are probably as many different ways of estimating ion radii as there are ways of estimating atom and ion electronegativities. One of the ways to arrive at a set of ion radii is the following. Many compounds are known for which variation of the cation has essentially no effect on the cation-anion internuclear distance. This phenomenon is interpreted to mean that for such compounds the cations just fit into or are somewhat smaller than the holes they occupy in the anion lattices. In other words, the distance between the anion nucleus and the hole center is fixed by the fact that the anions are just in contact with one another. According to this interpretation, the anion radius is simply half of the anion-anion internuclear distance. In this vein, "absolute" radii can be established for anions, and these in turn can be used to determine the cation radii in salts of those same anions where the cation-anion distance is *not* independent of the cation. By systematically considering various salts, one expands a list of cation and anion radii. The procedure inevitably produces some discrepancies, which have been studied from several points of view. Probably the most widely accepted ion radii are those of Pauling,[6] who introduced corrections for the greater internal compression forces in crystals with ions of high charge. These corrected radii are termed **crystal radii**, and some of them are presented in Table 2–7. A word of caution is in order regarding these values. Their absolute magnitudes are of questionable meaning; they have their greatest use in the operational sense of *comparing* relative ion sizes and internuclear distances in crystals.

TABLE 2–7

SOME IONIC RADII (pm)

								H^- 208
Li^+ 60	Be^{2+} 31			B^{3+} 20	C^{4+} 15	NH_4^+ 148	O^{2-} 140	F^- 136
Na^+ 95	Mg^{2+} 65			Al^{3+} 50	Si^{4+} 41		S^{2-} 184	Cl^- 181
K^+ 133	Ca^{2+} 99 *	Cu^+ 96	Zn^{2+} 74	Ga^{3+} 62	Ge^{4+} 53		Se^{2-} 198	Br^- 195
Rb^+ 148	Sr^{2+} 113	Ag^+ 126	Cd^{2+} 97	In^{3+} 86	Sn^{4+} 71		Te^{2-} 221	I^- 216
Cs^+ 169	Ba^{2+} 135	Au^+ 137	Hg^{2+} 110	Tl^{3+} 95	Pb^{4+} 84			
				Tl^+ 140	Pb^{2+} 121			

*Mn^{2+} (80), Fe^{2+} (76), Co^{2+} (74), and Ni^{2+} (69).

STUDY QUESTIONS

16. How many formula weights of MX_2 occur in the unit cell of the rutile structure? What is the coordination number of each cation and anion?

17. What is the ratio r_h/r_ℓ for the triangular holes in a layer of close packed ions? Were such a structure (with all trigonal holes filled) found for a compound, what would

be its empirical formula? Taking the lattice ion to be O^{2-}, are there any cations (except H^+) small enough to fit in the trigonal holes?

18. Calculate the volume of the unit cell of KCl and then the density of the salt (experimental value = 1.98 g/cm^3).

19. Which of the ionic structure types (CsCl or NaCl) is favored for a given compound MX if severe pressure is exerted on the crystal?

20. When the cation in the anion lattice holes is smaller than the hole, will the lattice energy be different from that predicted by the Born equation? How and why?

21. A simple binary compound, $M_m X_n$, is known to have a structure of face centered cubic packed anions with all T_d holes filled with M cations. What is the empirical formula of the compound?

22. Using sketch **10**, verify that the tetrahedral and octahedral hole/lattice radius ratios are 0.22 and 0.41, respectively.

23. Which would you predict would have higher *solid state* ionic conductivity, a salt MX with the NaCl structure or a salt MX with the zinc blende structure?

Layer Lattices and Incipient Covalency

Now that we have considered the most common structures for compounds that may be termed ionic, we must be careful to note that there are many common salts that have

TABLE 2-8

SOME EXPERIMENTAL AND CALCULATED LATTICE ENERGIES
(kJ/mole)

Compound	Calculated	Experimental	Difference	Structure
NaF	−904	−904	0	NaCl
NaCl	−757	−774	−17	NaCl
KCl	−690	−703	−13	NaCl
KI	−623	−649	−26	NaCl
CsF	−724	−732	−8	NaCl
CsCl	−623	−657	−34	CsCl
CaF$_2$	−2594	−2590	+4	CaF$_2$
SrCl$_2$	−2067	−2125	−58	CaF$_2$
BaF$_2$	−2339	−2314	+25	CaF$_2$
AgF		−954		NaCl
AgCl	−699	−904	−205	NaCl
MgBr$_2$	−2138	−2393	−255	Layer
MgI$_2$	−1983	−2314	−331	Layer
CaI$_2$	−1895	−2067	−172	Layer
MnF$_2$	−2761	−2774	−13	Rutile
MnBr$_2$	−2176	−2439	−263	Layer
MnI$_2$	−2008	−2381	−373	Layer
CdF$_2$	−2644	−2774	−130	CaF$_2$
CdI$_2$	−1987	−2435	−448	Layer
HgF$_2$	−2611	−2761	−150	CaF$_2$

lattice energies considerably in excess of those expected on the basis of the ionic model of crystalline materials. Some examples of compounds and the deviations between the theoretical and experimental lattice energies are cited in Table 2-8. A characteristic feature of those compounds for which the deviations are large is that they have either a small, highly charged cation or a large, polarizable cation combined with a relatively large, polarizable anion. These characteristics are consonant in a vague way with the idea of emerging covalency in the cation-anion attraction.

In some instances these deviations become particularly marked so that layer lattices are formed. For example, $CdCl_2$ possesses a lattice of face centered cubic Cl^- ions with half of the octahedral holes occupied, as required by the stoichiometry. The interesting feature, however, is that only *alternate layers* of holes are occupied by the Cd^{2+} ions.

It is exceedingly difficult to make and view a two-dimensional sketch of the lattice structure of such a material when we want to show clearly both the fcc structure of the Cl^- lattice *and* the layer structure. Careful inspection of the fcc lattice of anions shows that this structure can also be drawn as parallel sheets of anions.[7] Figure 2-5 shows two different perspectives of the fcc structure; the fcc structure is defined by anions from four different parallel sheets. Figure 2-6 gives presentations of one and two such sheets. Notice, in Figure 2-6(b), how the ions of the upper layer (full circles) rest in the cusps of the lower layer (broken circles). Such a "packing" arrangement is designated **cubic close packing** to distinguish it from simple **cubic packing**, in which the ions of the upper layer are situated directly above those of the lower layer.

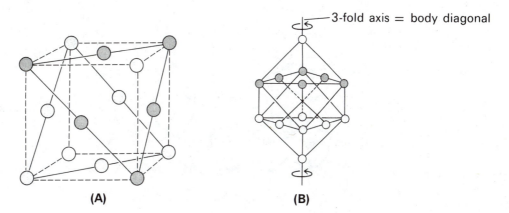

Figure 2–5. Two different perspectives of the fcc lattice to emphasize the parallel sheets of anions. Note O_h hole at unit cell center.

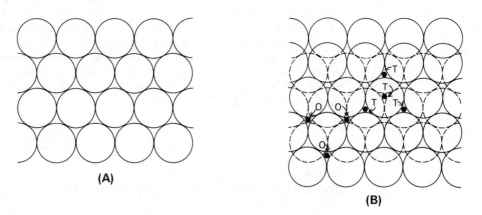

Figure 2–6. (a), One of the sheets of Figure 2–5 extended in two directions. (b), Two close packed layers with typical octahedral and tetrahedral holes identified.

In the previous discussion of the fcc lattice we were able to identify one octahedral hole per anion and two tetrahedral holes per anion. This, of course, can also be done for the partial fcc structure in Figure 2-6(b), where several such holes are identified by the labels O and T. Picking any ion in the upper layer, we can readily identify between the upper and lower layers four tetrahedral holes to which the chosen ion contributes. If a third layer of anions were to be laid on top of the second, the pattern would repeat itself so that, with any one ion contributing 1/4 to each of its tetrahedral holes, we conclude, as before, that there are 8/4 = 2 tetrahedral holes/anion. Similarly, any one ion contributes to the creation of three octahedral holes below it and to three above it, for a net of one octahedral hole/anion. In this way, you see that the octahedral holes themselves may be thought of as occurring in parallel sheets mid-way between the anion sheets. If all the octahedral holes are filled, we have the NaCl structure discussed earlier.

Returning to the $CdCl_2$ structure, note that the MX_2 stoichiometry requires that only half of the octahedral holes be filled by Cd^{2+} ions. This could be done by Cd^{2+} ions occupying half of the octahedral holes in each sheet of holes; but, in fact, $CdCl_2$ has the layer structure in which *all* the holes in alternate hole-sheets are occupied. This, of course, means that the structure may be thought of as a sheet of cations sandwiched between two sheets of anions (Figure 2-7), and this three-layer arrangement repeats itself perpendicular to the page. Consequently, we find adjacent layers of anions with no cations between them to hold them together. Such a structure is easily sheared, as the coulombic forces holding the sandwiches or layers together are those between the anions of one sandwich and the cations of another.

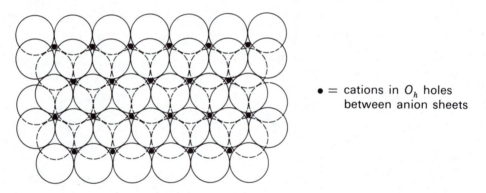

• = cations in O_h holes between anion sheets

Figure 2-7. The $CdCl_2$ layer structure.

Related to the $CdCl_2$ structure is that of $CrCl_3$, which, because of the stoichiometry, has only one-third of the octahedral holes occupied with Cr^{3+} ions. Again we encounter the layer structure, with only two-thirds of the holes in alternate sheets occupied.

As an alternative to the layer lattice type, there are a few examples in which the role of bridging halides is carried to the extreme of "one-dimensional" chains. Most notable among these are $BeCl_2$, $CuCl_2$, and $PdCl_2$. In Be^{2+} we find a small cation of high charge, and in Pd^{2+} a polarizable cation of high charge. The $BeCl_2$ chains may be viewed as edge-sharing *tetrahedra* of $BeCl_4$ sub-units

The $PdCl_2$ and $CuCl_2$ chains are similarly viewed as edge-sharing *squares* of MCl_4 sub-units.

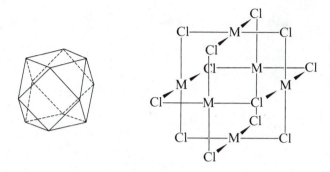

This structure of edge-sharing squares for $PdCl_2$ is called the α-form. Of greater thermo-dynamic stability, however, is a hexameric β-form of corner-shared squares,[8] as follows:

In this cuboctahedral structure, Cl appears at the polyhedral vertices and Pd at the face centers.

**STUDY
QUESTIONS**

24. Using ion radii and assuming hard spheres, predict structural types for $FeCl_2$, NaH, MnF_2, CuI, $NaBH_4$, and TlCl. Which salts might exhibit layer lattices?

25. Both $FeCl_2$ and $FeCl_3$ have layer structures based on close-packed anions. How do their structures differ? Aside from the anions, in what sense are their structures alike? Make sketches like that of Figure 2–7 to illustrate your points.

26. Which salt in each pair do you expect to show the larger deviation from the ideal ionic behavior?

$CaCl_2/ZnCl_2$ MgF_2/TiO_2 $MgCl_2/BeCl_2$
$CdCl_2/CdI_2$ Al_2O_3/Ga_2O_3
ZnO/ZnS $CaCl_2/CdCl_2$

27. Make a sketch, like that of Figure 2–7, for $CrCl_3$ (each Cr^{3+} has a C.N. = 6, each Cl^- has a C.N. = 2).

SOLID STATE ELECTROLYTES[9]

The electrical conductivity of solids is generally far less than that of solutions because the greater steric congestion in solids inhibits ion migration. Two materials with conductivities approaching those of aqueous electrolytes, Li_xTiS_2 ($x \leq 1$) and sodium beta alumina $[(Na_2O)_{1+x} \cdot 11Al_2O_3]$ ($x \leq 0.3$), have revolutionized solid state electro-chemistry.

Ionic conductivity depends on the formation of disordered crystals or of interstitial compounds. **Frenkel defects**, common for the cations of the alkaline-earth halides, involve displacement of a cation from its normal lattice hole to an energetically less favorable hole called an **interstitial site**. Usually it is the (smaller) cation which is respon-sible for the Frenkel defect (see Figure 2–8a). **Schottky defects** typically occur in alkaline-earth and alkali halides; they are created by cation/anion pair vacancies. The creation of these defects can be due to either (1) the disordering effect of entropy (in which case the defects are in equilibrium with the regular lattice and called *intrinsic*) or (2) the incorporation of impurity ions of charge different from that of the proper

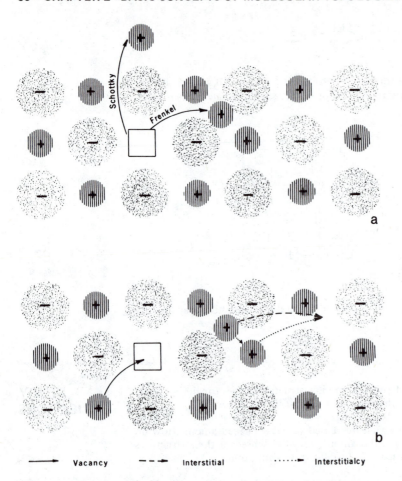

⟶	Vacancy	– – –► Interstitial	·······► Interstitialcy

Figure 2-8. (a) Frenkel (interstitial) and Schottky (vacancy) models of lattice defects that can result in ionic conductivity. (b) Three classical mechanisms for ionic conductivity in crystalline solids. [From G. C. Farrington and J. L. Briant, *Science*, **204**, 1371 (1979). Copyright 1979 by the American Association for the Advancement of Science.]

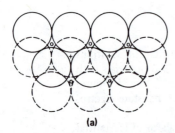

(a)

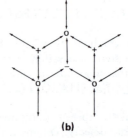

(b)

Figure 2-9. (a) A top view of the S^{2-} ions with O_h and T_d interstitial sites between the S^{2-} layers. The O_h holes, marked by small circles, lie in a plane mid-way between the S^{2-} layers. The T_d holes occur in one or the other of two planes: the holes marked + define a plane half-way between the O_h hole plane and the upper S^{2-} layer, and the holes marked − define a plane half-way between the O_h plane and the lower S^{2-} layer. (b) The up-down/zig-zag interstitial migration of Li^+ ions responsible for the electrolyte property of Li_xTiS_2.

cation. The presence of Ca^{2+} at a Na^+ site in NaCl, for example, requires a Na^+ vacancy elsewhere in the lattice in order to maintain electrical neutrality; these are called *extrinsic* defects. The intrinsic defects increase in importance at high temperature.

Three mechanisms for ionic transport in solids are vacancy, interstitial, and interstitialcy migrations (see Figure 2–8b). Vacancy migration involves counter-flow of cations and vacancies between the cation sites normally occupied. Interstitial migration entails ion migration between interstitial sites. The interstitialcy mechanism is the cooperative hop of an ion from a normal to an interstitial site, while an interstitial ion moves into the newly vacated normal site. The requirements for high conductivity are (1) a high concentration of potential charge carriers, (2) a high concentration of vacancies or interstitial sites, and (3) a low activation energy for ion site-hopping.

Both sodium beta alumina and $Li_x TiS_2$ have layer structures with the Na^+ and Li^+ ions confined to planes between Al/O and Ti/S layers, respectively. The $Li_x TiS_2$ structure is the easier to describe of the two.

Following the arguments on page 64 and Figure 2–6, there are one O_h and two T_d holes per S of TiS_2. Occupation of all O_h holes in alternate cation sheets by the Ti^{4+} ions gives a sandwich structure to TiS_2. There remain one sheet of interstitial O_h sites half-way between the S-Ti-S sandwiches and a sheet of interstitial T_d holes above and below the O_h sheet and half-way between it and the S layers (Figure 2–9). Ion or molecule occupation of the O_h or T_d interstices gives what are called **intercalation compounds**. Introduction of Li^+ ions into the O_h holes between TiS_2 sandwiches occurs when TiS_2 is reduced by Li dissolved in liquid ammonia, by reaction of TiS_2 with butyl-lithium, or by electrochemical reduction of TiS_2 in the presence of Li^+ ions. The migration of lithium ions in $Li_x TiS_2$ is limited to two dimensions, those parallel to the TiS_2 sandwiches. It is believed that the cation migration mechanism entails Li^+ hopping from $O_h \rightarrow T_d \rightarrow O_h \rightarrow T_d \rightarrow \cdots$ sites; thus, it presents an example of the interstitial mechanism, and the defect is extrinsic in nature.

The use of solid electrolytes holds great promise for the development of high energy-density batteries, by avoiding the use of aqueous electrolytes. Perhaps you have already heard of the Na/S battery discovered by the Ford Research Laboratories (Figure 2–10).

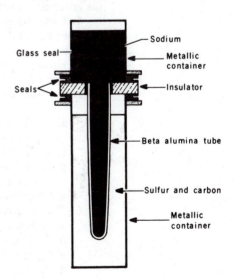

Figure 2–10. Diagram of Na/Na^+ beta alumina/S electrochemical cell. [From G. C. Farrington and J. L. Briant, *Science*, **204**, 1371 (1979). Copyright 1979 by the American Association for the Advancement of Science.]

A sodium anode and sulfur cathode are separated by the beta alumina solid electrolyte, and the battery operates at about 300°C. The cell reaction is

$$2Na + 3S \xrightleftharpoons{2e^-} Na_2 S_3$$

Practical cells of this type are one-fourth the size and half the weight of Pb/PbO_2 cells of the same energy and power.

STUDY QUESTIONS

28. The solid Ag_2HgI_4 (yellow) consists of cubic close packed I^- with Ag^+ and Hg^{2+} ions in tetrahedral holes. Are there sufficient cations to occupy all the T_d holes? In the range from 40° to 50°C, the color changes to red and the conductivity increases to a value ~1000 times greater than normal for an ionic solid at 50°, yet the anion lattice remains cubic close packed. What do you suppose happens within the crystal in the 40°–50° range?

29. Which should show the greater conductivity, $LiTiS_2$ or $NaTiS_2$? Explain.

30. Describe the operation of the galvanic cell consisting of a Li anode, a TiS_2 cathode,

and dioxolane $\left(\begin{array}{c} \end{array} \right)$ as the solvent medium.

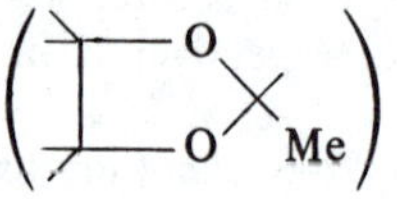

31. Why does entropy favor disordered crystals? Does this explain why raising the temperature enhances intrinsic defects?

CHECKPOINTS

1. Lewis structures are graphical depictions of shared and lone electron pairs.

2. Atoms tend to share electrons so as to achieve four valence pairs.

3. When the atom octet concept can be achieved in more than one way, write resonance structures to convey this idea. Generally, resonance structures arise to show a role intermediate between bond and lone pair characters for certain pairs.

4. Using the concept of electron-electron repulsion (VSEPR), the number of distinct electron pairs about the atoms allows you to predict the three-dimensional shape of the molecule.

5. Having arrived at the molecular shape, it is useful to delineate the symmetry equivalence of the various atoms in the molecule; to do this, the concept of a group of symmetry elements for each molecule is introduced. The principle elements/operations are:

    a. E, the identity (360° rotation about any axis).
    b. C_n, the proper rotation (rotation about an axis by $2\pi/n$).
    c. σ, the reflection plane (mirror image formation through the plane)
    d. S_n, the improper rotation (a C_n-like operation followed by reflection in a plane $\perp$ to the axis)
    e. i, the inversion center (image inversion through a point).

6. The key groups are:
    a. C_s (E only)
    b. C_n (E and C_n only)
    c. C_{nv} (E, C_n and $n\sigma$ containing C_n)
    d. D_n (E, C_n and $nC_2 \perp$ to C_n)
    e. D_{nh} (E, C_n, nC_2 and $\sigma \perp$ to C_n).

REFERENCES

[1] G. N. Lewis, *J. Amer. Chem. Soc.*, **38**, 762 (1916).
[2] R. J. Gillespie, "Molecular Geometry," Van Nostrand Reinhold Co., London (1972); *J. Chem. Educ.*, **51**, 367 (1974); *ibid.*, **47**, 18 (1970).

[3] It has been assumed by many that "Pauli forces" (the spin-correlation forces, p. 000) between the spin-bearing electrons determine the stereochemistry of valence electron pairs. This idea is under attack: J. L. Bills and R. L. Snow, *J. Amer. Chem. Soc.,* **97**, 6340 (1975): R. L. Snow and J. L. Bills, *J. Chem. Educ.,* **51**, 585 (1974). See also reference 6 at the end of Chapter 1.

[4] M. Orchin and H. H. Jaffe, *J. Chem. Educ.,* **47**, 246, 372, 510 (1970) is an excellent supplementary source for the topics of this section. Also, G. Davidson, "Introductory Group Theory for Chemists," Applied Science Publishers, London, 1971, is highly recommended at this level. Also, F. A. Cotton, "Chemical Applications of Group Theory," 2nd Ed., Wiley-Interscience, N.Y., 1971.

[5] The most comprehensive source book for the non-specialist on the subject of solid state structures is A. F. Wells, "Structural Inorganic Chemistry," 4th Edition, Oxford, London (1975).

[6] L. Pauling, "The Nature of the Chemical Bond," 3rd Edition, Cornell University Press, Ithaca (1960).

[7] S-M. Ho and B. E. Douglas, *J. Chem. Educ.,* **46**, 207 (1969) gives further good photos and drawings of ion packing and the lattice holes.

[8] K. Brodersen *et al., Z. Anorg. Allgem. Chem.,* **337**, 120 (1965); R. Mattes, *Z. Anorg. Chem.,* **364**, 290 (1969); U. Wiese *et al., Angew. Chem. Internat. Edn.,* **9**, 158 (1970).

[9] G. C. Farrington and J. L. Briant, *Science,* **204**, 1371 (1979).

3. THE DIRECTED ATOMIC ORBITAL VIEW OF CHEMICAL BONDS

In this chapter you will learn of one matter-wave conceptualization of electron pairs in molecules, the familiar hybrid atomic orbital model. Not only will you see that such orbitals have neither more nor less intrinsic significance than the Schrödinger atomic orbitals but, more importantly, you will learn of their application to the interpretation of some chemical experiments. Herein lies the **raison d'être** *for the hybrid concept: the interpretation of some molecular properties is more simply achieved from a directed atomic orbital view than from the equivalent Schrödinger, non-directed orbital view (the topic of the next chapter).*

DIRECTIONAL ATOMIC ORBITALS

To review our progress to this point in developing a physical description of electrons in molecules, we have utilized the following concepts: (1) perfect pairing of electrons between different atoms to form electron pair bonds, (2) π resonance electronic structures to allow for alternate designations of electron pairs as lone or π bond pairs, (3) repulsive forces between electron pairs as a guide to the basic structures of molecules, with refinements to allow more detailed descriptions of molecular structure and trends in structures of related molecules, and (4) a formal symbolism for the characterization of molecular structures according to the inherent symmetry of molecules. The question of great importance then is whether the wave model can be extended from atoms to molecules. Fortunately the answer is yes, and we will begin the transition from classical to wave models by introducing the concept of **hybrid atomic orbitals** (hao).

The principal feature of molecular structure with which the wave model must be consistent is the angular distribution of electron pairs about given atoms. Consequently, we will be primarily interested in the use of the angular parts of atomic orbital functions (as opposed to the radial parts).

We begin the examination of the wave model with the example of the linear triatomic molecule BeH_2. We will concentrate our attention on the Be atom. The VSEPR model has us thinking of the Be—H bond electron pair distributions as radially distributed from, and to each side of, the Be atom. The question, then, is whether we can use the (centrosymmetric) Schrödinger valence orbital functions for the isolated Be atom ($2s$ and $2p$ orbitals) to develop alternate Be atomic orbitals, which should have the characteristic that one is directed to one side, and the other to the other side, of the Be atom. Note that the centrosymmetry of the $2s$ and $2p$ orbitals means that each is equally directed to both sides of the Be atom. *You should clearly understand that such an "alternate orbital" model has no physical significance; the significance lies in a convenient conceptualization of bonding electron pair distributions.*

You may well ask at this point, "Why bother to create from the Schrödinger orbitals an equivalent set of hybrid atomic orbitals?" The answer is simple and direct. First, we achieve a unification of the concept of **localized** electron pairs (lone and bond pairs) with the wave model of electrons in atoms and molecules. Secondly, chemical interpretations of molecular properties are often particularly convenient when based on the ideas of localized lone and bond pairs of electrons.

The question now comes to, "which Be ao's are to be hybridized to mimic the desired properties of h_1 and h_2?"

$$H \xleftarrow{h_2} Be \xrightarrow{h_1} H$$

The answer is, "those valence orbitals that do not have angular nodes along the directions of h_1 and h_2." This eliminates P_x and P_y, which have zero electron probability along the bond axes. Furthermore, it should be evident to you that the electron density cloud for BeH_2 must have the same overall symmetry as the nuclear charges (positions). If this were not the case, electron density would be "piled up" at one end of the molecule at the expense of the other, and the Be—H distances would be different. Therefore, the electron density must be unchanged by rotating, inverting, or reflecting that density about the symmetry elements. This is just a way of saying that h_1 and h_2 are equivalent in all ways except direction.

Thus, we will form from the pair $(2s, 2p_z)$ two new atomic orbitals, to be called hybrid atomic orbitals (hao), by sum and difference. Let us represent h_1 by the combination

$$h_1 = a\phi_s + b\phi_z$$

and h_2 by the orthogonal combination†

$$h_2 = b\phi_s - a\phi_z$$

The equivalence of these new orbitals by symmetry (the inversion operation, located at the Be position, interconverts them) implies the following relation between h_1 and h_2:

$\hat{\imath}$ on $h_1 = h_2$ \hspace{2em} (from the equivalence of h_1 and h_2)

$\hat{\imath}$ on $(a\phi_s + b\phi_z) = b\phi_s - a\phi_z$

$a\phi_s - b\phi_z = b\phi_s - a\phi_z$ \hspace{1em} (from the inversion symmetry of ϕ_s and ϕ_z)

This last relation requires that

$$a = b$$

The result given for $\hat{\imath}$ on $(a\phi_s + b\phi_z)$ follows from the spherical shape of the Be $2s$ orbital and the tangent (at the Be nucleus) shape of the p orbital. Inversion of a sphere about its center leaves the sphere unchanged in all respects: we write $\hat{\imath} \cdot \phi_s = \phi_s$. Inversion of the double sphere of the p orbital about the center (point of tangency) only changes the signs of the spheres: we write $\hat{\imath} \cdot \phi_z = - \phi_z$

---

†To form the orthogonal partner to a two-orbital function, reverse the coefficients and change the sign of one of the latter. Then,

$$\int h_1 h_2 \, d\tau = ab \int \phi_s^2 \, d\tau \; - \; ab \int \phi_z^2 \, d\tau = ab - ab = 0.$$

Orthogonality is a property of atomic orbitals (i.e., $\int \phi_s \phi_z d\tau = 0$) which should also be retained in a hao theory.

If h_1 is to be normalized, then:

$$\int h_1^2 \, d\tau = a^2 \int \phi_s^2 \, d\tau + a^2 \int \phi_z^2 \, d\tau + 2a^2 \int \phi_s \phi_z \, d\tau = 1$$
$$= 2a^2 = 1$$
$$a = \sqrt{1/2}$$

This means that the analytical forms of the hao's are

and

$$h_1 = \sqrt{1/2}\,(\phi_s + \phi_z)$$

$$h_2 = \sqrt{1/2}\,(\phi_s - \phi_z)$$

By making sketches (Figure 3-1) of the angular parts of the hao wave functions, we readily see that h_1 and h_2 satisfy our imposed requirement that the hao represent directed (toward the hydrogen nuclei) electron functions for the Be atom. In making such sketches we use the fact that the superposition (algebraic addition) of two wave forms of opposite sign at some points in space leads to **destructive interference** of the waves and a reduction of the wave amplitude at those points. Similarly, superposition of two wave forms of the same sign at certain points in space leads to **constructive interference** at those points.

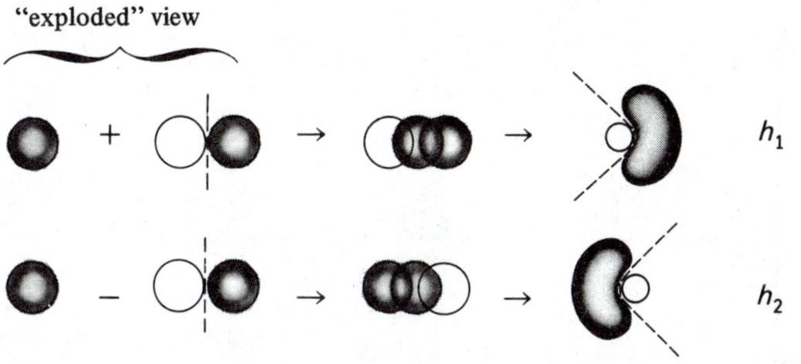

"exploded" view

Figure 3-1. Sketches of h_1 and h_2 and their origins from ϕ_s and ϕ_z.

To emphasize for you that there is no *physical* distinction to be made between the *pair* of functions (h_1, h_2) and the *pair* of functions (ϕ_s, ϕ_z), the angular probability function $h_1^2 + h_2^2$ is seen to be identical to that of $\phi_s^2 + \phi_z^2$:

$$h_1^2 = \tfrac{1}{2}\phi_s^2 + \phi_s\phi_z + \tfrac{1}{2}\phi_z^2$$
$$h_2^2 = \tfrac{1}{2}\phi_s^2 - \phi_s\phi_z + \tfrac{1}{2}\phi_z^2$$
$$h_1^2 + h_2^2 = \phi_s^2 + \phi_z^2$$

The "whole" of the electron distribution is the same from both points of view; only the "pieces" $[(h_1, h_2)$ and $(\phi_s, \phi_z)]$ are different. The middle terms in these expressions are quite important. For h_1, at those points in space where ϕ_s and ϕ_z have the same sign (the region of positive z coordinate), the product $\phi_s \cdot \phi_z$ is positive and the probability is increased by $\phi_s \cdot \phi_z$; whereas at points where ϕ_s and ϕ_z are opposite in sign (negative z), $\phi_s \cdot \phi_z$ is negative and the probability is diminished. One envisions displacement of wave amplitude from points of $z < 0$ to points of $z > 0$. Equal displacement in the opposite sense occurs for h_2. In relation to $h_1^2 + h_2^2$ vs. $\phi_s^2 + \phi_z^2$, the displacements offset each other so that there is *no* net change in *total* point density, i.e.,

$$h_1^2 + h_2^2 = \phi_s^2 + \phi_z^2$$

Finally, it is because of the equal contributions of $2s$ and $2p_z$ functions to the total probability that we call these particular hao's "sp" (or di, or digonal) hybrids (the $s:p$ ratio is $1:1$).

The hao schemes for other commonly encountered VSEPR structures are given in Appendix A at the end of this chapter. These can be worked out by the same basic procedure as was used for the sp hao of BeH_2 but, understandably, the complexity of the derivation increases as the number of directional orbitals increases. Table 3-1 is a summary, with examples, of the hybrid schemes for the σ pairs of each VSEPR structure.

TABLE 3-1

IDENTIFICATION OF STRUCTURAL AND HAO PATTERNS

Stereochemically Distinct Electron Pairs	Example	HAO of *atom	HAO Name
2	Me—Be*—Me	sp	di (digonal)
	O=C*=O	$sp + 2p\pi$	
3	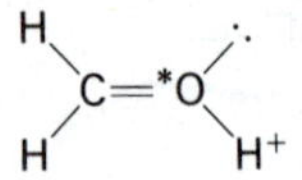 H₂C=*O (with H⁺)	$sp^2 + p\pi$	tr (trigonal)
	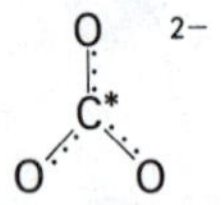 CO₃²⁻ (C*)	$sp^2 + p\pi$	
4	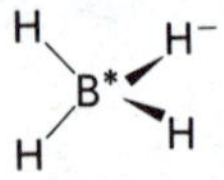 BH₄⁻ (B*)	sp^3	te (tetrahedral)
	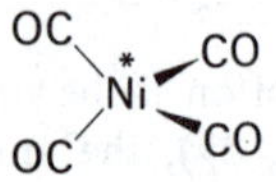 Ni*(CO)₄	sp^3	
5	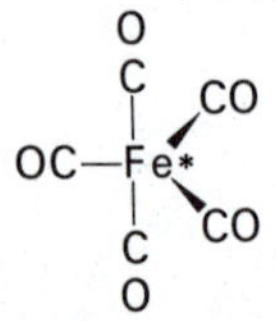 Fe*(CO)₅	dsp^3	di + tr / dp + sp^2
6	 SF₆ (S*)	$sp^3 d^2$	oc (octahedral)
	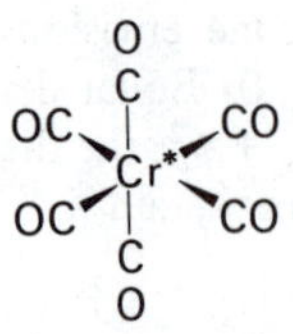 Cr*(CO)₆	$d^2 sp^3$	

In addition to the so-called sigma orbital framework formed by the hao, we need an orbital description for the double bond encountered with some resonance structures. In contrast to the sigma hao's, which have the property of a maximum in electron probability *along the internuclear axis*, the so-called pi electron pairs are characterized by electron probability maxima at points *off the internuclear axis*. The most common instances arise when the two atoms for which we accept a double bond resonance structure have either atomic p or d orbitals, as shown in Figure 3-2. The various possibilities are designated as $p-p$ π bonding, $d-p$ π bonding, and $d-d$ π bonding. Examples of the first two types are given.

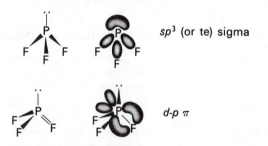

Figure 3-2. Sketches of σ, $p-p$ π, and $d-p$ π orbitals.

To summarize this description of hao's, you will find in Table 3-1 examples of molecules with two through six electron pair clouds about some atom and the *idealized* hybridization scheme for each.. Two points are important in examining this table. First, in cases where there is more than one pair shared between two atoms, one of these pairs is treated as arising from a combination of sigma hao's from each atom. Additional pairs are accounted for in terms of combinations of π type atomic orbitals from each atom. The second observation to be made is that the idealized hao's are only approximately correct descriptions. A case in point is the oxygen atom of H_2COH^+. Here, in an approximate sense, the oxygen atom is described as sp^2 hybridized even though the local symmetry about the oxygen is *not* D_{3h} (a similar comment applies to the carbon atom; the correct molecular symmetry is C_s).

1. Sketch the hybrids given by

   a) $\sqrt{\frac{1}{2}}(s + p_z)$

   b) $\sqrt{\frac{1}{2}}(d_{z^2} + p_z)$

   c) $\sqrt{\frac{1}{2}}(s + d_{x^2 - y^2})$

STUDY QUESTIONS

2. For a general hybrid ao given by

$$h = a\phi_s + b\phi_p \qquad\qquad \text{(where } a^2 + b^2 = 1\text{)}$$

plot the value of $2ab$ in the wave interference term $(2ab\phi_s\phi_p)$ as a function of the fraction s character (a^2). Remember, $a^2 + b^2 = 1$, and $0 \leqslant a^2 \leqslant 1$.

3. List the idealized central atom hybrid atomic orbitals of

a) $XeOF_4$ d) P in P_4
b) SO_3 e) ClF_3
c) $SnCl_3^-$ f) Al in Al_2Cl_6

MOLECULAR PROPERTIES CONVENIENTLY INTERPRETED WITH THE DIRECTED ATOMIC ORBITAL CONCEPT

In the preceding chapter we have developed a qualitative model for molecular structure, which has now led us to the use of directed or hybrid atomic orbitals. In the next few sections we will examine the utility of such an approach to conceptualization of bond formation, and several of the interpretative schemes for molecules properties that have been developed from the directed orbital point of view.

Bond Formation and Orbital Overlap

One of the properties of sp hybrid atomic orbitals discussed in the opening section was that the *centroid of electron density for each hybrid is shifted away from the nucleus relative to that of either the valence s or the valence p orbital.* The basic idea behind bond formation is much the same as that behind the hybridization concept. Just as hybridization is the mixing of two atomic orbitals (on the same center) *to enhance the orbital amplitude along some internuclear axis,* the combination or mixing of two orbitals from different atoms leads to constructive interference in the internuclear region. In terms of probability functions, this means that the greater the reinforcement, the greater is the probability of the electron pair in the internuclear region (where the probability must be large if the electron pair is to do a good job of holding the two nuclei together).

In analogy to the formation of hybrid atomic orbitals by superposition of pure atomic orbitals, we allow for the constructive interference of the two hybrid atomic orbitals on different atoms by writing the wavefunction for the bond orbital as

$$\psi_b = ah_1 + bh_2$$

where h_1 and h_2 are hybrids, on adjacent atoms, directed at each other.

As is necessary, the bond function must be normalized to unity, so

$$\int \psi_b^2 \, d\tau = a^2 \int h_1^2 \, d\tau + b^2 \int h_2^2 \, d\tau + 2ab \int h_1 h_2 \, d\tau = 1$$

As h_1 and h_2 are already normalized, the first two integrals on the right side of this equation have values of 1. The third integral is commonly called the **overlap integral** (symbol = S) and has the important physical significance that the greater it is, the greater is the probability that an electron in the bond orbital ψ_b will be found in the internuclear region. In other words, if *both* h_1 and h_2 have large amplitudes or probabilities in the internuclear region, then their product $h_1 \cdot h_2$ will also be large in this region, and a large overlap integral results when the two hybrid orbitals are strongly reinforcing. *Theoretically, then, a useful criterion for the strength of a bond is the magnitude of the overlap integral between the hybrid orbitals that define the bond.* The sketches of Figure 3-3 should help in giving a pictorial presentation of the reinforcement of overlapping atomic orbitals.

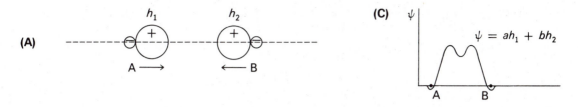

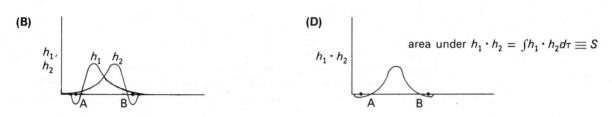

Figure 3-3. Bond formation as viewed from interfering hybrid waves: (A) hybrid angular sketches, (B) the hybrid radial functions, (C) the bonding orbital $ah_1 + bh_2$, (D) the product $h_1 \cdot h_2$ illustrating overlap of h_1 and h_2.

Bond Formation and Orbital Electronegativities

In the theory of bond formation, the strength of the chemical bond depends somewhat on the similarity in energy of the interacting orbitals; orbitals of similar energy tend to form stronger bonds than do others of dissimilar energies. The phrase **energy match** is used to describe this similarity.

The concepts of orbital energy and atom electronegativity are closely related; the relationship was developed by Mulliken[1] and amplified by Hinze and Jaffe.[2]

Mulliken began his argument for the definition of electronegativity by considering electron transfer between two atoms, A and B,

$$A^- B^+ \leftrightarrow A^+ B^-$$

He argued that if the electronegativities of A and B were identical, the two ionic pairs would be of equal energies. The energy of the first form depends on the ionization potential of B and the electron affinity of A, while the energy of the second form depends on the ionization potential of A and the electron affinity of B. The ionization potential and electron affinity of A are defined by the energies of the following:

$$I_A: A \rightarrow A^+ + e^-$$
$$E_A: A^- \rightarrow A + e^-$$

If the energies of the two ionic structures are equal, then $I_B - E_A = I_A - E_B$ or $I_A + E_A = I_B + E_B$ is the condition that specifies equal A and B electronegativities. If A is more electronegative than B, then $I_A + E_A > I_B + E_B$. Therefore, Mulliken defined the electronegativity of an atom to be the sum of its ionization potential and electron affinity: $\chi_A = I_A + E_A$. Current usage is to define $\chi = \frac{1}{2}(I_A + E_A)$, so as to work with smaller numbers for χ.

One of the more interesting ramifications of this definition of atomic electronegativity is that the latter depends on the configuration of electrons in the valence shell. For example, the ionization potential of a carbon atom is different for the configurations $2s^2 2p^2$ and $2s^1 2p^3$. As ionization potential and electron affinity data are generally available for ions, Mulliken's definition also allows for the determination of ion electronegativity.

The ground valence configuration of carbon is s^2p^2, for which two ionization potentials and one electron affinity may be defined:

$$I_s \text{ from } s^2p^2 \to s^1p^2 + e^-$$

$$I_p \text{ from } s^2p^2 \to s^2p^1 + e^-$$

$$E_p \text{ from } s^2p^3 \to s^2p^2 + e^-$$

$$\chi_{C_p} = \frac{1}{2}(I_p + E_p) = \frac{1}{2}(11.3 + 1.1) = 6.2 \text{ eV}$$

In this way, you see that the atom ionization potential is not a unique concept; you should distinguish between the ionization potentials of different orbitals just as you would distinguish between the effective nuclear charges and energies of different orbitals. The idea from Chapter 1 that Z^* is larger for the s orbital than for the p orbital is bearing fruit.

Similarly, for the configuration s^1p^3

$$I_s \text{ from } s^1p^3 \to s^0p^3 + e^- = 21.0 \text{ eV}$$
$$E_s \text{ from } s^2p^3 \to s^1p^3 + e^- = 8.9 \text{ eV} \qquad \chi_s = 15.0 \text{ eV}$$

$$I_p \text{ from } s^1p^3 \to s^1p^2 + e^- = 11.3 \text{ eV}$$
$$E_p \text{ from } s^1p^4 \to s^1p^3 + e^- = 0.3 \text{ eV} \qquad \chi_p = 5.8 \text{ eV}$$

Hinze and Jaffe have used a computer to convert atomic spectral and ionization data into three concepts that are of use to the chemist in characterizing the hybridized condition of the atom called the **valence state**. For carbon, using the abbreviation *te* to represent the tetrahedral hybrid orbital,

$$s^2p^2 \to te^4 \text{ defines } P_{te}, \text{ the \textbf{valence state promotion energy}}$$

$$te^4 \to te^3 + e^- \text{ defines } I_{te}$$
$$te^5 \to te^4 + e^- \text{ defines } E_{te} \qquad \chi_{te} = \frac{1}{2}(I_{te} + E_{te})$$

Here you find the valence state concepts of promotion energy, ionization potential, electron affinity, and electronegativity related to one another. Values of electronegativities for the representative elements, in various valence states corresponding to known structures, are reported in Table 3-2.

Just as orbital electronegativities were found to be different for the $2s$ and $2p$ orbitals of the carbon atom, *orbital electronegativities vary with the type of hybrid—that is, with the relative s and p character of the hybrid.* For the *sp* (or *di*gonally) hybridized nitrogen atom, as found in nitriles, the high s character of the hybrid yields a high electronegativity for the orbital. For sp^2 (or *tri*gonally) hybridized nitrogen, as found in pyridine, the hybrid is reduced in electronegativity because of the decrease in s character. The sp^3 (or *te*trahedrally) hybridized nitrogen shows the lowest electronegativity of all hybrids, since its s character is lowest.

For many chemical purposes, the detailed information in Table 3-2 is not necessary. You are already familiar, from your freshman course, with Pauling's scale of nominal atom electronegativities.[3] The Mulliken and Pauling scales are generally equivalent, with the Pauling scale values being given approximately by

$$\chi_P \approx \frac{1}{3}\chi_M - 0.2$$

A table of Pauling values is given in Appendix B, along with atom ionization potentials and electron affinities in Appendices C and D.

TABLE 3-2

SOME MULLIKEN ELECTRONEGATIVITIES (eV)

	H							
s	7.2							

	Li		Be		B		C		N		O		F
s	3.1	di^2	4.8	tr^3	6.4	$di^2\pi^2$	10.4, 5.7	$di^3\pi^2$	15.7, 7.9	$di^3\pi^3$	20.2, 10.3	s	31.3
p	2.1	te^2	3.9	te^3	6.0	$tr^3\pi$	8.8, 5.6	$tr^4\pi$	12.9, 8.0	$tr^5\pi$	17.1, 20.1	p	12.2
						te^4	8.0	te^5	11.6	te^6	15.3		

	Na		Mg		Al		Si		P		S		Cl
s	2.8	di^2	4.1	tr^3	5.5	$di^2\pi^2$	9.1, 5.7	$di^3\pi^2$	11.3, 6.7	$di^3\pi^3$	12.4, 7.9	s	19.3
p	1.6	te^2	3.3	te^3	5.4	$tr^3\pi$	7.9, 5.6	$tr^4\pi$	9.7, 6.7	$tr^5\pi$	10.9, 7.7	p	9.4
						te^4	7.3	te^5	8.9	te^6	10.2		

	K		Ca		Ga		Ge		As		Se		Br
s	2.9	di^2	3.4	tr^3	6.0	$di^2\pi^2$	9.8, 6.5	$di^3\pi^2$	9.0, 6.5	$tr^4\pi^2$	10.6	s	18.3
p	1.8	te^2	2.5	te^3	6.6	$tr^3\pi$	8.7, 6.4	$tr^4\pi$	8.6, 7.0	te^6	9.8	p	8.4
						te^4	8.0	te^5	8.3				

	Rb		Sr		In		Sn		Sb		Te		I
s	2.1	di^2	3.2	tr^3	5.3	$di^2\pi^2$	9.4, 6.5	$di^3\pi^2$	9.8, 6.3	$tr^4\pi^2$	10.5	s	15.7
p	2.2	te^2	2.2	te^3	5.1	$tr^3\pi$	8.4, 6.5	$tr^4\pi$	9.0, 6.7	te^6	9.7	p	8.1
								te^5	8.5				

Values can be computed only for orbitals holding 1 electron. For the carbon and nitrogen families it is possible to have both hybrid and π atomic orbitals half-filled. *di*gonal $\equiv sp$ hybrid, *tr*igonal $\equiv sp^2$ hybrid, *te*trahedral $\equiv sp^3$ hybrid.

---

4. Arrange the following list of atomic orbitals in ascending order of electronegativity.

   a) F(2s), O(2p), Na(3s), F(di)
   b) N(p), O(tr), N(te), C(p)

5. (a) Which orbital pair has the best energy match?
      i) C_{te} with N_{te}
      ii) C_{te} with N_{di}
      iii) C_{te} with O_{te}
   (b) Which orbital in each pair above lies higher in energy (the electron is less tightly held in the atom)?

6. (a) Which atom lone pair is most tightly held in its atom?
      i) the lone pair of N_{te}
      ii) the lone pair of N_{di}
      iii) the lone pair of O_{te}
      iv) the lone pair of O_{tr}
   (b) Which of these corresponds to the strongest nucleophile or Lewis base?

7. On the basis of electronegativities, state which member of each pair has the more polar bond, and indicate the direction of polarity. Use the nominal atom electronegativities in Appendix B.

   C–N or N–O
   P–S or S–Cl
   Sn–I or C–I

STUDY QUESTIONS

Bond Distances

At this point it is interesting to comment on the results of calculations by Maccoll[4] that show the variation of overlap with fractional valence orbital s character in overlapping sp hybrid orbitals. Figure 3-4 is a summary of the results, from which it may readily be seen that overlap of two atomic s orbitals is larger than that of two atomic p orbitals, and both are smaller than the overlaps of the customary sp^3, sp^2, and sp hybrids. It is important to note that the overlap (and presumably the bond strength) increases as the s/p character of the hybrids becomes better balanced (did you work study question 2?).

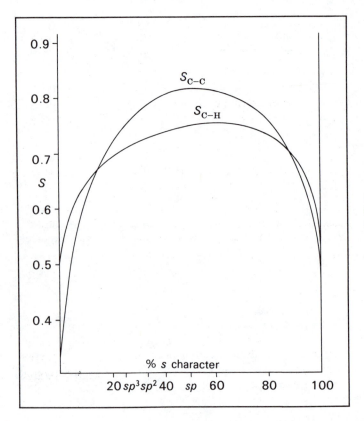

Figure 3-4. The overlap integral for C–C and C–H bonds as a function of % s character. [From A. Maccoll, *Trans. Faraday Soc.,* **46**, 369 (1950).]

Because the valence s ao's are more contracted than the valence p ao's, the s orbital overlap gains in importance (relative to the p overlap) as the internuclear distance shortens. Thus, at shorter bond distances we expect to find greater ϕ_s character in the hao's that best describe the bond pair density. That is, the shorter the chemical bond distance, the more s character there is in the best hao's. These developments markedly demonstrate one conceptual advantage to viewing chemical bond formation from an overlapping hao vantage point. The C–H distances cited below illustrate this simple correlation, as does the fact that the C–C bond of methylacetylene (sp^3/sp overlap) is shorter and stronger than that of ethane (sp^3/sp^3 overlap).

	r_{CH}	Csp^n
H—C≡C—H	106 pm	sp
H₂C=CH₂ (ethylene)	108 pm	sp^2
CH_4	109 pm	sp^3

We are now nearly in a position to illustrate the utility of the hybrid orbital model in the interpretation of other molecular properties. One last, very useful concept must be introduced. Up to this point in the discussion, we have idealized atom hybridization in the sense that all hybrids about a central atom are considered equivalent. In terms of s and p character, this means that all have been considered to have equal $s:p$ ratios. This is, of course, true when symmetry requires it. For example, the carbon hybrids in methane are all of the same $s:p$ character. Similarly, all four carbon hybrids have identical $s:p$ character in CF_4. What about a compound like CH_3F? Here symmetry does not require equivalence of all four carbon hybrids; rather, three (the carbon hybrids directed toward the hydrogens) must have identical $s:p$ ratios, but the hybrid toward fluorine is expected to be different. In what way? The concept of **isovalent hybridization** as promoted by Bent[5] clarifies the situation. The following statements are the basis for the isovalent hybridization idea:

a. The more electronegative an atom, the greater will be its demand for electron density from the atom(s) to which it is bound.
b. The greater the p character of a hybrid orbital, the less its electronegativity.
c. Accordingly, an atom will direct greater p character, and less s character, into its hybrid directed toward the more electronegative of its substituents.
d. Conversely, an atom will direct less p character, and greater s character, into the hybrids directed toward its less electronegative substitutents.

With respect to H_3CF, the C → H hao's should have $> 25\%$ s character, while the C → F hybrid should have $< 25\%$ s character.

Several important aspects of molecular electronic structures and geometries can be interpreted in terms of this concept of hybridization. The inductive transmission of charge withdrawal takes on greater meaning with this model. For example, replacing a hydrogen atom of CH_4 with a fluorine leads to loss of charge from carbon to fluorine, via increased p character in the carbon hybrid toward fluorine. The increased s character in the remaining hybrids toward hydrogen makes the carbon appear to have increased electronegativity from the point of view of the hydrogens. In this way, the hydrogens participate in the intramolecular charge flow toward fluorine. Additionally, since greater hybrid s character tends to lead to shorter bonds, the model is consistent with the observed shortening of the CH bonds.

Atoms containing lone pairs as well as bond pairs tend to direct greater s character into the hybrid identified with the lone pair. This also has a reasonable physical basis, since the lone pair (unlike the bond pair) moves under the influence of only a single nuclear charge and "needs" to experience greater *central* atom nuclear charge than the bond pairs. Within the hybrid orbital model, then, it is reasonable that the lone pair hybrid possesses greater s character. Given the greater angular diffuseness of s than p atomic orbitals, this greater s character of lone pair hybrids correlates well with the VSEPR idea that lone pair electrons subtend greater solid angles about an atom than do bond pairs.

A final structural example involving pentavalent phosphorus is appropriate. You should note that the axial and equatorial hybrids are not required by symmetry to be equivalent. The bond lengths of axial P—Cl bonds and equatorial P—Cl bonds of PCl_5 are different, the axial (219 pm) being greater than the equatorial (204 pm). This result is expected from application of the VSEPR model, for the axial pairs are acted upon by three equatorial pairs, whereas the equatorial pairs are acted upon by only the two axial pairs. In hybrid orbital terms, one describes the shorter equatorial bonds as greater in phosphorus s character than the axial bonds. This is an argument in favor of viewing the phosphorus hybrids as *principally* sp^2 (equatorial) and dp (axial). Similarly, PCl_3F_2 exhibits a structure in which both fluorines are axial. This is successfully interpreted with the VSEPR model; and again the isovalent hybridization model of (ideal) $sp^2 + dp$ central atom hybridization is used to describe the phosphorus hybrids. Comparing PCl_5 and PCl_3F_2, the axial hybrids of the latter require less s character than do those of the former.

Force Constants

The classical model of vibrating atoms may be viewed as assuming that Hooke's law of restoring forces is operative between atoms defining bonds and between atoms defining bond angles. This approach has great intuitive value for the chemist, and we will briefly discuss some aspects of this model for their value in characterizing bonds and changes in bonds incurred by chemical reaction.

Two oscillating nuclei whose motion is such that the center of gravity of the molecule remains fixed (no molecular translational motion) can be treated by the expedient of an equivalent system of a single atom of effective or reduced mass given by

$$\frac{1}{\mu} = \frac{1}{m_a} + \frac{1}{m_b}$$

oscillating about some equilibrium point; m_a and m_b are the masses of the two atoms defining the diatomic molecule. According to Hooke's law, the restoring force for a displacement, x, of this effective mass from its equilibrium value is given by $F = -kx$, where k is the force constant. Newton's first law ($F = ma$) leads to the following equation for the vibrational frequency, ν:

$$\nu = \frac{1}{2\pi} \sqrt{\frac{k}{\mu}}$$

Often the experimentalist expresses the frequency of the oscillation in units of reciprocal distance by using the relation $\lambda\nu = c$, where c is the speed of light, 3×10^{10} cm/second. That is,

$$\tilde{\nu} = \frac{1}{\lambda} = \frac{1}{2\pi c} \sqrt{\frac{k}{\mu}}$$

Two important generalizations emerge from this equation. First, the larger the force constant, the higher is the vibrational frequency. Second, the smaller the reduced mass of the oscillator (the lighter the atoms defining the chemical bond), the higher is the frequency. These trends are nicely illustrated by the following comparisons:†

	k, md/Å	μ, amu	$\sqrt{k/\mu}$	$\tilde{\nu}$
C−N	5.3	6.46	0.91	~1000 cm^{-1}
C=N	10.5	6.46	1.27	~1500 cm^{-1}
C≡N	17.2	6.46	1.63	~2000 cm^{-1}
C−C	~5	6.0	.91	~1000 cm^{-1}
O−H	~5	0.941	2.30	~3600 cm^{-1}

From comparison of these nominal values we find a regular, but of course non-linear, increase in vibrational frequency with increasing force constant and constant reduced mass, and in the last two instances, increased frequency with decreased reduced mass and constant force constant. Also interesting is the rough proportionality of CN force constant to bond order. An association such as this has led chemists to expect a rough correlation of bond energy with force constant.

In a chemical reaction, *the change in vibrational frequency for a bond may be diagnostic of the change in force constant for the bond and thus of the nature of the reaction.* For example, the coordination of the carbonyl oxygen of ethyl acetate by a Lewis acid such as BBr_3 should decrease the double bond character of the carbonyl link

---

†To make the actual computation, the following conversion factors are required: 1 amu = 1.660×10^{-27} kg; 1 md/Å = 100 N/m; and 1 N kg^{-1} m^{-1} = 1 s^{-2}.

and thereby reduce the C=O vibrational frequency because of a reduction in C=O bond order.

$$CH_3-C\overset{\ddot{O}:}{\underset{\ddot{O}-CH_3}{}} + BBr_3 \rightarrow \overset{CH_3}{\underset{\underset{CH_3}{\ddot{O}}}{}}C=\ddot{O}_{BBr_3}$$

The reduction in C=O bond order is predicted by consideration of the change in relative importance of the three important π resonance structures. That is, the single bond structures for the carbonyl oxygen are more important in coordinated than in "free" ester.

$$R-C\overset{\ddot{O}:}{\underset{\ddot{O}-R'}{}} \leftrightarrow R-\overset{\oplus}{C}\overset{\overset{\ominus}{\ddot{O}:}}{\underset{\ddot{O}-R'}{}} \leftrightarrow R-C\overset{\overset{\ominus}{\ddot{O}:}}{\underset{\overset{\oplus}{O}-R'}{}}$$

In fact, the C=O stretching frequency of ethyl acetate, at ~1600 cm^{-1}, is observed to decrease by about 190 cm^{-1} on adduct formation with BBr$_3$. It is useful to note, too, that the low energy shift strongly suggests that it is the *carbonyl oxygen* that acts as the donor atom in this reaction and not the *ester oxygen*. (This has been confirmed by x-ray structural analysis of the adduct.) The resonance structures given above suggest that *ester oxygen* coordination would incur an *increase in the carbonyl frequency*.

In using the arguments advanced above, one must be aware that the sigma orbitals have been ignored. Often this is done with good justification, since the sigma bonds are viewed as overlapping valence *sp* hybrids. Because of the *s* character (high effective nuclear charge) of the σ hybrids, σ electron pairs are less polarizable than $p\pi$ electron pairs. In short, the carbon $p\pi$ ao is less electronegative than the carbon "sp^2" sigma hao, so the polarization of the C=O electron density occurs mainly through the π bond pair as opposed to the σ bond pair.

A particularly interesting application[6] of these ideas can be made for the CN stretching frequencies of CN$^-$, HCN, and SCN$^-$. We will examine the changes in σ and π bonding between C and N as a result of addition of H$^+$ and S to the cyanide carbon. We expect stiffening of the C≡N σ bond in both cases, with a lesser change for S than H$^+$ (sulfur should be less electronegative than the *bare* proton). In both cases, the increase in carbon hybrid *s* character *and* polarization of the σ pair toward carbon act in concert to strengthen the bond. Considering the π resonance structures

$$:\overset{\ominus}{C}\equiv N: \leftrightarrow :C=\ddot{N}:^{\ominus}$$

$$H-C\equiv N: \leftrightarrow H-\overset{\oplus}{C}=\ddot{N}:^{\ominus}$$

$$\overset{\ominus}{:}\ddot{S}-C\equiv N: \leftrightarrow :S=C=\ddot{N}:^{\ominus}$$

we readily predict an increase in C≡N π bond order for HCN (the first structure, with no charge separation, is favored) but a decrease in π bond order for SCN$^-$ (the second structure, with the charge on N, is favored). Combining the σ and π bonding effects, it is straightforward that HCN should have a greater CN force constant than CN$^-$, but a prediction is not so easy for SCN$^-$ because the σ and π effects are oppositely directed. The experimental CN stretching frequencies are 2080 cm^{-1} (CN$^-$), 2097 cm^{-1} (HCN), and 2006 cm^{-1} (SCN$^-$). Clearly, the predicted increase is realized for HCN, and the π effect is seen to dominate the σ effects for SCN$^-$.

One further example of the use of vibrational frequency–bond strength correlations concerns the mode of coordination of dimethylsulfoxide (DMSO), $(CH_3)_2 SO$, which possesses two donor atoms. The two important π resonance structures for DMSO are

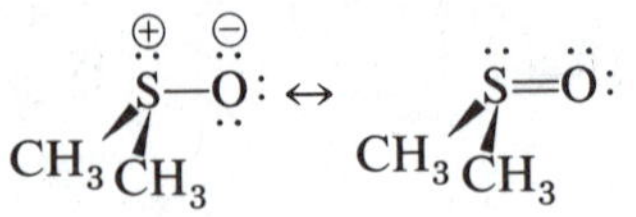

$$\overset{\oplus}{S}—\overset{\ominus}{\overset{..}{\underset{..}{O}}}: \leftrightarrow \quad S=\overset{..}{\underset{..}{O}}:$$

Coordination of the oxygen atom enhances the contribution of the first structure relative to the second, while sulfur coordination favors the second relative to the first. By analogy with $RCO_2 R'$, oxygen coordination should have a slight effect on the S—O σ bond, while sulfur coordination should strengthen the S—O σ bond (cf. CN^-). Consequently, the combined σ/π effects suggest a decreased SO frequency for oxygen coordination and increased $\tilde{\nu}_{SO}$ for sulfur coordination. Examples of ambidentate behavior are found[7] in $Cr(DMSO)_6^{2+}$, which exhibits oxygen coordination and an SO stretching frequency 27 cm^{-1} lower than that of the free DMSO, and in $PtCl_2 (DMSO)_2$, which exhibits sulfur coordination and an SO absorption band 61 cm^{-1} above that of the free ligand.

Nuclear Spin Coupling

Another experimental result that is easily interpreted from the hao view is that of the coupling of a hydrogen nuclear spin to the nuclear spin of an atom to which the hydrogen is bound. You are already aware that the proton spin property makes possible the nmr absorption experiment. In the presence of a laboratory magnetic field, the nuclear magnetic moments (intrinsic to the spin angular momentum) of a sample of protons will exhibit nearly equal numbers of aligned and anti-aligned nuclear magnets; the nmr experiment detects the aligned magnets by measuring the work required to "flip" them to an anti-aligned condition. The position of the absorption line on an energy scale represents the energy absorbed by the proton upon flipping. If the hydrogen is bound to an atom that also possesses a nuclear magnet, the pair of hydrogen/other-atom magnets gives rise to four spin states (an upward-pointing arrow represents a nuclear magnet aligned with the laboratory field):

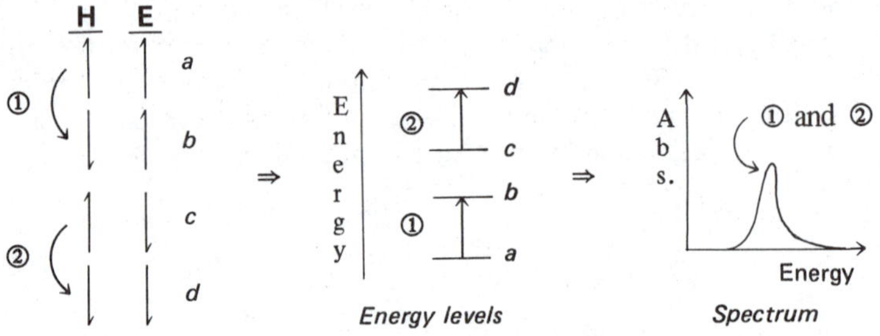

These four possibilities represent different energy states for the system of four magnets, and a real sample will have molecules nearly equally distributed over the four possibilities.

If there were no communication between the hydrogen and other-atom magnets, flipping the hydrogen spin as in ① would require the same energy as in ② (because the proton is acted upon by only the *external* magnetic field). If by some mechanism the *internal* field of the other-atom magnet augments the *external* field, as in ① say, and opposes the *external* field, as in ② for example, "flipping" the hydrogen spin in ① would require more energy than in ②. This will give rise to two different energy absorption phenomena for the protons.

It is the coupling mechanism in which we are interested here. The electron pair defining the H—E bond may be treated as two particles also possessing opposite magnetic moments. If neither H nor the other atom possessed a nuclear magnet, the α and β electrons would have equal probability of occurring at the hydrogen atom and at the other atom, say carbon:

$$H \quad C \quad = \quad H \quad C$$

$$\alpha \quad \beta \qquad \beta \quad \alpha$$

If carbon, as in the ^{13}C isotope, has a nuclear spin, this equal distribution of electrons will be upset with the result

$$H \quad \uparrow C \quad > \quad H \quad \uparrow C$$

because of the preference for alignment of electron/nuclear magnets. The ↑ and ↓ moment electrons no longer have the same probability functions (the orbitals are slightly different). Accordingly, the hydrogen experiences a net *internal* magnetic field from the unequal electron probabilities at H, and which is opposed to the external field. Molecules with the *carbon magnet aligned with the external field* exhibit an *internal field at hydrogen anti-aligned with the external field.*

Internal field at hydrogen	Carbon nuclear alignment
↓	C↑
↑	C↓

Including the electron moments, the net direction of the internal field at hydrogen is determined from

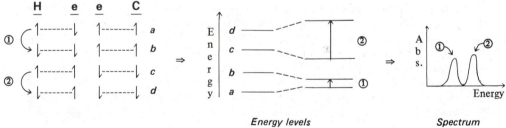

Energy levels   *Spectrum*

Now the "flip" ② requires more energy than the "flip" ① and the H absorption occurs at two photon frequencies instead of one. Clearly, the *internal* field at hydrogen depends on the electron probability at the proton. For hydrogen, the $1s$ ao has non-zero amplitude at the nucleus, and this enhances the *internal* field felt by the proton nuclear magnet. Similarly, the **spin polarization** of the C—H bond pair will increase as the $2s$ character of the carbon hybrid increases, because the carbon nuclear field felt by the electrons increases. This phenomenon is called **Fermi contact coupling** (of electron/nuclear magnets).

To illustrate the sensitivity of the difference in absorption processes ① and ② to the s character of the carbon hybrid, we list the following ^{13}CH coupling constants; J (defined as the energy difference between the two proton absorption events): CH_4 (sp^3, 125 Hz), C_2H_4 (sp^2, 156 Hz), C_2H_2 (sp, 249 Hz). Hz is the Herz frequency unit, sec^{-1}, and is proportional to energy through $E = h\nu$. Even more subtle changes in carbon s character can be detected by CH coupling, as illustrated by CH_4 (sp^3, 125 Hz) compared to HCF_3 (sp^3, 239 Hz). Here again we find application of Bent's isovalent hybridization scheme.

Table 3-3 summarizes typical data that are interpretable in this fashion.[8] Before leaving this topic we should pause to note that correlating J with hybrid %s character alone ignores what may be another important consideration. Electron withdrawal by a substituent descreens the central atom nucleus and enhances the valence s orbital amplitude at the nucleus (a contraction effect). This also may enhance J.[8]

Structure	J(Hz)	Structure	J(Hz)
CH_4	125	$HC(F)CH_2$	200
C_2H_4	156	H_2CO	172
C_2H_2	249		
HCF_3	239	$HC(O)(Me)$	172
$HC(OMe)_3$	186	(NMe_2)	191
		(OH)	222
$H_3C(SiH_3)$	122	$HC{\equiv}(P)$	211
(PH_2)	128	(CPh)	251
(NH_2)	133	(N)	269
(CN)	136	(NH^+)	320
(NH_3^+)	145		
(OH)	141	H_2P^-	139
(NO_2)	146	H_2PH	182
(F)	149	$H_2PH_2^+$	547
		H_2PCH_3	186
$H_3Si(H)$	203	H_2PCF_3	200
(SiH_3)	196		
H_3N	40		
H_3NH^+	54		

TABLE 3-3

SOME REPRESENTATIVE J_{EH} DATA

Bond Energies

From your general and organic chemistry courses, you will recall the utility of the concept of additivity of bond energies. That such a concept works in practical applications reaffirms the idea that it is possible to think of localized bond pairs of electrons as characteristic of the atoms defining the bond. Here we will pursue the concept of an experimental bond energy and its relation to atom hybridization.

The definition of a bond energy is not ambiguous at all for a diatomic molecule. Experimentally the bond energy is simply the internal energy change required to separate the molecule into its two constituent atoms (the molecule and atoms in the gas phase):

$$A{-}B_{(g)} \rightarrow A_{(g)} + B_{(g)} \qquad D_{AB} = \Delta E$$

This definition of a bond energy, applied to a polyatomic molecule, must be carefully considered in relation to the valence state concepts discussed earlier. For general usage, the *bond energies* of AB_n molecules are taken to be $1/n$ times the internal energy change for the process

$$AB_{n(g)} \rightarrow A_{(g)} + nB_{(g)}$$

$$E_{AB} = \frac{1}{n} \Delta E$$

and are to be distinguished from the concept of individual *bond dissociation energies,* D_{AB}.

The theoretical interpretation of bond energies requires the factorization of atom relaxation energies (these are simply the negatives of the valence state promotion energies, P, of p. 78) to get at the bond energies in which the products represent atoms not in their ground electronic states but with the valence state configurations appropriate to the molecule under consideration. In this way we arrive at more meaningful bond energies for interpretation in terms of the effects of multiple bond formation, sigma orbital hybridization, and the degree of overlap of atomic hybrids.

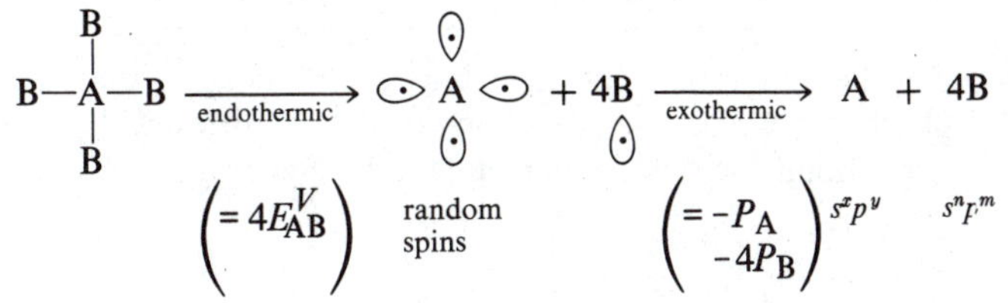

$$
\begin{array}{c}
B \\
| \\
B-A-B \xrightarrow[\text{endothermic}]{} \;\bigodot A \bigodot + 4B \xrightarrow[\text{exothermic}]{} \; A \;+\; 4B \\
| \\
B
\end{array}
$$

$$
\left(= 4E^V_{AB}\right) \qquad \begin{array}{c}\text{random}\\\text{spins}\end{array} \qquad \left(\begin{array}{c}= -P_A\\ -4P_B\end{array}\right) \; s^x p^y \qquad s^n_{\ } r^m
$$

As an interesting application of the ideas thus far discussed, the bond energies (kJ/mole) for the hydrides of some of the second and third row elements are given in the following table (taken from Table 6-3):

C 414	N 389	O 464
Si 318	P 326	S 368

Given that the heavy atom promotion energies to the tetrahedral valence states P_{te} are

C 628	N 954	O 812	F 502
Si 481	P 556	S 456	Cl 259

it is apparent that the bonds formed by the lighter elements in each group are truly stronger than those of the heavier elements. To illustrate the comparison of CH_4 with SiH_4, for example, we write

$$4E_{CH} = 4(414) = 4E^V_{CH} - P_C - 4P_H$$

$$4E_{SiH} = 4(318) = 4E^V_{SiH} - P_{Si} - 4P_H$$

$$1656 = 4E^V_{CH} - 628 - 0$$

$$1272 = 4E^V_{SiH} - 481 - 0$$

$$E^V_{CH} = \tfrac{1}{4}(2284) = 571$$

$$E^V_{SiH} = \tfrac{1}{4}(1753) = 438$$

and the CH bond is 133 kJ/mole stronger than the SiH bond, exclusive of atom relaxation energies. This is a ramificaion of the greater diffuseness of the atomic orbitals of the heavier elements. The diffuseness has its origin in the radial parts of the hybrid orbital functions, and it results in smaller overlap integrals, and consequently weaker bonds, for the orbitals of the heavier elements with hydrogen. For methyl compounds the same trend is found:

C 347	N 331	O 360
Si 305	P 264	S 259

For fluorides and oxides, however, the trend is often reversed, presumably because fluorine and oxygen have unshared electron pairs for multiple bond formation with the heavier elements. The bond energies are:

C 490	N 280	O 213	Fluorides
Si 598	P 498	S 285	
C 360	N 163	O 142	Oxides
Si 464	P 368	S —	

Correction for the promotion energies shows, for example, that the P—F bond in PF_3 is 85 kJ/mole stronger than the N—F bond in NF_3. This reversal is probably due not only to P—F multiple bond character enhancing the P—F energy:

$$\ddot{P}—\ddot{F}: \leftrightarrow \ddot{P}=F:$$

but also to greater lp/lp repulsions for N—F, acting to weaken that bond more:

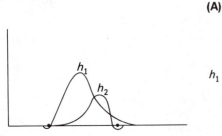

Another interesting example of the application of these ideas is to be found in the instability of the heavy elements of Groups III and IV with oxidation states of III and IV. Boron and carbon form stable compounds of formulae BX_3 and CX_4 with X = halide, whereas the elements Tl and Pb form as the most stable halides the compounds TlX and PbX_2. Drago[9] has discussed this problem thoroughly and finds that the energy gained by forming two more bonds in each case does not offset the promotion energy involved in achieving trigonal and tetrahedral valence states (tr^3 from s^2p^1 and te^4 from s^2p^2, respectively). It does not appear that the promotion energies are inordinately large for Tl and Pb, but rather that the bonds formed by trigonal Tl and tetrahedral Pb are not very strong. Again, the weakness of the bonds of Tl and Pb is to be attributed to the highly diffuse character of the atomic orbitals of these elements.

To help in visualizing the importance of orbital compactness in producing large overlap integrals, Figure 3-5 schematically illustrates the effect.

(A)

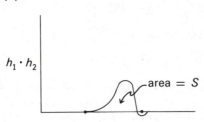

Figure 3-5. The effect of increasing radial diffuseness of an atomic hybrid on the overlap of that hybrid with another. (A) h_1 "compact," (B) h_1 "diffuse."

(B)

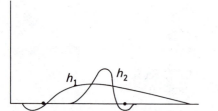

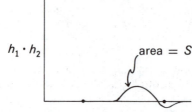

8. Use the VSEPR model to predict the stable structures for PCl_2F_3, $PCl_2(CF_3)_3$, and SF_4; then describe the "ideal" central atom hybrids for each electron pair. For each of these molecules, tell how the hybrids differ in s character in actuality ("non-ideal").

9. (a) Compare, on the basis of the hao model, the predicted bond strengths of the $P-F$ bonds in PF_3 and PF_5. (b) Rank the bonds in order of decreasing Fermi contact coupling (J_{PF}).

10. Predict a VSEPR structure for HCP. The nmr spectra of this compound reveal $J_{HC} = 211$ Hz, $J_{CP} = 54$ Hz, and $J_{HCP} = 44$ Hz. In what ways do these data (and the facts that the J_{CP} of PR_3 compounds are much smaller than 54 Hz and that J_{HCP} is typically 0 to 25 Hz) support the predicted structure?

11. Predict an order of vibrational frequencies for:

   a) $Xe-O$ in XeO_2, $XeOF_4$
   b) $I-Cl$ in ICl_3
   c) $M-H$ in CH_4, SiH_4

12. Consider the compound $Me_3N \cdot SnMe_3Cl$, derived from the Lewis acid-base addition of Me_3N to $SnMe_3Cl$.

   a) Sketch the structure of the compound, showing the geometry of the N, C, and Cl atoms about the Sn atom. (Be sure to tell *briefly* why you pick one isomer or another.)
   b) Predict a direction of change for the $Sn-Cl$ stretching frequency upon formation of the adduct and justify it, briefly.
   c) Do you expect J_{SnCH} to increase or decrease upon adduct formation?

13. Which should have the stiffer $C=O$ bond, CO_2 or cyanate?

14. Which group should show the greater σ orbital electronegativity toward X? Explain.

a) $\overset{\diagup}{B}-X$ or $\overset{\diagup}{C}-X$ b) ⬡$-X$ or H_3C-X c) $\overset{..}{\underset{..}{O}}=X$ or $H\overset{..}{\underset{..}{O}}-X$

15. Consider the molecule SOF_4.

   a) Draw Lewis structures for the two principal resonance forms.
   b) Estimate limits for the $S-O$ bond order.
   c) Predict the stable isomer.
   d) Are all $S-F$ bonds of equal length? If not, tell which are longer.
   e) Tell how the bond angles differ from ideal.
   f) Give the correct molecular point group.
   g) Rank the sulfur hybrids in order of decreasing s character.
   h) Which should have the stronger, shorter $S-O$ bond, SOF_4 or SOF_2?

16. For each of the following stoichiometries, give the "main group" group number(s) from which the central atom may come.

   a) EF_3 b) EF_4 c) EF_5 d) EF_6

17. For set EF_3 of question 16, answer each of the following questions for an example from *each* group:

   a) What are the *spd* notations for the central atom hybrids of your examples, assuming "ideal" VSEPR structures?
   b) Comparing bond pairs in the different EF_3 molecules, rank the bonds in order of predicted strength on the basis of E hybridization alone.
   c) Rank the EF_3 bonds in order of predicted magnitude of $E-F$ Fermi contact coupling (assume the F hybridization to be constant).

18. Consider the molecular fragments (i) $\overset{\ominus}{\overset{\cdot\cdot}{M}}-N\equiv C-R$ and (ii) $\overset{\ominus}{\overset{\cdot\cdot}{M}}-C\equiv N-R$, and determine for which of these the lone pair on M is more likely to become a bonding pair (M=N $\cdots$ or M=C $\cdots$). Which species, on the basis of this answer alone, will have higher ν_{CN}?

CHECKPOINTS

1. Two waves, both of non-zero amplitude at some point in space, may be in phase (both positive or both negative amplitudes) or out of phase (opposite sign amplitudes). The in-phase (or constructive) interference case leads to a new wave of amplitude greater than either component alone. Destructive interference implies wave cancellation.

2. From the probability density meaning of the squares of these new waves, constructive interference implies a build-up of electrons while destructive interference means lower electron probability.

3. Because s waves have greater effective nuclear charge than p waves, hybrid s-p atomic orbitals have energies which decrease, and electronegativities which increase, with %s character.

4. The angular parts of the s and p waves determine the solid angles subtended by the various hybrid orbitals and their orientation with respect to each other. This property of hybrids fits well the predictions of molecular shape from VSEPR.

5. Bonds formed from hybrids with high s character tend to be shorter and stronger, with greater electronegativity, higher vibrational frequencies, and greater coupling of nuclear spins.

REFERENCES

[1] R. S. Mulliken, *J. Chem. Phys.*, **2**, 782 (1934); **3**, 573 (1935).

[2] J. Hinze and H. H. Jaffe, *J. Am. Chem. Soc.*, **84**, 540 (1962); J. Hinze, M. A. Whitehead, and H. H. Jaffe, *J. Am. Chem. Soc.*, **85**, 148 (1963); J. Hinze and H. H. Jaffe, *J. Phys. Chem.*, **67**, 1501 (1963).

[3] L. Pauling, "The Nature of the Chemical Bond," Cornell University Press, Ithaca, N.Y. (1960, 3rd edition); A. L. Allred, *J. Inorg. Nucl. Chem.*, **17**, 215 (1961).

[4] A. Maccoll, *Trans. Faraday Soc.*, **46**, 369 (1950).

[5] H. A. Bent, *Chem. Rev.*, **61**, 275 (1961); also, C. A. Coulson, "Valence," Oxford University Press, New York (1952), pp. 198–201.

[6] K. F. Purcell, *J. Amer. Chem. Soc.*, **89**, 247, 6139 (1967); **91**, 3487 (1969).

[7] J. H. Price, A. N. Williamson, R. F. Schramm, and B. B. Wayland, *Inorg. Chem.*, **11**, 1280 (1972).

[8] G. E. Maciel, *et al.*, *J. Amer. Chem. Soc.*, **92**, 1 (1970); A. H. Cowley and W. D. White, *ibid.*, **91**, 1917 (1969).

[9] R. S. Drago, *J. Phys. Chem.*, **62**, 353 (1958).

Appendix to Chapter 3

Appendix A: Basic Hybrid Atomic Orbitals

sp^2 (D_{3h})

(tr_1, tr_2, tr_3)

Schrödinger: $s + p_x, p_y$

	ao character	
	s	p
$tr_1 = \dfrac{1}{\sqrt{3}}(\phi_s + \sqrt{2}\phi_x)$	1/3	2/3
$tr_2 = \dfrac{1}{\sqrt{3}}\left(\phi_s - \sqrt{\dfrac{1}{2}}\phi_x + \sqrt{\dfrac{3}{2}}\phi_y\right)$	1/3	2/3
$tr_3 = \dfrac{1}{\sqrt{3}}\left(\phi_s - \sqrt{\dfrac{1}{2}}\phi_x - \sqrt{\dfrac{3}{2}}\phi_y\right)$	1/3	2/3

Table continued on the following page.

Appendix A: Basic Hybrid Atomic Orbitals (*Continued*)

sp^3 (T_d)
$(te_1, \ldots, te_4)$
Schrödinger: $s + p_x, p_y, p_z$

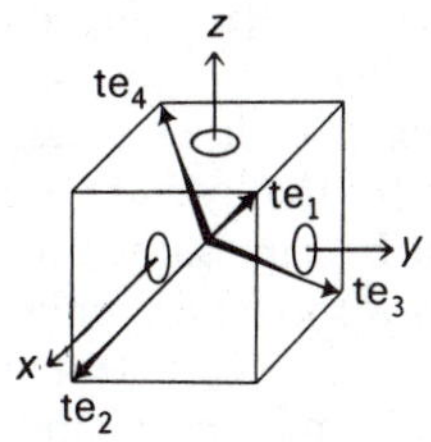

	ao character	
	s	p
$te_1 = \frac{1}{2}(\phi_s + \phi_x + \phi_y + \phi_z)$	1/4	3/4
$te_2 = \frac{1}{2}(\phi_s + \phi_x - \phi_y - \phi_z)$	1/4	3/4
$te_3 = \frac{1}{2}(\phi_s - \phi_x + \phi_y - \phi_z)$	1/4	3/4
$te_4 = \frac{1}{2}(\phi_s - \phi_x - \phi_y + \phi_z)$	1/4	3/4

dsp^3 (D_{3h})
(tr_1, tr_2, tr_3) (equatorial)
(di_4, di_5) (axial)
Schrödinger: $s, d_{z^2} + p_z + (p_x, p_y)$

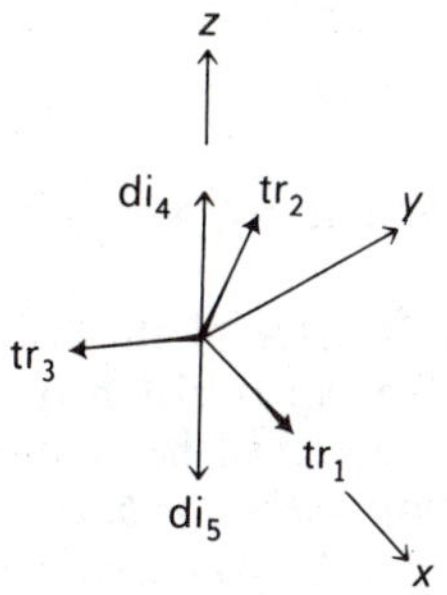

	ao character		
	s	p	d
$tr_1 = \sqrt{\frac{1}{3}}(\phi_s + \sqrt{2}\phi_x)$	1/3	2/3	
$tr_2 = \sqrt{\frac{1}{3}}\left(\phi_s - \sqrt{\frac{1}{2}}\phi_x + \sqrt{\frac{3}{2}}\phi_y\right)$	1/3	2/3	
$tr_3 = \sqrt{\frac{1}{3}}\left(\phi_s - \sqrt{\frac{1}{2}}\phi_x - \sqrt{\frac{3}{2}}\phi_y\right)$	1/3	2/3	
$di_4 = \sqrt{\frac{1}{2}}(\phi_z + \phi_{z^2})$		1/2	1/2
$di_5 = \sqrt{\frac{1}{2}}(\phi_z - \phi_{z^2})$		1/2	1/2

d^2sp^3
$(oc_1, \ldots, oc_6)$
Schrödinger: $s + (d_{z^2}, d_{z^2-y^2}) + (p_x, p_y, p_z)$

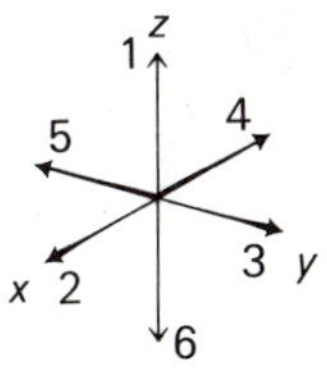

	ao character		
	s	d	p
$oc_1 = \frac{1}{\sqrt{6}}(\phi_s + \sqrt{2}\phi_{z^2} + \sqrt{3}\phi_z)$	1/6	1/3	1/2
$oc_2 = \frac{1}{\sqrt{6}}\left(\phi_s - \sqrt{\frac{1}{2}}\phi_{z^2} + \sqrt{\frac{3}{2}}\phi_{x^2-y^2} + \sqrt{3}\phi_x\right)$	1/6	1/3	1/2
$oc_3 = \frac{1}{\sqrt{6}}\left(\phi_s - \sqrt{\frac{1}{2}}\phi_{z^2} - \sqrt{\frac{3}{2}}\phi_{x^2-y^2} + \sqrt{3}\phi_y\right)$	1/6	1/3	1/2
$oc_4 = \frac{1}{\sqrt{6}}\left(\phi_s - \sqrt{\frac{1}{2}}\phi_{z^2} + \sqrt{\frac{3}{2}}\phi_{x^2-y^2} - \sqrt{3}\phi_x\right)$	1/6	1/3	1/2
$oc_5 = \frac{1}{\sqrt{6}}\left(\phi_s - \sqrt{\frac{1}{2}}\phi_{z^2} - \sqrt{\frac{3}{2}}\phi_{x^2-y^2} - \sqrt{3}\phi_y\right)$	1/6	1/3	1/2
$oc_6 = \frac{1}{\sqrt{6}}\left(\phi_s + \sqrt{2}\phi_{z^2} - \sqrt{3}\phi_z\right)$	1/6	1/3	1/2

In the upper left-hand corner you find the "*sp*" notation for the hao along with the point group symbol for the *set* of hybrids. Each hybrid is represented by an appropriately directed vector. This is followed by a listing of the *specific* Schrödinger ao's of the *identical* symmetry pattern. The linear combinations of these Schrödinger ao's that, by wave superposition, are directed along the hybrid "vectors" are given in the center; to the right of these you find the square of the s ao coefficient and the sum of the squares of the p and/or d ao coefficients. It is, of course, these numbers which determine the "$sp^x d^y$" designation of the hybrids. The fact that the sum of these numbers across a row equals 1 derives from the hao normalization. The fact that the sum of these numbers down a column equals 1 for s, "x" for p, and "y" for d derives from the complete distribution of 1 s, "x" p and "y" d Schrödinger ao's among the hybrids.

While the linear combinations of Schrödinger ao's look complicated in each case, the relative signs of the p/d ao coefficients are simply determined by the hybrid "vector" orientations. Realizing that the p_x and p_y ao's themselves can be represented as unit vectors along the x and y axes, the unit vector tr_1 of "sp^2" is seen to have only a $+x$ component and no y component; thus, ϕ_y does not appear and the coefficient of ϕ_x is positive. The vector tr_2, on the other hand, has a $-x$ component and a $+y$ component. Turning to te_3, you find x, y, and z components with $-$, $+$, and $-$ signs. The d ao's cannot be represented by simple vectors; the coefficient signs are, however, not difficult to explain. Taking oc_3 as an example, the phasing of all s, p, and d ao's in the direction of the $+y$ axis must lead to wave "build-up" in that direction:

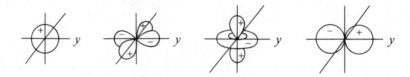

Thus, ϕ_x and ϕ_z do not contribute; $\phi_{x^2-y^2}$ and ϕ_{z^2} both contribute to oc_3 with negative coefficients.

Finally, we call your attention to the fact that the axial hybrids of the trigonal bipyramid (dsp^3) geometry are not symmetry-equivalent to the equatorial hao. The equatorial hybrids must involve ϕ_x and ϕ_y but not ϕ_z, while the axial hao's involve ϕ_z but not ϕ_x or ϕ_y. Each of these two sets requires, in addition, *either or both* of ϕ_s and ϕ_{z^2}. The ambiguity is not resolved by symmetry constraints. As you predicted in Chapter 2 (see the VSEPR discussion of ClF_3), the axial bonds are usually longer than the equatorial bonds. *For the hao model to be consistent with this topological feature,* the equatorial hybrids must have greater ϕ_s contribution, while the axial hao's have greater ϕ_{z^2} character. In this appendix we have presented the limiting case of ϕ_{z^2} in di_4 and di_5, and ϕ_s in tr_1, tr_2, and tr_3.

Appendix B: Nominal Electronegativity Values for the Elements

H 2.2																	
Li 1.0	Be 1.5											B 2.0	C 2.5	N 3.1	O 3.5	F 4.1
Na 1.0	Mg 1.2											Al 1.5	Si 1.7	P 2.1	S 2.4	Cl 2.8
K 0.9	Ca 1.0	Sc 1.2	Ti 1.3	V 1.5	Cr 1.6	Mn 1.6	Fe 1.6	Co 1.7	Ni 1.8	Cu 1.8	Zn 1.7	Ga 1.8	Ge 2.0	As 2.2	Se 2.5	Br 2.7
Rb 0.9	Sr 1.0	Y 1.1	Zr 1.2	Nb 1.2	Mo 1.3	Tc 1.4	Ru 1.4	Rh 1.4	Pd 1.4	Ag 1.4	Cd 1.5	In 1.5	Sn 1.7	Sb 1.8	Te 2.0	I 2.0
Cs 0.9	Ba 1.0	La 1.1	Hf 1.2	Ta 1.3	W 1.4	Re 1.5	Os 1.5	Ir 1.6	Pt 1.4	Au 1.4	Hg 1.4	Tl 1.4	Pb 1.6	Bi 1.7	Po 1.8	At 2.0
Fr 0.9	Ra 1.0															

Lanthanides ~ 1.1

Appendix C: First Ionization Potentials (eV) of the Main Group and Transition Elements

H 13.6																	He 24.6
Li 5.4	Be 9.3											B 8.3	C 11.3	N 14.5	O 13.6	F 17.4	Ne 21.6
Na 5.1	Mg 7.6											Al 6.0	Si 8.1	P 10.5	S 10.4	Cl 13.0	Ar 15.8
K 4.3	Ca 6.1	Sc 6.5	Ti 6.8	V 6.7	Cr 6.8	Mn 7.4	Fe 7.9	Co 7.9	Ni 7.6	Cu 7.7	Zn 9.4	Ga 6.0	Ge 7.9	As 9.8	Se 9.8	Br 11.8	Kr 14.0
Rb 4.2	Sr 5.7	Y 5.7	Zr 6.4	Nb 6.8	Mo 7.1	Tc 7.3	Ru 7.4	Rh 7.5	Pd 8.3	Ag 7.6	Cd 9.0	In 5.8	Sn 7.3	Sb 8.6	Te 9.0	I 10.5	Xe 12.1
Cs 3.9	Ba 5.2	La 5.2	Hf 7.0	Ta 7.9	W 8.0	Re 7.9	Os 8.7	Ir 9.0	Pt 9.0	Au 9.0	Hg 10.4	Tl 6.1	Pb 7.4	Bi 7.3	Po 8.4	At 9.5	Rn 10.7

Appendix D: Electron Affinity Values (eV) of the Elements (estimated values in parentheses)

H 0.75									He (−0.22)
Li 0.62	Be (−2.5)			B 0.24	C 1.27	N 0.0	O 1.47	F 3.34	Ne (−0.30)
Na 0.55	Mg (−2.4)		Cu 1.28	Al 0.46	Si 1.24	P 0.77	S 2.08	Cl 3.61	Ar (−0.36)
K 0.50	Ca (−1.62)		Ag 1.30	Ga (0.37)	Ge 1.20	As 0.80	Se 2.02	Br 3.36	Kr (−0.40)
Rb 0.49	Sr (−1.24)		Au 2.31	In (0.35)	Sn 1.25	Sb 1.05	Te 1.97	I 3.06	Xe (−0.42)
Cs 0.47	Ba (−0.54)			Tl (0.5)	Pb 1.05	Bi 1.05	Po (1.8)	At (2.8)	Rn (−0.42)
Fr (0.46)									

4. THE UNDIRECTED ORBITAL VIEW OF CHEMICAL BONDS

Although the LCAO technique was initially developed shortly after Schrödinger published his model of the atom, the chemical community is just now beginning to probe the power of this theoretical technique and is finding it to be tremendously useful in chemical interpretations of molecular spectra, structure, and reaction mechanisms. The LCAO-MO model is less restrictive at the outset than the localized pair approach by refusing to be predisposed on the ideas of valence s and p ao mixing (hybridization) and localization of electron pairs. Orbital symmetries play a very important role in this model, as do the ideas of matter wave interference and orbital nodal structure.

INTRODUCTORY COMMENTS

In this chapter you will learn how to apply the approximation known as the **linear combination of atomic orbitals (LCAO)** method,[1] which has been important to the chemist by yielding chemically useful wavefunctions for molecules. The basic assumption in this method is as follows: molecules can be thought of as aggregates of atoms close enough that they perturb one another; however, the atoms are not so perturbed that we have to forego the useful idea that the electrons, when in the vicinity of each nucleus, behave much like they do in the isolated atom. It is assumed, then, that the molecular wavefunction for each electron can be written as a linear combination of atomic orbital wavefunctions for the various atoms. Thus, the molecular orbital functions are assumed to be of the form

$$\psi_i = c_1\phi_1 + c_2\phi_2 + \cdots$$

where the ϕ's are the atomic orbitals of the atoms making up the molecule. The linear combination could include an infinity of atomic orbitals (the more terms carried, the greater the accuracy of the calculation) but, for our purpose, it suffices to include only the valence atomic orbitals of each atom.

In this text we will treat the results of such calculations as empirical findings with the intent of systematizing orbital patterns associated with a few basic nuclear geometries. There are quite a few systematic principles to be found in the results of the "number crunching" process of determining molecular orbitals. *Most important to us will be the concepts of symmetry, orbital probabilities given by ψ_i^2, and the phase relationships between the atomic orbitals as combined into LCAO molecular orbitals.*

MOLECULAR ORBITAL PROBABILITY FUNCTIONS

As was thoroughly discussed in the preceding chapter and in the atomic structure chapter, if the orbital functions are to have physical meaning, each function must be normalized to unity, corresponding to unit probability that the electron will be found in all of space. The mathematical condition is

$$\int \psi_i^2 \, d\tau = 1$$

where $\psi_i^2(x, y, z)$ is the point probability for an electron in ψ_i at the point (x, y, z).

Since the molecular orbital function ψ_i is given by

$$\psi_i = c_1 \phi_1 + c_2 \phi_2 + \cdots$$

the point probability density function is given by

(4–1)
$$\begin{aligned}\psi_i^2 &= c_1^2 \phi_1^2 + c_2^2 \phi_2^2 + 2c_1 c_2 \phi_1 \phi_2 + \cdots \\ &= c_1^2 P_1 + c_2^2 P_2 + 2c_1 c_2 P_{12} + \cdots\end{aligned}$$

where the P's are point probability density contributions from the various ao's and the integrated probability is

(4–2)
$$\int \psi_i^2 \, d\tau = c_1^2 + c_2^2 + 2c_1 c_2 S_{12} + \cdots$$

The last equation follows from the preceding one if the individual atomic orbitals are normalized ($\int \phi_m^2 \, d\tau = 1$), as usual, and S_{12} is the **overlap integral** between atomic orbitals ϕ_1 and ϕ_2, as discussed in the earlier section on overlapping hybrid ao's (see p. 76).

Mulliken[2] has given a very useful interpretation of this probability function that will form the basis for our later interpretation of molecular orbital functions.

The function in equation (4–1) is the point probability function. The region of space of special interest to us is that between the nuclei for ϕ_1 and ϕ_2—the bond region. The term $c_1 c_2 P_{12}$ is large only in this region because ϕ_1 and ϕ_2 are *simultaneously* large only in this region. If $c_1 c_2 P_{12}$ has a + sign at points in the bond region, then you see that there is *constructive* wave interference and the interference term *amplifies* the electron probability above that of $c_1^2 P_1 + c_2^2 P_2$ alone. Such a situation ($c_1 c_2 P_{12} > 0$ in the bond region) means that the molecular orbital is **bonding** between ϕ_1 and ϕ_2. On the other hand, a molecular orbital can be **anti-bonding** between ϕ_1 and ϕ_2; this will arise when P_{12} is opposite in sign to $c_1 \cdot c_2$ for points in the bond region. The waves ϕ_1 and ϕ_2 *destructively* interfere ($c_1 c_2 P_{12} < 0$) so as to *reduce* the electron probability below the value given by $c_1^2 P_1 + c_2^2 P_2$ at points in the bond region. An important consequence of $c_1 c_2 P_{12} < 0$ is that, for a series of points in the bond region, the interference term exactly cancels $c_1^2 P_1 + c_2^2 P_2$ so that $\psi_i = 0$. These points define a **nodal surface** in the bond region.

In equation (4–2), c_1^2 gives the ϕ_1 character of the mo, c_2^2 gives the ϕ_2 character, and $c_1 c_2 S_{12}$ (called the **overlap probability**) reflects the net build-up or depletion of electron density at all points in the bond region from interference of ϕ_1 and ϕ_2. This idea of ao character for an mo is just like that of s and p character for hybrid ao's (where ϕ_1 and ϕ_2 belong to the same nucleus). When $c_1^2 \gg c_2^2$, the mo appears more like ϕ_1 than ϕ_2.

The question remains, how do you tell the sign of S_{12}? Let us analyze two situations:

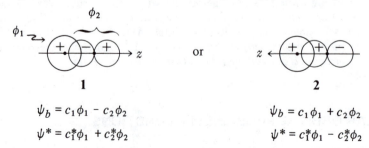

$$\psi_b = c_1 \phi_1 - c_2 \phi_2 \qquad\qquad \psi_b = c_1 \phi_1 + c_2 \phi_2$$
$$\psi^* = c_1^* \phi_1 + c_2^* \phi_2 \qquad\qquad \psi^* = c_1^* \phi_1 - c_2^* \phi_2$$

The difference between **1** (for which $S_{12} < 0$) and **2** (for which $S_{12} > 0$) is occasioned by a difference in choice of coordinate systems. The bonding mo for **1** is written $c_1 \phi_1 - c_2 \phi_2$

so that with $c_1 c_2 > 0$, and since $P_{12} < 0$ in the bond region, the interference term $-2c_1 c_2 P_{12} > 0$. For **2**, the bonding mo is written $c_1 \phi_1 + c_2 \phi_2$, and $c_1 c_2 P_{12} > 0$ also. Below the bonding wavefunctions are written the forms of the anti-bonding mo's for **1** and **2**; the superscript * is commonly used to denote anti-bonding functions.

This example illustrates the fact that the choice of coordinate system is physically immaterial, but mathematically affects the signs before c_1 and c_2. Given an mo function, you must know the coordinate system used to derive the mo's in order to describe the orbital as bonding or anti-bonding.

Based on these simple physical arguments of electron probability, you should expect an electron to be *stabilized* (have a lower energy) in the *bonding case* and *destabilized* in the *anti-bonding case.* The effect of in-phase and out-of-phase combinations of atomic orbitals on the energy of electrons "occupying" them is graphically characterized in terms of an mo energy level diagram (**3**) for a general two-orbital case. The energies of the interacting atomic orbitals are indicated by horizontal lines to the left and right of **3**, and the energies of electrons in bonding and anti-bonding molecular orbitals are given in the middle; note that the diagram conveys the information that the electron energy for the *bonding* orbital is *lower* than that of either atomic orbital, whereas the electron energy for the *anti-bonding* orbital is *higher* than that of either atomic orbital.

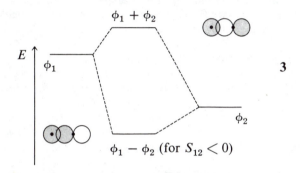

Table 4–1 summarizes the commonly encountered ao overlap types; the identification of each type is based on whether the overlap probability (P_{12}) arises from interference of just one lobe of each ao (σ), from two lobes (π), or from four lobes (δ).

TABLE 4–1

THE COMMON TYPES OF ATOMIC ORBITAL OVERLAP

Type	Examples	
Sigma	⬤⬤	*s,s*
	⬤⬤⬤	*s,p*
	⬤⬤⬤	*p,p*
	✿•✿	*d,d*
Pi	(p orbitals)	*p,p*
	(d orbitals)	*d,d*
	(p,d orbitals)	*p,d*
Delta	(d orbitals)	*d,d*

With these basic ideas of orbital probability and energy in mind, we will now proceed to examine and discuss the qualitative orbital patterns and energy level diagrams for some basic structural arrangements of nuclei.

PRINCIPLES OF MO CONSTRUCTION AND INTERPRETATION

"Linear" (One-Dimensional) Molecules

Thinking back to Chapter 1, one of the more important distinctions between ao's of different energy is the number of nodal surfaces. This is no less true for molecules. Realizing that each mo is characterized by a specific nodal pattern (number and placement), we draw the following key generalization from theoretical studies:

All molecular orbital nodes must be symmetrically disposed.

For linear arrangements of nuclei, the consequences of this statement are far-reaching in the development of an "intuition" of orbital shapes. The orbital nodal patterns calculated for the general cases of two through six atoms are collected in Figure 4–1 for future reference. You will note the following characteristics:

1. *Successively higher energy orbitals have mo nodes symmetrically placed and increasing in number by one, in order from most stable to least stable.*
2. *An mo node through a nucleus means that the ao from that atom does not contribute.*
3. *The mo energy increases with increasing numbers of nodes because the net number of bonding interactions (between neighboring atoms) decreases by two as the number of mo nodes increases by one.*

Furthermore, there is a unifying relationship between the placement of mo nodes in one orbital and that in the next higher orbital. A central mo node "splits" and shifts outward by $\sim d$ (d = bond distance), and this new pair of nodes continues to shift outward by $\sim d$ until placed just inside the terminal atoms (for the cases in which n is odd, the last shift, of necessity, is by $\sim d/2$). This pattern is continued for all successive central mo nodes (check the $n = 5$ and $n = 6$ cases). As the two- and three-orbital patterns are particularly common in chemistry, they should be committed to memory.

These mo nodal patterns are basic; they apply equally well to something as esoteric as a linear string of two to six hydrogen atoms and to the more practical example of π molecular orbitals of the unsaturated hydrocarbons C_2H_4 through C_6H_8. To be specific about the application of these general orbital patterns, we shall focus on the $p\pi$ ao's of a triatomic molecule (NO_2^+, CO_2, O_3, NO_2, C_3H_5, etc.). In all these cases the $p\pi$ mo's have the form **4**. The subscript n in π_n means **non-bonding**; there is no interaction between nearest-neighbor atoms.

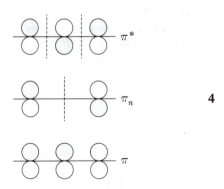

4

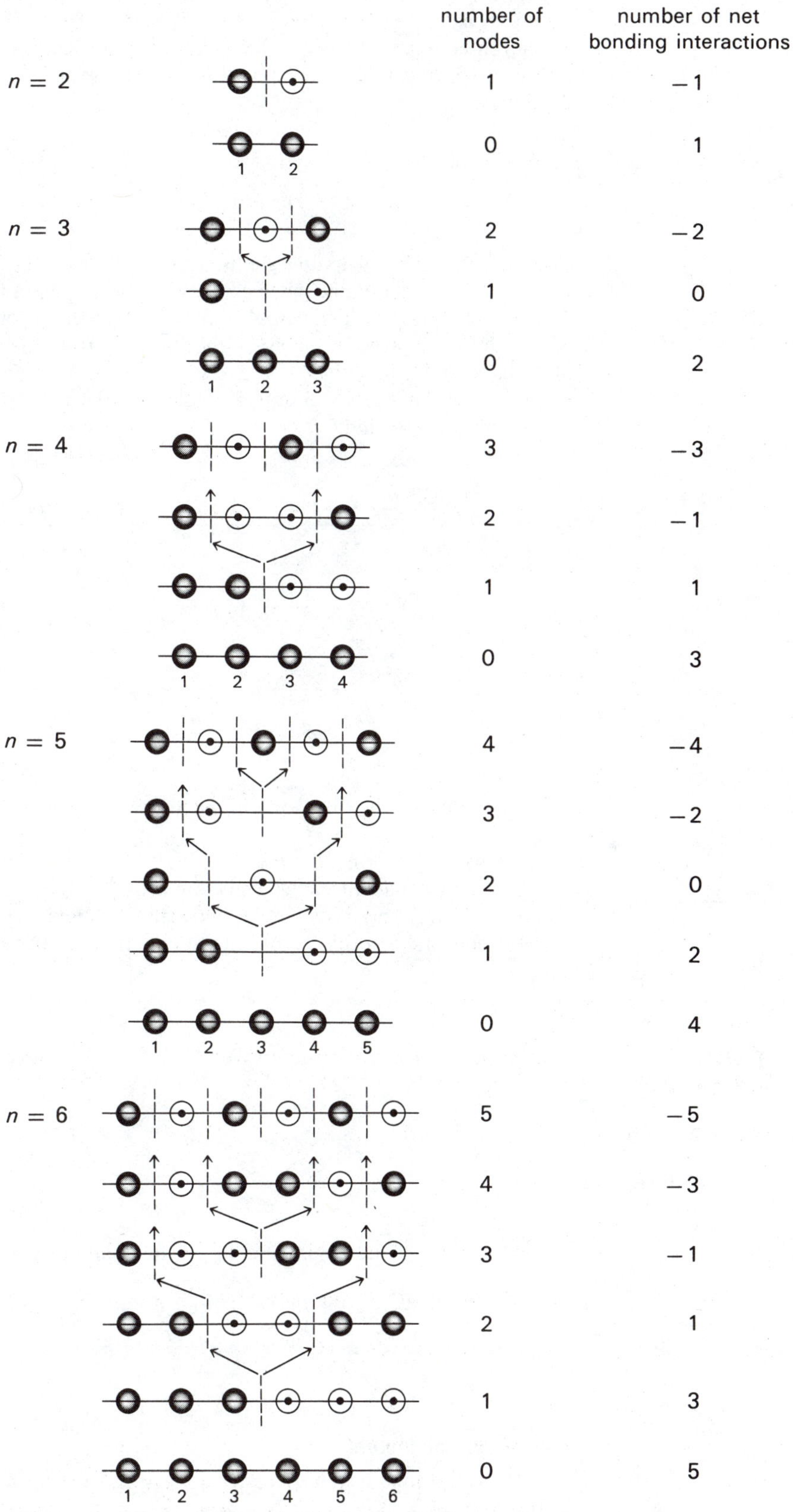

	number of nodes	number of net bonding interactions
$n = 2$	1	−1
	0	1
$n = 3$	2	−2
	1	0
	0	2
$n = 4$	3	−3
	2	−1
	1	1
	0	3
$n = 5$	4	−4
	3	−2
	2	0
	1	2
	0	4
$n = 6$	5	−5
	4	−3
	3	−1
	2	1
	1	3
	0	5

Figure 4-1. Molecular orbital nodal patterns for two to six atoms in a "linear" chain, where each atom contributes one atomic orbital to the LCAO mo's.

Electrons populating π and π^* are distributed over all three atoms, while electrons in π_n are constrained to the terminal atoms. The case of the allyl radical is interesting, for the π electron configuration is $(\pi)^2 (\pi_n)^1$ and the odd electron is concentrated at the terminal carbon atoms; it is important to note that the same description is achieved by π resonance structures:

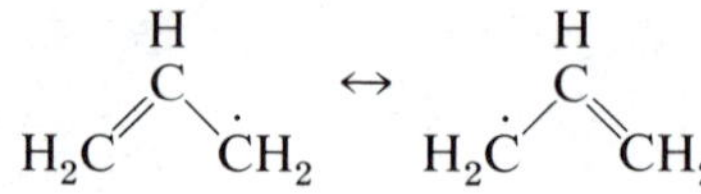

H
C   ·
H_2C CH_2 $\leftrightarrow$ H_2C CH_2

Note that the Lewis structures are used to decide that there are three π electrons.

A final note should be made concerning the ao phases for the terminal atoms. Inspection of the examples in Figure 4-1 reveals our convention of assigning to the left-most orbital the same phase (+) in every mo. Starting from the all-bonding mo for every value of n, the right-most ao phase alternates $+, -, +, -, \ldots$. This regularity is illustrated for the important case $n = 3$ by **5** and **6**, which are the appropriate patterns for the p and s sigma mo's of XeF_2 and CO_2.

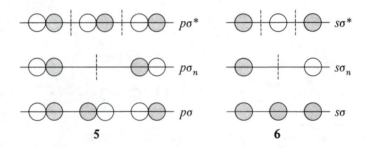

To summarize the rules for constructing any linear mo pattern:

1. Left-most ao + in all mo's.
2. Right-most ao alternates sign from lowest to highest mo.
3. The lowest mo has no nodes between neighbors.
4. Nodes increase by 1, and are symmetrically placed, in successively higher mo's.

STUDY QUESTIONS

1. From π resonance structures of NO_2^- you are aware that there are two π electron pairs:

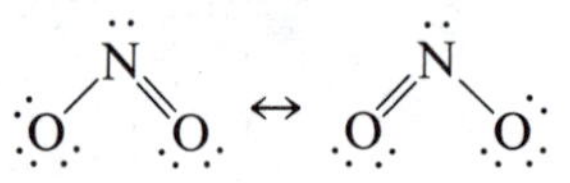

:Ö   N   Ö: $\leftrightarrow$ Ö   N   :Ö:

What does the mo model predict for the bond order of each N—O bond?

2. Sketch the $s\sigma, p\sigma$, and $p\pi$ mo's of I_3^-.

Solid State Conductors

It is appropriate at this point to consider the electronic structures of the metallic elements because the phenomenon of electron transport provides a good application of the mo model. Briefly, the three principal lattice types for metals are the **body centered cubic**, **face centered cubic**, and **hexagonal close packed**; sketches of these structures are given in Figure 4-2. Notice that the number of nearest neighbors of each atom in the bcc structure is 8, whereas it is 12 for both fcc and hcp structures.† With such high coordination numbers for each atom, we immediately encounter great difficulty in proposing a structure-predicting bonding model for metals according to the covalent, electron-pair bond concept.

Figure 4-2. The three common metal lattices: (a) face centered cubic, (b) hexagonal close packed, and (c) body centered cubic. In (a) and (b), atoms in the adjoining unit cell are shown to help in visualizing the coordination number.

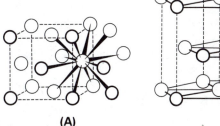
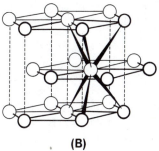
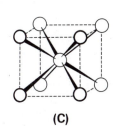

(A)   **(B)**   **(C)**

In many ways the most successful model of the electronic structures of metals (but this model is not particularly helpful with regard to lattice structure) is an extended form of molecular orbital theory that deals with physical properties of metals such as electrical conductivity. The theory is known as the **band theory** of metals, for reasons that will become obvious from the following brief outline.

The overlapping of orbitals on adjacent atoms leads to bonding, non-bonding, and anti-bonding orbitals. Within limits, the greater the number of overlapping orbitals, the greater the energy gap between bonding and anti-bonding orbitals. Applying these ideas to a crystal of a metallic element (a "molecule" of an astronomical number of atoms), we are confronted with so many lattice orbitals that the band of orbital energies for the valence electrons of the metal atoms approaches a continuum. Figure 4–3 depicts the

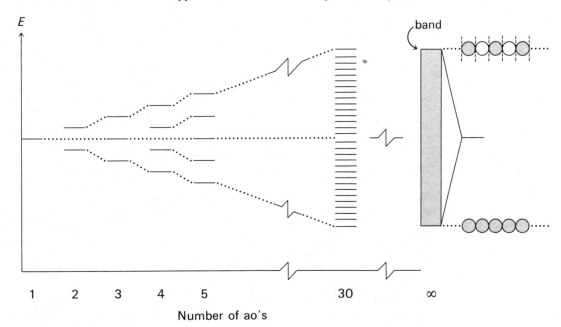

Figure 4-3. Correlation of one atom/one *s* ao with the energy band for a linear chain of *N* atoms/*N* *s* ao's.

---

†There is a strong similarity between the fcc and bcc structures, which appears on close inspection. Both structures present the central atom with 8 neighbors at the corners of a cube. In the bcc structure these are shown in Figure 4-2 (c) at unit cell corners. In the fcc structure, the 8 neighbors are at unit cell face mid-points. The additional 4 neighbors in the fcc cell are at the unit cell corners and are somewhat closer to the central atom.

creation of such a band from atomic orbitals (*s* ao's, say) for a linear chain of metal atoms. A key point is that as the number of atoms in the chain increases, an energy band develops. Because the band width for any particular type of atomic orbital depends on orbital overlap at the normal metal-metal distances of metallic lattices, the inner or core orbitals form fairly narrow bands separated by appreciable energy gaps from each other and from the valence orbital bands (Figure 4–4). Furthermore, since these atomic orbitals contain pairs of electrons, all lattice orbitals arising from core atomic orbitals are filled with electrons.

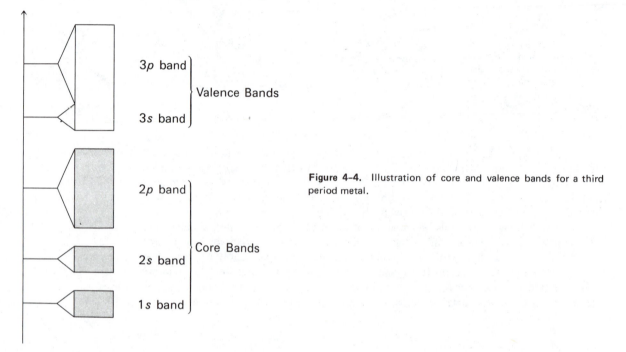

3*p* band ⎫
          ⎬ Valence Bands
3*s* band ⎭

2*p* band ⎫

Figure 4–4. Illustration of core and valence bands for a third period metal.

          ⎬ Core Bands
2*s* band

1*s* band ⎭

We will not digress on the mechanism for electric-field-induced electron migration in a crystal, but will simply outline a diagnostic test for the ease of this migration. Unoccupied lattice orbitals must be available for the transport of valence electrons throughout the crystal. An **insulator** is represented (Figure 4–5) as a case in which completely filled bands are separated by a considerable energy gap from unoccupied bands. **Conductors** are identified as cases in which a particular valence band is only partially filled or, if enough electrons are available to fill a particular band, two adjacent bands are contiguous so as to provide a composite partially filled band. In either case, application of an electric field (a voltage drop across the crystal) will induce conduction. Either description could apply to the alkali metals with only half-filled **s** orbital bands, while contiguous valence *s* and *p* bands are required to account for the conductance observed with alkaline earth metals. According to this model, **intrinsic semiconductors** are identified as elements for which sufficient electrons are available to occupy one band completely, but in which another band is to be found across an energy gap that is small in comparison with thermal energies. Silicon and germanium are familiar to you as belonging to this class. In this way we can account for the increase in conductivity with temperature, which is a characteristic property of semiconductors but not of metals. **Photoelectric** materials fit into this model as cases for which the gap between a filled valence band and an empty valence band is comparable to visible/ultraviolet photon energies. Here, electronic absorption of the photon energy causes an electron to jump from a filled to an empty band so that both higher and lower bands are partially filled. Another optical property of metals that fits into this model is their high reflectivity or luster. With a near continuum of unoccupied orbital energies in the range of visible radiation, a metal could absorb and subsequently emit radiation of nearly all wavelengths in the visible range, so as to appear "white" and highly reflective.

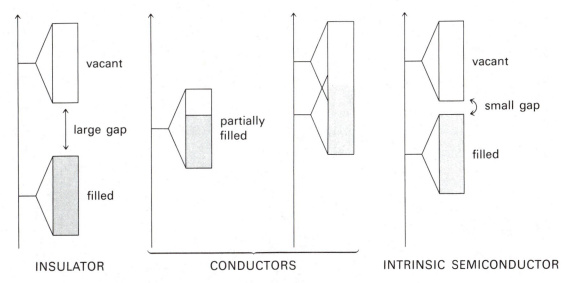

Figure 4-5. Schematics of the band concept for insulators, conductors, and intrinsic semiconductors.

The semiconductor properties of many materials can be controlled by "doping" semiconductors with elements having fewer or greater numbers of valence electrons than possessed by the semiconductor atoms (doped semiconductors are distinguished by the term "extrinsic"). In these ways, normally filled valence bands can be made less than filled (such a semiconductor is called a *p*-type, for positive or "hole" charge carrier) or normally unoccupied valence bands can become partially occupied (called an *n*-type, for negative charge carrier), and the conductivity is a direct function of crystal content of the doping element.

STUDY QUESTION

3. Do you expect GaAs to be an *ionic* solid? What do you expect would happen to the electrical conductivity of GaAs if Se were added? If Ga were added? If a crystal of GaAs is doped with Se at one end and with Ga at the other, in which direction would electric current (electrons) flow more readily? Would you anticipate greater or less discrimination in direction of electron flow from such an extrinsic semiconductor, compared to one doped at one end only?

Cyclic (Two-Dimensional) Molecules

A second important structural class of molecules is those with cyclic structures; you are aware of many from organic chemistry, and there are many inorganic types as well. The results given in Figure 4-1 for "linear" molecules provide a nice basis for introducing the forms of molecular orbitals of cyclic compounds. A convenient mnemonic device to use for the cyclic structures is to start with the orbital sketches of the linear molecule with the same number of atoms, and to "wrap" the linear arrangement into the cyclic arrangement:

This may cause separate nodal planes in certain of the linear mo's to coincide or shift. The key to the cyclic mo structures lies in the nodal structure. *The all-important symmetry requirement is that the nodal lines pass through the center of the polygonal arrangement of nuclei.* It is straightforward now to sketch the cyclic orbital patterns. This is done in Figure 4-6 for the cyclic cases of three through six atoms.

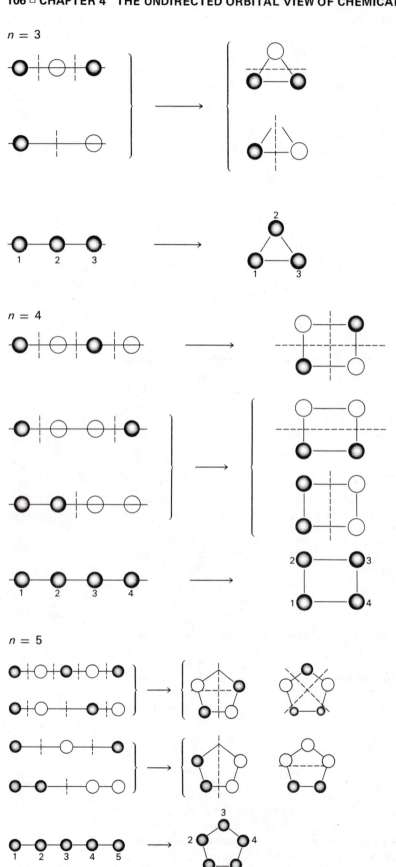

Figure 4-6. Cyclic mo nodal structures from linear counterparts.

Illustration continued.

$n = 6$

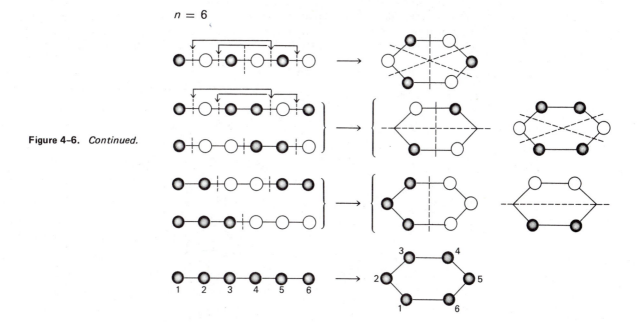

Figure 4-6. *Continued.*

Let us now consider each case in detail. For three atoms, the three linear-case mo's with 0 through 2 nodes transform into a single mo with no nodes and two mo's with one node each; these nodes are mutually perpendicular. As perhaps expected, the latter two cyclic mo's have the same energy (they have the same number of nodes) and are called degenerate orbitals—even though the sketches appear very different. (Trust your intuition!)

For the four-atom case, the non-degenerate mo's of the linear configuration transform into a most stable orbital with no nodes, a degenerate pair of orbitals with a single node each (which are mutually perpendicular), and a higher-lying orbital with two nodes. Pay particular attention to the fact that symmetry dictates that the outer nodes of the third and fourth linear mo's must coalesce into a single node (drawn horizontally) in the third and fourth cyclic mo's.

The five-atom case presents us with our first instance of nodes that, while passing through atoms in the linear case, shift off those atoms to adhere to the symmetry requirement that the nodes must pass through the center of the polygon. The most stable linear mo transforms into the most stable cyclic mo with no nodes. The next two linear mo's, with one and two nodes, transform into a degenerate pair with one node each; these nodes are mutually perpendicular. Note that the nodes at the second and fourth atoms in the third linear mo must shift off these atoms and coalesce in the third cyclic structure. The two highest linear mo's transform into a degenerate pair of cyclic mo's with two nodes each. Again note the node coalescence.

For the six-atom case, the no-node linear mo becomes the most stable no-node orbital of the hexagon. The one- and two-node linear orbitals become the degenerate, single node cyclic orbitals. The three- and four-node linear orbitals become a second degenerate pair at higher energy than the previous pair, owing to the presence of two nodes each. Finally, the five-node linear orbital becomes a three-node high-energy orbital of the hexagon. (The arrows to pairs of nodes in the fifth and sixth linear mo's identify those that coalesce.)

As the number of nodes is directly related to the orbital energy, it is quite straight-forward to rank the orbitals in order of increasing energy and thus to determine the electron configuration of the molecule. To illustrate, the molecule benzene, with six pi electrons, is predicted to have the π electron configuration $(\pi_1)^2 (\pi_2, \pi_3)^4$. The π_1 orbital is most stable, with six bonding interactions between nearest neighbors; while π_2 and π_3, each with a *net* of two bonding interactions, are next most stable, degenerate, and bonding. The orbitals π_4 and π_5 each have a *net* of two anti-bonding interactions and so are properly described as anti-bonding orbitals, while π_6 is most anti-bonding with six anti-bonding interactions. An interesting aside here is that $C_6H_6^-$ should be a powerful one-electron reducing agent while $C_6H_6^+$ should be a strong one-electron oxidant.

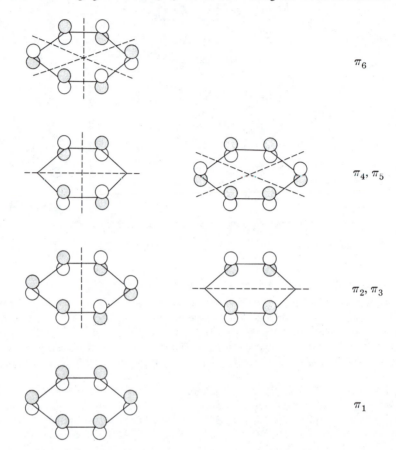

π_6

π_4, π_5

π_2, π_3

π_1

The cyclopentadiene molecule is predicted to be a radical with a $(\pi_1)^2(\pi_2, \pi_3)^3$ configuration; it should be an oxidant forming the stable cyclopentadienide anion, $C_5H_5^-$, and this is a known species. In fact, this anion will occur repeatedly throughout later sections of the text in connection with transition metal complexes, at which time the orbital sketches presented here will be used.

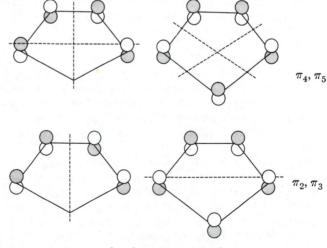

π_4, π_5

π_2, π_3

[π_1 shown on next page]

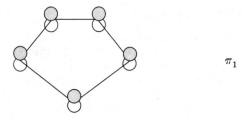

π_1

The symmetries of the cyclic molecular orbitals in Figure 4-6 can be summarized by writing down the phase (+ or −) of each atomic orbital in LCAO form. (We are interested only in the sign of the mo at each atom, not its amplitude there.) This is done in Table 4-2 for the triangle, square, and pentagon. *This is an important table to which we will frequently refer; familiarize yourself with its contents.*

TABLE 4-2

FORMS (RELATIVE AO PHASING) OF THE CYCLIC LCAO MOLECULAR ORBITALS

Triangle	$1 + 2 + 3$	
	$\left.\begin{array}{l} 1 \quad\;\; - 3 \\ 1 - 2 + 3 \end{array}\right\}$	degenerate
Square	$1 + 2 + 3 + 4$	
	$\left.\begin{array}{l} 1 + 2 - 3 - 4 \\ 1 - 2 - 3 + 4 \end{array}\right\}$	degenerate
	$1 - 2 + 3 - 4$	
Pentagon	$1 + 2 + 3 + 4 + 5$	
	$\left.\begin{array}{l} 1 + 2 \quad\;\; - 4 - 5 \\ 1 - 2 - 3 - 4 + 5 \end{array}\right\}$	degenerate
	$\left.\begin{array}{l} 1 - 2 \quad\;\; + 4 - 5 \\ 1 - 2 + 3 - 4 + 5 \end{array}\right\}$	degenerate

4. With which π mo's of cyclopentadienide do the s and p ao's of a silver ion overlap when the cation is positioned along the C_5 rotation axis?

STUDY QUESTION

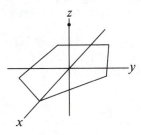

(Hint: Look for coincidence of Ag^+ p ao nodes with those of the cyclopentadienide mo's.)

Beryllium Hydride and Hydrogen Chloride

To escalate to a slightly more complex case, we shall consider the linear BeH_2 molecule discussed in the last chapter; this time we will use the centrosymmetric Schrödinger ao's to create molecular orbitals, rather than hao's for Be alone. A convenient view of BeH_2 is that of inserting a Be atom between two hydrogen atoms. The molecular point group is $D_{\infty h}$ as it is for H_2; the ψ and ψ^* wavefunctions (see $n = 2$ of Figure 4-1) for the H atoms are still good but, because the overlap of s_1 with s_2 is now so small, the ψ and ψ^* orbital energies are little different from that of an isolated H 1s ao. These mo's (ψ, ψ^*) are symmetrized atomic orbitals for terminal atoms; *we will refer to all such* **terminal atom symmetry orbitals** *in this chapter with the acronym* **TASO**.

The problem at hand is to *consider the valence ao's of the central atom* (Be) *for possible overlap (symmetry likeness) with the H_2 TASO's*. This can be done by inspection. To illustrate the inspection procedure, we will locate the molecule on a coordinate axis system, sketch the TASO's, sketch the central atom ao's, and compare for non-zero overlaps.

TASO's   Central ao's

From these sketches, it should be apparent that the Be s ao has $S > 0$ with ψ, that the Be p_z ao has $S < 0$ with ψ^*, and that all other comparisons yield $S = 0$. Construction of the mo energy diagram is straightforward. Recognizing first that the Be 2s and 2p ao's are less electronegative than the H 1s ao, the starting ao levels are "skewed" to reflect that difference in Figure 4-7. The Be 2s is to be taken in sum and difference combinations with ψ, and the same is true for the Be p_z with ψ^*. The combination $\phi_s + \psi$ accounts for a bonding mo because the Be—H overlaps are constructive; $\phi_s - \psi$, on the other hand, causes destructive Be—H overlaps and this corresponds to anti-bonding. We can distinguish these by calling the first $s\sigma$ and the second $s\sigma^*$. For the (ϕ_z, ψ^*) combinations it is $\phi_z - \psi^*$ that leads to bonding (call this mo $p\sigma$) and $\phi_z + \psi^*$ that leads to antibonding (call this mo $p\sigma^*$). Finally, note that the Be ϕ_x and ϕ_y ao's are unshifted and non-bonding (call them $p\pi_n$).

Chemical interpretations follow after you decide which mo's are occupied with electrons. Since BeH_2 has four valence electrons (two from Be and one each from the H atoms), the molecular electron configuration is

$$(s\sigma)^2(p\sigma)^2(p\pi_n)^0(s\sigma^*)^0(p\sigma^*)^0$$

There are two bonding electron pairs (a reassuring result!). *Both* contribute to defining *each* Be—H "bond," so that we might say, in rough terms, that each Be—H bond has a bond order of 1 ($\frac{1}{2}$ from each of $s\sigma$ and $p\sigma$). In short, we have arrived at the same qualitative point as was reached by the directed ao approach of the last chapter. Finally, we must introduce the important concepts of *Highest Occupied Molecular Orbital* (**HOMO**)

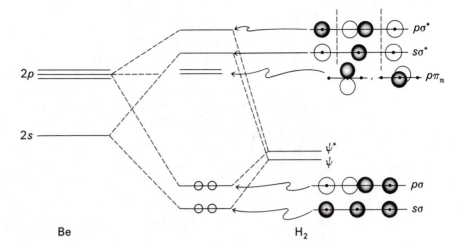

Figure 4-7. The energy diagram and orbital sketches for BeH_2.

and *Lowest Unoccupied Molecular Orbital* (**LUMO**). In BeH_2, the $p\sigma$ mo is the HOMO and the $p\pi_n$ pair of ao's is the LUMO. In later chapters you will see that the energies and shapes of HOMO and LUMO have an important bearing on the understanding of molecular reactions. Another important interpretation to be recognized is that, because H is more electronegative than Be, the electron pair probabilities for $s\sigma$ and $p\sigma$ should be greater at H than at Be. That is, writing

$$s\sigma = a\phi_s + b\psi$$
$$p\sigma = c\phi_z - d\psi^*$$

we expect $a^2 < b^2$ and $c^2 < d^2$. The $s\sigma$ and $p\sigma$ mo's are *polarized* toward the hydrogen atoms. Another way of saying this is that $s\sigma$ has greater ψ character than ϕ_s character, and similarly for $p\sigma$. A convenient idea here, to be used in the upcoming discussion of HCl, is that the lower-energy TASO's are contaminated with the Be ao's and move to lower energy, while the higher-energy Be ao's are contaminated by the TASO's and shift to higher energy. In short, *lower levels shift lower while higher levels shift higher.*[3]

As a final point, to which we shall refer later, note that the Be s ao has a symmetry entirely different from that of the Be p ao's; consequently, there is no s, p mixing in this case. The next example will show such mixing.

Hydrogen chloride is a simple heteronuclear diatomic molecule from which we may learn several useful simplifying concepts for later application to more complex molecules. The valence ao's for this molecule are simply the H $1s$ and the $3s$, $3p_x$, $3p_y$, and $3p_z$ ao's of chlorine. Making ao sketches on a coordinate axis system produces the following:

Matching H and Cl ao's for overlap, you find that the ϕ_x and ϕ_y ao's are again a degenerate non-bonding pair, while the H $1s$ ao overlaps both ϕ_s and ϕ_z of Cl.

How do we handle a problem with *three overlapping orbitals*? Three linear combinations must be taken, but how is it to be done? While a digital computer with the proper program could handle this situation directly, we will take a more controlled, stepwise approach by the brazen technique of simply ignoring one of these ao's *at the start*, and later we will dispense with such convenient shoddiness to arrive at the final answer.

The energy level diagram of Figure 4-8 is developed as follows. The chlorine ao's are placed at lower energy than that of hydrogen. The ϕ_x and ϕ_y ao's of chlorine are unshifted and non-bonding ($p\pi_n$). The three-orbital situation for bonding is simplified initially by ignoring the low energy ϕ_s ao of chlorine. The choice of the ϕ_s ao rather than the ϕ_z ao is dictated by the fact that *mixing of two ao's depends not only on their overlap integral* (greater overlap, greater mixing) *but also on their likeness in energy* (better energy match leads to greater mixing). This is a result we take without apology from the mathematics of molecular orbital calculations. Completion of the energy level diagram is now straightforward. The ϕ_s level is shown unshifted and non-bonding, while the H $1s$ ao is combined in-phase (to produce $p\sigma$) and out-of-phase (to produce $p\sigma^*$) with ϕ_z.

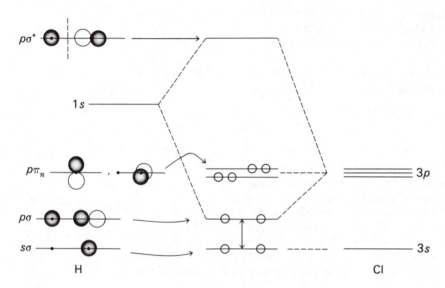

Figure 4-8. The "first approximation" energy diagram for HCl, with the second approximation mixing anticipated by the arrow between $s\sigma$ and $p\sigma$.

HCl has four pairs of valence electrons, yields the electronic configuration

$$(3s)^2(p\sigma)^2(p\pi_n)^4(p\sigma^*)^0$$

and shows three lone pairs (the $3s$ and $p\pi_n$ pairs) and one bond pair (from $p\sigma$). The bond order is qualitatively one, and HOMO = $p\pi_n$ while LUMO = $p\sigma^*$. The $p\sigma$ mo is polarized toward chlorine, while $p\sigma^*$ is polarized toward hydrogen.

What do we now do about ignoring the fact that ϕ_s and the H ao do overlap and should be mixed? Two guides we have used for mixing of ao's to form mo's are also applicable for mixing simple mo's to form better mo's. (i) The mo's ϕ_s, $p\sigma$, and $p\sigma^*$ all overlap, the fact we have been ignoring. *The overlap arises because ϕ_s overlaps the hydrogen ao appearing in $p\sigma$ and $p\sigma^*$*. (ii) The mixing of $p\sigma$ with ϕ_s will be greater than the $\phi_s/p\sigma^*$ mixing because of the better energy match. Now, what are the phasings of $p\sigma$ and $p\sigma^*$ when they are to be mixed into ϕ_s?

We know that the $3s$ ao must be stabilized by the mixing with $p\sigma$ and $p\sigma^*$ (lowest goes lower); to achieve this with $p\sigma$ and $p\sigma^*$ *as drawn* in Figure 4-8, both $p\sigma$ and $p\sigma^*$ must contaminate ϕ_s with positive phases ($3s \rightarrow 3s + p\sigma + p\sigma^*$) so as to introduce H $1s$/Cl $3s$ bonding. Interestingly, the ϕ_z contributions are out-of-phase in $p\sigma$ and $p\sigma^*$, and the new $3s$ mo has less ϕ_z character than might otherwise have been the case; the new ϕ_s mo is primarily H $1s$/Cl $3s$ in character, and the new mo is somewhat bonding between H and Cl. What about the new $p\sigma$ mo? Its contamination by ϕ_S must shift it to higher energy, a requirement met by $p\sigma \to p\sigma - \phi_S$ so as to introduce H $1s$/Cl $3s$ anti-bonding character into the $p\sigma$ mo. (Note that $p\sigma$ contaminates ϕ_S with a positive phase, and the converse is necessary for ϕ_S contamination of $p\sigma$.) Similarly, ϕ_S contaminates $p\sigma^*$ with a negative phase.

To illustrate these improvements in ϕ_S, $p\sigma$, and $p\sigma^*$ in an exaggerated way, we would sketch **7**.

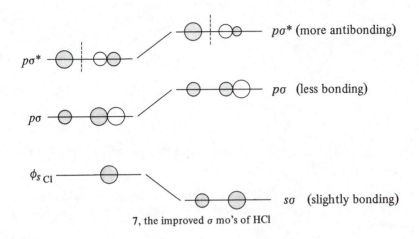

7, the improved σ mo's of HCl

As the sketches imply, ϕ_S is stabilized by new bonding character, while $p\sigma$ is destabilized by some anti-bonding character. In qualitative terms, bonding character has been transferred from $p\sigma$ to ϕ_S.

Before leaving this discussion of the mo's of HCl, we should mention the photo-electron spectrum of HCl, an experimental result that is straightforwardly interpreted by means of the mo level diagram of Figure 4-8. The photoelectron spectroscopy experiment is based on the idea that an electron may be ejected from a molecule by "striking" the molecule with a photon of energy in excess of that required for ionization.[4] After measuring the kinetic energy of the ejected electron, one calculates the work required for the ionization as the difference between the known photon energy and that kinetic energy: $I = (h\nu - \text{K.E.})$.

$$HCl + h\nu \to HCl^+ + e^-$$

In fact, the experiment should reveal several such ionization processes, for not all electrons in HCl feel the same effective nuclear charges.

$$HCl + h\nu \to HCl^+ + e^-$$
$$HCl + h\nu \to HCl^{+*} + e^-$$
$$\text{etc.}$$

} ejected e^- have different kinetic energies

Figure 4-9 shows that in the region between 12 and 18 eV there are in fact two ionization processes possible for HCl. Ignoring the fine structure in this high resolution spectrum, you may assign the band at $\sim$13 eV to removal of $p\pi_n$ electrons and the band at $\sim$16.5 eV to removal of a $p\sigma$ electron. The Cl $3s$ and core electrons of HCl can also be ionized in this fashion; but, as they experience greater effective nuclear charge, it requires photon energies much greater than 18 eV to eject them.

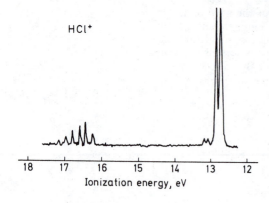

HCl⁺

Figure 4-9. The photoelectron spectrum of HCl in the region $12 < h\nu < 18$ eV. [From J. H. Eland, "Photoelectron Spectroscopy," Butterworths, London (1974).]

18   17   16   15   14   13   12

Ionization energy, eV

Summary

Now is a good time to stop and reflect on what has been learned from the mo formalisms of H_2, BeH_2, and HCl.

1. (From H_2) Two orbitals (ao's *or* mo's) that overlap (have $S \neq 0$) and that are of equal energy are taken in equal weights by sum and difference to form bonding and antibonding orbitals.

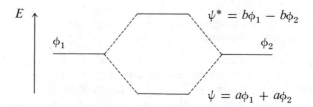

Electrons in ψ and ψ^* will be equally distributed between ϕ_1 and ϕ_2; ψ and ψ^* are *not* "polarized."

2. Two overlapping orbitals of unequal energy will *not* be combined with equal weights in sum and differences. *The more electronegative orbitals lie lower in energy.* The bonding combination (lower in energy) retains greater character of the orbital most like it in energy. The condition $a < b$ means that ψ is "polarized" toward the region of ϕ_2; electrons with orbital wave ψ have greater probability at ϕ_2 than at ϕ_1. The converse applies to ψ^*.

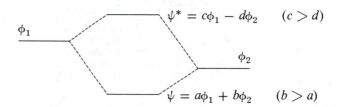

3. In both (1) and (2), ψ is lower in energy than ψ^* because ϕ_1 and ϕ_2 constructively interfere in the internuclear region; they destructively interfere in ψ^*.

4. When faced with three overlapping orbitals, at least one of which is of different energy, it is convenient *initially* to ignore one of them so as to reduce the problem to consideration of *two* overlapping orbitals.

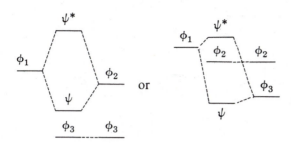

As a second step, we may remove this artificial convenience by recognizing that the ignored orbital overlaps ψ and ψ^*. Always, the lowest level moves lower, the highest moves higher, and the intermediate level shifts slightly up or down depending on its closer proximity to ψ or ψ^*, but remains intermediate.

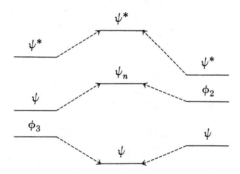

If worked to completion, the final answer is the same. Remember that *higher levels are mixed into lower levels in a bonding way, while lower levels are mixed into higher levels in an anti-bonding way.* The extent of mixing, as always, depends on the energy match and overlap.

MAIN GROUP DIATOMIC MOLECULES

Homonuclear Molecules

Staying with the basic structural unit of the diatomic molecule, we will now examine the slightly more complex case of two atoms of the same element, each of which has s and p atomic orbitals in the valence shell. Figure 4-10 gives sketches of the atomic orbital angular functions for the orbitals to be combined. It is quite easy to see in this case that the s orbitals overlap, the p_z orbitals overlap, and the p_x and p_y orbitals overlap. Recalling the $n = 2$ case from Figure 4-1, bonding and anti-bonding combinations of (s_1, s_2), (p_{z1}, p_{z2}), (p_{x1}, p_{x2}), and (p_{y1}, p_{y2}) are easily drawn, as in Figure 4-10.

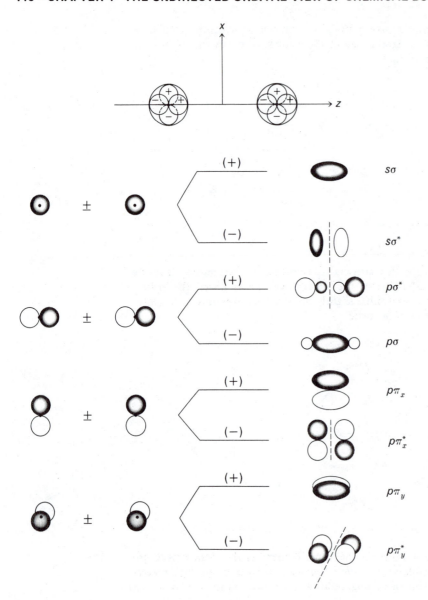

Figure 4-10. Sketches of the angular parts of the atomic and LCAO orbitals for the homonuclear diatomic molecule with valence s and p orbitals.

The "first appoximation" orbital energy diagram is given in Figure 4-11a. In sketching these diagrams we have made use of our knowledge that the atomic s orbital is more stable than the atomic p orbitals. The levels are labeled according to a common notation that identifies the molecular orbital according to type (σ or π) and according to origin in terms of atomic orbitals (s or p).

Another point necessary to bring up now is that the "first approximation" orbital treatment has completely ignored the fact that an s atomic orbital on *one* atom has a non-zero overlap with the p_z atomic orbital on the *other* atom. This deficiency is disconcerting, and its removal will establish a closer link between LCAO and hybrid orbital models.

To be specific, we have not allowed for interactions between the s_1 and p_{z2} atomic orbitals and between the s_2 and p_{z1} atomic orbitals, which are permitted by symmetry to have non-zero overlap integrals. To handle this by merely refining the simplistic orbital energy level diagram, we need only note that the *so* and *po molecular orbitals*

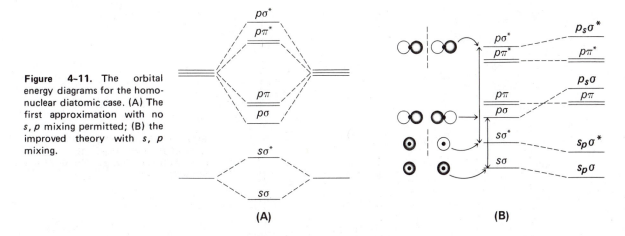

Figure 4-11. The orbital energy diagrams for the homonuclear diatomic case. (A) The first approximation with no s, p mixing permitted; (B) the improved theory with s, p mixing.

have the same symmetry and will have a non-zero overlap integral. Figure 4-12 illustrates this overlap with sketches (the crossed dashed lines draw your eye to the s/p overlap responsible for the mixing). The phases of $p\sigma$ and $s\sigma$ as drawn indicate that $s\sigma$ is stabilized by $+p\sigma$, while $p\sigma$ is destabilized by contamination with $-s\sigma$.

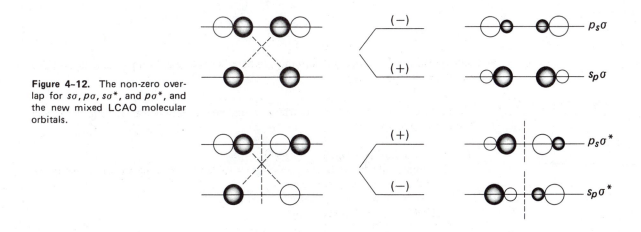

Figure 4-12. The non-zero overlap for $s\sigma, p\sigma, s\sigma^*$, and $p\sigma^*$, and the new mixed LCAO molecular orbitals.

The effect of this mixing on the $s\sigma$ and $p\sigma$ orbital energies is shown in Figure 4-11b. The $s_p\sigma$ orbital takes on a more stable energy (becomes more bonding), while the $p_s\sigma$ orbital is destabilized (loses bonding character). The renaming of $s\sigma$ and $p\sigma$ is meant to emphasize the mixing. A similar interaction between $s\sigma^*$ and $p\sigma^*$ is also possible, and this is indicated in Figures 4-11 and 4-12.

You will notice in Figure 4-11 that $p_s\sigma$ is shown at higher energy than $p\pi$. Both accurate calculations[5] and photoelectron spectra[6] indicate that $p_s\sigma$ can lie higher in energy than $p\pi$. Figure 4-13 is the photoelectron spectrum of N_2, which has a valence electron configuration

$$(s_p\sigma)^2 \ (s_p\sigma^*)^2 \ (p\pi)^4 \ (p_s\sigma)^2$$

With the orbital diagram thus confirmed for N_2, we can make some chemical interpretations. The LUMO is the degenerate pair of $p\pi^*$ mo's, and the HOMO is the bonding $p_s\sigma$ mo. There are two π bonding electron pairs ($p\pi$), and the σ "bond" is best described

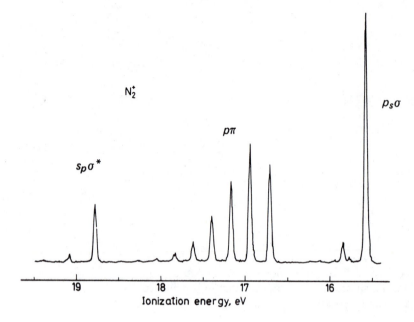

Figure 4-13. The photoelectron spectrum of N_2 in the region $15 < h\nu < 20$ eV. [With the kind permission of Professor W. C. Price.]

as a superposition of the $p_s\sigma$ and $s_p\sigma$ mo's (where the $s_p\sigma$* electron pair contributes in a somewhat anti-bonding way).

Continuing on to the other important second row homonuclear diatomics, O_2 and F_2, a summary is as follows:

O_2: 1. Configuration = $(s_p\sigma)^2 \cdots (p\pi$*$)^2$
2. Hund's rule predicts the ground term to have two unpaired electrons in $p\pi$*, and O_2 is, in fact, paramagnetic.
3. The bond order is ~2 (net $\sigma = 1$, net $\pi = 1$)
4. HOMO = $p\pi$*, LUMO = $p_s\sigma$*

F_2: 1. Configuration = $(s_p\sigma)^2 \cdots (p\pi$*$)^4$
2. All electrons paired.
3. The net bond order is 1 (no net π bonding)
4. HOMO = $p\pi$*, LUMO = $p_s\sigma$*

Heteronuclear Molecules

In inorganic chemistry perhaps the most important heteronuclear diatomic molecules (aside from the hydrogen halides) are CO, CN⁻, and NO. All appear as important ligands in transition metal chemistry, and we need to pause here to discuss their mo's, using CO as an example. The *first approximation* will be invoked initially, and the development closely follows that for the homonuclear diatomic molecule. The primary difference now is the polarization of the mo's toward oxygen and toward carbon. Because the oxygen atom ao's are lower in energy than the carbon atom ao's, the "skewing" results in the $p\sigma$ mo being lower than the $s\sigma$* (see **8**, p. 119). The orbital sketches to the right in **8** are meant to reflect the mo polarizations through the relative ao sizes drawn. To proceed to

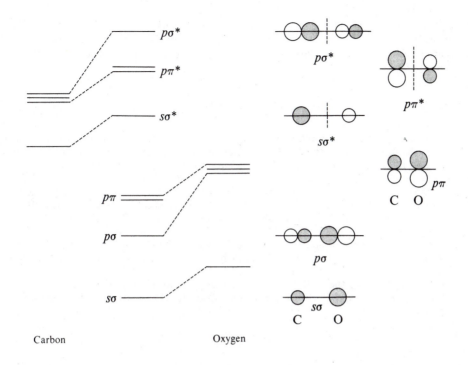

8, first approximation mo's for CO

the second level of approximation, we need only mix $s\sigma/p\sigma$ and $s\sigma^*/p\sigma^*$.† The refined mo's now take on the appearances (exaggerated for clarity) shown in **9**. The phasing relationships are:

$s\sigma^*$ into $p\sigma^*$: $+$ $\quad p_s\sigma^*$

$p\sigma^*$ into $s\sigma^*$: $-$ $\quad s_p\sigma^*$ **9, improved σ mo's for CO**

$s\sigma$ into $p\sigma$: $-$ $\quad p_s\sigma$

$p\sigma$ into $s\sigma$: $+$ $\quad s_p\sigma$

Carbon monoxide possesses 10 valence electrons and therefore has a valence configuration

$$(s_p\sigma)^2\ (p_s\sigma)^2\ (p\pi)^4\ (s_p\sigma^*)^2$$

The HOMO is the "carbon lone pair orbital" $s_p\sigma^*$, while the $p\pi^*$ mo's are the LUMO. You will later see that the polarizations of these mo's are important to the Lewis acid/base chemistry of CO (and NO^+ and CN^-, which are isoelectronic with CO). There are roughly three net bonding pairs (two π and one σ), describing the usual formulation $C\equiv O$. Similarly, NO has an electron configuration

$$(s_p\sigma)^2\ (p_s\sigma)^2\ (p\pi)^4\ (s_p\sigma^*)^2\ (p\pi^*)^1$$

---

†While it is correct also to mix $s\sigma^*$ with $p\sigma$ and $p\sigma^*$ with $s\sigma$, the poor energy match in the latter case means that the $p\sigma^*/s\sigma$ mixing will be minimal, while we shall simply ignore for convenience sake the $s\sigma^*/p\sigma$ mixing in this discussion.

and so should be paramagnetic with a single unpaired electron. A familiar aspect of NO is its ready oxidation to NO^+. This corresponds to loss of the $p\pi^*$ electron, an associated increase in NO bond order (from $2\frac{1}{2}$ to 3), and a decreased NO distance. The HOMO of NO ($p\pi^*$) plays an important role in the structures of its compounds.

STUDY QUESTIONS

5. The mixing of $s\sigma^*$ into $p\sigma$ for CO was ignored in the text. Comment on the effect of such mixing, after making mo sketches to reflect changes in mo shape and polarization. [Hint: note that the (C_s, O_z) overlap is greater than that of (O_s, C_z).]

6. Compare the HOMO's of CO and N_2 and draw a conclusion about the relative Lewis basicities of the C and N atoms of these molecules (consider not only the HOMO amplitudes at C and N, but also the Z^* experienced by the HOMO pairs).

7. From the polarization and shape of the HOMO of CO, which "end" (C or O) should be more basic in the Lewis sense? Is this mo predominately C_{2s} or C_{2p} in character?

8. From the form of the HOMO of NO, deduce the more likely point of attack (N or O) by an F atom upon NO to form nitrosyl fluoride. Is formation of the other isomer possible?

9. Sketch a photoelectron spectrum of BeH_2.

10. Describe the LUMO of I_2. Describe, by giving the ground and excited electronic configurations, the lowest-energy electronic transition responsible for the color of I_2.

11. Using the mo scheme (both level ordering and orbital sketches) of the text, account for the three ionization bands in the photoelectron spectrum of CO.

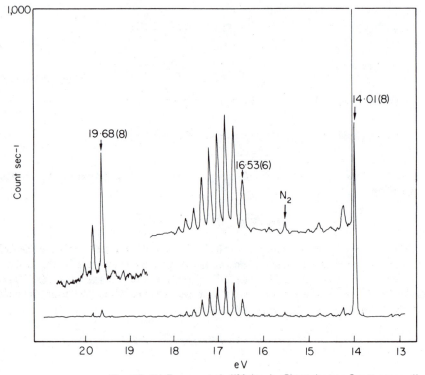

[From D. W. Turner, *et al.*, "Molecular Photoelectron Spectroscopy," John Wiley and Sons, New York (1970).]

CHECKPOINTS

1. Using the wave interference model for electron waves, molecular electron waves (orbitals) can be derived from unhybridized ao's.

2. These mo's are delocalized because each unhybridized ao is symmetrically distributed about its nucleus.

3. Electrons in mo's are distributed in space so as to be compatible with the symmetry of distribution of nuclear charges.

4. Examination of mo's shows increasing electron energy with increasing number of nodes (as is true for ao's).

5. These nodes are symmetrically positioned.

6. When two or more ao's interact to form mo's, the lowest energy mo is lower than any of the ao's; conversely, the highest energy mo is higher than any of the ao's.

7. In simple cases where the statement has meaning, an mo appears most like (is polarized toward) the ao most like it in energy.

8. Mo's for linear atom chains are a convenient basis for generating the mo's of atom rings.

9. The highest occupied mo (HOMO) in a molecule usually determines its Lewis base and nucleophilic properties; the lowest unoccupied mo (LUMO) usually determines its Lewis acid and electrophilic nature.

REFERENCES

[1] R. S. Mulliken, *Phys. Rev.,* **32**, 186, 761 (1928); **41**, 49 (1932); F. Hund, *Z. Physik,* **51**, 759 (1928); **73**, 1 (1931); S. R. La Paglia, "Introductory Quantum Chemistry," Harper & Row, New York (1971), Chapter 5.

[2] R. S. Mulliken, *J. Chem. Phys.,* **23**, 1833, 1841, 2338, 2343 (1955).

[3] R. W. Jotham, *J. Chem. Educ.,* **52**, 377 (1975).

[4] An excellent little book on this subject has been written by J. H. Eland, "Photoelectron Spectroscopy," Wiley, New York (1974).

[5] C. W. Scherr, *J. Chem. Phys.,* **23**, 569 (1955).

[6] See reference 4.

Part 2

DESCRIPTIVE NON-METAL CHEMISTRY WITH INTERPRETATIONS

5. THE DONOR/ ACCEPTOR CONCEPT

The impact of Lewis's definition of the acid and base and the concept of the electron pair bond has been felt throughout all of chemistry. Often an understanding of nucleophilic and electrophilic reaction steps depends on knowledge of "adduct-like" formation of intermediates or activated complexes, and in favorable instances the intermediates are even sufficiently stable to be isolated. In fact, proposed mechanisms are often judged reasonable or unreasonable against a background of donor/acceptor chemistry. Another area of great practical impact by the donor/acceptor concept is compound solubility and the choice of a solvent for chemical synthesis.

INTRODUCTION

Few concepts have been so enduring as the concept of acids and bases. The development of the concept of intermolecular interactions from Arrhenius' restrictive original idea [acids are sources of protons (H^+) and bases are sources of hydroxide ion (OH^-)] through Brönsted's generalization[1] [acids are proton sources (HA) and bases are proton consumers (A^-, the conjugate base of HA)] and Lewis' idea[2] of the electron pair bond (acids are electron pair acceptors, while bases are electron pair donors) to the modern concept (acids are molecules with a relatively low-energy LUMO and bases are molecules with a fairly high-lying HOMO) illustrates most dramatically how chemical concepts evolve over a period of years. Nowadays we recognize that Brönsted's acid is itself an adduct (for example, HCN) and generally can act as a Lewis acid via hydrogen bonding or by proton transfer to the conjugate base of a weaker acid:

$$H\!-\!C\!\equiv\!N\!: \ + \ Et_2O\!: \ \rightleftarrows \ Et_2O \ \cdots \ HCN$$

HCN also happens to be an example of a Lewis adduct that can act as a Lewis base:

$$H\!-\!C\!\equiv\!N\!: \ + \ SbF_5 \ \rightarrow \ HCN \ \cdots \ SbF_5$$

To help introduce you to the scope of the acid-base concept and how the detection of new chemical species by modern instrumental techniques has added even greater depth to its basis, we will begin this chapter with a sampling of examples of acids, bases, and adducts from the various groups of the periodic table. We forecast what is to be found with the generalizations contained in the following chart.

Type	Acid	Base
hydrogen	protonic (characteristically bound to electronegative atoms)	hydridic (characteristically bound to electropositive atoms)
non-metals	coordinatively unsaturated (may exhibit valence shell expansion)	generally non-bonding electron pairs
metals	coordinatively unsaturated (may exhibit valence shell expansion)	generally non-bonding electron pairs (low oxidation states, organometals)

As you examine each of the samples to follow, ask yourself the following questions:

Which atom(s) is the acceptor and which is the donor?
In which category in the chart does each example fit?
Are the structures of the acid molecular fragment and the donor fragment in the adduct much different from their structures as separate entities?
Is the presence of an acceptor orbital (LUMO) in the acid (or the donor orbital, HOMO, in the base) readily apparent for the separated fragment or is that orbital more apparent when the fragment assumes the structure it is to have in the adduct?

SURVEY OF ADDUCT TYPES

Acidic and Basic Hydrogen[3]

In addition to "complete" proton transfer reactions such as

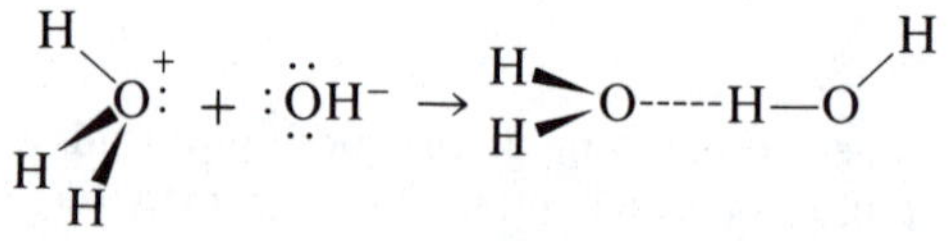

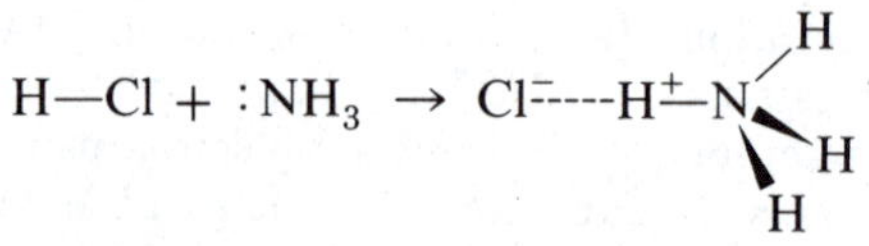

$$H_2O^+ + :OH^- \rightarrow H_2O\text{-----}H\text{--}O\text{--}H$$

$$H\text{--}Cl + :NH_3 \rightarrow Cl^{=}\text{-----}H^{\pm}N(H)(H)(H)$$

we must also include the hydrogen bonding interaction further exemplified by

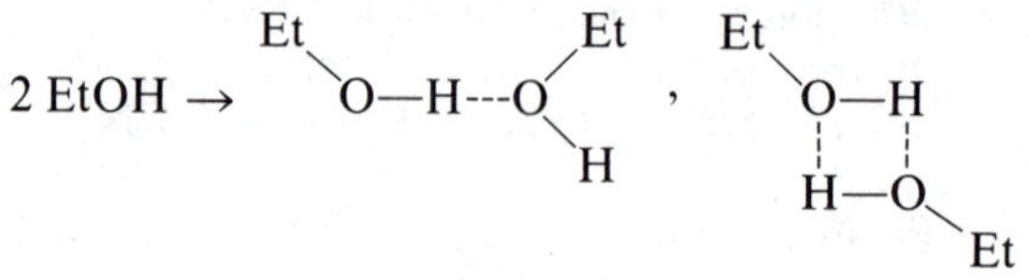

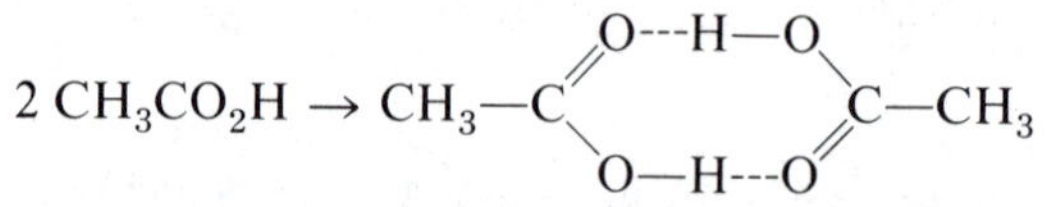

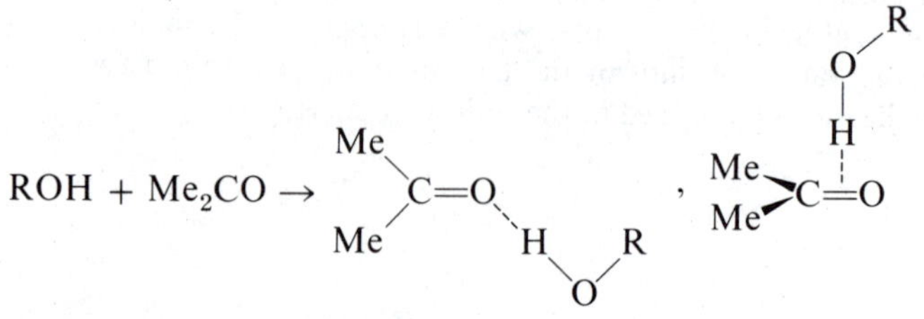

$$2\ EtOH \rightarrow$$

$$2\ CH_3CO_2H \rightarrow CH_3\text{--}C$$

$$ROH + Me_2CO \rightarrow$$

Hydrogen bonds are weak as bonds go; typical values for the enthalpy of hydrogen bond formation fall in the range from 4 to 60 kJ/mole. Because of the limited covalent energy of many hydrogen bonds, and the fact that the AH bond and donor lone pair dipoles are large, it is generally important to recognize that dipole-dipole attractions make a relatively large contribution to the hydrogen bond energy. A second important property of hydrogen bonded adducts is the greatly decreased AH stretching frequency (and longer bond length) in the adduct AHD than in free AH. This is expected because the electron pair in the A—H bond becomes more localized on A through σ resonance structures like

$$A{-}H \cdots \overset{\ominus}{:}D \leftrightarrow \overset{}{A}\overset{}{:} \cdots \overset{\oplus}{H}{-}D$$

Finally, you might expect to find that the hydrogen bond is not highly directional. In fact, alcohols dimerize with anti-parallel OH dipoles favored over the "head-to-tail" structure. In the ethanol dimer example,

$$\begin{array}{c} Et \\ \diagdown \\ O{-}H \\ \vdots \\ H{-}O \\ \diagdown \\ Et \end{array}$$

we find a bent hydrogen bond resulting from the structural constraint imposed on the formation of two hydrogen bonds.

The occurrence of such dimers (and even higher oligomers) of alcohols is made interesting by the fact that alcohol viscosities and boiling points are related to the extent of such "self-association." In seeming contradiction to the usual effect of electronegativity on the strength of the hydrogen bond, greater electronegativity of the alkyl group causes *less* self-association. With normal bases, greater electronegativity of the alkyl group causes greater association between the alcohol and the donor; but for *self-association*, where the donor atom is the oxygen atom of another alcohol molecule, the donor character of the oxygen atom is reduced as the substituent increases in electronegativity, causing a decrease in self-association.[4] This follows from the fact that the oxygen is directly bound to the substituent. The withdrawal of charge from the oxygen by the substituent serves to reduce the basicity of oxygen lone pairs by increasing their effective nuclear charge.

In the acetic acid dimer

$$CH_3{-}C\begin{array}{c} O{\cdots}H{-}O \\ \diagup \qquad \diagdown \\ \diagdown \qquad \diagup \\ O{-}H{\cdots}O \end{array}C{-}CH_3$$

we find two linear, asymmetric hydrogen bonds between carboxylic acid functional groups. The linearity of the O—H———O structure is constrained by the structure of the carboxyl groups. In a series of RCO_2H, there is again an inverse correlation between acidity and self-association because of the greater sensitivity of the oxygen basicity than the hydrogen acidity to the alkyl group electronegativity. A noteworthy feature of the carboxylic acid dimer hydrogen bond is that its stability is greater than that of the corresponding alcohol dimer (for example, ethanol *vs.* acetic acid). The acetyl substituent on the OH group of acetic acid is certainly more electronegative than the ethyl group of ethanol, and this causes the OH group of the acid to function as a stronger acid. That the OH oxygen atom of the acid is a poorer donor than its alcohol counterpart is of little consequence, since in the acid it is the carbonyl oxygen group that functions as the donor atom.

The $(ROH + Me_2CO)$ example of hydrogen bonding above is interesting for several reasons. First, there are no structural constraints on the angle of the oxygens about the acid proton; second, using a ketone as an example of a Lewis base, we have an opportunity to note that evidence has been cited[5] for the Lewis base character of the carbonyl π bonding pair of electrons. In this sense, and at least toward hydrogen-bonding acids, the carbonyl group can behave as a π donor as well as a σ donor (by using one of the lone pairs about the carbonyl oxygen). It was not too many years ago that such an idea would have been viewed with skepticism, but now such behavior is more readily accepted (we will see more examples of π donors when we consider aromatic organic molecules) and points up the necessity of dropping Lewis' original concept of Lewis bases as acting through lone pair electrons only. We see here the first instance of what are normally considered to be *bond pair electrons acting in a Lewis base capacity* (more examples will follow).

Hydridic hydrogen, meaning hydrogen bound to an electropositive or low electronegativity element, quite frequently acts as a donor atom in intermolecular interactions. In this capacity hydrides may be added to that class of molecules for which bond pair electrons may be involved in adduct formation. A very important example arises in the dimerization of borane,[6] wherein we see the BH_3 molecule acting both as an acid (through the boron atom) and as a base (through the hydrogen atom).

diborane(6)
colorless gas
spontaneously flammable
b.p. = −93°C

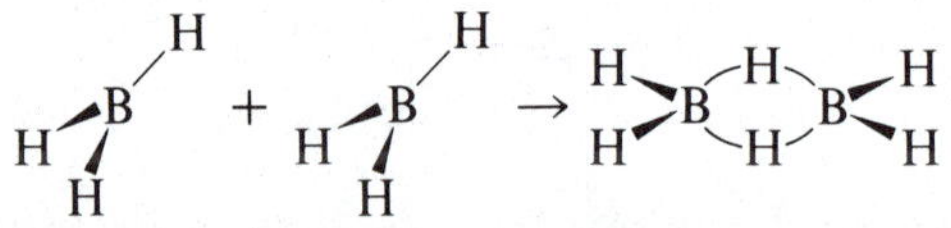

B_2H_6 could therefore be considered an adduct. As a simple extension of these ideas, we may view BH_4^- as an adduct with the capacity to act as a donor, as the following examples[7] illustrate.

$Al(BH_4)_3$ ≡

liquid
m.p. = −65°
b.p. = 45°

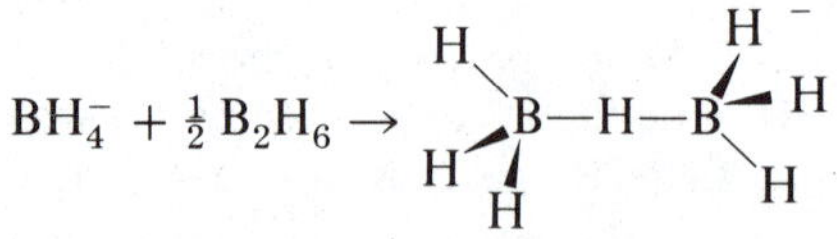

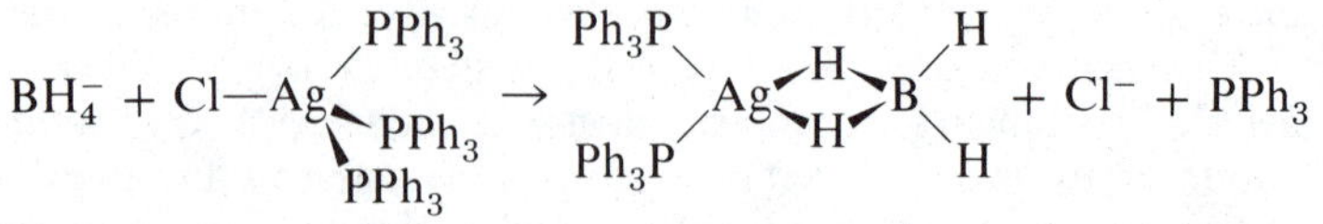

$Al(BH_4)_3$ and $(Ph_3P)_2AgBH_4$ are especially interesting and worth remembering because they illustrate the dibasic or bidentate ability of BH_4^-.

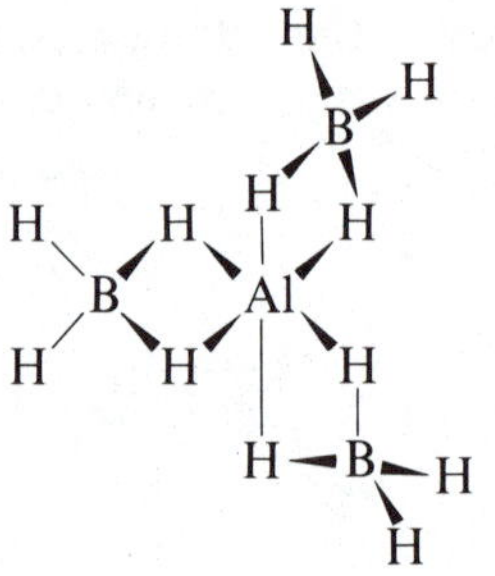

In these examples we encounter the so-called three-center bridge bond, a recurring unit in stable inorganic structures. In a molecule like $B_2H_7^-$ the bridging hydrogen atom is symmetrically positioned and the molecular point group is D_{3d}; that is, the boron atoms are symmetry equivalent. To describe the $B-H_b-B$ bonding we may proceed as follows. Each $B(H_t)_3$ unit has mo's much like those of NH_3. The H_b 1s ao principally overlaps the BH_3 LUMO's, which individually look a great deal like the lone pair (HOMO) of NH_3.

$$H_3B \circ \,) \qquad (H) \qquad (\, \circ BH_3$$

This problem is one of the "three orbital linear chain" type (see Figure 4-1). Accordingly, the $B-H_b-B$ mo's appear as

$$\text{———} \qquad H_3B \circ \,) \mid \bigcirc \mid (\, \circ BH_3$$

$$\text{———} \qquad H_3B \circ \,) \mid \qquad (\, \circ BH_3$$

$$\text{—oo—} \qquad H_3B \circ \,) \quad \bigcirc \quad (\, \circ BH_3$$

After accounting for the six $B-H_t$ bond electron pairs, there remains a single pair to occupy the bonding mo of the $B-H_b-B$ unit. Such a situation is called "three-center/two-electron" bonding. Note that the "bridge bond" corresponds to a bond order of $\frac{1}{2}$ for each boron. This model also applies to *each* bridge bond of B_2H_6.

Non-Metal Acids and Bases

BORON AND ALUMINUM

Probably the most familiar examples of Lewis acids come from the compounds formed by these elements. Normally tricovalent, these elements do not thereby satisfy the "octet rule," which in modern terminology means that a valence shell orbital is unoccupied and of fairly low energy (a stable LUMO). This orbital plays the role of the acceptor mo or ao. The acid-base concept nicely accommodates many of the chemical and structural features of this class of molecules. In the next chapter we will examine more closely the tendency of organo- and hydrido-boranes and the lighter organo- and hydrido-alanes to dimerize with the characteristic of bridging H and C atoms.† Similarly, the heavier halides of aluminum (all solids) dimerize with bridging halogens§ but, interestingly, AlF_3 is a polymeric solid (m.p. = 1290°) with six fluorides about each aluminum (see also the next chapter).

[spontaneously flammable for R = Me
m.p. = 15°
b.p. = 126°]

[sublimes at 178°]

---

†The dimerization enthalpy of $AlMe_3$ has been determined to be –84 kJ/mole; C. H. Henrickson and D. P. Eyman, *Inorg. Chem.*, **6**, 1461 (1967). With CH_3 bridging groups, one still may envision three-center/two-electron bonding; comparing CH_3 with hydrogen, an $\sim sp^3$ carbon hybrid takes the place of the 1s ao of hydrogen. See also J. C. Huffman and W. E. Streib, *Chem. Comm.*, 911 (1971).

§Unlike H and CH_3, the halides possess additional lone pair electrons. Thus, a three-center mo description is not *required* here.

As the boron halides do not dimerize like those of aluminum, an intramolecular acid/base interaction, described as π bonding between boron and halogen, seems to inhibit formation of the B_2X_6 dimer.

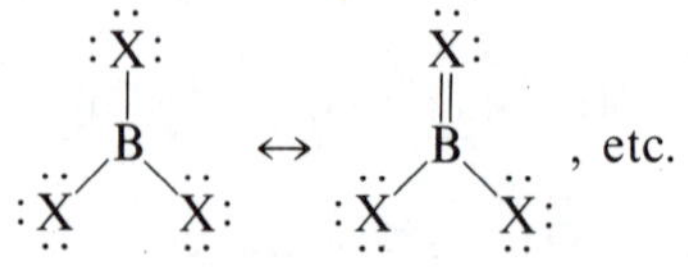

$$ \text{(structures)} \quad , \text{etc.} $$

This intramolecular "adduct" formation is of great importance to the ability of the boron halides to function as Lewis acids in general. Nevertheless, it appears safe to say that these dimerization and intramolecular donor/acceptor interactions are not generally effective in achieving "saturation" of the chemical valency of boron and aluminum; the trivalent compounds of both elements are known to be readily involved in adduct formation with other Lewis base molecules.

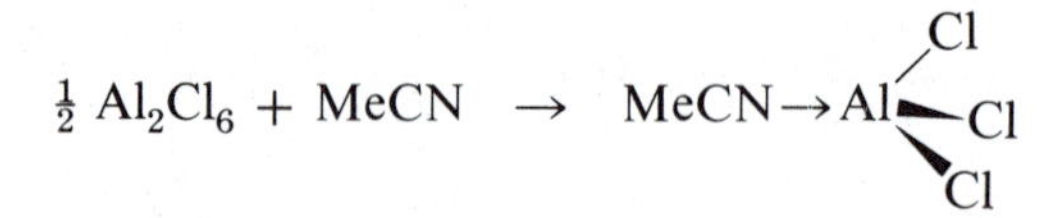

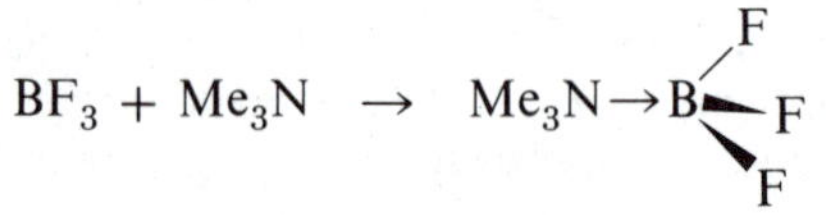

$$ \tfrac{1}{2} Al_2Cl_6 + MeCN \rightarrow MeCN \rightarrow AlCl_3 $$

$$ BF_3 + Me_3N \rightarrow Me_3N \rightarrow BF_3 $$

CARBON AND SILICON

Among the strongest known Lewis acids are the carbonium ions so important in organic chemistry. You will recognize a resemblance of many of these compounds to isoelectronic boron acids, but of course the greater nuclear charge of carbon renders the carbonium ions particularly effective as acids (but very poor as donors; to wit, CH_3BH_2 dimerizes but $CH_3CH_2^+$ apparently does not). The high acidity of carbonium ions is illustrated by the reaction

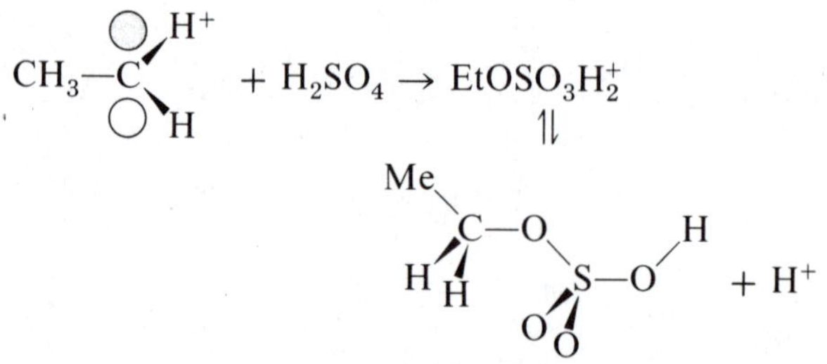

$$ CH_3-C(OH^+)(OH) + H_2SO_4 \rightarrow EtOSO_3H_2^+ $$

$$ \updownarrow $$

$$ \text{(structure)} + H^+ $$

Normally, the oxygens of H_2SO_4 are weakly basic, but the carbonium ion is such an avid acid that complexation occurs.

Perhaps more startling than carbonium ions as acids are the acids and bases of conjugated carbon compounds. These are the π acids and bases referred to earlier in the outline of hydrogen bond interactions. For example, the aromatic phenyl ring of anisole is known[8] to perturb the OH function by acting as a π donor.

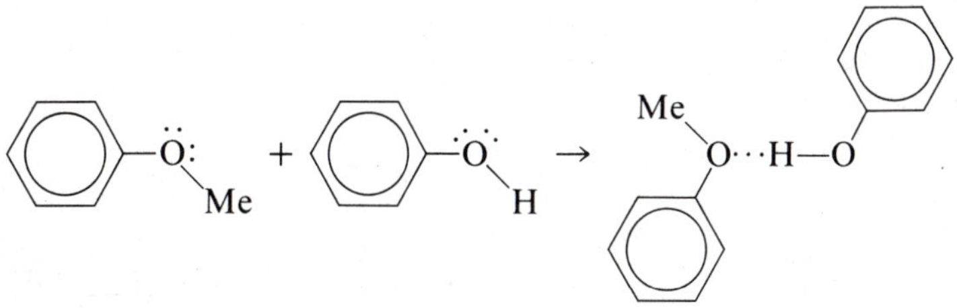

and

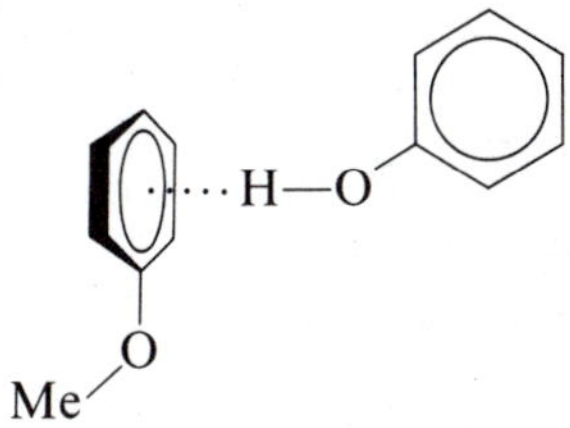

Benzene itself is known to form weak adducts with π acceptors like tetracyano-ethylene. The structure of the adduct is believed to be that of parallel benzene and TCNE planes with C_{2v} symmetry

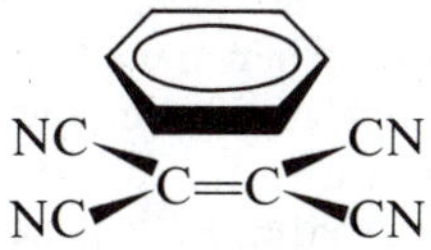

Study of the electronic structures of such organic π-acid/π-base adducts has aroused considerable interest.[9] A characteristic feature of the stronger adducts is that they can be isolated as solids, with the acid-base molecules alternately layered upon one another in chains. These solids have the intriguing property of being anisotropic conductors of electricity, current flowing easily normal to the molecular planes but negligibly in directions parallel to the aromatic planes (in this sense, they are one-dimensional organic "metals"). That the compounds are often highly colored means that a low-energy electronic transition (in the visible region) is possible and suggests their use in photo-electric devices.

The π HOMO of benzene is depicted as a pair of degenerate mo's possessing a single node each (see Figure 4-2). LCAO calculations for TCNE reveal a reasonably stable LUMO with similar nodal properties. The interaction between these as donor–acceptor mo's is held responsible for the formation of the adduct.

top view of C_6H_6 HOMO's

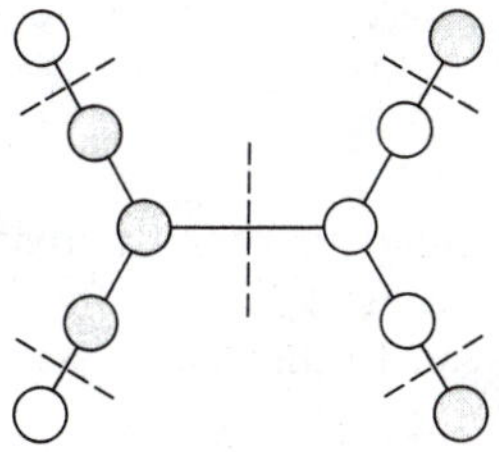

top view of TCNE LUMO

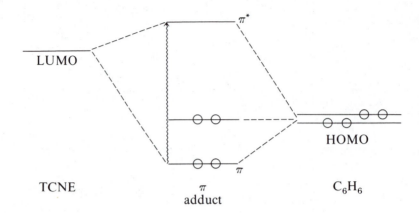

As reflected by the diagram, the delocalization of HOMO electrons into LUMO is weak by virtue of the fact that LUMO energy is higher than HOMO energy.

A photon-induced electronic excitation is indicated by the vertical wavy line. Because the absorption process causes *transfer of an electron from an mo with greater amplitude on the donor to an mo with greater amplitude on the acceptor, the transition is called a* **charge transfer** *absorption—conveying the idea of photoinduced transfer of charge from one fragment to another.* When the energy gap between LUMO and HOMO is not too large, the absorption band appears in the visible region, in which case the adduct will be colored. With deference to this occurrence of a "low" energy electronic transition, such adducts are often referred to as "charge transfer" complexes.

Before leaving the survey of carbon donors, we must recognize that carbanions in general should be very effective donors. Probably the most stable carbanion is cyanide, and it is known to function as a donor in many reactions[10] such as

$$BH_3 \cdot THF + CN^- \rightarrow BH_3CN^- \xrightarrow{BH_3 \cdot THF} BH_3CNBH_3^-$$

$$2Me_3SiBr + 2CN^- \rightarrow Me_3SiCN + Me_3SiNC + 2Br^-$$

and particularly with metal ions (you will see such complexes throughout Chapters 10 to 14 on transition metals).

Although much weaker in its donor ability than CN^- (because oxygen has greater nuclear charge than nitrogen), CO is known to form a wide variety of adducts, in particular with organometallic species; some examples are

<table>
<tr><td>OC
  Ni—CO
OC
OC
[colorless liquid
b.p. = 43°]</td><td>O
C
OC Fe—CO
OC
C
O
[yellow liquid
b.p. = 103°]</td><td>O
C
OC Cr CO
OC CO
C
O
[colorless solid
sublimes *in vacuo*]</td></tr>
</table>

For many years, it was puzzling to inorganic chemists that metal atoms (which did not seem likely candidates for good Lewis acids) should form adducts with CO (recognized to be a rather poor donor toward most other Lewis acids; *e.g.*, BF_3 and SbF_5 do not form adducts with CO).

The dilemma is rationalized by recognizing that the metal valence electrons are available for **back-bonding** to the CO. This concept is summarized by writing, for example,

$$\overset{\ominus}{Ni} \leftarrow C \equiv O: \quad \longleftrightarrow \quad \overset{\ominus}{Ni} \leftarrow C = \overset{\oplus}{O}: \quad \longleftrightarrow \quad Ni \rightleftarrows C = \overset{..}{O}:$$

In this sense you find the five valence pairs of Ni acting as Lewis base pairs; the metal acts as an acid *and* a base while CO simultaneously displays donor *and* acceptor character.

Another important example of CO as a donor occurs in its reaction with diborane.

$$: CO: + \tfrac{1}{2} B_2H_6 \rightarrow \begin{array}{c} H \\ \diagdown \\ H \diagup B-CO \\ H \end{array} \quad \begin{bmatrix} \text{gas} \\ \text{m.p.} = -137° \\ \text{b.p.} = -64° \end{bmatrix}$$

Here you find another example of back bonding with CO. The B—C sigma bond is supplemented by delocalization of the B—H bond pairs onto CO. Since σ bond pairs (rather than lone or π bond pairs) are involved in conjugation to CO, the phenomenon is called **hyperconjugation.**

$$H_3 \overset{\ominus}{\equiv} B \leftarrow C = \overset{\oplus}{O}: \quad \longleftrightarrow \quad H_3 \overset{\oplus}{=} B = C = \overset{\ominus}{O}:$$

The other elements of the fourth group also exhibit acid and base properties. As with third and later row elements of the main groups, we find for silicon the ability to expand the "octet" to form five- and six-coordinate adducts. This, of course, is a primary feature of the reactions of silicon with nucleophiles. Significant in this regard has been the synthesis and isolation of a 1:1 adduct of SiF_4 (b.p. = $-86°$) and F^-. The anion SiF_5^- was unknown until it was stabilized[11] in crystalline form with a quaternary ammonium cation of like size and charge.

$$SiF_4 + [R_4N]F \rightarrow [R_4N][SiF_5]$$

Another element from this group that has been studied extensively is tin. A large variety of adducts in which tetravalent Sn becomes five-coordinate are known; the trigonal bipyramidal structure[12] of Me_3SnCl adducts is believed to feature the Cl and donor atoms in apical positions, on the basis of VSEPR and the high electronegativity of Cl and O.

$$Me_3SnCl + Me_2CO \rightarrow \begin{array}{c} \text{Me} \quad \text{Cl} \\ \diagdown \, | \\ \text{Me} \diagup \text{Sn} - \text{Me} \\ \uparrow \\ : O \diagdown \quad \text{Me} \\ \diagdown C \diagup \\ | \\ \text{Me} \end{array}$$

As examples of Si and Sn species with basic properties, we can cite the anions R_3M^-. R_3Si^- is capable of displacing Cl^- from alkyl chlorides, which makes it very useful as a nucleophile for synthesis of unsymmetrically substituted organo-silanes. The trichlorostannate ion is also a useful nucleophilic base[13] and the following reaction, featuring formation of a metal-metal adduct bond,

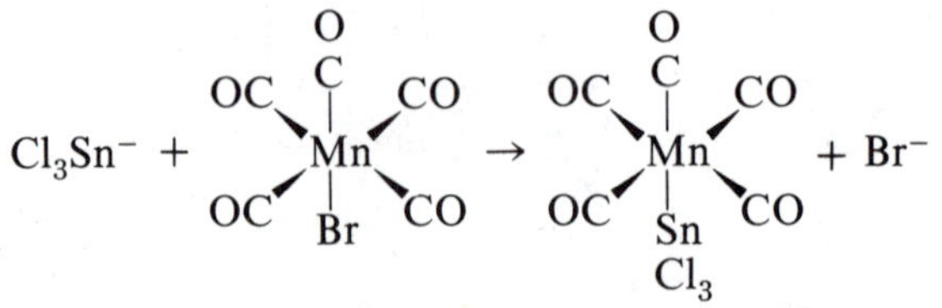

$$Cl_3Sn^- + \underset{\underset{Br}{\overset{OC}{\underset{|}{}}}{\overset{\overset{O}{\overset{\|}{C}}}{\underset{OC}{\overset{OC}{}}}}Mn\overset{CO}{\underset{CO}{}} \rightarrow \underset{\underset{\underset{Cl_3}{Sn}}{\overset{OC}{\underset{|}{}}}{\overset{\overset{O}{\overset{\|}{C}}}{\underset{OC}{\overset{OC}{}}}}Mn\overset{CO}{\underset{CO}{}} + Br^-$$

may be simply viewed as a displacement of one donor (Br^-) from an adduct ($Mn(CO)_5Br$) by another. This interesting reaction is one of many leading to metal-metal bonds.

The trichlorostannate ion may itself be regarded as an acid-base adduct between Cl^- and $SnCl_2$. Diamagnetic stannous chloride is formally electron deficient, in the same sense as the boron trihalides, and could act as an acid or base (draw an electron dot structure for $SnCl_2$). $SnCl_3^-$ is one example of the many adducts in which the Sn atom acts as an acid. With organic solvents, 1:1 adducts of the general formula $SnCl_2 \cdot S$ are common.[14] Unfortunately, no examples can be cited at this time of $SnCl_2$ acting solely as a Lewis base. At any rate, we can cite[15] one example of $SnCl_2$ acting as a donor and acceptor in the same compound, via an "insertion" reaction:

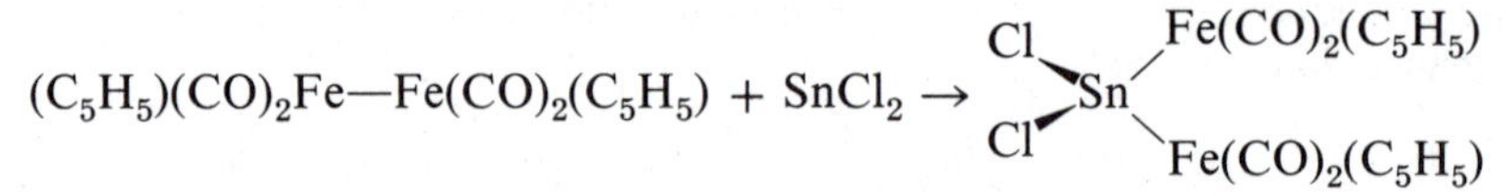

$$(C_5H_5)(CO)_2Fe\text{—}Fe(CO)_2(C_5H_5) + SnCl_2 \rightarrow \underset{Cl}{\overset{Cl}{}}Sn\underset{Fe(CO)_2(C_5H_5)}{\overset{Fe(CO)_2(C_5H_5)}{}}$$

Here the $SnCl_2$ "inserts" into the Fe-Fe linkage, acting, in effect, as an acid toward an anion, $(C_5H_5)Fe(CO_2)^-$, and as a base toward a cation, $(C_5H_5)Fe(CO_2)^+$. This behavior is reminiscent, with good reason, of that of the isoelectronic carbenes (R_2C) of organic chemistry. Many more examples of this reaction type are explored in Chapter 17.

NITROGEN AND PHOSPHORUS

This family affords many common examples of Lewis donors (the trivalent compounds of nitrogen and phosphorus) and a few less widely appreciated examples of Lewis acids (the pentavalent compounds of phosphorus). The amines and ammonia are among the most commonly recognized and generally the strongest of the nitrogen donors. It was noted in the introduction that HCN, like the organic nitriles, is known also to act as a weak donor. As a class, the pyridines are of intermediate basicity and complete the correlation of lone pair hybridization with donor strength, a correlation to be more fully explored in the next section. As a striking example of a substituent effect on basicity, we might note that NF_3 (b.p. = $-129°$) is a notoriously weak amine donor but that it does, like other amines, participate in an oxidation reaction with oxygen atoms that is reminiscent of the Lewis interaction:[16]

$$F_3N + \frac{1}{2}O_2 \xrightarrow[\text{dischg.}]{\text{elec.}} F_3N\text{—}\ddot{\text{O}}:$$

Similarly, trivalent phosphines undergo this reaction to form the highly stable (in both the thermodynamic and kinetic senses) phosphine oxide link (see Chapter 6).

A particularly interesting phosphine donor, to be encountered again in the study of organometallic compounds in Chapters 16 and 17, is PF_3. While the reaction with diborane is more or less conventional, the ability of gaseous PF_3 to convert solid nickel metal to the gaseous tetrakis(trifluorophosphine) nickel complex

$$PF_{3(g)} + \frac{1}{2} B_2H_{6(g)} \rightarrow F_3P\text{---}BH_{3(g)}$$

$$4PF_{3(g)} + Ni_{(c)} \rightarrow \underset{F_3P}{\overset{F_3P}{\diagup}}Ni\underset{PF_3}{\overset{PF_3}{\diagdown}} \quad (g)$$

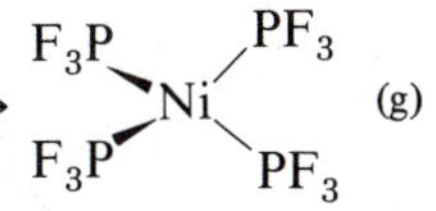

is quite remarkable. Normally one might not think of nickel atoms as being very good Lewis acids. In fact, CO (as noted in the previous section) and PF_3 share this unusual ability to form MD_n complexes.[17]

An interesting aspect of PCl_5 chemistry is that, while the vapor is monomeric, the solid is a crystalline salt (sublimes at $162°$) and thereby illustrates an important class of acid-base behavior—the halide transfer reaction. It is interesting to speculate that lattice energies have a great deal to do with this property of PCl_5, because PF_5 (b.p. = $-75°$) does not form an ionic solid (but PF_6^- salts are known). Complementary to this aspect of PCl_5 as a Lewis acid is the occurrence[18] of the simple 1:1 adduct $PCl_5 \cdot py$. Important examples of halide ion transfer arise particularly with pentavalent antimony.

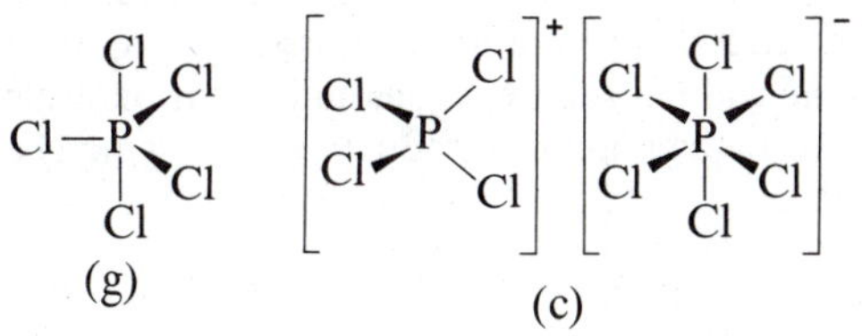

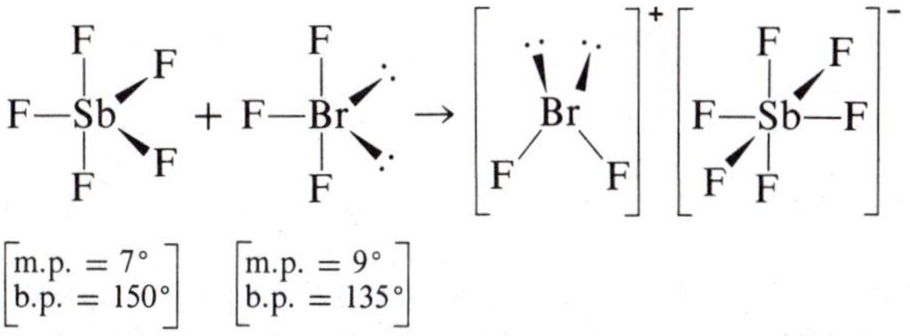

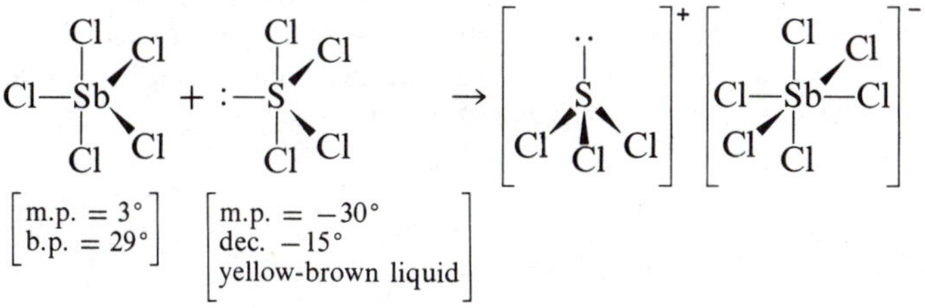

With other reagents, such as arsenic trifluoride, the fluoride transfer is incomplete and bridging fluoride structures often result. AsF_2^+ should be, particularly as an electron "deficient" cation, more highly acidic than SbF_5 (compare AsF_2^+ with BrF_2^+ and SCl_3^+ with regard to the "octet" rule).

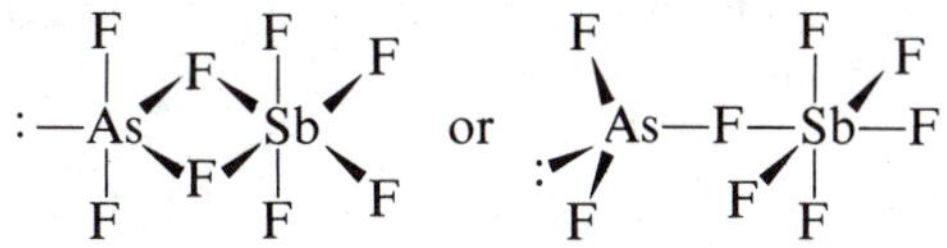

Both complete halide transfer and association via bridging halide are also encountered with the halogen fluorides. It is well to be alert to the fact that this *halide transfer/association behavior is characteristic of halides of heavier elements and plays an important role in the chemistries of such compounds.*

Our final example of pentavalent phosphorus concerns its acid character. The formation of PCl_6^- with a properly sized cation and of $PCl_5 \cdot py$ have already been mentioned. A less conventional example[19] is the reaction of phosphoryl trichloride with an organic base like Et_3N:

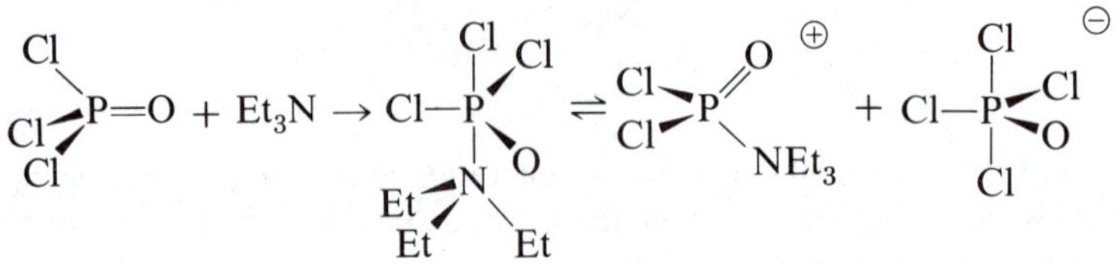

$$\begin{bmatrix} \text{m.p.} = 2° \\ \text{b.p.} = 105° \end{bmatrix}$$

leading to conducting solutions with $POCl_3$ as the solvent.

OXYGEN AND SULFUR

While many examples of oxygen as a donor can be cited [water, ethers, ketones, amine and phosphine oxides, and thionyl ($>SO$) and sulfuryl ($>SO_2$) compounds are a few of the important classes], we are hard pressed to find examples in which oxygen acts as an acid. The amine and phosphine oxides and sulfur oxides mentioned above could be considered, in fact, to be examples of adducts with the oxygen atom playing the role of the acid:

$$\left.\begin{matrix} R_3N: \\ X_3P: \\ R_2\ddot{S}: \\ R_2SO \end{matrix}\right\} \xrightarrow[\text{transfer}]{\text{oxygen atom}} \left\{\begin{matrix} R_3N\ddot{O}: \\ X_3PO \\ R_2SO \\ R_2SO_2 \end{matrix}\right.$$

Such oxygen atom transfer reactions play an important role in phosphorus and sulfur chemistry.

A contemporary example of dioxygen adduct formation arises with the so-called "oxygen carrier" transition metal complexes. Complexes with the ability to reversibly add O_2 are of great interest as analogs to the biological carriers such as hemoglobin and myoglobin. Recent studies in inorganic laboratories[20] have confirmed that such carrier complexes are characterized by an angular $M-O-O$ structure.

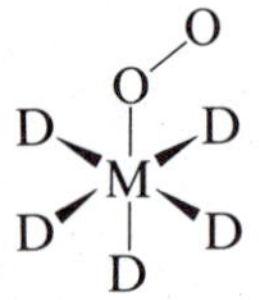

The formation of the $M-O$ bond has been of some interest. The MD_5 unit, when $M = Co^{2+}$, has a single unpaired electron directed along what is to become the $M-O$ bond axis (this statement will be justified in Chapter 9). Recalling that O_2 has two unpaired electrons in π^* mo's, the stage is set for bond formation by the overlap of the Co and one of the O_2 π^* orbitals. The result then is a $Co-O$ bond pair and a single unpaired electron residing in the other O_2 π^* mo.

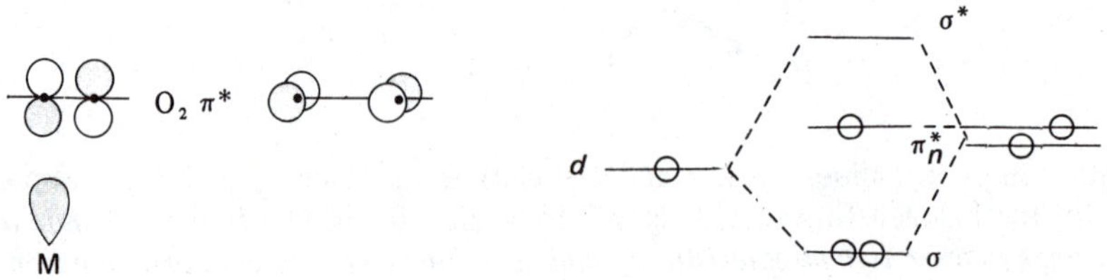

Whether Co^{2+} is oxidized to Co^{3+} (and O_2 reduced to O_2^-) depends on whether and to what extent the π^* mo's of O_2 lie below the metal d ao in energy.[21]

Yet another reaction of dioxygen with certain square planar metal complexes is known to give a symmetrically bonded O_2 group.[22] Called "oxidative addition," the reaction produces an adduct that *could* be viewed as an adduct of O_2^{2-},

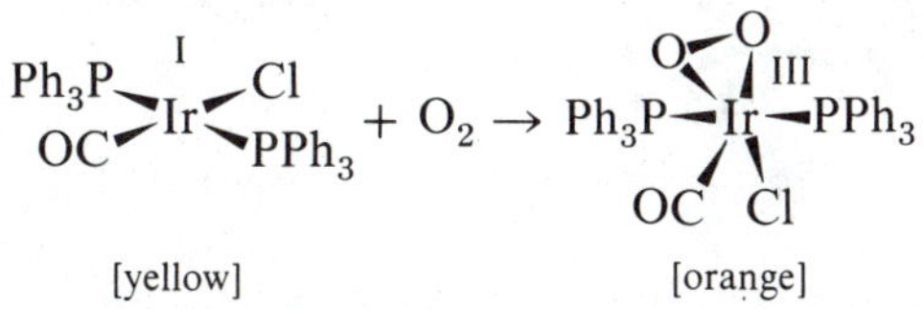

[yellow]   [orange]

From an mo view of O_2^{2-}, the O_2^{2-} donor mo (HOMO) is a π^* mo having non-zero overlap with an Ir d ao (d_{xy}, for example).

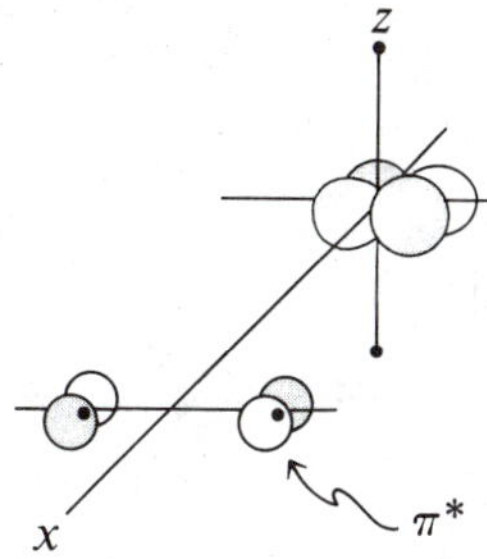

The delocalization of these O—O anti-bonding electrons onto the metal causes an O—O linkage strengthening, as evidenced by an O—O distance of 130 pm. This bond distance is very close to that in superoxide, $[O_2]^-$, where there are only three π^* electrons.

Sulfur has a richer acid-base chemistry than does oxygen. Not only do we find examples of sulfides† and phosphine sulfides as donors

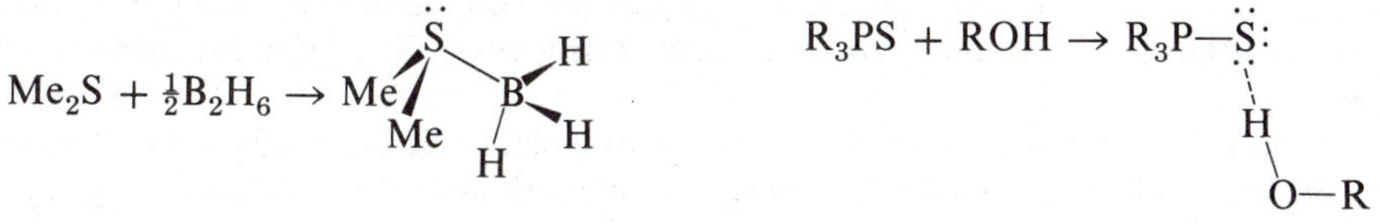

but the sulfoxides can also act as Lewis bases (at either sulfur or oxygen).[23]

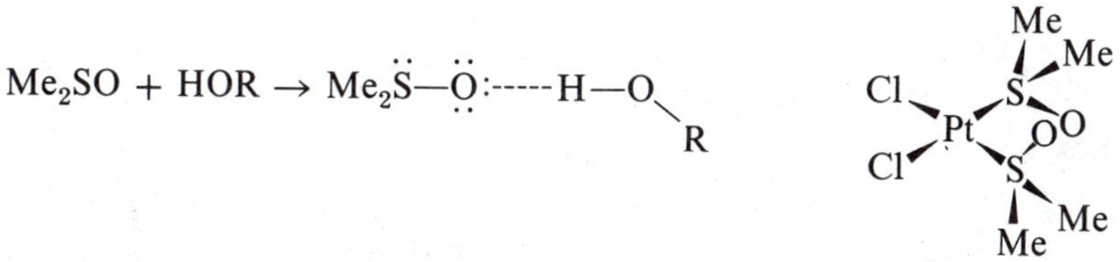

Two of the more interesting sulfur compounds exhibiting acid-base properties are the binary oxides, SO_3 and SO_2. The trioxide is a voracious acid but a notoriously weak (oxygen) donor. Accordingly, SO_3 is often encountered as a vapor at room temperature (b.p. = 45°) but, interestingly, has polymeric structures as a solid. The trimeric ring and

---

†$Me_2S \cdot BH_3$ (m.p. = $-38°C$, b.p. = $97°C$) is thought to present a particularly convenient source of BH_3 for use in the laboratory because it is 18% by weight BH_3, is not a gas (like B_2H_6), and avoids the hazards associated with ethereal BH_3 (THF/BH_3 solutions, only 1.5 per cent by weight BH_3, are commercially available). J. Beres, *et al.*, *Inorg. Chem.*, **10**, 2072 (1971).

chain structures of two different crystalline forms suggest that an acid-base self-interaction takes place in the solid.

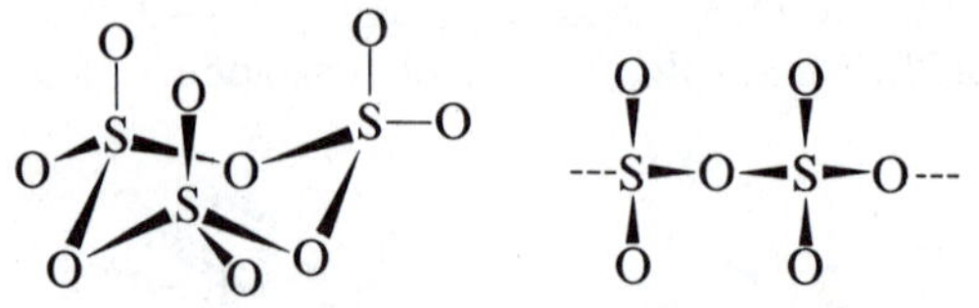

[m.p. = 17°]   [m.p. = 33°]

More conventional examples involve SO_3 acting as an acid toward donors like trimethylamine, alcohols, and thioalcohols:

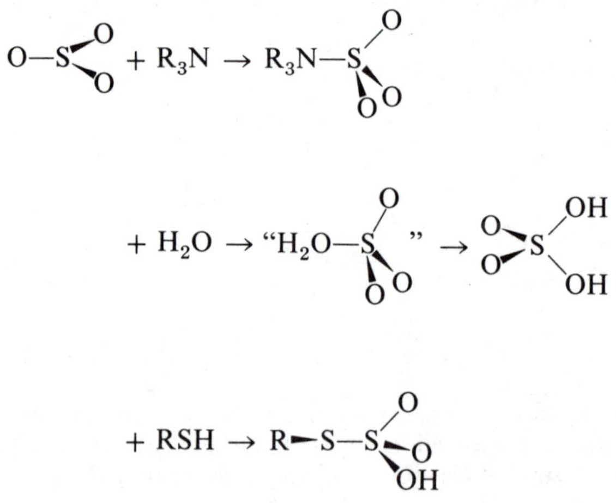

Simple amine and dioxane adducts are particularly useful as sulfonating reagents, for they are solids and afford less vigorous, more easily controlled reactions than does SO_3 itself.[24]

Even more interesting than SO_3 is sulfur dioxide. Its acid-base character is weak but unusual; not only can the sulfur act as an acid site, but both the sulfur and oxygen atoms can act as donors.[25]

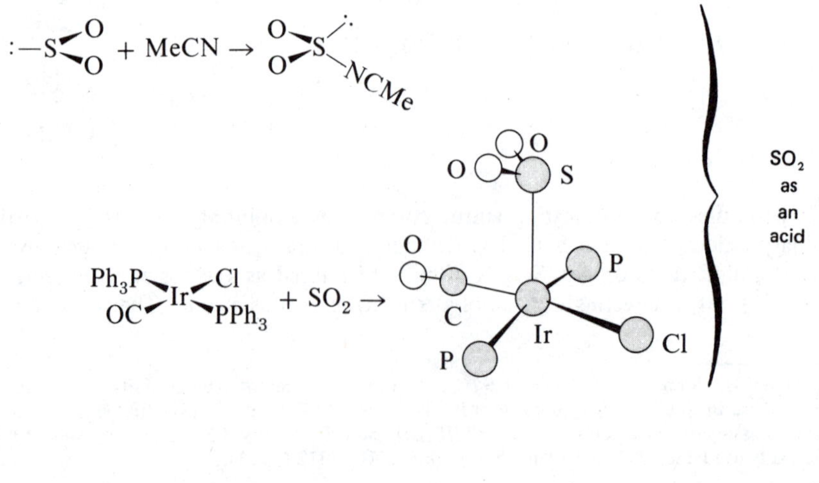

SO_2
as
an
acid

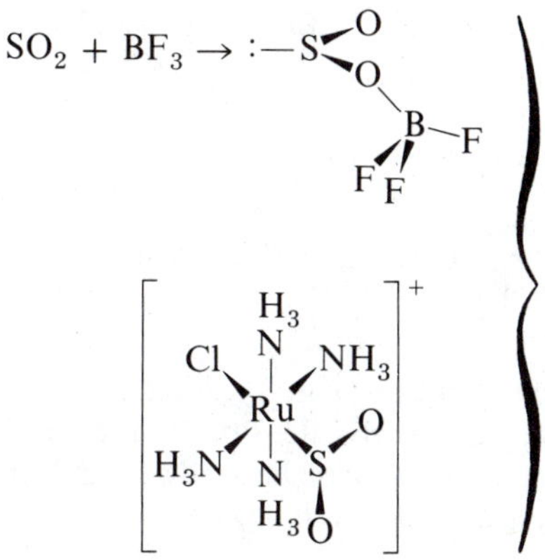

$$SO_2 + BF_3 \rightarrow \quad :-S \overset{O}{\underset{O}{\diagup}} \quad B-F \atop F \; F$$

$$\left[\begin{array}{c} H_3 \\ Cl \underset{Ru}{\overset{N}{\diagdown}} NH_3 \\ H_3N \underset{N}{\diagup} \overset{O}{\underset{S}{\diagdown}} \\ H_3 \; O \end{array} \right]^+$$

$$\left. \begin{array}{c} \\ \\ \\ \\ \\ \end{array} \right\} \quad \begin{array}{c} SO_2 \\ as \\ a \\ base \end{array}$$

To help you tie things together, notice that the SO_2 valence shell is isoelectronic with $SnCl_2$ and that they form structurally similar adducts. The analogy extends to the insertion reaction also, for SO_2 is known to insert into metal-carbon and metal-metal bonds (*cf.* p. 134).[26]

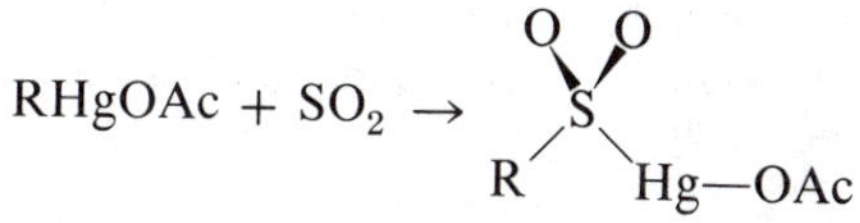

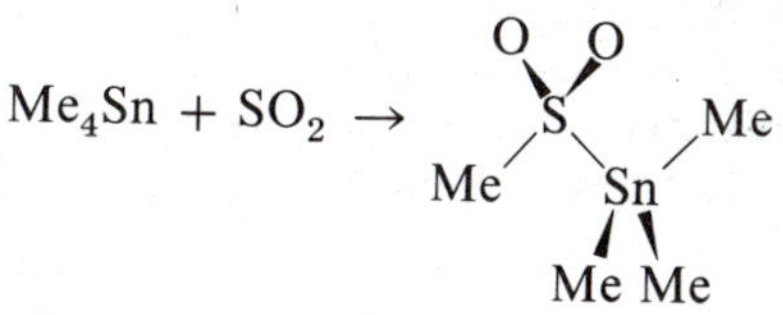

$$RHgOAc + SO_2 \rightarrow \quad \overset{O \; O}{\underset{R \quad Hg-OAc}{S}}$$

$$Me_4Sn + SO_2 \rightarrow \quad \overset{O \; O}{\underset{\underset{Me \; Me}{Sn}}{\underset{Me \quad \quad Me}{S}}}$$

Because SO_2 is highly polar and weakly acidic-basic, and has a usefully high boiling point of $-10°C$, it is an important non-aqueous solvent, either alone or in combination with SbF_5 and HF as a "super acid" solvent.

Finally, we should mention that SF_4 has been reported[27] to act as a fluoride ion acceptor in the reaction with quaternary fluoride salts

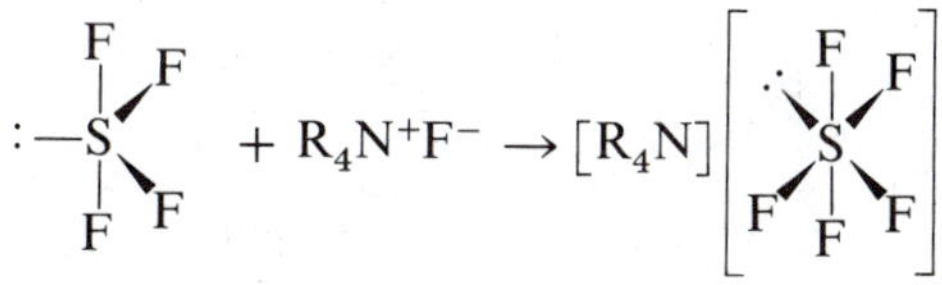

$$:-S \overset{F \; F}{\underset{F \; F}{\diagup}} + R_4N^+F^- \rightarrow [R_4N] \left[\overset{F \; F}{\underset{F \; F}{\underset{F}{S}}} \right]$$

$$\left[\begin{array}{c} colorless \; gas \\ b.p. = -40° \end{array} \right]$$

where again (refer to the SiF_5^- and PCl_6^-, PCl_4^+ examples given earlier) lattice stabilization of the adduct is of importance.

HALOGENS

The halide ions are well known as Lewis bases, particularly toward metal ions but also in the ion transfer and bridging capacities noted above with the Group V elements and the halides of aluminum. The elemental forms of the halogens are known to have primarily acidic properties, but one example of Cl_2 acting as a base has been reported[28] in the reaction

$$ClF + Cl_2 + SbF_5 \rightarrow [Cl_3][SbF_6]$$

which may be viewed as F^- transfer from one acid (Cl^+) to another (SbF_5) with stabilization of Cl^+ by adduct formation with Cl_2.

The acid functions of the halogens are largely restricted to iodine for the simple reason that the other elements, particularly fluorine and chlorine, are such strong oxidants that their reactions with donors carry on beyond adduct formation to oxidation products. One example of Cl_2 acting as an acid can be cited:[29]

$$[Ph_4As][Cl] + Cl_2 \xrightarrow{-78°} [Ph_4As][Cl_3]$$

The product anion is linear, as expected from the VSEPR model.

One of the more colorful halogens, iodine (purple in the gas phase), gives solutions of varying shades of color when dissolved in organic solvents. In the simplest terms, such solutions contain both "free" I_2 and complexed iodine in equilibrium with each other, and the color of the solution is simply a superposition of the different colors of I_2 and $I_2 \cdot D$. A well-known example of the Lewis acid character of I_2 is its reaction with iodide ion to form the yellow triiodide ion, I_3^-. (This reaction is made use of when aqueous solutions of I_2, normally susceptible to air oxidation, are stabilized by addition of I^-.)

To account for the role of I_2 as an acid, and the colors of I_2 and $I_2 \cdot D$, we need to identify the LUMO of I_2 and examine its overlap with the donor orbital of a general base. I_2 is a homonuclear diatomic molecule with seven valence pairs; thus, the lowest energy acceptor orbital of I_2 is $p\sigma^*$ (see Chapter 4) directed along the I-I axis.

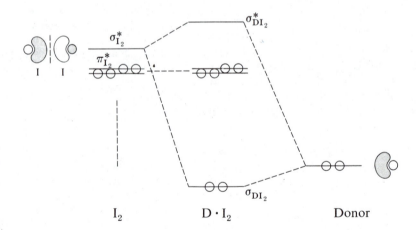

A very important consequence of the adduct bond formation is that while the occupied donor orbital is stabilized by the constructive mixing-in of a little of the I_2 σ^* mo, the σ^* mo of I_2 is destabilized by the destructive mixing-in of a little of the donor orbital (remember that the destabilization is associated with anti-bonding character in the *adduct bond* region).

These ideas, which follow from the LCAO model of adduct formation, have important consequences for the spectral properties of I_2 complexes. To begin with, I_2 in the gas phase has a purple color owing to absorption of photons in the middle of the visible range of radiation (~500 nm). The electronic transition responsible for this absorption of light involves excitation of an electron from the I_2 π^* mo into the I_2 σ^* mo.

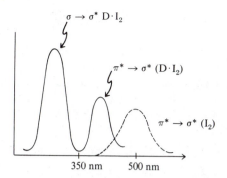

Adduct formation, as you have seen, results in a destabilization of the I_2 $\sigma*$ mo, with the consequence of making the $\pi* \to \sigma*$ absorption process more energetic. That is, radiation of shorter wavelength is now required for the electronic transition. We expect, and find, that the stronger the adduct bond formed with iodine, the more shifted to the blue is the absorbed photon.

Another interesting feature of the spectrum of I_2 adducts is the occurrence of a strong absorption in the near ultraviolet region, which is not present in the spectrum of I_2 alone. This new absorption band is assigned to excitation of an electron from the adduct bond orbital (mainly the donor lone pair orbital) into the adduct anti-bonding mo (mainly the I_2 $\sigma*$ mo). Because charge density within the adduct shifts from the donor toward the I_2 region, this transition is referred to as a **charge-transfer** (sometimes abbreviated C-T) **transition**. As we might expect, there is also a correlation[30] between the energy of this electronic transition and the energy of the adduct bond between I_2 and donors.

Still other examples of acid behavior are to be found for the interhalogen compounds. ICl, for example, is normally a stronger acid than I_2, with the acid site being the iodine atom.

$$:\ddot{I}-\ddot{C}l: \quad + \text{MeCN:} \to \text{MeCN}-I-Cl:$$

$$:\ddot{I}-\ddot{C}l: \quad + \quad Cl^- \to [Cl-I-Cl]^-$$

$$\begin{bmatrix} \text{red solid/liquid} \\ \text{m.p.} = 27° \end{bmatrix} \qquad\qquad\qquad \text{[yellow]}$$

Similarly, ICl_3 is a good chloride ion acceptor,

$$\begin{array}{c} \text{Cl} \\ | \\ :\!\!\diagdown I\!-\!Cl \\ | \\ \text{Cl} \end{array} \quad + Cl^- \to \begin{bmatrix} Cl\diagdown | \diagup Cl \\ I \\ Cl\diagup | \diagdown Cl \end{bmatrix}^-$$

$$\begin{bmatrix} \text{yellow-brown solid} \\ \text{red liquid} \\ \text{dec. } 77° \end{bmatrix} \qquad\qquad \begin{bmatrix} \text{yellow} \end{bmatrix}$$

a property also revealed by its dimerization (I_2Cl_6, D_{2h}) in the solid. Further, BrF_3 self-associates and appears to auto-ionize (the liquid is conducting)

$$\begin{array}{ccc} F\diagdown\!\underset{|}{Br}\!\diagup F\!\diagdown\!\underset{|}{Br}\!\diagup F & & F \\ F\diagup | \quad F\diagup | \quad F & \text{or} & \diagdown\!\underset{|}{Br}\!-\!F\!-\!\underset{|}{Br}\!-\!F \end{array}$$

$$\begin{bmatrix} \text{yellow-green liquid} \\ \text{m.p.} = 9° \\ \text{b.p.} = 126° \end{bmatrix}$$

and acts as a fluoride ion acceptor and fluoride ion donor (p. 135):

$$ClF_3 + BrF_3 \rightarrow [ClF_2][BrF_4]$$

XENON

Of the three fluorides of Xe(XeF_2, XeF_4, and XeF_6 are all colorless solids of m.p. 129°, 117°, and 50°), all are known to act as fluoride donors, and the hexafluoride is also known to function in an acid capacity.[31] From the difluoride, two different salts have been obtained, depending on whether an excess of XeF_2 or of fluoride acceptor is used:

$$XeF_2 + 2SbF_5 \rightarrow XeF_2 \cdot 2SbF_5$$
$$2XeF_2 + SbF_5 \rightarrow 2XeF_2 \cdot SbF_5$$

In the first (1:2) adduct we find a dinuclear anion with F^- bridging two SbF_5 groups, and in the (2:1) adduct there is a dinuclear cation (of C_{2v} symmetry; cf. the VSEPR prediction with F^- bridging, in effect, two XeF^+ moieties.

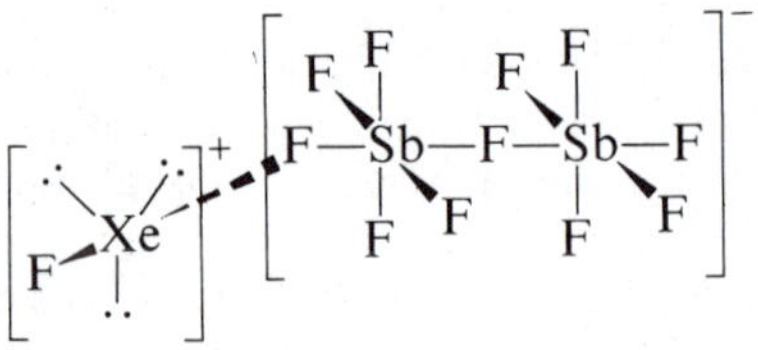

Two adducts of the tetrafluoride, both containing the XeF_3^+ cation, have been reported:

$$XeF_4 + SbF_5 \rightarrow [XeF_3][SbF_6]$$
$$XeF_4 + 2SbF_5 \rightarrow [XeF_3][Sb_2F_{11}]$$

In a similar way, XeF_6 and antimony pentafluoride form a compound with the composition $2XeF_6 \cdot SbF_5$, which presumably also contains a dinuclear cation $[Xe_2F_{11}]^+$. That the hexafluoride can also act as an acid is evidenced by the reactions

$$XeF_6 + F^- \rightarrow [XeF_7]^- \underset{\lambda}{\rightarrow} \tfrac{1}{2} XeF_6 + \tfrac{1}{2} [XeF_8]^{2-}$$

The crystalline form of XeF_6 itself reflects the dual acid-base nature of the compound, for in the solid phase one finds fluorines bridging xenon atoms in tetrameric ring and hexameric cage structures.

At least one oxide of xenon has been shown to exhibit Lewis acid behavior, in that XeO_3 will react with alkali fluorides to form the complex anion XeO_3F^-:

$$XeO_3 + KF \rightarrow K[XeO_3F]$$

As you might well expect by now, the anions are linked, by sharing F atoms, into chains.[32]

Metals

To complete this survey of the utility of the acid-base concept, we will emphasize a few more unusual properties of metals. Metal cations are widely known to act as acids—

this is the basis for a great deal of transition metal ion chemistry, and we will give this topic considerable attention in subsequent chapters. You have seen (p. 132) that with the right donors (*e.g.,* CO and PF_3) even metal *atoms* appear to act as Lewis acids and bases (σ acceptor, π donor). A closely related aspect of metal acid-base chemistry is their function as σ donor Lewis bases. The low-oxidation-state organometallic complexes have received considerable study recently with a view to understanding their chemistry in terms of Lewis base properties.[33] A few examples will suffice at this point. The manganese pentacarbonyl anion exhibits base character in the reactions

$$Mn(CO)_5^- + H^+ \rightleftharpoons H{-}Mn(CO)_5 \qquad (pK_a \sim 7)$$
$$+ Me_3SiCl \rightarrow Me_3Si{-}Mn(CO)_5 + Cl^-$$
$$+ Cr(CO)_6 \rightarrow [(OC)_5Cr{-}Mn(CO)_5]^- + CO$$

The last example might be viewed as another example of a low-oxidation-state metal atom (Cr) acting as an acid; the last two examples further illustrate (see p. 134) the concept of metal-metal bonds.

To illustrate that this base behavior of organometal compounds is not just a peculiarity of $Mn(CO)_5^-$, we note that tetracarbonylcobaltate($1-$)[isoelectronic with $Ni(CO)_4$] is capable of reacting in a conventional Lewis manner with indium tribromide.[34]

$$Co(CO)_4^- + InBr_3 \rightarrow \left[\begin{array}{c} \overset{\displaystyle O}{\underset{\displaystyle \|}{C}} \\ OC{-}\!\!\overset{\displaystyle |}{\underset{\displaystyle |}{Co}}\!\!{-}In{\Big\langle}^{\displaystyle Br}_{\displaystyle Br}{\,}^{Br} \\ OC \quad C \\ \underset{\displaystyle O}{} \end{array} \right]^{\ominus}$$

In $Co(CO)_4^-$ the metal atom has a filled valence shell with four pairs defining the Co–C σ bonds and five non-stereochemically active pairs involved in Co–C π bonding. Structural rearrangement to accommodate the presence of the indium atom about the cobalt indicates that one of the Co–C π pairs is easily diverted into a sigma valence role.

---

1. Compare RCH_2^+ and RCO^+ with the isoelectronic CH_3BH_2 and RBO, and explain why the latter do, but the former don't, dimerize and trimerize, respectively.

2. Describe the SO_2 LUMO which acts as the acceptor orbital on approach of a donor.

3. Interpret the finding that the boiling points of the following compounds are in the order shown.

   a) $H_2O > NH_3$
   b) $CH_3OH > CF_3CH_2OH > (CF_3)_2CHOH > (CF_3)_3COH$
   c) $CH_3CO_2H > ClCH_2CO_2H > CF_3CO_2H$

4. Which acid should show the largest shift to lower energy in AH stretching frequency (largest AH bond weakening) upon hydrogen bond formation with a Lewis base?

   a) H_2O c) NH_3
   b) EtOH   d) CF_3CH_2OH

5. What is the predicted effect of metal-CO retro-bonding on the CO bond strength? Would ν_{CO} (the C≡O stretching frequency) be larger, the same or smaller in free CO?

**STUDY
QUESTIONS**

6. Compute the atom formal charges in the valence structures below and suggest a preferred site for attack by a nucleophile (M has zero formal charge in the third structure).

$$\ddot{M}-C=\ddot{O}: \leftrightarrow \ddot{M}-C\equiv O: \leftrightarrow M=C=\ddot{O}:$$

7. Given that $[AlMe_2Ph]_2$ has a structure with bridging phenyl groups and the phenyl groups are perpendicular to the four-membered ring of Al and bridging carbons (see p. 129), comment on the similar roles of the phenyl groups and the Cl atoms in Al_2Cl_6 (specifically, describe the phenyl and Cl orbitals used for bridge bonding).

8. Predict products for the reactions

$$2ClF + AsF_5 \rightarrow$$
$$ClF_3 + AsF_5 \rightarrow$$
$$IF_5 + CsF \rightarrow$$
$$ClF_3 + CsF \rightarrow$$
$$ClF + CsF \rightarrow$$

9. Predict structures for

a) B_2H_5Me
b) $Al_2Me_2Cl_4$
c) $(R_2AlH)_3$
d) $(RBO)_3$
e) $Al_2Cl_7^-$
f) $AlH_3(NMe_3)_2$
g) $Sn_2F_5^-$

10. Dissect each of the adducts below into two fragments: an "easily recognized" acid and base. Within each fragment, which are the acceptor and donor atoms? As the criterion for "easily recognized," at least one of the fragments must have a reasonable chemical identity.

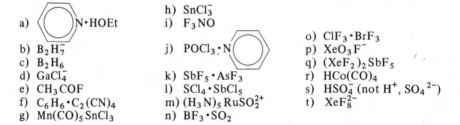

a) (structure) $N \cdot HOEt$
b) $B_2H_7^-$
c) B_2H_6
d) $GaCl_4^-$
e) CH_3COF
f) $C_6H_6 \cdot C_2(CN)_4$
g) $Mn(CO)_5SnCl_3$

h) $SnCl_3^-$
i) F_3NO
j) $POCl_3 \cdot N$ (structure)
k) $SbF_5 \cdot AsF_3$
l) $SCl_4 \cdot SbCl_5$
m) $(H_3N)_5RuSO_2^{2+}$
n) $BF_3 \cdot SO_2$

o) $ClF_3 \cdot BrF_3$
p) XeO_3F^-
q) $(XeF_2)_2SbF_5$
r) $HCo(CO)_4$
s) HSO_4^- (not H^+, SO_4^{2-})
t) XeF_8^{2-}

11. In the presence of ketones, aldehydes, and aryl ethers, alcohols often exhibit three OH stretching vibrational absorption bands. (One of these, that at highest energy, corresponds to the vibration in the free molecule.) Discuss the origins of the two lower energy bands and assign structures to the species responsible for them.

ACID-BASE STRENGTHS

The Thermodynamic Definition[35]

The preceding survey of acid-base compounds should have convinced you that a large number of compounds are characterized by conveniently high-energy occupied orbitals that can serve as the source of Lewis basicity, and conveniently low-energy unoccupied mo's that can serve as the basis for Lewis acid character. As a qualitative guide to intermolecular interactions, the concept of acids and bases is certainly useful. Nevertheless, we soon reach the point of asking quantitative questions such as, "If I place two donors in competition for a given acid, which will preferentially form the adduct?" or "Just how strong is the adduct formed by a given acid-base pair and what factors determine the strength of the interaction?"

These questions are not easy to answer in general, and research continues as experimental and theoretical answers are sought. One of the first problems we face is how to measure the "strength" of the adduct bond. A thermodynamic approach seems most reasonable, but even after agreeing on this we are faced with the question, "Which thermodynamic property of the reaction

$$A \cdot D \rightarrow A + D$$

are we to use—the internal energy change or the free energy change?" The latter seems to have more direct bearing on practical experience but is more difficult to interpret than the former, because of the inclusion of entropy changes. To give one example in which different conclusions are reached regarding the strengths of adducts, we give below some data[36] for dissociation of the $Me_3N \cdot BMe_3$ and $Me_3P \cdot BMe_3$ adducts, all species in the gas phase.

	$Me_3N \cdot BMe_3$	$Me_3P \cdot BMe_3$
$K_d^{100°C}$	0.47	0.13
$\Delta G_d^{100°C}$ (kJ/mole)	+2.5	+6.3
$\Delta H_d^{100°C}$ (kJ/mole)	+73.6	+69.0

According to the gas phase equilibrium constants (free energies) for adduct dissociation, the phosphine adduct is more than three times more stable—yet the bond energies of the adducts (more properly, the enthalpies) suggest that the N—B bond is 7 per cent stronger than the P—B bond. In both of these examples, the entropy change upon adduct dissociation (at 100°C) is nearly as important to ΔG as is ΔH.

In addition, there remains the difficult question of the phase of the reaction. An entirely gas phase reaction will be simplest to interpret because the complication of condensation energies (heats or free energies of solution, vaporization, and/or crystallization) is avoided. Consequently, we are best able to interpret gas phase enthalpies.

Illustrative Interpretations

Adduct formation involves heterolytic bond formation but, like the usual bond energy concept, depends on promotion energies (of molecules this time) and the formation of a "dative chemical bond." In this section you will see a rationale for adduct formation and its qualitative application to selected adducts. The rationale is to be developed according to the following outline for a reaction

$$AX_m + :DY_n \rightarrow X_mA—DY_n:$$

1. Structural change in the acid and base generally
   a. occurs at the expense of bonding between the acid or base atom and its substituents (A and X, and D and Y): **bond weakening**
   b. gives rise to increased VSEPR ($A—X \leftrightarrow A—X'$) and steric interaction between substituents ($X \leftrightarrow X'$): **back strain**

2. Donor-acceptor interaction (bond formation)
   a. depends on the relative energies of donor (HOMO) and acceptor (LUMO) orbitals and their overlap (often substituent dependent): **bond making**
   b. is inhibited by non-bonded (steric) effects ($X \leftrightarrow Y$) and VSEPR effects between the adduct bond and acid-substituent bond pairs ($A—D \leftrightarrow A—X$): **front strain**

The importance of these points can be best visualized in terms of an energy diagram.

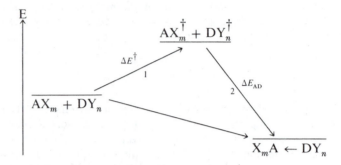

In preparing the acid and base for interaction with each other, we may visualize the distortion of each molecule into the structural configuration that it will adopt in the adduct. This will require the expenditure of energy and usually is greater for the acid because one of its atoms is to experience valence shell expansion. Generally, this means greater bond pair/bond pair and non-bonded repulsions between the substitutent atoms directly bound to the donor or the acceptor atoms (such a steric interaction is often referred to as **back-strain** or B-strain). Even when the back-strain is negligible, the distorted acid or base molecular structure cannot be of the lowest energy because of weakened bonding between the donor or acceptor atoms and their substituents. The second step of the energy cycle corresponds to the release of energy as the fragments are brought together. To analyze this we must, as a minimum, estimate *how well the donor and acceptor orbitals overlap one another and approach each other in energy.* Additionally, we may have to decide whether there will be steric interaction (**front-strain** or F-strain) between the substituents on the acid and base sites. There is always the added VSEPR about the acid as the donor pair enters the acid atom valence shell. Finally, the interaction may be facilitated or hindered by any special bonding interactions between the substituents and the acid or base atom.

PROTON AFFINITIES

The simplest acid-base reaction possible is that of donor protonation in the gas phase,

$$D + H^+ \rightarrow DH^+$$

We begin with these reactions because they allow us to analyze the donor-acceptor interaction unencumbered by acid reorganization and front-strain. Furthermore, data have recently been collected[37] that clearly reveal the effects of donor lone pair hybridization and structural reorganization on basicity. For example, the **proton affinities** of pyridine and piperidine are 940 and 962 kJ/mole, respectively. The sp^3 nitrogen of piperidine is the stronger donor. Recalling the discussions of Chapter 3 on the relationship between bond energy and hybrid percentage s character, you may be perplexed that the greater lone pair s character occurs with the weaker $D-H^+$ bond. In fact, there is a fundamental difference between these two bond cleavage processes: the process of Chapter 3 requires homolytic bond scission, while the proton affinity definition requires heterolytic cleavage.

$$D-H^+ \rightarrow D\cdot{}^+ + H \qquad \textbf{HA} \text{ (hydrogen affinity)}$$
$$D-H^+ \rightarrow D: \; + H^+ \qquad \textbf{PA} \text{ (proton affinity)}$$

In the present context, the energy of the homolytic scission is called the **hydrogen affinity**, HA. The two energies are related by the ionization potential difference between H and D.

$$PA = HA - (IP_D - IP_H)$$

Accordingly, the HA's of pyridine and piperidine are 565 and 485 kJ/mole, respectively, and thus are in the order expected on the basis of N hybrid s character $(sp^2 > sp^3)$. While the IP_H correction to HA is the same for all donors, the IP_D factor varies with lone pair Z^*; the expected order pyridine $>$ piperidine $(sp^2 > sp^3)$ is found (937 $vs.$ 837 kJ/mole). In short, the difference between pyridine and piperidine PA's depends not only on the difference of HA's (primarily reflecting HOMO/LUMO overlap differences) but also on the difference in IP's (reflecting the difference in HOMO/LUMO energy match):

$$(PA_{py} - PA_{pip}) = (HA_{py} - HA_{pip}) - (IP_{py} - IP_{pip})$$
$$-21 \quad = \quad 79 \quad - \quad 100$$

A brief physical interpretation of the difference between PA and HA is simply that with PA there is greater transfer of lone pair density from D to H^+ than is the case for HA, for which there is much less net electron transfer between $D^{.+}$ and H. Both δHA and δIP depend on the difference in lone pair s character, and δIP is more important.

A change in donor atom from nitrogen to oxygen should also cause a change in PA through a change in lone pair Z^* (HOMO/LUMO energy match). Accordingly, Et_2O has a smaller PA than Et_2NH (828 $vs.$ 962 kJ/mole), where both donors are "sp^3."

Given such Z^* effects on donor basicity, the lower PA of manxine than $(i\text{-}Pr)_3N$ (983 $vs.$ 996 kJ/mole) is a good illustration of the donor reorganization contribution to basicity. Manxine is a "strained" molecule for an "sp^3" nitrogen because of transannular CH_2 repulsions; consequently, the bridgehead CCC and CNC angles are $\sim115°$. The nitrogen lone pair should be predominately p ao in character and thus exhibit greater

manxine

basicity than that of $(i\text{-}Pr)_3N$. On the other hand, protonation of manxine should incur significant reorganization about the nitrogen as the p lone pair is polarized to form the $N-H^+$ bond.

That is, VSEPR forces should strongly favor a pyramidal nitrogen, in opposition to CH_2 repulsions constraining the nitrogen to be planar. This, then, is an (extreme) example of how donor reorganization is an influential factor in donor basicity. For comparison, quinuclidine and Et_3N exhibit the same PA of 987 kJ/mole, and space-filling models show that no strain energy is to be expected for quinuclidine.

quinuclidine

$(CH_3)_3N, (SiH_3)_3N$

As the next example of analyzing donor abilities of related donors, we will consider the adducts formed by trimethylamine and trisilylamine with boron trimethyl. With the same acid in both adducts, your attention should focus on the differences in the donors. As a first guess you might figure that the lower electronegativity of silicon than of carbon would lead to a stronger adduct by trisilylamine, because the greater shielding of nitrogen nuclear charge should yield a higher energy, less tightly held HOMO pair (favors bond making). A check of the adduct dissociation enthalpies reveals[38] this expectation to be wrong—the silylamine adduct has a bond enthalpy of less than 4 kJ/mole, while that of the normal amine is on the order of 75 kJ/mole. A clue to the flaw in the prediction is that trisilylamine has a planar structure[39] while the normal amine is pyramidal—the structural reorganization of the silylamine will be greater than that of the normal amine, and this will inhibit the base character of the silylamine (bond weakening and back strain).

The origin of the planar Si_3N grouping is a matter of some debate; most agree that the silicon valence d ao's act in a Lewis acid capacity so as to involve the nitrogen lone pair in π bonding, and this would be maximized by a planar structure. An additional

$$Si-\ddot{N}\diagdown \begin{matrix} Si \\ Si \end{matrix} \leftrightarrow \overset{\ominus}{Si}=\overset{\oplus}{N}\diagdown \begin{matrix} Si \\ Si \end{matrix}, \text{etc.}$$

factor might be that the electron releasing character of the silicon, with regard to the $Si-N$ σ bond, tends to flatten the molecule (this idea derives from increased bond-bond pair repulsion energy). In any event, the distortion of the planar molecule to the pyramidal shape needed for the adduct represents an energy cost apparently not made up by adduct bond formation. The effect of $N-Si$ π delocalization of the lone pair is double-edged. Once the molecule is in the pyramidal configuration, there would still be competition for the nitrogen lone pair by the silicon d orbitals; the silicon atoms compete with the Lewis acid for the amine lone pair. We expect all points from 1a through 2b (p. 145) to favor Me_3N as a donor.

PICOLINES

Many examples[40] exist that illustrate rather clearly the operation of steric effects. A fairly straightforward one is the comparison of pyridine, 2-picoline, and 4-picoline as donors toward the common acid BMe_3. The adduct "bond energies" are 71, 42, and

2-picoline   4-picoline

75 kJ/mole, respectively. Clearly, the reorganization energies (bond weakening and back strain) of the donors are small and should be very similar. The inductive effect of the methyl group should, by increased screening of the N nuclear charge, make the *ortho* isomer the stronger donor (bond making favored by better energy match). This example, then, is one showing the effects of F-strain with 2-picoline.

Et₃N, HC(C₂H₄)₃N

An example involving both F- and B-strain is to be found in the comparison of triethylamine and quinuclidine (1-aza[2.2.2]bicyclooctane). You have already seen that these donors have the same PA. Construction of a molecular model of triethylamine reveals considerable steric congestion (B-strain) if all three ethyl groups are to be "folded

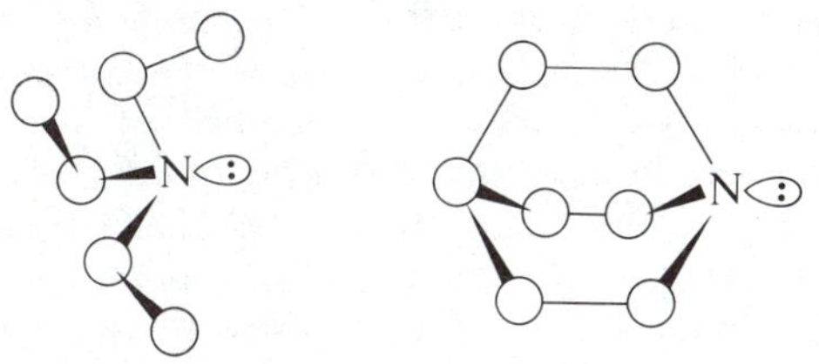

back" behind the nitrogen atom. In quinuclidine, the ethyl groups are constrained in the back position, so the idea of B-strain is not applicable. Toward boron trimethyl as the acid, quinuclidine is the stronger donor[36] (84 kJ/mole compared with 42 kJ/mole). The difference is ascribed to the steric congestion encountered by the third terminal methyl group of triethylamine as it attempts to avoid the methyls of BMe₃ (F-strain) and the terminal methyls of the other two ethyl groups (B-strain).

π RESONANCE EFFECTS

Pi resonance effects can be important in determining a molecule's donor or acceptor ability. It is generally observed that *N,N*-dimethylacetamide is a better donor than is acetone and that it is the oxygen atom, rather than the nitrogen atom, of the amide that acts as the site of basicity. Both observations naturally arise from the involvement of the single lone pair at nitrogen in three-center π bonding. (Oxygen, of course, has two in-plane lone pairs, in addition to the π pair of electrons involved in the conjugation).

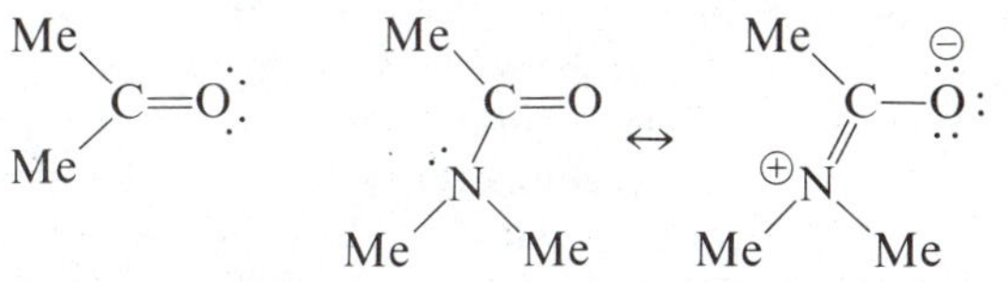

A very similar explanation is proposed for the observation that tris(dimethylamino)-phosphine is a stronger donor than is trimethyl phosphine[41] and that $(Me_2N)_xPMe_{3-x}$ (x = 1 to 3) all coordinate BH₃ through phosphorus, whereas $(Et_2NCH_2)_3P$ will easily add up to three BH₃ ligands, all at nitrogen.[42] In all these examples, the NR₂ function should be more strongly σ electron withdrawing than is the methyl group (bond making favored by better energy match for Me₂CO) but more strongly π donating (bond making favored by better energy match for the amide). Bond weakening (stretching the C=O bond) also favors the amide. There should be little distinction in the other points. It is apparently the case that the π donation or conjugation is more important than the σ induction.[40a]

On reflecting upon the nature of these substituent effects, it is worth noticing that the π conjugation effects just mentioned do not primarily involve the donor orbitals but indirectly affect their Z* and energies. This is to be contrasted with the direct effects of π bonding in the example of trisilylamine given earlier.

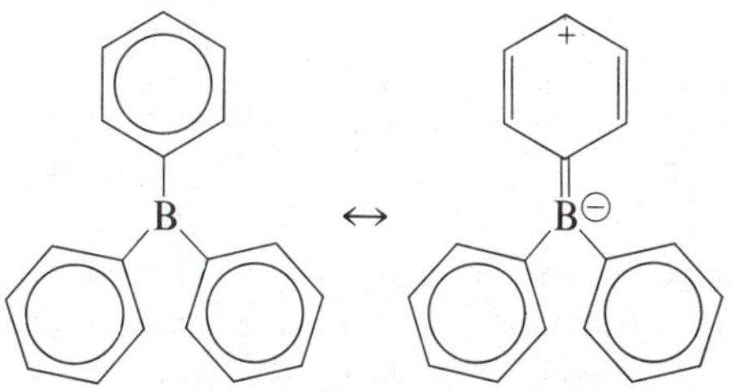

, etc.

Similar π factors are known to operate in Lewis acids. Direct conjugation of the boron acceptor orbital into the phenyl aromatic π system in triphenyl boron lowers the acidity of BPh_3 relative to BMe_3 (where only hyperconjugation is operative). This π interaction between acid site and substituent will persist even in the adduct pyramidal configuration [cf. $(H_3Si)_3N$]. Furthermore, the change from planar to pyramidal geometry is opposed by the B—C interaction. Were it not for these π factors (as well as F- and B-strain), you might have predicted that BPh_3 would be a stronger acid than BMe_3 on the basis of σ effects (the sp^2 hybrid of C_6H_5 is more electronegative than the sp^3 of CH_3 and so more effectively descreens the boron nucleus).

Similar effects are found in the rather surprising trend[43] of acidities $BF_3 < BCl_3 < BBr_3 \sim BI_3$. Simple substituent inductive effects (descreening of the boron nuclear charge via withdrawal of σ electron density) would lead you to expect that BF_3 would be the strongest acid of the series and the iodide the weakest. Rather, the boron-halogen π interaction is greatest for fluorine and decreases toward iodine. This translates into a "stiffer" molecule for BF_3 (bond weakening and perhaps the VSEPR back strain; more energy would be consumed in folding BF_3 to the extent of BI_3 or, alternately, BF_3 folds less). The degree of folding is important because, as before, the greater the folding the less anti-bonding, more localized, and more electronegative the boron acceptor mo becomes (bond making, both energy match and overlap). A general statement implicating all these factors would be that iodine less strongly competes against the donor for the boron acceptor orbital than does fluorine.

RETRODATIVE BONDING

One of the more interesting series of donors to compare consists of the trimethyl and trifluoro amines and phosphines. Toward BH_3, the order of donor strengths is found to be $Me_3P > Me_3N > F_3P > F_3N$. (Recall from the previous discussions of NF_3 that it exhibits negligible basicity, F_3NO being the only stable compound of NF_3 that could be considered an adduct.) For each of phosphorus and nitrogen, it is not surprising that the trimethyl compounds are more basic than the trifluoro compounds. What is perplexing is that the phosphine is a stronger donor than the correspondingly substituted amine. You have already seen that Me_3N forms a stronger bond with BMe_3, even with the steric interaction, than does Me_3P. Similarly, with boron trifluoride Me_3N forms a more stable adduct (>160 kJ/mole) than does Me_3P (79 kJ/mole), and F_3P shows negligible basicity.[44] Consequently, there seems to be some special property of BH_3 as an acid that makes Me_3P and F_3P appear to be better donors than expected.

This interesting reversal in donor strengths is also apparent in the adducts of BH_3 with organo sulfides and ethers (the sulfides forming the stronger adducts with BH_3, but weaker adducts with BF_3). Even carbon monoxide (CO and NF_3 have similar PA's of ~580 kJ/mole,[45] not very different from that of CH_4!) is known to form a moderately stable adduct with BH_3 and shares with PF_3 an ability to coordinate metal atoms, as we mentioned earlier in the survey of acid-base interactions.

Earlier in this chapter you learned that borane hydrogens exhibit some tendency to act in a Lewis base capacity. Coupled with the idea that phosphorus (but not nitrogen) and sulfur (but not oxygen) have vacant d valence ao's that could act in an acceptor (LUMO) capacity, it could be that back-bonding or retro-bonding occurs between the B—H bond density and the heavy atoms; such an effect would help stabilize an adduct directly and further stimulate greater σ donor character in the base. In resonance structure terminology, we would write

Now, a little thought about the fact that Me_3P forms a more stable adduct than F_3P with BH_3 should alert you not to overestimate the importance of this retro-bonding to the energy of the adduct bond.[46] The normal σ inductive effects seem to dominate a comparison of Me_3P with F_3P. In addition, there is evidence that OR and NR_2 substituents enhance the phosphorus donor ability by $p\pi$-$d\pi$ interaction of their lone pairs with phosphorus [cf. $(Me_2N)_3P$ in the preceding section].

12. SO_3 and BF_3 are (valence shell) isoelectronic and have analogous mo structures, but are emphatically different in acid strengths. Comment on the important difference in the number of S and B valence atomic orbitals and the role of π bonding, O to S and F to B, as a factor in the magnitude of $\Delta E^\dagger$ (see p. 146) for adduct formation.

13. Rationalize the linear SiNCS structure of H_3SiNCS, knowing that CH_3NCS is angular about N. Similarly, draw a conclusion about the difference in Si–P and Si–N bonding in $(H_3Si)_3P$ and $(H_3Si)_3N$ from the fact that (ignoring the hydrogens) the former has C_{3v} symmetry while the latter has D_{3h} symmetry.

14. Predict relative donor characters of Me_2O and $MeOSiH_3$.

15. "π bonding" is thought to be of prime importance to the order of adduct stabilities: $BI_3 \sim BBr_3 > BCl_3 > BF_3$. For each of the four contributions to adduct dissociation energy, tell what trend is expected from the trend in π bonding: $BF_3 > BCl_3 > BBr_3 > BI_3$.

16. For boranes and alanes ($= AlR_3$) that dimerize

$$2A \rightarrow A_2$$

the enthalpy of dimerization is in the range from 80 to 120 kJ/mole of dimer. Compute the range of bridge adduct bond energies; would you say that the (B, Al)–(H, R) bond pairs are weakly, moderately, or strongly basic? How does this "self-association" affect adduct stabilities in terms of ΔH for the reaction

$$D \cdot A \rightarrow D + \frac{1}{2}A_2?$$

17. The dissociation enthalpies for py·BMe_3 and py·BCl_3 are 71 and 159 kJ/mole. Rationalize the order.

18. Interpret the order of adduct stabilities (based on enthalpy of dissociation, ΔH_d) of methyl amines with BMe_3, which is

$$NH_3 < MeNH_2 < Me_3N < Me_2NH$$

19. Rationalize the following order of adduct stabilities (based on enthalpy of dissociation, ΔH_d):

$$Me_3N \cdot BF_3 \gg Me_3P \cdot BF_3 > Me_3N \cdot BMe_3 > Me_3P \cdot BMe_3 \gg H_3P \cdot BF_3 \gg H_3P \cdot BMe_3$$

20. Explain why ΔH_d decreases as x decreases in the series $Me_3N \cdot BF_xMe_{3-x}$ ($x = 3$, 2, 1, 0).

21. Knowing that BH_3 and BF_3 bond to different donor atoms in Me_2NPF_2, predict which acid bonds to which donor atom.

22. The reaction

$$M(CO)_n + PR_3 \rightarrow M(CO)_{n-1}PR_3 + CO$$

tends to be disfavored by R groups of good π donor character (like NR_2). Discuss this fact in terms of P σ donor and π acceptor (with respect to the metal atom) ability.

23. Account for the basicity order

$$
\underset{\substack{\text{NMe}_2}}{\overset{\substack{\text{Me}}}{>}} \text{C=O} \quad > \quad \underset{\substack{\text{Me}}}{\overset{\substack{\text{Me}}}{>}} \text{C=O} \quad > \quad \underset{\substack{\text{OMe}}}{\overset{\substack{\text{Me}}}{>}} \text{C=O}
$$

24. Predict whether NH_2^- or PH_2^- has the greater proton affinity.

AN ACID-BASE VIEW OF SOLVATION PHENOMENA[47]

By far the overwhelming majority of chemical studies are performed in the solution phase, so it behooves us to understand the role of the solvent in determining the species present in the solutions we study. It is difficult to predict quantitatively the effects of various solvent properties on solution behavior of compounds, but some generalizations—derived from experience—are possible that provide useful guidelines. In a general way, we appreciate that solubilization of a solute will occur when the attraction forces between solvent molecules and solute molecules are greater than those between the solvent molecules themselves and between the solute molecules themselves. These solvent-solute attractions may be **non-specific** (London and dipolar) or **specific** (with reference to an identifiable acid-base interaction between solvent and solute) in origin. The terms specific or non-specific thereby suggest solvation by a specific donor orbital-acceptor orbital interaction or by more general intermolecular attractions, respectively. Heterolytic cleavage of solute bonds to form ionic species in solution is another important aspect of solute-solvent interactions and requires efficient ion separation and solvation. These ideas are implied by equations such as

$$ MX_{n(c)} \overset{S}{\rightarrow} [MX_n]_{(s)} \overset{S}{\rightarrow} [MX_{n-1}^+]_{(s)} + [X^-]_{(s)} $$

where S is the solvent and [   ]$_{(s)}$ symbolizes solvation of the species in brackets.

To give a preliminary illustration of these ideas, it is known that $FeCl_3$ (an ionic solid that easily sublimes to form Fe_2Cl_6 dimers in the gas phase) gives solutions of Fe_2Cl_6 in some solvents, of $FeCl_3 \cdot S$ adducts in others, of $[FeCl_{3-x}S_{3+x}]^{x+} + [FeCl_4]^-$ in yet others, or $[FeCl_{3-x}S_{3+x}]^{x+} + [Cl]^-$ in still others. In the series of equilibria

$$ FeCl_{3(c)} \rightleftharpoons [Fe_2Cl_6] \rightleftharpoons [FeCl_3 \cdot S] \rightleftharpoons [FeCl_{3-x}S_{3+x}]^{x+} + [FeCl_4]^- $$

$$ \rightleftharpoons [FeCl_{3-x}S_{3+x}]^{x+} + [Cl]^- $$

solvent properties play an understandable role in determining what species actually arise in solution.

From a thermodynamic viewpoint, an important aspect of the solution process is the entropy change associated with the randomization of the solute/solvent phases. The net entropy change for dissolution (and ionization) of a solute need not, however, always be positive, for on a molecular level the association between solute and solvent might be more constrained than that between solute molecules on the one hand and that between solvent molecules on the other. Whether the enthalpy change accompanying a solution process is net exo- or endothermic similarly depends on the balance between solute interaction energies and solvent interaction energies relative to the solvent-solute interaction energies. This is where the acid-base concept makes an important contribution. If we recognize the solute to be a Lewis acid (or base) and consider its solution in a basic (or acidic) solvent, it stands to reason that a sufficiently strong solute-solvent interaction (adduct formation) might be possible to effect the solution process and may even inhibit further solute reactions. Examples of this will be examined later.

Another important consideration with many potential solutes is whether solute ionization can occur in a solvent. In more general terms, what species do we expect to find in a solution? The heterolytic cleavage of solute bonds to form cationic and anionic species will depend on the ability of the solvent to stabilize the incipient cation and anion, and ion solvation becomes an important consideration. Cations are often good Lewis acids and anions are often good Lewis bases so that, again, it is convenient to consider solute ionization and ion solvation in terms of the acid-base properties of the solvent. Specific examples will follow.

One solvent characteristic of great importance in causing ion formation by a solute is the dielectric constant. The dielectric constant of a continuous medium (and it *is* inherently *inconsistent* to view a liquid phase as a uniform continuum if we subscribe to the atomic and molecular concept of matter) is defined in terms of the medium's ability to attenuate the force between two ions:

$$F = (q_1 q_2/r^2)(1/4\pi\epsilon)$$

where ϵ is the dielectric constant of the medium. The generalization we need to recognize is that a high solvent dielectric constant ($\gtrsim 15$ to 20) tends to support a high degree of solute ionization. As an empirical observation, high dielectric solvents often possess high dipole moments and/or the combination of acidic hydrogens and donor atoms in the same molecule; in either case, considerable self-association of solvent molecules is possible. Such solvent attributes should be important to ion separation, since the first few layers of polar solvent molecules about the ion would rapidly attenuate its electric field.

Table 5-1 illustrates the basis for the expectation that a high solvent dielectric supports solute ionization. The solvents in this table are listed in order of decreasing dielectric constant; the ion association constants for the simple salts listed are seen generally to increase with decreasing dielectric constant of the solvent. The ion association constant is the equilibrium constant for the reaction

$$[M^+]_{(s)} + [X^-]_{(s)} \rightleftharpoons [M^+X^-]_{(s)}$$

and the quaternary ammonium salts are used in order to eliminate specific cation-solvent interactions as a contribution to the K_{as}.

To give a quick illustration of the application of these ideas, we will compare the solvents H_2O, NH_3, HF, and Me_2SO. Of these, water is often the best solvent, for it has a high dielectric constant (the dielectric constants of the four solvents are 78, 23, 175, and 47 in the order above), possesses a moderately high dipole moment (the dipole moments are 1.85, 1.47, 1.82, and 3.96 Debyes), and has both acidic protons and lone pairs on oxygen so that it solvates bases and acids well. While ammonia is probably a

TABLE 5-1

DATA ILLUSTRATING THE RELATIONSHIP BETWEEN DIELECTRIC AND SOLUTE IONIZATION

Solvent	Salt	ϵ	K_{as}
$HC(O)N(H)CH_3$	Et_4NBr	182	0
$CH_3C(O)N(H)CH_3$	Et_4NBr	166	0
$(CH_3)_2SO$	Bu_4NI	47	0
$(CH_2)_4SO$	Ph_4AsCl	44	0
$CH_3C(O)N(CH_3)_2$	Et_4NBr	38	20
$HC(O)N(CH_3)_2$	Me_4NBr	37	37
	Me_4NI		15
CH_3CN	Me_4NCl	36	78
	Me_4NBr		41
	Me_4NI		28
CH_3OH	Bu_4NBr	33	26
$(CH_3)_2CO$	Bu_4NI	21	164
$NOCl$	$[NO][FeCl_4]$	20	268
$POCl_3$	Et_4NCl	14	1390
	Et_4NBr		530
	Et_4NI		260
$(C_2H_5O)_3PO$	Et_4N Picrate	13	1600

better donor than water, it is also a poorer acid (less acidic hydrogens). Solute ionization is often not as great in ammonia as in water because the dielectric constant is lower and the anion solvation (via hydrogen bonding) is poorer. In situations where solvent basicity is the important property, NH_3 may be the better solvent. HF should be an exceptionally acidic solvent and should facilitate solution of donor molecules. Its high dielectric constant would favor appreciable solute ionization, but the fluorine atom is a very weak donor, making cation solvation not as favorable as with water or ammonia. Me_2SO possesses a sufficiently high dielectric constant to lead to solute ionization, and is a fairly good Lewis base. It should solvate cations and acidic solutes well and the high dipole moment helps in solvating anions (nevertheless, it lacks the acidic hydrogens of the other solvents).

To show how specific and non-specific solvation properties combine to determine the species in solution, we return to the $FeCl_3$ case brought up earlier. Of the solvents $(CH_3)_2SO$, pyridine, and acetonitrile, we expect the sulfoxide to show good donor ability toward the metal ion (a specific interaction), to show good anion solvation (a non-specific interaction of the ion-dipole type), and to be generally supportive of ionization. Accordingly, $FeCl_4^-$ and $FeCl_3$ solutes are dissociated into cationic chloro complexes $[FeCl_{3-x}S_{3+x}]^{x+}$ and chloride ion. Pyridine is generally a better donor than the sulfoxide but is much poorer at non-specific ion solvation. Accordingly, $FeCl_3$ forms the simple adduct $FeCl_3 \cdot$pyridine, and $FeCl_4^-$ is stable (does not dissociate to $FeCl_3 \cdot py + Cl^-$). Acetonitrile is a generally weak donor of moderate ion solvating ability, with the result that dissolution of $FeCl_3$ in acetonitrile effects ionization only to the extent of $FeCl_2S_4^+$ and $FeCl_4^-$ (the $FeCl_4^-$ anion does not appear to dissociate appreciably into cationic chloro complexes and Cl^- ion).

Thus, the choice of a particular solvent for a chemical reaction is dictated by its chemical properties, including but sometimes going beyond, the simple acid-base properties of the individual solvent molecules. A few examples will show how important the acid-base properties are in determining the feasibility of a chemical reaction. A good illustration is the well-known reaction in water

$$BaCl_2 + 2AgNO_3 \rightarrow 2AgCl_{(c)} + Ba(NO_3)_2$$

which occurs in the *reverse* direction in liquid ammonia

$$2AgCl + Ba(NO_3)_2 \rightarrow BaCl_{2(c)} + 2AgNO_3$$

In a qualitative sense, $Ba(NO_3)_2$ is "soluble" in both solvents, as is $AgNO_3$. The dramatic solvent effect is due to the difference that $BaCl_2$, but not $AgCl$, is soluble in water, whereas $AgCl$ is moderately soluble in ammonia while $BaCl_2$ is not. In short, the relative solubilities of $AgCl$ and $BaCl_2$ reverse on changing from H_2O to NH_3. Water is expected to solvate Cl^- more energetically than ammonia, but ammonia should solvate Ag^+ better than water. Obviously, the Ag^+ and Cl^- ion *solvation energies* are less than the $AgCl$ *lattice energy* in water but not in ammonia. This would seem to be due to a greater difference between H_2O and NH_3 cation solvations than anion solvations. For $BaCl_2$ and $AgCl$ the anion solvation order is, of course, the same $(H_2O > NH_3)$; but the cation solvation appears to reverse, as the $Ba^{2+} + 2Cl^-$ ion solvation energy in water is greater than the lattice energy, whereas it is not in ammonia.

If one wished to synthesize an ether adduct with boron trimethyl, he would be foolish to use a basic solvent like pyridine, which is a stronger donor than ether. The pyridine solvent would complex the boron trimethyl, and the ether molecules would be unable to displace pyridine. Conversely, an ether solvent would be a good choice for synthesis of the pyridine adduct. Acetonitrile (a less basic donor than ether) would be a logical choice for solvent in synthesis of the ether adduct.

Similarly, experience has shown[48] that HCl is not a good choice for synthesis of $GeCl_6^{2-}$ from $GeCl_4$ and Cl^- because the highly acidic HCl solvent competes with $GeCl_4$ for Cl^- through formation of the hydrogen-bonded complex HCl_2^-. $MeNO_2$, which is sufficiently polar to dissolve many quaternary ammonium chlorides but not basic enough to strongly complex the $GeCl_4$, does not solvate Cl^- well enough to inhibit formation of the hexachlorogermanate anion.

In an entirely analogous fashion, water is too basic to permit a reaction between phosphine and hydrogen ion to form PH_4^+ salts, whereas the very weakly basic solvent HCl makes synthesis of $PH_4^+BCl_4^-$ straightforward:[48]

$$PH_3 \xrightarrow{\text{HCl}} PH_4^+ + HCl_2^- \xrightarrow{\text{BCl}_3} [PH_4][BCl_4]_{(c)}$$

A reaction unknown in mixed aqueous/organic solutions is

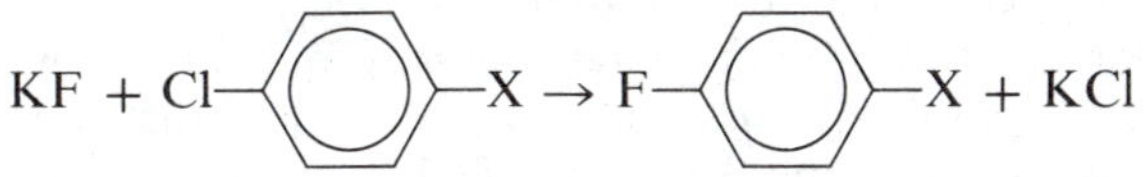

$$KF + Cl\!-\!\!\bigcirc\!\!-\!X \rightarrow F\!-\!\!\bigcirc\!\!-\!X + KCl$$

However, in Me_2SO we find the reaction possible,[49] probably because of poorer solvation of F^- by Me_2SO than by water.

As a final interesting example of the solvent effect on the apparent relative basicities of donors (earlier in this chapter you saw that solvation or condensation energies could play an important role in such equilibria), the pK_b's (defined by the following reaction) of amines fall in the unexpected order $NH_3 > Me_3N > MeNH_2 > Me_2NH$

$$\text{amine} + H_2O \xrightarrow{\text{H}_2\text{O}} \text{amine} \cdot H^+ + OH^-$$

the smaller pK_b, the greater the basicity), whereas the PA's (p. 146) fall in the order[50] $NH_3 < MeNH_2 < Me_2NH < Me_3N$ for the protonation reaction

$$\text{amine}_{(g)} + H^+_{(g)} \rightarrow \text{amine} \cdot H^+_{(g)}$$

The key to understanding the aqueous system ordering lies first in realizing that the proton basicity of the alkyl amines as a group is greater than that of ammonia; this dominates the comparison of R_xNH_{3-x} with NH_3. Within the alkyl amine series, alkyl substitution does lead to enhanced basicity in the gas phase, so we look to the difference in amine/ammonium ion solvation energies to account for the relatively large pK_b of Me_3N. The solvation (via $N \cdots H_2O$ hydrogen bonding) of the amines should *inhibit* ammonium ion formation most for Me_3N and least for $MeNH_2$. Solvation of the ammonium cations (via $N—H \cdots OH_2$) should *favor* ammonium ion formation most for $MeNH_3^+$ and least for Me_3NH^+. Thus, the change in solvation energies in the reaction amine $\rightarrow$ ammonium ion would tend to make Me_3N appear the weakest base and $MeNH_2$ the strongest base. Increasing methyl substitution increases the nitrogen atom gas phase basicity but decreases the aqueous basicity through solvation energies. The anomalous position of Me_3N in the aqueous series arises from the solvation inhibition of its basicity; this is an excellent example of the point that solvation energies may confuse interpretation of adduct bond energies.

For protonic and halide solvent systems, the ideas developed above for solvent-solute interaction can be expressed in terms of the concept of acid and base **leveling** by the solvent.[51] With water, any acid HA that extensively reacts with H_2O to form the hydronium ion is said to be too strong and is leveled to the acidity of H_3O^+. This is nothing more than a statement that, in water as solvent, H_2O is a stronger Brönsted base than is the anion A^-. Similarly, all bases stronger than OH^- are leveled in water to OH^-.

$$HA + H_2O \rightarrow H_3O^+ + A^-$$
$$B + H_2O \rightarrow BH^+ + OH^-$$

For ammonia, NH_4^+ is the strongest acid and NH_2^- is the strongest base. In liquid HF, HF_2^+ and HF_2^- are the strongest acid and base permitted.

$$2NH_3 \rightleftarrows NH_4^+ + NH_2^-$$
$$2HF \rightleftarrows H_2F^+ + HF_2^-$$

This concept leads to the idea that acetic acid is a strong acid in liquid ammonia (much as HCl is a strong acid in water) and NH_4OAc is a strong acid in ammonia (NH_4^+ is the strongest acid possible in NH_3). These ideas were of course, implied in the examples of the preceding paragraphs, where we found PH_3 to act like a strong Brönsted donor in liquid HCl but a weak Brönsted donor in H_2O. Cl^- is a strong base in HCl and accordingly is leveled to the basicity of HCl_2^-, making difficult the synthesis of $GeCl_6^{2-}$ (which must accordingly be considered a strong base, too) in liquid HCl.

The application of these ideas to solvent choice for synthesis work can be further illustrated by selecting a solvent for preparation of a salt of the urea anion. The choice could be approached in two equivalent ways: find a solvent that makes urea appear to be a strong acid,

$$\underset{\substack{\|\\ O}}{H_2N\overset{O}{\overset{\|}{C}}NH_2} + S \rightarrow SH^+ + HN\overset{O}{\overset{\|}{C}}NH_2^-$$

or find a solvent that will tolerate the urea anion as not a very strong base.

$$HN\overset{O}{\overset{\|}{C}}NH_2^- + S \not\rightarrow H_2N\overset{O}{\overset{\|}{C}}NH_2 + (S - H)^-$$

Liquid ammonia is a good choice, for it is highly basic; acids that are weak in water appear much stronger in ammonia. Similarly, we could predict that $(H_2NCO)NH^-$ would be a weaker base than $(H)NH^-$. Reaction of the amide ion with urea in ammonia would effect the desired synthesis of H_2NCONH^-.

$$NH_2^- + H_2N\overset{O}{\overset{\|}{C}}NH_2 \xrightarrow{NH_3} NH_3 + HN\overset{O}{\overset{\|}{C}}NH_2^-$$

This same concept can be applied to the choice of solvents for working with oxidizing and reducing agents.[52] While there is sometimes difficulty in defining a set of oxidized and reduced forms for a particular solvent, it stands to reason that some solutes will be too oxidizing or reducing for use in a particular solvent. We can, of course, set rough upper limits for each solvent, based on its most highly oxidized and reduced forms. In aqueous chemistry we are accustomed to the concept that oxidants more oxidizing than O_2 will oxidize water solvent, and that those more reducing than H_2 will reduce water (often, however, we can seemingly violate such dictates of thermodynamics because the gas overvoltages, a kinetic feature, are often appreciable). By analogy, one can work in HCl solvent with oxidizing agents up to the power of Cl_2 but only with reducing agents less powerful than H_2.

Solvent	"Reduced Form"	"Oxidized Form"
H_2O	H_2	O_2
HCl	H_2	Cl_2
HSO_3F	H_2	$S_2O_6F_2$
$AsCl_3$	As	Cl_2

Liquid Ammonia As A Solvent For Synthesis[53]

Many liquids (H_2SO_4, HSO_3F, SO_2, HF) offer novel media for a variety of "unusual" reactions. Space limitations permit an examination here only of NH_3. This solvent is quite useful for syntheses requiring very basic and/or strongly reducing reagents. Its physical properties are:

M.P. (°C)	-78
B.P. (°C)	-33
η (centipoise)	$0.25\ (-33°C)$
ρ (g cm^{-3})	$0.69\ (-40°C)$
ε	23
Λ (ohm^{-1} cm^{-1})	10^{-11}
K_{ion}	10^{-27}

Liquid ammonia is a colorless fluid only 70% as dense as water and only $\frac{1}{4}$ as viscous (an obvious advantage in filtration). In spite of the low boiling point of liquid ammonia, the liquid has a sufficiently high heat of vaporization ($\sim$20 kJ/mole) that it can be handled in unsilvered vacuum dewars at room temperature with tolerable loss to evaporation. Further comparing ammonia with water, the auto-ionization constant is considerably smaller (that for water is $\sim$5 $\times$ 10^{-14})

$$2NH_3 \rightleftharpoons NH_4^+ + NH_2^-$$

and the specific conductance is much smaller (that for water is $\sim 10^{-7}$). The Grotthuss (or proton "switching") mechanism for the high conductance of acidic or basic water

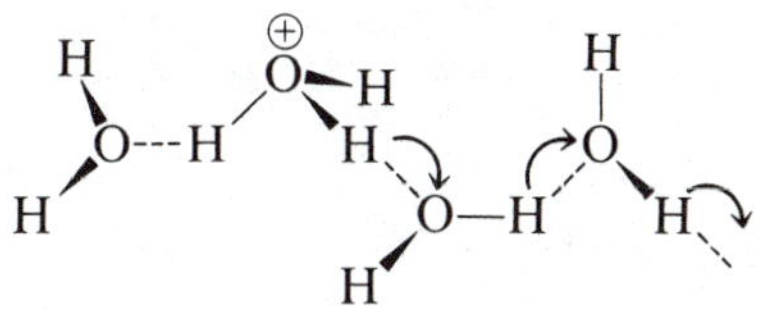

solutions is clearly not applicable to liquid ammonia solutions of NH$_4^+$ and NH$_2^-$. The data below reveal the anomalously high mobilities of H$^+$ and OH$^-$ in aqueous solvent and the

	H$_2$O	NH$_3$
Ion	λ_0	λ_0
H$_3$O$^+$	350	
NH$_4^+$	73	142
Na$^+$	50	158
OH$^-$	198	
NH$_2^-$		166
NO$_3^-$	71	177

perfectly normal mobilities of NH$_4^+$ and NH$_2^-$ in liquid ammonia. This marked difference in acid and base mobilities is, of course, expected from the point of view that NH$_4^+$ is a weaker hydrogen-bonding acid than is H$_3$O$^+$, in spite of the fact that NH$_3$ is a stronger donor than H$_2$O. Further evidence (in addition to the difference in H$_2$O and NH$_3$ boiling points) for the weaker hydrogen-bonding interactions and more difficult proton transfer in NH$_3$ than in H$_2$O comes from nmr studies of the exchange rates

$$NH_2^- + NH_3 \rightleftharpoons NH_3 + NH_2^-$$
$$NH_4^+ + NH_3 \rightleftharpoons NH_3 + NH_4^+$$

$k \sim 10^8$ M^{-1} sec^{-1}

$$OH^- + H_2O \rightleftharpoons H_2O + OH^-$$
$$H_3O^+ + H_2O \rightleftharpoons H_2O + H_3O^+$$

$k \sim 10^{10}$ M^{-1} sec^{-1}

where those for H$_2$O fall in the range of diffusion-controlled reaction velocities at 25°C.

Solubilities in liquid ammonia are interesting to contrast with those in water. It has been estimated that the non-specific solvation possible with H_2O is dominated by the high polarity (dipole moment) of the molecule, with the polarizability and London contributions being some five times less effective. With ammonia there is a more even balance between dipole and London forces, with the dipolar contribution being some two times smaller than that for water but the London contribution being two times greater. Consequently, it is not surprising to learn that liquid ammonia is the better solvent for solutes that have a large number of electrons or that are highly polarizable. This is illustrated by the fact that organic compounds such as hydrocarbons tend to have higher solubilities in ammonia than in water. Data that reflect this tendency of ammonia to best solvate species with high polarizabilities are given in Table 5-2.

As the solubility data suggest, salts of divalent ions are often much less soluble in ammonia than in water, but exceptions are known. Particularly for heavy transition metal ion salts (note the silver halides above), appreciable solubility can be traced to the energetic coordination of the metal ion by ammonia. This is an example of the importance of the donor strength and polarizability of ammonia.

In many respects the most intriguing solutes for liquid ammonia are the *alkali metals*. Since they are strong reducing agents, we might expect the alkali metals to cause H_2 evolution, as happens very rapidly with H_2O. Quite often, however, there is a kinetic inhibition of the reduction reaction in ammonia, so that alkali metal solutions are kinetically (but not thermodynamically) stable. On the other hand, insoluble metals do not generally exhibit an overvoltage effect. Before continuing with a summary of the properties of such solutions, an examination of the thermodynamics of redox behavior in ammonia from a general point of view is in order.

The H_2/NH_3 and NH_3/N_2 half-cell potentials for liquid ammonia show very little difference under standard state acid (1 M NH_4^+) and base (1 M NH_2^-) conditions. These

$$NH_3 + \frac{1}{2}H_2 \rightarrow NH_4^+ + e^- \qquad E^\circ = 0$$

$$4NH_3 \rightarrow 3NH_4^+ + \frac{1}{2}N_2 + 3e^- \qquad E^\circ = -0.04$$

$$NH_2^- + \frac{1}{2}H_2 \rightarrow NH_3 + e^- \qquad E^\circ = 1.59$$

$$3NH_2^- \rightarrow 2NH_3 + \frac{1}{2}N_2 + 3e^- \qquad E^\circ = 1.55$$

pH-dependent E°'s are diagrammed as solid lines in Figure 5-1; the narrow span in E°'s of only 0.04 V indicates that it should be very difficult to find solutes that would *not* be capable of oxidizing or reducing ammonia. For a chemical species to be stable in liquid ammonia, its redox half-cell potential would have to fall between the parallel solid

Cation	Cl$^-$	Br$^-$	I$^-$	NO$_3^-$
Li$^+$	1.43	—	—	—
Na$^+$	11.37	29.00	56.88	56.05
K$^+$	0.132	21.18	64.81	9.52
Rb$^+$	0.289	18.23	68.15	—
Cs$^+$	0.381	4.38	60.28	—
Ag$^+$	0.280	2.35	84.15	—
NH$_4^+$	39.91	57.96	76.99	—
Mg^{++}	—	0.004	0.156	—
Ca^{++}	—	0.009	3.85	45.13
Sr^{++}	—	0.008	0.308	28.77
Ba^{++}	—	0.017	0.231	17.88

TABLE 5-2

SOLUBILITIES OF METAL HALIDES IN LIQUID AMMONIA AT 0°C (g/100 g soln.)

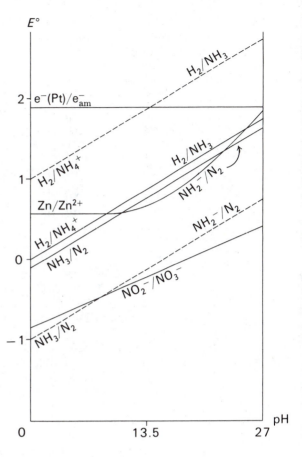

Figure 5-1. The oxidation potential diagram for liquid ammonia, illustrating the "overvoltage" range and pH dependence. e_{am}^- represents the solvated electron.

lines at the center of Figure 5-1. Reducing agents stronger than H_2 should reduce NH_4^+ and ammonia, and oxidizing agents stronger than N_2 should oxidize NH_2^- and ammonia. Fortunately, the H_2 and N_2 couples commonly, but not always, exhibit overvoltages of about 1 volt so that, as the dotted lines in Figure 5-1 indicate, the practical working range for liquid ammonia is considerably extended beyond that thermodynamically possible. Species that we would normally characterize as strong oxidizing agents (MnO_4^-, O_3, O_2^-) and reducing agents (Na, Cs) can accordingly be handled in liquid ammonia.

TABLE 5-3

STANDARD (1 M ACID) ELECTRODE POTENTIALS

	H_2O (25°C)	NH_3 (−35°C)
Cs/Cs$^+$	2.92	1.95
Rb/Rb$^+$	2.93	1.93
K/K$^+$	2.93	1.98
Na/Na$^+$	2.71	1.85
Li/Li$^+$	3.05	2.24
Zn/Zn^{2+}	0.76	0.53
Cd/Cd^{2+}	0.40	0.20
Pb/Pb^{2+}	0.13	−0.32
H$_2$/H$^+$	0	0
Cu/Cu^{2+}	−0.34	−0.43
Cu/Cu$^+$	−0.52	−0.41
Ag/Ag$^+$	−0.80	−0.83
I$_2$/I$^-$	1.06	
Br$_2$/Br$^-$	1.60	
Cl$_2$/Cl$^-$	1.97	

	H_2O	NH_3
Cl^-	−314	−285
Br^-	−289	−276
I^-	−255	−251
H^+	−1075	−1163
Li^+	−490	−506
Na^+	−393	−406
K^+	−322	−322
Rb^+	−305	−297
Cs^+	−264	−264
Ag^+	−464	−552
Zn^{2+}	−2063	−2218
Cd^{2+}	−1791	−1912

TABLE 5-4

FREE ENERGIES OF
SOLVATION (KJ/MOLE)

Some reagents exhibit a very different potential dependence on pH than do the H_2/NH_3 and N_2/NH_3 couples and can be utilized only in basic or acidic solutions. For example, the solvated electron is stable in solutions at least a little more basic than neutral (pH = 13.5), nitrate ion is stable to the acid side of neutral solutions, and Zn metal (insoluble) reduces ammonia at all but two narrow pH ranges.

Table 5-3 presents some electrode potential data[54] for metals in liquid ammonia and, for comparison, similar data for aqueous solution. The data for liquid ammonia, of course, constitute a self-consistent set of data for making comparisons within that solvent. While it is not possible to make a precise thermodynamic measurement of the relative H_2 electrode potentials in the two solvents, useful estimates place the reference potential in ammonia approximately 1 volt more negative than that in water. The data as presented allow us to conclude that the degree of solvation of the alkali metal ions is nearly the same for water and ammonia. For the transition metal ions, however, ammonia solvation is greater than water solvation (recall the previous discussion of solubilities), and this is to be expected from the recognized greater basicity of ammonia. Ion solvation energy values derived from thermodynamic data and shown in Table 5-4 are consistent with these conclusions.

Now we can return to the topic of metal solutions. The metal properties that favor their dissolution are those of low metal lattice energy, high cation solvation energy, and —interestingly enough—low metal ionization potential. Consequently, it is the alkali metals, the heavy alkaline earth metals, and divalent rare earth metals that show greatest solubility. Solubilities of the alkali metals fall in the range from 10 to 20 moles of metal/ 1000 grams of NH_3, so that quite concentrated solutions can be prepared (Table 5-5).

The three most striking physical properties of these solutions are their color (dilute solutions are blue, concentrated solutions are bronze), their conductivity (dilute solutions have conductivities an order of magnitude greater than those of salts in water, while the concentrated solutions have conductivities approaching those of pure metals), and their

	gram-atoms metal/kg NH_3	ΔH (kJ/mole)	
Li	15.7	−41	(0.07 M)
Na	10.9	+5.9	(0.4 M)
K	11.9	0.0	(0.07 M)
Rb	—	0.0	(0.13 M)
Cs	25.1 (−50°C)	0.0	(0.19 M)
Ca	—	−82	(0.02 M)
Sr	—	−87	
Ba	—	−79	

TABLE 5-5

SOLUBILITIES AND HEATS OF
SOLUTION OF METALS IN LIQUID
AMMONIA (−33°C)

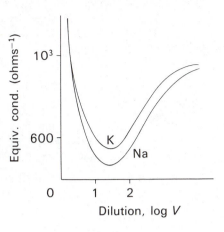

Figure 5–2. Equivalent conductance of metal-ammonia solutions at –33.5°. V = liters of ammonia (of density 0.674) in which one gram-atom of metal is dissolved.

magnetic properties (at infinite dilution the magnetic susceptibilities approach one equivalent of unpaired electrons, and the solutions become progressively less paramagnetic as the concentration of metal increases). The conductivity/concentration curves for Na and K are shown in Figure 5-2. While the full nature of the species present in these solutions is not understood at present, the conductivity and magnetic data seem best interpreted in terms of the presence of five species (M, M_2, M^+, M^-, and e_{am}^-) related according to the equilibria.

$$M \rightleftharpoons M^+ + e_{am}^- \qquad K \sim 10^{-2}$$
$$2M \rightleftharpoons M_2 \qquad K \sim 10^{-4}$$
$$e_{am}^- + M \rightleftharpoons M^- \qquad K \sim 10^{3}$$

The symbol e_{am}^- denotes the intriguing concept of an electron solvated by ammonia solvent. While the identity of this species is still subject to considerable research, agreement has centered on the qualitative correctness of a "cavity" (on the order of 300 pm in diameter) of NH_3 molecules in which the electron is trapped. The quantitative aspects of the conductance and magnetic data seem to indicate that the solvated electron behaves in a normal fashion for an anionic species in liquid ammonia, in that it is extensively ion-paired with a cationic species.

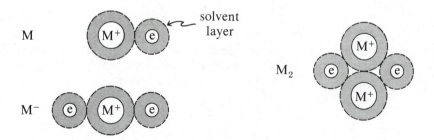

Liquid ammonia is a very versatile solvent for many reactions that are impossible in aqueous media. In particular, it is a very useful solvent for working with materials that are too strongly basic for aqueous media—examples are alkoxides, sulfides, and organic amide anions. Just as in aqueous systems, many reactions can be effected in NH_3 by simple metathesis, if the solubility relationships are right. Owing to the general high solubility of NO_3^- in ammonia, we find reactions like

$$Ba(NO_3)_2 + 2AgCl \rightarrow BaCl_2\downarrow + 2AgNO_3$$
$$2KOR + M(NO_3)_2 \rightarrow M(OR)_2\downarrow + 2KNO_3$$

In further analogy with aqueous chemistry, wherein metals or basic materials react with hydronium ion (H_3O^+) to evolve H_2 or form "hydrides," we find reactions in ammonia like

$$Co + 2NH_4NO_3 \rightarrow Co(NH_3)_6^{2+} + H_2 + 2NO_3^-$$
$$\text{"}Mg_xSi_y\text{"} + NH_4NO_3 \rightarrow SiH_4$$

The counterpart of OH^- in water chemistry is played, in ammonia chemistry, by NH_2^-, often with novel results (generally KNH_2 has sufficient solubility to be useful—$LiNH_2$ and $NaNH_2$ do not):

$$C_2H_2 + KNH_2 \rightarrow KC_2H + NH_3$$
$$NaNH_2 + 2KNH_2 \rightarrow K_2[Na(NH_2)_3]$$
$$AgNH_2 + KNH_2 \rightarrow K[Ag(NH_2)_2]$$
$$Zn + 2KNH_2 + 2NH_3 \rightarrow K_2[Zn(NH_2)_4] + H_2$$

Similarly, many non-metal halides undergo so-called solvolytic reactions in water to eliminate HX and form a hydroxy derivative of the non-metal halide; in liquid ammonia we find many analogous reactions leading to amide derivatives in place of the hydroxide derivatives:

$$SOCl_2 + 4NH_3 \rightarrow SO_2(NH_2)_2 + 2NH_4X$$
$$SO_3 + 2NH_3 \rightarrow [NH_4][SO_3NH_2]$$

In this last reaction the initial adduct $SO_3 \cdot NH_3$ is too strong an acid for NH_3 solvent (compare the solvation of SO_3 by H_2O) and is "leveled" to the acidity of NH_4^+.

Among the more interesting reactions possible in liquid ammonia are those of alkali metal solutions. These solutions are strongly reducing and make possible the formation of interesting reduced species. Among the simple electron transfer reactions we find the reaction with O_2

$$e_{am}^- + O_2 \rightarrow O_2^-$$

to form superoxides that are stable in NH_3, you recall, because the overvoltage effect is large. The production of complexes of metal *atoms* from complexes of cationic metals is particularly intriguing:

$$2e_{am}^- + Pt(NH_3)_4^{2+} \rightarrow Pt(NH_3)_4$$
$$+ Ni(CN)_4^{2-} \rightarrow Ni(CN)_4^{4-}$$

The large majority of reduction reactions by metals in ammonia are not simple electron transfer reactions but entail further reaction of the initially produced reduction product. Here we find reduction of non-metal hydrogen bonds,

$$e_{am}^- + AH \rightarrow A^- + \frac{1}{2}H_2 \qquad A = OR, NH_2, PH_2$$

and of carbon-sulfur bonds,

$$e_{am}^- + R_2S \rightarrow \frac{1}{2}R_2 + RS^-$$

More generally, the electron transfer reactions lead to heterolytic bond cleavage by net two-electron processes. A few examples are

$$2e_{am}^- + Ge_2H_6 \rightarrow 2GeH_3^-$$
$$2e_{am}^- + RX \rightarrow R^- + X^- \xrightarrow{NH_3} RH + X^- + NH_2^-$$

(the occurrence of the last step depends on R^- being more basic than NH_2^-).

$$2e_{am}^- + PhHN{-}NH_2 \rightarrow PhNH^- + NH_2^-$$

$$2e_{am}^- + NNO \rightarrow N_2 + O^{2-} \xrightarrow{NH_3} OH^- + NH_2^-$$

(O^{2-} is too strong a base for ammonia and is leveled to OH^- and NH_2^-.)

$$2e_{am}^- + RCH{=}CH_2 \rightarrow [R\bar{C}H{-}\bar{C}H_2] \xrightarrow{2NH_3} RCH_2CH_3 + 2NH_2^-$$

Of some interest to those developing techniques for forming catenated species are the reactions of Li, Na, and K with elemental sulfur to open and cleave the eight-membered sulfur ring, thereby forming various anionic polysulfides

$$\frac{16}{n} e_{am}^- + S_8 \rightarrow \frac{8}{n} S_n^{2-} \qquad n = 1 \text{ to } 7$$

Even more interesting is the reduction of diorganotin dihalides to form the diorgano-stannate anion. The possibility of R_2Sn^{2-} being a good nucleophile accounts for the succeeding reactions with RX and R_3SnX to give unique Sn polymers:

$$8e_{am}^- + 2Me_2SnBr_2 \rightarrow 2Me_2Sn^{2-} + 4Br^-$$

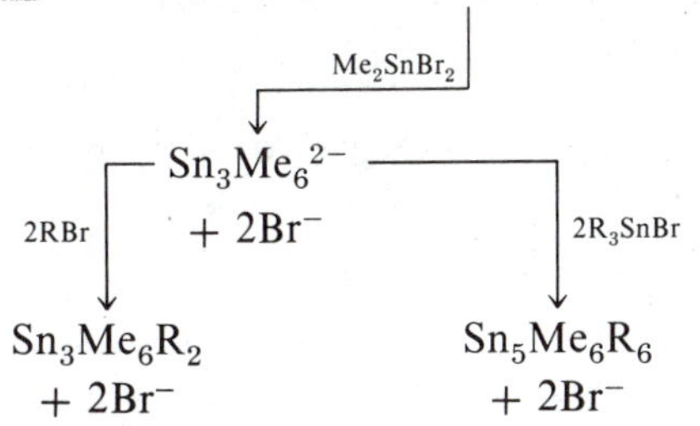

25. Describe the solvent properties that

   a) would give ClO_4^- strong apparent Brönsted base character;
   b) would give $B(OH)_3$ strong apparent Brönsted acid character;
   c) would give HNO_3 and HI different apparent (Brönsted) acid strengths;
   d) would distinguish apparent basicities of NH_2^- and PH_2^-;
   e) would permit amphoteric behavior of AsF_5.

26. $FeCl_4^-$ ion is yellow in color, whereas Fe_2Cl_6 has a reddish color; 0.1 M solutions of $FeCl_3$ in $POCl_3$ *and* in $PO(OR)_3$ are red and *both* turn yellow on dilution. Titration of red $FeCl_3/POCl_3$ solutions with colorless $Et_4NCl/POCl_3$ solutions causes the color to change from red to yellow with a sharp end-point at 1:1 $FeCl_3$ to Et_4NCl. Furthermore, as far as is known, oxychloride solvents form adducts with Lewis acids wherein the oxygen, rather than chlorine, is the donor atom.

   a) Are these observations best represented by the equilibria
      a) $Fe_2Cl_6 + 2POCl_3 \rightleftarrows 2FeCl_4^- + 2POCl_2^+$
      b) (a) shifted to right on dilution
      c) $FeCl_4^- + POCl_2^+ + Et_4N^+Cl^- \rightleftharpoons Et_4N^+ + FeCl_4^- + POCl_3$
      or by
      a) $Fe_2Cl_6 + 4POCl_3 \rightleftharpoons FeCl_2(OPCl_3)_4^{2+} + FeCl_4^-$
      b) (a) shifted to right on dilution
      c) $FeCl_2S_4^{2+} + 2Cl^- \rightleftharpoons FeCl_4^-$

   b) Does your "intuition" suggest $FeCl_3$ or $POCl_2^+$ to be the stronger acid? In other words, do you believe that $FeCl_3$ is a strong acid in $POCl_3$, so as to be leveled to $POCl_2^+$?

c) Vibrational spectra of $AlCl_3 \cdot OPCl_3$ and $GaCl_3 \cdot OPCl_3$ as solids and as melts give no evidence for $POCl_2^+$. How could one distinguish $POCl_2^+$ from O-coordinated $POCl_3 \cdot$ acid on the basis of vibrational spectra?

27. Do you think it would be possible to determine the end-point of a titration such as

$$BCl_3 + Et_3N \rightarrow Et_3N \cdot BCl_3$$

in CCl_4, using a "normal" organic indicator like phenolphthalein?

28. Account for the facts that 1:1 and 2:1 ratios of $FeCl_3$ to $ZnCl_2$ in $POCl_3$ lead to quantitative formation of $FeCl_4^-$.

29. Account for the high conductivity of $AlCl_3$ in CH_3CN.

CHECKPOINTS

1. Lewis acid-base interaction is the basis for conceptualization of intermolecular interactions, and therefore is intimately tied to reaction mechanisms (excepting radical formation).

2. Donor orbitals can be lone pairs, σ bond pairs, and π bond pairs.

3. Lewis acids generally suffer a drastic change in molecular structure on adduct formation and are termed coordinately unsaturated.

4. The Lewis acid structure change gives rise to a relatively "low energy" vacant orbital for overlap with the electron pair of the base.

5. The ideas of bond breaking, bond making, and steric interactions are the components of analysis of the interaction between an acid and base.

a. Steric interactions are described as front and back strain.

b. Bond making depends on the overlap of the donor/acceptor orbitals and on the degree of matching of orbital energies.

c. The concept of conjugation (normal or hyper) can be important to understanding the acidity or basicity of a molecule.

REFERENCES

[1] J. N. Brönsted, *Recl. Trav. Chim. Pays-Bas*, **42**, 718 (1923).

[2] G. N. Lewis, "Valence and the Structure of Molecules," The Chemical Catalogue Co., New York (1923).

[3] Good general references on the subject of hydrogen bonds are: G. C. Pimentel and A. L. McClellan, "The Hydrogen Bond," Freeman, San Francisco (1960). W. C. Hamilton and J. A. Ibers, "Hydrogen Bonding in Solids," Benjamin, New York (1968).

[4] A. D. Sherry and K. F. Purcell, *J. Amer. Chem. Soc.,* **94**, 1853 (1972).

[5] E. M. Arnett, *et al., J. Amer. Chem. Soc.,* **92**, 2365 (1970).

[6] The enthalpy for the dimerization is ~ −146 kJ/mole; K. Wade, "Electron Deficient Compounds," Nelson, London (1971).

[7] B. D. James and M. G. H. Wallbridge, *Progr. Inorg. Chem.,* **11**, 99 (1970); S. J. Lippard and K. M. Melmed, *Inorg. Chem.,* **6**, 2223 (1967); E. B. Baker, *et al., J. Inorg. Nucl. Chem.,* **23**, 41 (1961).

[8] B. B. Wayland and R. S. Drago, *J. Amer. Chem. Soc.,* **86**, 5240 (1964).

[9] R. Foster, "Organic Charge-Transfer Complexes," Academic Press, New York (1969).

[10] J. R. Berschied, Jr., and K. F. Purcell, *Inorg. Chem.,* **9**, 624 (1970); R. C. Wade, *et al., ibid.,* **9**, 2146 (1970); E. C. Evers, *et al., J. Amer. Chem. Soc.,* **81**, 4493 (1959).

[11] H. C. Clark, *et al., Inorg. Chem.,* **8**, 450 (1969); F. Klanberg and E. L. Muetterties, *Inorg. Chem.,* **7**, 155 (1968); J. J. Harris and B. Rudnor, *J. Amer. Chem. Soc.,* **90**, 575 (1968).

[12] T. F. Bolles and R. S. Drago, *J. Amer. Chem. Soc.,* **88**, 3921, 5730 (1968).

[13] J. F. Young, *Adv. Inorg. Chem. Radiochem.,* **11**, 91 (1968).

[14] R. J. H. Clark, *et al., J. Chem. Soc. (A),* 2687 (1970); J. D. Donaldson, *et al., J. Chem. Soc. (A),* 2928 (1968).

[15] D. E. Fenton and J. J. Zuckerman, *Inorg. Chem.,* **8**, 1771 (1969); F. Bonati and G. Wilkinson, *J. Chem. Soc.,* 179, 1964; R. D. Garisch, *J. Amer. Chem. Soc.,* **84**, 2486 (1962); J. F. Young, *Adv. Inorg. Chem. Radiochem.,* **11**, 91 (1968).

[16] W. B. Fox, *et al., J. Amer. Chem. Soc.,* **92**, 9240 (1970).

[17] H. J. Plastas, *et al., J. Amer. Chem. Soc.,* **91**, 4326 (1969); J. C. Green, *et al., Chem. Comm.,* 1121 (1970); I. H. Hillier, *et al., Chem. Comm.,* 1316 (1970).

[18] I. R. Beattie and M. Webster, *J. Chem. Soc.,* 1730 (1961); I. R. Beattie, *et al., J. Chem. Soc. (A),* 2772 (1968).

[19] M. Baaz and V. Gutmann, *Monatsch. Chem.,* **90**, 276 (1959).

[20] F. Basolo, B. M. Hoffman, and J. A. Ibers, *Acc. Chem. Res.,* **8**, 384 (1975); D. A. Summerville, R. D. Jones, B. M. Hoffman, and F. Basolo, *J. Chem. Educ.,* **56**, 157 (1979).

[21] B. S. Tovrog, D. J. Kitko, and R. S. Drago, *J. Amer. Chem. Soc.,* **98**, 5144 (1976).

[22] A current brief review of all O_2 complexes is given by L. Vaska, *Acc. Chem. Res.,* **9**, 175 (1976).

[23] J. H. Price, *et al., Inorg. Chem.,* **11**, 1280 (1972).

[24] H. H. Sisler and L. F. Audrieth, "Inorganic Syntheses," Vol. II, 173 (1946).

[25] S. J. LaPlaca and J. A. Ibers, *Inorg. Chem.,* **5**, 405 (1966); L. H. Voight, J. L. Katz, and S. E. Wiberley, *ibid,* **4**, 1157 (1965); J. W. Moore, *et al., J. Amer. Chem. Soc.,* **90**, 1358 (1968).

[26] C. W. Fong and W. Kitching, *J. Amer. Chem. Soc.,* **93**, 3791 (1971); M. R. Churchill and J. Warmald, *Inorg. Chem.,* **10**, 572 (1971); A. Wojcicki, *et al., ibid,* **10**, 2130 (1971); *J. Amer. Chem. Soc.,* **93**, 2535 (1971).

[27] K. O. Christie, *et al., Inorg. Chem.,* **11**, 1679 (1972).

[28] R. J. Gillespie and M. J. Morton, *Quart. Rev.,* **25**, 553 (1971); O. Glemser and A. Sinali, *Angew. Chem. Internat. Edn.,* **8**, 517 (1969).

[29] E. F. Riedel and R. D. Willet, *J. Amer. Chem. Soc.,* **97**, 701 (1975).

[30] R. S. Mulliken and W. B. Person, "Molecular Complexes," Wiley, New York (1969); ref. 9.

[31] R. J. Gillespie, *et al., Chem. Comm.,* 1543 (1971); G. R. Jones, *et al., Inorg. Chem.,* **9**, 2264 (1970); J. H. Holloway and J. G. Knowles, *J. Chem. Soc. (A),* 756 (1969); F. O. Sladky, *et al., ibid,* 2179, 2188 (1969); V. M. McRae, *et al., Chem. Comm.,* 62 (1969).

[32] D. J. Hodgson and J. A. Ibers, *Inorg. Chem.,* **8**, 326 (1969).

[33] More attention will be given to this important property of organometal compounds in Chapters 16 and 17. J. C. Kotz and D. G. Pedrotty, *Organometal. Chem. Rev. A,* **4**, 479 (1969); D. F. Shriver, *Accts. Chem. Res.,* **3**, 231 (1970); R. B. King, *ibid,* **3**, 417 (1970).

[34] J. K. Ruff, *Inorg. Chem.,* **7**, 1499 (1968).

[35] See, for example, T. D. Coyle and F. G. A. Stone, Some Aspects of the Coordination Chemistry of Boron, in "Progress in Boron Chemistry," H. Steinberg and A. L. McCloskey, Eds., Macmillan, New York (1964).

[36] H. C. Brown, *J. Chem. Soc.,* 1248 (1956).

[37] D. H. Aue, *et al., J. Amer. Chem. Soc.,* **97**, 4136, 4137 (1975); **98**, 311, 318 (1976).

[38] S. Sujiski and S. Witz, *J. Amer. Chem. Soc.,* **76**, 4631 (1954); A. B. Burg and E. S. Kuljian, *ibid,* **72**, 3103 (1950).

[39] B. Beagley and A. Conrad, *Trans. Faraday Soc.,* **66**, 2740 (1967).

[40] A good compilation of amine·BR_3 adduct stabilities is ref. 36.

[40a] K. F. Purcell and J. P. Zapata, *J. Amer. Chem. Soc.,* **100**, 2314 (1978).

[41] See E. Fluck, *Topics in Phosphorus Chemistry,* **4**, 291 (1967).

[42] C. Jouany, *et al., J. Chem. Soc. (Dalton),* 1510 (1974).

[43] D. G. Brown, *et al., J. Amer. Chem. Soc.,* **90**, 5706 (1968); D. F. Shriver and B. Swanson, *Inorg. Chem.,* **10**, 1354 (1971).

[44] See ref. 35. Also W. A. G. Graham and F. G. A. Stone, *J. Inorg. Nucl. Chem.,* **3**, 164 (1956).

[45] M. A. Haney and J. L. Franklin, *Trans. Faraday Soc.,* **65**, 1794 (1969); W. L. Jolly and D. N. Hendrickson, *J. Amer. Chem. Soc.,* **92**, 1863 (1970); D. Holtz, *et al., Inorg. Chem.,* **10**, 201 (1971).

[46] E. L. Lines and L. F. Centofanti, *Inorg. Chem.,* **13**, 2796 (1974); A. H. Cowley and M. C. Damasco, *J. Amer. Chem. Soc.,* **93**, 6815 (1971); R. W. Rudolph and C. W. Schultz, *ibid,* **93**, 6821, (1971); R. W. Parry, *et al., Inorg. Chem.,* **11**, 1, 1237 (1972).

[47] R. S. Drago and K. F. Purcell, *Progr. Inorg. Chem.,* **6**, 217, Wiley, New York (1964); R. S. Drago and K. F. Purcell, "Non-Aqueous Solvent Systems," T. C. Waddington, Ed., p. 211, Academic Press, New York (1965); D. W. Meek, "The Chemistry of Non-Aqueous Solvents," Vol. I, J. J. Lagowski, Ed., Academic Press, New York (1966).

[48] T. C. Waddington, "Non-Aqueous Solvents," Appleton–Century–Crofts, New York (1969).

[49] G. C. Finger and C. W. Kruse, *J. Amer. Chem. Soc.,* **78**, 6034 (1956).

[50] M. S. B. Munson, *J. Amer. Chem. Soc.,* **87**, 2332 (1965); J. I. Brauman and L. K. Blair, *ibid,* **90**, 6561 (1968); J. I. Brauman, *et al., ibid,* **93**, 3914 (1971).

[51] H. H. Sisler, "Chemistry in Non-Aqueous Solvents," Reinhold, New York (1961).

[52] See ref. 48.

[53] J. J. Lagowski and G. A. Moczygemba, in "The Chemistry of Non-Aqueous Solvents," Vol. II, J. J. Lagowski, Ed., Academic Press, New York (1967); W. L. Jolly and C. J. Hallada, in "Non-Aqueous Solvent Systems," T. C. Waddington, Ed., Academic Press, New York (1965).

[54] H. Strehlow, in "The Chemistry of Non-Aqueous Solvents," Vol. I, J. J. Lagowski, Ed., Academic Press, New York (1966).

6. NON-METAL COMPOUNDS: IMPORTANT FUNCTIONAL GROUPS

In this chapter we will look into the more (and some less) highly developed aspects of main-group element compounds. Against a background of bond energies, we will summarize and compare the known classes of compounds formed by the elements primarily from Groups III through VIII and from the second and third periods of Mendeleev's Periodic Table in its long form. Our interest here is in surveying briefly the structural units on which the chemistries of these elements are based and in focusing your attention on the similarities and dissimilarities between these units from element to element.

FREE ENERGY, REACTION POTENTIAL, AND EQUILIBRIUM

Review

The thermodynamic quantity that has the most direct bearing on the course of a chemical reaction is the Gibbs free energy change for the reaction, ΔG. But you are familiar with other quantities that describe the stability of a chemical system and that are related to ΔG. Of great practical use are the standard reaction potential†, E°, and the equilibrium constant, K;

$$\Delta G^\circ = - nFE^\circ$$

and

$$\Delta G^\circ = - RT \ln K$$

where n is the number of Faradays of electricity involved in the reaction and $F = 96{,}500$ coul = 96.48 kJ volt^{-1} (gram equivalent)$^{-1}$.

Estimation of Reaction Spontaneity

In the estimation of the relative stabilities of reactants and products (these stabilities always being measured in terms of ΔG), we find it most important to consider the relative enthalpies and entropies of reactants and products:

$$\begin{aligned}\Delta G &= \Delta H - T\Delta S \\ &= \Delta E + \Delta(PV) - T\Delta S\end{aligned}$$

†The standard reaction potential, E°, is derived from the weighted difference of the standard reduction potentials.

$$\mathrm{A} + ne^- \rightarrow \mathrm{A}^{n-} \quad :E_A^\circ$$

$$\mathrm{B} + me^- \rightarrow \mathrm{B}^{m-} \quad :E_B^\circ$$

$$m\mathrm{A} + n\mathrm{B}^{m-} \rightarrow m\mathrm{A}^{n-} + n\mathrm{B} \quad :E^\circ = E_A^\circ - E_B^\circ$$

If we are ever to develop an intuition for the spontaneity of reactions we must, explicitly or implicitly, develop a "feel" for the relative importance of each of these terms and how each favors or disfavors the reaction as written.

Perhaps the quantity most difficult for which to develop an intuition is the entropy change, ΔS. Because ΔS is a measure of the relative "randomness" of products and reactants, we can predict that a reaction that increases the number of gaseous molecules is favored by $T\Delta S$. Similarly, in condensed phases, entropy generally favors an increase in the number of molecules.

The $\Delta(PV)$ term is negligible for condensed phase reactions and usually for reactions involving gases; in the cases of reactions involving gases, $\Delta(PV) \cong \Delta(nRT)$, which amounts to a contribution (at $300°K$) of less than 2.5 kJ for each mole of gas produced or consumed. Because of this, the practicing chemist often interprets ΔH and ΔE as equivalent quantities.

With regard to ΔE, it is obvious that the arrangement of atoms to form molecules with the greatest binding energies will be favored. It is also likely that the arrangement with the greatest binding energies will be the one most constrained, so that often ΔE and $T\Delta S$ are of opposite sign.

On occasion, the spontaneity of a reaction is controlled by $T\Delta S$ (whenever $|T\Delta S| > |\Delta H|$); but most often it is found that, at temperatures near $300°K$, the $T\Delta S$ term amounts to only a few kJ/mole whereas ΔH is one to two orders of magnitude larger. Some data[1] are given in Table 6-1 to illustrate the point that ΔH is often dominant over $T\Delta S$.

Compound	ΔH_f°	ΔG_f°
$SO_{2(g)}$	-297	-301
$H_2O_{(l)}$	-243	-230
$H_2S_{(g)}$	-21	-33
$NH_{3(g)}$	-46	-17
$PH_{3(g)}$	$+8$	$+17$
$KNO_{3(c)}$	-494	-393

TABLE 6-1

STANDARD ENTHALPIES AND FREE ENERGIES OF FORMATION (kJ/mole)

Table 6-2 presents some data (derivable from free energies and enthalpies of formation) with which we can make some additional, useful points. First of all, the first two reactions involve an appreciable change in the number of molecules and exhibit reasonably large entropy changes (nevertheless, ΔH dominates $T\Delta S$). For the next two reactions, with no change in numbers of gas molecules, the entropy change at $300°K$ is quite small. The final three reactions exhibit entropy contributions $(T\Delta S)$ on the order of 40 kJ/mole. The last of these reactions makes the interesting point that when the enthalpy change for a reaction is less than 40 kJ/mole, the entropy contribution to the free energy may determine the sign of ΔG. Finally, only the third, fifth, and sixth reactions proceed, unassisted, at a significant rate at $300°K$. This, of course, warns you that *thermodynamic spontaneity is a necessary but not sufficient condition for the observation of a chemical reaction.* The requirement that a low energy pathway must exist for the reaction to occur will be discussed in the next chapter. Moreover, the data in Table 6-2 serve to illustrate the commonly made assumption that spontaneity of a reaction can be judged by the enthalpy change and that entropy contributions must be allowed for when ΔH is small.

TABLE 6-2

SOME REACTION FREE ENERGIES AND
ENTHALPIES AT 300°K

	$\Delta H°$ (kJ/mole)	$\Delta G°$ (kJ/mole)
$SF_{6(g)} + 3H_2O_{(g)} \rightarrow SO_{3(g)} + 6HF_{(g)}$	-188	-318
$Xe_{(g)} + 3F_{2(g)} \rightarrow XeF_{6(g)}$	-402	-280
$PCl_{3(g)} + AlBr_{3(c)} \rightarrow PBr_{3(g)} + AlCl_{3(c)}$	-13	-17
$CH_{4(g)} + 2O_{2(g)} \rightarrow CO_{2(g)} + 2H_2O_{(g)}$	-803	-799
$CaO_{(c)} + CO_{2(g)} \rightarrow CaCO_{3(c)}$	-180	-130
$AsCl_{3(l)} + 3NaF_{(c)} \rightarrow AsF_{3(g)} + 3NaCl_{(c)}$	-105	-134
$CH_3OH_{(l)} + NH_{3(g)} \rightarrow CH_3NH_{2(g)} + H_2O_{(g)}$	$+17$	-17

HEATS OF REACTIONS FROM BOND ENERGIES

The semantic/conceptual distinction to be drawn between bond dissociation energies and bond energies derives from the usage that **bond dissociation energy** (D) refers to the energy involved in homolytically breaking one mole of a particular bond in a molecule, whereas **bond energies** (E) are defined so that their sum (one term for each bond in the molecule) equals the atomization energy of a mole of the molecules. For example, each step of the reaction

$$CH_4 \rightarrow CH_3 + H \rightarrow CH_2 + 2H \rightarrow CH + 3H \rightarrow C + 4H$$

involves a CH bond dissociation energy (D_{CH} values for this example are 435, 444, 444, and 339 kJ/mole, respectively), whereas the C—H bond energy for methane is the average of these, $E_{CH} = 415$ kJ/mole. A bond dissociation energy value depends on the valence states of both atoms defining the bond and on the substituents about them. It is the bond energies with which you are most familiar from your organic chemistry course; for organic molecules in particular the bond energies are often additive, making possible an effective estimate of the heat of formation of a molecule.

Table 6-3 contains a listing of bond energies,[2] which will be of use to you in the later discussion of the stabilities of various non-metal compounds. To make the use of bond energies easier for the prediction of the spontaneity of reactions in which pairs of atoms in different molecules exchange partners, some of the data in Table 6-3 are presented in a useful form in Figure 6-1. A couple of examples will show how easy it is to use the bond energy terms presented in this way.†

---

† Be alert to the fact that the bond energies are defined to be positive. If you use the rule "final state energy minus initial state energy" to estimate the reaction energy, you must change the sign of the numerical result.

Bond	E (kJ/mole)	Compounds
O=O	498	O_2
H—H	435	H_2
O—H	464	H_2O
O—O	142	H_2O_2
F—F	159	F_2
F—O	213	F_2O
F—H	569	HF
Cl—Cl	243	Cl_2
Cl—O	205	Cl_2O
Cl—H	431	HCl
Cl—F	251	ClF
Br—Br	192	Br_2
Br—H	368	HBr
Br—F	251	BrF
I—I	151	I_2
I—H	297	HI
I—F	243	IF
I—Cl	209	ICl
I—Br	176	IBr
S=S	431	S_2
S=O	523	SO
S—H	368	H_2S
S—S	264	S_8, H_2S_n
S—Cl	272	SCl_2
Se—Se	159	Se_2Cl_2, Se_2Br_2
Se—H	305	H_2Se
Se—Cl	243	$SeCl_2$
N≡N	946	N_2
N≝O	632	NO
N≡O$^+$	1054	NO$^+$
N—H	389	NH_3
N—O	163	NH_2OH
N—F	280	NF_3
N—Cl	188	NCl_3
N—N	159	N_2H_4, N_2F_4
N=O	594	NOF, NOCl
P≡P	523	P_2
P—H	326	PH_3
P—P	209	P_4, P_2H_4
P—F	498	PF_3
P—Cl	331	PCl_3
P—Br	268	PBr_3
P—O	368	P_4O_6
As—As	180	As_4
As—O	331	As_4O_6
As—F	485	AsF_3
As—Cl	310	$AsCl_3$
As—Br	255	$AsBr_3$
As—H	297	AsH_3
Sb—Sb	142	Sb_4
Sb—Cl	314	$SbCl_3$
Sb—Br	264	$SbBr_3$
Sb—H	255	SbH_3
C—C	347	Organic
C=C	611	Organic
C≡C	837	Organic
C—H	414	Organic
C—N	305	Organic
C—O	360	Organic

TABLE 6-3

SOME BOND ENERGIES

TABLE 6-3 (Continued)

Bond	E (kJ/mole)	Compounds
C=O	745	Organic
C=O	803	CO_2
C≡O	1075	CO
C—F	490	CF_4
C—Cl	326	CCl_4
C—Br	272	CBr_4
Si—H	318	Si_nH_{2n+2}
Si—Si	226	$Si_{(c)}$, Si_nH_{2n+2}
Si—O	464	$SiO_{2(c)}$
Si=O	640	$SiO_{2(g)}$
Si≡O	803	$SiO_{(g)}$
Si—F	598	SiF_4
Si—Cl	402	$SiCl_4$
Si—Br	331	$SiBr_4$
Si—C	305	$Si(CH_3)_4$, $SiC_{(c)}$
Si—N	335	$Si_3N_{4(c)}$
Ge—Ge	188	$Ge_{(c)}$, Ge_nH_{2n+2}
Ge—H	285	Ge_nH_{2n+2}
Ge—O	360	$GeO_{2(c)}$
Ge—F	473	GeF_4
Ge—Cl	339	$GeCl_4$
Ge—Br	280	$GeBr_4$
Ge—I	213	GeI_4
Ge—N	255	$Ge_3N_{4(c)}$
Sn—Sn	151	$Sn_{(c)}$
Sn—H	251	SnH_4
Sn—C	209	$Sn(CH_3)_4$, $Sn(CH_3)_3H$
Sn—Cl	314	$SnCl_4$
Sn—Br	268	$SnBr_4$
B—F	644	BF_3
B—Cl	444	BCl_3
B—Br	368	BBr_3
B—I	272	BI_3
B—B	301	B_2F_4
B—O	523	Boron esters, ester chlorides and hydroxides

You may be interested in whether HF and BrF might react to produce HBr and F_2.

$$F—H + Br—F \rightarrow Br—H + F—F$$

Locate the F—H and Br—H terms in column 1 (the H column) and the Br—F and F—F terms in the second column. The conversion F—H → Br—H incurs a loss of stability (a positive ΔE), as does the conversion Br—F → F—F. Consequently, you predict—correctly—that H—F and Br—F will not react to form significant amounts of HBr and F_2.

Another example involves the H and OR exchange between an amine and a phosphate ester:

$$N—H + P—OR \rightarrow P—H + N—OR$$

This sort of reaction is quickly seen not to be favorable, largely because of the greater P—OR than N—OR functional group stability. On the other hand, the conversion (reduction) of a fluorophosphine to a phosphine with R_3SiH is a possibility:[3]

$$Si—H + P—F \rightarrow P—H + Si—F$$

owing to the greater stability of the Si—F than P—F function.

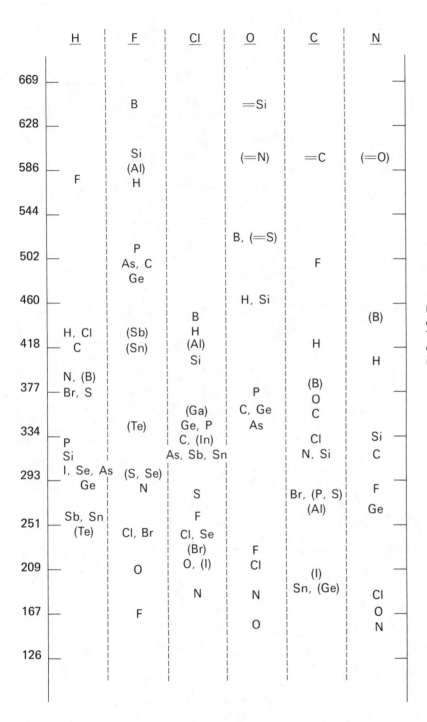

Figure 6-1. A graphical presentation of "nominal bond energies" from Table 6-3. Values in parentheses are either dissociation energies or estimated values.

A Caveat Concerning Condensed Phase Reactions

In the preceding section we implicitly introduced the idea of using a Haber cycle at the molecular level to interpret gas phase heats of reaction in terms of a two-step process (bond breaking *vs.* bond making). More generally, we deal with reactions in condensed phases, and this complicates our use of bond energies to predict the feasibility of new reactions or to understand known chemical systems. Haber cycles, which are practical applications of Hess' Law, help us to understand such reactions. As exemplified by the following general reaction, the Haber cycle shows us in what way we must allow for the "condensation" energies of reactants and products.

$$A_{(g)} + B_{(g)} \xrightarrow{\Delta G_{(g)}} C_{(g)} + D_{(g)}$$

$$\Delta G^A_{(cond)} \downarrow \quad \Delta G^B_{(cond)} \downarrow \quad \Delta G^C_{(cond)} \downarrow \quad \Delta G^D_{(cond)} \downarrow$$

$$A_{(cond)} + B_{(cond)} \xrightarrow{\Delta G_{(cond)}} C_{(cond)} + D_{(cond)}$$

Using Hess' Law,

$$\Delta G_{(cond)} = \Delta G_{(g)} + \left\{ \sum^{products} \Delta G_{(cond)} - \sum^{reactants} \Delta G_{(cond)} \right\}$$

and an entirely analogous equation can be written for enthalpies. In this formalism the condensed phase for any one compound could be any of the following: solid, pure liquid, or solution. Correspondingly, the $\Delta H_{(cond)}$ are the lattice energy (ignoring ΔPV), the heat of condensation (or the negative of the heat of vaporization), and the heat of solution. The *corrective term* in brackets (so called because the term corrects the gas phase free energy or enthalpy to that for the condensed phase) is often small because it is the difference between two quantities that frequently are of similar magnitude. Nevertheless, the corrective term must be considered whenever the $\Delta G_{(g)}$ we predict is small. Thus the corrective term, typically up to 40 kJ/mole, could have a critical effect on the existence of compounds with $|\Delta G_{(g)}|$ estimated at ~40 kJ/mole. Such a circumstance is most likely when molecules self-associate by hydrogen bonding or other acid-base interaction or by non-specific interactions (dipole-dipole or London attractions), as discussed in Chapter 5.

While we will encounter many examples later in this chapter, it will help to illustrate how the corrective term can be of importance. The ΔE_f° of the alkali metal fluorides is an interesting case. The Haber cycle can be written as:

$$M_{(g)} + \tfrac{1}{2}F_{2(g)} \xrightarrow{\Delta E^\circ_{(g)}} M^+_{(g)} + F^-_{(g)}$$

$$-S \downarrow \qquad \qquad \swarrow E_0$$

$$M_{(c)} + \tfrac{1}{2}F_{2(g)} \xrightarrow{\Delta E_f^\circ} MF_{(c)}$$

Here S is the sublimation energy of the pure metal, while E_0 is the lattice energy of the salt MF. Hess' Law tells us that

$$\Delta E_f^\circ = \Delta E^\circ_{(g)} + (E_0 + S)$$

The $\Delta E^\circ_{(g)}$ can itself be broken down, using a Haber cycle, to give

$$\Delta E^\circ_{(g)} = I_M + \frac{1}{2}D_{F_2} - A_F$$

where I_M is the ionization potential of the metal and A_F is the electron affinity of F.

Thus,

$$\Delta E^{\circ}_{(g)} = I_M + 79 - 335 = I_M - 256 \text{ kJ/mole}$$

Consequently we have

$$\Delta E^{\circ}_f = (I_M - 256) + (E_0 + S) \text{ kJ/mole}$$

The following table shows these two terms for Li through Cs:

	$I_M - 256$	$(E_0 + S)$	ΔE°_f
Li	+264	-870	-606
Na	+238	-808	-570
K	+163	-724	-561
Rb	+146	-695	-131
Cs	+121	-669	-548

Here is a good example of the importance of the "corrective" condensation energy term. The formation of $M^+ + F^-$ in the gas phase is thermodynamically unfavored (assuming that the sign of $\Delta G^{\circ}_{(g)}$ is determined by that of $\Delta E^{\circ}_{(g)}$) and it falls to the lattice energy of the product to alter this state of affairs.

STUDY QUESTIONS

1. How large must be K for the reaction $A^- + B^+ \rightarrow C^+ + D^-$ if the products are to be contaminated with no more than 0.1% of each reactant? What value of ΔG° does this condition require? How large a change in $\ln K$ is incurred by a change in ΔG° of 5.77 kJ/mole? Reflecting on the influence of "condensation energies," would you say that choice of solvent could be critical to reaction efficiency?

2. Referring to the reaction in problem 1, what is the minimum difference in half-cell E°'s required to meet the product purity criterion?

3. For a reaction $A + B \rightarrow C$, for which $K = 10^2 \text{ M}^{-1}$, what ratio of A_0 to B_0 is required to produce a product contaminated by no more than 0.1% of B? (A_0 and B_0 are the initial concentrations of A and B). If $B_0 = 1$ M, what must A_0 be?

4. Is the reaction

$$R_3P{=}O + Si_2Cl_6 \rightarrow R_3P + (Cl_3Si_2)O$$

exothermic? (See p. 196 for an estimate of the P=O bond energy.)

5. With reference to Figure 6-1, predict whether

$${>}B{-}F + HOR \rightarrow {>}BOR + HF$$

or

$${>}B{-}Cl + HOR \rightarrow {>}BOR + HCl$$

is more exothermic. To what do you ascribe the greater driving force in your choice?

6. Would

$$\begin{array}{ccccc} S_2Cl_2 & + & 2H_2S_n & \rightarrow & H_2S_{2n+2} & + & 2HCl \\ Cl{-}S_2{-}Cl & & H{-}S_n{-}H & & H{-}S_{2n+2}{-}H \end{array}$$

be a reasonable (from a thermodynamic view) scheme for synthesizing sulfanes?

7. Is the reaction

$$R_3SiOR + BX_3 \rightarrow R_3SiX + BX_2(OR)$$

a potentially useful way to convert a silyl ether to a silyl fluoride or chloride?

8. Do entropy and solvent polarity factors favor scrambling reactions like

$$BX_3 + BY_3 \rightleftharpoons BX_2Y + BY_2X \quad ?$$

NON-METAL FUNCTIONAL GROUPS

In these remaining sections, our emphasis will be on covalently bonded elements. Of special interest are the tendencies of the elements to catenate via both saturated and unsaturated linkages, and properties of their bonds to hydrogen, carbon, oxygen, and the halogens. The mechanisms of the reactions of these functional groups are compared and contrasted in the next chapter. Here we wish to use bond energy ideas to account for the existence (or non-existence) of the important non-metal functions.

Boron and Aluminum

Bond Energies for Reference			
B—B	301	Al—H	(272)
B—H	(389)	Al—C	(251)
B—C	(372)	Al—F	(582)
B—F	644	Al—Cl	(418)
B—Cl	444		
B—O	523		

Boron is a most interesting element, in that it tends to form trivalent compounds typical of most non-metallic elements but, in doing so, is left in a state of valency unsaturation (more valence ao's than electron pairs). This is a "mild" form of the extreme valency unsaturation found in metals. Thus, you should not be surprised to find some boron compounds with rather bizarre structures, at least for a non-metallic element. In many ways, boron exhibits compounds that are in the ill-defined region between metallic and "normal" covalent compounds. This is no more strikingly shown than by the element and its binary compounds with metals, the so-called **borides**.

Elemental boron exists in three different crystallographic forms, but all are characterized by icosahedral B_{12} clusters linked together in a three-dimensional network.[4] The icosahedron is a regular polyhedron with 12 vertices and 20 faces. There are two high-symmetry polyhedra with this same number of vertices; the cuboctahedron has six square faces and eight triangular faces, while the icosahedron has only equilateral triangular faces.†

THE B—B FUNCTION

---

†The cuboctahedron has 24 edges, while the icosahedron has 30. The difference of 6 edges affords a convenient view of the relation between the two structures. Notice the 6 square faces of the cuboctahedron. By connecting a pair of diagonally opposite vertices in each square face, each square is converted to two triangles. "Relaxation," so that all edges are of equal length, produces the icosahedron.

icosahedron   *cuboctahedron*

The boron atoms are located at the vertices of the icosahedron, and this leads us to note that each atom has *five* nearest neighbor atoms. Quite obviously we cannot identify an electron pair bond with each edge of the icosahedron (there are 30 edges), so it is difficult (but not impossible) to rationalize the bonding of the boron atoms in such a structure in terms of a "localized" orbital model.

It would appear that formation of such a polyhedral structure with extensive delocalization of valence electrons is highly effective in stabilizing the element. Even though the more common, impure, amorphous boron is highly reactive, truly crystalline boron is extremely hard and resistant to chemical attack, even by hot, concentrated solutions of most oxidizing agents.

THE B–H FUNCTION

With hydrogen,[4,5] boron could ostensibly form a simple borane, BH_3. This compound demonstrates dramatically the effect of valency unsaturation in trivalent boron compounds. BH_3 is a quite unstable species and readily dimerizes into B_2H_6, known as diborane [or more frequently as diborane(6), where the (6) signifies the number of hydrogens in the molecule]. This simplest member of the borane series exhibits the general feature of all **neutral boranes**, the bridging hydrogen atom.

$$\text{H}\diagdown_{\text{H}}\text{B}\diagup^{\text{H}}\diagdown_{\text{H}}\text{B}\diagup^{\text{H}}_{\diagdown\text{H}} \qquad \begin{bmatrix} \text{m.p. } -165° \\ \text{b.p. } \;\;-93° \\ \text{pyrophoric} \end{bmatrix}$$

The enthalpy change for the dimerization reaction

$$2BH_3 \rightarrow B_2H_6$$

is about 146 kJ exothermic.[6] As cleavage of a terminal B–H bond is endothermic by about 377 kJ/mole (estimated), the B–H–B bridge has a somewhat higher energy of about 450 kJ/mole.† Hydrogen in a bridging role is not really all that unique in chemistry. The unique feature here, of course, is the $2e^-$ (as opposed to $4e^-$) nature of the structural unit (see p. 129).

A few examples are given now to impress on you that boron catenation, hydrogen bridge bonding, and polyhedral structures are the characteristic features of the hydrides of boron.

After diborane, the next most complex neutral borane is **tetraborane(10)**:

$$\begin{matrix} & & \text{H} & & \\ & & | & & \\ \text{H}\diagdown & \text{H}\;\;\text{B}\;\;\text{H} & & \text{H} \\ \;\;\;\;\text{B} & & & \text{B}\diagup \\ \text{H}\diagup & \text{H}\;\;\;\;\;\;\text{H} & & \diagdown\text{H} \\ & & \text{B} & & \\ & & | & & \\ & & \text{H} & & \end{matrix} \qquad \begin{bmatrix} \text{m.p. } -120° \\ \text{b.p. } \;\;\;\;16° \\ \text{pyrophoric} \end{bmatrix}$$

---

†This value is derived from the energy cycle:

$$2BH_3 \xrightarrow[754]{} 2BH_2 + 2H \longrightarrow B_2H_6$$

with -146 over the arc from $2BH_3$ to B_2H_6.

Note the bicyclo (C_{2v}) structure of four borons and four bridging hydrogens, the B—B bond, and the presence of two borons with two terminal hydrogens.

Pentaborane(9), the most thermally stable[7] of the lower boranes, has the regular (C_{4v}) structure of a square-based pyramid of boron atoms, each with a single terminal hydrogen, and four bridging hydrogens between basal borons. The apical boron seems to be bound to the basal borons through five-center, six-electron bonding (the $6e^-$ character of the apical/basal boron binding is determined from the fact that 24 valence electrons are present, with two required for each terminal and bridging H link).

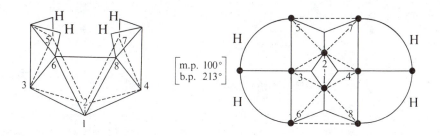

Theoretical studies[8] reveal the occurrence of a three-center BBB electron pair within the triangular faces. However, because the molecule has C_{4v} symmetry and not the C_s symmetry appropriate to the structure **1**, we must actually write down four equivalent resonance structures.

An interesting example of the polyhedral boron structure arises for **decaborane(14)**. This structure (where each vertex represents a B—H unit) is seen to be an incomplete icosahedron (p. 176) of boron atoms, each with a terminal hydrogen. The missing eleventh and twelfth B—H units are effectively replaced with four hydrogens of the bridging type. In the localized electron pair scheme at the right, we find two kinds of three-center BBB "bonds." Four of the so-called "closed" type (as in B_5H_9 above) are located in faces outlined by dashed lines. Two "open" three-center BBB "bonds" (reminiscent of the BHB bridge) are defined by the 5,3,6 and 7,4,8 edges.

The neutral boranes, as a class of compounds, have no simple stability characteristics. Using Wade's[9] suggested bond energies (BH = 372, BB = 310, BHB = 452, BBB = 381), you may readily verify that the B_nH_m are thermodynamically unstable with respect to the elements, to air oxidation (products = B_2O_3, H_2O), and to hydrolysis (products = $B(OH)_3$, H_2). However, the rates of such reactions are highly variable, owing to the feasibility (or lack of it) of mechanistic pathways for such reactions. The only significant generalization to be made is that the neutral boranes owe their existence to kinetic factors.

Boron also forms binary, anionic compounds with hydrogen. Some examples would be BH_4^-, $B_2H_7^-$ (again with bridging hydrogen, as the structure [C_{3v} more likely than D_{3h}] is $BH_3-H-BH_3^-$), and the **closo† boranes** of generic formula $(BH)_n^{2-}$. These latter compounds all display polyhedral structures, with known examples having $n = 6$ to 12. The BH units of these anions are arranged in as regular a fashion as possible so that the polyhedra have triangular faces.

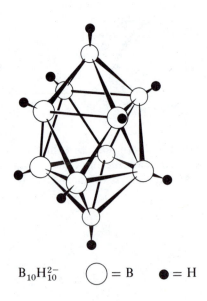

$$B_{10}H_{10}^{2-} \qquad \bigcirc = B \qquad \bullet = H$$

The simplest hydride anion of boron, BH_4^-, is very important in chemical synthesis, where it finds use as a reducing agent and has the striking property of being "stable" in water (unlike B_2H_6!), as well as soluble in ether-type solvents. (As will be discussed in the next chapter, this stability is kinetic in origin. The facts that the speed of hydrolysis depends on [H^+] and that solutions of BH_4^- become basic on hydrolysis,

$$BH_4^- + 4H_2O \rightarrow 4H_2 + B(OH)_3 + OH^-$$

translate into a sharply decreasing rate of hydrolysis after only a slight degree of decomposition.) Readily available commercially, salts of BH_4^- are convenient sources of B_2H_6 through treatment with acids such as 85% H_3PO_4 and BF_3.

$$NaBH_4 + H_3PO_4 \rightarrow \frac{1}{2}B_2H_6 + H_2 + NaH_2PO_4$$

$$3NaBH_4 + 4BF_3 \xrightarrow{\text{ether}} 3NaBF_4 + 2B_2H_6$$

THE B–C FUNCTION

One of the most important chemical properties of the B–H function is its ability to add to olefin and other unsaturated linkages in a reaction known as *hydroboration*.

---

†The word "closo" designates a closed, polyhedral boron skeleton.

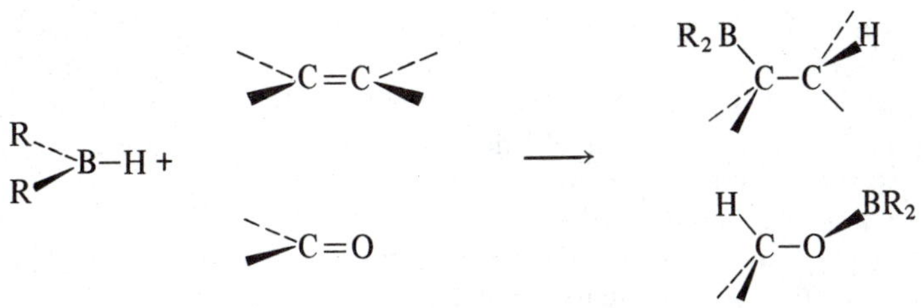

Not only is this a convenient route to organoboranes and borate esters, but hydrolysis of the initial product will yield alkanes or alcohols as products:

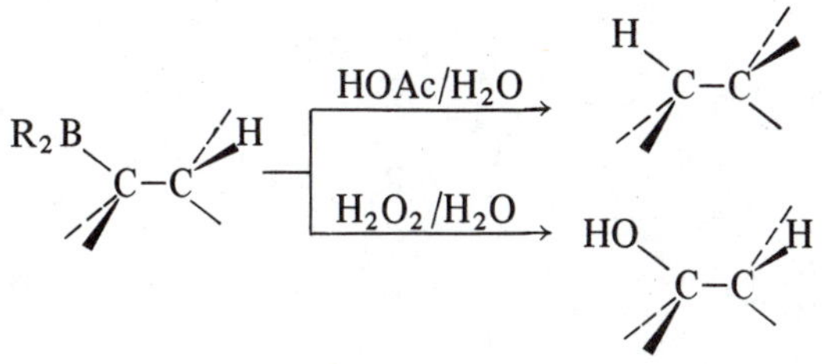

It is interesting that B_2H_6 is ineffective in this reaction; only monomeric BR_xH_{3-x} (x = 0, 1, 2) are effective, presumably because the mechanism requires addition of the boron Lewis acid to the π bond of the olefin as a first step. The role of ether solvents in promoting the dissociation of B_2H_6 is treated in Chapter 7, p. 234.

Given that BH_3 dimerizes to relieve the unsaturation at boron and that CH_3 is known to exhibit a tendency to bridge like hydrogen in many organometallic compounds, it is somewhat surprising that $B(CH_3)_3$, and **triorganoboranes** in general, do not dimerize. It has been suggested that steric compression (particularly for branched hydrocarbons) is at least partly responsible for this, but it also may be correct that hyperconjugation of the alkyl groups with the boron may also be responsible for lowering the acidity of boron.

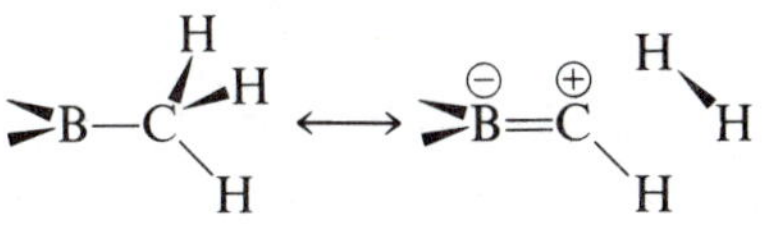

This seems to be particularly true for the triarylboranes:

Consistent with such a view of internal valency saturation, the triarylboranes are less susceptible to nucleophilic attack and are generally non-flammable compounds, whereas the lighter alkylboranes are pyrophoric.

Analogous steric and intramolecular valency saturation ideas apply to the monomeric boron trihalides, but significant differences in B—X bond strengths appear on comparison of hydrolytic stabilities. BF_3 is stable in aqueous solution and renders the solutions acidic by the reaction

THE B—halogen FUNCTION AND SOLVOLYSIS

$$BF_3 \cdot OH_2 \rightleftharpoons BF_3OH^- + H^+$$

THE B–N FUNCTION

and it is known to form solids with the composition $BF_3 \cdot OH_2$ and $BF_3 \cdot 2OH_2$; the other trihalides of boron rapidly hydrolyze to form HX and boric acid $(B(OH)_3)$. These hydrolysis reactions have a negative enthalpy change for all halogens except fluorine, and appear to be general in that alcohols and amines effect similar solvolyses. Such reactions constitute a basis for polymerization reactions.

To preview some polymerization reactions (their mechanisms are to be discussed in the next chapter, and a more comprehensive view will be taken in Chapter 8), boron trihalides react with many secondary amines with loss of HX to give species of the type R_2N-BX_2. These **aminoboranes** may then oligomerize

$$BX_3 + R_2NH \; \rightleftharpoons \; R_2NH \cdot BX_3 \; \rightleftharpoons \; R_2N=BX_2 + HX \; \rightleftharpoons \; \frac{1}{n}(R_2NBX_2)_n + HX$$

to give species such as

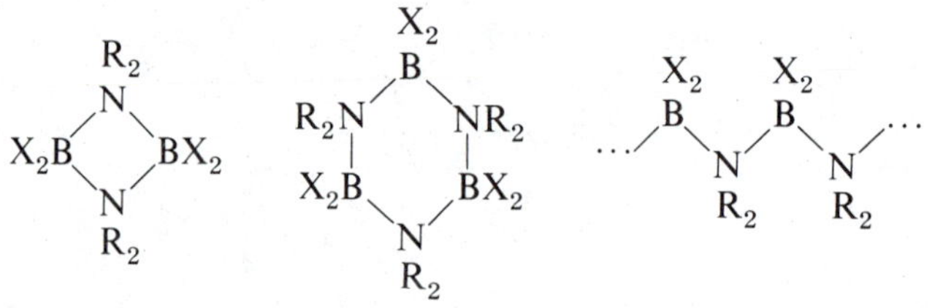

depending on steric and kinetic factors. Again, R/X combinations having incompatible steric requirements disfavor the acid/base condensation of the R_2NBX_2 monomers, which can be stabilized by intramolecular dative bonding. Given[10] that the dimerization enthalpies are typically no larger than 80 kJ/mole dimer (exothermic), while $T\Delta S \approx +40$ kJ/mole dimer, it is not surprising that steric congestion (see Et_3NBMe_3 in Chapter 5, p. 149) has an observable effect on the magnitude of ΔG for dimerization.

An important reaction of this type is also observed to occur on pyrolysis† of $NH_3 \cdot BH_3$ (where H_2 instead of HX is evolved).

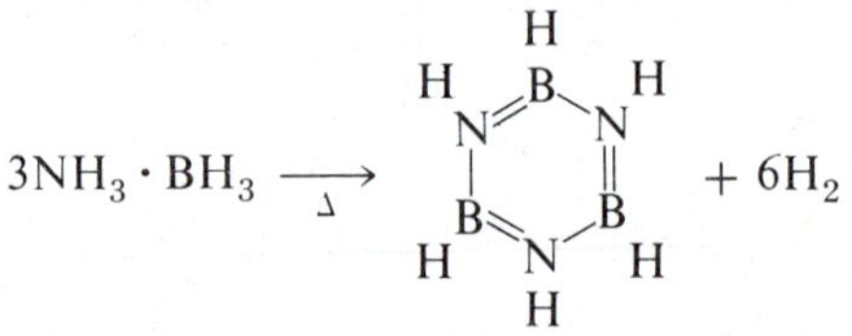

$$3NH_3 \cdot BH_3 \; \xrightarrow{\Delta} \; \text{(borazine ring)} \; + \; 6H_2$$

This interesting reaction produces the inorganic analog of benzene, $(HBNH)_3$, a compound commonly called **borazine**. In contrast to benzene, borazine is much more reactive; this is generally attributed to the more localized nature of borazine π electrons:

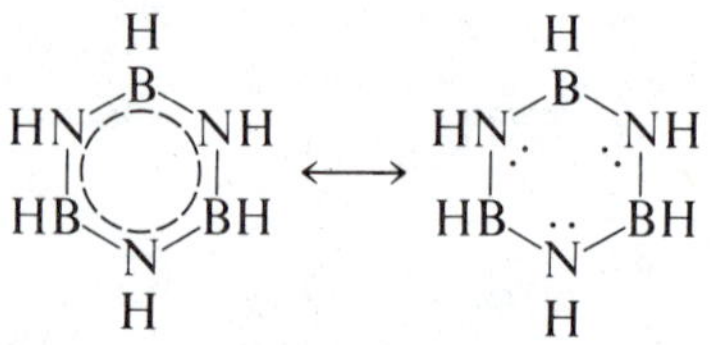

Reaction of NH_4Cl with BCl_3 at elevated temperatures

$$3NH_4Cl + 3BCl_3 \; \xrightarrow{\Delta} \; (HNBCl)_3 + 9HCl$$

---

†This too is a solvolysis reaction, but one which requires high temperature to effect elimination of H_2 from the 1:1 adduct.

leads to the borazine analog with Cl replacing H at boron. This product is useful for introducing three organo groups at boron by Grignard reactions (see Chapter 8).

The other reasonably well known halides of boron are the **diboron tetrahalides** B_2X_4, which exhibit boron catenation.[11] As an interesting example of how the structure of a molecule can change when packed into a crystal lattice, both the fluoride (m.p. $-56°$) and the chloride (m.p. $-98°$) are known to have mutually perpendicular BX_2 groups (D_{2d} symmetry) in the gas phase but are planar molecules (D_{2h}) in the solid.

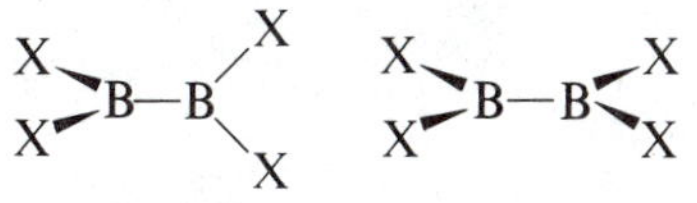

gas (D_{2d})   solid (D_{2h})

While it is not surprising that the B—B bond suffers cleavage in air (B_2F_4 is pyrophoric) so as to form B_2O_3 (see below), it is interesting that the B—B bond of B_2Cl_4 survives solvolytic reaction with liquid H_2O and alcohols to yield $B_2(OH)_4$ and its esters $B_2(OR)_4$, in analogy with the trihalides.

A dominant feature of boron chemistry is the strength of its bonds to oxygen.[12] Given that the dissociation energy of BO is only about 795 kJ/mole for an estimated (from molecular orbital theory) bond order of 2.5, while the B—O "single" bond has an energy of about 523 kJ/mole, it is not to be expected that boron-oxygen compounds will exhibit —B=O as a structural feature. Thus, the —B=O link would be expected to undergo dimerization. The high "single" bond energy is an important consequence of some dative p-$p\pi$ interaction between boron and oxygen. Consequently, the stable (solid) **boric oxide**, B_2O_3, has a structure based on chains of BO_3 units (triangular), rather than the simpler molecular arrangement of O=B—O—B=O.† Boric oxide is a major constituent (14%) of borosilicate glass, a type of glass sold under the trade names Pyrex (Corning Glass Works) and Kimax (Owens-Illinois). In fact, the triangular BO_3 unit is a structural feature common to almost all boron-oxygen compounds, including a large number of anionic borate salts that have chain, ring, and three-dimensional structure. The structures **2** through **6** illustrate the chain and ring structures possible with 3- and 4-coordinate boron.

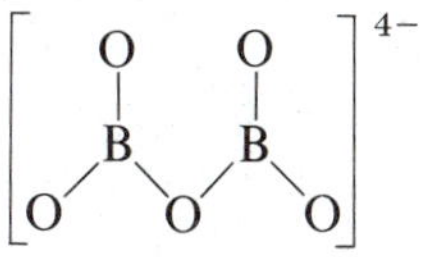

THE B—O FUNCTION

as in $Mg_2B_2O_5$

2

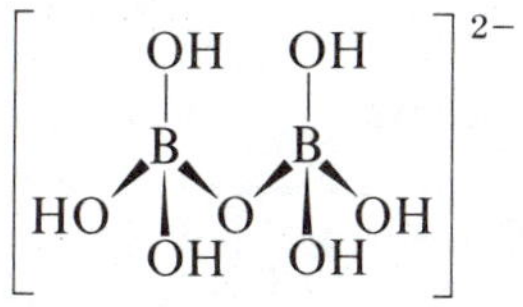

$B_2O(OH)_6^{2-}$

3

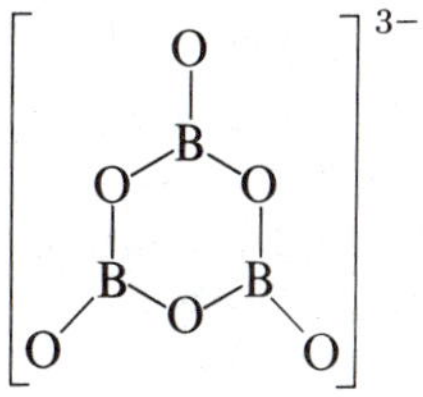

as in $Na_3B_3O_6$

4

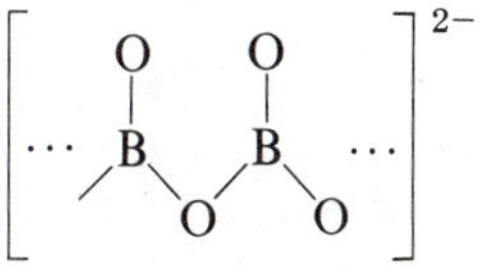

as in CaB_2O_4

5

---

†The chains may be most easily viewed as association by the action of the oxygen atom of the terminal B=O units of the hypothetical O=B—O—B=O monomers as a Lewis base on the boron atom (as an acid) of another monomer.

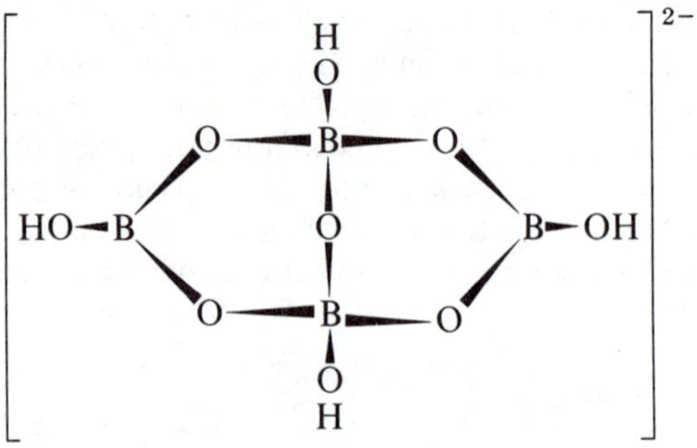

as in
"$Na_2B_4O_7 \cdot 10H_2O$", Borax
6

Compound **6** is **borax**, easily the most important polyborate in both dollar sales and tonnage. Perhaps you are already familiar with its use as a household cleaner; a few years ago, when environmentalists objected to the use of phosphates in laundry detergent, some consumers returned to their earlier use of borax cleaners. Borax is found in arid regions of the world, especially the desert Southwest of the United States. These were apparently regions of volcanic activity in the past, and boron was transported to the surface by hot springs or steam in the form of boric acid; there the acid combined with surface rocks such as carbonates or salines and was converted into alkali or alkaline earth metal borates. If this activity occurred in an arid region, the water soon evaporated after borate formation (or after the solution accumulated in a lake), leaving immense beds of borate salts.

Boric acid is soluble in water, where it behaves as a weak acid. Interestingly, the acidic behavior is not due to the simple ionization

$$B(OH)_3 \not\rightleftharpoons (HO)_2BO^- + H^+$$

but rather to ionization of the acid-base adduct $(HO)_3B \cdot OH_2$

$$B(OH)_3 + 2H_2O \rightleftharpoons B(OH)_4^- + H_3O^+ \qquad pK_a = 9.2$$

in analogy with $BF_3 \cdot OH_2$. Consistent with such a formulation is the fact that ΔS is much more negative (−0.130 kJ/K·mole) than usual for simple proton ionization.[13] An important medicinal application of this mild acid is in eye-washes.

The final examples of boron oxyacids are those with one or two OH groups of boric acid replaced by organo groups.[14] The former, $RB(OH)_2$, are referred to as **boronic acids** while the latter, R_2BOH, are called **borinic acids**. They can be prepared by hydrolysis of the corresponding diorgano boron halide, while the boronic acids have the interesting property of self-dehydrating, often quite easily.

$$3RB(OH)_2 \rightarrow (RBO)_3 + 3H_2O$$

The molecular species RBO is not expected to be stable, however, so that trimerization to form the alkyl **boroxine** $(RBO)_3$ occurs (note the analogy with bor*azine*, p. 180).

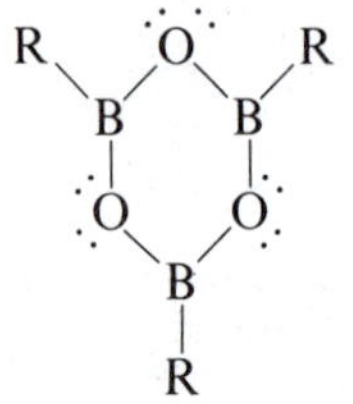

As an element of the third row of the Periodic Table, aluminum has d orbitals in its valence shell, and it is a common practice to expect valence shell expansion for these elements. Thus, the structural chemistry of aluminum is based not only on tetrahedral structures but also on octahedral geometries. Whether such Al d ao involvement is critical to the formation of 6-coordinate Al^{3+} structures is far from established. Rather, nothing more esoteric than the difference between B^{3+} and Al^{3+} sizes (VSEPR) and electronegativity may account for the 6-coordinate structures of many Al compounds.

As a point of contrast with boron, aluminum is not known[15] to form catenated polyhedral hydride species. Instead we find the simple **alane**, AlH_3, to be a highly poly-merized solid with bridging hydrogen atoms[16] (its heat of formation[17] is only ~ -12 kJ/mole). Thus, we find here a good example of aluminum "going one better" than boron on valency saturation. The structure of the solid is indicated by the following figure, where you find each Al in an octahedral environment with each hydrogen serving in a bridging capacity.

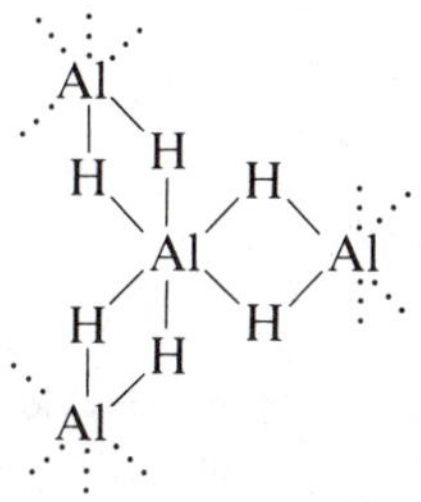

THE Al–H FUNCTION

It is an interesting fact that treatment[18] of $LiAlH_4$ in THF (tetrahydrofuran) with anhydrous H_2SO_4 can be controlled so as to produce the solid AlH_3.

$$H_2SO_4 + 2LiAlH_4 \xrightarrow{\text{THF}} 2AlH_{3(c)} + Li_2SO_4 + 2H_{2(g)}$$

With water and alcohols, as you could expect for a metal hydride, AlH_3 solvolyzes to form "hydrous oxides" or aluminum alkoxides, $Al(OR)_x H_{3-x}$, respectively.

The simple hydridic anion AlH_4^- is known in combination with several cations, and, like BH_4^-, it is a *very* useful reducing agent for inorganic and organic materials. The structures of tetrahydridoaluminate salts of second row cations (Li^+, Be^{2+}, and B^{3+}) all feature bridging hydrogen structures, and this has some important consequences for the particularly useful Li^+ salt. The association of Li^+ with AlH_4^- is thought to yield "molecular" species in ether and aromatic hydrocarbon solvents, and this is the structural feature that accounts for the solubility. It is an interesting observation that *tertiary* amines react with $LiAlH_4$ to produce amine adducts of alane, R_3NAlH_3, and Li_3AlH_6. While Na_3AlH_6 is known† (made simply by reaction of MH with $MAlH_4$), only $LiAlH_4$ reacts with tertiary amines to form a 1:1 adduct. Thus, it must be that the high degree of asso-ciation of Li^+ with the hydrogens of AlH_4^- is important kinetically and thermodynamic-ally for the displacement of H^- from AlH_4^-.

$$3MAlH_4 + 2R_3N \rightarrow 2R_3N \cdot AlH_3 + M_3AlH_6 \qquad (M = Li, \neq Na)$$

Finally, we should note that there is a marked difference in the hydrolytic stabilities of BH_4^- and AlH_4^-. Whereas BH_4^- is kinetically stable in aqueous solution as explained earlier, AlH_4^- violently hydrolyzes. Clearly, a difference in mechanisms for the hydrolyses of tetrahydridoborate and tetrahydridoaluminate is indicated.

Another useful class of alanes consists of the **organoalanes**. These are most often considered to be organometallic compounds. Again calling to mind the size difference of B and Al, the organoalanes are *dimeric* in condensed phases but somewhat dissociated in the gas phase ($\Delta H \sim 80$ kJ/mole).[19] Even $Al(C_6H_5)_3$ possesses the dimeric structure,

THE Al–C FUNCTION

---

†Note that M_3BH_6 is unknown, and considered not likely to exist.

and the mixed compound $Al_2(CH_3)_4(C_6H_5)_2$ exhibits bridging phenyl groups rather than bridging methyls.

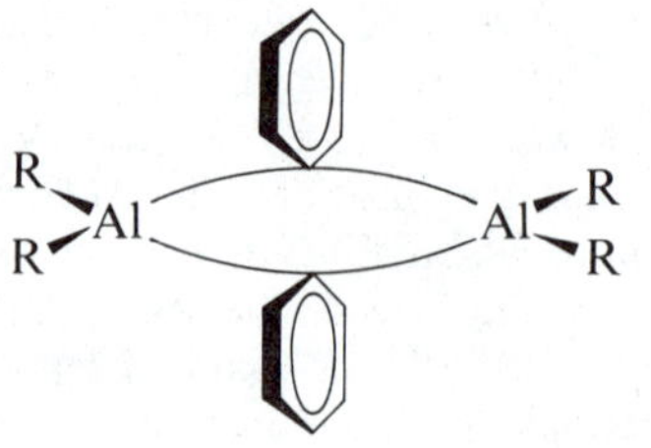

An interesting idea to ponder at this point is the role of $C_6H_5^-$ as an intra- or intermolecular π donor. In BPh_3, you saw this action in the form

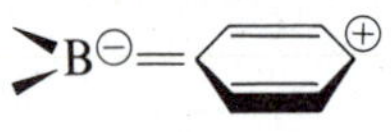

and steric congestion also seems to inhibit dimerization. In the $Al_2R_4Ph_2$ structure above it seems that Ph, in contrast with CH_3, may act much like a bridging *halogen*, involving both the carbon sp^2 ao and an occupied π mo in bridge bonding (cf. p. 129).

THE Al–N FUNCTION

With monofunctional bases, the organoalanes form simple 1:1 adducts like the borane analogs; but a point of demarcation, illustrating valence shell expansion for Al, is reached with alane adducts.[20] For example, trimethyl amine forms a 2:1 adduct with AlH_3; as you could predict (from the electronegativities of N and H), the H atoms take up equatorial sites and the amine nitrogens take up axial sites in the trigonal bipyramid.

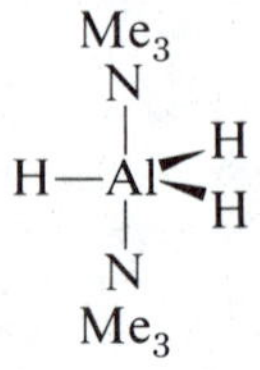

This behavior is not found with primary and secondary amines because "solvolytic" reactions eliminate alkane and form **amidoalanes**, R'_2NAlR_2,

$$R'_2HN-AlR_3 \rightarrow R'_2NAlR_2 + RH$$

and **imidoalanes**, $R'NAlR$. As we found for the boron analogs, these monomeric structures (with, formally, $\ddot{N}$–Al and $\dot{N}$=Al linkages) are unstable with respect to condensation, where the amide and imide groups function to bridge Al atoms as in **A** and **B**, below.[21] With primary amines, from which solvolysis would ostensibly lead to monomers with the empirical formula $R'N$=AlR, nature prefers condensation into three-dimensional tetrameric "cubane-like" imido structures of D_{2d} symmetry.

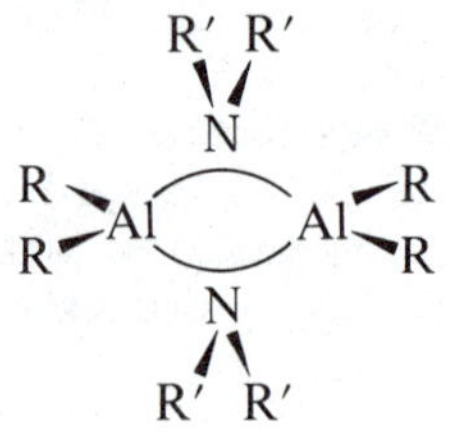

A 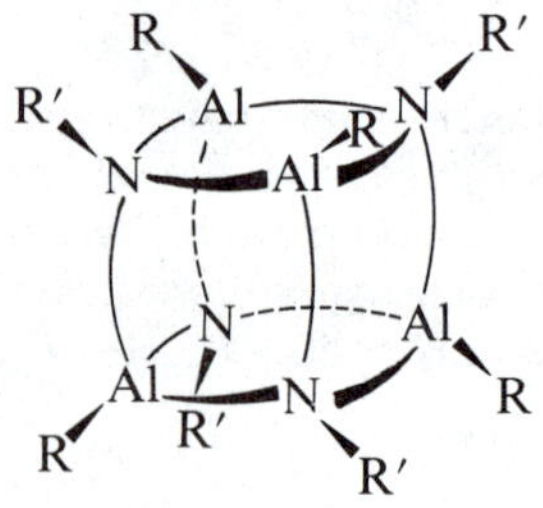   **B**

Again you are reminded of less severe steric problems in Al compounds than in their boron analogs.

Having seen the theme of bridging roles played by hydrogen, carbon, and nitrogen, we fully expect to find the halides serving a similar function. All four halides of empirical formula AlX_3 are known. The fluoride is thermally the most stable member of the series and exhibits the polymeric bridged structure like AlH_3. The chloride possesses a layer structure similar to $CrCl_3$ (p. 64), but it is easily vaporized (below 200°C) into gaseous dimers. The bromide and iodide exclusively exhibit the D_{2h} dimeric structure and are easily vaporized.† The dimers are reasonably stable but have dissociation enthalpies sufficiently small ($\sim$ 120 kJ/mole)[22] that D_{3h} symmetry monomers can be observed at high temperatures.

THE Al–halogen
FUNCTION

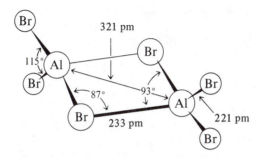

Solution of the aluminum halides in hydroxylic solvents leads to displacement of halide ions in a manner typical of metal halide salts. Partly owing to factors related to the formation of $Al(OH_2)_6^{3+}$, these solutions are not as acidic ($pK_a \approx 5$) as those of boron halides (which, except for BF_3, completely hydrolyze to $B(OH)_3$ and HX). Even hydrated forms of the aluminum halides are known, such as $AlF_3(OH_2)_3$, $Al(OH_2)_6F_3$, and $Al(OH_2)_6Cl_3$. It is suggestive of a low polarizing power of the 6-coordinate hydrated aluminum cation that the trihydrate of aluminum fluoride is dehydrated, rather than solvolyzed, by the application of heat, even though the last water is driven out only at red heat. In solution, the hexaquo cation and its derivatives from successive replacement of H_2O by fluoride are present, as well as the anion AlF_4^-. In solutions with high concentrations of added F^-, the hexafluoroaluminate anion§ makes an appearance and completes the structural analogy with the three hydride species (AlH_3, AlH_4^-, and AlH_6^{3-}).

It is the Lewis acid properties of the aluminum halides, like those of the boron halides, that make them such useful synthesis reagents. They are best known as catalysts in Friedel-Crafts reactions of organic acid chlorides. The role of $AlCl_3$ and $AlBr_3$ in these reactions is to associate strongly with or abstract the halides from the acid chlorides, forming *in situ* the tetrahaloaluminate anion and the highly reactive carbonium ion. While the carbonium ion salts are not usually isolated, the salt $CH_3CO^+Al_2Cl_7^-$ has been isolated.[23] With one bridging chloride, the $Al_2Cl_7^-$ anion has a C_{3v} geometry.

While only one oxide of aluminum is known, it exists in two important forms. The so-called γ-form of Al_2O_3 is man-made and has a somewhat disordered arrangement of aluminum ions surrounded tetrahedrally and octahedrally by oxide ions. Prepared by thermally dehydrating "hydrous oxides" of aluminum, **alumina** is characterized by small particle size and a fairly reactive nature. It readily absorbs water, is soluble in strong acids and bases, and is widely used as "chromatographic alumina" in laboratories. Doping γ-Al_2O_3 with iron and titanium oxides produces synthetic blue sapphire.

THE Al–O
FUNCTION

---

†The r_+/r_- ratios are (0.37, Al/F), (0.28, Al/Cl), (0.26, Al/Br) and (0.23, Al/I). The ideal r_h/r_ϱ for cubic close packed anions is 0.41 at O_h sites. This ideal is met well by Al^{3+} in a F^- sub-lattice, not too well for Cl^-, and not at all for Br^- and I^-. Thus, it is much easier for $AlCl_3$ than AlF_3 to pass into the gas phase as Al_2Cl_6. Again, in the trend from 6- to 4-coordination of Al^{3+}, we see a feature of main group chemistry determined by VSEPR/steric forces.

§This anion occurs naturally as cryolite (Na_3AlF_6), so important to the large scale electrolytic manufacture of Al by the Hall process.

Heating the γ-form above $1000°C$ leads to the more compact structure of the α-form. Here all aluminum ions are octahedrally surrounded by oxide ions, and each oxide bridges four aluminum ions. That there are too many cations per anion for a $CdCl_2$ type layer lattice is consistent with the hardness of Al_2O_3. Rather, two out of every three octahedral holes between oxide layers contain Al^{3+} in the fashion

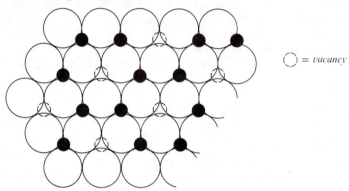

$\bigcirc = vacancy$

The α-form, as expected, is much less reactive than the γ-form and is very hard (you are probably familiar with the use of **corundum** as an abrasive material for grinding and polishing; corundum is the α-form of Al_2O_3).†

Solid solutions of Cr_2O_3/corundum constitute the gem ruby, which has the well-known red color at less than 8 per cent Cr_2O_3 but a green color at higher concentrations (Cr_2O_3 is green, while Al_2O_3 is white). The red color at low Cr^{3+} concentration seems to arise from the size difference between Al^{3+} and Cr^{3+}, the latter having a radius about 15 pm greater. In essence, Cr^{3+} ions are forced to occupy the smaller Al^{3+} holes in the Al_2O_3 lattice (Cr_2O_3 has an expanded corundum-like lattice). This causes the electronic transitions within the Cr^{3+} ion that are responsible for its green color to shift to higher energy and allow red light to be transmitted by the crystal (Chapter 9 explores the basis for the colors of transition metal ions).

STUDY QUESTIONS

9. The "closo" borane anion $(BH)_{12}^{2-}$ has an icosahedral structure. From a valence electron count, does it seem reasonable to you that the carborane $B_{10}C_2H_{12}$ exists? Of the three geometric isomers [(1,2), (1,7), and (1,12), where the numbers give the positions of the carbon atoms in the scheme below] possible for the carborane, two may be readily interconverted at $450°C$. Starting with the icosahedron below, discover which pair [(1,2 ⇌ 1,7), (1,2 ⇌ 1,12) or (1,7 ⇌ 1,12)] are interconverted simply by expansion of the icosahedron into a cuboctahedron and then relaxation back to an icosahedron. (Hint: see footnote, p. 175.)

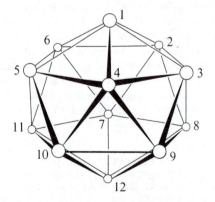

---

†The surface oxide responsible for rendering metallic aluminum unreactive is yet another crystal form, similar to that of NaCl, but with (as required by the 2:3 stoichiometry) every third cation missing.

10. Knowing that B_2H_6 and BH_4^- hydrolyze in aqueous acid, why do you suppose 85% H_3PO_4 (1:1 mole ratio H_2O/H_3PO_4) can be used in the synthesis of diborane? What is the likely experimental arrangement and order of mixing of the reagents in this synthesis?

11. Knowing that $Al_2Cl_7^-$ has a C_{3v} geometry, what do you conclude about the Al−Cl−Al angle? Is this expected for the central Cl^- sharing only two electron pairs with the two Al ions? Suggest a feature of the Al−Cl−Al bonding to account for that angle.

12. By analogy with carbon allotropes (diamond and graphite), sketch two structures for boron nitride, $(BN)_x$.

13. Compare the thermal yields of

$$CH_{4(g)} + 2O_{2(g)} \rightarrow CO_{2(g)} + 2H_2O_{(g)}$$

and

$$B_2H_{6(g)} + 3O_{2(g)} \rightarrow B_2O_{3(g)} + 3H_2O_{(g)}$$

on the basis of kJ/(g hydride). Would use of B_2H_6 as a substitute for natural gas be practical? (What are the physical properties of the products?)

14. Predict products for the following reactions.

a) $2NaBH_4$ + I_2 $\rightarrow 2NaI + ?$
   (reducing   (oxidizing
   Lewis base)   Lewis acid)

b) $B_2Cl_4 + C_2H_4 \rightarrow ?$ (the intermediate species derives from B_2Cl_4 acting as a π acid and C_2H_4 acting as a π base, after a simple structural change in B_2Cl_4)

c) $Me_2PH + \frac{3}{2}B_2H_6 \xrightarrow{\Delta} H_2 + ?$

d) $R_2BH + C_2H_4 \xrightarrow{\Delta} ?$ Would steric congestion be a factor in the molecular form of the *final* product?

e) the product from (d) + $H_2O \xrightarrow{H^+} C_2H_6 + ?$

15. Predict products for the following reactions.

a) $LiAlH_4 + 4RCHCH_2 \rightarrow ?$ (a Li^+ salt of 4-coordinate Al; related to question 14[d])

b) $LiAlH_4 + Me_3NH^+Cl^- \rightarrow H_2 + ?$

c) $LiAlH_4 + 4ROH \rightarrow 2H_2 + ?$

d) $4LiH + AlCl_3 \xrightarrow{Et_2O} 3LiCl + ?$

e) $Al_2X_6 + Al_2R_6 \rightarrow ?$ (product has bridging X)

Carbon and Silicon

Bond Energies for Reference

Si=Si	318	C≡C	837
Si−Si	226	C=C	611
Si−H	318	C−C	347
Si−C	305	C−H	414
Si−O	464	C−O	360
Si−F	598	C−F	490
Si−Cl	402	C−Cl	326

Having studied organic chemistry in some depth, you are already well aware of carbon chemistry. Our emphasis in this section will be on silicon, but at the end of this section you will find study questions on some less familiar "inorganic" carbon compounds. These questions will introduce those compounds by asking you to make predictions, comparisons, and interpretations based on structural, bonding, and thermodynamic ideas.

THE Si—Si FUNCTION

One of the important features of carbon chemistry is the propensity of the element to undergo catenation; therefore, an important question is why silicon does not seem to exhibit the same property, or at least why catenated silicon compounds are so much more reactive. In large measure, the reactivity of catenated silicon compounds with other compounds is due to the fact that silicon may experience valence shell expansion, a fact believed to have an important effect on the rates of nucleophilic attack on silicon compounds. This idea will be more fully developed in the next chapter; right now, we want to establish the range of useful Si functionalities.

The list of catenated silicon compounds is not particularly extensive at the moment. **Silanes** of formula Si_nH_{2n+2} are known[24] for $n \leq 8$, and the larger values of n correspond to greater branching of the Si backbone. For lower values of n, hydrosilanes, chlorosilanes, and organosilanes have been well characterized. Among the non-cyclic chlorosilanes[25] are the compounds $SiCl_4$, Si_2Cl_6, Si_5Cl_{12}, and Si_6Cl_{14}. Pyrolysis (at 1150°) of SiF_4 in the presence of Si leads to a polymeric solid $(SiF_2)_n$ via condensation of the "carbene-like" SiF_2 at lower temperature.[26] Controlled pyrolysis of this chain polymer provides a route to chain perfluorosilanes, Si_nF_{2n+2}, up to $n = 14$.

The thermal stability of the hydrosilanes, as a series, can be analyzed in terms of a Haber cycle (p. 173) for the reaction

$$Si_{n-1}H_{2n(g)} + Si_{(c)} + H_{2(g)} \rightarrow Si_nH_{2n+2(g)}$$

Elemental silicon possesses the diamond structure, in which each silicon atom is tetrahedrally bound to four other silicon atoms. The rupture of four Si—Si bonds to liberate one silicon atom, for insertion into an Si—Si chain, has a net effect of breaking only two Si—Si bonds. One such bond is to be formed when $Si_{n-1}H_{2n}$ expands to Si_nH_{2n+2}. Therefore, to a first approximation there is a net endothermicity to the above reaction of 226 kJ/mole from the breaking and making of Si—Si bonds. The other bond energy changes associated with this reaction are the endothermic breaking of the H_2 bond and the exothermic making of two Si—H bonds (a net change of -201 kJ/mole). For silane, SiH_4, this balance of bond breaking and bond making amounts to an endothermicity of 8 kJ/mole. The endothermicity increases by some 25 kJ/mole for each SiH_2 unit added. Consequently, all silanes, like neutral boranes, are unstable with respect to the elements; but SiH_4 and Si_2H_6 seem to be capable of prolonged stability at room temperature (a kinetic effect). By comparison, CH_4 has a heat of formation of -92 kJ/mole and a net *exothermicity* of 46 kJ/mole for each CH_2 unit of expansion. The low heats of formation of hydrosilanes are easily traced to the relative weakness of the Si—H bonds ($3sp$/H overlap not as good as $2sp$/H overlap) and not to the weakness of the Si—Si bond. Even though the Si—Si bond energy is some 120 kJ/mole less than that of carbon, relative to the hydrocarbons this actually favors the formation of the hydrosilanes. Nevertheless, an important conclusion is that the atomization of each Si_nH_{2n+2} requires much less energy than atomization of each C_nH_{2n+2} so that conversion of Si—H and Si—Si bonds to Si—X (X = halogen) bonds tends to be easier than the corresponding conversions in C_nH_{2n+2}.

An important aspect of silicon chemistry is the behavior of the hydrosilanes in oxidizing media. For example, combustion of Si_nH_{2n+2} homologs is exceedingly exothermic (and kinetically rapid, so that the hydrosilanes are pyrophoric); this is attributable to the high Si—O bond energy on one hand, and low Si—H and Si—Si energies on the other. Furthermore, using bond energies from Table 6-3 (p. 170), the hydrolysis reactions

$$MH_{4(g)} + 2H_2O_{(g)} \rightarrow MO_{2(g)} + 4H_{2(g)}$$

for carbon and silicon are, respectively, *endothermic* by 167 kJ/mole and 109 kJ/mole. As $SiO_{2(g)}$ is much less stable than quartz [the $Si=O$ double bond energy (640 kJ/mole) is much less than twice the $Si-O$ bond energy (464 kJ/mole), presumably in this case because the $Si-O$ "single" bond in quartz still has appreciable double bond character, so $SiO_{2(g)} \rightarrow SiO_{2(c)}$ is *exothermic* by 577 kJ/mole], the hydrolysis reaction to produce quartz is thermodynamically spontaneous with an estimated $\Delta E°$ of -469 kJ/mole.

Judging from the bond energies cited at the beginning of this section, you might surmise that halosilanes (except, as usual, for fluorides) and organosilanes tend to be thermodynamically quite unstable in combination with O_2 and hydroxylic solvents. While this is true in principle, in practice the reactions of halo- and organosilanes with O_2 are kinetically slow so that both can often be handled in air; other than the fluorosilanes, the halosilanes vigorously solvolyze in hydroxylic solvents, e.g.,

THE Si–halogen FUNCTION AND SOLVOLYSIS

$$SiCl_4 + 4HOR \rightarrow Si(OR)_4 + 4HCl$$

The high stability of the $Si-O$ bond is responsible for the utility of **organosiloxane** polymer (silicones), which are prepared by hydrolysis of the organohalosilanes:

THE Si–O FUNCTION

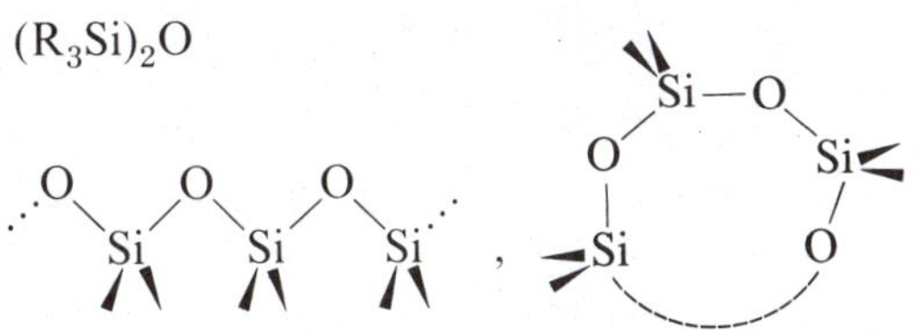

(chain)   *(cyclic)*

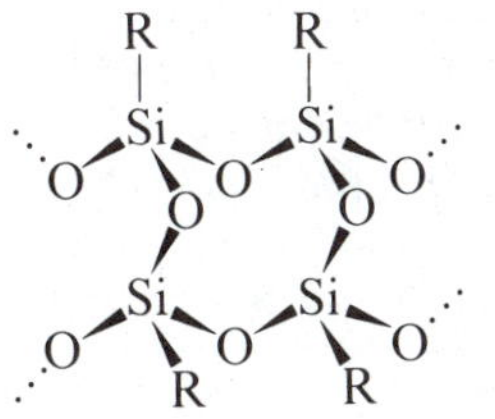

The dichlorosilanes are useful for forming linear chains of $Si-O$ backbones, the trichlorosilanes make possible the cross-linking of linear chains, and the monochlorosilanes are chain terminators. Consequently, hydrolysis of mixtures of all three produce polymers with properties (degree of cross-linking) dependent on the starting mixture.[27]

The polymerization reactions, of course, owe their occurrence to the instability of **silanols** (*cf.* boronic acids) presumably initially formed from the hydrolysis.

$$2R_3SiOH \rightarrow (R_3Si)_2O + H_2O$$

Note that this reaction is approximately thermally neutral and largely entropy-driven.

Nature achieves unsurpassed variety in its structural variations of binary combinations of silicon and oxygen. Such natural polymer chemistry achieves heights at which the synthetic chemist can only marvel. In the interest of correlating the structures of these polymers, we will limit your exposure to a few of the more important structural identities. The simplest natural binary combination of silicon and oxygen is, of course, SiO_2; in the gas phase, it has a structure like that of its congener, CO_2. However, owing to the low energy of two $Si=O$ bonds (640 kJ/mole each) relative to four $Si-O$ "single" bonds (464 kJ/mole each), SiO_2 undergoes self-association to form two important structural arrangements. **Cristobalite (7a)** is an orderly structure derived from the diamond structure of elemental silicon.† In ordinary hydrocarbon chemistry, adamantane **(7b)** exhibits the same 10-atom skeletal cage. Insertion of an oxygen atom between each pair of silicon atoms considerably expands (~50 percent) the Si frame, creating a cage with a large hole at the center. This openness is a characteristic feature of silicate materials and, as you shall see in a moment, it is important to the highly useful ion-exchange and molecular sieve (chromatographic) properties of these materials. A second structural feature you will see again and again is the six-membered ring of silicon atoms (where oxygen atoms are imposed between each Si—Si pair).

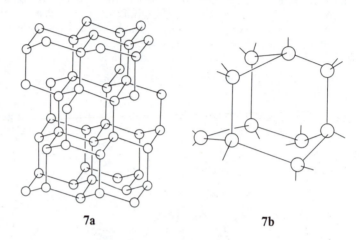

7a   7b

The more stable form of silica, **quartz**, has the intriguingly useful property of being an optically active polymer. This physico-chemical characteristic is a direct consequence of its structure. The very complicated arrangement of silicon and oxygen atoms (though each silicon is surrounded by a tetrahedral arrangement of oxygen atoms) is difficult to show without the use of three-dimensional models. The most direct way to describe the structure here is to imagine stacking springs so as to form the hexagonal arrangement shown in the center of **8**:

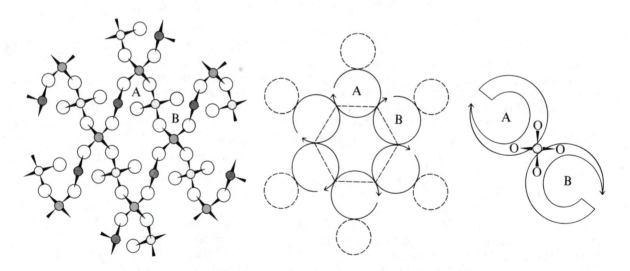

8   Views down the helical axis of quartz. Shaded circles represent Si atoms at three depths, with the darkest ones closest to you.

---

† See the unit cell perspective of zinc blende, structure **6** on p. 58.

The points of tangency of the springs define, in projection, a hexagon and locate the silicon atoms along the coils.† These are separated by oxygen atoms. Notice that three silicon and three oxygen atoms complete a 360° turn along a coil. These coils *must* all turn in the same direction in order for the silicon atoms to find themselves in a tetrahedral environment of oxygen atoms, and this imposes the optical activity feature upon the structure. Notice that the SiO_2 units manage to create rings of six silicons bridged by oxygen atoms, in spite of an obviously more intricate scheme of self-association than found in cristobalite.

An interesting situation arises for materials in which the silicon atoms of silicates are replaced by aluminum atoms. We may think of this either as replacement of Si^{4+} ions by Al^{3+} ions or as replacement of Si atoms by Al^- ions—the views are equivalent, for they maintain the number of valence electrons fixed. Because of the excess negative charges so generated, counter-ions must also be present to maintain electrical neutrality. Presumably any number of Si atoms may be replaced by Al, but only in certain materials containing equal numbers of Si and Al atoms are the structures related§ to those of silica. The quartz structure is so compact that the channels between helices can accommodate only counter-ions as small as Li^+, Be^{2+}, and B^{3+}. $LiAlSiO_4$ is an example. To accommodate larger cations, a more open structure is needed; this can be provided by the cristobalite structure, an example of which is $NaAlSiO_4$. Note in both of these examples, and in the ones to follow, that the ratio of (Al + Si) atoms to oxygen is 1:2. This is dictated by the fact that each derives from replacement of Si in an SiO_2 polymer by Al.

Of much more practical significance are minerals with more open structures. There are a great variety of these materials presenting a number of complicated structures. We will mention here one of the more symmetric structures found in **chazabite**, (Ca, Na_2)(Al_2Si_4)O_{12}·$6H_2O$. Remembering that an oxygen atom bridges each Al and Si atom, we can most simply describe the structure by drawing lines to connect the Al and Si atoms, with the basket-like result shown below.

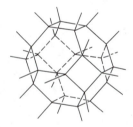

By counting the vertices (Al, Si) and edges (O) of this polyhedron, you can determine that the unit contains 24 (Al + Si) atoms and 48 oxygen atoms, and is simply an expanded form of the cuboctahedron (p. 176).

The linking together of these cuboctahedra to form a three-dimensional lattice is achieved by the 24 bridging oxygen atoms *exo* to the surface of the polyhedron. Considering the two most symmetric ways in which these "inter-cuboctahedra bridging" oxygens can be grouped leads to two well-known **zeolite** structures. Grouping the 24 bridging oxygens into four mutually tetrahedrally disposed hexagons leads to the **faujasite** structure {$NaCa_{0.5}$($Al_2Si_5O_{14}$)·$10H_2O$}, where the cuboctahedra take up a diamond-like structure (see **7** for a different perspective).

---

†Be careful to realize that the silicon atoms at these points are not, in fact, coplanar.

§Note that some distortion of the O−Si−O skeleton is inevitable because of the differing sizes of Al^{3+} and Si^{4+}.

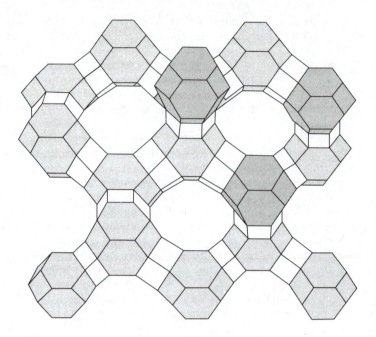

By considering the 24 bridging oxygens to be grouped into six octahedrally disposed squares, we are led to the structure of the synthetic zeolite called **Zeolite A** $\{Na_{12}(Al_{12}Si_{12}O_{48})\cdot27H_2O\}$.

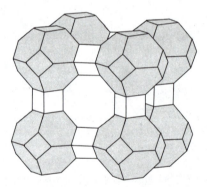

The baskets, as well as the prominent channels in these networks, provide for the chromatographic and ion exchange properties characteristic of this class of minerals. By now you should understand more clearly the idea behind calling such materials "molecular sieves."

STUDY QUESTIONS

16. Recall or look up the structure of graphite and predict the C−C bond order. Does your result fit with the fact that the C−C distance in graphite is a little greater than that in benzene?

17. a) Is the fluorination of graphite to produce poly(carbon fluoride) exothermic?

$$(C)_n + \frac{n}{2}F_2 \rightarrow (CF)_n$$

Which compound, graphite or $(CF)_n$, should be more stable to air oxidation to CO_2?

b) To be of advantage as a substitute for graphite lubricant, $(CF)_n$ must have what structural characteristic? Knowing that the hexagons of $(CF)_n$ have a "boat" structure, make a sketch of several fused "boats" (show the CF bonds, too) so as to compare the lamellar structure directly with that of graphite.

18. The *simpler* metal carbides are of three types with distinctly different hydrolytic behavior:

$$Be_2C \text{ or } Al_4C_3 \xrightarrow{H_2O} CH_4 + \text{metal hydroxide}$$

$$Na_2C_2 \text{ or } CaC_2 \xrightarrow{H_2O} C_2H_2 + \text{metal hydroxide}$$

$$Mg_2C_3 \xrightarrow{H_2O} C_3H_4 + Mg(OH)_2$$

$$SiC \xrightarrow{H_2O} \text{inert}$$

One of these is a covalent solid. Which? In the ionic solid cases, infer a structure for the carbon anion present. Carborundum (SiC) is an extremely hard material used as an abrasive. Identify a sketch in this chapter that describes its structure.

19. Construct π mo diagrams for the linear molecules C_nO_2 for $n = 1,3,5$ and for $n = 2,4$. (See Chapter 4, p. 101, for the linear chain mo pattern.) Which set yields free radicals? Only one of these sets (n even or odd) is known; which is it?

20. Account for the large $Si-O-Si$ angle of $\sim 150°$ in $(H_3Si)_2O$.

21. Is it surprising to you to learn that Me_3SiOH is as good a hydrogen-bonding acid as t-BuOH? If so, why and how can you rationalize this finding? If not, why not? Do you expect a large $Si-O-H$ angle?

22. Ph_3SiX reacts with bipyridine 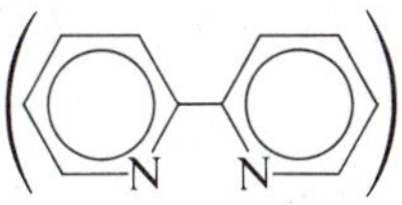 to form an ionic solid of formula $Ph_3SiX \cdot bipy$. Propose a geometry for the silicon cation; note and comment on any unusual aspect regarding the positions of the carbon and nitrogen atoms about silicon.

23. Make a "bed-spring" sketch of hypothetical racemic quartz (*i.e.*, similar to 8). With silicon ions at the points of tangency, satisfy yourself that the silicon is in a *planar* environment of four oxygens in racemic quartz, whereas coils of like symmetry produce a tetrahedral environment for silicon.

Nitrogen and Phosphorus

Bond Energies for Reference			
$N \equiv N$	946	$P \equiv P$	523
$N = N$	418	$P - P$	209
$N - N$	159	$P - H$	326
$N - H$	389	$P - C$	268
$N - C$	305	$P - O$	368
$N - O$	163	$P - F$	498
$N - F$	280	$P - Cl$	331
$N - Cl$	188		

Whereas the order of single bond energies for the Group IV elements is $C-C > Si-Si$, the converse appears to apply for the Group V elements, with $E_{N-N} < E_{P-P}$. A nominal

THE N−N
FUNCTION

value for the N−N single bond is 159 kJ/mole, while that for P−P is 209 kJ/mole.† Interestingly, this ordering of light and heavy atom single bond energies is maintained for the remaining groups of the representative elements. It has been generally associated with the first appearance at Group V of lone pairs on the neighboring atoms and may be attributed to lone pair-lone pair repulsions on the neighboring atoms. Whatever the reason, the ordering of single bond energies has an important connection with the number and types of catenated N−N and P−P compounds known to date. Hydrazine, the simplest catenated N−N compound, is unstable with respect to its elements, and the higher homologs become increasingly unstable. This, of course, derives primarily from the fact that the N−N single bond energy is only about one-sixth of the N≡N triple bond energy (946 kJ/mole).

In addition to N_2H_4, nitrogen is known to form the analogous fluoride N_2F_4, a gas with the interesting property of dissociating ($\Delta H \approx 84$ kJ/mole) into NF_2 radicals. The remaining examples of catenated nitrogen involve oxygen as a substituent: N_2O_4 is a dimer of NO_2 (N−N bond dissociation energy of ~63 kJ/mole; N_2O_2 is a dimer found in solid NO; and N_2O_3 may be thought of as the association product of NO + NO_2 (N−N bond dissociation energy of ~42 kJ/mole). All have weak N−N bonds and are highly dissociated at room temperature. A little later, we shall consider these oxides further.

THE P−P FUNCTION

These characteristics of catenated N−N compounds contrast sharply with the apparently greater stability of analogous P−P compounds. A good case in point is that elemental nitrogen has only one allotropic form, the N_2 molecule. Elemental phosphorus, on the other hand, exists in no fewer than three allotropic forms—all of which are solids at room temperature. All three forms are characterized by P−P single bonds. **White phosphorus (9)** consists of P_4 tetrahedra such that each P forms single bonds to the other three P atoms of the tetrahedron.

$$\begin{bmatrix} \text{waxy} \\ \text{m.p. } 44° \\ \text{b.p. } 287° \end{bmatrix}$$

P−−−|−−−P

P
P
9

Red phosphorus has an intriguing double layer structure[28] consisting of mutually perpendicular, connected, pentagonal tubes.

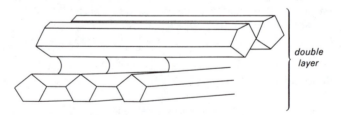

double layer

The pentagonal shape arises from the repeating pattern of fused bicyclo(3.3)P_8 units, as in **10**. Notice that the fused three-membered rings of P_4 have opened to a fused five-membered ring system. In **10** you notice also the insertion of a P atom at alternate bridgehead positions; these provide the "connectors" to the tubes of the partner layer.

---

† More striking is the comparison

B−B (301) < C−C (347) ≫ N−N (159) ~ O−O (142) ~ F−F (159)

Si−Si (226) ~ P−P (209) < S−S (264) ~ Cl−Cl (243).

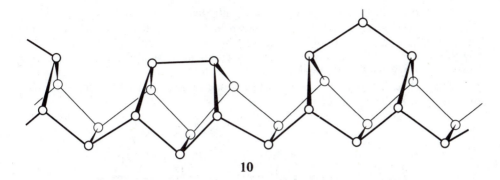

10

The most stable allotrope of the element, a flaky **black** solid, has a layer lattice, **11** (notice the occurrence of the "chair" six-membered ring, as in carbon and silicon), where each P is connected to three other P atoms via single bonds (the internal strain of the white and red allotropes is fully removed in this structure).

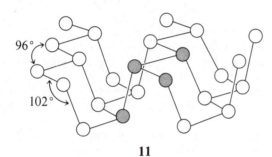

11

With a bond energy of only 523 kJ/mole for P_2 ($P\equiv P$), the $\Delta E°$ for the reaction

$$2P_{2(g)} \rightarrow P_{4(g)}$$

is nominally −209 kJ/mole while that for

$$2N_{2(g)} \rightarrow N_{4(g)}$$

is estimated to be +937 kJ/mole (with no allowance for internal strain), because $E\equiv E > 3E-E$ for nitrogen but $E\equiv E < 3E-E$ for phosphorus.

With regard to the relative stabilities of N=N and P=P bonds, we find a trend like that for C=C *vs.* Si=Si. You may have learned in organic chemistry that some (often explosively dangerous) $(-N=N-)_x$ chain and cyclic compounds are known. There is one inorganic analog of these, N_2F_2, called difluorodiazene. The stable azide ion (N_3^-) and unstable hydrazoic acid (HN_3) are more familiar to you; phosphorus analogs of all these materials are unknown.

Owing in part to the reversal (relative to the carbon/silicon trend) in order of single bond energies, phosphorus has yielded more catenated compounds than has nitrogen.[29] Thus, the chain molecules P_2H_4, P_3H_6, and P_4H_6 have been reported, as has P_2F_4. Several organopolyphosphines with chain $[P_2R_4, P_3(C_6H_5)_5, HP(CF_3)-P(CF_3)-PH(CF_3)]$ and cyclic $[P_4R_4, P_5(CF_3)_5,$ and $(PCF_3)_x$ for $x > 5]$ structures are known. A few diphosphines with oxygen substituents,[30] such as $HP_2O_5^{3-}$ (**12**) and $P_2O_6^{4-}$, add to the list of examples of the P−P link.

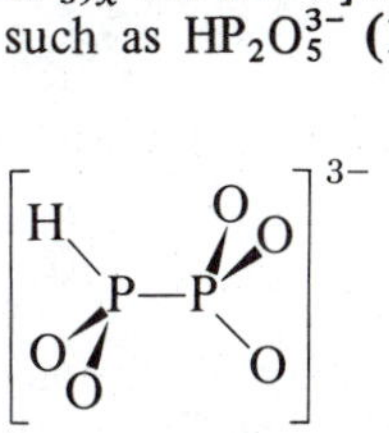

$$\left[\begin{array}{c} H \\ O \end{array} \underset{O}{\overset{O}{P}}-P \underset{O}{\overset{O}{}} \right]^{3-}$$

12

THE P–O FUNCTION

One further striking difference between nitrogen and phosphorus concerns their compounds with oxygen. Phosphorus is known to form two stable binary oxygen compounds: P_4O_6 and P_4O_{10} (more commonly written as P_2O_3 and P_2O_5), both of which have structures based on insertion of oxygen atoms into the P–P bonds of the tetrahedral P_4 unit **9** (and reminiscent of the Si ↔ cristobalite relation). As both structures below suggest, each phosphorus atom is single bound to three (in P_4O_6) or four (in P_4O_{10}) oxygen atoms in these structures.

As was mentioned in Chapter 3 in the discussion of bond energies, and also in the preceding discussions of Si–O bonds, these P–O linkages are most likely only formally "single" bonds. At this stage in your development as a chemist you know that no such oxides of nitrogen have ever been prepared; rather unusual reagents or conditions (and precautions!) will be necessary to do so, as the bond energy data allow you to estimate the heat of formation of N_4O_6 (with a structure analogous to that of P_4O_6) to be greater than +1450 kJ/mole.

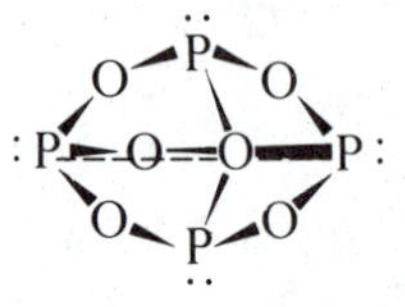

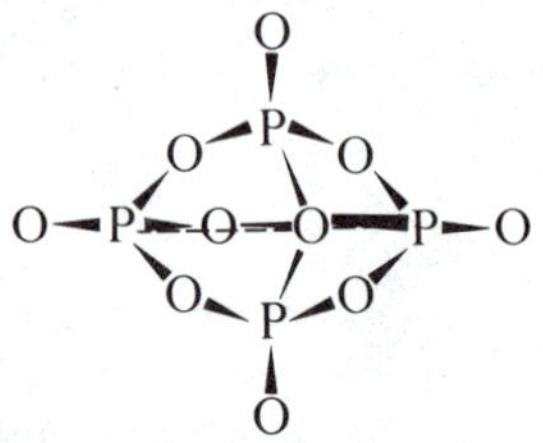

Related to these binary oxide structures for phosphorus are the structures of the phosphorus oxyacids and their esters. H_3PO_4 and its ester, R_3PO_4, present the phosphorus atom in a C_{3v} environment of oxygen atoms, one of which is the terminal type found in P_4O_{10}. In one sense, P_4O_6 is somewhat unique because phosphorous acid, H_3PO_3, and most of its esters do not have the simple C_{3v} structures $P(OR)_3$; rather, they also contain a terminal PO linkage so that their formulae are better written as $RPO(OR)_2$ **(13)**. This tendency extends to the lower phosphorus acid H_3PO_2 **(14)** (hypophosphorus

acid), which is only a monobasic acid. (The P–H hydrogens in H_3PO_3 and H_3PO_2 are not acidic.) In this connection it is interesting to note that the hypothetical compound $P(OH)_3$ has an estimated heat of formation (from Table 6-3) of -761 kJ/mole. Consequently, with a net endothermicity of 138 kJ/mole estimated for the breaking of an OH bond and formation of a PH bond, we can judge that the energy gained by the P=O dative bonding for the terminal oxygen must be in excess of 138 kJ/mole, a not unreasonable lower limit. In fact, one estimate of the PO_t bond energy is 544 kJ/mole, about 176 kJ/mole in excess of the P–O– bond energy. This tendency for terminal P=O formation

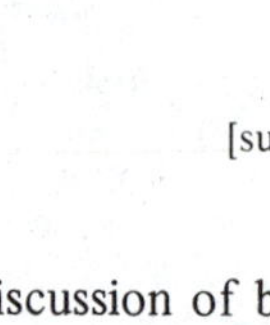

is further evident in the condensed (poly) phosphoric acids, $H\left(O-\overset{\overset{\displaystyle O}{|}}{\underset{\underset{\displaystyle OH}{|}}{P}}-OH\right)_n$ and phos-

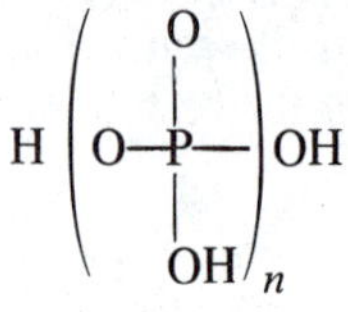

phoryl halides, $X_3P=O$.

We will complete this preliminary discussion of basic phosphorus compounds by noting that all the phosphorus trihalides, pentahalides, and oxyhalides vigorously solvolyze in hydroxylic solvents to form phosphoric and phosphorous esters, except for the fluorides. Refer to the bond energy terms at the beginning of this section and you will see that this is not unexpected. The unique role of fluorine among the halogens probably stems from a tendency toward P=F character, as with oxygen. Given this character for oxygen and fluorine, you might wonder whether PN bonds tend to show the same characteristic. In fact, the phosphorus halides and oxyhalides do solvolyze in secondary amines to form PN bonds,

$$X_3P{=}O + 6R_2NH \rightarrow (R_2N)_3PO + 3(R_2NH_2)Cl$$

and an interesting structural feature of these compounds is the planarity of the PNR_2 groups. This strongly implicates P=N bond character.[31]

Unlike phosphorus, which forms only two binary oxides, nitrogen forms many. The simplest, of course, is the colorless diatomic π^* radical NO, **nitric oxide**, which tends to dimerize weakly ($\Delta H \approx 17$ kJ/mole) in the liquid (b.p. $-152°$) and solid (m.p. $-164°$) states to $(NO)_2$ and is readily oxidized to the nitrosyl cation $(NO)^+$. The structure of the colorless[32] dimer is planar and cyclic (C_{2v}) with long (220 pm), weak bonds between the nitrogens. In the dimerization, the weak "N—N bond" energy and the planar structure are easily understood to arise from the half-occupied HOMO of NO. The constructive overlap of both the nitrogen p and oxygen p ao components of the π^* HOMO introduces both N—N and (to a lesser extent) O···O bonding and an ONN angle of ~100°.

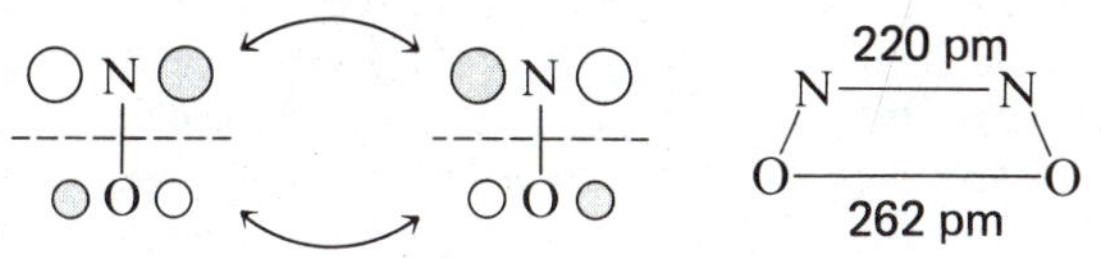

NO_2 (m.p. $-11°$, b.p. $21°$) is another familiar oxide of nitrogen. Like NO, **nitrogen dioxide** is a free radical (but a σ, not π, radical) with a slight tendency to dimerize to N_2O_4 and thereby lose its brown color. The N—N bond in this compound is weak (distance = 175 pm and bond energy ≈ 63 kJ/mole). The most stable structure of the dimer is planar.

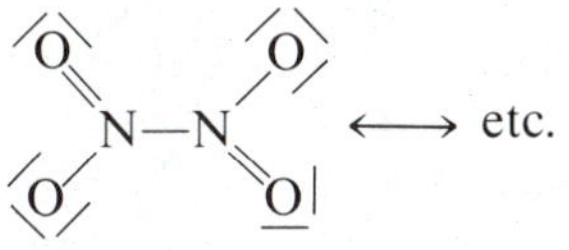

The planarity of N_2O_4, according to a full mo study,[33]

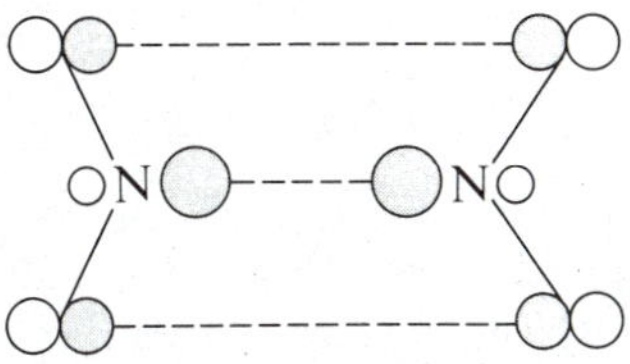

arises from the fact that the in-plane HOMO of NO_2 has appreciable O character, affording some long range bonding between *syn*-oxygens in the N_2O_4 mo otherwise characterizing the N—N bond.

Dinitrogen trioxide, which may be thought of as an association product of NO and NO_2, exists in a pure form only as a blue solid (m.p. -102°). In solution and as a liquid, extensive reversible dissociation into NO and NO_2 occurs (dissociation energy $\approx$42 kJ/ mole). The stable form[34] of this molecule has a weak N—N bond (C_s symmetry, distance $\approx$186 pm), but the interesting fact that nitrogen isotope studies[35] reveal rapid scrambling of nitrogen atoms

$$*NO + NO_2 \rightleftharpoons NO + *NO_2$$

may indicate the presence in fluid phases of a species (perhaps an intermediate) with an ONO—NO structure (*i.e.*, ONO—*NO $\rightleftharpoons$ ON—O*NO).

In addition to NO and NO_2 as oxides with an identifiable existence at room temperature, **nitrous oxide** (or dinitrogen oxide) is a well-known gas (b.p. -89°) with a $C_{\infty v}$ structure. Until recently, this and N_3^- were the only clear examples of stable compounds of N_2. N_2O seems to be the least reactive of the nitrogen oxides and has uses as an anesthetic (laughing gas) and as the propellant in aerosol cans of dessert toppings.

The least stable (tends to explode) of the nitrogen oxides is the white solid (sublimes 33°), **dinitrogen pentoxide**. An interesting property of this material is that its structure varies with phase. The solid is actually ionic, consisting of **nitronium** (NO_2^+) cations and **nitrate** anions, whereas the gas phase structure is that of a "dinitro ether" $[O(NO_2)_2]$.

A few oxyhalides of nitrogen are known; they have structures derived from the radical binary nitrogen oxides, NO and NO_2, or (alternately) from the cations NO^+ and NO_2^+. For example, the gases XNO (the **nitrosyl halides**, where X = F, Cl, Br) have structures of C_s symmetry as expected from overlap of an ao from X (or X^-) with the $\pi*$ HOMO of NO (or the $\pi*$ LUMO of NO^+).† The nitrosyl halides are very reactive oxidizing agents and, as the acid halides of HNO_2, vigorously solvolyze in water to produce **nitrous acid** (HONO). Similarly, the gases **nitryl fluoride** and **chloride** are conveniently viewed as combinations of X and NO_2 or of X^- and NO_2^+. Considering them as acid halides of HNO_3, the idea of heterolytic cleavage of the X—N bond most directly accounts for the solvolysis reaction:

$$XNO_2 + H_2O \rightarrow HX + HNO_3$$

One nitrogen oxyhalide seems to stand out as being related to the phosphorus oxyhalides: **trifluoramine oxide**, F_3NO, has a C_{3v} structure. It happens to be a kinetically stable, though toxic, gas (b.p. -88°) that is not easily hydrolyzed and is probably the most unreactive member of the class of amine oxide compounds (R_3NO). It shares with hydroxylamine the unusual characteristic of being an oxide of a tetrahedral nitrogen. (Note that while the terminal oxygen of F_3NO possesses lone pairs, the nitrogen atom does not; the Nlp—Olp interaction is not present to weaken the N—O link (see below).) The stability of F_3NO to nucleophilic attack is thought to derive from the steric shielding of the attack site (N) by the F and O atoms. Additionally, NO multiple bonding might confer extra stability to the bond:

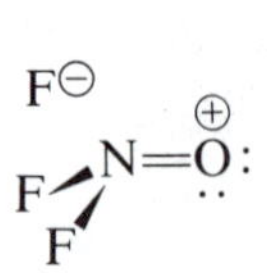

A consistent structural thread runs through all the N/O compounds discussed thus far: with the exception of R_3NO and R_2NOH compounds, at least one oxygen is bound to nitrogen via multiple bonds. This feature is no surprise when we compare the strength

---

†Both views are supported by chemical evidence; neat ClNO and BrNO are slightly dissociated into X_2 + NO, while in solution they often undergo reaction via X^- + NO^+.

of the N=O bond to that of the N—O single bond (~586 $vs.$ ~167 kJ/mole). In comparison with C—O (~356 kJ/mole) and C=O (~753 kJ/mole) bonds, however, it appears that both N=O and N—O bonds are ~175 kJ/mole weaker.

To understand the "low" N/O bond energies, we must inspect more closely the compounds from which these bond energies are derived. As is suspected to be the case with hydrazine, the N—O lp/lp repulsions of **hydroxyl amine** could appreciably lower the bond energy in comparison to the CO bond of, say, methanol.

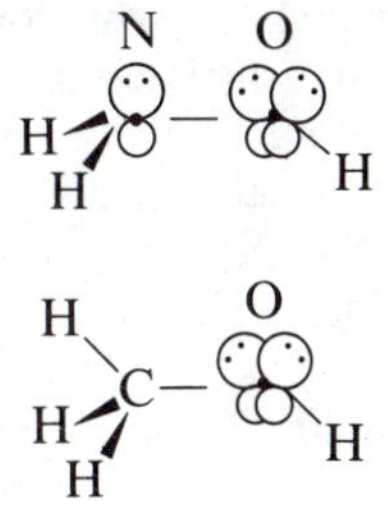

An analogous lp/lp versus bp/lp relation arises for XNO and the organo carbonyl function:

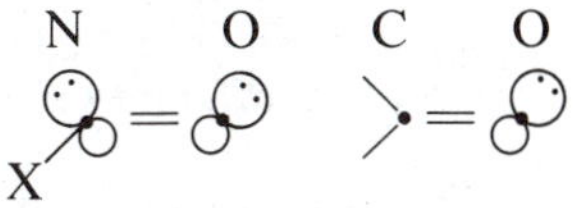

The characteristic ⟍N=O and ⟩N=O structural features for the oxides and oxyhalides are, of course, also observed in the oxyacids formed by nitrogen. The two most commonly known oxyacids of nitrogen are **nitrous** (HNO_2) and **nitric** (HNO_3). The admittedly unusual view of HNO_3 as an adduct of OH^- and NO_2^+ is of some value in systematizing the nitrating power of HNO_3 as a liquid or in strongly acidic and/or ionizing media. Sulfuric acid apparently dehydrates HNO_3 to form NO_2^+ according to the reaction

$$HNO_3 \xrightarrow{H_2SO_4} NO_2^+ + HSO_4^- + H_2O_{(solvated)}$$

At this point it is well to recall from p. 198 that N_2O_5 is the anhydride of HNO_3 and is known to exist in the ionic form $NO_2^+NO_3^-$.

Nitrous acid is considerably less stable than nitric acid, in the sense that acid solutions tend to disproportionate into NO and HNO_3 (note that there is no change in the number of OH bonds, but an increase in N=O π bonding—and an associated loss of effective N lp/O lp repulsions—is achieved in this reaction)

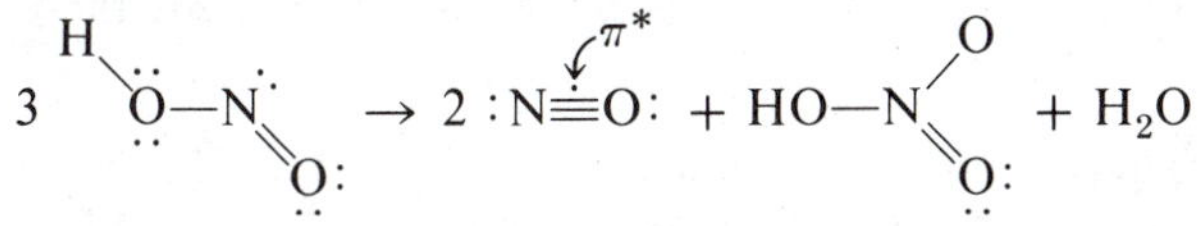

$$3\pi_{NO} \rightarrow 4\pi_{NO}$$

$$3 \text{ lp/lp} \rightarrow 0 \text{ lp/lp}$$

Consistent with this behavior is the fact that HNO_2 is both a good oxidizing agent and a good reducing agent, forming NO in the first instance and NO_3^- in the second.

The structure of HNO_2 shown in the preceding equation is easily related to that of the nitrosyl halides if we take the point of view that HNO_2 is an adduct between OH^- and NO^+. This means, of course, that the acidic proton is bound to oxygen, not to nitrogen, and the molecular symmetry is C_s. (While an N—H structure would eliminate N lp/O lp repulsion, such stabilization occurs at the expense of converting O—H to N—H.) In contrast with this structure, it is an interesting property of NO_2^- that it can act as a Lewis base toward metal ions via either N-coordination or O-coordination. Analogy is found, then, with organic nitro and nitrite compounds.†

THE N—halogen AND P—halogen FUNCTIONS

Our survey of comparative nitrogen and phosphorus structures is not complete until we note that both elements form **trihalides**, while only phosphorus forms **pentahalides** (except with iodine). The nitrogen trihalides are rather unimportant as *common* laboratory materials because of their unpredictable nature. The least dangerous to handle (with regard to explosion hazards) is the gas NF_3. The infamous solid, "nitrogen triiodide," is actually formulated as $NI_3 \cdot NH_3$; it is safe to handle in aqueous solution but is shock sensitive as a solid. The black color of the crystals forecasts a rather unusual structure:[36] chains of NI_4 tetrahedra built by bridging iodine atoms. The presence of NH_3 molecules serves to link (and stabilize?) the chains via hydrogen bonding.

STUDY QUESTIONS

24. Using resonance structures and bond order arguments, predict which, NaN_3 or HN_3, has an N_3 structure more distorted toward N_2 and therefore which is more hazardous with respect to decomposition to produce N_2. Does knowing that both r_{NN} are the same (= 116 pm) in N_3^- but different (124 pm, 113 pm) in HN_3 relate to your theoretical arguments?

    Which in each pair is more likely to detonate?

    $$RbN_3, \ AgN_3$$
    $$Hg(N_3)_2, \ Ba(N_3)_2$$
    $$KN_3, \ CuN_3$$

25. The problem of chemical conversion (nitrogen "fixation") of N_2 to NH_3 is a significant one that illustrates the importance of thermodynamic concepts. Consider the relative ease of the steps

    $$N_2 \xrightarrow{H_2} N_2H_2 \xrightarrow{H_2} N_2H_4 \xrightarrow{H_2} 2NH_3$$

    Using NN bond energies, which step is most costly (in terms of energy), and which of the species N_2, N_2H_2, or N_2H_4 therefore requires the greatest chemical activation (either by extreme temperature conditions or by catalytic combination) so as to weaken the NN link? The first three species exhibit a lone pair on each nitrogen. How does the lp/lp repulsion increase for $N_2 \rightarrow N_2H_2$ compare with that of $N_2H_2 \rightarrow N_2H_4$?

26. Which compound, NF_3 or NCl_3, is more unstable with respect to reversion to the elements? Are both unstable in this sense? Which presents the greater hazard in handling?

27. How might you explain why the fumes generated by HNO_3 oxidation of Cu are colorless for a few millimeters above the solution surface, but brown thereafter?

28. With regard to the structure of $N_2O_{5(g)}$, would a linear N—O—N geometry favor a D_{2h} or D_{2d} (twisted NO_2 groups) structure?

---

†Were there any evidence for a C_{2v} structure for HNO_2, we could point out an analogy with phosphorous acid.

29. Is nitric oxide thermodynamically stable with respect to nitrous oxide and nitrogen dioxide?

	NO	N_2O	NO_2
ΔH_f°	+92	+79	+33

30. Since there is no difference in the number of P−P bonds per phosphorus atom in the white and black allotropes of the element, why is the former so much more (a) volatile and (b) reactive than the latter?

31. Estimate the enthalpy of

$$2P_2R_2 \rightarrow (PR)_4$$

in order to identify whether the cyclophosphine may be produced during attempts to synthesize compounds containing −P=P−.

32. Diphosphine, a pyrophoric volatile liquid, is photosensitive, yielding P_4 and phosphine

$$3P_2H_4 \xrightarrow{h\nu} 4PH_3 + \frac{1}{2}P_4$$

Argue that this reaction is approximately thermally "neutral" ($\Delta E \approx 0$) but free-energy spontaneous.

33. Draw resonance structures for HPO_3 (obtained from thermal dehydration of H_3PO_4 at $\gtrsim 300°$) and the conjugate base PO_3^-.

Oxygen and Sulfur

Bond Energies for Reference

O=O	498	S=S	431
O−O	142	S−S	264
O−H	464	S−H	368
O−C	360	S−C	(272)
O−F	213	S−F	(285)
O−Cl	205	S−Cl	272
		S−O	?
		S=O	523

The trend in relative second and third row element single bond energies is carried on by the members of Group VI: O−O, 142 kJ/mole, and S−S, 264 kJ/mole. On the other hand, the (formal) double bond energies for O=O and S=S are 498 kJ/mole and 431 kJ/mole, respectively, an order you should expect by now from the properties of the preceding groups. These data in themselves reflect the general instability of catenated oxygen compounds and the much wider occurrence of catenated sulfur compounds. For example, even the elemental forms mirror this different propensity for multiple bond formation by oxygen and catenation by sulfur. Oxygen exists in two allotropic forms, O_2 (dioxygen) and O_3 (ozone), but not as cyclic O_n. Both allotropes are characterized by O−O multiple bonding. Sulfur, on the other hand, is found naturally in the form of S_8 rings of a D_{4d} crown structure (15) but S_2 is found only at high temperatures in the gas phase.

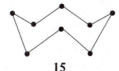

15

THE O—halogen FUNCTION

The simple **halides** of oxygen include the fluoride, OF_2, and chloride, OCl_2. The former is a highly toxic compound of marginal stability and high oxidizing power,

$$\frac{1}{2} O_{2(g)} + F_{2(g)} \rightarrow OF_{2(g)} \qquad \Delta E^\circ = -17 \text{ kJ/mole}$$

$$OF_2 + 2H^+ + 4e^- = H_2O + 2F^- \qquad E^\circ = 2.1 \text{ v}$$

while the chloride violently decomposes on heating:

$$\frac{1}{2} O_{2(g)} + Cl_{2(g)} \rightarrow OCl_{2(g)} \qquad \Delta E^\circ = 84 \text{ kJ/mole}$$

The high reactivities of both compounds could be attributed to oxygen-halogen lone pair repulsions, to the stability of O_2 (where it is a product), and to the stability of oxygen and fluorine or chlorine bonds to other elements.

THE O—O FUNCTION

An understanding of the terminal group properties desirable for stabilizing oxygen catenation is emerging. Certainly, the terminal groups must form strong bonds to oxygen, as do H and R, and as F probably would were it not for lone pairs. Based on Bent's idea of bond strength and s orbital distribution among hybrids (Chapter 3, p. 81), high terminal group electronegativity should help strengthen the O—O bond. You should also expect that oxygen lone pair involvement in π bonding with the substituent would be of primary value to strengthening the O—O bond.

Organic peroxides are known to be generally dangerous compounds. Particularly hazardous are the hydroperoxides (HO_2R), commonly found in the laboratory in bottles of ethers possessing α H atoms (Et_2O, C_4H_8O). It is an interesting observation, however, that fluorocarbon peroxides tend to be more stable than their hydrocarbon analogs. This observation has led to the preparation of a few trioxides [$(CF_3)_2O_3$, $(C_2F_5)_2O_3$, and $CF_3O_3C_2F_5$] and even a mixed dioxide-trioxide ($CF_3—O_3—CF_2—O_2—CF_3$). One must appreciate, however, that these still are not routine reagents; they most often are highly unstable (thermodynamically and kinetically) with respect to reactions with oxidizable compounds and sometimes to thermal and shock-initiated decomposition. They are best handled in small quantities at low temperatures.

In addition to "stabilization" by perfluoro groups as termini, terminal groups such as SO_3, FSO_2, and SF_5 seem to make possible the preparation of peroxy compounds. Specific examples are

$CF_3O_2—SF_5$

$HO_2—SF_5$

$HO_2—SO_3H$ (commonly called Caro's acid)

$FSO_2—O_2—SO_2F$ (commonly called peroxydisulfuryldifluoride)

$CF_3—O_2—SO_2F$

$SO_3—O_2—SO_3^{2-}$ (commonly called peroxydisulfate)

While the relation between CF_3, for example, and sulfur in the +6 oxidation state may seem obscure, it is reasonable to view the "stabilization" of the O—O link by these groups as arising from their conjugation with the oxygen lone pairs so as to confer some bonding character to the lone pair. Perhaps more important may be the fact that, with reduced lp/lp repulsion, better advantage may be taken of the O—O σ orbital overlap, which improves with electronegative termini.

$$-\ddot{\text{O}}-\ddot{\text{O}}-\text{SF}_5 \longleftrightarrow -\ddot{\text{O}}-\overset{\oplus}{\ddot{\text{O}}}=\overset{\ominus}{\text{SF}}_5$$

THE S—S FUNCTION

In contrast to the exciting (!) chemistry associated with oxygen catenation, sulfur has a rich, mild chemistry of catenation. As there is no change in the number of H—S and S—S bonds for the reaction

$$\text{H}_2\text{S}_{(g)} + \frac{n-1}{8}\,\text{S}_{8(g)} \rightarrow \text{H}_2\text{S}_{n(g)}$$

it should not surprise you that the hydrogen persulfides (**sulfanes**) have been well characterized.[37] Similarly, the **polysulfide anions** have been studied in aqueous solutions and the compounds from Na_2S through Na_2S_5 have been identified. Again, because there is no net change in the number of S—S bonds, reactions such as the following are nearly thermally neutral:

$$\text{S}_4^{2-}{}_{(aq)} + \text{S}_{(c)} \rightarrow \text{S}_5^{2-}{}_{(aq)}$$

In addition to the anionic polysulfides, recent studies have shown the existence of **cationic polysulfides**. Working with highly acidic, non-reducing solvents (this use of HF and HSO_3F as solvents was forecast in Chapter 5), Gillespie[38] has succeeded in preparing and characterizing the species S_8^{2+}, S_{16}^{2+}, and S_4^{2+}.

Two-electron oxidation of S_8 by the peroxide $\text{S}_2\text{O}_6\text{F}_2$ apparently leads to some degree of bond formation between S atoms opposite each other in the crown ring (recall the allotropes of P, with which S^+ is isoelectronic). Because of ring strain, the formation of the "bicyclo" structure necessitates the conversion of the crown structure (15) to the *endo-exo* form (16) observed for S_8^{2+}.

16

Incomplete oxidation of S_8 apparently leads to the formation of paramagnetic S_8^+, which tends to dimerize to S_{16}^{2+}. The structure of this dication may be simply that of a dimer of two S_8 crowns connected by an S—S bond, $(\text{S}_8)_2^{2+}$. Strong oxidation of S_8^{2+} can lead to isolation of the S_4^{2+} cation, which has a planar structure and is diamagnetic. Interestingly, this molecule can be described in terms of S=S bonding. The π out-of-plane mo's of this species are, of course, those characteristic of the cyclic four-atom case (Chapter 4), with a net, single π-bonding electron pair.

In summary, one of the striking features of the chemistry of elemental sulfur is its propensity for forming cyclic species.† Whereas the S_8 ring can be opened on reduction

---

†Other all-sulfur ring structures are known (S_{6-10} and S_{12}); all are unstable at room temperature (the decomposition reactions are favored by entropy and internal ring strain energy) and can be retained only at low temperatures.

of sulfur to form a variety of chain-like polyanions, oxidation of S_8 causes oligomerization of the rings in the formation of S_{16}^{2+}, introduces cross-ring bonding in the formation of S_8^{2+}, and results in symmetrical cleavage of the ring to form the S_4^{2+} cation.

The examples of sulfur catenation go far beyond those mentioned thus far. Fluorine is known to form only the simpler catenated species such as the gas S_2F_2 and the liquid **disulfurdecafluoride**, the latter featuring sulfur in the +5 oxidation state. The structure[39] of S_2F_{10} is, as expected, based on the octahedron, with the sixth position about each sulfur being occupied by the other sulfur atom.

An intriguing situation arises for **disulfurdifluoride**. The more stable isomer of this compound has the C_s structure[40] **17** while the less stable isomer, and probably the one most of us would have "intuitively" expected to be the more stable, has the normal (C_2) peroxide structure **18**.

17

18

The thermodynamics of the isomerization reaction

$$XSSX \rightarrow X_2SS$$

do not depend on the magnitude of E_{XS} (if we assume E_{XS} to be the same in both structures). That is,†

$$\Delta E = (E_{S-S}^{XSSX} - E_{S=S}^{X_2SS}) = -167$$

Thus, it seems reasonably consistent for nature to prefer the "sulfoxide" structure **17**. Why, then, H_2S_2 and Cl_2S_2 have never been reported to exhibit the "sulfoxide" structure is unexplained.

In the previous discussions of peroxide compounds we learned that complex oxysulfur(VI) groups apparently act as stabilizers of the peroxide linkage. A similar characteristic is found for polysulfur linkages. Thus, the **sulfanemonosulfuric** acids and their salts have been prepared; however, HS_nSO_3H and salts of $S_nSO_3^{2-}$ are not very stable at room temperature. The simplest member of this series is well known as **thiosulfate**; while its conjugate acid is not particularly stable, you are well aware that sodium thiosulfate is available at any store dealing in photographic supplies.

A simple extension of the ideas in the previous paragraph should lead you to expect the existence of **polythionate** anions, and indeed many ions of the type $O_3S-S_n-SO_3^{2-}$ are known in alkaline solutions (the acids are unstable). The best characterized members of this series are those with $n = 0$ through 4 (dithionate, trithionate, etc.); the simplest, dithionate, is best known and least reactive of the series.

---

†The use of 167 kJ/mole from $(E_{S=S} - E_{S-S})$ is an approximation, since the S=S in X_2SS is somewhat different from that in S_2

Both the S–S σ and π bonds should be weaker than the nominal values used to arrive at 167 kJ/mole, making it difficult to estimate by how much $(E_{S=S} - E_{S-S})$ may be in error.

At least formally related to the dithionate ion is the **dithionite** ion, $S_2O_4^{2-}$ (the conjugate base of dithionous acid, $H_2S_2O_4$). While aqueous solutions of the anion are not very stable for extended periods,† when fresh they make good scavengers for O_2 impurities in gases (Fieser's reagent is a solution of dithionite and 2-anthraquinone). The ease with which $S_2O_4^{2-}$ can be oxidized is probably related to the fact that the ion dissociates into SO_2^- radicals (solutions of $S_2O_4^{2-}$ are paramagnetic). The C_{2v} structure[41] **(19)** of the dithionite ion is intriguing, especially since the O—S—O planes deviate from parallelity by only $30°$, compared to $70°$ expected for tetrahedral sulfur atoms. (Study question 40 leads you through an accounting of this structure.)

$$\left[\begin{array}{c} S\!\!-\!\!S \\ \diagdown\;\;\diagdown \\ O\;\;O\;\;O\;\;O \end{array} \right]^{2-} \qquad 19$$

The final example of catenated sulfur is a hybrid of the dithionite and dithionate ions. **Disulfite** ion, $S_2O_5^{2-}$, plays an important role in the aqueous chemistry of sulfurous acid, arising as it does in concentrated solutions of bisulfite by elimination of H_2O from two molecules of HSO_3^-. Similarly, thermal dehydration of bisulfite salts produces disulfite salts. A convenient view of this derives from the known donor/acceptor reaction

$$SO_2 + SO_3^{2-} \rightarrow S_2O_5^{2-}$$

Easily the most important acid of sulfur is H_2SO_4. The existence of sulfurous acid (H_2SO_3), on the other hand, has never been proved; rather, SO_2, which is highly soluble in water, simply forms a hydrated species much as CO_2 does. The bisulfite ion (HSO_3^-), however, is well known and exhibits an interesting tautomerism like H_3PO_3. Two species are believed to be in equilibrium in dilute aqueous media:

THE S—O
FUNCTION

$$\left[\begin{array}{c} \ddot{S} \\ O\;\;O\;\;O\!\!-\!\!H \end{array} \right]^{-} \rightleftharpoons \left[\begin{array}{c} H \\ | \\ S \\ O\;\;O\;\;O \end{array} \right]^{-}$$

As mentioned in the preceding paragraph, the formation of detectable amounts of disulfite ion results from an equilibrium ($pK \sim 1$) requiring $2HSO_3^-$, but the mechanism of the reaction seems not to have been determined.

Judging from examples given already, the S—O bond is an important feature of sulfur chemistry. The two best known simple oxides are SO_2 and SO_3. Both molecules are believed to involve strong $p\pi$–$p\pi$ multiple bonding between sulfur and oxygen, which is necessary to avoid valency unsaturation at sulfur; the persistence of strong, short S—O bonds in **thionyl** ($>\ddot{S}{=}O$) and **sulfuryl** ($>SO_2$) compounds is generally taken to implicate some degree of multiple bond character in these bonds, too. In fact, the remarkably small variation in S—O distance (143-150 pm) for a wide variety of SO compounds seems to imply a characteristic constant multiple bond nature for those bonds.

Sulfur trioxide is an extremely strong Lewis acid, a fact behind its self-association in the solid state. Three crystalline forms of SO_3 are known. Rapid condensation of the gas at low temperatures ($< -80°C$) leads to cyclic trimers with a melting point of about $16°C$ (γ-form). The more stable β-form consists of helical chains of —O—S—O—S backbones. The form with the highest melting point (α, $60°C$) has an unknown structure, but, like

---

†The products are $S_2O_3^{2-}$ and HSO_3^-.

THE S—halogen FUNCTION

the β-form, has an asbestos-like appearance and so may consist of cross-linked chains. (The γ-form melts to a liquid of monomers and trimers and boils [45°C] at temperatures below the melting point of the α-form.)

For the simple **halides** formed by sulfur, we find that the red liquids SCl_2 (b.p. 59°) and SCl_4 (m.p. −31°) and the gases SF_4 (b.p. −40°) and SF_6 (sublimes −64°) are known (S_2F_{10} and S_nX_2 were mentioned earlier). The initially surprising non-existence of SF_2 is often rationalized in thermodynamic terms by a highly favorably energy for the disproportionation reaction,†

$$3SF_{2(g)} \rightarrow \frac{1}{4}\,S_{8(g)} + SF_{6(g)} \qquad \Delta E \approx -544\text{ kJ/mole}$$

even though SF_2 should be quite stable with respect to the elements:

$$\frac{1}{8}\,S_{8(g)} + F_{2(g)} \rightarrow SF_{2(g)} \qquad \Delta E \approx -146\text{ kJ/mole}$$

It is interesting to note, after this, that SF_4 does not disproportionate into S and SF_6, though the reaction should be thermodynamically spontaneous. Similarly, SCl_4 does not disproportionate into S and SCl_6 §, but then neither does SCl_2. The failure of both SF_4 and SCl_4 to disproportionate must be kinetic in origin, although for chlorine there could be a general thermodynamic barrier to formation of SCl_6. It is supposed that steric congestion about S in SCl_6 may be too severe. All of this should still leave some questions in your mind. How does SF_2 disproportionate to SF_6 without also producing SF_4? Why does SCl_2 not decompose to S_8 and SCl_4? An interesting complication arises from the observation that SCl_2 does slowly decompose to S_2Cl_2:

$$2SCl_2 \quad \rightarrow \quad S_2Cl_2 \quad + Cl_2$$
$$\text{[red liquid]} \qquad \text{[yellow liquid; b.p. 137°]}$$

At this point we can only conclude that some interesting mechanistic constraints are operating.

The remarkable property of SF_6 is its chemical inertness. The compound does not experience acid or base hydrolysis near room temperature, even though the reaction is expected to be strongly spontaneous from the thermodynamic point of view. SF_6 gas is so safe to handle and so stable that it is used as an insulator in high-voltage electrical equipment. One facile reaction of SF_6 is electron transfer; metallic sodium, in glycolic solvents, effects the reduction of SF_6 to sodium fluoride and sodium sulfide.

Of greater chemical utility is SF_4. While the hydrolytic stability of SF_4 is low, as expected, it is a very useful fluorinating agent. Generally speaking, the compound can be used to convert metal and non-metal oxide compounds into the corresponding fluorides without destroying other groups or oxidizing the metals and non-metals. Particularly useful is the fluorination of keto ($>C=O$) and phosphoryl ($>P=O$) groups.

---

†Note that this reaction is exothermic, no matter what the value of E_{S-F}, unless that value should be greatly different for SF_2 and SF_6.

§In this regard, SCl_4 is quite a bit less stable than its analog, SF_4. The tetrachloride is synthesized by the low temperature (−80°C) chlorination of SCl_2 but decomposes to the lower chlorides at $\sim$ −31°C.

As we found to be the case for phosphorus, sulfur is known to form mixed oxy-halide compounds. The simple molecular compounds are the **thionyl halides** [(F, Cl, or Br)$_2$SO, with C_s symmetry] and the **sulfuryl halides** [(F or Cl)$_2$SO$_2$. with C_{2v} symmetry]. In striking analogy with other non-metal halides and oxyhalides, the thionyl and sulfuryl fluorides are relatively stable to hydrolysis, whereas thionyl chloride and bromide, and sulfuryl chloride, are vigorously solvolyzed by water and other hydroxylic solvents. The terminal SO bonds in these compounds are remarkably similar to SO bonds in other compounds and are believed to possess considerable multiple bond character. All are stable to decomposition at normal temperatures.

A derivative chemistry of SF$_4$ is beginning to develop, with the existence of OSF$_4$ (C_{2v} symmetry) and SF$_4$(OSF$_5$)$_2$ having been established. It has been confirmed, by the formation of R$_4$N$^+$SF$_5^-$ and SF$_3^+$BF$_4^-$, that SF$_4$ has acid and base properties. In this connection, note that OSF$_4$ may be thought of as an adduct of SF$_4$ and O, much like the amine and phosphine oxides from Group V.

To close this section on Group VI compounds, we might point out a further analogy with the chemistries of silicon and, particularly, phosphorus. **Esters** of sulfurous and sulfuric acids are readily achieved by solvolysis of thionyl chloride and sulfuryl chloride:

$$X_2SO_n + 2ROH \rightarrow (RO)_2SO_n + 2HCl \qquad \begin{matrix} n = 1 \text{ thionyl} \\ n = 2 \text{ sulfuryl} \end{matrix}$$

In the case of sulfurous esters, the tautomerization reaction (seen before for HSO$_3^-$) to produce a second terminal SO bond is found

$$(RO)_2SO \rightarrow (RO)SO_2(R)$$

(Trifluoromethyl)sulfuric acid, with properties much like those of HSO$_3$F, possesses such a structure.

An unfortunate situation in the nomenclature of sulfur-oxygen compounds has arisen over the years, and it seems best to pause for a moment to identify the different usages. The $>\ddot{S}-O$ group is variously referred to as **thionyl**, **sulfoxide**, and **sulfinyl**, while $>SO_2$ goes by the names **sulfuryl**, **sulfone**, and **sulfonyl**. All are so commonly used that they need to be committed to memory.

---

34. Compare the approximate energies for

$$O_2F_2 \rightarrow O_2 + F_2$$

$$S_2F_2 \rightarrow \frac{1}{4}S_8 + F_2$$

Is a distinction to be made in the care with which the difluorides must be handled? Which of the two difluorides is potentially more useful for *selective* fluorination reactions?

35. Generalize in equation form the energy of

$$XO_nX \rightarrow XO_2X + \frac{n-2}{2}O_2$$

What is the prognosis for preparation of polyoxide materials?

STUDY QUESTIONS

36. Would decomposition of RO_2H more likely produce $(RH + O_2)$ or $(ROH + \frac{1}{2}O_2)$?

37. Comment on the order of thermochemical bond energies $E_{O-O} < E_{O-F}$ with respect to lp/lp repulsions.

38. Why should S_8^{2+} prefer the bicyclo(3.3) rather than bicyclo(1.5) or bicyclo(2.4) structures?

39. a) Use Pauling's defined relation between bond energy and atom electronegativity (see Chap. 3, App. B) to estimate E_{S-Br} and E_{S-I}:

$$\Delta\chi = \frac{1}{23}[E_{S-X} - (E_{S-S} \cdot E_{X-X})^{1/2}]$$

From these, estimate the enthalpy change for

$$\frac{1}{8} S_{8(g)} + X_{2(g)} \rightarrow SX_{2(g)}$$

b) Were you to attempt synthesis of SBr_2 or SI_2, would there be cause to anticipate product disproportionation to S_8 and SX_6?

40. SO_2^- is a π radical; make a sketch of the half-occupied π HOMO. Now use the HOMO sketch to analyze the dimerization of SO_2^- to account for the following structural characteristics of dithionite: long S—S bond of 240 pm, nearly parallel SO_2 planes, and eclipsed (C_{2v}) geometry. Is there an analogy here with $(NO)_2$ and N_2O_4 (p. 197)?

41. From donor/acceptor concepts predict possible structures for $S_2O_5^{2-}$. Identify the donor and acceptor atoms in each case.

42. By analogy with the hydrolysis of SO_3, propose a synthesis method for HS_nSO_3H.

43. Propose a mechanism for the incorporation

$$S_2O_4^{2-} + {}^*SO_2 \rightleftarrows O_2S^*SO_2^{2-} + SO_2$$

of labeled sulfur dioxide into dithionite.

Halogens

Bond Energies for Reference

F—F	159	Cl—Cl	243	Br—Br	192	I—I	151
F—H	569	Cl—H	431	Br—H	368	I—H	297
F—C	490	Cl—C	326	Br—C	272	I—C	213
F—O	213	Cl—O	205	Br—O	(201)	I—O	(201)
F—Cl	251	Cl—F	251	Br—F	251	I—F	243
				Br—Cl	218	I—Cl	209
						I—Br	176

The question of the X=X double bond as a primary functional unit does not arise for the Group VII elements, since single bond formation completes the octet of each atom in the pair. (From the molecular orbital description of the diatomics, recall that there are predicted to be two fully occupied π bonding orbitals and two fully occupied π^* anti-bonding orbitals—a situation giving no net double bonding.) The order of single bond energies of F_2 and Cl_2 is as expected from the trend started by the earlier groups: F—F, 159 kJ/mole and Cl—Cl, 243 kJ/mole. This factor in itself should tend to make compounds of the type EF_x more stable with respect to reversion to the elements than those of the type ECl_x with respect to reversion to their elements. While this is what is generally found, the relative bond energies of the halogens are only partly responsible for the observed greater stabilities of the covalent fluorides.

A second factor of considerable importance is that, quite generally (see Figure 6-1), the E—F bond energies are greater than the E—Cl bond energies. This observation is believed to be associated with a variety of factors such as (1) smaller size for F when steric congestion may be a factor, (2) better σ ao overlap when E of EX_n is a second period element, (3) the occurrence of the E=F resonance form for third period elements, and (4) the advantage of high Z_X^* (effective nuclear charge) when "electron rich" species like XeX_n are considered. This is the thermodynamic analysis of the existence of compounds such as NF_3, OF_2, and XeF_2 on the one hand, but the "non-existence" or instability of NCl_3 (easily detonated), OCl_2 (explodes easily at room temperature), and $XeCl_2$ on the other.

Since fluorine, and sometimes chlorine, promotes formation of compounds with the central element in several oxidation states, it is pertinent to inquire as to the relative stabilities of the higher and lower halides in such cases:

$$EX_{x(g)} \rightarrow EX_{y(g)} + \frac{x-y}{2} X_{2(g)}$$

The energy change for such a reaction can be analyzed in terms of the different E—X bond energies and the X—X bond energy:

$$\Delta E = \{xE_{EX}(EX_x) - yE_{EX}(EX_y)\} - \frac{x-y}{2} E_{XX}$$

E_{EX} in the higher-valent compound will often be less than E_{EX} in the lower-valent compound; but the term in brackets above will still be positive (for example, the Cl—F bond energies in ClF and ClF_3 are 251 and 176 kJ/mole, respectively). That this first term in the expression for ΔE is positive and larger than the second term is the thermodynamic requirement for the stability of the higher-valent compound. Furthermore, in comparing F with Cl, it is generally true that E_{EX} for either valency of E is greater for fluorine than for chlorine, which means that the first term in brackets will be a larger positive number when X = F than when X = Cl. This tends to make covalent fluorides more stable toward loss of halogen than chlorides. Finally, the second term is smaller for fluorine than for chlorine, a situation that amplifies the difference between the thermodynamic stabilities of fluorides and chlorides toward the decomposition reaction.

The heavier halogens are known to form several well-defined **catenated cations**[42] and **anions**. The element iodine, as perhaps expected, has yielded the greatest number of cationic species (I_2^+, I_3^+, and I_5^+); but even bromine is known to form Br_2^+ and Br_3^+, while the cation Cl_3^+ has been identified for chlorine. The iodine species I_2^+ (blue) and I_3^+ (red-brown) have been observed in fluorosulfuric acid (HSO_3F) upon treatment of I_2 with the oxidant peroxydisulfuryldifluoride:

$$2I_2 + S_2O_6F_2 \rightarrow 2I_2^+ + 2SO_3F^-$$
$$3I_2 + S_2O_6F_2 \rightarrow 2I_3^+ + 2SO_3F^-$$

CATENATED
HALOGEN
FUNCTIONS

The cation I_3^+ has a bent triatomic structure, as expected from the VSEPR concept. The bromine cations Br_2^+ and Br_3^+ have also been prepared by oxidation with $F_2S_2O_6$ in the so-called "superacid" $HSO_3F/SbF_5/3SO_2$ as solvent. The structure of the triatomic ion is presumed to be angular like that of I_3^+.

Thus far, attempts to prepare Cl_3^+ by oxidation of Cl_2 in fluorosulfuric acid have been unsuccessful; rather unusual is its synthesis by the displacement of F^- from ClF by Cl_2 (at $-78°C$) in the presence of the strong fluoride ion acceptor AsF_5:

$$Cl_2 + ClF + AsF_5 \rightarrow [Cl_3][AsF_6] \quad [43]$$

With respect to catenated anionic halides, fluorine (the light element of the group) is not known to form any, chlorine appears to form small amounts of Cl_3^- in saturated aqueous solutions of Cl_2/Cl^-, and iodine is known[44] to form several species: I_3^- is the best known (recall from analytical chemistry that I_2 is more highly soluble in KI solutions than in H_2O alone, and that such solutions are useful as a way to stabilize I_2 solutions against air oxidation), and I_n^- anions with $n = 5, 7, 8$, and 9 have also been reported. Whereas I_3^- has $D_{\infty h}$ symmetry, the higher **polyiodides** are best thought of as iodide or triiodide anion weakly bound to one or more I_2 molecules.

An interesting point about I_5^- in particular is that its structure is not the D_{4h} geometry found for ICl_4^-. To what extent this is due to lattice forces and to what extent to anticipated intra-anion steric congestion in a D_{4h} structure is uncertain.

THE halogen−F FUNCTION

The heavier halides are known to form highly reactive compounds with fluorine. Chlorine and bromine exhibit the first three positive oxidation states with fluorine by forming EF, EF_3, and EF_5, whereas the more reactive element iodine apparently does not form a stable monofluoride but readily forms IF_3, IF_5, and IF_7. It is most interesting that the thermal stabilities of the **halogen fluorides** increase with the number of fluorines, a fact that may have both kinetic and thermodynamic roots (refer to the generalizations on page 209). While these substances are certainly reactive enough to be used as fluorinating agents, they are also such strong oxidizers that they tend to degradatively fluorinate most other compounds, in an often violent way. Their use as fluorinating agents for inorganic materials is often possible but, unless diluted with an inert carrier or solvent, they are dangerous to handle in the presence of organic compounds.

An important, characteristic feature of these interhalogen fluorides, brought out in the survey section of Chapter 5, is their simple Lewis acid action toward the fluoride ion; examples of adducts are ClF_2^-, ClF_4^-, BrF_4^-, BrF_6^-, and IF_6^- (these ions are all isoelectronic with the inert gas compounds XeF_n). Alternately, when in contact with compounds like BF_3, AsF_5, and SbF_5 (good F^- acceptors), the halogen fluorides readily form cationic species by loss of F^-. Examples of this behavior are ClF_2^+, ClF_4^+, BrF_2^+, BrF_4^+, IF_4^+, and IF_6^+. An interesting example is ClF, which, much like the preparation of Cl_3^+ given above, loses F^-; but the Cl^+ so generated is readily coordinated by a ClF molecule to form Cl_2F^+. Knowing that such fluoride transfer reactions are a characteristic feature of the chemistries of these materials, it comes as no surprise that most of them exhibit conductivity to some degree by virtue of equilibria such as

$$IF_5 \rightleftarrows \frac{1}{2} IF_4^+ + \frac{1}{2} IF_6^-$$

Even though we implied above that F^- transfer leads to ion formation, at least in the solid state the situation is far from ideal. With the "salt" $ClF_2^+SbF_6^-$, for example, the cations and anions pack in such a way as to leave the Cl atom in a distorted square planar environment of fluorides. That is, two of the fluorines from a nearby SbF_6^- appear[45] to be shared with the Cl.

As with the catenated halogen ions and the interhalogen compounds, the **oxides** of the halogens are highly reactive, often violently temperamental materials of limited use to anyone other than the specialist skilled in handling them. The known oxides of fluorine, OF_2 and O_2F_2, have been mentioned in the section on oxygen compounds. Of all the chlorine oxides (OCl_2, ClO_2, Cl_2O_3, Cl_2O_4, Cl_2O_6, and Cl_2O_7), perhaps it is not unexpected that the heptoxide, with the chlorines of highest oxidation state but most shielded from nucleophilic attack, is the most stable (note the similarity with SF_6 and F_3NO "stabilities"). The structure[46] of this oily compound (b.p. $82°$) is similar to that of an ether $[(ClO_3)_2O$, where the $Cl-O-Cl$ unit is not linear] and is reminiscent of the isoelectronic species $S_2O_7^{2-}$. An interesting relationship exists between the structure of ClO_2 (a yellow gas, b.p. $10°$) and Cl_2O_4 (a liquid, b.p. $45°$). Reasoning by analogy, we might suppose that **dichlorine tetroxide** is simply the dimer of chlorine dioxide (recall nitrogen dioxide and, in particular, SO_2^-), but no such dimer is known; rather, the structure of the tetroxide can be described as that of **chlorine perchlorate** ($Cl-OClO_3$). This appears to be a not uncommon arrangement, for both $FClO_4$ and $BrClO_4$ are known to have similar structures.

The halides form quite a few useful oxyacids. With the recent preparation[47] of HOF, all halogens can be included in the list of **hypohalous** acids. All but fluorine form **halic** acids, HXO_3, and, with the recent synthesis[48] of perbromic acid, all but fluorine form **perhalic** acids, HXO_4.

Typical of the hypohalous acids, HOF is quite unstable and decomposes, at room temperature, into O_2 and HF with a half-life of something less than an hour. The acid rapidly decomposes in aqueous media by oxidation of H_2O to H_2O_2 with generation of HF (refer to the next chapter for notes on the mechanism of this reaction). Similarly, the other hypohalous acids are not best prepared by oxidation of H_2O with X_2. The standard potentials for the reactions

THE halogen–O FUNCTION

$$\frac{1}{2} X_2 + H_2O \rightarrow H^+ + HOX + e^- \qquad \begin{matrix} X = Cl & E° = & -1.63 \text{ v} \\ Br & & -1.59 \text{ v} \\ I & & -1.45 \text{ v} \end{matrix}$$

$$\frac{1}{2} X_2 + e^- \rightarrow X^- \qquad \begin{matrix} X = F & E° = & 1.44 \text{ v} \\ Cl & & 1.36 \text{ v} \\ Br & & 1.07 \text{ v} \\ I & & 0.54 \text{ v} \end{matrix}$$

$$H_2O \rightarrow \frac{1}{2} O_2 + 2H^+ + 2e^- \qquad\qquad E° = -1.23 \text{ v}$$

are pH dependent and can be used to show that the equilibrium constants for formation of HOX from X_2 and H_2O are all less than 10^{-4}, while those for oxidation of water to O_2 range from 10^7 for F_2 down to 10^{-23} for I_2. A better method of synthesis seems to be to use the strong, insoluble base HgO to remove the halide ion (formed as insoluble HgX_2) and to avoid the presence of H^+ that naturally arises in the reaction

$$X_2 + H_2O \rightarrow HOX + X^- + H^+$$

$$2X_2 + H_2O + HgO \rightarrow 2HOX + HgX_2$$

Of all the hypohalous acids, only HOF actually effects the oxidation of water to H_2O_2 or O_2 at an appreciable rate and accounts for the difficulties in past synthesis attempts. An additional interesting point is that the pH dependence of the $E°$'s for the formation of HOX suggests that it should be possible to stabilize the hypohalite ions in basic solution. This, in fact, is the basis for preparation of chlorine laundry bleaches by electrolysis of brine (this produces OCl^- in a basic solution). With Br and I the chemistry is more interesting and complex. Hypobromite and hypoiodite are both unstable with respect to disproportionation into halide and halate, and the disproportionation reaction

$$3OX^- \rightarrow 2X^- + XO_3^-$$

at room temperature is fast for Br and I. Thus, OCl^- is relatively long-lived at room temperature, OBr^- solutions need to be cooled, and IO^- simply decomposes rapidly at temperatures down to the freezing point of its solutions. The instability of the OX^- anion is not unexpected in view of the reactivity of peroxides and halogens with which the hypohalites are (valence shell) isoelectronic.

Disproportionation of the halate ions

$$4XO_3^- \rightarrow X^- + 3XO_4^-$$

into halides and perhalate ions is thermodynamically favored to a high degree for chlorine, but the reaction is very slow. For bromate and iodate, the disproportionation is highly unfavorable in a thermodynamic sense. Thus, the halate ions are quite stable in aqueous solutions, but for different reasons. In acid solution the perhalates are all strong oxidizing agents, but it is interesting that the reduction potentials for the XO_4^-/XO_3^- half-cells are in the order Br(1.76) > Cl(1.23), again reflecting a relatively high thermodynamic stability for ClO_4^-. Nevertheless, all the halate anions and acids in concentrated form can be dangerous in contact with organic materials, so that such contact should be effected, if necessary, under highly controlled and protective conditions.

The aqueous chemistry of periodic acid is easily the most intriguing of the series. HIO_4 rapidly aquates to form the so-called **orthoperiodic** acid, H_5IO_6, by stepwise addition (accompanied by proton transfer between oxygens) of two water molecules to the iodine atom to leave the structure $(HO)_5IO$. Again, we note a tendency for iodine to exhibit high oxidation states with highly electronegative substituents.

In closing this section on the oxyhalides, we should mention that it is generally believed that multiple bonding between the halogen and oxygen is an important feature of the electronic structures of the higher oxidation states. Supposed failure of fluorine to support multiple bonds with oxygen in fluorate and perfluorate is often mentioned as a reason for believing that they cannot be synthesized. This argument is based on the assumption that halogen valence d orbitals are necessary for multiple bonding to oxygen, but it should be noted that molecules such as PF_5 and F_3NO are calculated[49] to be stable without invoking participation of central atom d orbitals. A more reasonable hypothesis for postulating the "non-existence" of FO_4^- and FO_3^- would be the thermo-dynamic VSEPR/steric congestion about the small F atom.

44. Compare the series of $S_xO_y^{n-}$ and Cl_xO_y compounds for correspondence. As the structures of Cl_2O_3 and Cl_2O_6 are not known, do you feel confident in assuming, by analogy, they have the structures of $S_2O_3^{2-}$ and $S_2O_6^{2-}$? What other reasonable structures for Cl_2O_3 and Cl_2O_6 can you suggest?

45. Why do you suppose that oxygen, but not fluorine, is found combined with chlorine in the +7 oxidation state? Is the reason for the chemical inertness of SF_6 related to the basis for your answer?

46. Contrast the structures of Cl_2O_7, $B_2H_7^-$, $Al_2Cl_7^-$, $S_2O_7^{2-}$, and $(Cl_3Si)_2O$.

Noble Gases[50]

One of the more exciting developments in inorganic chemistry in recent years has been the delineation of xenon chemistry. All compounds that are currently well characterized contain the highly electronegative fluorine and/or oxygen atoms. Less well characterized and presently less useful compounds are KrF_2[51] ($E_{Kr-F} \approx 50$ kJ/mole) and $XeCl_2$.[52] Very recently there has been a report[53] of $FXeN(SO_2F)_2$, which is claimed to possess a Xe–N bond (no structural or thermodynamic data are available, however) and which therefore extends to nitrogen the elements known to form bonds to a rare gas atom.

The Xe–F bond energy[54] is estimated to be about 126 kJ/mole and that [55] of Xe–O is lower yet, about 84 kJ/mole. Thus, the simple fluorides XeF_2, XeF_4, and XeF_6 are all thermodynamically stable with respect to the elements. On the other hand, the simple oxides XeO_3 (solid) and XeO_4 (gas) explosively revert to the elements.

$$XeF_n \rightarrow Xe + \frac{n}{2}F_2 \qquad \Delta E \cong +46n$$

$$XeO_n \rightarrow Xe + \frac{n}{2}O_2 \qquad \Delta E \cong -167n$$

$$XeCl_n \rightarrow Xe + \frac{n}{2}Cl_2 \qquad \Delta E < 4n$$

This difference in XeO_n/XeF_n stabilities is due not so much to the slightly stronger XeF bonds as to the much lower bond energy of F_2 than O_2. Thus the oxygen compounds are seen to benefit from some degree of kinetic stability. The fact that E–Cl bonds are normally weaker than E–F bonds suggests that $XeCl_n$ compounds should also be less thermodynamically stable than XeF_n, but more stable than XeO_n. Given this background on XeF_n and XeO_n, you should not be surprised to learn that $XeOF_4$ (liquid, m.p. -46°), $XeOF_2$ (solid, m.p. 31°), XeO_2F_2, and XeO_3F_2 have been synthesized, in spite of their expected instability† toward loss of O_2.

---

†Stability here is used in an operational sense; it would appear that kinetic factors may be important for these compounds. On the other hand, the Xe–F bonds might enhance the Xe–O bond strength as you have seen for F_3NO [the Xe–O bond stretching force constants are 710 ($XeOF_4$) and 580 (XeO_3, XeO_4) N/m].

XeF_6 is an interesting species since its structure in the solid state is that of tetrameric and hexameric rings formed by square pyramidal XeF_5^+ units bridged by F^- ions. This tendency for XeF_6 to expand its coordination shell and to act as a F^- acceptor is also evident in the fact that it forms salts such as $M^+XeF_7^-$ (M must be a large cation like Rb or Cs). Even more interesting is the fact that these salts decompose on mild heating to evolve XeF_6 and form the *eight*-coordinate species M_2XeF_8.

A similar consequence of the Lewis acid nature of XeO_3 is that aqueous OH^- coordinates XeO_3 ($K \approx 10^{-3}$) to form solutions of the anion, $HXeO_4^-$. It is the base-induced disproportionation of this ion that leads to the **perxenate** anion, XeO_6^{4-}:

$$OH^- + HXeO_4^- \rightarrow \frac{1}{2}O_2 + \frac{1}{2}Xe + \frac{1}{2}XeO_6^{4-} + H_2O$$

In aqueous solution this reaction equation actually conceals the fact that perxenate ion is a fairly strong base, with $HXeO_6^{3-}$ being the principal species, even when the pH is as high as 13. Sodium and barium salts of XeO_6^{4-} have been isolated, and all exhibit O_h symmetry for the anion. Protonation and dehydration of the barium salt with concentrated sulfuric acid is the technique used to synthesize the dangerously explosive gas XeO_4 (T_d geometry). The use of concentrated H_2SO_4 for this reaction may be contrasted with the fact that dilute acid solutions effect an alternate rapid decomposition mode of XeO_6^{4-}, leading to $HXeO_4^-$, O_2, and H_2O as products.

$$HXeO_6^{3-} + 2H^+ \rightarrow HXeO_4^- + \frac{1}{2}O_2 + H_2O$$

Such reactions, as opposed to complete decomposition, should reinforce in your thinking the idea that Xe—O compounds seem to enjoy a certain degree of mechanistic stability.

Mechanistic considerations must also be important in the hydrolytic behavior of the higher oxidation state fluorides. XeF_4 and XeF_6 are rapidly hydrolyzed in water to form relatively stable solutions of XeO_3 (HF is the only other product in XeF_6 hydrolysis, but the hydrolysis of XeF_4 is more complex in that it involves disproportionation with Xe and O_2 as products along with XeO_3 and HF):

$$3XeF_4 + 6H_2O \rightarrow XeO_3 + 2Xe + \frac{3}{2}O_2 + 12HF$$

Controlled hydrolysis of XeF_6 with the stoichiometric quantity of H_2O leads to $XeOF_4$ and 2HF.

To close this section on noble gas compounds, let us note that reactions of KrF_2 are under development. In analogy with XeF_2, ionic materials of stoichiometry $(KrF_2)_2SbF_5$ (actually $[Kr_2F_3][SbF_6]$) and $KrF_2(SbF_5)_2$ (actually $[KrF][Sb_2F_{11}]$) have been prepared.[56]

STUDY QUESTIONS

47. Estimate the energy for the reactions

$$FXeN(SO_3F)_2 \rightarrow Xe + FN(SO_3F)_2$$
$$\rightarrow Xe + \frac{1}{2}F_2 + \frac{1}{2}((FSO_3)_2N_2)$$

in order to predict the thermodynamically more favored decomposition products (the XeN bond energy is unknown). Assuming the favored reaction to be exothermic, place an upper limit on E_{XeN}.

48. Predict structures and point group symbols for $XeOF_2$, $XeOF_4$, XeO_2F_2, and XeO_3F_2.

49. Would the synthesis of $XeOF_4$ by XeF_6 hydrolysis in the gas phase probably be exothermic?

50. Using Pauling's thermochemical method (see problem 39, p. 208) for estimating atom electronegativities, insert the XeF and XeO thermochemical bond energies to estimate χ_{Xe}. Xe is like what halogens in its χ value?

51. To which neutral and ionic halogen compounds are the known XeF_n and XeO_n compounds similar in structure?

CHECKPOINTS

1. Non-metal to halogen bonds tend to be reactive.

2. Bonds to F, O, and N from third row elements and from boron are fairly stable and feature partial multiple bond character.

3. Therefore, a characteristic of non-metal chlorides (of B and third row elements) is solvolysis by amines and alcohols to eliminate HCl and form $E\overset{\cdot\cdot}{=}N$ and $E\overset{\cdot\cdot}{=}O$ functions.

4. The $E=E$ and $E=E'$ functional groups tend to be stable when E and E' are from the second period but unstable with respect to oligomer formation (rings, chains) when E and E' are from the third period.

5. Bonds to F, O, and N from the second row elements N, O, and F tend to have low energies, probably because of lone pair/lone pair repulsions. Their reactions are sometimes of the free radical type.

6. Catenated N, O, and halogen species tend to be highly reactive for the reason given in point 5.

7. The hydridic functions (BH, AlH, and to some extent SiH) are useful for addition to unsaturated linkages ($C=C$, $C\equiv C$, $C=N$, $C=O$, $P=O$, $S=O$, etc.). Hydroboration is the best established of these.

REFERENCES

[1] There are many collections of such data, some newer than others and having more reliability for difficult-to-handle compounds. To be internally consistent, we have relied heavily on D. A. Johnson's "Some Thermodynamic Aspects of Inorganic Chemistry," Cambridge University Press, London (1968). This little book is highly recommended. In addition to the sources used by Johnson, supplementary data may be found in: L. Brewer, *et al.*, *Chem. Rev.*, **61**, 425 (1961); *ibid.*, **63**, 11 (1963); R. C. Feber, Los Alamos Report LA-3164 (1965); B. Darwent, *Nat. Stand. Ref. Data Ser., Nat. Bur. Stand.*, 1970. NSRDS-NBS31.

[2] See ref. 1, p. 267 of Johnson's book.

[3] This exchange has actually been used to prepare some R_xPH_y compounds: J. W. Gilje, *et al.*, *Chem. Comm.*, 813 (1973).

[4] E. L. Muetterties, Ed., "The Chemistry of Boron and Its Compounds," Wiley, New York (1967); R. M. Adams, Ed., "Boron, Metallo-Boron Compounds, and Boranes," Interscience, New York (1964); K. Wade, "Electron Deficient Compounds," Nelson, London (1971).

[5] See ref. 4; W. N. Lipscomb, "Boron Hydrides," W. A. Benjamin, New York (1963); R. J. Brotherton and H. Steinberg, "Progress in Boron Chemistry," Vols. 2 and 3, Pergamon Press, New York (1970); H. C. Brown, "Boranes in Organic Chemistry," Cornell University Press, Ithaca (1972).

[6] L. H. Long, *Prog. Inorg. Chem.,* **15**, 1 (1972).

[7] All the lower B_nH_m are thermodynamically unstable with respect to elemental B and H_2. Their "stability" derives from mechanistic factors. See K. Wade (ref 4). For a review of B_5H_9, see D. F. Gaines, *Accts. Chem. Res.,* **6**, 416 (1973).

[8] E. Switkes, *et al., J. Amer. Chem. Soc.,* **92**, 3847 (1970).

[9] See K. Wade (ref. 4).

[10] H. Nöth and H. Vahrenkamp, *Chem. Ber.,* **100**, 3353 (1967).

[11] Less well characterized are $(BX)_4$ (T_d), $(BCl)_8$ $(D_{2d}$, dodecahedron), $(BCl)_9$, and $(BCl)_{12}$; A. G. Massey, *et al., J. Inorg. Nuclear Chem.,* **33**, 1195, 1569 (1971).

[12] See ref. 4.

[13] R. P. Bell, *et al.,* in "The Chemistry of Boron and Its Compounds," E. L. Muetterties, Ed., Wiley, New York (1967).

[14] K. Torssell, in "Progress in Boron Chemistry," Vol. I, H. Steinberg and A. L. McCloskey, Ed., Pergamon Press, New York (1964).

[15] S. Cucinella, *et al., Inorg. Chem. Acta Revs.,* **2**, 51 (1970); E. C. Ashby, *Adv. Inorg. Chem. Radiochem.,* **8**, 283 (1966).

[16] J. W. Turley and H. W. Rinn, *Inorg. Chem.,* **8**, 18 (1969).

[17] G. C. Sinke, *et al., J. Chem. Phys.,* **47**, 2759 (1967).

[18] N. M. Yoon and H. C. Brown, *J. Amer. Chem. Soc.,* **90**, 2926 (1968).

[19] C. H. Henrickson and D. P. Eyman, *Inorg. Chem.,* **6**, 1461 (1967).

[20] K. M. MacKay, "Hydrogen Compounds of the Metallic Elements," Spon, London (1966).

[21] J. K. Ruff, in "Developments in Inorganic Nitrogen Chemistry," Vol. 1, Elsevier, New York (1968).

[22] K. Wade, *J. Chem. Educ.,* **49**, 502 (1972).

[23] M. E. Peach, *et al., J. Chem. Soc., A.,* 366 (1969).

[24] A classical writing from many points of view, you should read A. Stock, "Hydrides of Boron and Silicon," Cornell University Press, Ithaca (1933); K. Borer and C. S. G. Phillips, *Proc. Chem. Soc.,* 189 (1959); B. J. Aylett, *Adv. Inorg. Chem. Radiochem.,* **11**, 249 (1968).

[25] G. Urry, *J. Inorg. Nucl. Chem.,* **26**, 409 (1964).

[26] P. L. Timms, *et al., J. Amer. Chem. Soc.,* **87**, 2824 (1965).

[27] An interesting, short exposé of the utility and preparation of "silicone liquid rubbers" is by J. A. C. Watt, *Chem. Britain,* **6**, 519 (1970).

[28] H. Thurn and H. Krebs, *Acta Cryst.,* **B25**, 125 (1969).

[29] A. H. Cowley, "Compounds Containing Phosphorus-Phosphorus Bonds," Dowden, Hutchinson and Ross, Stroudsburg, Pa. (1973).

[30] J. E. Huheey, *J. Chem. Educ.,* **40**, 153 (1963).

[31] A. H. Cowley and J. R. Schweiger, *J. Amer. Chem. Soc.,* **95**, 4179 (1973) reviews evidence (preferred donor atoms and rotational barriers for OP—N compounds) indicating P=N character: I. G. Csizmadia, *et al., J. Chem. Soc., Chem. Comm.,* 1147 (1972), however, make a case for the importance of σ release (P → N) as a factor in the planarity of N in H_2PNH_2.

[32] In one of the "Textbook Errors" features of *J. Chem. Educ.,* J. Mason [52, 445 (1975)] notes that $(NO)_2$ is *not* blue and does *not* have a C_{2h} structure, as is often written. Her article also contains structural and thermodynamic data references for $(NO)_2$.

[33] R. Alrichs and F. Keil, *J. Amer. Chem. Soc.,* **96**, 7615 (1974); J. M. Howell and J. R. Van Wazer, *ibid.,* **96**, 7902 (1974).

[34] A. H. Brattain, *et al., Trans. Faraday Soc.,* **65**, 1963 (1969).

[35] E. Leifer, *J. Chem. Phys.,* **8**, 301 (1940); W. Spindel in "Inorganic Isotope Synthesis," R. H. Herber, Ed., W. A. Benjamin, New York (1962).

[36] Ammonia is present because the synthesis of NI_3 is by I_2 oxidation of aqueous ammonia. The structure is reported by H. Hartl, *Z. Anorg. Chem.,* **375**, 225 (1968).

[37] In fact, while this reaction is thermally neutral, entropy factors favor $H_2S + S_8$. The real reason the sulfanes have been well characterized lies in the fact that no suitable *low* energy pathway connecting H_2S_n to H_2S and S_8 is available: T. K. Wiewioroski, *Endeavor,* **29**, 9 (1970); E. Muller and J. B. Hyne, *Can. J. Chem.,* **46**, 2341 (1968).

[38] R. J. Gillespie and J. Passmore, *Accts. Chem. Res.,* **4**, 413 (1971).

[39] R. B. Harvey and S. H. Bauer, *J. Amer. Chem. Soc.,* **75**, 2840 (1953).

[40] R. D. Brown, *et al., Australian J. Chem.,* **18**, 627 (1965).

[41] J. D. Dunitz, *Acta Cryst.,* **9**, 579 (1950).

[42] R. J. Gillespie, and J. Passmore, *Adv. Inorg. Chem. Radiochem.,* **17**, 49 (1975).

[43] The salt reverts to reactants at room temperature; R. J. Gillespie and M. J. Morton, *Inorg. Chem.,* **9**, 811 (1970).

[44] L. E. Topol, *Inorg. Chem.,* **10**, 736 (1971); A. E. Wells, "Structural Inorganic Chemistry," 4th Edition, Oxford, London (1975).

[45] K. O. Christie and C. J. Schack, *Inorg. Chem.,* **9**, 2296 (1970); A. J. Edwards and R. J. C. Sills, *J. Chem. Soc. A.,* 2697 (1970).

[46] J. D. Witt and R. M. Hammaker, *J. Chem. Phys.,* **58**, 303 (1973).

[47] M. H. Studier and E. H. Appleman, *J. Amer. Chem. Soc.,* **93**, 2349 (1971).

[48] G. K. Johnson, *et al., Inorg. Chem.,* **9**, 119 (1970); J. R. Brand and S. A. Bunck, *J. Amer. Chem. Soc.,* **91**, 6500 (1969).

[49] F. Keil and W. Kutzelnigg, *J. Amer. Chem. Soc.,* **97**, 3623 (1975).

[50] J. H. Holloway, "Noble Gas Chemistry," Methuen, London (1968).

[51] S. R. Gunn, *J. Phys. Chem.,* **71**, 2934 (1967); D. R. MacKenzie and I. Fajer, *Inorg. Chem.,* **5**, 669 (1966); J. J. Turner and G. C. Pimentel, *Science,* **140**, 974 (1963).

[52] L. Y. Nelson and G. C. Pimentel, *Inorg. Chem.,* **6**, 1758 (1967).

[53] R. D. LeBlond and D. D. DesMarteau, *Chem. Comm.,* 555 (1974).

[54] B. Weinstock, *et al., Inorg. Chem.,* **5**, 2189 (1966); M. Karplus, *et al., J. Chem. Phys.,* **40**, 3738 (1964).

[55] S. R. Gunn, *J. Amer. Chem. Soc.,* **87**, 2290 (1965).

[56] H. Selig and R. D. Peacock, *J. Amer. Chem. Soc.,* **86**, 3895 (1964); R. J. Gillespie and G. J. Schrobilgen, *J. Chem. Soc., Chem. Comm.,* 90 (1974).

7. NON-METAL COMPOUNDS: FUNCTIONAL GROUP REACTION MECHANISMS

This chapter is dedicated to the question of what happens along the route from reactants to products in a chemical transformation. The ideas about "how" molecular species interact in one or more steps, and about what their subsequent fates are, define the concept of a mechanism and are often highly speculative. In this chapter your understanding of the concepts of intermolecular interaction will be given free rein. An important aspect of all this will be to learn the currently accepted semantics and concepts surrounding the ideas of chemical activation and reaction step types.

BASIC CONCEPTS

STOICHIOMETRIC MECHANISMS AND ACTIVATION STEPS

Chemical reactions are often grouped for convenience into the following types: substitution, exchange (both intermolecular and intramolecular), addition or elimination (perhaps involving free radical species), solvolysis, and redox.

$$EX + D \rightarrow ED + X \quad \text{(substitution)}$$
$$EX + DY \rightarrow EY + DX \quad \text{(exchange)}$$
$$EX + DY \rightleftharpoons DEXY \quad \text{(addition/elimination)}$$
$$EX + HA \rightarrow EA + HX \quad \text{(solvolysis)}$$
$$Ox_1 + Red_1 \rightarrow Red_2 + Ox_2 \quad \text{(redox)}$$

Since such labels are meant to characterize the structural relationship between the compositions of products and reactants, they usually reflect little of the mechanisms of such reactions. It is often found, as you shall see in this chapter, that mechanistically these ostensibly different reaction types have a great deal in common. Consequently, only a few mechanistic concepts carry us a long way in understanding reaction types. It is hoped that, by approaching mechanism studies in a systematic way, sufficient understanding of how molecules react can be developed to make generalizations possible for rational syntheses of new materials.

In this preliminary section we will examine the mechanisms commonly invoked to explain reactions in which one group replaces or substitutes for another. In the survey section to follow, you will find the concepts given here to be applicable also to the so-called exchange, addition, elimination, and solvolysis reactions, and even to some redox reactions.

As the majority of chemical reactions are carried out in solutions, we will confine our attention to the concepts developed for such reaction conditions. In a general reaction between R and R' to form P and P', we recognize that many steps may be required to complete the transformation from reactants to products and that some of these steps will be much faster than others. It appears that often, but not always, a single step is so much slower than all others that the speed of that step effectively determines the form of the rate law expression. *The rate law will then contain information as to the nature of all steps preceding and including the slowest one.* In the simplest terms we can envision any of the following as the slowest step in a reaction sequence:

(a) R and R′ diffuse together to be trapped momentarily in a solvent cavity or "cage" as an *encounter* or *cage complex*, R:R′.

(b) R and R′ smoothly transform into P and P′ (in the form of a cage complex P:P′).

(c) or R and R′ transform first into a new species, R—R′, which subsequently evolves into P and P′.

Situation (b) as the slowest step characterizes what we will call an **interchange** (or **I**) **mechanism**, and (c) as the slowest step characterizes what we will refer to as an **associative** (or **A**) **mechanism**, distinguished by the formation of a bond between R and R′ to produce an **intermediate**, R—R′. Situation (a) as the slow step characterizes a *diffusion controlled reaction*.

Yet another possibility exists. Perhaps the cage complex R:R′ does not lead to products. Rather, R unimolecularly dissociates into species P′ and X. The *intermediate* X can then recombine with P′ (to reform R or some stereochemical isomer of R) or be long-lived enough to diffuse into a solvent cavity with R′ and undergo reaction to form P. This situation, with the R → X + P′ step slowest, characterizes what we mean by a **dissociative** (or **D**) **mechanism**.

These possibilities for a **stoichiometric mechanism**[1] can be summarized by the equations (examples you have seen in organic chemistry are also given):

$$(I) \quad R + R′ \rightarrow \boxed{R:R′ \rightarrow P:P′} \rightarrow P + P′ \qquad\qquad H_3CCl:OH^- \rightarrow CH_3OH + Cl^-$$

$$(A) \quad R + R′ \rightarrow \boxed{R:R′ \rightarrow R—R′} \rightarrow P:P′ \rightarrow P + P′$$

$$R_2CO:H_2O^* \rightarrow R_2C\begin{smallmatrix}OH\\ \\ ^*OH\end{smallmatrix}$$

$$\downarrow$$

$$R_2CO^* + H_2O$$

$$(D) \quad \boxed{R \rightarrow X:P′} \rightarrow X + P′ \qquad\qquad (t\text{-Bu})_3CBr \rightarrow (t\text{-Bu})_3C^+ :Br^-$$

$$X + R′ \rightarrow X:R′ \rightarrow P \qquad\qquad (t\text{-Bu})_3C^+ + OH^- \rightarrow (t\text{-Bu})_3COH$$

The **transition state structure** for the *I* mechanism is envisioned to be

$$\left[\begin{array}{c} \overset{H}{} \overset{H}{} \\ HO\text{--}C\text{--}Cl \\ \underset{H}{} \end{array} \right]^-$$

and you will note there is no net change in the number of bond pairs in the carbon valence shell (the HO—C—Cl structure is a typical three-center/four-electron mo problem†). Both partial bond-making and partial bond-breaking characterize this transition state. Net bond-making is the dominant characteristic of the *A* mechanism intermediate; in the ketone example above, the CO π bond is supplanted by a CO σ bond in the intermediate.

---

†Formally, there are five electron pairs about carbon, but one of these pairs is predominantly localized on the oxygen and chlorine atoms.

$$\left[\begin{array}{c} R \overset{\textstyle R}{\underset{\textstyle \underset{\textstyle H_2}{O}}{\overset{\textstyle |}{C}}}{-}O \end{array} \right]$$

The intermediate for the D mechanism is characterized by net bond breaking:

$$\left[\begin{array}{c} t\text{-Bu} \diagdown \quad \diagup t\text{-Bu} \\ \underset{\textstyle \underset{\textstyle t\text{-Bu}}{|}}{C} \end{array} \right]^{+}$$

These concepts are probably best understood in terms of what is called an **energy profile diagram** of the change in energy of the chemical system as the nuclei of the reacting parts execute the motion leading to the substitution (this motion, even though often not quantitatively known, is referred to as the **reaction coordinate**).

Figure 7–1 summarizes the differences between D, A, and I processes. The dashed curve illustrates the I process, in which reactants are smoothly transformed into products by the formation of a single, unstable "high" energy chemical species referred to as the **transition state** or **activated complex**. The A and D processes (solid line) involve two steps showing the appearance of two activated complexes, separated along the reaction coordinate by a minimum that is characteristic of the **intermediate**. The A mechanism implies the formation of the first activated complex by an **associative activation mode**, a, in which the reactant R:R$'$ forms a bond to create the intermediate R–R$'$. The second step requires a much smaller activation energy for the conversion of R–R$'$ into product P:P$'$. The D mechanism, on the other hand, has an energy profile similar to that of the A process, but the initial high activation energy step involves a **dissociative activation mode**, d, leading to complete fragmentation of R into one of the products, say P$'$, and a species (intermediate X) of kinetically significant stability. The intermediate then rapidly reacts with R$'$ to form the final product P. In contrast to the A mechanism, R$'$ does *not* play a kinetically significant role in the D mechanism. With regard to the intimate aspects of an I mechanism, one can envision a transition state complex involving considerable fragmentation of R with little progress being made in the formation of the new bond between R and R$'$, or the converse. Such activation processes are then distinguished on the intimate level as I_d and I_a, respectively.

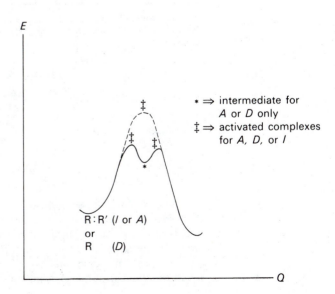

Figure 7–1. A reaction energy profile diagram illustrating the formation of an intermediate, characteristic of A and D processes (———), and the formation of an activated complex, characteristic of an I process (———).

In summary, the symbols D and A convey not only the concepts of unimolecular and bimolecular reaction steps, but also the idea that these steps are the *slow* ones and proceed by d (bond-breaking) or a (bond-making or "adduct" formation) so as to produce reactive intermediates. The symbol I conveys the concept of a slow bimolecular reaction step that proceeds with predominantly bond-breaking (I_d) or bond-making (I_a) character so that no reactive intermediate is formed. With reference to the D and A reaction sequences on p. 220, the "boxed" steps are implied to be the slow ones. Were these steps not the slow ones, the A mechanism would become D in $R-R'$ ($\rightarrow P:P'$) and the D mechanism would become A in $X + R'$ ($\rightarrow X:R'$).

CHEMICAL VALENCY IDEAS

If R is a structure representing valency unsaturation for the central atom, an I or A mode is anticipated. Compounds of elements for which valence shell expansion is established by the existence of "stable" adducts are prime candidates for the A mechanism. If the reagent R is known to be valency saturated, the most reasonable pathways are D and I.

RATE LAW EXPRESSIONS

The details of rate law interpretation are saved for a more advanced text; the following charts express the correlation of stoichiometric mechanism with rate laws.

Known Mechanism	Predicted Rate Law
I for R + R$'$	Rate = k[R][R$'$]
A for R + R$'$	Same as I
D for R	Rate = k[R]

Known Rate Law	Implied Mechanism
Rate = k[R][R$'$]	(i) I or A in R + R$'$
= (Kk)[R][R$'$]†	(ii) D in R$-$R$'$
Rate = k[R]	(i) D in R
	(ii) Any of the second order mechanisms where [R$'$] is so large as to be approximately time independent

†In a two-step reaction, the first step of which is fast in forward and reverse directions, the slowness of the second step means that the species in the first step are nearly at equilibrium. To a good approximation:

$$R + R' \underset{}{\overset{fast}{\rightleftharpoons}} R - R' \qquad K = [R-R']/[R][R']$$

$$R - R' \xrightarrow{slow} P + P' \qquad k \equiv \text{rate constant}$$

Rate = k[R $-$ R$'$] $\doteq Kk$[R][R$'$].

Very seldom are we in a position of knowing the mechanism and wanting to predict a rate law. It is important to realize that the mechanism → rate law correlation is unique but that the rate law → mechanism translation is rarely so. Ambiguities in interpretation arise for the second order rate law in particular, but also for the first order law when [R$'$] is essentially constant throughout the reaction; that is, k[R$'$] is constant during a kinetic measurement and so is indistinguishable from a rate constant alone.

SUBSTITUENT EFFECTS

Because of anticipated effects on bond-making and bond-breaking energies, variation of substituents can help complete the description of a mechanism. By implication, I_d and, particularly, D mechanisms should have rate constants that are sensitive to substituent changes that affect the ease of bond-breaking. On the other hand, I_a and, particularly, A mechanisms are sensitive to bond-making influences.

Mechanism	Rate Constant Leaving Group Dependence	Rate Constant Entering Group Dependence
D	large	none
A	small	large
I_d	significant	less significant
I_a	less significant	significant

STUDY QUESTION

1. Sketch a reaction coordinate diagram for a reaction exhibiting an induction period (*i.e.*, one that *initially* shows a faster disappearance of reactants than formation of products). Hint: Is the intermediate more or less stable than the reactants?

SURVEY OF REACTIONS

In the remainder of this chapter we will present some proposed "inorganic" reaction mechanisms, supported by kinetic studies where possible, broadly organized on the basis of the number of electron pairs in the valence shell of the reactive atom. This is done to establish, so far as possible, a relationship between reactant structure and proposed mechanism.

As you examine the reactions of functional groups, look for the utility of the following concepts:

(a) the estimated strengths of bonds to entering and leaving groups,
(b) the "stability" of leaving groups,
(c) the degree of "unsaturation" at the reactive centers, *e.g.*, the availability of reasonably low lying unoccupied orbitals (atomic or molecular) in the valence shells, and
(d) the steric bulk of entering and leaving groups.

By way of generalization, we may begin by anticipating the following. For those molecules with only *three bonding pairs* of electrons in the valence shell (those that are electron deficient in the sense of the octet rule) you may expect many reactions certainly to involve an associative step somewhere along the reaction coordinate, but this does not always mean the rate controlling step or activation mode is necessarily *a*. Those atoms with *four electron pairs* in the valence shell are by far the most numerous and can be divided into sub-classes according to whether they have none, one, two, three, or four lone pairs of electrons (corresponding to Groups IV through VIII). You will see examples of attack (via the lone pairs present) *by* these atoms *upon* others, and, conversely, attack (via valence shell expansion) *upon* the heavier atoms from these groups. The ideas developed for the four-electron-pair cases can often be carried over with simple modification to those molecules whose reactive centers have *five* and *six pairs* of valence electrons (the trigonal bipyramidal and octahedral structures).

One Valence Pair

HYDROGEN

**THE ACIDIC
H–E FUNCTION**

Perhaps the most commonly encountered reactions involving attack upon a hydrogen atom are those belonging to the **solvolysis class**. Nucleophilic attack on hydrogen is encountered when solvents with a more or less acidic hydrogen atom (protonic solvents) are brought into contact with materials that are strongly basic toward the proton. Such solvents form hydrogen-bonded adducts with weaker Lewis base materials; with stronger bases, the proton transfer is often complete and leads to solvolysis. Many of these reactions are exceedingly fast—often diffusion controlled—and presumably involve simple, rapid transfer of the proton. Typical reactions are:

$$NaH + NH_3 \rightleftharpoons H_2 + NH_2^- + Na^+$$

$$\left.\begin{array}{l} CaS \\ CaO \\ CaC_2 \\ Ca_3N_2 \\ Ca_3P_2 \end{array}\right\} + H_2O \rightleftharpoons \left.\begin{array}{l} HS^- \\ OH^- \\ C_2H_2 \\ NH_3 \\ PH_3 \end{array}\right\} + OH^-$$

Such reactions provide for convenient *preparations of deuterated hydrides* if D_2O is used. The reaction of CaC_2 is typical of most metal carbides, and is particularly interesting in that the reaction product suggests that the carbide is actually an acetylide. Be_2C and Al_4C_3 are unusual in that they produce methane on hydrolysis and could be called methides. Other examples of fast ($k \gtrsim 10^8$ M^{-1} sec^{-1}, diffusion controlled) proton transfer reactions are those involving simple exchange such as

$$NH_3 + NH_4^+ \rightleftharpoons NH_4^+ + NH_3$$

$$NH_4^+ + H_2O \rightleftharpoons NH_3 + H_3O^+$$

**THE NON-POLAR
AND HYDRIC
H–E
FUNCTIONS**

Exchange reactions for non-acidic hydrogen seem to feature H in a bridging capacity.

$$BCl_3 + MePhNpSiH \rightleftharpoons MePhNpSiCl + BCl_2H$$

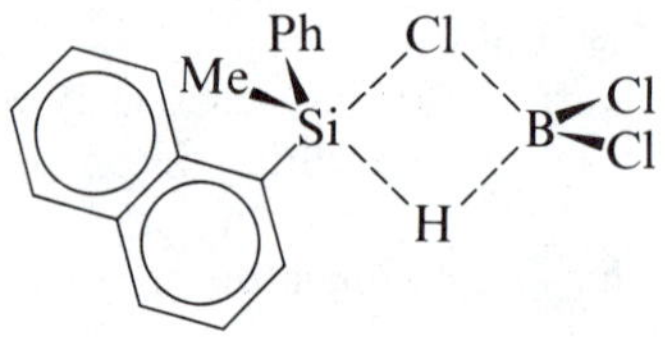

Because of the observed configuration retention by the chiral silane, it is believed[2] that the four-center structure shown for the exchange of H and Cl is correct. The concept of such a structure is not foreign to you if you reflect that many ground state structures are known in which H, and also Cl, serve as bridging functions (Chapters 5 and 6). It is most likely that such a species is an intermediate.

Particularly interesting are the **catalytic hydrogenation reactions** of organic compounds (see Chapter 18 for more examples). Some of these reactions are thought[3] to proceed through a four-center species such as

$$\begin{array}{c} H{\rightarrow}M \\ | \qquad \longrightarrow B{-}H^+ + H{-}M^- \\ B{:}{\rightarrow}H \end{array}$$

for activation of molecular hydrogen (where base = pyridine, a carboxylate anion, or water; and metal = the monovalent silver, mercury, and copper ions, or Pt^{2+}). All of these metals are classified as Lewis acids, and the "push-pull" base-acid action favors heterolytic cleavage of the H_2 molecule to form a metal hydride and a protonated base. It is the metal hydride that then acts as the catalyst (reactive intermediate) in the hydrogenation reaction. A strictly inorganic reaction of this type is the oxidation of H_2 by aqueous Cr(VI), Fe(III), etc.,

$$H_2 + \frac{2}{3}Cr(VI) \xrightarrow{Cu^{2+}} 2H^+ + \frac{2}{3}Cr(III)$$

where the reaction steps are believed to be

$$Cu^{2+} + H_2 \underset{k_{-1}}{\overset{k_1}{\rightleftharpoons}} CuH^+ + H^+$$

$$CuH^+ + Cu^{2+} \xrightarrow{k_2} 2Cu^+ + H^+$$

$$2Cu^+ + \frac{2}{3}Cr(VI) \longrightarrow 2Cu^{2+} + \frac{2}{3}Cr(III)$$

which express the importance of the catalytic role of Cu^{2+} in electrophilic cleavage of the H_2 bond (the possible role of the solvent is omitted).

The hydrides, in which hydrogen is bound to fairly electropositive metals, tend to be extremely reactive compounds. In addition to the simple binary hydrides (such as NaH noted above), which make excellent desiccants, you are aware of the class of complex anion species such as the hydrides of Group III elements (BH_4^-, AlH_4^-, $Al(BH_4)_3$, etc.), which are very important reagents as sources of H^- and for reduction reactions. The ions AlH_4^- and GaH_4^- are far more reactive (even explosively) with H_2O than is BH_4^-. The reasons for this have not been demonstrated† at this time, since mechanism studies appear to have been conducted only for BH_4^-.

The tetrahydroborate ion, for which the hydrolysis is drastically affected by solution pH, has been studied[4] in considerable detail. It is interesting that both hydrogen *exchange* and *hydrolysis* reactions can be observed.

$$BH_4^- + {}^*HOH \rightleftharpoons BH_3^*H^- + HOH$$

$$BH_4^- + 4H_2O \rightarrow B(OH)_4^- + 4H_2$$

The BH_4^- *hydrolysis* rate law expression indicates both acid independent and acid catalyzed pathways. The rate constant for the acid dependent path is $\sim 2 \times 10^6$ M^{-1} sec^{-1}, so that elimination of H_2 is about 15 times faster than hydrogen exchange. A very important result, namely that mainly HD (less than 4% H_2 and less than 1% D_2) is evolved from reactions of BH_4^- in D_2O and BD_4^- in H_2O, respectively, strongly militates against formation of reactive *intermediates* with the structures **1** and **2**.

    1   2

---

†One could certainly argue greater nucleophilicity of the hydrogens in AlH_4^- and GaH_4^- and/or more ready nucleophilic attack upon Al and Ga because of their greater ease of valence shell expansion.

A mechanism based on loss of HD from equatorial positions in **1** would also permit loss of H_2 with equal facility. Axial/equatorial HD loss would also require equal loss of H_2. Analogous comments apply to **2**. Furthermore, assuming that the exchange and hydrolysis reactions proceed from the same species, this species must have a structure (such as **3**) in which two hydrogens are equivalent and different from the other three bound to boron:

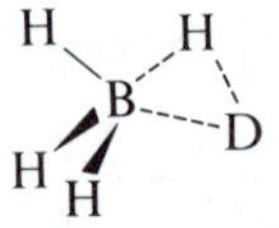

3

STUDY QUESTION

2. Sketch an energy profile for

$$BH_4^- + {}^*H^+ \begin{cases} BH_3 + {}^*H_2 \\ B{}^*H_4^- + H^+ \end{cases}$$

showing both reactions to proceed from a common intermediate. If formation of this intermediate were overwhelmingly the dominant activation process, would there by any detectable difference in the rates of formation of H_2 and H^+ exchange?

Three Valence Pairs

BORON

THE B–X FUNCTION

The most important examples of three valence shell pairs come from Group III of the Periodic Table, and boron has been the most widely studied. A very general reaction for the boron halides is **solvolysis**. With protonic solvents the reaction products are hydrogen halide and a substituted borane. The reaction of the boron halides with water is rapid and vigorous when the halide is Cl, Br, or I, and it probably involves coordination of H_2O as a rapid first step, to be followed by elimination of HX to leave BX_2OH. The latter continues to react, forming $BX(OH)_2$ and finally $B(OH)_3$. The fluoride is unusual in that it can form (below 20°C) a relatively stable adduct with H_2O. This adduct melts at 10°C and, as could be expected, is a strong acid:

$$BF_3 \cdot OH_2 + H_2O \rightleftharpoons H_3O^+ + BF_3OH^-$$

Studies[5] of the **elimination** of HX from the **solvolysis** reactions of amines with BX_3 and $PhBX_2$ in weakly basic aromatic solvents have revealed some interesting aspects of the solvolysis of B–X bonds. First we must note some generalizations about the relationship between reactants and products of these reactions. With tertiary amines, no solvolyses are observed so that only simple 1:1 adduct formation occurs:

$$R_3N + BX_3 \rightarrow R_3N \cdot BX_3$$

With secondary amines, one equivalent of HX is evolved per amine

$$R_2HN + BX_3 \rightarrow R_2NBX_2 + HX$$

and experimental control of the stoichiometry of reactants leads to the aminoboranes: BX_2NR_2, $BX(NR_2)_2$, or $B(NR_2)_3$. With less sterically hindered *primary* amines, much more interesting reactions occur, which are highly dependent on the nature of the amine substituents. When R of RNH_2 is an aryl group such as phenyl or *o*-tolyl, the first step of the reaction is the formation of a 1:1 adduct as an insoluble intermediate, which subsequently undergoes loss of HX upon heating. In both of these cases, as well as with most primary alkyl amines, the final reaction product is the borazine **4** resulting from loss of two equivalents of HX per adduct.

$$RH_2N + BX_3 \rightarrow \frac{1}{3}[RNBX]_3 + 2HX$$

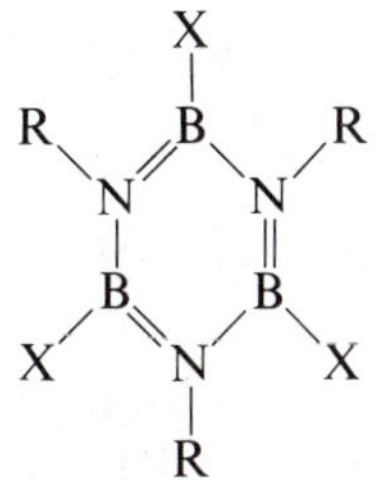

4

Characteristic **substitution** reactions of Group III halides of some importance are the following:

$$3H^- + BF_3 \xrightarrow{R_2O} BH_3 + 3F^-$$

$$4H^- + AlCl_3 \xrightarrow{R_2O} AlH_4^- + 3Cl^-$$

$$MeO^- + BCl_3 \longrightarrow MeOBCl_2 + Cl^-$$

The first two are very important reactions, as they constitute methods for synthesis of the important hydrides B_2H_6 and lithium aluminumhydride. The detailed mechanisms of these reactions have not been clarified, but the idea that the strong base H^- initially coordinates the boron and aluminum is very likely correct.† Subsequent reactions (*I* mechanism) with H^- to displace the less basic bases F^- and Cl^- from $[HEX_3]^-$ appear appropriate.

A reaction closely related to HX elimination from amine boranes is H_2 elimination in the thermal decomposition of H_3NBH_3 (actually performed by heating $LiBH_4$ and NH_4Cl above 250°C) to give borazine, **5**. For possible mechanistic aspects of this reaction you should note the analogy with the HX elimination reactions of aniline- and toluidine-BH_3 adducts to form the N- and B-substituted borazines (**4**). Given that

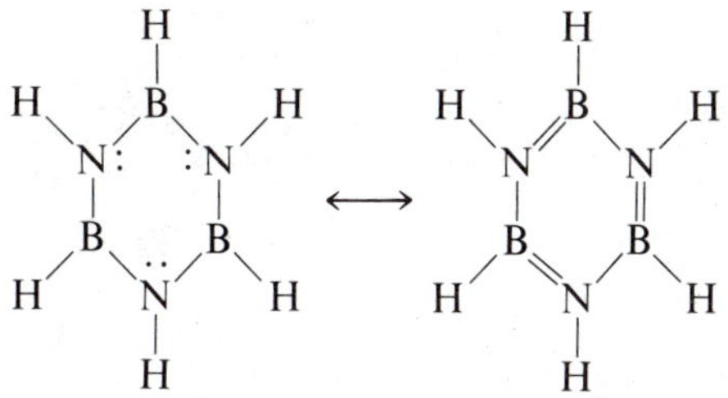

THE B–H FUNCTION

5a   **5b**

---

†In fact, reactions of X^- with EX_3 will lead to EX_4^- as products.

resonance structure **5a** should be quite important, you should not be surprised to learn that borazine has been found[6] to react with the hydrogen halides to form the analog of 1,3,5-trihalocyclohexane (the borons are the 1,3,5 atoms in borazine). This is a cleaner route than synthesis of $H_3N \cdot BHCl_2$, followed by thermal evolution of one equivalent of HCl.

In closely related thermal decomposition reactions, secondary amine adducts of borane (at temperatures above 100°C) lose H_2 and apparently dimerize† to generate cyclobutane analogs **6** (note that dimers form when X = H in R_2NBX_2, but when X is the larger halogens only monomers can form).

$$R_2NHBH_3 \rightarrow H_2 + \frac{1}{2}[R_2NBH_2]_2$$

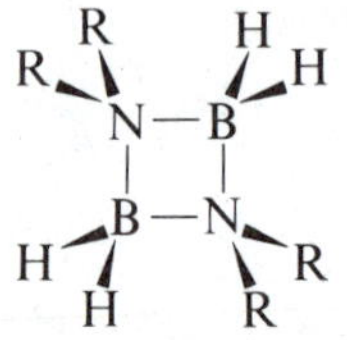

6

Certainly one of the most important reactions of three coordinate boron is the **addition** of BH_3 to olefins (alternately, an insertion of C=C into the BH bond). The so-called "hydroboration" reaction[9] is used to introduce one, two, or three alkyl groups at boron.

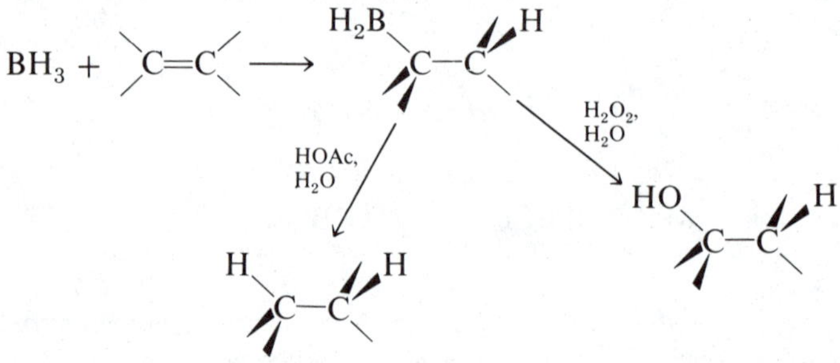

Several aspects of this highly important hydroboration reaction have been carefully determined. First of all, the reaction is strongly anti-Markownikoff, perhaps because the more highly substituted carbon better supports development of positive charge during the later stages of the reaction. An example is 2-methyl-1-butene (**7**), for which 99% of the product has boron attached to the terminal carbon. With 4-methyl-2-pentene (**8**) there is no significant preference for boron attachment at one or the other carbon.

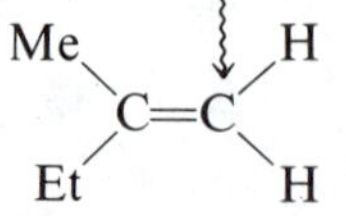

7 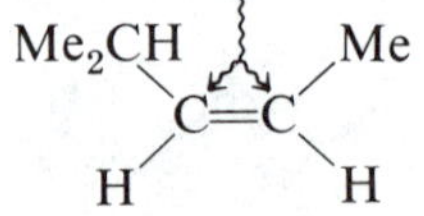   **8**

---

†Trimerization is apparently inhibited by steric interactions between R groups in axial positions of what would be a cyclohexane structure.

The details of the hydroboration reaction mechanism are not known, but an educated guess would have a weak olefin π-donor complex formed early in the reaction. While such an adduct probably would be symmetrical for symmetrical olefins, it would not be so for unsymmetrical olefins. The primary activation step should then be the shifting of the boron toward one carbon while the hydrogen is transferred to the other carbon.

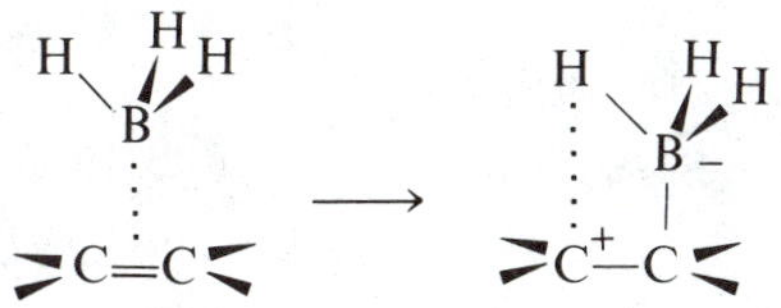

Given the quite different steric requirements of the methyl and isopropyl groups, it appears that steric factors are not of great importance to the stereochemistry of the hydroboration reaction. A steric effect can be observed, however, on the *extent* of the hydroboration reaction. With 2-methyl-2-butene (**9**) the reaction stops after formation of *bis*(3-methyl-2-butyl)borane, while 2,4,4-trimethyl-2-pentene (**10**) yields only 2,2,4-trimethyl-3-pentylborane.

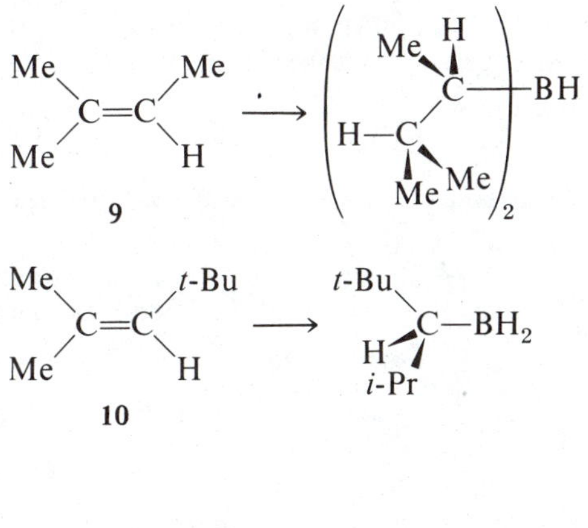

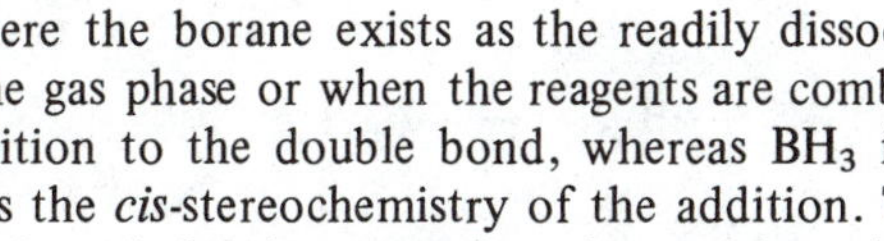

Another important aspect of the reaction is that it proceeds smoothly and quickly in ether solvents, where the borane exists as the readily dissociated adduct ether·BH_3, but is very slow in the gas phase or when the reagents are combined neat. Clearly, B_2H_6 is ineffective for addition to the double bond, whereas BH_3 is much more reactive. A third feature concerns the *cis*-stereochemistry of the addition. This is deduced from the production of *trans*-2-methylcyclopentanol and *trans*-2-methylcyclohexanol from 1-methylcyclopentene and 1-methylcyclohexene, respectively.

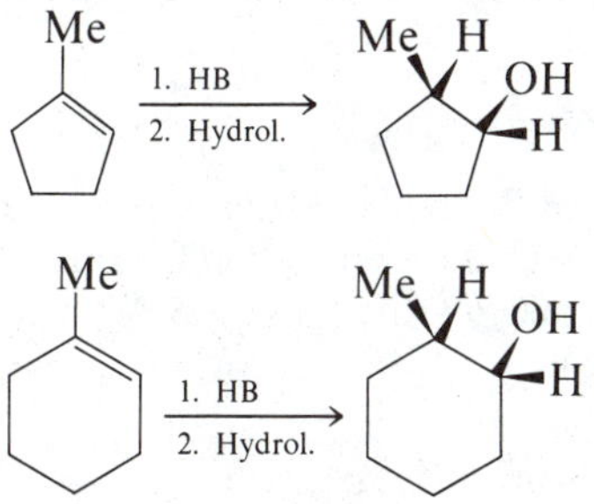

THE B—organo
FUNCTION

Solvolysis of alkyl boranes,† BR_3, interestingly, is accelerated by the presence of aqueous carboxylic acids[7] (mineral acids are not nearly so effective); and hydroperoxide ion, OOH^-, is an excellent catalyst[8] for decomposition of organoboranes. The facility of both reactions in relation to the ineffectiveness of the mineral acids is believed to be a consequence of initial adduct formation by the carbonyl (or peroxo) oxygen atom followed by transfer of an alkyl group to the hydroxy groups (a mechanism for the disproportionation of peroxides bears a good likeness to the second of these; see p. 248):

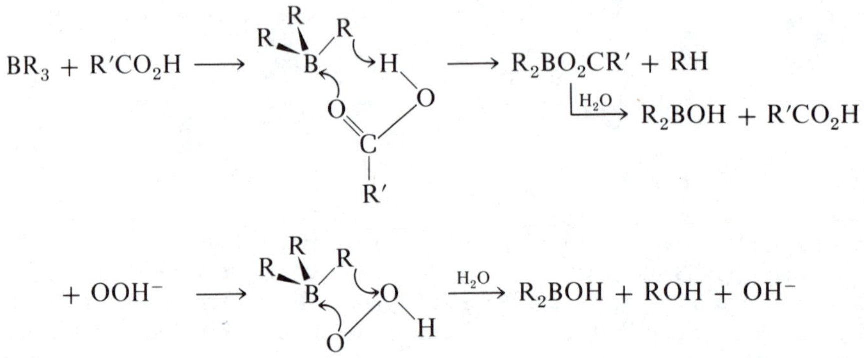

$$BR_3 + R'CO_2H \longrightarrow \text{[adduct]} \longrightarrow R_2BO_2CR' + RH \xrightarrow{H_2O} R_2BOH + R'CO_2H$$

$$+ OOH^- \longrightarrow \text{[adduct]} \xrightarrow{H_2O} R_2BOH + ROH + OH^-$$

The relative strengths of C—O and O—O bonds appear to be as important to the nature of the final products as are the relative strengths (acidities) of the O—H bonds. Both mechanisms (and we must be frank in saying that they are not the only ones consistent with the nature of the products) bear a certain similarity to the "push-pull" concept mentioned earlier for the activation of H_2.

STUDY
QUESTIONS

3. Analyze the fact that $BF_3 \cdot OH_2$ eliminates HF much more slowly than $BCl_3 \cdot OH_2$ eliminates HCl, in terms of bond-breaking and bond-making influences on the activation energy.

4. In a synthesis of B_2H_6 by hydride reduction of BF_3, the reaction products are actually not consistent with

$$3MH + BF_3 \rightarrow \frac{1}{2} B_2H_6 + 3MF$$

but with

$$3MH + 4BF_3 \rightarrow \frac{1}{2} B_2H_6 + 3MBF_4$$

Viewing the reaction as proceeding through steps of H^- attack upon BF_xH_{3-x} ($x = 3,2,1$), what is a possible role for the excess BF_3 required?

Four Valence Pairs

The examples that could be cited in this category are numerous indeed; in fact, examples can be drawn from all the elements in Groups IV through VII. In addition, the adducts of Group III elements fall into this class. We will consider some representative reactions from all these groups.

---

†Note the use of these reactions on p. 228.

BORON

Several four-coordinate boron compounds have been carefully studied for nucleophilic displacement at the boron as the reactive center. The **exchange** reaction of excess Me_3N for that in Me_3NBMe_3 has been shown[10] to proceed with a rate law that is first order in adduct concentration. The enthalpy of activation (+71 kJ/mole) is indistinguishable from the adduct dissociation enthalpy, and the entropy of activation is positive and fairly large (+63 J/K mole). These findings strongly support the view that the activation step in the exchange reaction is unimolecular dissociation of the adduct to give free BMe_3 (a D mechanism), which subsequently rapidly combines with free Me_3N to form the adduct.

It is particularly interesting to contrast this mechanism for donor exchange of Me_3NBMe_3 with those for $Me_3E \cdot NMe_xH_{3-x}$ (E = Ga, In), for which the exchange reactions are so rapid that attempts[11] to determine the rate laws have been unsuccessful. To be so rapid the reactions proceed without adduct dissociation (contrast the BMe_3 case) since the adduct dissociation enthalpies are all about 80 kJ/mole. The somewhat larger sizes of Ga and In atoms presumably are responsible for the differences in displacement mechanisms of these Group III adducts (it may be that the availability of $4d$ and $5d$ atomic orbitals in the valence shells of Ga and In also facilitate the formation of a five-coordinate transition state for Ga and In). Arguments could be made for either I or A mechanisms.

Even more interesting are the results of studies[11] into the mechanism for the electrophilic acid-exchange reactions:

$$Me_3Ga + Me_3Ga \cdot NMe_xH_{3-x}$$

$$Me_3In + Me_3In \cdot NMe_xH_{3-x}$$

For both Lewis acids, the reactions are of the D type *when the amine is tertiary or secondary* [the activation energies are quite close to the adduct bond energies (40 to 80 kJ/mole) and the entropies of activation are positive and large (40 to 120 J/K mole)]. With the primary amine and ammonia, however, the reactions are either A or I according to the rate laws, have activation energies of only half the adduct bond energies, and have negative entropies of activation (–40 J/K mole). The dependence of this mechanism for exchange on the number of methyl groups about nitrogen is seen as a steric phenomenon (refer to Figure 7-2).

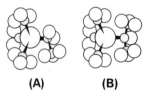

Figure 7–2. (A) $(CH_3)_3GaN(CH_3)H_2$, assuming tetrahedral symmetry of the gallium and nitrogen atoms, shown with van der Waals radii for all protons. (B) $(CH_3)_3GaN(CH_3)_2H$, assuming tetrahedral symmetry of the gallium and nitrogen atoms, shown with van der Waals radii for all protons.

(A)   (B)

A study[12] of the *hydrolysis* of substituted pyridine adducts of BCl_3 resulted in the conclusion that the hydrolysis reactions proceed by a dissociative mechanism

$$py \cdot BCl_3 + 3H_2O \rightarrow py + 3HCl + B(OH)_3$$

in which the activation step seems to involve heterolytic cleavage of a B–Cl bond. H_2O solvation by hydrogen bonding is an important contributor to the selection of the B–Cl bond, rather than the B–N bond, for heterolytic cleavage. The activation energies and rate constants for the first order rate expression suggest the reactivity order

If the process were D in Am·BCl_3, this order would be the exact reverse of that expected for rate determining cleavage of the pyridine bond to boron (there should be a steric enhancement of the rate, relative to 4-Mepy, for the 2-methylpyridine adduct). A truly bimolecular rate step (I path) in which H_2O displaces pyridine or Cl^- is also inconsistent with this order, as the electrophilic character of the boron is expected to follow the order pyridine adduct $>$ 4-methylpyridine adduct. The order is that expected for cleavage of a B—Cl bond. The low rate constant for the 2-methylpyridine adduct is interpreted to mean that the transition state species for this adduct does not experience the expected degree of resonance stabilization (**11**) for a planar boron species

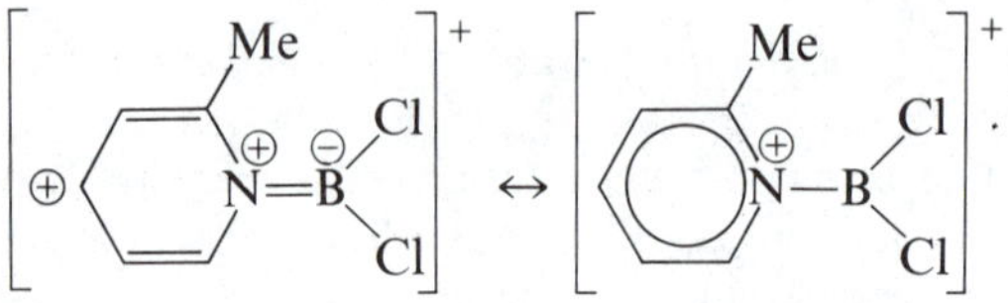

11

because the steric interaction between the 2-methyl group and the remaining chlorines prevents formation of the planar structure **11** by the incipient cation:

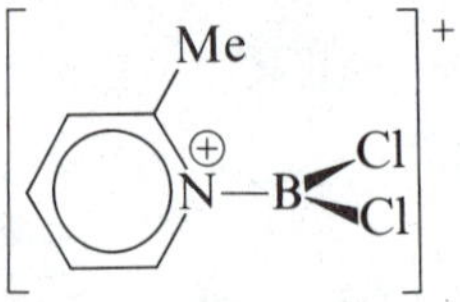

In comparing hydrolysis (in H_2O) of py·BX_3 with aminolysis (in MeCN) of MeCN·BX_3, one finds a drastic change in reaction channel: B—X $vs.$ MeCN—B cleavage. The greater basicity of py than MeCN makes D—B bond cleavage harder, while simultaneously making X more anionic, with greater $X\cdots H_2O$ solvation to aid B—X cleavage.

THE B—H FUNCTION

Hydrolysis of amine adducts of BH_2(aryl) and BH_3 seem always to involve a process that is dependent on H_3O^+. When pyridine adducts of BH_2(aryl) were studied,[13] it was found that the rate constant increases with the electron releasing ability of the substituent at the 4 position of the phenyl ring. Thus, the relative rate constants were X = Cl (0.5), H (1.0), CH_3 (4.3), and CH_3O (22.3).

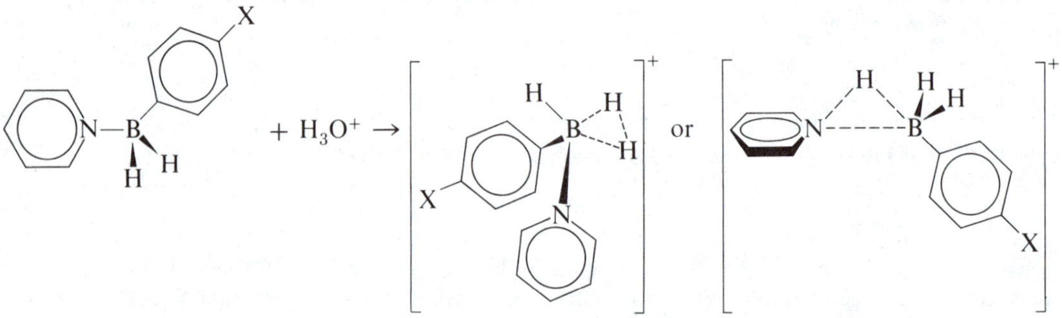

12   **13**

This order fits nicely with the concept (an A path) of the transition state for these reactions as attack of the electrophile H^+ either upon the hydridic hydrogens of the borane unit (**12**) (compare the earlier discussion of BH_4^- hydrolysis in acid media) *or upon the pyridine nitrogen* (**13**).

An extensive series of studies[14] into the mechanisms of *hydrolysis* of amineboranes has led to the conclusion that these acid catalyzed reactions proceed by *attack at the amine nitrogen*. The BH_3 adducts of the tertiary amines Me_3N, Et_3N, and quinuclidine

($\langle\hspace{1.5em}N\rangle$) all hydrolyze according to a second order rate law: first order in each of

amineborane and H_3O^+. The activation energies are all closely the same (100 kJ/mole) and the activation entropies are quite small. The similarity in behavior of Et_3N and quinuclidine adducts seems to strongly militate against rearward attack by H_3O^+ at nitrogen (*cf.* Me_3Ga attack at nitrogen of $Me_3N\cdot GaMe_3$ on page 231). Substituted aniline adducts of BH_3 also exhibit[15] a second order hydrolysis pathway (k_2), but a term that is first order in adduct concentration (k_1) is also observed. Given the following information, the first order mechanism appears to involve adduct dissociation (a *D* path) while the second order term involves frontal H_3O^+ attack at nitrogen (an *I* or *A* path):

(a) electron-releasing phenyl substituents (of aniline) reduce k_1 and increase k_2, while withdrawing groups have the opposite effects;
(b) replacement of aniline hydrogens by alkyl groups increases k_1 but reduces k_2;
(c) the BD_3 adducts exhibit rates of hydrolysis no different from those of the isotopic BH_3 adducts (this seems to eliminate a BH_5-like intermediate);
(d) increasing alkyl substitution in the series RNH_2, R_2NH, R_3N results in decreasing k_2 (k_2 should increase if a BH_5-like intermediate were formed, but would decrease from steric congestion about nitrogen for H^+ attack at the N—B bond).

Presumably, an explanation for the appearance of a first order pathway for the anilines relative to the tertiary amine donors is their lower base strength. In contrasting the greater hydrolytic stability of the amineboranes with that of BH_4^-, we might cite the greater electronegativity of the amine nitrogen than of the hydride ion as rendering the borane hydrogens less hydridic (recall that with BH_4^-, rapid reaction of H_3O^+ with the borane hydrogens occurs). Most interesting is the general observation that phosphine borane adducts are stable toward hydrolysis in *hot* mineral acids. At present there appears to be no real understanding of this property of the phosphine boranes (but see study question 7).

Particularly intriguing are the *solvolyses* by ammonia of simple borane adducts. For example, the simple ethane analog H_3NBH_3 can be formed by the reaction

$$R_2OBH_3 + NH_3 \xrightarrow{\text{ether}} H_3NBH_3 + R_2O$$

or by

$$LiBH_4 + NH_4Cl \xrightarrow{\text{ether}} LiCl + H_2 + H_2NBH_3$$

In the former reaction it is not known whether the borane substrate is the dialkyl ether adduct or free (monomeric) borane, but it seems unlikely that the reaction could proceed through attack on B_2H_6 by NH_3 (see the next paragraph). The stoichiometric mechanism for the second reaction is most likely to involve attack upon BH_4^- by NH_4^+ or H^+ to liberate BH_3 or the ether adduct (recall the acid catalyzed hydrolysis reaction of BH_4^-), either of which could be attacked by NH_3 as in the first reaction.

As an example of kinetic (as opposed to thermodynamic) control of reaction products, the reaction of B_2H_6 with liquid ammonia at temperatures lower than $-78°C$ produces[16] a mono-hydrogen bridged adduct by partial cleavage of the borane dimer (note the structural analogy with $B_2H_7^-$):

Warming the reaction mixture above $-78°C$ to remove the solvent ammonia stimulates further reaction to form the "diammoniate" of diborane with a salt structure

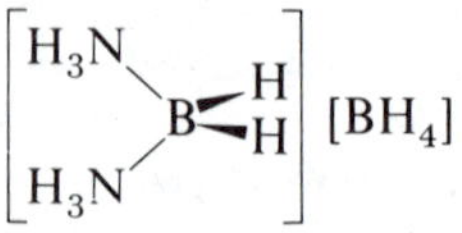

$$\left[\begin{array}{c} H_3N \\ \\ H_3N \end{array} \searrow B \begin{array}{c} H \\ H \end{array} \right] [BH_4]$$

Note the *unsymmetrical* cleavage of diborane in the absence of ether solvent (whose role seems to be to cleave B_2H_6 *symmetrically*).

Whether a reaction of some donor with B_2H_6 will lead to symmetrical or unsymmetrical cleavage of the diborane cannot be predicted with great certainty yet. Review[17] of the known examples of both kinds of cleavage does bring out a pattern and suggests a few important concepts. All such reactions are presumed to proceed by a two-step mechanism:

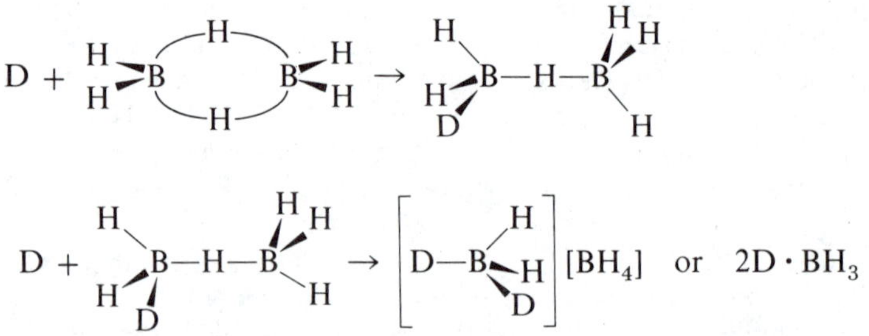

Selection of one or the other of the two boron atoms as the point of attack by the second donor atom seems to determine the course of the reaction. By virtue of the greater electronegativity of most donor atoms than of H^-, the relative partial charges of the two borons would tend to favor unsymmetrical cleavage. If the donor molecules are large, however, steric congestion during attempted attack at the same boron would tend to direct the second donor to the other boron to produce $2 D·BH_3$. Such a pattern has been established for 1,2-tetramethylene diborane and 1,2-*bis*(tetramethylene) diborane:

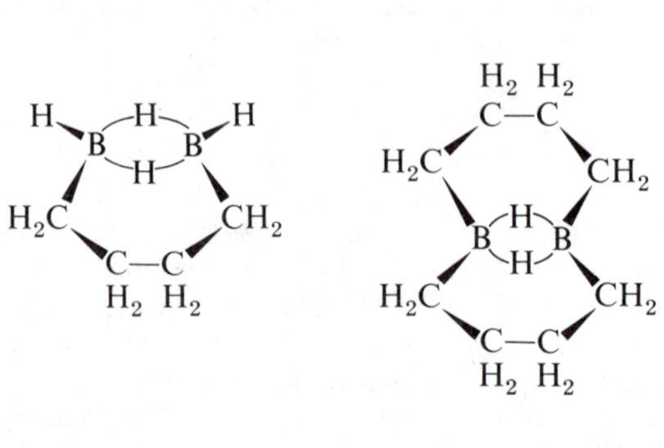

Since the latter borane is more sterically hindered than the first, which, in turn, is more hindered than diborane, the tendencies noted in Table 7-1 for the $NH_{3-x}Me_x$ series of donors make good sense. Steric factors may also play a role in the formation of symmetrical cleavage products for phosphines and sulfides, although the low electronegativities of the donor atoms in these Lewis bases may be an additional factor. As noted earlier, ethers seem to effect symmetrical cleavage of diborane; this might be explained by steric congestion, but the relatively high electronegativity of the oxygen atom in these donors would seem to favor unsymmetrical cleavage. Quite possibly the weaknesses of

Borane	Donor	Cleavage
$B_2C_8H_{18}$	NH_3	unsymmetrical
	$NMeH_2$	symmetrical
	NMe_2H	symmetrical
	NMe_3	symmetrical
$B_2C_4H_{12}$	NH_3	unsymmetrical
	$NMeH_2$	unsymmetrical
	NMe_2H	symmetrical
	NMe_3	symmetrical
B_2H_6	NH_3	unsymmetrical
	$NMeH_2$	unsymmetrical $\gg$ symmetrical
	NMe_2H	symmetrical $>$ unsymmetrical
	NMe_3	symmetrical
	R_3P	symmetrical
	R_2S	symmetrical
	R_2O	symmetrical
	H_2O	(unsymmetrical)

TABLE 7-1

SUMMARY OF BORANE CLEAVAGE PRODUCTS

the ether oxygen-boron bonds in the $BH_2(OR_2)_2^+$, if formed, make possible a rapid hydride transfer from BH_4^- to the cation with the equilibrium favoring the simple ether·borane adduct. Suuport for these ideas comes from the recent synthesis[18] of $BH_2(OH_2)_2^+BH_4^-$ at $-130°C$, where the donor is the unhindered water molecule. The possibility of rapid hydride displacement of one of the H_2O molecules from the cation could not be examined, of course, owing to rapid H_2 evolution at the higher temperatures apparently necessary for such a displacement reaction.

STUDY QUESTIONS

5. Can you draw a structual and chemical analogy between frontal attack of H^+ upon the $N-B$ bonds of amine boranes and upon the $B-H$ bond of BH_4^-?

6. Why could attack of H^+ upon the nitrogen atom of py·BH_2(aryl) and py·BH_3 be reasonable when it is known that such attack does not occur for py·BX_3?

7. Could hyperconjugation in $H_3P·BH_3$ account for the stability of this adduct in hot mineral acids?

SILICON[2]

Not surprisingly, one of the most widely studied classes of reagents is the silanes, in particular the organosilanes. A very important aspect of the reactions of silanes is the absence of firm evidence for the formation of a siliconium ion, the silicon counterpart of the carbenium ion known to be important in many reactions of carbon. On the contrary, the evidence for a wide range of silanes is that substitution and solvolysis reactions involve an associative step (often the activation step) between a nucleophile and the silane substrate. Presumably, the ease with which silicon can expand its valence shell is of great importance to this general aspect of silane reactions. *The ability of third row atoms to expand their valence shells is widely held to be responsible for many of the reactivity differences between second and third row elements.* Comparing carbon and silicon, for example, R_3CCl is known to hydrolyze slowly by either an I or D mechanism. For R_3SiCl, however, the solvolysis reactions are generally fast and are found to obey second order rate expressions; considerable data exist that support the a mode of activation for many of these reactions.

Evidence[19] for an A mechanism comes from studies of the class of silanes symbolized by the formula Ar_3SiO_2CR (Ar is a substituted aryl group). In the alcoholysis reactions (formation of Ar_3SiOR')

$$Ar_3Si(O_2CR) + R'OH \rightarrow Ar_3SiOR' + HO_2CR$$

of these compounds, it has been determined that electron-withdrawing substituents on the aryl ring and R groups enhance the rate of reaction, whereas electron-releasing substituents on the aryl moieties lower the rate constants. The effect of aryl substituents is, of course, opposite to that expected for an activation step of dissociation of RCO_2^-. Providing further characterization of the transition state for the solvolysis reactions, studies[20] of relative rate constants for the hydrolyses of the Si—H bonds in the compounds Et_3SiH and the bicyclosilanes **14** and **15** have yielded the ratios $1:10:10^3$.

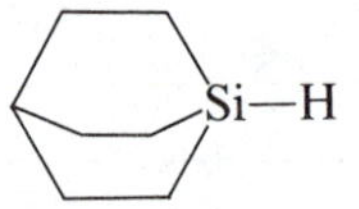

14 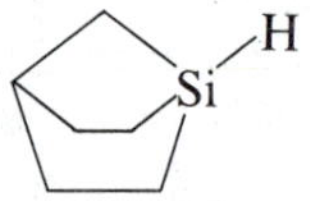   **15**

$$(14 \text{ or } 15) + H_2O \xrightarrow{OH^-} R_3SiOH + H_2$$

For the bicyclo compounds, front side attack is required and little difference is found between the triethyl and bicyclooctyl structures. The norbornane analog **15** has a structure of a tetrahedron highly distorted toward a geometry appropriate to a five-coordinate transition state (the hydrogen and ring carbons are "equatorial," the bridgehead carbon is "axial").[21] Hence the larger rate constant.

In tetrahedral carbon chemistry, where five-coordinate intermediates are uncommon, retention of configuration usually implies a D mechanism while inversion implies an I mechanism. Extension of this generalization to silicon is not possible because the A mechanism (with formation of a trigonal bipyramidal intermediate) is likely; subsequent steps with retention *or* inversion are possible, as shown below.

In generalizing[22] the conditions for a change in silicon chirality, it appears that *good leaving groups, X, in combination with nucleophiles more basic than X*, favor inversion (a good leaving group is defined in this context as one with a pK_b greater than 8, such as OAc^-), whereas poor leaving groups usually lead to retention of configuration but sometimes exhibit inversion (a poor leaving group is one with a pK_b less than 4, such as H^- and OR^-). Thus, R_3SiCl and R_3SiO_2CR' (usually) change configuration in substitution reactions, whereas R_3SiOR' and $R_3SiOSiR'_3$ frequently retain their configurations. By way of examples we can take the following as characteristic (Si^* implies an optically active Si):

$$R_3Si^*OCH_3 + AlH_4^- \rightarrow R_3Si^*\!\!\overset{\displaystyle Me}{\underset{\displaystyle H}{\overset{\displaystyle O}{\diagup\!\!\diagdown}}}\!\!AlH_3 \xrightarrow[Et_2O]{retention} R_3Si^*H + AlH_3(OMe)^-$$

$$R_3Si^*OAc + MeOH \xrightarrow[xylene]{inversion} R_3Si^*OMe + HOAc$$

The racemization reaction[23]

$$R_3Si^*F \xrightarrow{MeOH} racemization$$

is a striking example that not only makes us entertain the idea of a five-coordinate intermediate but also leads us to the important concept of *pseudorotation* as a pathway for intramolecular rearrangements.† For silicon compounds this concept provides (i) a mechanism for racemization *without* the necessity of siliconium ion formation and (ii) a mechanistic route for a rearward attack *with retention of configuration*.

While several pseudorotation mechanisms have been proposed, we will confine our attention to just one that is most often believed to be correct—the so-called **Berry mechanism**. The Berry mechanism[24] is a *pseudorotation process for simultaneously interchanging two equatorial groups with the two apical groups, while the third equatorial group* (called the **pivot group**) *remains an equatorial group*. The whole process is represented by the symbol ψ_i, where i is the pivot group. The transition state for the pseudorotation is (ideally) a square-based pyramidal structure.

The mechanistic significance of the trigonal bipyramidal *intermediate* and its pseudorotation is illustrated in the chart on the following page.

The methanol-catalyzed racemization of R_3Si^*F follows the left-most path. Methanolysis of R_3Si^*OAc seems to follow the left-center path (or perhaps simple Walden inversion prior to ψ_2). Whether one finds inversion or retention depends on the relative rates of ψ_N and ψ_X.§ Whether substitution occurs depends on the ratio of the rate of ψ_1 to the rate of loss of X or of N. The sufficient conditions for Walden inversion are that ψ_2 is slow and X is a good leaving group.

Relative electronegativities and π-bonding properties of the groups bound to the central atom may influence the relative energies of the isomers. For known ground state trigonal bipyramidal molecules, more electronegative and weakly π-bonding groups are directed to an apical positon; pseudorotamers with the least electronegative and most strongly π-bonding groups in equatorial positions should be lower in energy than those that place such groups axially. Thus, we can often identify (via the preferred pivot atom) the preferred pseudorotations on the basis of electronegativity and π-bonding arguments.

STUDY QUESTIONS

8. Propose a mechanism for substitution at tetrahedral silicon that could lead to a mixture of products, some exhibiting retention and others exhibiting inversion of the original chirality.

9. Predict whether retention or inversion of chirality about silicon is expected for

   a) $R_3SiOR' + KOH \rightarrow R_3SiOH + KOR'$

   b) $R_3SiO_2CR' + LiAlH_4 \xrightarrow{Et_2O} R_3SiH$

   c) $R_3SiCl + MeNO_2 \rightarrow R_3SiCl + MeNO_2$ (assume $\psi_{Cl} > \psi_{MeNO_2}$)

   d) $R_3SiCl + KOH \xrightarrow{xylene} R_3SiOH + KCl$ (assume $\psi_{Cl} < \psi_{OH}$)

---

†By this is meant any process internal to a molecule for permuting the positions of groups about a central atom. A recent discussion of the more general aspects is given by J. I. Musher, *J. Chem. Educ.*, **51**, 94 (1974).

§An often useful predictor is that in the trigonal bipyramidal intermediate, the structure with the more electron withdrawing of X and N at an axial site is preferred (see Study Question 9).

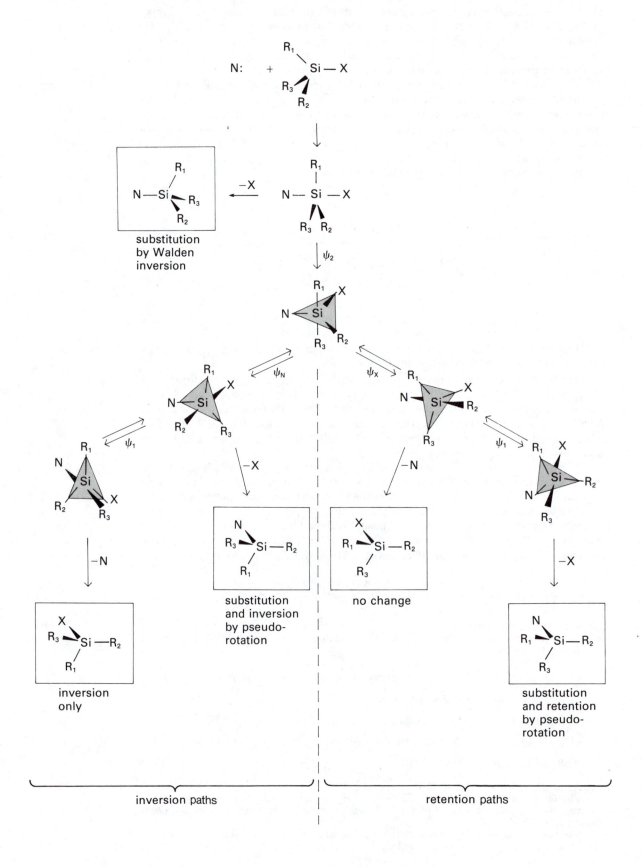

10. Is it possible for a nucleophile to displace X, with inversion of silicon chirality,

$$R_3SiX + Nuc \rightarrow R_3SiNuc + X$$

by attacking at an $R-Si-X$ edge?

11. Compare an I displacement of X from R_3SiX, where inversion accompanies displacement, and the A mechanism for the same reaction.

NITROGEN AND PHOSPHORUS

Trivalent nitrogen provides for many interesting reaction pathways because of the great variety of compounds and structures known. Simple *electrophilic* **substitution** of coordinated BMe_3 by free BMe_3 in the adduct Me_3NBMe_3 has been shown[10] to proceed by the same d activation you discovered (p. 231) for exchange of coordinated and free Me_3N in the adduct. On the other hand, exchange of free and coordinated $GaMe_3$ and $InMe_3$ in adducts with primary amines and ammonia has been shown[11] to occur by an I mechanism, presumably through the transition state structure **16**.

THE N−B, Ga, In FUNCTIONS

16

With secondary and tertiary amine adducts, exchange of free and coordinated $GaMe_3$ and $InMe_3$ proceeds by a D mechanism like that of BMe_3.

Displacement reactions by *nucleophilic attack* upon substituted amines, NH_2X, have received considerable attention because they are involved in synthesis of hydrazine from chloramine and of hydroxylamine from nitrous acid, to name just two examples. Quite generally the rate laws are second order—that is, first order in both the nitrogen species and the nucleophile. Even so, as you learned in the introductory section of this chapter, different mechanistic pathways are consistent with such a rate law:

a) $NH_2X + Nuc \rightarrow NucNH_2^+ + X^-$
b) $NH_2X + Nuc \rightarrow NucX^+ + NH_2^-$
c) $NH_2X + Base \rightarrow NHX^- + BaseH^+$

then

$$NHX^- + Nuc \rightarrow NucNH + X^-$$

or (when Base = Nuc)

$$NHX^- \rightarrow products$$

Mechanism (a) is meant to imply attack by Nuc upon the nitrogen atom, (b) is meant to imply attack upon X, and (c) is called the **conjugate base (CB) mechanism**, for the first step is rapid generation of the conjugate base of the substrate. Which of these mechanisms is followed by a given system depends on the detailed nature of X and Nuc; generally speaking, pathways (a) and (b) are most commonly found and are of the I variety—that is, the rates depend on both the bond-making (Nuc attacking the substrate) and bond-breaking (X departing from the substrate) features of attaining the transition state.

A substituted amine that exhibits many of the characteristic features of the displacement reaction (a) for trivalent nitrogen is the hydroxylamine-O-sulfonate anion[25] ($H_2NOSO_3^-$). At low acidities the *anion* reacts with various nucleophiles such as triphenyl-

THE N−O FUNCTION

phosphine (to form $Ph_3PNH_2^+$), triethylamine (to form $Et_3NNH_2^+$), and I^- (to form $\dot{N}H_3$ and I_2). The reaction with I^- is ostensibly an oxidation-reduction reaction, but all three reactions embody the steps characteristic of displacement reactions. That is, all have in common a second order rate expression that is first order in the nucleophile and first order in hydroxylamine-O-sulfonate. The second order rate constant is independent of H^+ concentration until the pH becomes low enough to cause appreciable concentration of the conjugate acid to build up. That the second order rate constant is sensitive to the nature of the nucleophile and exhibits a large negative activation entropy (~ -120 J/K mole) emphasizes the importance of bond-making in the transition state. That Ph_3P and I^-

$$Ph_3P(2.0 \text{ M}^{-1} \text{ sec}^{-1}) > I^-(0.07) > Et_3N(0.02) \gg Br^-(< 10^{-4}) > Cl^-$$

react fastest with the anion is believed to reflect an important contribution from retro-bonding between the nitrogen lone pair and P and I orbitals so as to lower the activation energy of attack upon nitrogen. A further indication that the activation step is of the *I*

$$H_3P\text{------}NH_2^+\text{------}OSO_3^-$$

type stems from the observation that *N*-alkyl substitution reduces the rate constants. The general mechanism then depends heavily on attack at the nitrogen with simultaneous displacement of the SO_4^{2-} group.

The fact that the rate constants for these reactions *decrease* as the pH is lowered to a point where the anion is appreciably converted to its conjugate acid is an important clue. A decrease in rate is what is to be expected if the conjugate acid

$$H_2NOSO_3^- + H^+ \rightleftharpoons H_2NOSO_3H$$

is less reactive than the anion. Independent experiments, under such acid conditions that the anion is fully protonated, reveal no change in the form of the rate law (where the acid simply replaces the anion) and so no drastic change in mechanism is expected. The effect of anion protonation is confined to a reduction (by nearly an order of magnitude) in velocity of the reaction—the activation energy is simply slightly higher for attack on the acid than on the anion. Of considerable importance is the site of anion protonation.[26] Were an oxygen of the sulfato group to be protonated (H_2NOSO_3H), you are likely to be thinking that the rate constant should *increase;* for not only should HSO_4^- be a better leaving group (weaker base) than SO_4^{2-}, but also the more electronegative OSO_3H group should render nitrogen more susceptible to nucleophilic attack. Were protonation to occur at the nitrogen (to form the zwitterion form $H_3N^+OSO_3^-$), you should still predict that the increased nitrogen formal charge would facilitate the attack, but *hinder* the N—O bond-breaking as SO_4^{2-} is displaced from NH_3^{2+}. Clearly, this view is consistent with the experimental observation of a lower rate constant for the acid than for the anion. Consequently, it appears that protonation occurs (in solution) at the nitrogen, and this incurs a slightly greater hindrance to bond-breaking than enhancement of bond-making (I_d).

In a related reaction, this time with H_2NOH, organophosphines undergo a general acid catalyzed (H^+, BF_3) oxidation to phosphine oxides.

$$R_3P + NH_2OH \xrightarrow{\text{acid}} R_3PO + NH_3 \cdot \text{Acid}$$

In contrasting this reaction with those given above in which the phosphine attacks the *nitrogen* of $H_2NOSO_3^-$ and the *nitrogen* of H_3NOSO_3, note the difference in apparent electrophilicities of the bridging oxygens of H_3NOH^+ and H_3NOSO_3. Thus, Ph_3P appears to favor attack at oxygen of H_3NOH^+ but at the nitrogen of H_3NOSO_3. The dilemma of the fickle behavior of phosphorus is resolved by recognizing the considerable steric congestion in the transition state for attack at the bridging oxygen of hydroxylamine-O-sulfonate.

The *hydrolysis* of nitramide, H_2NNO_2, is most interesting in that the strongly conjugating and electron withdrawing NO_2 group alters the mechanism for hydrolysis from type (a) for $X = OSO_3^-$ to the CB pathway[27] [type (c)]. The hydrolysis of nitramide is general-base catalyzed (not limited to OH^-), and the product is N_2O. [With OH^-, (a) or (b) would have led to $H_2NOH + NO_2^-$ or $NH_3 + HNO_3$ as products.]

$$H_2NNO_2 + \text{base} \rightleftharpoons HNNO_2^- + \text{base} \cdot H^+$$

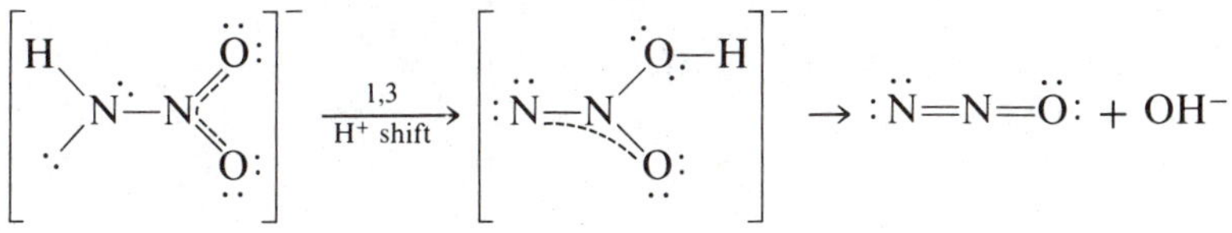

The general rate expression for the hydrolysis is first order in nitramide and first order in each base present:

$$\text{Rate} = [H_2NNO_2]\left\{ \sum k_B [B] \right\}$$

It is expected that the activation step is the *D* loss of OH^- from $N_2O_2H^-$, as both proton transfer steps are expected to be relatively fast. (Note the analogy of $NN(O)OH^-$ to $ON(O)OH$ and of NNO to ONO^+.)

Many of the redox reactions of nitrous acid are analogous in that OH^- is displaced by nucleophiles. For example, the nitrosation reactions of amines at low concentrations of HNO_2 follow a third order rate law:[32]

$$HNO_2 + R_2NH \rightarrow R_2NN{=}O + H_2O$$

$$\text{Rate} = k\,[\text{amine}][H^+][HNO_2]$$

Judging from the rate expressions, one mechanistic possibility is that a rapid protonation of HNO_2 establishes a steady-state concentration of $H_2NO_2^+$ (although it is not known which oxygen is protonated, **17** is more likely the kinetically important one), which

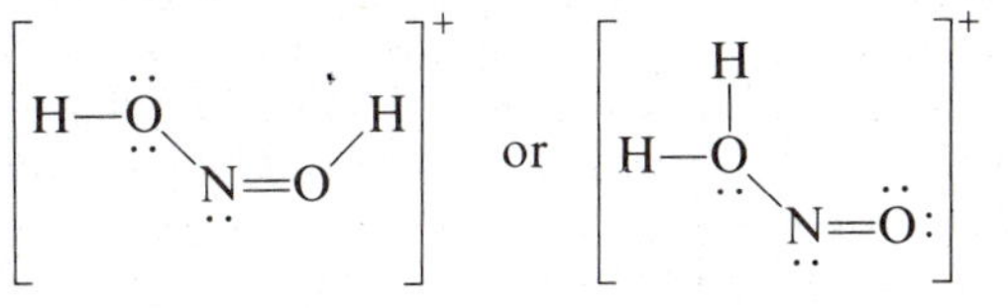

17

experiences a rate controlling attack by the amine to form, after rapid loss of a proton, the product nitrosamine:

$$H_2NO_2^+ + \text{Amine} \xrightarrow{\text{slow}} \text{Amine}{-}NO^+ + H_2O$$

$$R_2HN{-}NO^+ \xrightarrow{\text{fast}} R_2N{-}N{=}O + H^+$$

Interestingly, the rate law expression changes drastically at high concentrations of HNO_2. The third order expression

$$\text{Rate} = k\,[\text{amine}][HNO_2]^2$$

most likely arises from disproportionation of HNO_2 into N_2O_3 (NO_2^- attack upon $H_2NO_2^+$ or HNO_2 upon HNO_2),

$$2HNO_2 \rightleftharpoons H_2O + \underset{}{\overset{}{\text{O}_2\text{N}-\text{N}\text{O}}}$$

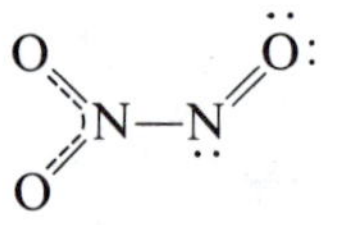

followed by rate controlling attack at the nitrosyl nitrogen of N_2O_3 to form, after elimination of H^+, the product nitrosamine:

$$\text{Amine} + N_2O_3 \xrightarrow{\text{slow}} \text{Amine}-NO^+ + NO_2^-$$

$$R_2HNNO^+ \xrightarrow{\text{fast}} R_2N-N{=}O + H^+$$

With secondary amines the reactions cease at production of the *N,N*-dialkyl nitrosamine. With primary amines, however, the initial product (*N*-alkyl nitrosamine) is unstable and decomposes, after a 1,3-tautomerization to form a hydroxyimide (**18**), into N_2 and ROH.

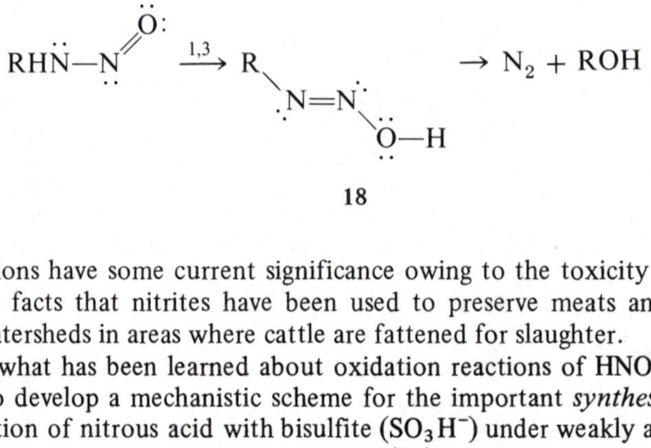

$$RH\ddot{N}-N{=}\ddot{O} \xrightarrow{1,3} R-N{=}N-\ddot{O}-H \rightarrow N_2 + ROH$$

18

These reactions have some current significance owing to the toxicity of nitrosamine to man and the facts that nitrites have been used to preserve meats and are found in abundance in watersheds in areas where cattle are fattened for slaughter.

THE N–S FUNCTION

In view of what has been learned about oxidation reactions of HNO_2, you are in a good position to develop a mechanistic scheme for the important *synthesis of hydroxyl amine* by reduction of nitrous acid with bisulfite (SO_3H^-) under weakly acid conditions. An intermediate hydroxylaminedisulfonate acid (**19**) can be isolated in this reaction

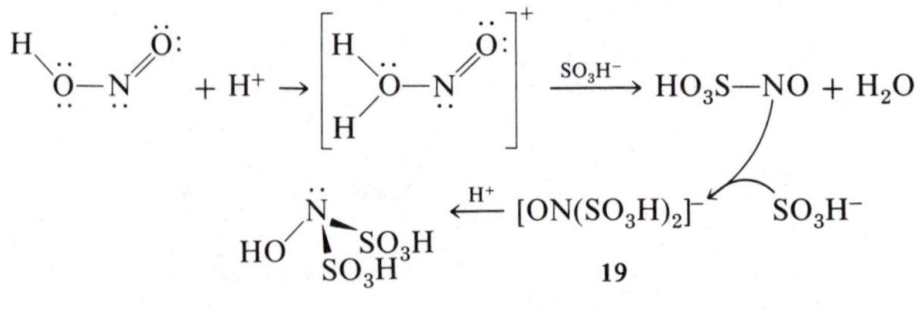

$$H-\ddot{O}-N{=}\ddot{O} + H^+ \rightarrow \left[\overset{H}{\underset{H}{\ddot{O}}}-N{=}\ddot{O} \right]^+ \xrightarrow{SO_3H^-} HO_3S-NO + H_2O$$

$$\underset{\text{HO}}{\overset{SO_3H}{N}}{-}SO_3H \xleftarrow{H^+} [ON(SO_3H)_2]^- \xleftarrow{SO_3H^-}$$

19

so the scheme seems entirely reasonable. (Compare the nitrosation reactions of amines.) As in the hydrolysis of amine$-$O$-$sulfonates,† formation of the zwitterionic form of the acid (20), or an assist by the acidic conditions, could lead to solvolysis in successive steps. This time a monosulfonic acid intermediate (21) can be isolated.

$$(HO)N(SO_3H)_2 \rightleftharpoons \underset{\mathbf{20}}{\overset{\displaystyle H}{\underset{\displaystyle}{\,}}} \quad \xrightarrow{H_2O} [(HO)NH(SO_3H)] + H_2SO_4$$

$$HONH_2 + H_2SO_4 \xleftarrow[H_2O]{}$$

$$\mathbf{21}$$

In application, this method of preparation of hydroxylamine suffers if one is careless and uses an excess of bisulfite, for hydroxylamine itself will oxidize bisulfite and form ammonia.

$$NH_2OH + SO_3H^- \rightarrow NH_3 + H_2SO_4$$

In view of the preceding mechanisms for H_2NX as an oxidizing agent, we might expect bisulfite to attack at nitrogen, first to displace OH^-; this could be followed by H_2O attack upon the H_2N-SO_3H form of H_3NSO_3. On the other hand, recalling the hydroxyl amine oxidation of Ph_3P, attack by sulfur could also occur directly at the OH oxygen after protonation of N, with displacement of NH_3.

The chloramines (NH_2Cl, $NHCl_2$, and NCl_3) provide us with some interesting, and often very complex, reactions.[28] The detailed kinetic information available for their reactions is somewhat limited by the difficulties in handling them. A few courageous chemists, with the requisite confidence in their laboratory technique, have provided what data is available. All these compounds are best handled in solution, where their detonation hazard is somewhat reduced. Their synthesis is achieved by chlorine *oxidation* of ammonia in aqueous media, where the active oxidizing species is actually hypochlorite or its conjugate acid.

THE N$-$Cl FUNCTION

$$NH_3 + OCl^- \rightarrow NH_2Cl + OH^-$$

$$\underset{(8 < pH < 12)}{2NH_2Cl} + H^+ \rightleftharpoons \underset{(3 < pH < 5)}{NHCl_2} + NH_4^+$$

$$\underset{(3 < pH < 5)}{2NHCl_2} + H^+ \rightleftharpoons \underset{(pH < 3)}{NCl_3} + NH_4^+$$

All three compounds are related by strongly pH dependent equilibria, so pH control is critical to successful synthesis of one or the other. Of the three chloramines, only NCl_3 is sufficiently insoluble in water that it soon separates as a yellowish oil phase.§ This is dangerous, for the "pure" liquid is shock- and light-sensitive. Fortunately, dilute CCl_4

---

†Note, however, the occurrence of an N$-$S, not N$-$O, bond in these compounds.
§Note that the low solubility is expected for a very weakly basic amine.

solutions are not too unstable, so extraction[29] of NCl_3 into the CCl_4 layer of a two-phase H_2O/CCl_4 synthesis medium provides a way to avoid many of the dangers of NCl_3 synthesis.

Of the three chloramines, NH_2Cl is the easiest to work with and undergoes an important *reductive* **solvolysis**[30] by *liquid ammonia*. Attack by NH_3 at the chlorine would lead to no net reaction (although an exchange of N atoms would be effected, this appears not to have been studied), whereas attack by NH_3 on the nitrogen of NH_2Cl produces hydrazinium chloride (with the excess NH_3 that is present, the hydrazinium ion is converted to hydrazine). This reaction constitutes the basis for the **Raschig**[31] **synthesis of hydrazine**.

$$NH_3 + NH_2Cl \rightarrow N_2H_5^+ + Cl^- \xrightarrow{NH_3} N_2H_4 + NH_4^+ + Cl^-$$

Since the preparation of chloramine itself is conveniently carried out by the reaction of NH_3 with Cl_2, a useful synthesis of hydrazine entails the chlorination of liquid ammonia.†

Throughout this chapter you can find numerous examples of **trivalent phosphorus** in the role of a nucleophile. One of the less conventional but more important of these reactions might be termed *oxidative addition*. A particularly simple example involves the chlorination of PCl_3:

$$PCl_3 + Cl_2 \rightarrow PCl_5$$

This reaction is reversible to a slight extent ($\Delta G°$ only ~38 kJ/mole for the *gas phase* chlorination reaction). Though no mechanistic studies of this reaction have been performed, in the gas phase, at least, an asymmetric bimolecular **addition** seems plausible.

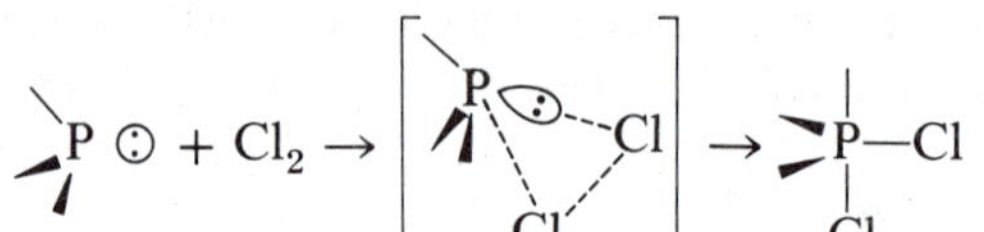

$$\searrow\!\!\!\overset{}{P} \odot + Cl_2 \rightarrow \left[\begin{array}{c} \overset{P}{\diagup}\!\!\!\cdots\!\!\overset{\cdot\cdot}{}\!\!\!Cl \\ Cl \end{array} \right] \rightarrow \searrow\!\!\overset{|}{\underset{|}{P}}\!\!-Cl$$

THE P—OR FUNCTION

Just such a mechanism was proposed by Arbusov[33] in his studies of the reactions of alkyl phosphites with alkyl chlorides:

$$(RO)_3P + R'Cl \rightarrow (RO)_2P(O)R' + RCl$$

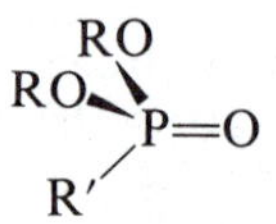

$$\begin{array}{c} RO \\ RO\!\!\searrow\!\!\!P{=}O \\ R' \end{array}$$

----

†Note that the formation of HOCl in water is analogous to the chlorination of NH_3 in ammonia. Also, the fact that neither $NHCl_2$ nor NCl_3 is found in liquid NH_3 means either that NH_3 attack upon them to form NH_2Cl is very rapid or that NH_2Cl is not sufficiently nucleophilic toward Cl_2 to compete with NH_3 attack on Cl_2.

Presumably the first step in this reaction is addition of R'X, followed by loss of X⁻, followed by X⁻ attack upon R.

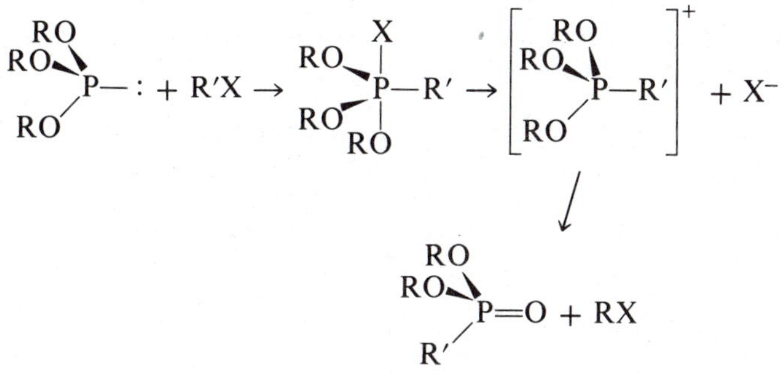

In support of just this mechanism, recent studies[34] of the chlorination of a cyclic catechol phosphite proved the formation of the five-coordinate intermediate

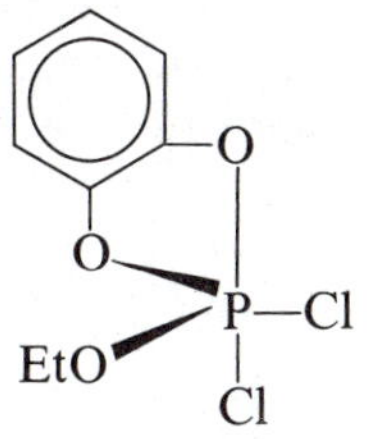

at −85°C. On warming to −40°C, the Arbusov product

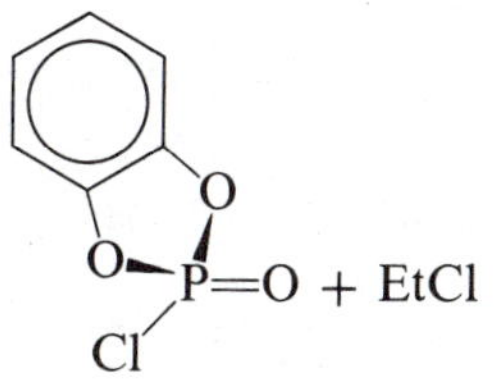

formed.

As a less conventional example[35] of the oxidative addition step, n-BuLi exchanges organo groups with the optically active phosphine MePhP(CH₂Ph) and changes the phosphine chirality:

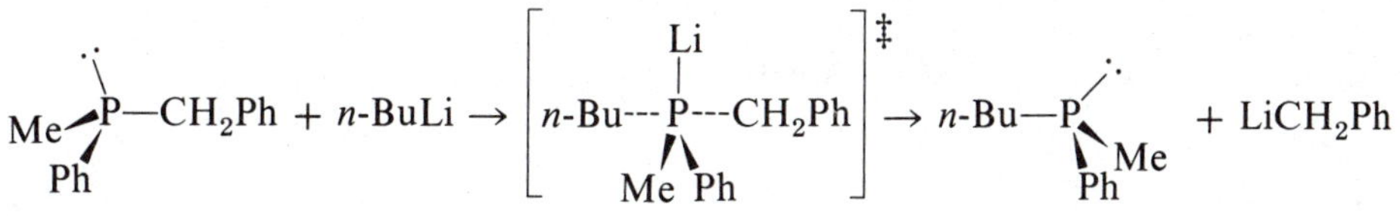

THE P–R FUNCTION

The special features of this reaction are the need for the lone pair to take an equatorial position in the transition state and/or intermediate, the stabilization of the lone pair by Li⁺, and the lack of the two successive pseudorotations (ψ_{Li}, then ψ_{benzyl}) required for retention. Though ψ_{Li} might be facile, pseudorotation about benzyl would cause Li to take an axial position prior to loss of benzyl.

STUDY QUESTIONS

12. Propose a mechanism for the decomposition of hyponitrous acid, via $HN_2O_2^-$, and compare it with the base catalyzed decomposition of the isomeric nitramide.

13. What analogy can you find between the dehydration reaction of HNO_2 at low pH and the nitrosation of amines by HNO_2?

14. Why do you suppose chlorination of liquid ammonia fails to yield $NHCl_2$ and NCl_3 as products?

15. Write a reaction scheme for the non-redox conversion of HNF_2 into N_2F_2 by OH^-.

OXYGEN AND SULFUR

THE O–O FUNCTION

For examples of reactions involving central atoms with four electron pairs, two of which are lone pairs, we look to the reactions of two-coordinate oxygen and sulfur compounds. In addition to the many examples encountered earlier in which oxygen acts as a nucleophile, there are other interesting examples in which the oxygen may appear electrophilic (refer to the last section for rections of phosphines with NH_2OH). Reactions in which hydrogen peroxide acts as an **atom transfer oxidant**† fit this category. With oxygen acting as an electrophile, we find a ready explanation for the oxidation of CN^- to cyanate, OCN^-, of sulfides to sulfoxides, of iodide to iodine, and of carbanions to ethers:

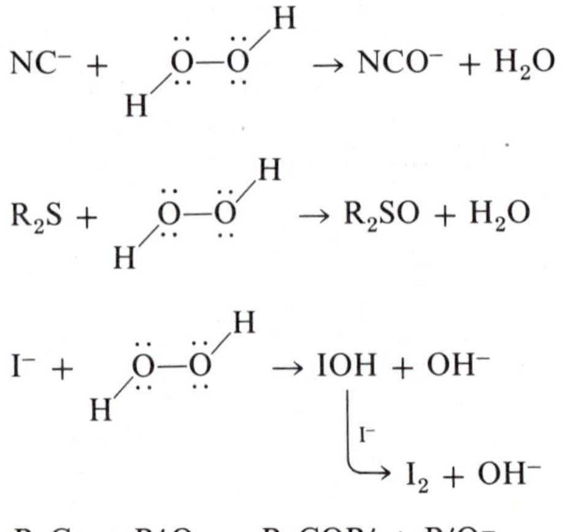

$$NC^- + \underset{H}{\overset{\ddot{O}-\ddot{O}\diagup^H}{\diagup}} \rightarrow NCO^- + H_2O$$

$$R_2S + \underset{H}{\overset{\ddot{O}-\ddot{O}\diagup^H}{\diagup}} \rightarrow R_2SO + H_2O$$

$$I^- + \underset{H}{\overset{\ddot{O}-\ddot{O}\diagup^H}{\diagup}} \rightarrow IOH + OH^-$$
$$\overset{I^-}{\longrightarrow} I_2 + OH^-$$

$$R_3C^- + R_2'O_2 \rightarrow R_3COR' + R'O^-$$

By way of generalization,[36] it has been determined that many peroxides react with nucleophiles according to a second order rate law:

$$\text{Rate} = k[\text{peroxide}][\text{nucleophile}]$$

In the case of hydrogen peroxide it has been shown that the second order rate constant exhibits a pH dependence given by

$$k = k_2 + k_H [H^+]$$

---

†There is more discussion of this reaction type in Chapter 12.

with the implication that $H_3O_2^+$ can act as an electrophile with attack occurring at the OH oxygen to displace H_2O (compare this with the acid catalyzed reactions of HNO_2 and H_2NOH). From studies of the activation parameters for nucleophilic displacements from peroxide oxygen [*viz.*, reactions of H_2O_2, $(t\text{-butyl})O_2H$, HSO_5^-, and CH_3CO_3H] it appears that the $\Delta H^{\ddagger}$ for these reactions falls in the range from 40 to 60 kJ/mole and $\Delta S^{\ddagger}$ in the range from -67 to -163 J/K mole; both are consistent with an a activation step, most reasonably I_a. The fairly large discrimination† between nucleophiles is also a good indication that bond formation is an important part of the activation step. Similarly, the reaction rates of various peroxy acids show a good correlation of rate constant with pK_b of the displaced group; for example, the order

$$H_3O_2^+(H_2O) > H_3PO_5(H_2PO_4^-) > HSO_5^-(SO_4^{2-}) > CH_3CO_3H(CH_3CO_2^-) > H_2O_2(OH^-)$$

is found for the oxidation of Br^- (the species in parentheses are the leaving groups). Whether this correlation should be taken to imply that bond-breaking is also important to the activation step or simply implies that the basicity of the leaving group determines the ease with which the new bond is formed is unclear.

By means of oxygen tracer studies it has been determined[37] that, in the oxidation of NO_2^- by peroxyacids, the NO_3^- product contains one oxygen atom derived from the peroxy acid. This, of course, lends credence to the hypothesis that the stoichiometric mechanism involves direct attack of the nucleophile upon a peroxide oxygen. Interestingly, H_2O and OH^- (the latter is generally a reasonably strong nucleophile) have been shown by isotope tracer studies to exchange oxygen atoms with peroxides at *immeasurably slow rates*. Comparison of rate constants for second order nucleophilic attacks at saturated carbon and peroxide oxygen reveals that oxygen is much more discriminating than carbon in these reactions. This has been taken to reflect the greater importance of lone-pair/lone-pair repulsions (between oxygen and nucleophile, 22) in the peroxide reactions.

22

This is consistent with slow reactions for second row nucleophiles (H_2O, OH^-) and faster reactions with the more "polarizable" nucleophiles from the third and later periods. (See p. 240.) This concept is the kinetic equivalent (for a mechanisms) of the vicinal lone pair repulsion as a factor in bond energies.

Another aspect of peroxide chemistry that is important from a kinetic view is **disproportionation**. At constant pH, the rate laws appear to be generally valid far to either side of pH = pK_a of the peroxide.

$$2RO_2H \rightarrow 2ROH + O_2$$

$$\text{Rate} = k_a[RO_2H]^2 \quad \text{and} \quad \text{Rate} = k_b[RO_2^-]^2$$

(in acidic solution)   *(in basic solution)*

The rate constants k_a and k_b show a pH dependence such that at pH values below the pK_a of the peroxy acid, k_a is *inverse* first order in $[H^+]$, while at pH values above the

---

†Remember that a variation in $\Delta G^{\ddagger}$ of ~20 kJ/mole causes a change in k of ~10^4.

$p\dot{K}_a$ of the peroxy acid, k_b is first order in $[H^+]$. *Both* rate expressions derive (see Study Question 19) from the ionization equilibrium of H_2O and

$$\text{Rate} = k_2[RO_2H][RO_2^-]$$

and fit the general concept of nucleophilic attack (RO_2^-) at peroxide oxygen (RO_2H). Furthermore, taking Caro's acid (H_2SO_5) as an example, the $\Delta H^\ddagger$ for the auto-redox reaction is 50 kJ/mole and $\Delta S^\ddagger$ is –84 J/K mole, values in the "normal" range for nucleophilic attack at peroxide oxygen. An intriguing question arises regarding the nature of the transition state species in such auto-redox reactions. In general, it is expected that the attacking nucleophile would select the OH oxygen of the peroxy acid so as to displace an RO^- group in the activation step.

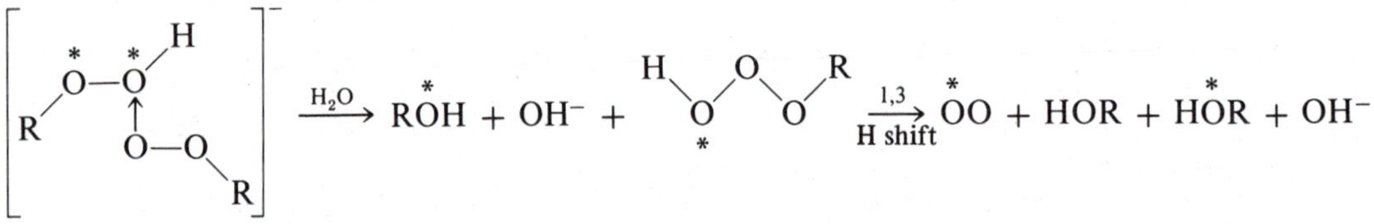

By isotope (^{18}O) labeling both peroxide oxygen atoms of Caro's acid (O_3SOOH) and carrying out the reaction with unlabeled acid, it has been determined that the oxygen atom isotopes in the product O_2 are highly "scrambled." The really interesting result of *unscrambled* oxygen arises for peroxyacetic acid. Owing to the trigonal planar structure about the carboxyl carbon and its expected greater electrophilicity than the peroxy oxygen, it seems reasonable that the entering nucleophile prefers to attack at carbon. With an assist from proton transfer, the following mechanism accounts for the fact that the evolved O_2 consists of oxygen atoms originally bound to each other.

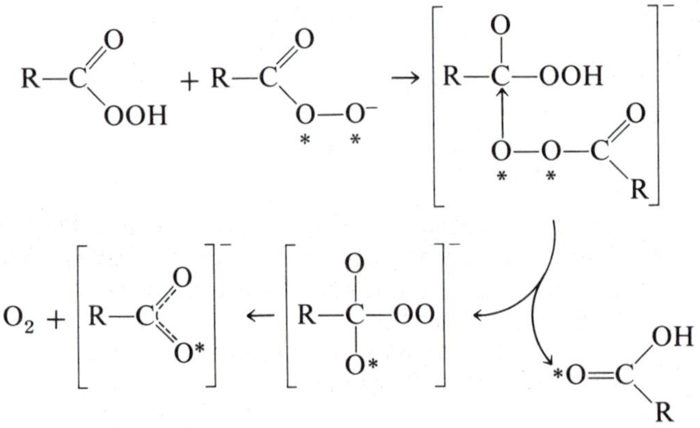

Be careful to note that this mechanism does *not* suggest that *all* peroxy linkages remain intact throughout the reaction. You see that half of them are ruptured, with those oxygen atoms appearing only in carboxylic acid groups.

THE O—halogen FUNCTION   In some respects you might expect the hypohalous acids (HOX) to react in the fashion of peroxy acids and hydroxylamine; the analogy is clear if the X group is viewed as a substituent like OR of the peroxy acid or NH_2 of hydroxylamine. In this case, the oxygen of HOX could act as an electrophile or the oxygen of OX^- could act as a nucleophile. There are times, however, when it is more appropriate to view OX^- as a "pseudo" halogen. The peroxide analogy should be valid for HOF and OF^-, but when X = Cl, Br, or I, the electrophilicity of X may exceed that of oxygen and reactions occur by nucleophilic attack at the *halogen* (compare the syntheses of the chloramines and most reactions of hydroxylamine). In these cases, the site of attack chosen by the nucleophile may depend strongly on the relative "leaving" group stabilities and on the relative electrophilicities of O and X.

$$\text{Nuc} + \text{HOX} \rightarrow \text{Nuc---O} \Big\langle \begin{array}{c} \text{H} \\ \text{X} \end{array} \rightarrow \text{Nuc---OH}^+ + \text{X}^-$$

or

$$\text{Nuc---X---OH} \rightarrow \text{Nuc---X}^+ + \text{OH}^-$$

The decomposition of HOF and HOCl in aqueous media appear to be good cases in point. The product O_2 could arise by a mechanism analogous to that discussed above for peroxide disproportionation,

$$\text{OX}^- + \text{HOX} \rightarrow O_2 + \text{HX} + \text{X}^-$$

or by the disproportionation of H_2O_2 formed as an intermediate

$$\text{HOX} + H_2O \rightarrow H_3O_2^+ + \text{X}^- \rightarrow \text{etc.}$$

HOF has been so recently prepared that its reactions with other materials are not well documented. For the better known hypochlorous acid, more reactions have been studied. In addition to the decomposition reaction just mentioned, oxygen *atom transfer oxidations* of SO_3^{2-}, IO_3^-, and NO_2^- proceed[40] according to second order rate laws, and in the case of nitrite the use of tracer oxygen confirms[41] that the third oxygen of the product nitrate derives from the hypochlorous acid.

$$\text{HOCl} + :SO_3^{2-} \rightarrow \text{HSO}_4^- + \text{Cl}^-$$
$$\text{HOCl} + :IO_3^- \rightarrow \text{HIO}_4 + \text{Cl}^-$$
$$\text{HO}\overset{*}{\text{C}}\text{l} + :NO_2^- \rightarrow \text{HO}\overset{*}{\text{N}}\text{O}_2 + \text{Cl}^-$$

In the next section on halogen reactions we will consider the evidence for nucleophilic attack upon the halogen of HOX. Before leaving the subject of nucleophilic attack at peroxy and hypohalous oxygen, we should note that not all such reactions are of the polar, displacement type. In the presence of trace metal ions these oxidants achieve reaction through free radical intermediates, where the role of the metal ion is to catalyze the free radical decomposition of R_2O_2 and HOX.[36]

Reactions involving *sulfenyl* sulfur (−S−) seem not to be much different in mechanism from those at oxygen, except that they are often much faster.[42] Generally speaking, reactions with nucleophiles follow a second order rate law

$$\text{Rate} = k[\text{RSX}][\text{Nuc}]$$

THE S–S FUNCTION

where Nuc is some nucleophile. It is often the case that such reactions are acid catalyzed, and a particularly interesting example is the acid catalyzed **exchange** of disulfides:

$$2\text{RSSR}' \rightarrow \text{RSSR} + \text{R}'\text{SSR}'$$

The mechanism of these exchange reactions has been clarified by studies of the related exchange of sulfur between Ph*SH and PhSSPh. This reaction is also catalyzed by acid, and the remarkable finding that the *rate constant* depends on the acid *anion* is an important clue to the pathway.[43] Specifically, the relative rate constants are $HClO_4$

(0), HCl (1.0), HBr (10^2), and HI (10^4). Thus, the mechanism must account for *anion* catalysis:

$$\text{PhSSPh} + \text{HX} \rightleftharpoons \text{PhS}\overset{\overset{\text{H}}{|}}{\text{S}}\text{Ph}^+ + \text{X}^-$$

$$\text{PhS}\overset{\overset{\text{H}}{|}}{\text{S}}\text{Ph}^+ + \text{X}^- \underset{\text{slow}}{\rightleftharpoons} \text{PhSX} + \text{HSPh}$$

$$\text{PhSX} + \text{Ph*SH} \underset{\text{slow}}{\rightleftharpoons} \left[\text{PhS}(\overset{\overset{\text{H}}{|}}{\text{*S}})\text{Ph}\right]^+ + \text{X}^-$$

$$\left[\text{PhS}(\overset{\overset{\text{H}}{|}}{\text{*S}})\text{Ph}\right]^+ + \text{X}^- \rightleftharpoons \text{PhS*SPh} + \text{HX}$$

Here we have written all steps so you can see that this mechanism satisfies the randomization requirement that the chemical identities of PhSH and Ph*SH are lost at the equilibrium for intermediates (PhSX + PhSH).†

The HCl catalyzed exchange[44] of the organo groups of RSSR′ is now seen to be actually an exchange of RS groups proceeding through steps like

$$\text{Cl}^- + \left[\text{---S}\overset{\overset{\text{H}}{|}}{\text{---S---}}\right]^+ \rightleftharpoons \text{---SCl} + \text{---SH}$$

and

$$\text{---SCl} + \text{---SS---} \rightleftharpoons \left[\text{---S}\overset{|}{\text{S}}\text{S---}\right]^+ + \text{Cl}^-$$

Because of the multiplicity of species present in the details of this scheme, we have avoided writing down all possible steps where the terminal organo groups (R, R′) are specified. In fact, the shortened version above should satisfy you that scrambling is achieved by proper choices of R and R′ in the second step (for example, R—SCl + R′—SS—R′).

At the present time there is insufficient information to prove whether nucleophilic substitutions at sulfenyl sulfur occur by an *I* mechanism or by an *A* mechanism. As mentioned in the introduction to this chapter, a proof is hard to come by. The steric inhibition by R observed[45] for the exchange reactions

$$\text{*SO}_3^{2-} + \text{RSSO}_3^- \rightarrow \text{RS*SO}_3^- + \text{SO}_3^{2-}$$

parallels almost exactly that for Br⁻ exchange with RCH_2Br, a known *I* process. It would seem that bond formation is of some importance in the activation step, but the data

---

†In the second step we presume that X⁻ attacks the unprotonated sulfur because HSPh is a better leaving group (less basic) than PhS⁻.

cannot distinguish between I and A mechanisms. Relative rates for R = (Me, Et, i-Pr, t-Bu) are $(2, 1, 10^{-2}, 10^{-5})$ and are characteristic of displacement reactions at sulfenyl sulfur.

16. Refer to the mechanism suggested for the auto-decomposition of Caro's acid and predict the ratios $*O_2 : *OO : O_2$, assuming that there are no isotope effects on the rates of any steps. Do assume rapid H^+ exchange between ROOH and ROO$^-$, and base your answer on the relative equilibrated amounts of the electrophile/nucleophile pairs: (labeled/unlabeled), (labeled/labeled), and (unlabeled/unlabeled).

17. Discuss the similarity between the acid catalyzed action of H_2O_2 as an electrophile and acid catalyzed oxidations by HNO_2 by identifying a common leaving group.

18. How might the enhanced nucleophilicities of third period, as contrasted with second period, elements toward peroxide oxygen be accounted for? Do you see a relation to their nucleophilicities toward $NH_2SO_4^-$ (p. 240)?

19. Verify that

$$\text{Rate} = k_2[RO_2H][RO_2^-]$$

may be converted into

$$\text{Rate} = k_a[RO_2H]^2 \text{ at low pH,}$$
$$\text{Rate} = k_b[RO_2^-]^2 \text{ at high pH.}$$

(Hint: See the footnote to the table on p. 222.)

HALOGENS

The reactions of the halogens bring us to the case of four electron pairs, three of which are lone pairs. Perhaps the most interesting reactions here are the **solvolysis** reactions (which could also be considered as redox reactions). We have already mentioned the solvolysis of Cl_2 by NH_3 (in liquid ammonia) leading to the formation of hydrazine and chloride ion. In aqueous systems the halogens initially yield halide ion and hypohalous acid.

$$X_2 + H_2O \rightarrow X^- + HOX + H^+$$

The kinetics of these reactions have been studied[46] in detail, with the finding that the major pathway for hydrolysis involves attack of H_2O and/or OH^- at the X_2 molecule to displace X^- and form HOX (note that these reactions could also be considered as simple substitution or displacement reactions). For the pathway involving attack of H_2O upon the halogen, the rate controlling step is the formation of the intermediate X_2OH^-; whereas for attack on X_2 by OH^-, the rate controlling step appears to be the dissociation of X^- from X_2OH^-.

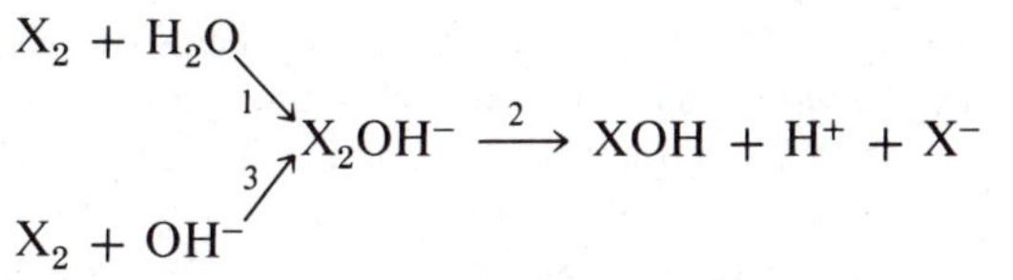

$$X_2 + H_2O \xrightarrow{\;1\;} X_2OH^- \xrightarrow{\;2\;} XOH + H^+ + X^-$$
$$X_2 + OH^- \xrightarrow{\;3\;}$$

Rate constants for the three steps for Cl, Br, and I are reported to be:

	Cl	Br	I
k_1 (sec^{-1})	11.0	110	2.1
k_2 (sec^{-1})	—	5×10^9	3×10^7
k_3 (M^{-1} sec^{-1})	10^{10}	10^{10}	10^{10}

With H_2O solvent, k_1 is of necessity pseudo-first order; k_2 is truly first order, while k_3 is second order. Judging from the magnitude of k_3, path 3 is diffusion controlled.

THE O–halogen FUNCTION REVISITED

An interesting feature of these reactions is that the HOX acids exhibit different kinetic stabilities in aqueous solution. As mentioned in the preceding section, HOF is apparently rapidly attacked by H_2O (or OH^-), and the final product O_2 results from decomposition of H_2O_2.

$$HOF + H_2O \rightarrow F^- + H^+ + H_2O_2$$

With X = Cl, the reaction of HOCl with H_2O is less rapid and the production of O_2 is much slower. HOBr solutions seem to be more unstable with respect to **disproportionation** (oxygen transfer) to BrO_3^- and Br^- than to *solvolysis*. It appears that one HOBr reacts with another (Br attack at oxygen) to form Br^- and $HBrO_2$, while the latter reacts further with HOBr to yield BrO_3^- and more Br^-.

$$HOBr + *BrO^- \rightarrow (HO)*BrO + Br^-$$

or

$$HOBr + *BrOH \rightarrow (HO)*BrO + Br^- + H^+$$

then

$$HOBr + *BrO(OH) \rightarrow *BrO_2(OH) + Br^- + H^+$$

Tracer studies[41] have revealed that, in addition to H_2O attack upon HOX to form H_2O_2, oxygen exchange between H_2O and HOCl and between H_2O and HOBr occurs in hypohalite solutions, illustrating the point made in the last section that attack on the heavier hypohalites can occur at either O or X.

Five Valence Pairs

PHOSPHORUS AND SULFUR

For examples involving reaction of molecules with five valence electron pairs, penta-valent phosphorus and tetravalent sulfur should come to your mind. As is well illustrated by silicon, a characteristic feature of such higher valency third row atoms is their ability to expand their valence shells to form six-valence-pair structures; therefore, it is not surprising that most chemists think of solvolysis and substitution reactions as proceeding through the formation of intermediates with six valence pairs about P and S. For example, SF_4 forms, in its reaction with $Me_4N^+F^-$, an adduct salt $Me_4N^+SF_5^-$ and also may be oxidized to the five-coordinate OSF_4. This behavior for sulfur(IV) contrasts with the uncertainty regarding a sulfenyl intermediate (p. 250). As examples of the tendency of phosphoranes to form six-coordinate species, PCl_5 forms adducts with amines, ether, and Cl^-.

THE P–halogen AND S–halogen FUNCTIONS

The *trivalent halides* of phosphorus undergo **solvolysis** reactions with acidic amines, alcohols, and water to produce HX and the trivalent substituted phosphines. These reactions involve nucleophilic attack by nitrogen or oxygen at the phosphorus atom. With *pentavalent phosphorus halides* the reactions are much more exothermic and proceed with greater speed, so that care is sometimes required in bringing the PX_5 compounds into contact with acidic solvents. The phosphoryl halides (X_3PO) are similarly reactive; an important reaction for them is alcoholysis to form esters of phosphoric acid, $(RO)_3PO$.

The only qualification to make concerning these comments is with reference to fluorides: in spite of the fact that solvolysis of fluorides can be thermodynamically spontaneous, such reactions are usually much slower than is the case with the heavier halides.

In highly related reactions, SF_4 and thionyl chloride ($SOCl_2$) hydrolyze to form SO_2 and HX (recall that sulfurous acid is actually the SO_2 solvate with formula $SO_2 \cdot H_2O$), and sulfuryl chloride (SO_2Cl_2) vigorously hydrolyzes to form sulfuric acid and HCl. Following the established pattern for fluorides, thionyl and sulfuryl fluorides are kinetically resistive to hydrolysis, in spite of a high thermodynamic spontaneity for formation of H_2SO_4 and SO_2, respectively. This resistivity to nucleophilic attack is most pronounced for SF_6. While the electrophilicity of the central atom in these fluorides should be at its highest, the short E—F bonds and high charge density of the fluorines seem to interfere seriously with nucleophiles (O, N, etc.) crowding into the coordination sphere of the central atom.

A few reflections on the relation of these reactions to those of the chloramines are valuable at this point. The chlorines of NCl_3 are subject to nucleophilic attack, while with PX_3 it is the phosphorus atom that experiences nucleophilic attack, to the exclusion of the halogens. This is attributed to (i) the much greater electronegativity of the Cl_2N group than of the Cl_2P group, (ii) the more "open" PX_3 structure, and (iii) the ability of phosphorus to expand its valence shell. It might seem to you that in the pentavalent halides, PX_5, the PX_4 group could render the "fifth" halogen sufficiently electrophilic to experience attack by nucleophiles. In fact, there are five halogens present to support the (+5) oxidation state of phosphorus in PX_5 and three to support the (+3) oxidation state in PX_3. Viewed in this way, it is difficult to see how the PX_5 halogens could be significantly more electrophilic than their PX_3 analogs. Rather, it is the central atom that should have a markedly increased electrophilicity, as is found.

Formation of a 5 valence pair species as an intermediate can confer nucleophilic activity to a reagent. For example, SF_4 will not react with chlorine in the absence of CsF. A low nucleophilicity expected for SF_4 should be markedly enhanced by formation of SF_5^- as an intermediate. Attack upon Cl_2 to displace Cl^- now appears feasible.

$$SF_4 + Cl_2 + CsF \rightarrow Cl_2 + Cs^+ SF_5^- \rightarrow SF_5Cl + CsCl$$

Many pentavalent phosphorus compounds have been found to be stereochemically non-rigid, as might be anticipated from our earlier inspection of five-coordinate silicon intermediates. It is becoming clear that axial fluorines of PF_5, for example, rapidly interconvert with equatorial fluorines via the Berry mechanism of *pseudorotation*.[47] The first order rate constant for this intramolecular isomerization is so large that it cannot be measured, even at $-100°C$. The theoretical estimate[48] of 20 kJ/mole for the activation enthalpy of the pseudorotation appears quite reasonable in that, assuming a negligible activation entropy, it leads to a predicted rate constant at $173°K$ of $\sim 10^6$ sec^{-1}.

Compounds such as CH_3PF_4 and HPF_4 have been studied,[49-52] with the finding that rapid pseudorotation about phosphorus takes place. In all three of these RPF_4 compounds ($R = Me_2N$, Me, H), the R groups are equatorially positioned in the most stable isomeric forms and pseudorotation presumably occurs with these groups as pivots. Interestingly, the compounds $(CH_3)_3PF_2$, $(CH_3)_2PF_3$, and H_2PF_3 do not undergo a rapid pseudorotation (presumably because the presence of two or more R groups raises the barrier to pseudorotation by forcing at least one of the R's into an axial position in the pseudorotamer), whereas PCl_2F_3 and PBr_2F_3 exhibit[53] quite low intramolecular exchange barriers (~ 25 kJ/mole). Rather than *intra*molecular exchange, the R_3PF_2 and R_2PF_3 compounds undergo (in sufficiently concentrated solutions) rapid exchanges of fluorine atoms between phosphorus atoms by dimeric, six-coordinate phosphorus intermediates or transition states (*inter*molecular exchange).

The activation parameters of these *inter*molecular exchange reactions[54] do not vary much and are consistent with an *a* activation step: $\Delta H^{\ddagger} \approx 17$ kJ/mole and $\Delta S^{\ddagger} \approx -160$ J/K mole. Other compounds believed to undergo such bimolecular intermolecular fluorine exchange are ClF_3 and BrF_3. Note the striking contrast in exchange behavior for molecules with two or more electron pairs (either lone pairs or R—P bond pairs) strongly preferring equatorial positions (ClF_3, BrF_3, R_2PF_3, R_3PF_2) and those with less than two such pairs.

On the basis of this limited series of phosphorus compounds, it would appear that we could generalize that facile pseudorotation usually occurs if there is present only one group that strongly prefers (because of π-bonding with P or low electronegativity) an equatorial position. When two or more such groups occur, there tends to be no exchange or halogen exchange by the intermolecular pathway. Note that the phrase "strongly prefers" is relative to the other groups present. For example, Cl_2PF_3 has a much lower barrier to pseudorotation than does Me_2PF_3. Such general guidelines are just that, for the scientific community has yet to establish the role of steric effects and the effect of synergic interaction between rotating and pivot groups on the barrier to pseudorotation. Nor is it fully understood how the central atom may affect the barrier magnitudes of inter- and intramolecular exchange processes.[55]

The reactions of *esters* and *acids* of phosphorus and sulfur make some interesting points that we need to review. While these compounds are properly thought of as examples with five valence electron pairs, one of the pairs participates in double bonding to a terminal oxygen; thus, the stereochemistry is approximately tetrahedral and much like that of the silicon compounds discussed previously.

The phosphate esters, and particularly the fluorophosphate esters, have recently attracted much attention because of their biochemical activity. Fluorophosphate esters of general formula

$$RO\underset{F}{\overset{\overset{\displaystyle R}{\overset{\displaystyle |}{O}}}{\!\!\diagdown\!\!P\!\!}}=O$$

are exceedingly toxic to insects and animals because of a nucleophilic displacement reaction, *in vivo*, of the fluoride by the acetylcholinesterase enzyme. When phosphorylated in this way, the enzyme is rendered inactive and can no longer serve its normal function to assist the solvolysis of acetylcholine (**23**) in muscle tissue (the enzyme reacts

$$\underset{Me}{\overset{\displaystyle O}{\overset{\displaystyle \|}{\underset{\diagdown}{C}}}}\overset{}{\diagup}\;OC_2H_4NMe_3^+$$

23

with acetylcholine by nucleophilic attack at the acetyl group, which is subsequently hydrolyzed and the enzyme regenerated).

Studies of the reactions of the toxin Sarin (**24**) with various nucleophiles show[56]

$$\underset{\underset{\displaystyle Me}{F}}{\overset{\displaystyle (i\text{-}Pr)-O}{\diagdown}}P=O\;+\;Nuc\;\rightarrow$$

24

that the second order rate constants are sensitive to the nature of the nucleophile; this implies either that a rapid pre-equilibrium forms a five-coordinate phosphorus intermediate or that bond-making is important in reaching the transition state in the activation step. Some illustrative data are (k_2 values given in parentheses): F^- (highly reactive) $> O_2H^-$ (10^5) $> OH^-$ (10^3) $> S_2O_3^-$, Cl^-, Br^-, I^- (unreactive). Furthermore, *hydrolysis* of organophosphates exhibits a pH dependence indicative of alternative mechanisms in different pH ranges. For example diisopropylfluorophosphate exhibits a dependence of the pseudo-first order rate constant on pH, as illustrated in Figure 7–3. In the low pH region the rate constant decreases as the H^+ concentration decreases, a behavior that suggests that an acid catalyzed reaction occurs at low pH:

$$\underset{RO}{\overset{RO}{>}}\underset{F}{\overset{|}{P}}{=}O + H^+ \rightleftharpoons (RO)_2POF \cdot H^+ \xrightarrow[\text{slow}]{H_2O} (RO)_2P(O)OH + H^+ + F^-$$

Since protonation of the OR oxygen would actually facilitate ROH displacement, it appears that the most likely donor atom is the terminal oxygen, as you expect from the behavior of phosphate esters as Lewis bases. In the high pH region we find the rate constant increasing with increasing OH^- concentration, which suggests direct attack on the fluorophosphate by OH^-.

$$(RO)_2POF + OH^- \rightarrow (RO)_2PO_2H + F^-$$

At moderate pH's, where neither the H^+ nor the OH^- concentration is sufficiently high to make the previous two pathways kinetically significant, we find k to be independent of pH; this is consistent with a relatively slower, uncatalyzed attack by the H_2O molecule.

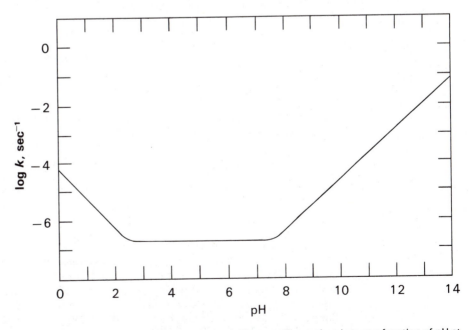

Figure 7–3. The rate of hydrolysis of diisopropylfluorophosphate as a function of pH at 25°C. [From J. O. Edwards, "Inorganic Reaction Mechanisms," W. A. Benjamin, New York, 1965.]

In addition to the sensitivity of second order rate constants to the nature of the entering group (as in the reactions of Sarin), studies of the nucleophilic displacement reactions

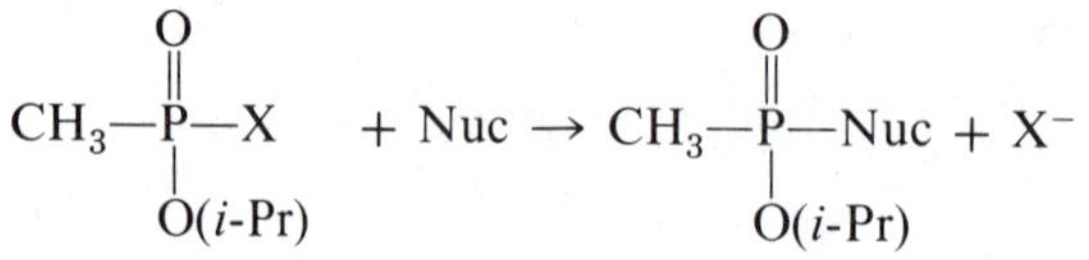

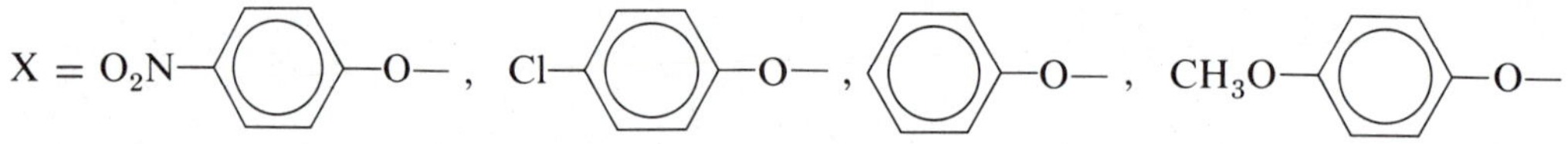

$$CH_3-\overset{\overset{O}{\|}}{\underset{\underset{O(i\text{-}Pr)}{|}}{P}}-X \;\;+ Nuc \rightarrow CH_3-\overset{\overset{O}{\|}}{\underset{\underset{O(i\text{-}Pr)}{|}}{P}}-Nuc + X^-$$

$$X = O_2N-\!\!\left\langle \bigcirc \right\rangle\!\!-O-\;,\;\; Cl-\!\!\left\langle \bigcirc \right\rangle\!\!-O-\;,\;\; \left\langle \bigcirc \right\rangle\!\!-O-\;,\;\; CH_3O-\!\!\left\langle \bigcirc \right\rangle\!\!-O-$$

have shown that the rate constants decrease in the order of this listing of leaving groups. Whether this is best interpreted to mean that significant bond-breaking occurs in the transition states of the reactions (the *I* mechanism) or simply that the electron withdrawing character of the leaving group influences the ease of the bond-making process (either in the transition state or via a pre-equilibrium formation of five-coordinate phosphorus) is uncertain at this time.

THE P–OR FUNCTION   Studies of trimethylphosphate using labeled (with ^{18}O) OH^- and H_2O have yielded the interesting result[57] that when OH^- is the nucleophile, the tagged oxygen appears primarily in the dimethyl phosphate product and not in the methanol product:

$$(MeO)_3PO + {}^*OH^- \rightarrow (MeO)_2PO^*O^- + MeOH$$

With $H_2^{18}O$ as a nucleophile, however, the tagged oxygen appears in the methanol product and not in the phosphate product:

$$(MeO)_3PO + H_2{}^*O \rightarrow (MeO)_2PO(OH) + Me^*OH$$

OH^- thus attacks the phosphorus atom much more readily than the carbon, while the converse holds for water. For aryl phosphates, which are not susceptible to nucleophilic attack by OH^- or H_2O at carbon, the ratio of relative second order rate constants for attack at phosphorus[58] is $k_{OH^-}/k_{H_2O} \approx 10^4$, a result further demonstrating the general distinction you have seen with other substrates between OH^- and H_2O as nucleophiles. Similarly, F^- attacks the phosphorus of alkyl phosphates, while the heavier halides tend to attack the carbon to produce methyl halides.

In the reaction of triphenylphosphate with OH^- it has been ascertained[57,59] that the activation enthalpy is only 40 kJ/mole and that the activation entropy is large and negative (–159 J/K mole). The small activation energy distinctly implies considerable bond formation in the transition state and the negative $\Delta S^{\ddagger}$ adds strength to this argument: an A activation step to form a highly organized five-coordinate intermediate is indicated, as opposed to an I mechanism, because the latter mechanism is generally associated with a larger $\Delta H^{\ddagger}$ and, particularly, a smaller $|\Delta S^{\ddagger}|$.

In addition to the results given above, two findings are particularly germane to the question of intermediate formation and the detailed pathway of phosphate ester hydrolysis.[60] First of all, five-membered ring esters (25)

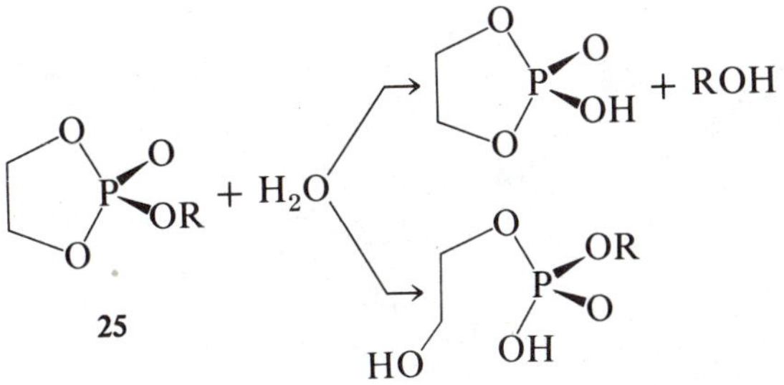

hydrolyze by loss of OR *and* by ring opening millions of times faster than do their acyclic analogs. Secondly, replacement of a cyclic phosphate ring oxygen with CH_2 gives a phostonate (26),

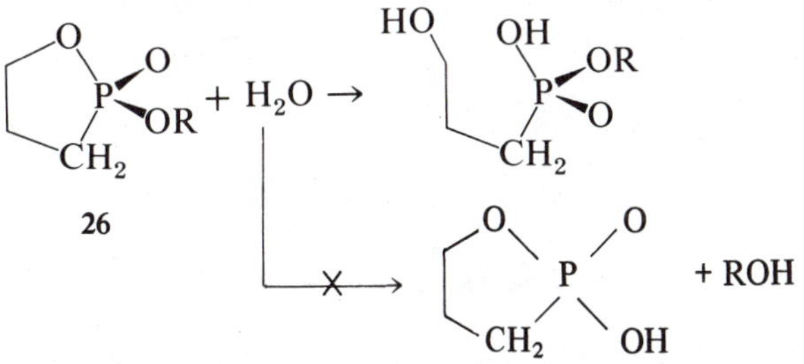

which hydrolyzes almost *exclusively* by ring opening, and the ring opening still proceeds at velocities many orders of magnitude faster than the analogous reaction of the acyclic analog. You might be thinking at this point that ring strain would account for the rate enhancement of the ring opening, but the important question is, "How does the ring feature so markedly enhance OR loss in the phosphate and so markedly inhibit OR cleavage in the phostonate?" The answer must lie in the structure of the intermediate and the position of the departing group within that structure.

Before passing on to consideration of the intermediate, let us briefly note that the ring O–P–O angle of 90° to 100° would make very unlikely a Walden inversion (I mechanism) path for OR loss. In such a transition state the O–P–O angle would become extremely strained (this translates into a high activation energy) with both ring oxygens in equatorial positions of a trigonal bipyramidal structure.

The analysis and elimination of mechanistic possibilities for the hydrolysis reactions of phosphates and phostonates are somewhat lengthy and cerebral. The important conclusions are summarized in the following scheme for the pathway for phosphate (E = oxygen) and phostonate (E = carbon) hydrolysis.

Special features of this path are summarized as follows:

a) The initial attack by H_2O is upon the face opposite the ring oxygen.

b) There is rapid protonation of the ring- or OR-oxygen.

c) Dissociation of the axial P—O bond in the ring leads to ring opening of both phosphate and phostonate and a change in their chirality.

d) Pseudorotation about the terminal P—O group places the HOR group axially for rapid loss and net solvolysis of that group.

e) Pseudorotation is not likely for E = C, so phostonates experience only ring-opening.

f) Distortion by the ring of the tetrahedral substrate (as in **15**, p. 236) facilitates attack upon the (RO, E, O) face,

making the *cyclic* substrate reactions faster than those of the *acyclic* analogs.

20. Why is it reasonable [see discussion of MePO(*i*-Pr)X] that O_2N—⟨◯⟩—O^- would

be a better leaving group than ⟨◯⟩—O^-?

21. Suggest a series of steps for the reduction of chlorate by sulfite:

$$ClO_3^- + 3SO_3^{2-} \rightarrow Cl^- + 3SO_4^{2-}$$

22. Does it surprise you that the five-membered-ring phosphinate

does not hydrolyze any faster than its acyclic analog, in marked contrast to the hydrolyses of phosphate and phostonate esters?

CHECKPOINTS

1. Rate laws provide information of the stoichiometric composition of the transition state species or intermediate. From this information, one can often determine whether the mechanism is
  a) *D* (dissociative) or
  b) *I* (interchange) or *A* (associative)
2. Mechanisms feature
  a) Bond breaking in the activation step if *D* applies;
  b) Bond making, with the appearance of an intermediate, if *A* applies;
  c) Dominant bond breaking (I_d) or bond making (I_a) if *I* applies.
3. A feature of many reactions is rapid formation of a reactive precursor; this situation results in an observed rate constant which is the product of the pre-equilibrium constant and the rate constant for the activation step: $k_{obs} = Kk$.
4. Leaving-group stability and leaving-group solvation are helpful predictors of the course of displacement reactions.
5. Acid and base catalysis, particularly the conjugate base mechanism, are helpful concepts.
6. Pseudorotation of five-coordinate intermediates is an important feature of reactions at silicon and phosphorus.

[1] Note that the stoichiometric mechanism notation reflects the composition (stoichiometry) of the transition state species. Good references with examples from transition metal reactions are: C. H. Langford and H. B. Gray, "Ligand Substitution Processes," W. A. Benjamin, New York, 1965; C. H. Langford and V. S. Sastri in "Inorganic Chemistry, Series One," Vol. 9, M. L. Tobe (Ed.), University Park Press, Baltimore, 1972; M. L. Tobe, "Inorganic Reaction Mechanisms," Nelson, London, 1972.
[2] R. H. Prince, *Nucleophilic Displacement at Some Main Group Elements*, in "Inorganic Chemistry, Series One," Vol. 9, pp. 353ff, M. L. Tobe (Ed.), University Park Press, Baltimore, 1972.
[3] J. Halpern, *Quart. Rev.* (Chem. Soc. London), **10**, 463 (1965); *Ann. Rev. Phys. Chem.*, **16**, 103 (1965); J. Chem. Educ., **45**, 372 (1968).
[4] W. L. Jolly and R. E. Mesmer, *J. Amer. Chem. Soc.*, **83**, 4470 (1961); *Inorg. Chem.*, **1**, 608 (1962); R. E. Davis, *et al.*, *J. Amer. Chem. Soc.*, **84**, 885 (1962); J. A. Gardiner, *ibid.*, **86**, 3165 (1964); B. S. Meeks, Jr., and M. M. Kreevoy, *ibid.*, **101**, 4918 (1979).

[5] H. S. Turner and R. J. Warne, *J. Chem. Soc.*, 6421 (1965); R. K. Bartlett, *et al.*, *J. Chem. Soc. (A)*, 479 (1966); J. C. Lockhart, *ibid.*, 809 (1966); J. R. Blackborow, *et al.*, *J. Chem. Soc. (A)*, 49 (1971).

[6] A. W. Laubengayer, *et al.*, *Inorg. Chem.*, **4**, 578 (1965).

[7] H. C. Brown and K. Murray, *J. Amer. Chem. Soc.*, **81**, 4108 (1959).

[8] H. C. Brown, *et al.*, *J. Amer. Chem. Soc.*, **78**, 5694 (1956); *ibid.*, **83**, 1001 (1961); H. G. Kuivila, *ibid.*, **76**, 870 (1954).

[9] H. C. Brown's recounting ("Boranes in Organic Chemistry," Cornell University Press, Ithaca, 1972) of his and his students' experiences in this area is not only informative but highly entertaining reading. See also K. Wade, "Electron Deficient Compounds," Nelson, London, 1971.

[10] A. H. Cowley and J. H. Mills, *J. Amer. Chem. Soc.*, **91**, 2910 (1969).

[11] J. B. DeRoos and J. P. Oliver, *Inorg. Chem.*, **4**, 1741 (1965); *J. Amer. Chem. Soc.*, **89**, 3970 (1967).

[12] G. S. Heaton and P. N. K. Riley, *J. Chem. Soc., (A)*, 952 (1966).

[13] R. E. Davis and R. E. Kenson, *J. Amer. Chem. Soc.*, **89**, 1384 (1967).

[14] H. C. Kelly and J. A. Underwood, III, *Inorg. Chem.*, **8**, 1202 (1969).

[15] H. C. Kelly, *et al.*, *Inorg. Chem.*, **3**, 431 (1964).

[16] S. G. Shore and C. L. Hall, *J. Amer. Chem. Soc.*, **88**, 5346 (1966); D. R. Schultz and R. W. Parry, *ibid.*, **80**, 4 (1958); S. G. Shore and R. W. Parry, *ibid.*, **80**, 8 (1958).

[17] D. E. Young and S. G. Shore, *J. Amer. Chem. Soc.*, **91**, 3497 (1960).

[18] P. Finn and W. L. Jolly, *Inorg. Chem.*, **11**, 1941 (1972).

[19] G. Schott, *et al.*, *Chem. Ber.*, **99**, 291, 301, (1966); **100**, 1773, (1967).

[20] L. H. Sommer, *et al.*, *J. Amer. Chem. Soc.*, **79**, 3295 (1957); **81**, 251 (1959).

[21] R. L. Hilderbrandt, *et al.*, *J. Amer. Chem. Soc.*, **98**, 7476 (1976).

[22] L. H. Sommer, "Stereochemistry, Mechanism, and Silicon," McGraw–Hill, New York, 1965.

[23] L. H. Sommer and P. G. Rodewald, *J. Amer. Chem. Soc.*, **85**, 3898 (1963).

[24] R. S. Berry, *J. Chem. Phys.*, **32**, 933, (1960).

[25] P. A. S. Smith, *et al.*, *J. Amer. Chem. Soc.*, **86**, 1139 (1964); J. H. Krueger, *et al.*, *Inorg. Chem.*, **12**, 2714 (1973); *ibid.*, **13**, 1736 (1974).

[26] In the solid state it is the nitrogen that bears the proton: R. E. Richards and R. W. Yorke, *J. Chem. Soc.*, 2821 (1959).

[27] K. J. Laidler, "Chemical Kinetics," McGraw–Hill, New York, 1965; A. A. Frost and R. G. Pearson, "Kinetics and Mechanism," Wiley, New York, 1961; I. R. Wilson in "Comprehensive Chemical Kinetics," Vol. 6, C. H. Bramford and C. F. H. Tipper (Eds.), Elsevier, New York, 1972.

[28] A thorough review of the preparation and reactions of these compounds is given by J. Jander, *Nitrogen Compounds of Chlorine, Bromine, and Iodine*, in "Developments in Inorganic Nitrogen Chemistry," Vol. 2, C. B. Colburn (Ed.), Elsevier, New York, 1973.

[29] R. M. Chapin, *J. Amer. Chem. Soc.*, **51**, 2112 (1929).

[30] F. N. Collier, *et al.*, *J. Amer. Chem. Soc.*, **81**, 6177 (1959).

[31] F. Raschig, "Schwefel und Stickstoffstudien," Verlag Chemie G.m.b.H., Berlin, 1924.

[32] A good, critical review of the kinetics and controversial mechanisms proposed for nitrosation reactions is given by J. H. Ridd, *Quart. Rev. (London)*, **15**, 418 (1961).

[33] A. E. Arbusov, *J. Russ. Phys. Chem. Soc.*, **42**, 395 (1910); R. G. Harvey and E. R. Somber in Vol. 1 and B. Miller in Vol. 2 of "Topics in Phosphorus Chemistry," M. Grayson and E. J. Griffith (Eds.), Wiley, New York, 1964.

[34] A. Skowronska, *et al.*, *J. Chem. Soc., Chem. Comm.*, 791 (1975).

[35] E. P. Kyba, *J. Amer. Chem. Soc.*, **97**, 2554 (1975). See also J. Omelanczuk and M. Mikolájczyk, *Chem. Comm.*, 1025 (1976).

[36] An excellent analysis of peroxide reaction kinetics and mechanisms is given by J. O. Edwards, "Inorganic Reaction Mechanisms," W. A. Benjamin, New York, 1964.

[37] M. Anbar and H. Taube, *J. Amer. Chem. Soc.*, **76**, 6243 (1954).

[38] See Ref. 36; J. F. Goodman and P. Robson, *J. Chem. Soc.*, 2871 (1963).

[39] E. Koubek, *et al.*, *J. Amer. Chem. Soc.*, **85**, 2263 (1963); J. F. Goodman, *et al.*, *Trans. Faraday Soc.*, **58**, 1846 (1962).

[40] M. W. Lister and P. Rosenblum, *Can. J. Chem.*, **39**, 1645 (1961).

[41] M. Anbar and H. Taube, *J. Amer. Chem. Soc.*, **80**, 1073 (1958).

[42] A good summary of nucleophilic substitution at sulfur is given by J. L. Kice in "Inorganic Reaction Mechanisms," Part II, J. O. Edwards (Ed.), Interscience, New York, 1972.

[43] A. Fava and G. Reichenbach, cited by E. Cuiffarin and A. Fava in "Progress in Physical Organic Chemistry," Vol. 6, A. Streitwieser and R. W. Taft (Eds.), Interscience, New York, 1968.

[44] R. E. Benesch and R. Benesch, *J. Amer. Chem. Soc.*, **80**, 1666 (1958). Note that the mechanism proposed in this reference is in error in that it does not recognize the acid anion effect treated by A. Fava, *et al.*, *J. Amer. Chem. Soc.*, **79**, 833 (1957).

[45] A. Fava and A. Iliceto, *J. Amer. Chem. Soc.*, **80**, 3478 (1958).

[46] M. Eigen and K. Kustin, *J. Amer. Chem. Soc.*, **84**, 1355 (1962).

[47] P. Russegger and J. Brickmann, *Chem. Phys. Letters*, **30**, 276 (1975); *J. Chem. Phys.*, **62**, 1086 (1975).

[48] A. Strick and A. Veillard, *J. Amer. Chem. Soc.*, **95**, 5574 (1973).

[49] G. M. Whitesides and H. L. Mitchell, *J. Amer. Chem. Soc.*, **91**, 5374 (1969).

[50] E. L. Muetterties, *et al.*, *J. Amer. Chem. Soc.*, **94**, 5674 (1972).

[51] L. S. Bartell and K. W. Hansen, *Inorg. Chem.*, **4**, 1777 (1965); P. Russegger and J. Brickmann, *J. Chem. Phys.*, **62**, 1086 (1975); R. R. Holmes, *Accts. Chem. Res.*, **5**, 296 (1972).

[52] P. M. Treichel, *et al., J. Amer. Chem. Soc.,* **89**, 2017 (1967).

[53] E. L. Muetterties, *et al., Inorg. Chem.,* **2**, 613 (1963); **3**, 1298 (1964); **4**, 1520 (1965).

[54] T. A. Furtsch, *et al., J. Amer. Chem. Soc.,* **92**, 5759 (1970).

[55] W. G. Klemperer, *et al., J. Amer. Chem. Soc.,* **97**, 7023 (1975); E. L. Muetterties, *et al., Inorg. Chem.,* **3**, 1298 (1964).

[56] A. L. Green, *et al., J. Chem. Soc.,* 1583 (1958); also Ref. 36.

[57] P. W. C. Barnard, *et al., J. Chem. Soc.,* 2670 (1961).

[58] C. A. Bunton, *et al., J. Org. Chem.,* **33**, 29 (1968).

[59] C. A. Bunton, *Accts. Chem. Res.,* **3**, 257 (1970).

[60] F. H. Westheimer, *Accts. Chem. Res.,* **1**, 70 (1968).

8. NON-METAL COMPOUNDS: SYSTEMATICS OF FUNCTIONAL GROUP TRANSFORMATION

In this chapter we will present, insofar as is possible, a general view of the syntheses of the important classes of non-metal compounds. Because non-metal halides are by-and-large readily obtained and are quite reactive, they are usually the reagents of choice for synthesis of a wide variety of derivatives. In particular, you will discover general routes to hydrides, organo, oxo, and amino compounds. Solvolysis reactions provide entry to a fascinating variety of inorganic polymers.

SPECIAL TECHNIQUES

Your experiences in organic laboratories have taught you many important techniques for synthesis and handling of compounds, and these find direct application in an "inorganic" laboratory. There are, however, several techniques commonly used in the manipulation of air- or moisture-sensitive compounds with which you may be unfamiliar. Furthermore, the inorganic chemist is sometimes called upon to deal with electrolytic techniques and high-energy electrical discharge methods for producing reactive species, and it is these methods we wish to introduce to you before you see their use in synthesis. Our overview will be limited in scope, for all we wish to achieve is your awareness of the equipment and fundamental techniques. For further details you should consult the original literature and the several excellent texts for use in inorganic laboratory courses.[1]

The Chemical Vacuum Line

The vacuum line, a very basic version of which is shown in Figure 8-1, is one of the most useful apparatuses available to the synthetic chemist. Using such an apparatus, one can measure small quantities of volatile liquids and gases,† carry out reactions, and separate and measure products. Working on the millimolar scale, measurements of quantities can be made quite accurately and, since everything can be handled within the line, a complete material balance can often be achieved, even in the most complex reactions. Although these advantages are available to many synthesis apparatuses, the obvious advantage of the vacuum line is that one is able to handle air-sensitive liquids and gases such as the boron hydrides or halides and aluminum alkyls.

---

†This is achieved with a mercury or oil pump, driven by atmospheric pressure, to move a volatile sample into a bulb of pre-calibrated volume (not shown in Figure 8-1). There the gas pressure is measured and the ideal gas equation is used to compute the number of moles of gas at hand. The sample subsequently can be returned to the synthesis section of the line.

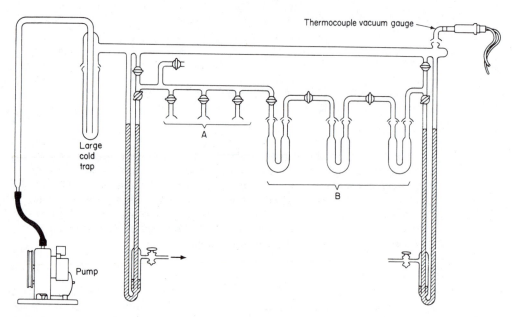

Figure 8-1. A basic chemical vacuum line. [From W. L. Jolly, "The Synthesis and Characterization of Inorganic Compounds," Prentice-Hall, Inc., Englewood Cliffs, N. J., 1970.]

The pressure in a vacuum line is usually maintained at about 10^{-6} torr (= 10^{-6} mm Hg) by a combination of a mechanical pump (as shown in Figure 8-1) and an oil or mercury diffusion pump (though not shown in the figure, it is placed between the mechanical pump and the large trap†). Pressure is monitored with a thermocouple vacuum gauge§ or mercury or oil manometers. In order to introduce material into the line, one can attach a gas cylinder or flask to one of the standard taper joints in section A. These are isolated from other sections of the line by "key" type stopcocks coated with special greases or by Teflon needle valves.

Section B of the line consists of a series of U-traps often called a "fractionation train." Together they compose a crude distilling column, with which one can separate compounds having sufficiently different volatilities. (This usually means that two compounds must have vapor pressures that differ by a factor of at least 100 at a given temperature.) For example, in order to purify BCl_3 (*i.e.*, separate other volatiles) one must remove from it the most likely impurity, HCl, and other possible compounds such as $SiCl_4$. In order to effect the purification, one passes the impure mixture through a first trap held at $-78°$ (by immersing the first U-trap in a Dewar flask containing a Dry Ice-acetone mixture) and then into a second held at $-196°$ (liquid nitrogen). The Dry Ice trap condenses $SiCl_4$ (m.p. = $-70°$) and allows the BCl_3 (v.p. = 2.4 torr at $-78°$) and HCl to pass through to be collected at $-196°$. If the mixture of HCl and BCl_3 is then passed through a $-112°$ trap‡, the BCl_3 is retained at this low temperature, while the HCl

---

†This trap is normally held at $-196°C$ by immersion in a Dewar flask containing liquid nitrogen; its purpose is to protect the pumps from contamination.

§The temperature of a length of resistance wire depends on thermal conduction by gases to the line walls. The higher the pressure, the cooler the wire. A thermocouple measures the wire temperature.

‡In order to achieve a temperature of $-112°$, one uses a CS_2 "slush bath." This is made by adding liquid nitrogen to CS_2 until the liquid just turns to a "slush." That is, the temperature of CS_2 is lowered to its normal freezing point. When contained in a Dewar flask, such a bath will maintain a trap at this temperature for 30 minutes or more. Obviously, other cold temperatures can be achieved in the same way, that is, by making a slush of an appropriate solvent. The most common are C_6H_{12}, $+7°$; CCl_4, $-23°$; $CHCl_3$, $-63°$; CH_2Cl_2, $-97°$; CS_2, $-112°$; and methylcyclohexane, $-126°$.

is allowed to pass (v.p. of HCl is 124 torr at $-112°$) to be collected in a final trap at $-196°$.

Plasmas

The preparation of materials of marginal or poor thermodynamic stability but reasonable kinetic stability usually requires that we have techniques available for producing "unstable" reactants. Two common techniques are those that utilize an electron flux or electric/magnetic fields to fragment stable molecules. Examples of such apparatuses are shown in Figure 8-2. An "electrode" discharge tube (A) is characterized by an alternating (60 Hz) voltage of several keV applied to the exposed electrodes, thereby maintaining a glowing discharge through a stream of one reactant flowing through the U-tube. The effluent stream of atoms and ions intercepts a flux of the other reactant a short distance downstream. To maintain the discharge and to minimize recombination of plasma particles, this experiment is conducted at plasma pressures of only a few mm Hg. A characteristic problem with these bare electrode techniques is that the electrode material often enters into reaction with the plasma. This can be turned to advantage in certain cases by using an electrode material that efficiently scavenges unwanted fragments.

One way to avoid altogether the bare electrode problem is to utilize an "ozonizer" as shown in Figure 8-2(B). Here the electrodes are separated from the plasma by thin glass walls. This arrangement requires voltages on the order of several tens of keV and sometimes of higher frequency than 60 Hz. Notice that a larger electrode surface is designed into the apparatus than that used in Figure 8-2(A).

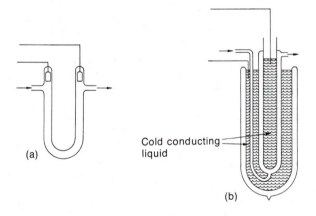

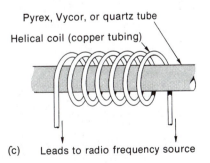

Figure 8-2. (a) Electrode discharge tube. (b) Ozonizer. (c) Inductive discharge tube. [From W. L. Jolly, in "Technique of Inorganic Chemistry," vol. 1, H. B. Jonassen and A. Weissberger (Eds.), Wiley–Interscience, New York, 1963.]

A technique that avoids the use of electrodes altogether is the microwave or radio-frequency discharge tube. In its simplest form, Figure 8-2(C), a reactant gas is converted to a plasma by passing it through a region of high-frequency electric/magnetic fields, generated by a high-power radiofrequency source.

All these techniques are carried out in conjunction with a chemical vacuum line, for the control of gas pressures, flow rates, and sample collections is easily accommodated within such a system. Very often these high-energy devices are very inefficient in terms of product yield and so place great demands on one's patience and vacuum line technique for collecting and isolating the desired product.

Photochemical Apparatus

Yet another way to "destabilize" stable molecules is to supply them with internal energy by photon absorption.† This method is somewhat more selective than the plasma methods, for it requires that a reactant molecule be capable of experiencing electronic excitation by photon capture. In this process a normally bonding or non-bonding electron is excited into an anti-bonding orbital, thereby weakening some bonds; fragmentation may occur directly, or an insurmountable activation barrier for a ground electronic state molecule may be more easily overcome. A difficulty with the method is that an excited state molecule often returns quickly to the ground state, before it encounters a reaction partner or experiences fragmentation. Therefore, in contrast to the plasma methods, a static system rather than a flow system is photolyzed with improvement in yield. Two photosynthesis arrangements are shown in Figure 8-3; (A) is typical for solution work, and (B) is characteristic of the set-up for gases§. The UV lamp generates a lot of energy not used in the reaction *per se*, which can cause severe heating problems. Thus, in (A) you see a water-cooled jacket serving the purpose of cooling both lamp and reaction mixture; a purging stream of some inert gas such as nitrogen stirs the mixture and also helps to remove any product gases. In (B) a rapid stream of air serves as a coolant. (A) is inherently more efficient than (B) because in the former configuration the reaction mixture completely surrounds the lamp. A final point: Pyrex strongly absorbs certain frequencies of radiation in the UV region, whereas quartz does not, so quartz containers are more generally useful.

Electrolysis

This is a "high-energy" technique you may remember from your freshman chemistry laboratory. As with the plasma techniques discussed earlier, energy is supplied to the reactants in the form of an electric current; but there the similarity stops. The electrolysis technique can be much more selective about which reactant is energized; it can be used in both static and flow configurations; it is limited to much smaller voltages, and therefore requires a conducting medium to support current flow. The arrangement shown in Figure 8-4 is used in the synthesis of peroxydisulfate by oxidation of bisulfate. Notice the use of electrical energy (2.1 eV) to generate the *radical* $HSO_4 \cdot$ by the reactions

$$HSO_4^- \rightarrow HSO_4 \cdot + e^-$$

$$2HSO_4 \cdot \rightarrow S_2O_8^{2-} + 2H^+$$

---

†The energy equivalents of 200, 300, and 400 nm photons are 598, 399, and 299 kJ/mole, respectively. Generally this energy is not completely isolated in one "bond," but you see that the energy content is sufficient to break "single" chemical bonds. Which bonds break is a dynamical problem determined only in part by the molecular orbital configuration and the location of the weakest bond.

§Total sample pressures much less than 1 atm are used, to allow for sample expansion by heating and/or a change in the number of moles of gases during reaction.

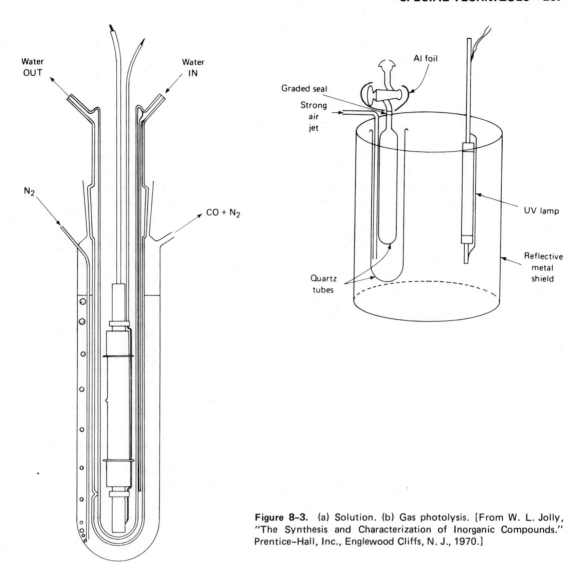

Figure 8-3. (a) Solution. (b) Gas photolysis. [From W. L. Jolly, "The Synthesis and Characterization of Inorganic Compounds." Prentice–Hall, Inc., Englewood Cliffs, N. J., 1970.]

Figure 8-4. Electrolytic synthesis of peroxydisulfate. [From R. J. Angelici, "Synthesis and Technique in Inorganic Chemistry," 2nd Ed., W. B. Saunders Company, Philadelphia, 1977.]

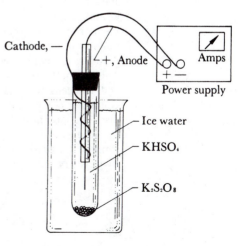

An important part of the apparatus is the temperature bath, in this case used to cool the reaction mixture and facilitate the precipitation of $K_2S_2O_8$. Other applications to be encountered later in this chapter involve both high (molten salts) and low (liquid HF) temperature electrolysis.

AN OVERVIEW OF STRATEGY

Probably the single most important class of non-metal compounds, at least in a synthetic sense, is that of the halides. The fluorides are useful for introducing fluorine as a substituent in other compounds, and the chlorides find great use through substitution of chlorine by other groups (for such reactions the chlorides tend to be more reactive than the fluorides; in fact, *a general approach to synthesis of non-metal fluorides is to replace chlorine by fluorine, using metal fluoride salts*).

The great utility of the non-metal chlorides is summarized by the following reaction types, which forecast the organization of this chapter (E and M are general symbols for non-metal and metal elements, and R and A carry their usual representations of organo and electronegative groups).

$$E{-}Cl + MR \rightarrow E{-}R \qquad \text{alkylation or arylation}$$
$$+ MH \rightarrow E{-}H \qquad \text{hydrogenation}$$
$$+ M \;\; \rightarrow \frac{1}{x}[E{-}E]_x \qquad \text{coupling}$$
$$+ HA \rightarrow E{-}A + HCl \qquad \text{solvolysis}$$

These are the primary reaction types with which we will be concerned. Other, perhaps less general, synthetic strategies will be mentioned as appropriate. Reactions subsequent to those above also will be examined when appropriate.

SYNTHESIS OF HYDRIDO AND ORGANO DERIVATIVES

Before we proceed to examine the application of chlorine substitutions, we need to note two important exceptions. First of all, the difficulty of handling and performing reactions with OCl_2 and NCl_3 means that these reagents are not useful precursors to H_2O, NH_3, ethers, and amines! The other important exception is that high oxidation state compounds of third row elements such as SR_4, SR_6, PR_5 (where R = H or alkyl group) have never been synthesized. The instability of SCl_4 and the non-existence of SCl_6 seem to preclude their use as precursors for SR_4 and SR_6. While PCl_5 is a stable compound of synthetic utility, it is likely that hydrogen and alkyl groups are generally too reducing to confer high stability on the higher oxidation states of phosphorus and sulfur. This is not to say that such compounds, particularly mixed forms containing hydrogen and halogens or organic groups and halogens, are not possible. In fact, you have encountered compounds such as $ArSF_3$ and PR_nF_{5-n} in previous chapters. Others will be discussed later in this chapter.

The replacement of chlorine by hydrogen or organic groups may be represented in simplest terms as an **"anion exchange" reaction** *using highly reactive forms of hydridic hydrogen or metal alkyls/aryls.*

$$ECl_n + \text{"MH"} \rightarrow EH_n + n\text{"MCl"}$$
$$+ \text{"MR"} \rightarrow ER_n + n\text{"MCl"}$$

The most convenient metal hydrides for this purpose are **lithium aluminum hydride** ($LiAlH_4$) and **sodium borohydride** ($NaBH_4$), but in certain cases binary hydrides like LiH and NaH are used. For the alkylation and arylation reactions, **Grignard reagents** (RMgX), **lithium alkyls** (LiR), and **organo mercury compounds** (HgR_2) are effective and commonly used. Many of these "anion exchange" reactions can be performed with pure reagents; but a more common technique, allowing better control of the reactions, is to

use a weakly basic/acidic ethereal solvent [tetrahydrofuran ; diethyl ether;

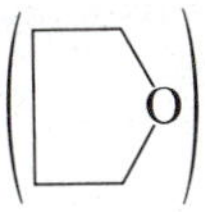

monoglyme ($CH_3OC_2H_4OCH_3$); or diglyme ($CH_3OC_2H_4OC_2H_4OCH_3$)] .

In the survey of methods used to produce the non-metal hydrides, you will encounter the technique of **acid hydrolysis of an anionic form of the non-metal**; this method is included for its historical interest and also because it can be particularly convenient for the introduction of deuterium in place of normal hydrogen.

$$E^{n-} + nHA \rightarrow EH_n + nA^-$$

An alternate method for synthesis of organo derivatives will also be mentioned for those hydrides of more hydridic nature (the hydrides of boron, aluminum, and silicon). This technique is one of *addition of a non-metal–hydrogen bond across an olefinic linkage* and is called **hydrometallation**:

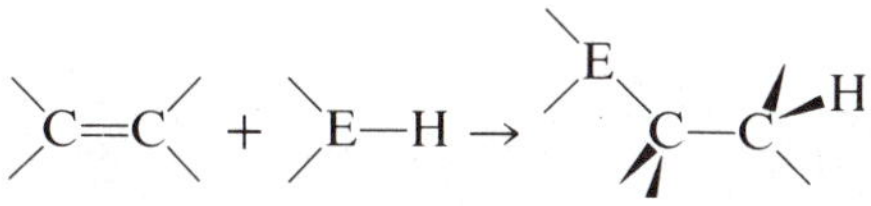

To close this overview, be aware that the production of mixed organohydrides of the non-metals is easily achieved in most cases by tandem application of the methods used to introduce organic functionalities and hydrogen in place of chlorine. The reactions of non-metal chlorides with organometals can usually be controlled by stoichiometrically limiting the amount of organometal used, so as to replace chlorine only partially with the organic moiety. Reduction of the mixed organochloride compound with, say, $LiAlH_4$ produces the mixed organohydride.

$$ECl_n + \text{“MR”} \rightarrow ER_xCl_{n-x}$$
$$\xrightarrow{\text{“MH”}} ER_xH_{n-x}$$

Organometal Reagents†

To add a measure of continuity to the ensuing descriptive chemistry, we should pause briefly to mention the source of the alkylating/arylating reagents used in some of the substitution reactions. As the preparation of these reagents and their properties can be more systematically approached in a general treatment of organometal synthesis, it is there (Chapter 16) that you will find a more complete treatment.

Grignard reagents are already familiar to you from organic chemistry. There you learned of their synthesis from Mg turnings and the appropriate organohalide:

$$Mg + RX \xrightarrow{\text{“ether”}} RMgX$$

---

†A more comprehensive discussion of organometal synthesis is given in Chapters 16 and 17.

In an entirely analogous way, one obtains **organolithium** reagents:

$$2Li + RX \xrightarrow{B} RLi + LiX\downarrow \quad (B = benzene\ or\ hydrocarbon)$$

(As organolithium reagents tend to be reactive with ethers, such solvents are usually avoided.) An alternate route is afforded by use of a **diorganomercury** with Li,

$$R_2Hg + 2Li \rightarrow 2RLi + Hg\downarrow$$

which is also useful for preparing **diorganomagnesium** compounds by a similar reaction.

$$R_2Hg + \underset{(excess)}{Mg} \rightarrow R_2Mg + Hg$$

This reaction is usually conducted in the absence of a solvent, followed by extraction of R_2Mg with the desired solvent. To come full circle, a source of R_2Hg is the reaction

$$HgCl_2 + 2RMgCl \rightarrow R_2Hg + MgCl_2\downarrow$$

Boron and Aluminum

Of all the metallic hydrides, the two MH_4^- anions from this group are the most important. For aluminum, the important compound is $LiAlH_4$ (dec. $\sim150°$) prepared from LiH and $AlCl_3$:

(8-1)
$$8LiH + Al_2Cl_6 \xrightarrow{\text{"ether"}} 2LiAlH_4 + 6LiCl\downarrow$$

in an ether solvent (in which LiCl is insoluble). The tetrahydroaluminate anion is important both as a powerful reducing agent (this solid hydride becomes *incandescent* upon hydrolysis by a small amount of H_2O!) and as a precursor to the production of other aluminum hydride species.

The tetrahydroborate anion is an important reducing agent, particularly since it is a less vigorous reducing agent than AlH_4^- and so presents the synthetic chemist with an alternate, easily handled reductant. Its production as a lithium salt from LiH and BF_3 is analogous to the formation of $LiAlH_4$:

(8-2)
$$4LiH + BF_3 \xrightarrow{\text{"ether"}} 3LiF\downarrow + LiBH_4$$

The use of the fluoride instead of boron trichloride is not required, for combination of almost any strongly hydridic reagent with either BF_3 or BCl_3 can be recommended. It is useful, but not really unexpected, to note that control of the reacting stoichiometries leads to production of the gas B_2H_6 (diborane).

(8-3)
$$6NaH + 8BF_3 \xrightarrow{\text{"ether"}} 6NaBF_4 + B_2H_6\uparrow$$

This binary hydride of boron is highly soluble in tetrahydrofuran, and such solutions are very convenient sources of borane for synthetic work. (As noted in Chapter 7, p. 234, ethers symmetrically cleave B_2H_6, so that such solutions contain the weak adduct $BH_3 \cdot THF$ rather than B_2H_6.)

Of some historical importance is the reaction of metal borides with acid to produce a variety of binary boron/hydrogen compounds. First discovered by Alfred Stock[2] (the developer, for his work with the gaseous products of this reaction, of the modern vacuum line and associated techniques), the mixture of products can be thermally degraded to the simplest member of the boron hydride series, B_2H_6.

$$MgB_2 \xrightarrow{\text{acid}} B_nH_m \xrightarrow{\Delta} B_2H_6 \tag{8-4}$$

Silicon

The known binary hydrides of silicon, Si_nH_{2n+2} ($n \leq 8$), are highly pyrophoric gases and liquids easily prepared from the corresponding chlorosilanes, Si_nCl_{2n+2}, by **reduction** with $LiAlH_4$ [3]

$$Si_nCl_{2n+2} + LiAlH_4 \rightarrow Si_nH_{2n+2}$$
$$\begin{bmatrix} n = 1: \text{ b.p. } -112° \\ n = 4: \text{ b.p. } 109° \end{bmatrix} \tag{8-5}$$

and by reduction, under more strenuous conditions, of SiO_2 (!)

$$SiO_2 + LiAlH_4 \rightarrow Si_nH_{2n+2} \text{ mixture}$$

In analogy with his work with boron, Stock found it possible to synthesize many members of the silane series by **acid hydrolysis** of magnesium silicide

$$Mg_2Si \xrightarrow{\text{acid}} Si_nH_{2n+2} \text{ mixture} \tag{8-6}$$

The interesting feature of this reaction is that it points up rather emphatically that the silanes are not readily susceptible to acid hydrolysis; it suggests that the hydrogens of the silanes are not nearly as hydridic in nature as those of the boron analogs. (However, be alert to the fact that silicon compounds are highly susceptible to attack by base, so the silanes are easily hydrolyzed in aqueous base.)

Nitrogen and Phosphorus

Trichloramine (p. 243) is not a very useful precursor for production of ammonia or amines. Probably the most convenient synthesis of ammonia or amines in the laboratory is from neutralization of NH_4^+ or $NH_xR_{4-x}^+$ salts with strong base.

$$NH_4X + OH^- \rightarrow NH_3 + X^- + H_2O \tag{8-7}$$

Use of ammonium salts does presuppose the prior formation of NH_3, and a more elementary reaction, the **acidification** of magnesium nitride, is convenient for this purpose. This reaction of the anionic form of a non-metallic element is useful† for the production of ND_3 in the laboratory (at this point you might wish to review study question 6-18 on the similar hydrolysis of Be_2C and Al_2C_3).

$$Mg_3N_2 \xrightarrow{D_2O} ND_3 + Mg(OD)_2 \tag{8-8}$$

---

†The alkali and alkaline earth metals all form nitrides from N_2 at elevated temperatures (Li does so at room temperature). For the preparation of ND_3 in this manner, see G. H. Payn, in "Mellors Comprehensive Treatise on Inorganic and Theoretical Chemistry," Vol. 8, Suppl. I, *Nitrogen*, Part I, Section V, Longmans, London, 1964.

The mechanistic details of the Raschig method for generation of hydrazine, the other important hydride of nitrogen, were discussed in Chapter 7. The reaction of hypochlorite with ammonia proceeds through chloramine as an intermediate, and care must be taken to minimize the degradation of hydrazine by its counter-productive reaction with chloramine:

(8-9)
$$2NH_3 + OCl^- \rightarrow N_2H_4 + Cl^- + H_2O$$
$$\xrightarrow{2NH_2Cl} N_2 + 2NH_4Cl$$

Phosphorus trichloride is much more easily synthesized and handled than its nitrogen analog and can be used with a reducing agent such as $LiAlH_4$ to produce phosphine[4]

(8-10)
$$PCl_3 \xrightarrow{LiAlH_4} PH_3$$
$$\begin{bmatrix} \text{air stable} \\ \text{b.p. } -88° \end{bmatrix}$$

As mentioned in the introduction to this section on chloride replacements, PCl_5 reduction does not lead to PH_5 (an as yet unprepared compound) but rather produces PH_3. Mixed H and F compounds of P(V) are known, however. Use of SnH/PF "exchange" has been reported[5] for the syntheses of HPF_4 and H_2PF_3:

(8-11)
$$PF_5 + Me_3SnH \rightarrow HPF_4, H_2PF_3 + [Me_3Sn][PF_6]$$

These P(V) hydride species are unstable and tend to eliminate HF slowly according to

(8-12)
$$HPF_4 \rightarrow HF + PF_3$$

Three pentaorgano derivatives of P(V) are known at present. The first prepared was Ph_5P (by Wittig[6] in 1948) according to the multi-step scheme:

(8-13)
$$Ph_3PO \xrightarrow[\text{2. HCl}]{\text{1. PhLi}} [Ph_4P]Cl \xrightarrow{HI} [Ph_4P]I \xrightarrow{PhLi} Ph_5P$$
$$[\text{dec. } 124°]$$

The starting material, Ph_3PO, can be obtained by Ph substitution of Cl in Cl_3PO.

It should come as no surprise that alkali and alkaline earth phosphides can be **hydrolyzed** to form phosphine.[7] This alternate route to PH_3 suffers from the simultaneous production of some contaminating diphosphine.

(8-14)
$$Na_3P + 3H_2O \rightarrow PH_3 + 3NaOH \ (+ \text{ some } P_2H_4, H_2)$$

This is important to remember since, unlike PH_3, P_2H_4 is a low boiling (52°) pyrophoric liquid, and the spontaneous combustion of its vapor in air is sufficient to cause ignition of PH_3. The low volatility of P_2H_4 at its m.p. (−99°) does provide, however, for its ready separation from PH_3 in a vacuum fractionation train.

As you could anticipate by extrapolation from the weakly hydridic character of silanes, the phosphines exhibit even less hydridic character. PH_3 in *aqueous* solution is neither acidic nor basic but does show reducing character toward the heavier transition metal ions in aqueous systems, where formation of stable phosphorus oxyacids thermodynamically assists the oxidation of PH_3.

Oxygen and Sulfur

It is highly unlikely that you would be faced with preparing H_2O in the laboratory. At the risk of sounding ridiculous, reduction of OCl_2 would be a most undesirable way of making H_2O. As a reminder of the general scheme exemplified by (8-14), (8-8), (8-6), and (8-4), H_2O (and, more importantly, D_2O) can be synthesized by treatment of a simple metal oxide with (H,D)X (but even D_2O is readily available from commercial sources). In a similar vein, the strongly basic character of ionic metal oxides is of further use to the practicing chemist for removal of trace amounts of H_2O from organic solvents:

$$BaO \xrightarrow{H_2O} Ba(OH)_2$$

The synthesis of hydrogen peroxide can be approached in either of two ways. As sodium and barium readily combust in air to form appreciable amounts of the corresponding peroxide salts, their **hydrolysis** affords a useful synthesis of H_2O_2. The separation of H_2O_2 from the aqueous reaction mixtures requires, in both cases, distillation with utmost care to avoid a disastrous explosion.

$$Na_2O_2 + 2H_2O \xrightarrow{H_2O} H_2O_2 + 2NaOH$$

$$BaO_2 + H_2SO_4 \xrightarrow{H_2O} H_2O_2 + BaSO_4\downarrow$$

(8-15)

The use of Na_2O_2 requires greater care and technique than use of BaO_2 because the production of OH^- in the Na_2O_2 hydrolysis means that the pH dependent decomposition of H_2O_2 proceeds more rapidly than under the acid conditions of BaO_2 hydrolysis. The convenience of being able to separate the insoluble $BaSO_4$ by filtration prior to distillation should not go unnoticed.[8]

The second widely known technique for synthesizing hydrogen peroxide is related to those of (8-15), but with the distinction that a *non-metal* peroxide is hydrolyzed. Peroxydisulfate ion is electrolytically generated[9] from aqueous $KHSO_4$ at a voltage of -2.2 v at a platinum electrode (see Figure 8-4). (The O_2 overvoltage prevents oxidation of H_2O.)

$$K^+ + HSO_4^- \xrightarrow{H_2O} \frac{1}{2}K_2S_2O_{8(c)} + H^+ + e^-$$

The potassium salt is preferred because of its insolubility at the temperature (maintained at about $5°C$) of the electrolysis. This defers the formation of $H_2S_2O_8$ and its hydrolysis to form Caro's acid (H_2SO_5), which would subsequently hydrolyze to form the desired hydrogen peroxide.

$$H_2S_2O_8 \xrightarrow{H_2O} H_2SO_5 + H_2SO_4$$
$$\xrightarrow{H_2O} H_2O_2 + H_2SO_4$$

(8-16)

Postponement of this hydrolysis is necessary to avoid leaving the hydrogen peroxide in contact with Caro's acid, for the two react to produce O_2 (the mechanism was discussed in Chapter 7, p. 248).

$$H_2O_2 + H_2SO_5 \rightarrow O_2 + H_2O + H_2SO_4$$

Thus, it is preferred that the isolated salt $K_2S_2O_8$ be hydrolyzed with dilute H_2SO_4 and the hydrogen peroxide quickly removed by *vacuum* distillation (the boiling point of H_2O_2 is $150°C$ at 1 atmosphere pressure).

Halogens[10]

The hydrogen halides, in principle, could be synthesized by the technique of **hydrolysis** of the element in the anionic form. Thus, acidification of metal fluorides and chlorides or hydrolysis of non-metal fluorides and chlorides leads directly to the hydrogen halide.

$$MF_n \xrightarrow{\text{acid}} HF$$

(8-17)

$$SOCl_2 \xrightarrow{H_2O} HCl + SO_2$$

To isolate the pure HX gas, you must avoid H_2O as a solvent; sulfuric acid would seem a good choice, for not only does it provide a source of hydrogen ion but it also acts as a desiccant. This technique works well for HF and HCl, but H_2SO_4 oxidizes HBr and HI to Br_2 and I_2, respectively. A better method for synthesis of HBr is careful hydrolysis of a non-metal bromide such as PBr_3, keeping the latter in excess.

(8-18)

$$PBr_3 + 3H_2O \rightarrow H_3PO_3 + 3HBr$$

The best methods for production of HI (interestingly enough, gaseous HI is thermo-dynamically unstable relative to the elements) involve reduction of elemental iodine:

$$I_2 + H_2S \xrightarrow{H_2O} 2HI + \frac{1}{8}S_{8(c)} \quad \text{(for aqueous solutions of HI)}$$

(8-19)

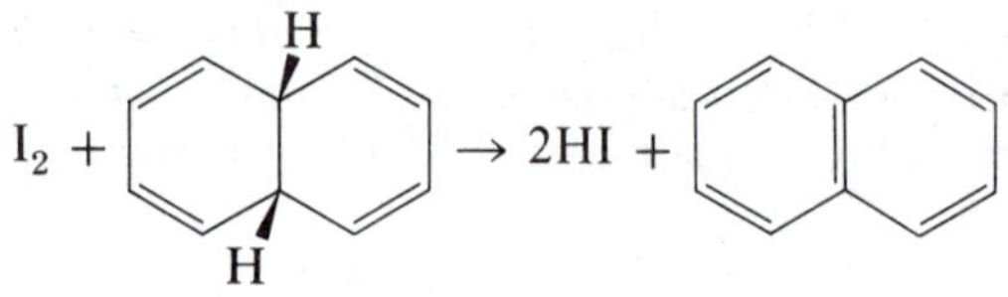

STUDY QUESTIONS

1. Give steps for syntheses of the following from halides and any special reagents identified in parentheses.

   a) $Al(BH_4)_3$
   b) $Al_2Me_3Cl_3$ (any alkyl halide)
   c) $AlH_3(NMe_3)_2$ (NMe_3)
   d) Me_2SiH_2
   e) $[PCl_3R][AlCl_4]$ (any alkyl chloride; the only phosphine cation to appear does so as the product)
   f) Me_2PH (any methyl halide)
   g) $LiNH_2$ (NH_3)
   h) $Et_2PO(OH)$
   i) D_2O
   j) $[R_3PNH_2]Cl$ (NH_3)

2. Match reactants with products.

   a) $[MeNBCl]_3 + NaBH_4 \rightarrow$ 1) Al_2Et_6
   b) $[MeNBCl]_3 + PhMgCl \rightarrow$ 2) $n\text{-BuLi}$
   c) $2Al + 3Et_2Hg \rightarrow$ 3) $Si(C_2H_5)_4$
   d) $SiCl_4 + 2ZnEt_2 \rightarrow$ 4) $[(CH_3)_3Si]_3N$
   e) $Me_3SiCl + Na[N(SiMe_3)_2] \rightarrow$ 5) $[MeNBH]_3$
   f) $ClNF_2 + R_2Hg \rightarrow$ 6) $EtCl + EtNF_2$
   g) $Li + n\text{-BuCl} \rightarrow$ 7) $[MeNBPh]_3$

Hydrometalation and Others

As anticipated in the introduction to this section, organoboranes, alanes, and silanes (under proper conditions) can be synthesized by the addition of E—H to olefins. These reactions can be utilized, in the cases of boron and aluminum, to introduce one, two, or three alkyl groups in place of hydrogen about the Group III atoms:

hydroboration:[11]

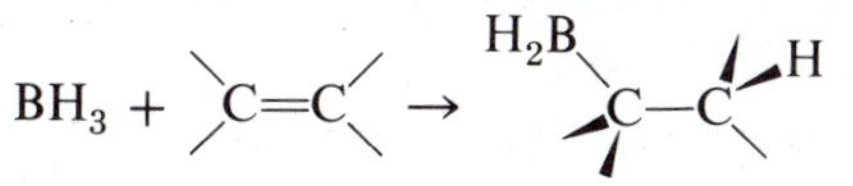

$$BH_3 + \ \text{>C=C<} \ \rightarrow \ \text{H}_2\text{B—C—C—H} \tag{8-20}$$

hydroalumination:

$$LiAlH_4 + 4RCH{=}CH_2 \rightarrow Li[Al(CH_2CH_2R)_4] \tag{8-21}$$

(1) In hydroboration it is BH_3, not B_2H_6, that is reactive, a fact that dictates the use of weakly basic ethers as solvent (see p. 234).

(2) The addition is *anti*-Markownikoff, as in

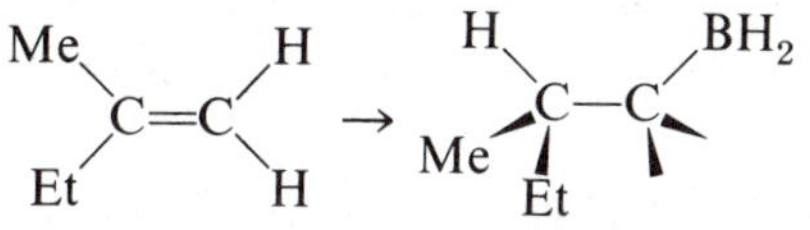

$$\text{Me, Et \ C=C \ H, H} \rightarrow \text{Me, Et \ C—C \ BH}_2$$

1

(3) Hydrolysis of the addition product **1** with aqueous mineral acid introduces H in place of BH_2 (**2a**)

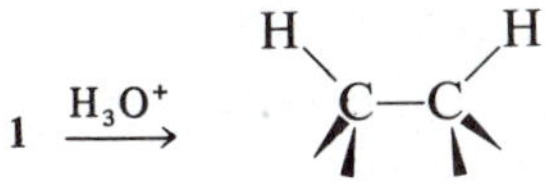

$$\textbf{1} \ \xrightarrow{H_3O^+} \ \text{H—C—C—H} \qquad \textbf{1} \ \xrightarrow{H_2O_2} \ \text{H—C—C—OH}$$

2a 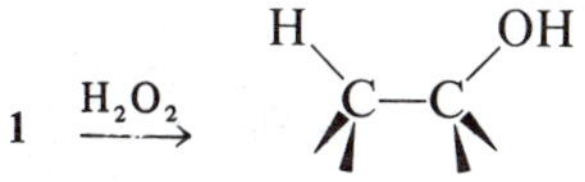   **2b**

whereas the more facile H_2O_2-assisted hydrolysis produces the alcohol **2b** (see p. 228).

(4) Addition is *cis-*, as exemplified by

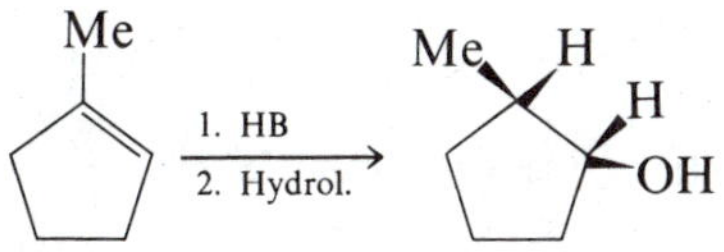

$$\text{(Me-cyclopentene)} \ \xrightarrow[\text{2. Hydrol.}]{\text{1. HB}} \ \text{(Me, H, H, OH cyclopentane)}$$

As we mentioned earlier in discussing the synthesis of silane, the Si—H bond appears to be considerably less hydridic than B—H and Al—H bonds. Accordingly, triorganosilanes can be caused to add to olefin functions, *if* a so-called "coordinatively unsaturated" transition metal complex is used as a catalyst. The Si—H bond is believed to undergo cleavage with formation of M—H and M—Si bonds (oxidative addition to M [see Chap. 17 for more on this]):

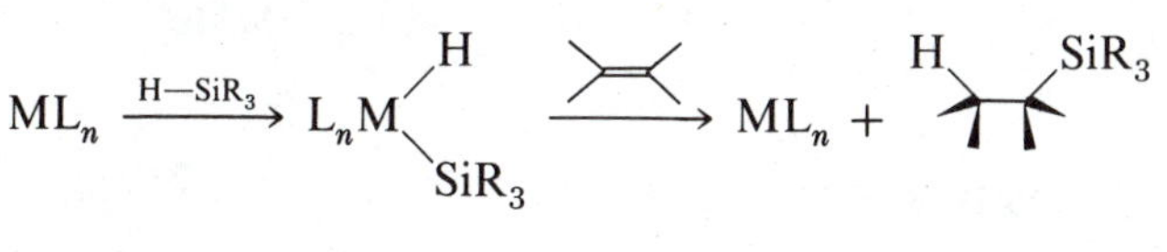

$$ML_n \ \xrightarrow{H{-}SiR_3} \ L_n\text{M}\big\langle {\ \ \text{H} \atop \text{SiR}_3} \ \xrightarrow{} \ ML_n + \ \text{H—SiR}_3 \tag{8-22}$$

You will note that "ML_n" behaves in a general way like an unsaturated group, which easily adds and transfers the H and SiR_3 fragments. Clearly, the effect of ML_n is a kinetic one.

No similar techniques have been reported yet for aiding the addition of P—H† and S—H bonds to olefins; presumably the unaided addition is even less likely than for SiH because these bonds are not sufficiently polar or have the wrong polarity to effect the addition.

While on the subject of organosilanes, we should mention the interesting reduction of organodisilanes by sodium to form triorganosilane anions:[12]

(8-23)
$$Si_2R_6 \xrightarrow{\text{Na}} NaSiR_3 \xrightarrow{\text{R'—Cl}} R_3SiR' + NaCl$$

These silicon anions are good nucleophiles and readily displace chloride from organo-chlorides, thereby affording a useful means of producing unsymmetric organosilanes.

Entirely analogous reactions for the isoelectronic organosulfides and organophosphines can be cited:[13]

(8-24)
$$R_{2,3}E + R—I \rightarrow [R_{3,4}E]I$$

The reaction of organophosphines to produce phosphonium salts is very important in the case of the triorganomethyl phosphonium cation because of its reaction with n-butyl lithium to form **3**, the phosphorus **ylid** (alkylidene phosphorane).[14]

(8-25)
$$R_3P—CH_3^+ \xrightarrow{\text{BuLi}} C_4H_{10} + \left[R_3P{=}CH_2 \leftrightarrow R_3 \overset{\oplus}{P} {-} \overset{\ominus}{\ddot{C}} H_2 \right]$$
$$\mathbf{3}$$

The polar nature of the phosphorus-methylene bond renders the ylid very reactive, and its use as an *in situ* reagent for replacing the carbonyl oxygen of a ketone with methylene is very important. We are reminded here of the affinity of phosphorus for oxygen in forming the phosphoryl linkage, P=O.

$$R_2C{=}O + Ph_3P{=}CH_2 \rightarrow Ph_3PO + R_2CCH_2$$

(8-26)

$$
\begin{array}{c}
Ph \\
| \\
Ph {>} P{-}CH_2 \\
Ph \quad | \qquad | \\
O{-}CR_2
\end{array}
$$

(the intermediate)

CATENATION BY COUPLING OF HALIDES

The mixed organochlorides of the non-metals make useful reagents for preparing various organo compounds containing other substituents. Among the more interesting utilizations of these reagents are the coupling reactions, which have been extensively studied, particularly for silicon and phosphorus. Some examples will serve to illustrate the versatility of these reagents. Synthesis of asymmetric disilanes, presumably via a "Grignard-like" intermediate, can be achieved[15] by reacting different triorganosilyl-chlorides in the presence of Mg:

(8-27)
$$Ph_3SiCl + Me_3SiCl \xrightarrow{\text{Mg}} Ph_3SiSiMe_3$$

---

†See, however, study question 4.

Both silicon and phosphorus monochlorides undergo a Wurtz-type coupling on reduction with Na [compare with (8-23)]:

$$2R_{3,2}ECl \xrightarrow{Na} R_{6,4}E_2 + 2NaCl \tag{8-28}$$

An extension of this reaction, using the dichloro compounds, has made possible the synthesis of the penta (**4a**), hexa, and heptasilyl compounds:[16]

$$Me_2SiCl_2 + 2Na \rightarrow \frac{1}{n}\{Me_2Si\}_n + 2NaCl$$

$$\tag{8-29}$$

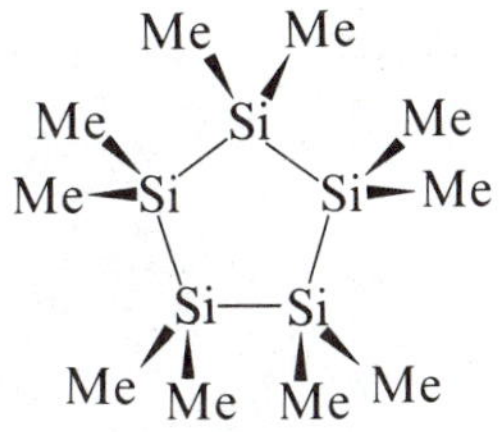

4a 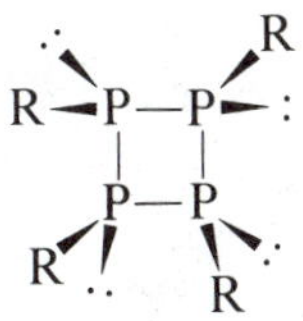   **4b**

Mercury, too, has been found to be an effective reductant for this type of reaction, particularly for the synthesis of cyclic polyphosphines[17] (**4b**).

$$RPI_2 \xrightarrow{Hg} \frac{1}{n}\{RP\}_n + HgI_2 \tag{8-30}$$

In yet another reaction type, mixtures of monochloro and monohydrido phosphines may be caused to eliminate HCl to form the corresponding asymmetric diphosphine[18]

$$R_2PCl + R'_2PH \xrightarrow[\Delta]{} R_2PPR'_2 + HCl \tag{8-31}$$

while use of dichloro and dihydrido phosphines leads to the expected unsymmetric polycyclic analogs of **4b**:

$$RPCl_2 + R'PH_2 \rightarrow \frac{1}{n}\{RPPR'\}_n + 2HCl \tag{8-32}$$

An interesting extension of this type of reaction has led to formation of a chain triphosphine:

$$2R_2PCl + R'PH_2 \xrightarrow[\text{base}]{\text{organic}} R_2P\overset{\overset{\displaystyle R'}{|}}{-}P-PR_2 \tag{8-33}$$

STUDY
QUESTIONS

3. Give steps for the syntheses of the following from the halides and any special reagents given in parentheses.

a) $Na[B(OCH_2R)_4]$ (any carbonyl compound)
b) $1,2\text{-}C_2H_4(BCl_2)_2$ and $1,2\text{-}C_2H_4(BMe_2)_2$ (any hydrocarbon and methyl halide)

c) $Li[Al(C_2H_4R)_4]$ (any hydrocarbon)
d) $Cl_3SiC_2H_4R$ (do *not* use an MR alkylating reagent, but any hydrocarbon)
e) $Me_3Si_2(n\text{-}Pr)_3$

f) 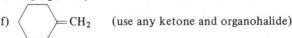$=CH_2$ (use any ketone and organohalide)

g) $[Me_3S]I$

4. Match reactants and products.

a) $NaBH_4 + \frac{1}{2}I_2 \rightarrow$

b) $\frac{1}{2}B_2H_6 + 1,5\text{-}C_8H_{12} \rightarrow$ not isolated $\xrightarrow{H_2O_2}$

c) $B_2H_6 + 2C_4H_6 \rightarrow$ polymer $\xrightarrow{\text{distill}}$

d) $BMe_3 + O_2 \xrightarrow[\text{flow system}]{N_2}$

e) $\frac{1}{2}Al_2Me_6 + C_2H_4 \rightarrow$

f) $KSiH_3 + MeCl \rightarrow$

g) $2PF_2I + Hg \rightarrow$

h) $PH_3 + HCl + 4R_2CO \rightarrow$

i) $Me_3SiNR_2 + Et_3SiH \rightarrow$

1) Me_2BO_2Me

2) $Al_2Me_4(n\text{-}Pr)_2$

3) P_2F_4

4) $MeSiH_3$

5) $\frac{1}{2}B_2H_6$

6) $Me_3Si_2Et_3$

7)

8) $[P(CH_2OH)_4]Cl$

9) 1,5- and 1,6-$C_8H_{14}(OH)_2$

INORGANIC POLYMERS BY SOLVOLYSIS REACTIONS[19]

The so-called solvolysis reaction is a general reaction for non-metal halides and organohydrides (at least the hydridic ones). This reaction is analogous to the coupling reactions noted for phosphorus in the preceding section, but it leads to replacement of chloride or hydrogen about the non-metal atom by oxygen or nitrogen. *The reactions are simple in form;* **elimination of HCl or H_2** *by reaction of $E-Cl$ or $E-H$* (in those cases where the hydrogen is sufficiently hydridic) *with $H-O$ or $H-N$* leads to formation of the corresponding oxyacid, its ester, or the amide of the parent non-metal compound. In those instances of ECl_2 functions reacting with H_2O or NH_3 or RNH_2, a common process is the elimination of two or more moles of HCl with formation of multiple bonds between E and O or N. It is often the case that such multiple bonds are less stable than an equivalent number of single bonds so that one finds **polymerization** *to form chains, rings, and polyhedra to be a common feature of non-metal halide solvolysis reactions.*

The reaction types to be discussed are given by the following general equations:

$$R_xECl_n + nHA \rightarrow R_xEA_n + nHCl$$

$$+ \frac{n}{2}H_2A \rightarrow \underbrace{\frac{1}{3}(R_xEA)_3, \frac{1}{4}(R_xEA)_4,}_{\substack{\text{cyclic or} \\ \text{polyhedral}}} \quad \underbrace{(R_xEA)_\infty}_{\substack{1\text{-},2\text{-, and }3\text{-} \\ \text{dimensional chains}}} \quad + nHCl$$

Boron and Aluminum

The boron halides react with alcohols to form the simple esters of boric acid. Particularly interesting among this class of reactions are those of the monoorgano and diorgano chlorides and hydrides with water:

$$RBCl_2 \xrightarrow{H_2O} \underset{5}{RB(OH)_2} \xrightarrow[\Delta]{} \frac{1}{3}\underset{7}{(RBO)_3} + H_2O \qquad (8\text{-}34)$$

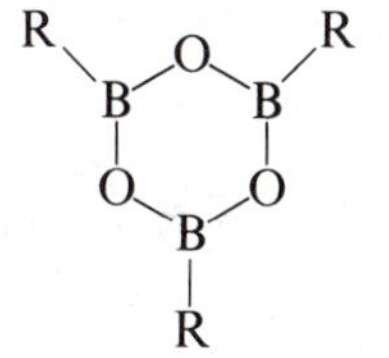

$$\begin{bmatrix} R = Me \\ b.p. \quad 79° \\ m.p. \quad -68° \end{bmatrix}$$

$$R_4B_2H_2 \xrightarrow{H_2O} 2\underset{6}{R_2BOH} \xrightarrow[\Delta]{} \underset{8}{R_2BOBR_2} + H_2O \qquad (8\text{-}35)$$

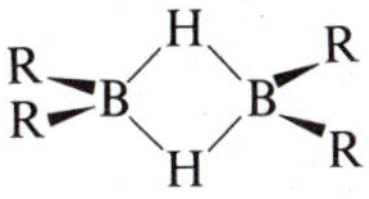

The products, boronic (5) and borinic (6) acids, are acids in the Brönsted sense and have the interesting property (like the parent H_3BO_3) that they can be thermally dehydrated to form the anhydrides[20] 7 and 8. The anhydrides of boronic acids are cyclic trimers called **boroxines**, with a distinct similarity in structure (but not chemistry) to benzene.

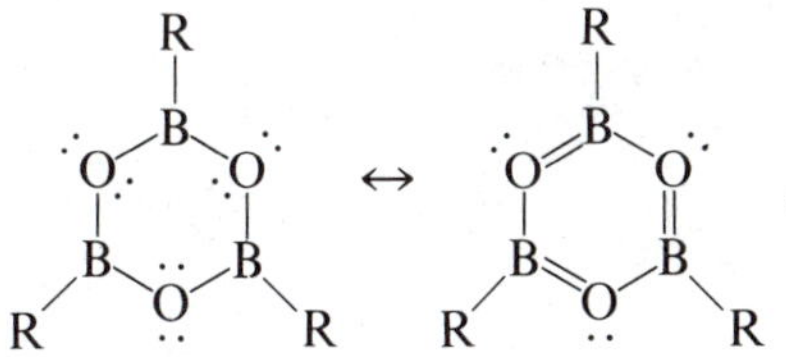

, etc.

[As an aside, note the nomenclature system that designates $(RBO)_3$ as bor*oxines* and $(RBNR')$ as bor*azines*.]

Analogous reactions are known for amines, although it is often the case that more forcing conditions (presence of a tertiary amine base or elevated temperature) are required to effect HCl or H_2 elimination. The nature of the products (monomers, dimers, trimers) is highly dependent upon the steric bulk of groups bound to the boron and to the amine nitrogen. Both the rate of polymerization and the thermodynamic stability of the condensation reaction product depend on steric and electronic factors associated with the substituents. The following points are important:

(i) You can assume that tertiary amines lead to formation of 1:1 adducts called **amineboranes**.

$$BX_3 + NR_3 \rightarrow X_3B \cdot NR_3 \qquad (8\text{-}36)$$

(ii) Secondary amines may lead to either monomeric or dimeric products, depending on the nature of the boron and nitrogen substituents through their influence on the stability of the $(B=N \leftrightarrow B-\ddot{N})$ linkage in the monomer and on steric interactions between "eclipsed" X and R groups in the nearly planar dimer (9).

$$BX_3 + NHR_2 \rightarrow BX_3 \cdot NHR_2$$

$$-HX$$

$$(8\text{-}37)$$

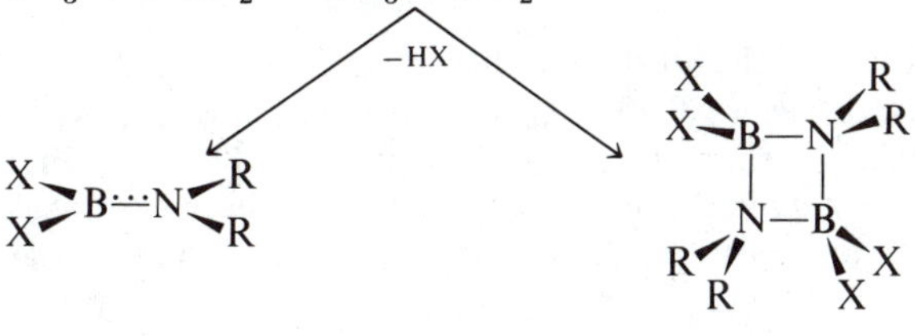

9

Trimerization of the **aminoboranes** (X_2BNR_2) appears uncommon, for the steric interaction between R and X may become too severe when $X \neq H$. When $X = H$, trimerization to the cyclohexane analog is apparently no problem.

(iii) Primary amines can lead to trimeric structures (**borazines**) analogous to the boroxines.

(8-38)

$$BX_3 + NH_2R \rightarrow X_3B \cdot NH_2R \xrightarrow{\Delta} (XBNR)_3$$

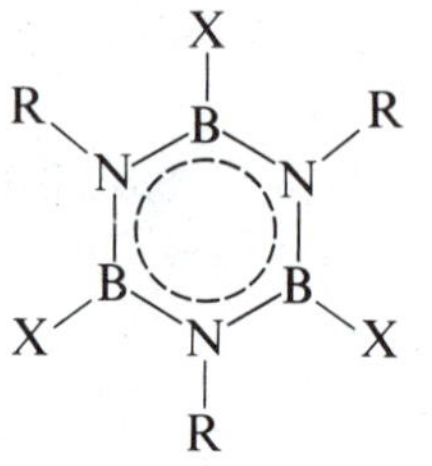

(iv) That stoichiometry can have a great deal to do with the products formed in unhindered cases is illustrated by the complete solvolysis of BCl_3 by liquid ammonia†

(8-39)

$$BCl_3 + 3NH_3 \xrightarrow{NH_3} B(NH_2)_3 + 3[NH_4]Cl$$

and the reaction of BCl_3 with excess aniline

(8-40)

$$BCl_3 + \underset{\text{excess}}{PhNH_2} \rightarrow B(NHPh)_3 + 3[PhNH_3]Cl$$

leading not to dimeric or trimeric products, but to the fully substituted amides.

An interesting point of demarcation of the reaction pathways of BCl_3 and BH_3 (actually B_2H_6) arises in their reactions with liquid ammonia and derives from the difference (monomer *vs.* dimer) in structure of these acids. As noted above, BCl_3 leads to the fully substituted amide whereas diborane leads to borazine via unsymmetric cleavage of diborane to produce initially the salt $[H_2B(NH_3)_2][BH_4]$ (see p. 234).

As you might have anticipated by now, phosphine adducts such as

$$2Me_2PH + B_2H_6 \xrightarrow{\Delta} 2Me_2PH \cdot BH_3$$

may be pyrolyzed to eliminate H_2 and thereby produce not a monomeric "boraphosphene," but rather the cyclic **phosphinoborane** trimer with a "cyclohexane" chair conformation (similar to the "saturated" aminoborane trimers).[21]

(8-41)

$$Me_2PH \cdot BH_3 \xrightarrow{\Delta}$$

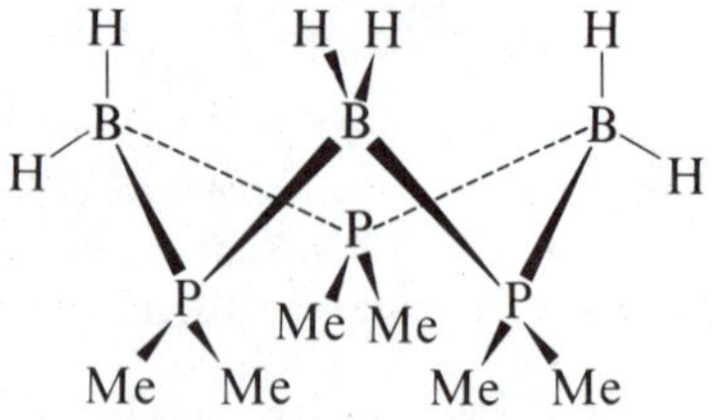

Note that the larger size of phosphorus than of nitrogen should facilitate the formation of trimers, as should weaker $P=B$ character in the monomer (H_2BPMe_2). The thermal and chemical stability of this trimeric structure is unusually great. The origin of the inertness is not completely understood, although arguments have been made that hyperconjugation of B—H bond density with the phosphorus d atomic orbitals (**10**) may be an

---

†$B(NH_2)_3$ is unstable and has never been characterized.

important contributor to reduced B—H hydridic character.[22] (You will encounter this phosphorus acceptor concept again, in the section after next, where an NR group occurs in the place of BH_2.) It would be well to recall at this point the analogous stability of H_3PBH_3 to acid hydrolysis (p. 233).

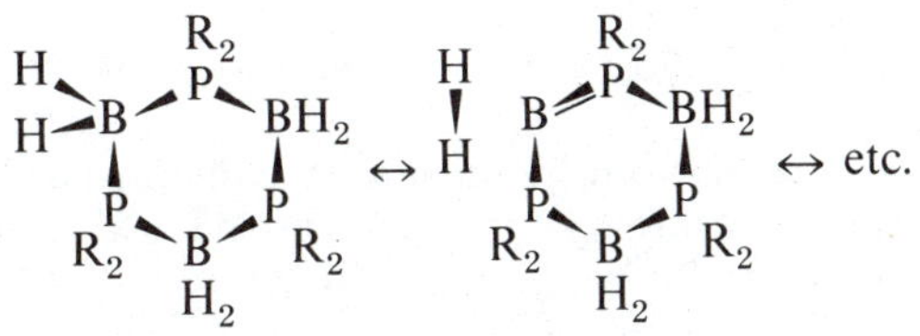

10

Whatever the reason, the high chemical and thermal stability of the phosphinoborane trimer has spurred interest in parlaying this tendency into development of phosphino-borane polymers for various commercial uses. These efforts have been only partially successful, and the study of this class of compounds continues.

Many of the solvolysis products of boron halides and hydrides find counterparts when aluminum is the central atom. An important difference between boron and aluminum is the propensity of aluminum to increase its coordination number. Thus, the boroxine and borazine structures are not found in Al^{3+} chemistry. Rather, the **alkoxides** of aluminum, for example, are dimeric and even tetrameric. A sterically hindered alkoxide like aluminum *t*-butoxide achieves only the dimeric structure

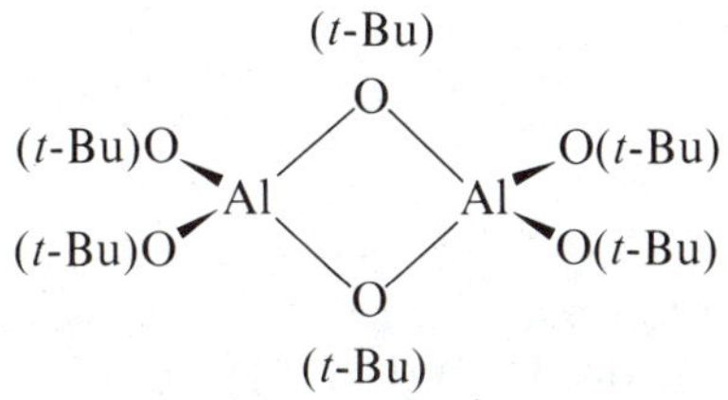

whereas the important aluminum isopropoxide exhibits both trimeric and tetrameric (**11**) structures, the latter containing both four-coordinate and chiral six-coordinate aluminum centers.[23]

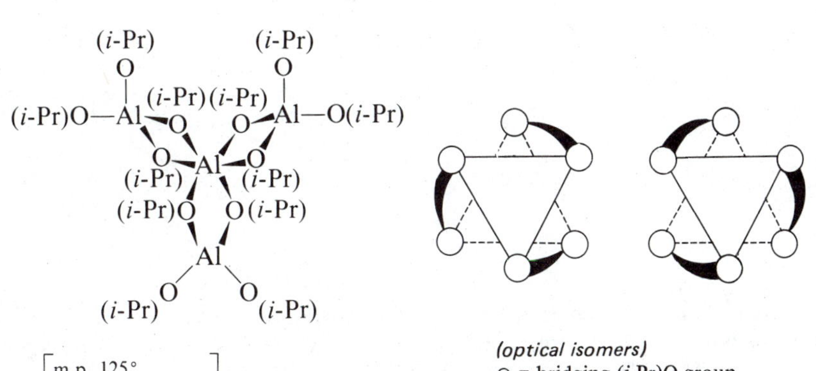

11 [m.p. 125°
b.p. 242°/10 torr]

(optical isomers)
○ = bridging (*i*-Pr)O group

The best methods for synthesizing the alkoxides appear to be the reaction of the chloride with the appropriate sodium alkoxide or alcohol (with NH_3 present) or even the reduction of the corresponding alcohol with Al (aluminum being a very active metal) using $HgCl_2$ as a catalyst:

(8-42)

$$AlCl_3 + 3NaOR \rightarrow \frac{1}{n}\{Al(OR)_3\}_n + 3NaCl$$

$$Al + 3ROH \xrightarrow{HgCl_2} \frac{1}{n}\{Al(OR)_3\}_n + \frac{3}{2}H_2$$

There is a marked difference between boron and aluminum halides in regard to ease of solvolysis in general. This is manifest in the fact that more forcing conditions (use of NaOR or ROH/NH$_3$) are needed for synthesis of Al(OR)$_3$.

These ideas presented for the alkoxides of aluminum pertain to amides also. Amines solvolyze aluminum halides even less readily than do alcohols, so that alkali metal amides are useful reagents, with stoichiometry control, for preparing the polymeric aluminum amides.[24] As you might expect from the earlier discussion concerning use of LiAlH$_4$ as a very active reducing agent, amines and phosphines tend to be reduced (with H$_2$ evolution) when brought in contact with lithium aluminum hydride,[25] but note that formation of polymeric materials is difficult to achieve in this way since four moles of NR$_2^-$ (in the case of amines) are formed per mole of Al^{3+}.

$$LiAlH_4 \xrightarrow{R_2NH} LiAl(NR_2)_4 + 4H_2$$

(8-43)

$$\xrightarrow{PH_3} LiAl(PH_2)_4 + 4H_2$$

$$\xrightarrow{ROH} LiAlH(OR)_3 + 3H_2$$

One of the more intriguing examples of aminoalane or phosphinoalane condensations results from thermolyzing the *primary* amine and phosphine adducts of organoalanes.[26] The highly condensed tetrameric structure **12** sometimes encountered for these compounds is reminiscent of the hydrocarbon cubane.

$$R'NH_2 \cdot AlR_3 \rightarrow \frac{1}{4}(R'NAlR)_4 + 2RH$$

(8-44)

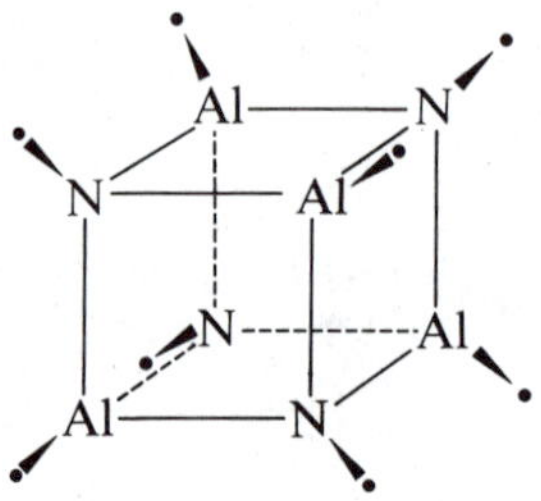

12

Quite generally, these elimination reactions are much more facile than the corresponding reactions of boron. This is dramatically illustrated by the fact that borazine is produced by pyrolysis of H$_3$BNH$_3$ or [(H$_3$N)$_2$BH$_2$][BH$_4$] at 250°, while the following scheme[27] reveals the more facile and more extensive elimination of H$_2$ from H$_3$AlNH$_3$:

(8-45)

$$AlH_3 + NH_3 \xrightarrow[-80°]{ether} AlH_3 \cdot NH_3$$

$$H_2 + (H_2Al-NH_2)_3 \xleftarrow{-40°}$$

$$\xrightarrow{20°} H_2 + (HAlNH)_x$$

The polymer obtained at room temperature is a non-volatile white solid that is insoluble in most solvents. Although the empirical formula is formally analogous to borazine, the unsaturation of $(HAlNH)_3$ is avoided by condensation (as in the tetramer **12**) into a polymeric solid, perhaps resembling **13**, for which $(CF)_n$ is a model.[28]

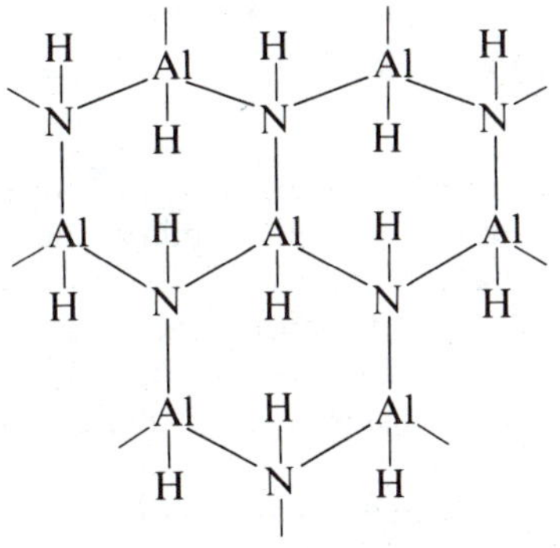

13

The reactions above illustrate a fundamental fact of Al−N adduct chemistry: *if both the aluminum and nitrogen have hydrogens attached, H_2 can be eliminated as above;* or, if either has an alkyl or aryl group and the other a hydrogen, then *elimination of hydrocarbon* can be observed. Some specific examples are as follows:[29]

$$Me_3Al \cdot NH_3 \xrightarrow{70°} CH_4 + (Me_2AlNH_2)_3 \xrightarrow{200°} CH_4 + (MeAlNH)_x$$
$$[m.p. 56.7°] \qquad\qquad [m.p. 134.2°] \qquad\qquad [glassy\ solid]$$

$$Et_2AlCl \cdot NH_2Me \xrightarrow{72-76°} C_2H_6 + (EtAlCl \cdot NHMe)_{2,3} \xrightarrow[54\ hr.]{210°} C_2H_6 + (ClAlNMe)_x$$
$$[m.p. -11.5°] \qquad\qquad [m.p. 91°] \qquad\qquad \begin{bmatrix} non\text{-}volatile \\ white\ solid \end{bmatrix}$$

(8-46)

As implied in (8-45) and (8-46), even unsaturated Al−N compounds such as the aminoalanes (R_2Al-NR_2') are very prone to association to dimers and trimers, unlike the aminoboranes, many derivatives of which are monomeric or only weakly associated (pp. 180 and 228). As pointed out above, this is due, in part at least, to the fact that B−N π bonding more efficiently relieves the coordinative unsaturation of the boron atom (better $p\text{-}p\pi$ overlap) and also to the fact that there is less steric hindrance to trimerization of (R_2Al-NR_2').

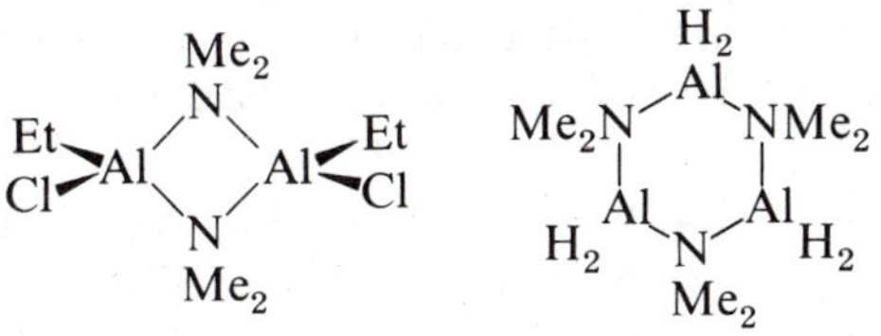

Silicon

Generally speaking, the Si−C bond is reasonably stable (the order of bond energies B−C > Si−C > Al−C is inversely related to the order of polarities and susceptibility to attack); because of this, certain organosilanes and their derivatives have found commercial usage. Probably the most important class of organosilanes consists of the **organosiloxanes**, commonly known as silicones. These materials are polymers, with highly stable Si—O backbones, formed by hydrolyses of $R_n SiX_{4-n}$. For example:

$$R_3SiX + H_2O \rightarrow (R_3Si)_2O \quad \textit{(disilyl ethers)}$$

(8-47)
$$R_2SiX_2 + H_2O \rightarrow (R_2SiO)_n \quad \textit{(ring and chain polymers)}$$

$$RSiX_3 + H_2O \rightarrow (RSiO_2)_n \quad \textit{(cross-linked chains)}$$

All these reactions proceed through formation of the corresponding silanol and, even though some diols and triols have moderate stability, they tend to eliminate H_2O and condense (as with the boronic and borinic acids, the E=O linkage affords less stability than oligomer formation). While somewhat of an art, it is possible to control the nature of the polymer formed by hydrolyzing mixtures of mono-, di-, and trihalosilanes; the dihalides tend to establish linear chains, the trihalides cross-link the chains, and the monohalides act as chain terminators.

In reactions analogous to the hydrolysis reactions, the organosilicon halides react with amines to form **silyl amines**, which bear a *formal* resemblance to alkyl amines. The compound trisilyl amine has the structurally interesting feature of a planar Si_3N frame and is anomalously weak as a Lewis base relative to Me_3N (see Chapter 5, p. 148).

(8-48)
$$3SiH_3X + NH_3 \rightarrow (SiH_3)_3N + 3HX$$

With respect to the VSEPR model of structures, you might initially think that trisilyl amine would have a pyramidal geometry about nitrogen. This conclusion is predicated on the assumption that the nitrogen atom is considered to have a *lone* pair. Many believe that the planarity of this compound is good evidence for the existence of $p\pi$-$d\pi$ interaction of the lone pair with the d orbitals of the silicon (recall the near planarity of the nitrogen in organic amides). Others have speculated

$$H_3Si\overset{..}{-}N\overset{SiH_3}{\underset{SiH_3}{\diagdown}} \leftrightarrow H_3Si=N\overset{SiH_3}{\underset{SiH_3}{\diagdown}} \text{, etc.}$$

that the electron-releasing nature of the silane facilitates the flattening of the nitrogen atom. Both the π overlap and σ factors act in the same direction, so an experimental answer as to which is more important is unlikely. In fact, the σ/π factors are synergically coupled (greater σ release to nitrogen facilitates retro-bonding from N to Si and *vice versa*). Given the larger size and lower electronegativity of phosphorus, the pyramidal structure of $P(SiH_3)_3$ can be rationalized by less σ bp/bp repulsion and correspondingly less retro-π interaction.

The compound **hexamethyldisilazane** or *bis*(trimethylsilyl)amine has attracted some attention as a "silylating" reagent. Reactions of a large variety of polar E—H groups (E = C, O, N, S) introduce the E—$SiMe_3$ linkage. For example:

(8-49)
$$(Me_3Si)_2NH + \left.\begin{array}{l} ROH \\ RCO_2H \\ RSH \\ RCONH_2 \\ HCN \\ H_2SO_4 \end{array}\right\} \rightarrow \left\{\begin{array}{l} ROSiMe_3 + Me_3SiNH_2 \\ RCO_2SiMe_3 \\ RSSiMe_3 \\ RCONH(SiMe_3) \\ Me_3SiCN \\ HSO_3(SiMe_3) \end{array}\right.$$

[b.p. 126°]

The facility of these reactions is due in large part to the Si—N π bonding in Me_3SiNH^- as a leaving group (from the other $SiMe_3$), for otherwise the amide group would not cleave very easily.

The reactions of dichlorosilanes with amines produce products generally in line with what you anticipate by now [note that NH is isoelectronic with O, and refer to (8-47)].

$$RR'SiCl_2 + NH_3 \rightarrow [RR'SiNH]_x$$

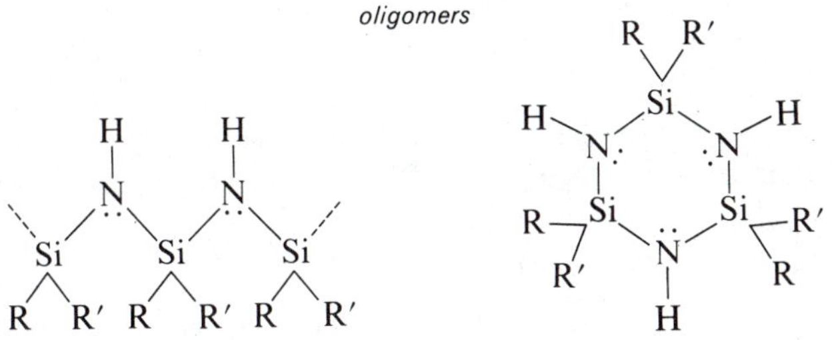

(8-50)

The progression of increased steric bulk represented by (R = R' = H; R' = H, R = Me, Et, Ph; R, R' = organic) results in less tendency for chain polymers and increasing yields of cyclic **tri-** and **tetrasilazanes**, $\{RR'SiNH\}_{3,4}$. When primary amines are used in place of NH_3, steric congestion worsens and **diaminosilanes** are produced:

$$Me_2SiCl_2 + 2NH_2R \rightarrow Me_2Si(NHR)_2$$

(8-51)

Deamination of these materials (compare the dehydration of silanols) by NH_4^+ catalysis is effective when R = Me, but the reaction becomes less facile as the bulk of R increases:

$$Me_2Si(NHMe)_2 \xrightarrow{NH_4^+} \{Me_2SiNMe\}_3$$

(8-52)

Like the aluminum analogs, but unlike the borazines, the cyclotriazanes exhibit nonplanar boat and chair structures; nevertheless, the Si—N bond distances and large Si—N—Si angles (up to 130°) suggest that some N → Si p-$d\pi$ bonding still takes place.

In parallel with the mild reactivity of Si—C, noted in the opening of this section, the silanes are marginally hydridic and do not rapidly hydrolyze in acidic media. This is dramatically illustrated by the fact that hydrogen halides will not react with silanes in the absence of aluminum halide (or other halide acceptor) as a catalyst[30]

$$SiH_4 + nHX \xrightarrow{Al_2X_6} SiX_nH_{4-n} + nH_2$$

(8-53)

($HAlCl_4$ should be a potent acid). Reaction of silane with alcohols occurs in the presence of alkoxide ion; when R = H the product $Si(OH)_4$ eliminates H_2O to form silicon dioxide.

$$SiH_4 + 4ROH \xrightarrow{OR^-} Si(OR)_4 + 4H_2$$

(8-54)

This behavior is nothing more than base catalyzed hydrolysis of silanes, where you are again reminded of the characteristic ease of nucleophilic attack upon silicon.

Nitrogen and Phosphorus

The bonds to hydrogen of nitrogen and of phosphorus (though much less so) are weakly acidic and consequently do not experience solvolytic reactions as do the hydrides of the preceding elements in their periods. With regard to the reactions of the halides, the only useful haloamine is chloramine ("useful" meaning that NH_2Cl is not an inordinately dangerous chemical to handle and thus might be used by chemists as a general synthetic material). As detailed in Chapter 7, chloramine experiences reactions with fairly strong nucleophiles through displacement of Cl^-. Thus, base hydrolysis and treatment with ammonia illustrate normal solvolysis behavior:

(8-55)

$$NH_2Cl + OH^- \rightarrow NH_2OH + Cl^-$$

$$NH_2Cl + 2NH_3 \rightarrow N_2H_4 + NH_4^+Cl^-$$

The halides of phosphorus solvolyze in the customary fashion, and several of these reactions are worthy of special comment. The trivalent halides undergo alcoholysis (vigorous) and aminolysis (less vigorous) to form the corresponding alkoxides and amides.

(8-56)

$$PCl_3 + 3H_2O \rightarrow H_3PO_3 + 3HCl$$

$$+ 3NH_3 \rightarrow P(NH_2)_3 + 3HCl$$

A characteristic of phosphorous acid and its trimethyl ester is a "keto-enol" type of rearrangement to produce a phosphoryl linkage:

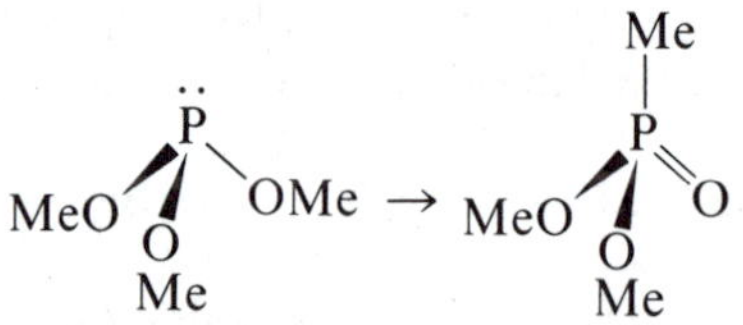

Taking a cue from the Arbusov reactions of phosphite esters with alkyl halides (see p. 244, Chapter 7):

(8-57)

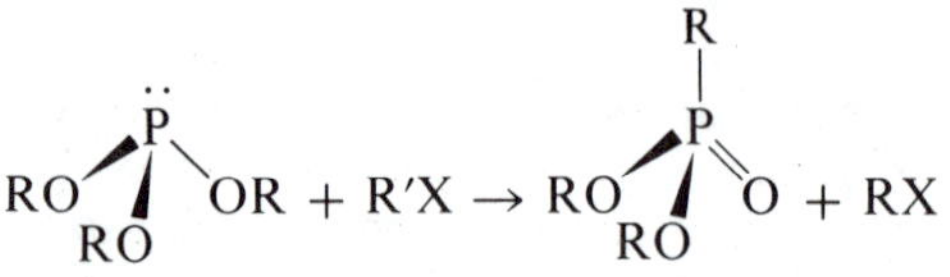

and from the fact that, in the alcoholysis of PCl_3, failure to remove the HCl produced by the addition of some base (like an amine) leads to $(RO)_2PHO$ as a product, the "keto-enol" rearrangement is bi-molecular.[31] In a most interesting example of substituent effects at phosphorus, the *bis*(pentafluorophenyl)phosphinous acid exhibits a "keto-enol" *equilibrium:*[32]

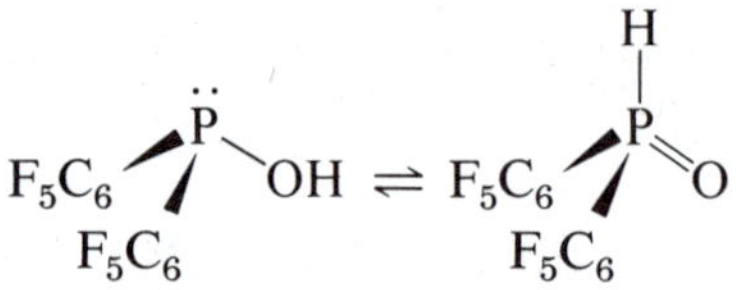

It appears that the electronegativity of C_6F_5 lowers the nucleophilicity of the P lone pair, and thereby reduces the stability of the "keto" form.

The solvolysis products from the reactions of phosphorus halides with amines define an extensive class of polymeric materials. In a moment we will consider the reactions of PCl_5, but first consider the reactions of PCl_3 and $POCl_3$. The latter reacts with NH_3 and with primary and secondary amines to form stable **triamides**[33] wherein the nitrogen atoms adopt a planar or nearly planar geometry:

(8-58)

$$POCl_3 + 3NH_3 \rightarrow PO(NH_2)_3$$

This structural feature played an important role in the basicity of $(Me_2N)_3P$ (Chapter 5, p. 149).

PCl_3 presents an aminolysis chemistry more in line with your expectations, as the following scheme shows:[34]

$$PCl_3 + 6NH_3 \xrightarrow[-78°]{CHCl_3} P(NH_2)_3 + 3NH_4Cl$$

$$\downarrow \text{deamination on warming}$$

$$PCl_3 + 5NH_3 \xrightarrow[-20°]{R_2O} \frac{1}{n}\{HN{=}PNH_2\}_n + 3NH_4Cl$$

$$\text{[unstable at R.T.]}$$

$$PCl_3 + 4NH_3 \xrightarrow{\text{liq. } NH_3} \frac{1}{n}(PN)_n + 3NH_4Cl$$

$$4PCl_3 + 18MeNH_2 \rightarrow P_4(NMe)_6 + 12NMeH_3Cl$$

(8-59)

This last compound, a crystalline material, bears a striking resemblance to P_4O_6 in structural and some chemical properties. To fit its formation into the scheme of NH_3 solvolysis, you may find it helpful to *imagine*† formation of $4P(NHMe)_3$ with subsequent deamination (intramolecular, four times, and intermolecular, two times) to yield§

$$2\{MeN{=}P{-}\overset{\displaystyle Me}{\overset{|}{N}}{-}P{=}NMe\}.$$ The four P=N links are less stable than four P—N links which arise from condensation into $P_4(NMe)_6$. This behavior leads us directly to the behavior of PCl_5 in contact with amines.

Following a pattern established by boron and aluminum, pentavalent phosphorus halides form a series of polymeric **phosphazenes**[35] with ammonia (the class of compounds commonly referred to as phosphonitrilic compounds).

$$PCl_5 + NH_4Cl \xrightarrow[C_2H_2Cl_4]{146°} \frac{1}{x}[Cl_2PN]_x + 4HCl$$

(8-60)

Perhaps best prepared by treating PCl_5 with NH_4Cl in a chlorinated hydrocarbon solvent, primarily trimeric (**14**) and tetrameric (**15**) dichlorophosphazene, $(NPCl_2)_n$, are formed.[36] These are useful for synthesis of various substituted phosphazenes, since the chloride substitution reactions common for non-metal chlorides work for this particular phosphonitrile.

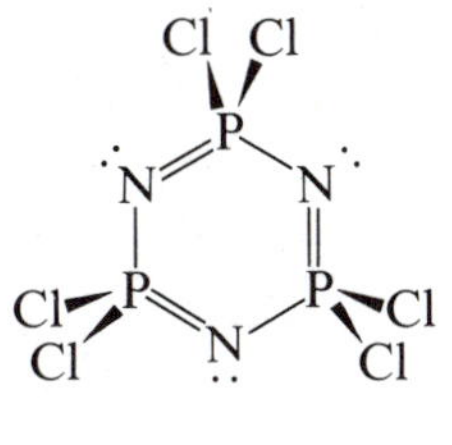

14 $\begin{bmatrix} \text{m.p. } 114° \\ \text{b.p. } 127°/13 \text{ torr} \end{bmatrix}$ **15** $\begin{bmatrix} \text{m.p. } 124° \\ \text{b.p. } 188°/13 \text{ torr} \end{bmatrix}$

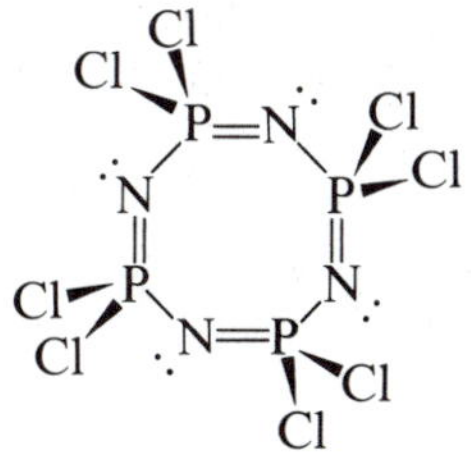

The reason for the condensation reaction is most apparent if we inquire into the structure of the monomeric unit, Cl_2PN.

---

†The mechanism of the reaction has not been determined.
§Compare this monomer to a P_2O_3 monomer: $O{=}P{-}O{-}P{=}O$.

$$\overset{Cl}{\underset{Cl}{>}}P\overset{\oplus}{=}\overset{\ominus}{N}\colon\;\leftrightarrow\;\overset{Cl}{\underset{Cl}{>}}P\equiv N\colon$$

This structure is not so bizarre as it may first seem to you, for a good structural analog is to be found in carbon chemistry. Phosgene (Cl_2CO) is valence shell isoelectronic with this monomer, but, of course (and this is an important difference), the nuclear charges are distributed quite differently in CO and PN. The polarity of the PN bond means that the phosphorus atom will exhibit significant acid character, and the nitrogen will exhibit significant base character; polymerization seems inevitable. In **14** and **15** you still find the necessity of PN double bond character, and the bond distances are observed[37] to be shorter than "normal" (158 pm instead of 178 pm). Furthermore, the trimer structure appears to have D_{3h} symmetry, so all PN bonds are equivalent. The ring electron pairs are fully delocalized. You might note that, even in the ring structure, there *may* also be residual delocalization of the in-plane nitrogen lone pairs onto the phosphorus atoms.

When the phosphorus is symmetrically substituted (as in **14**) it seems that a planar structure is preferred, but unsymmetric phosphorus substitution could well introduce a tendency for the ring to pucker, and this can be accommodated by the π electron system. An example of puckering is found with $[NP(OMe)_2]_3$, which exhibits a planar structure as formed (as a solid) but which undergoes a thermal rearrangement (an extension of the "keto-enol" migration discussed on p. 286) to form the boat structure[38] **16**:

(8-61)

$$[NP(OMe)_2]_3 \xrightarrow{\Delta} \left[MeNP(OMe)\right]_3$$
* like **14**

16

Apparently the larger tetrameric ring **15** introduces enough strain under D_{4h} symmetry that $(NPCl_2)_4$ and $[NP(NMe_2)_2]_4$ exhibit nonplanar structures,[37] without serious deterioration of the PN π bonding.†

The chloro-derivatives are useful for introducing other substituents at phosphorus atoms in the ring. Here we will mention one recently discovered unusual reaction illustrating bicyclic ring formation.[39] Treatment of the *bis*(ethylamino) derivative of **15** with excess Me_2NH in $CHCl_3$ not only resulted in full amination of both (PCl_2) groups and one of the (PCl) groups but also caused one of the NHEt groups to bridge to the opposite (PCl) group, with HCl elimination.

---

†In the context of these remarks on the N–P π interaction, you may wish to review the earlier comments about the possibility for B=P hyperconjugation in $(H_2BPH_2)_3$ and N=Si character in the cyclosilazanes.

When drawn in stereo projection, the bicyclo structure is reminiscent of adamantane

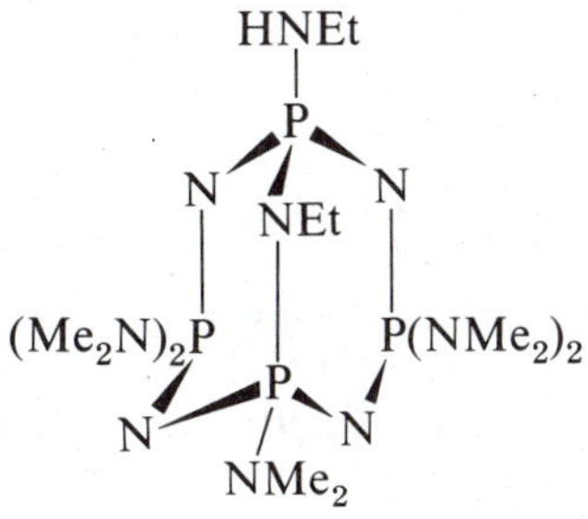

and of that of $P_4(NMe)_6$, p. 287.

It is useful that one can stoichiometrically control the hydrolysis of the penta-halides (X = F, Cl, Br) of phosphorus to form the important **phosphoryl halides**

$$PCl_5 + H_2O \rightarrow Cl_3PO + 2HCl$$

From these reagents one can form organophosphine oxides (R_3PO), esters of phosphoric acid [$(RO)_3PO$], phosphoryl amides [$(R_2N)_3PO$], and so on. Complete hydrolysis, of course, leads to phosphoric acid, indicating that $P(OH)_5$ is not a stable compound.† Similarly, esters of $P(OH)_5$ are not known to arise from alcoholysis of PCl_5.

You may be puzzled at this point, wondering whether hydrolysis of phosphorus halides, in analogy with the aminolysis reactions, produces *polymeric* P—O compounds. The answer is emphatically "yes" in that a large number of linear and cyclic phosphorus-oxygen compounds are known.[40] Rather than embark on a lengthy discussion of these compounds, let us note that the simplest species of this type are **diphosphoric acid (17)** (obtained by dehydration of phosphoric acid at $>200°$ or by its reaction with phosphoryl chloride)

$$2H_3PO_4 \rightarrow H_4P_2O_7 + H_2O$$
$$5H_3PO_4 + POCl_3 \rightarrow 3HCl + 3H_4P_2O_7$$

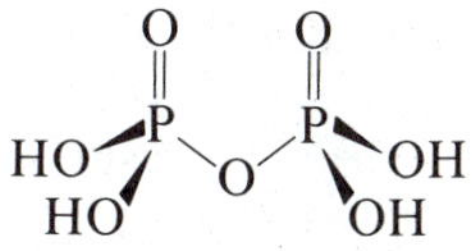

17
[m.p. 61°]

(8-62)

and the polymers of **metaphosphoric acid**, with the empirical formula HPO_3. Best characterized are the salts of metaphosphoric acid, which contain a non-planar trimeric anion **(18)** [the acid itself is found either as an amorphous (glassy) solid or as a viscous liquid].

$$nH_3PO_4 \xrightarrow{\text{dehyd.}} \frac{1}{n}(HPO_3)_n + nH_2O$$

(8-63)

$$nNaH_2PO_4 \xrightarrow{\text{dehyd.}} \frac{1}{n}(NaPO_3)_n + nH_2O$$

---

†Note that deaquation here, $P(OH)_5 \rightarrow PO(OH)_3 + H_2O$, bears a strong resemblance to the deamination reactions of P(III) amides as in equation (8-59) and to ylid formation (p. 276), and achieves the very stable P=O link.

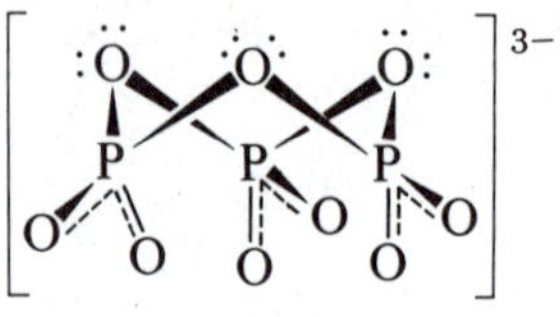

18

The chain polyphosphates, of which diphosphate is the simplest example, are of great significance in biosystems. In fact, the exothermic hydrolysis of adenosine triphosphate (ATP, **19**) to adenosine diphosphate (ADP, **20**) is of fundamental significance as a source of energy in living muscle cells.

(8-64)

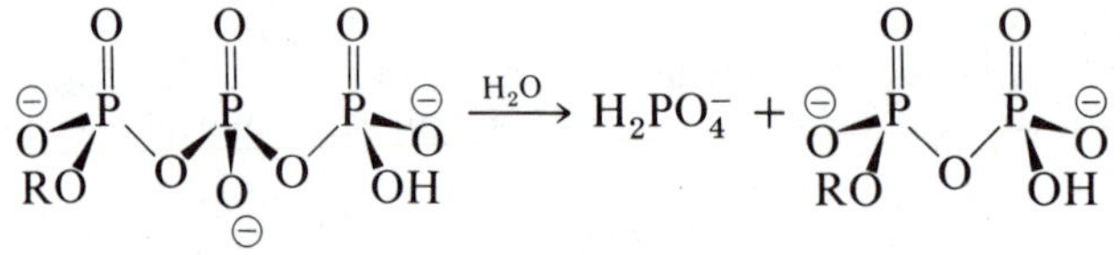

19   **20**

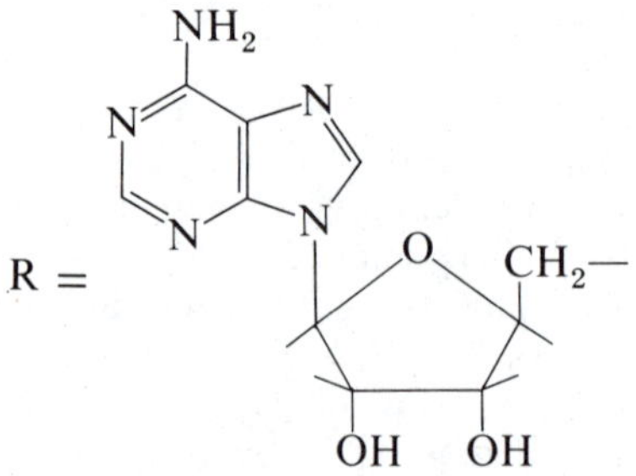

Note that this reaction is only mildly exothermic ($\Delta G° \approx -29$ kJ/mole) and is thermally neutral on an "additive bond energy" basis. Solvation energy and entropy changes provide considerable driving force for the reaction. The condensed phosphate best known to the general public is sodium **triphosphate** ($Na_5P_3O_{10}$)—used to control water hardness and pH in laundry detergents and one compound responsible for phosphate pollution of natural waters (the other main offender in this category is $NH_4H_2PO_4$, the field fertilizer).

The ultimate in condensed phosphate polyhedra is found in the anhydrides of phosphorus and phosphoric acids, commonly called phosphorus trioxide (**21**) (phosphorus oxide or tetraphosphorus hexoxide) and phosphorus pentoxide (**22**) (phosphoric oxide or tetraphosphorus decoxide). The molecular forms of these anhydrides (P_4O_{10} is also found in several polymeric forms) have the composition P_4O_6 and P_4O_{10}, as you will

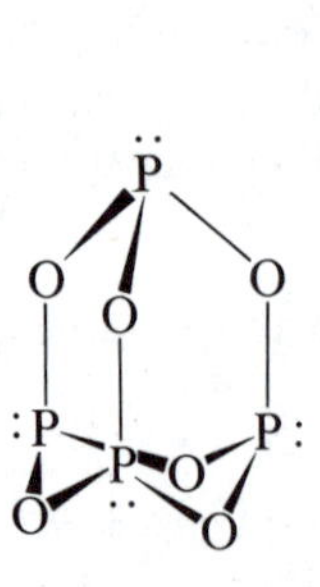

21 [m.p.   24° / b.p.   174°] 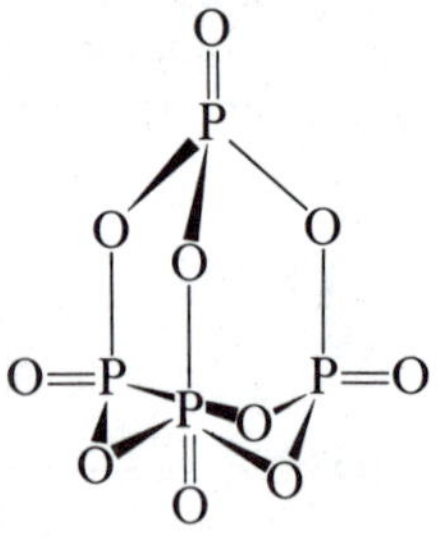   **22** [subl. 360°]

recall from the discussion of their structures in Chapter 6. Their synthesis, since they are *very* strong desiccants, is not best achieved by dehydration of H_3PO_3 and H_3PO_4 but by direct combination of elemental (white P_4) phosphorus and oxygen. Use of excess O_2 easily results in formation of P_4O_{10} but the yield of P_4O_6 from use of limited O_2 is only about half of the theoretical yield. As P_4O_6 and P_4 are soluble in organic solvents but polymeric red phosphorus is not, conversion of unreacted white phosphorus to red by UV radiation† permits the isolation of the trioxide.

$$P_4 + 5O_2 \rightarrow P_4O_{10} \tag{8-65}$$

$$P_4 + 3O_2 \rightarrow P_4O_6 + (P_4O_7, P_4O_8, P_4O_9) \, \S$$

As mentioned above, a most important property of P_4O_{10} is its strong desiccating ability, making this oxide useful in producing the anhydrides of other oxyacids (for example, by dehydration of $HClO_4$, HNO_3, and even H_2SO_4). These reactions with P_4O_{10} are typical of compounds with acidic hydrogens. We can illustrate this general reaction by mentioning that HF reacts to form both difluorophosphoric and fluorophosphoric acids.[41]

$$P_4O_{10} + 6HF \rightarrow 2F_2PO(OH) + 2FPO(OH)_2 \tag{8-66}$$

(You might recognize that a popular "fluoride" toothpaste uses the fluorophosphate anion as its anti-caries agent.)

Oxygen and Sulfur

As with nitrogen, the use of most oxygen halides as starting materials for synthesis of other materials is not to be recommended for other than the most experienced and careful chemist. Even so, most reactions of O_mX_n are not typical solvolysis reactions, in the sense of preceding examples in this section. The appearance of oxygen in solvolytic products is therefore essentially restricted to the roles of H_2O and ROH as the solvolyzing agents. Even the reactions of the most important oxygen halide, OCl^- (again in analogy with nitrogen chemistry), are usually not considered solvolysis reactions. Here you should be thinking of the oxygen transfer (redox) reactions like

$$NO_2^- + OCl^- \rightarrow NO_3^- + Cl^-$$

$$SO_3^{2-} + OCl^- \rightarrow SO_4^{2-} + Cl^-$$

Consequently, our attention will focus primarily on sulfur compound solvolysis.

For sulfur we would be somewhat restricted in considering only binary halides, for SF_6 is highly inert, SF_4 is useful primarily as a fluorinating agent, and SF_2 is not attainable in useful quantities. Of the chlorides, only S_2Cl_2 and SCl_2 are common (SCl_4 is known, but it decomposes at $\sim -50°C$ into SCl_2 and S_2Cl_2). Of some interest is the product of the reaction[42] of S_2Cl_2 with NH_4Cl or NH_3:

$$6S_2Cl_2 + 4NH_4Cl \rightarrow (SN)_4 + S_8 + 16HCl \tag{8-67}$$

Notice the parallel between this reaction and that [equation (8-59)] of PCl_3 with liquid ammonia. Furthermore, by noting that SN is valence shell isoelectronic with NO you are confronted once again with the characteristic condensation of "unsaturated" species containing third period elements. That is, NO only *weakly* dimerizes in the solid state.

---

†Remember that UV radiation (200 to 300 nm wavelength) supplies 600 to 400 kJ/mole to a molecule that absorbs a photon in this range.

§These products are simply intermediates between P_4O_6 and P_4O_{10} where 1, 2, and 3 phosphorus atoms have terminal oxygens.

The compactness of the $(SN)_4$ structure is the intriguing feature of this compound; if you were to assume a close analogy of S_4N_4 with the phosphonitrilic compounds, you might devise a structural/bonding model involving (as a single resonance structure) alternate S—N double and single bonds within an eight-membered ring. In doing so, however, you come up with an electronic structure with an unpaired electron on each sulfur (sulfur has one more valence electron than phosphorus).†

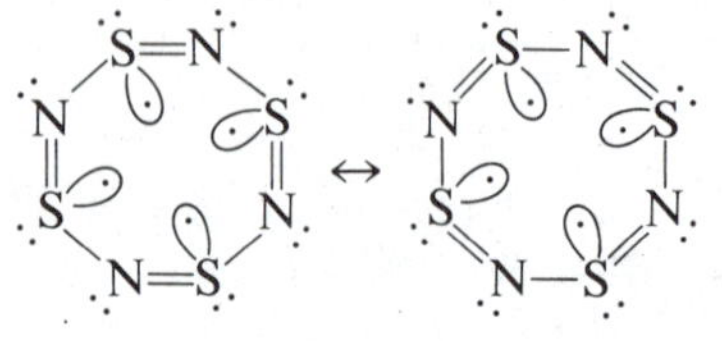

Your intuition should tell you that such a structure is probably not very stable with respect to further condensation; the experimental finding[43] that samples of the compound are *essentially* diamagnetic at room temperature confirms this suspicion. By analogy with the bicyclic structure of the S_8^{2+} cation (as discussed in Chapter 6), a reasonable choice for the structure would be the cage **23** (the structure actually found by X-ray analysis of the crystalline material[44]):

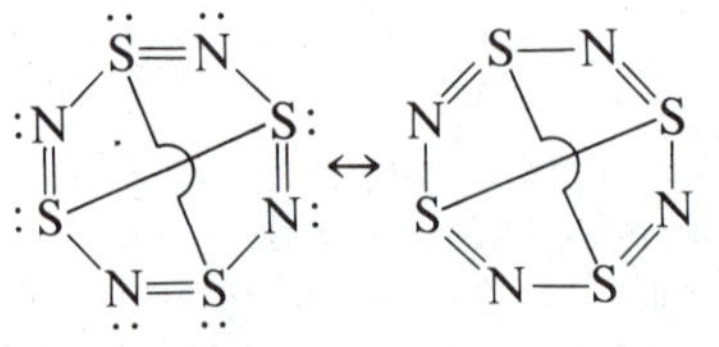

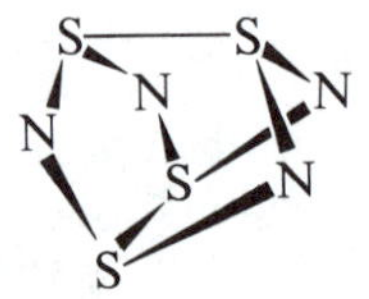

23

$$\begin{bmatrix} \text{orange, shock} \\ \text{sensitive} \\ \text{m.p. } 178° \end{bmatrix}$$

The S—S bond distances (258 pm) are long relative to the nominal values for S—S bonds ($\sim$210 pm) but considerably shorter than the non-bonded, van der Waals distance for two sulfur atoms ($\sim$360 pm). Consistent with the idea that the S—S bonding is weak is the thermochromic property of the solid: it is colorless at $-190°$C, orange at room temperature, and deep red at $100°$C. This color behavior is reminiscent of elemental sulfur, which, when heated, undergoes ring rupture to produce colored (reddish) paramagnetic chains of sulfur atoms. Thus, heating of S_4N_4 is likely to lead to some opening of the cage, making possible the electronic transitions responsible for absorption of light in the visible region. To be consistent with the C_{2v} symmetry of **23** you must describe the four "π" electron pairs as fully delocalized over the "cage" framework (the situation is not unlike that of the phosphonitriles).

One of the most interesting reactions[45] of $(SN)_4$ is the controlled thermolysis to produce cage opening and eventually the chain oligomer $(SN)_x$ called "polythiazyl." Using the apparatus shown in Figure 8-5, the orange $(SN)_4$ crystals (at A) are gently warmed ($85°$) so that the vapors pass through Ag wool (heated to $220°$ at B) and are condensed onto the liquid nitrogen cold finger (C). The colorless solid obtained at this point is S_2N_2 (this material is shock sensitive). Careful sublimation of the impure $(SN)_2$ crystals onto an ice-water cold finger (D) produces well formed crystals that eventually

---

†To be consistent with a D_{4h} molecular symmetry you must include both π resonance structures, as in the Kekule structures of benzene (the situation is not unlike that of the phosphonitriles). *Assuming* the ring to be planar, develop the cyclic out-of-plane π mo pattern to show that such a structure would actually have only two unpaired electrons (see p. 105, Chapter 4).

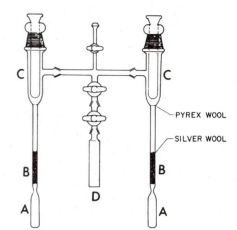

Figure 8-5. Apparatus for synthesis of $(SN)_x$ crystals. [Reproduced with permission from C. M. Mikulski, *et al., J. Amer. Chem. Soc.,* **97**, 6358 (1975). Copyright by the American Chemical Society.]

become reflective and *bronze-colored* on annealing at room temperature. The chain oligomer $(SN)_x$ can be safely handled, for it is neither shock nor thermally sensitive.

$$(SN)_4 \xrightarrow[\Delta]{\text{Ag wool}} (SN)_2 \xrightarrow[\Delta]{} (SN)_x \qquad \text{(8-68)}$$

The exciting property of this polymer is its high electrical conductivity (approaching that of Hg). In color and conductivity, the non-metal polythiazyl has "metallic" character. To understand the origin of the metallic "band" of orbital levels (Fig. 4-3) we first need to examine the bonding forces in the polymer.

Notice that these resonance structures imply all equal N—S bond distances, as is, in fact, found. The origin of the metallic conduction band can now be described by the following. The chain consists of repeating $(sp^2 + p\pi)$ hybridized atoms. The simplest repeating unit is $(-\ddot{N}=\ddot{S}=\ddot{N}-\ddot{S}-)$, for which there are four $p\pi$ ao's with three $p\pi$ electron pairs. In a chain of n of these units there is an electronic band derived from $4n$ ao's incompletely occupied by $3n$ electron pairs. A partially filled band normally leads to high electrical conductivity (see Chapter 4, p. 102).

This description provides for conductivity along the chain direction. In fact, the conductivity is not restricted to just this direction, so there must be "communication" between chains (in the form of π mo overlap) to permit electron motion between chains as well.

Tetrasulfur tetranitride is the starting reagent for preparation of a host of substituted, cyclic S—N compounds; we mention here only three interesting reactions. For example, oxidative fluorination of $(SN)_4$ with AgF_2 results in cage opening by fluorination of the sulfur atoms† (consistent with the ideas that the S—S bonds of **23** are the weak "links" and that S—F bonds are stronger than N—F bonds):

---

†It is a highly fascinating kinetic effect that monomeric FSN is known and that it trimerizes, rather than forms $\{FSN\}_4$!

(8-69)

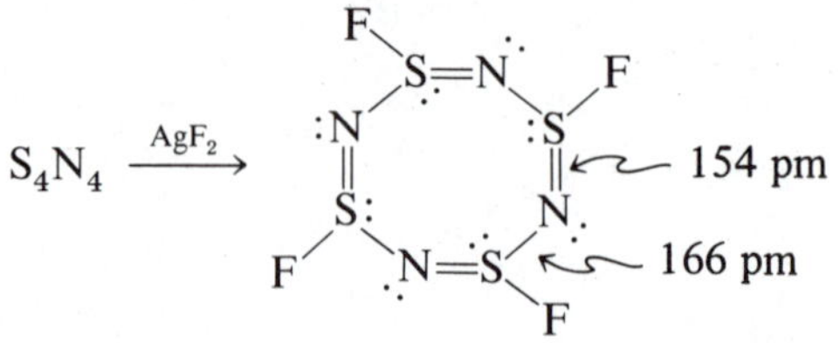

$$S_4N_4 \xrightarrow{\text{AgF}_2}$$

— 154 pm
— 166 pm

The analogy with $(F_2PN)_4$ seems obvious enough, yet the C_2 symmetry reported† for $(FSN)_4$ precludes the possibility of fully equivalent NS bonds. The tetramer of NSF does not have a planar structure [in analogy with an *unsymmetrically* substituted phosphonitrile tetramer $(NPXX')_4$] and is most simply described as SF groups lying along four edges of a "tetrahedron" of N atoms (**24**).

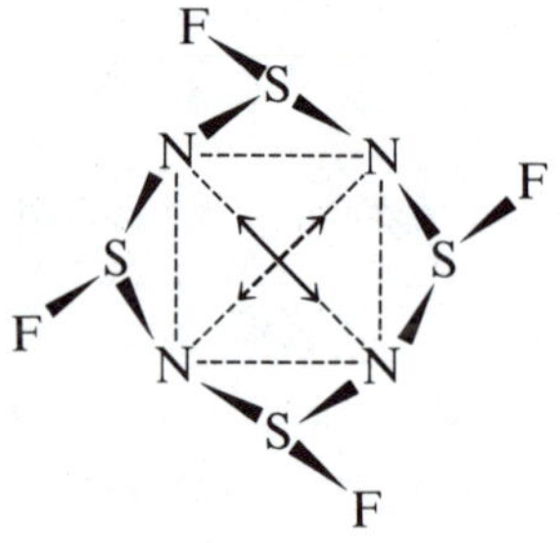

24

Reduction of tetrasulfur tetranitride with ethanolic stannous chloride disrupts the N—S multiple bonding altogether by creating NH bonds and giving a product $(SNH)_4$ isoelectronic with S_8 and having the analogous "tiara" structure[46] (**25**).

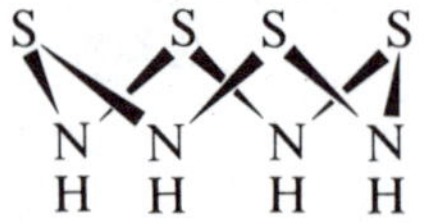

25

Oxidative chlorination of S_4N_4 cleaves the eight-membered ring to produce the (planar, D_{3h}) trimeric form of NSCl (**26**).

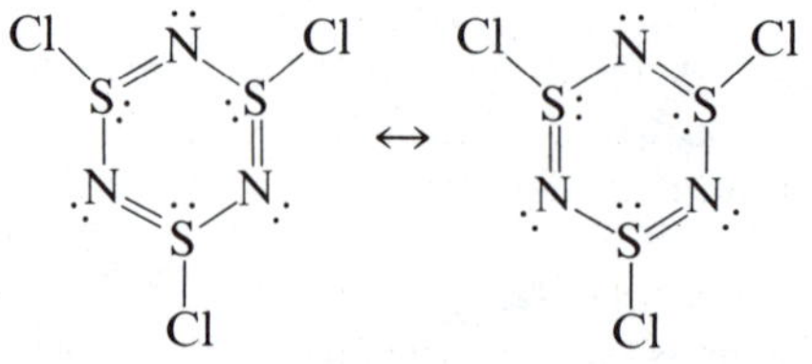

26

---

†C. A. Wiegers and A. Vos, *Acta Cryst.*, **16**, 152 (1963). This symmetry lowering would be most perplexing were the sulfur atoms to possess only a single acceptor orbital like boron in the borazines. In fact, the N—S "single bond" in {FSN}$_4$ is estimated to have a bond order of ~1.5. As with the phosphonitriles, ring distortions are not likely to greatly inhibit multiple bond formation so that intramolecular steric factors, sulfur substituent asymmetry, and even intermolecular steric effects in the solid state may cause ring distortions.

It is intriguing (and unexplained) that chlorination produces the ClSN trimer, while fluorination with AgF_2 retains the tetramer unit.†

Of great interest are the oxychlorides, since they provide an entry to a large number of important, naturally occurring sulfur compounds. While both thionyl and sulfuryl chlorides ($SOCl_2$ and SO_2Cl_2) are prepared from SO_2 by oxidative chlorination, the lower oxidation state compound can also be prepared by the exchange chlorination

$$SO_2 + PCl_5 \rightarrow SOCl_2 + POCl_3 \tag{8-70}$$

The S—Cl bonds of thionyl and sulfuryl chlorides undergo most of the reactions we have identified as being typical of non-metal chloride bonds. The compounds are quite reactive in solvolytic reactions and so provide convenient reagents for synthesis of **sulfonamides** $[(R_2N)_2SO]$ and **sulfamides** $[(R_2N)_2SO_2]$, **esters** of sulfurous and sulfuric acids $[(RO)_2SO$ and $(RO)_2SO_2]$, and **sulfoxides** $[R_2SO]$ and **sulfones** $[R_2SO_2]$.

Following the pattern of aminolysis behavior typical of the other non-metal chlorides, the sulfamide obtained from sulfuryl chloride and ammonia[47] can be thermally deaminated to yield the trimer[42] of NSO_2^- (27):

$$SO_2Cl_2 + 4NH_3 \rightarrow SO_2(NH_2)_2 + 2NH_4Cl \tag{8-71}$$

$$\xrightarrow[200°]{} (NH_4)_3(NSO_2)_3$$

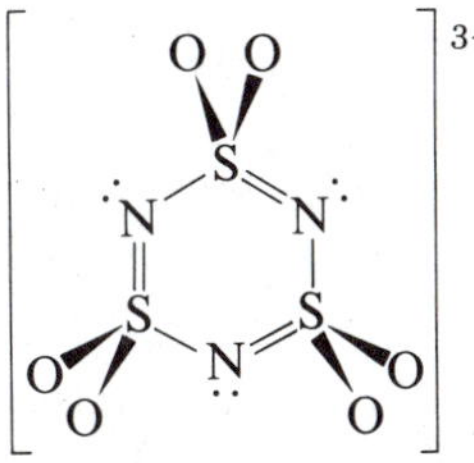

27

By noting that N^- is isoelectronic with O, you see that the NSO_2^- trimer fits nicely into the structural pattern for one form of solid SO_3 (*cf.* Chapter 6). Various salts of this anion are known, perhaps the most useful being that with silver; from it, various nitrogen-substituted derivatives can be formed by the ingenious strategy of simple metathesis with a halide of the groups to be added to the nitrogen atoms[42]

$$Ag_3(NSO_2)_3 + 3RX \rightarrow 3AgX_{(c)} + \quad\quad\quad \tag{8-72}$$

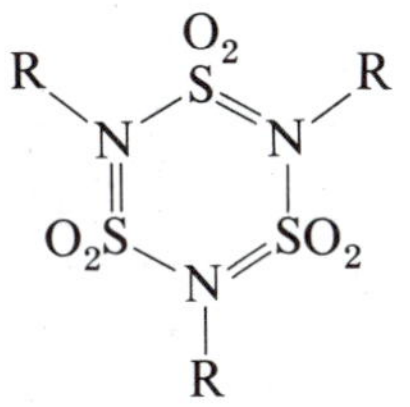

R = H, alkyl, etc.

These reactions expand the range of accessible trimeric S—N compounds beyond those of sulfur-substituted derivatives of $(ClSN)_3$, **26**.

---

† M. Becke-Goehring, "Ergebnisse und Probleme der Chemie der Schwefel-stickstoff-verbindungen," Academie-Verlag, Berlin, 1957. Here's an opportunity to use your imagination. Is it possible (based on your understanding of ring size effects for earlier elements) that steric interactions and/or weaker N → S bonds in the chloro substituted tetramer make its ring opening more facile? Perhaps you will find it important that {NSCl}₃ passes into the gas phase as monomers at 110°C.

The hydrolysis and alcoholysis reactions of thionyl and sulfuryl chlorides are extremely vigorous. The useful aspect of controlled thionyl hydrolysis is that SO_2 and HCl are products

$$SOCl_2 + 2H_2O \rightarrow SO_2 + 2HCl$$

and, being gases, are easily eliminated from a reaction mixture. For this reason, anhydrous metal chlorides are frequently prepared[48] from their hydrated forms by reaction with $SOCl_2$:

(8-73)
$$MCl_n \cdot xH_2O + xSOCl_2 \rightarrow MCl_n + 2xHCl + xSO_2$$

Recall also that thionyl chloride hydrolysis is a good reaction for synthesis of HCl (see the earlier section on HX synthesis).

Hydrolysis of $SOCl_2$ in excess water produces, of course, solutions of "sulfurous acid." "H_2SO_3" is not known in the form $(HO)_2SO$ but rather the formulation $H_2O \cdot SO_2$ seems more appropriate. Thus, the anhydride, SO_2, of H_2SO_3 is so easily obtained that

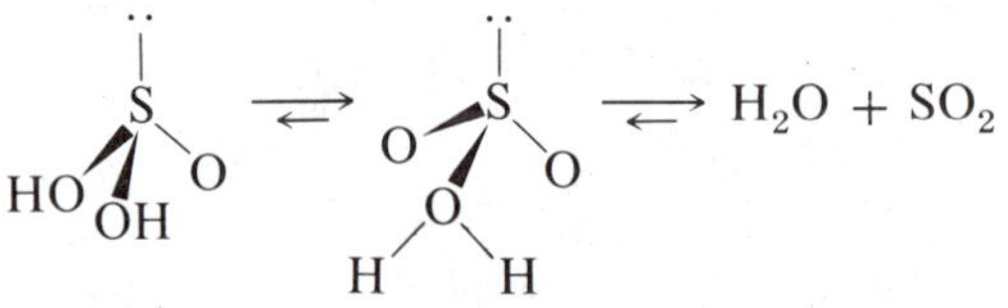

H_2SO_3 has never been isolated. Nevertheless, hydrolysis of $OSCl_2$ in an *excess* of the stronger (than H_2O) base OH^- forms bisulfite, which is well known in salt form. In solution HSO_3^- exhibits a 1,2 tautomerization in analogy with phosphorous acid.

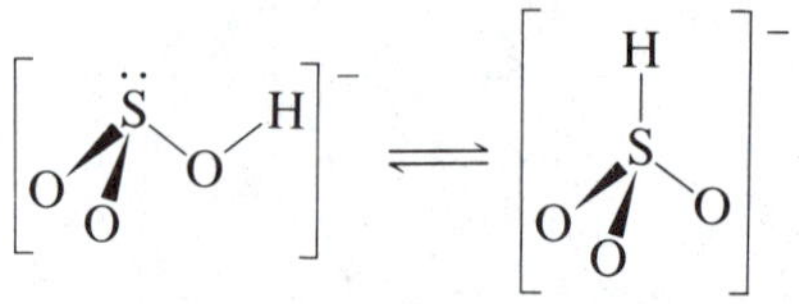

As with some of the triesters of phosphorous acid, dialkylsulfites also exhibit tautomeric forms:

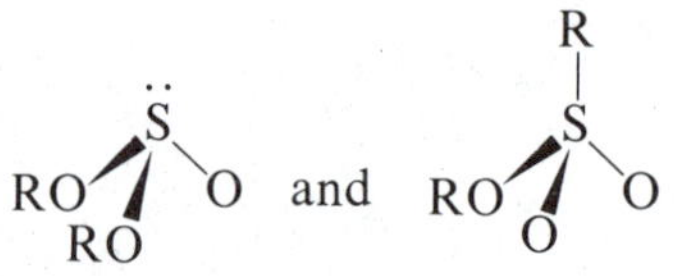

While it is possible to alkylate thionyl chloride to generate sulfoxides, an often-used synthesis relies on oxygenation (oxygen atom transfer) of the appropriate dialkylsulfide. Hydrogen peroxide is a useful reagent for this kind of oxidation, and again you find an analogy with organophosphorus chemistry (organophosphines and sulfides possess heavy atom lone pairs and their nucleophilic character renders oxygen atom transfer an effective reaction mode, as discussed in the last chapter, pp. 240 and 246).

Given the "solvate" nature of "H_2SO_3," you should not be surprised to learn that deamination of $SO(NH_2)_2$ (actually unknown) must be facile [also not particularly surprising in view of the behavior of $P(NH_2)_3$ as noted by reactions (8-59)]:

(8-74)
$$SOCl_2 + 2NH_3 \rightarrow [NH_4][NSO]^\dagger + 2HCl$$

---

†Note that N^- and O are isoelectronic.

The conjugate acid, $HN{=}S{=}O$, may be isolated ($<-94°C$) from a vapor phase reaction; on warming to $-70°C$ the monomer polymerizes, and then depolymerizes on further heating![49] An interesting change in the $SOCl_2/NH_3$ reaction channel arises when the reaction is performed in organic solvents ($CHCl_3$) with CaO present:[50]

$$SOCl_2 + NH_3 + CaO \rightarrow (HO)S{\equiv}N + CaCl_2 + H_2O \qquad (8\text{-}75)$$

This fascinating isomerization of HNSO has not been fully characterized; whether H_2O alone or in conjunction with unreacted CaO is responsible for the 1,3 proton shift can only be guessed.

Sulfuryl chloride reacts with water in a fashion analogous to thionyl chloride. If an excess of water is used, the SO_3 liberated dissolves to form a sulfuric acid solution. Again, alkylation reactions are possible to produce sulfones, although these can better be obtained by oxygenation of sulfides or sulfoxides.

$$SO_2Cl_2 + H_2O \rightarrow SO_3 + 2HCl \qquad \text{(but see } HSO_3Cl \text{ below)}$$

$$SO_2Cl_2 \xrightarrow{H_2O} H_2SO_4 + 2HCl \qquad (8\text{-}76)$$

$$R_2S \xrightarrow{O} R_2SO \xrightarrow{O} R_2SO_2$$

For the higher oxidation state of sulfur, both **fluorosulfuric** and **chlorosulfuric acid** are known. Both are prepared by the simple treatment of SO_3 with HX. You recognize HSO_3Cl as the reagent useful for sulfonation of aromatic compounds. HSO_3F is particularly useful as a solvent for high oxidation state species. Other important properties in its practical application as a solvent are that it can be removed from a reaction mixture by vacuum distillation (b.p. $163°C$), and it does not strongly *auto*-ionize via

$$2HSO_3F \rightarrow H_2SO_3F^+ + SO_3F^-$$

Halogens and Xenon

The solvolysis reactions of the interhalogens and rare gas halides that are worth discussing at this point are their hydrolysis reactions. These reagents are strongly oxidizing, and it is not recommended that one casually combine them with oxidizable reagents (in particular, organo compounds).

It appears that one can be very general in viewing the hydrolysis reactions of the interhalogen compounds; *the lighter, terminal halogen atoms are displaced as simple anions while the heavier, central halogen becomes oxygenated to the corresponding (same oxidation state) oxyacid.* Examples are

$$ClF + H_2O \rightarrow HOCl + H^+ + F^-$$

$$BrF_5 + 3H_2O \rightarrow HBrO_3 + 5H^+ + 5F^- \qquad (8\text{-}77)$$

Halogen oxyacids of intermediate oxidation state rapidly disproportionate into higher and lower oxidation state forms under acid conditions. For example,

$$ClF \xrightarrow{H_2O} [HOCl] \xrightarrow{acid} Cl^- + HClO_3$$

$$BrF_3 \xrightarrow{H_2O} [HBrO_2] \xrightarrow{acid} Br^- + HBrO_3 \qquad (8\text{-}78)$$

Attempts to prepare esters and amides of halogen oxyacids by reaction of a higher oxidation state interhalogen with a normal alcohol or amine will be very inefficient (and dangerous) because of degradation of the alcohol or amine. This appears to be a problem with XF_n in particular, because of reactions catalyzed by HF. Simple solvolysis reactions may be possible with perfluoroalcohols and amines and certain inorganic acids like SF_5OOH and $HN(SO_2F)_2$, but generally one is limited to the preparation of substituted hypohalites and haloamines.

Of the rare gas halides, only those of xenon have a moderately developed chemistry. The three fluorides of xenon do hydrolyze, but only the *hexafluoride* (surprisingly enough, from a thermodynamic view!) does so without immediate accompanying oxidation of the water (even in this case, though, the kinetically "stable" intermediate product of complete hydrolysis, xenon trioxide, can be dangerously explosive).[51]

$$XeF_6 + H_2O \rightarrow XeOF_4 + 2HF$$

$$XeF_6 + 3H_2O \rightarrow XeO_3 + 6HF$$

The simple product $Xe(OH)_6$ appears to dehydrate easily to form XeO_3, just as $Xe(OH)F_5$ apparently eliminates HF. Both xenon difluoride and xenon tetrafluoride oxidize water.[52] However, the difluoride undergoes only slow hydrolysis in acid media, while the reaction is much faster when base is present.

$$XeF_2 + 2OH^- \rightarrow Xe + \frac{1}{2}O_2 + 2F^- + H_2O$$

$$XeF_4 + 2H_2O \rightarrow \frac{1}{3}XeO_3 + \frac{2}{3}Xe + \frac{1}{2}O_2 + 4HF$$

The tetrafluoride does *not* produce the *unknown* dioxide (just as the difluoride does not produce the unknown monoxide[53]) on hydrolysis. The mechanisms of these reactions are not known, so one may only surmise that the monoxide and dioxide, if formed, rapidly disproportionate under the hydrolysis conditions to effect an oxidation reaction. The products from XeF_4 are suggestive of this view, but the stoichiometric relationship of XeO_3 and Xe suggests that a fairly complex mechanism prevails.

Finally, we might note that a compound with a Xe–N bond has been reported[54] from the "solvolysis" reaction

$$XeF_2 + HN(SO_2F)_2 \rightarrow FXe(NSO_2F)_2 + HF$$

XeF_2 is a good fluorinating agent, particularly in the presence of HF; it is considered that the fluorocarbon solvent used in this reaction plays the vital role of separating the HF produced as an immiscible layer so as to inhibit the degradation of $HN(SO_3F)_2$ [a side reaction that seems to dominate the reaction between XeF_2 and $HN(SO_2F)_2$ when the pure reagents are used].

STUDY QUESTIONS

5. Give steps for syntheses of the following from the halides, any normal organic compound, and any special reagents given in parentheses.

a) $Na[HB(OR)_3]$ (NaH)
b) $[MeNBCl]_3$
c) $[ROBO]_3$ (H_2O)
d) $[R_2AlPR'_2]$ (what are likely R, R'?)
e) $[Me_2AlNMe_2]_2$, $[Me_2AlNHMe]_3$ (why is one a dimer, the other a trimer?)
f) $[PhAlNMe]_4$
g) $[H_2SiO]_n$ (H_2O)
h) SiH_3NH_2 (NH_3)
i) $MeOSiEt_3$

j) $[NPF_2]_3$ (NH_3)

k) $[ClBNH]_3$, and from it, $[Cl_2BNH_2]_3$ (NH_3)

l) $MeCO(O_2H)$ (use an anhydride)

6. Match reactants and products.

a) $\frac{1}{2}B_2H_6 + 3ROH \rightarrow$ 1) $Al_2Et_4H_2$

b) $\frac{1}{2}Al_2Ph_6 + H_2NNMe_2 \rightarrow$ 2) $Si(NH)_2$

c) $Al_2Et_6 + HCl \rightarrow$ 3) $Li[Al(PMe_2)_4]$

d) $Al_2Et_6 \xrightarrow{\Delta}$ 4) $(Cl_3Si)_2O$

e) $LiAlH_4 + 4HPMe_2 \rightarrow$ 5) Me_3SiOEt

f) $2H_3SiNH_2 \xrightarrow{\Delta}$ 6) $P(OR)_3$

g) $2SiCl_4 + H_2O \xrightarrow[-80°]{\text{"ether"}}$ 7) $B(OR)_3$

h) $Si(NH_2)_4 \xrightarrow{\Delta}$ 8) $LiNMe_2$

i) $PCl_3 + 3ROH \xrightarrow{NR_3}$ 9) Al_2Et_5Cl

j) $PCl_3 + 3ROH \rightarrow$ 10) $Na[N_2H_3]$

k) $n\text{-BuLi} + Me_2NH \rightarrow$ 11) $(H_3Si)_2NH$

l) $Me_3SiNR_2 + EtOH \rightarrow$ 12) $HPO(OR)_2$

m) $EtNa + N_2H_4 \rightarrow$ 13) $[PhAlN_2Me_2]_4$

CHECKPOINTS

1. Non-metal halide bonds are reactive under conditions of attack upon the non-metal atom by nucleophiles:

a) H^- (MH, BH_4^-, AlH_4^-)

b) "R^-" ($RMgX$, R_2Hg, MR)

c) ROH

d) R_nNH_{3-n}

e) R_nPH_{3-n}

f) R_nSH_{2-n}

2. When the nucleophile and substrate produce an adduct with the element pairs (H,X) or (H,R) on adjacent atoms, HX and RH eliminations are possible.

3. The elimination reaction for HX may be base catalyzed or induced by heating.

4. The resulting $E=E'$ pi bond is generally thermodynamically unstable relative to EE' sigma bonds, so polymerization to join chains, rings, and polyhedra is common.

5. An alternate route to organoderivatives, the $E-H$ addition to olefins and alkynes, is possible for $E = B$, Al, and Si (with transition metal catalysis).

[1] R. J. Angelici, "Synthesis and Technique in Inorganic Chemistry," 2nd Ed., W. B. Saunders, Philadelphia, 1977; D. F. Shriver, "The Manipulation of Air-sensitive Compounds," McGraw-Hill, New York, 1969; W. L. Jolly, "The Synthesis and Characterization of Inorganic Compounds," Prentice–Hall, Englewood Cliffs, 1970.

[2] A. Stock, "Hydrides of Boron and Silicon," Cornell University Press, Ithaca, 1933.

[3] A. E. Finholt, et al., J. Amer. Chem. Soc., 69, 2692 (1947).

[4] S. R. Gunn and L. G. Green, J. Phys. Chem., 65, 779 (1961).

[5] P. M. Treichel, et al., J. Amer. Chem. Soc., 89, 2017 (1967).

[6] G. Wittig and M. Rieber, Annalen, 562, 187 (1955); Naturwiss., 35, 345 (1948).

[7] M. Baudler, et al., Z. anorg. allgem. Chem., 353, 122 (1967).

[8] This reaction is the one used by L. J. Thenard, Ann. chim. phys., 8, 306 (1818) when he reported the discovery of H_2O_2. A more general reference to hydrogen peroxide chemistry is W. L. Schumb, et al., "Hydrogen Peroxide," Reinhold, New York, 1955.

[9] W. L. Jolly, "The Synthesis and Characterization of Inorganic Compounds," Prentice–Hall, Englewood Cliffs, 1970.

REFERENCES

[10] "Inorganic Syntheses," Vols. 1, 3, and 7, McGraw–Hill, New York, 1939–63.

[11] H. C. Brown's recounting ("Boranes in Organic Chemistry," Cornell University Press, Ithaca, 1972) of his and his students' experiences in this area is not only informative but highly entertaining reading. See also K. Wade, "Electron Deficient Compounds," Nelson, London, 1971.

[12] A. G. MacDiarmid, *Adv. Inorg. Chem. Radiochem.,* **3,** 207 (1961).

[13] D. H. Chadwick and R. S. Watt, in "Phosphorus and its Compounds," Vol. 2, J. R. van Wazer (Ed.), Interscience, New York, 1961.

[14] H. Schmidbauer and W. Tronich, *Chem. Ber.,* **101,** 595, 604 (1968); A. Schmidt, *ibid.,* **101,** 4015 (1968); J. C. J. Bart, *J. Chem. Soc. (B),* 350 (1969). Note that *n*-Bu⁻ should initially attack the electrophilic phosphorus, followed by reductive elimination of C_4H_{10}.

[15] N. Duffant, *et al., Compt. Rend.,* **268C,** 967 (1969).

[16] F. S. Kipping and J. E. Sands, *J. Chem. Soc.,* 830 (1921); E. Carberry and R. W. West, *J. Amer. Chem. Soc.,* **91,** 5440 (1969).

[17] W. Mahler and A. B. Burg, *J. Amer. Chem. Soc.,* **80,** 6161 (1958).

[18] E. Fluck, *Compounds Containing P–P Bonds,* in "Preparative Inorganic Reactions," Vol. 5, W. L. Jolly (Ed.), Interscience, New York, 1968.

[19] As many of the non-metal solvolyses yield polymeric products (chains, rings, cages), this has been a generally active field. We recommend D. A. Armitage, "Inorganic Rings and Cages," Edward Arnold, London, 1972 for a more general survey of this field than we can give here. Armitage's survey includes not only more compound classes but also preparative schemes other than solvolysis.

[20] H. Steinberg, "Organoboron Chemistry," Vol. 1, Interscience, New York, 1964.

[21] A. B. Burg and G. Brendel, *J. Amer. Chem. Soc.,* **80,** 3198 (1958); R. I. Wagner and C. O. Wilson, Jr., *Inorg. Chem.,* **5,** 1009 (1966); W. Gee, *et al., J. Chem. Soc.,* 3171 (1965).

[22] See ref. 19; relevant structural data are given by W. C. Hamilton, *Acta Cryst.,* **8,** 199 (1955); G. J. Bullen and P. R. Mallinson, *Chem. Comm.,* 132 (1969). It is most interesting that P → B π bonding does not stabilize the monomer R_2PBH_2, relative to P → B σ bonding and oligimer formation, so that BH_2 → P π interaction becomes possible in the polymer.

[23] J. G. Oliver, *et al., J. Inorg. Nucl. Chem.,* **31,** 1609 (1969); *Chem. Comm.,* 918 (1968); *Inorg. Nucl. Chem. Letters,* **5,** 749 (1969).

[24] These are actually reductions by alkali metals in liquid ammonia. R. Brec and J. Rouxel, *Bull. Soc. Chim. France,* 2721 (1968); *Compt. Rend. Ser. C,* **270,** 491 (1970).

[25] These and other related reactions are cited and discussed by K. Wade and A. J. Banister in "Comprehensive Inorganic Chemistry," Vol. 1, pp. 1007ff, 1039ff, A. F. Trotman-Dickenson, *et al.* (Eds.), Pergamon Press, New York, 1973.

[26] S. Amirkhalili, P. B. Hitchcock, and J. D. Smith, *J. Chem. Soc., Dalton,* 1206 (1979); J. K. Ruff, in "Developments in Inorganic Nitrogen Chemistry," Vol. 1, C. B. Colburn (Ed.), Elsevier, New York, 1968.

[26] J. K. Ruff, in "Developments in Inorganic Nitrogen Chemistry," Vol. 1, C. B. Colburn (Ed.), Elsevier, New York, 1968.

[27] E. Wiberg and A. May, *Z. Naturforsch.,* **10B,** 229 (1955).

[28] L. B. Ebert, *et al., J. Amer. Chem. Soc.,* **96,** 7841 (1974).

[29] A. W. Laubengayer, *et al., J. Amer. Chem. Soc.,* **83,** 542 (1961).

[30] See ref. 2, and F. G. A. Stone, "Hydrogen Compounds of the Group IV Elements," Prentice–Hall, New York, 1962.

[31] A. D. F. Toy, in "Comprehensive Inorganic Chemistry," Vol. 2, A. F. Trotman-Dickenson, *et al.* (Eds.), Pergamon Press, New York, 1973.

[32] D. D. Magnelli, *et al., Inorg. Chem.,* **5,** 457 (1966). This compound is rather unusual among phosphinous acids in that it does not disproportionate into $(F_5C_6)_2PH$ and $(F_5C_6)_2PO(OH)$.

[33] R. Klement and O. Koch, *Chem. Ber.,* **87,** 333 (1954); "Inorganic Syntheses," Vol. 6, E. G. Rochow (Ed.), McGraw–Hill, New York, 1960.

[34] E. Fluck, *Topics in Phosphorus Chemistry,* **4,** 293 (1967); ref. 31.

[35] A recent short review of phosphazenes and their technological applications is H. R. Allcock, *Science,* **193,** 1214 (1976).

[36] H. R. Allcock, "Phosphorus-Nitrogen Compounds," Academic Press, New York, 1972; *Scientific American,* **230,** 66 (1974). A good survey is also given in ref. 31. A more recent source, with photographs of products made from polyphosphazenes, is H. R. Allcock, *Science,* **193,** 1214 (1976).

[37] G. J. Bullen and P. A. Tucker, *Chem. Comm.,* 1185 (1970).

[38] G. B. Ansell and G. J. Bullen, *J. Chem. Soc. (A),* 3026 (1968).

[39] T. S. Cameron and Kh. Mannan, *J. Chem. Soc., Chem. Comm.,* 975 (1975).

[40] A good summary may be found in ref. 31 (pp. 469–534).

[41] W. Lange, in "Inorganic Syntheses," Vol. 2, C. W. Fernelius (Ed.), McGraw–Hill, New York, 1946, gives a more convenient synthesis based on substitution of F for OH by $HF + H_3PO_4$.

[42] The product tetrasulfur tetranitride is actually formed in ~25% yield. Consult M. Schmidt and W. Siebert in "Comprehensive Inorganic Chemistry," Vol. 2, A. F. Trotman-Dickenson, *et al.* (Eds.), Pergamon Press, New York, 1973, for entries to the literature of this fascinating reaction and its products.

[43] R. C. Brasted, *J. Chem. Soc.,* 2297 (1965).

[44] D. Clark, *J. Chem. Soc.,* 1615 (1952).

[45] C. M. Mikulski, *et al., J. Amer. Chem. Soc.,* **97**, 6358 (1975); M. J. Cohen, *ibid.,* **98**, 3844 (1976); D. R. Salahub and R. P. Messmer, *J. Chem. Phys.,* **64**, 2039 (1976); R. H. Baughman, *et al., ibid.,* **64**, 1869 (1976).

[46] D. Chapman and T. D. Waddington, *Trans. Faraday Soc.,* **58**, 1679 (1962).

[47] W. Traube, *Chem. Ber.,* **25**, 2472 (1892).

[48] A. R. Pray, "Inorganic Syntheses," Vol. 5, T. Moeller (Ed.), McGraw-Hill, New York, 1957.

[49] W. P. Schrenk and E. Krone, *Angew. Chem.,* **73**, 762 (1961).

[50] M. Becke-Goehring, *et al., Z. anorg. Allgem. Chem.,* **293**, 294 (1957).

[51] B. Jaselskis, *et al., J. Amer. Chem. Soc.,* **88**, 2149 (1966); D. F. Smith, *Science,* **140**, 899 (1963); J. Shamir, *et al., J. Amer. Chem. Soc.,* **87**, 2359 (1965).

[52] S. M. Williamson and C. W. Koch, *Science,* **139**, 1046 (1963); E. H. Appelman and J. G. Malm, *J. Amer. Chem. Soc.,* **86**, 2297 (1964).

[53] C. D. Cooper, *et al., J. Mol. Spec.,* **7**, 223 (1961) detected the diatomic molecule in the gas phase, and the data suggest a bond energy < 38 kJ/mole.

[54] R. D. LeBond and D. D. DesMarteau, *Chem. Comm.,* **1974**, 555.

Part 3

DESCRIPTIVE TRANSITION METAL CHEMISTRY WITH INTERPRETATIONS

9. FUNDAMENTAL CONCEPTS FOR TRANSITION METAL COMPLEXES

In this chapter you will explore concepts helpful in understanding the properties of color, paramagnetism, and structure which form the basis for understanding the reactions of transition metal complexes to be discussed in subsequent chapters.

A SAMPLER

This last half of the text addresses the most significant aspects of the chemistry of the *d*-block or transition elements. Until recently, progress in this area has occurred across two broad fronts—classical coordination chemistry and organometallic chemistry—which often have been tended by two separate schools of chemists. It now seems clear that further significant progress in transition metal chemistry will require intimate knowledge of both areas.

By classical coordination chemistry is meant the chemistry of adducts formed by metals in their higher oxidation states (formally $\gtrsim$ +2) bonded to inorganic or organic ions or molecules. A companion characteristic is that the ligands bound to the metal ion are predominately sigma donors with moderate to weak π-acceptor or π-donor tendencies. Therefore, ions or molecules such as $[NiCl_4]^{-2}$, $[Co(H_2N-C_2H_4-NH_2)_2Cl_2]^+$, or $[(C_6H_5)_3P]_2NiCl_2$ can properly be called classical coordination compounds. Both basic and technological research with these materials continue at a rapid pace. Synthetic and structural determination studies are certainly important and have been spurred by progress on the theory of electronic structure. The latter has provided insight into the spectroscopic, magnetic, structural, thermodynamic, and kinetic properties of complexes.

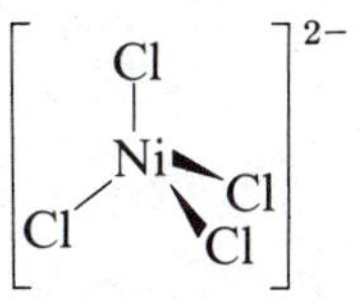

tetrachloronickelate
(2−)

1

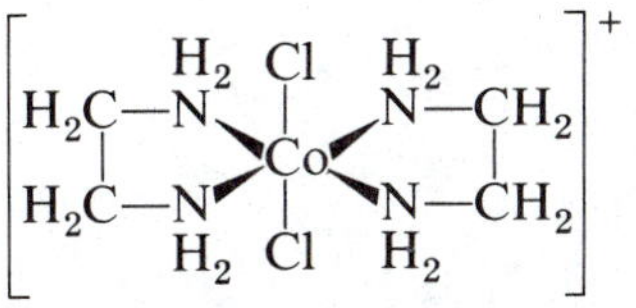

dichlorobis(ethylenediamine)
cobalt (+)

2

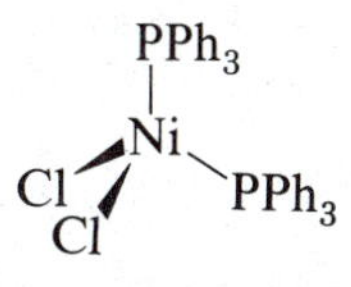

dichlorobis(triphenyl
phosphine)nickel

3

Paralleling the progress in classical coordination chemistry have been the rapid advances made in the coordination chemistry of the transition metals in their lower valence states, particularly in the area of organometallic chemistry. Impetus for these developments was provided by the synthesis of ferrocene (**4**) in 1951,[1] and homogeneous catalysts such as Vaska's compound, $IrCl(CO)[P(C_6H_5)_3]_2$ (**5**), in the 1960's.[2] Progress in the field has been so rapid and complete that generalizations concerning reactivity and stereochemistry similar to those found in organic chemistry are beginning to emerge.[3] This fascinating subject looms so large on the horizon of inorganic chemistry that Chapters 15–18 are devoted to it.

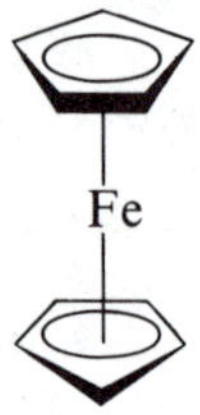

Fe

Ph_3P — Ir — Cl
OC — PPh₃

bis(*pentahapto*cyclopentadienyl)iron

4

trans-carbonylchlorobis(triphenylphosphine) iridium

5

The further development of inorganic chemistry requires that new areas be explored and new directions taken. One such area of transition metal chemistry under development is the synthesis and study of compounds containing metal-metal bonds, particularly metal cluster compounds; these are exemplified by structures **6** and **7**.[4]

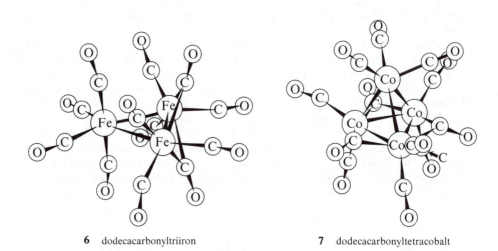

6   dodecacarbonyltriiron

7   dodecacarbonyltetracobalt

Another very important field for new development is the study of inorganic compounds of biological interest (see Chapter 19). For example, the coenzyme of vitamin B_{12} (**8**) and a model compound of this vitamin, methylbis(dimethylglyoximato)(pyridine) cobalt (**9**), have been the subject of extensive research.[5]

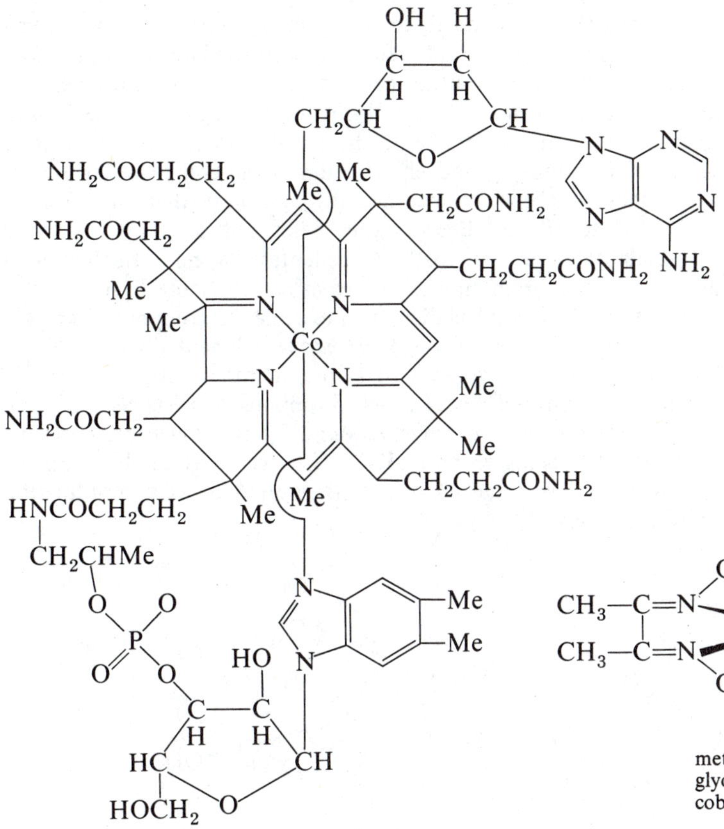

8

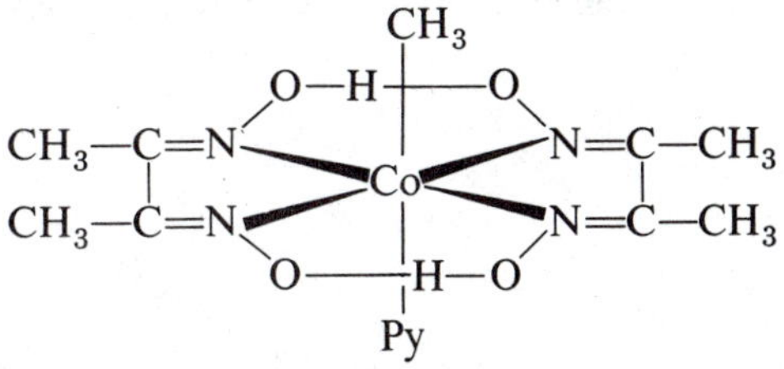

methylbis(dimethyl-
glyoximato) (pyridine)
cobalt

9

It is clear that, in order to understand the chemistry of vitamin B_{12} coenzyme, metal clusters, and other systems of future importance, chemists should be as knowledgeable as possible in both of the historically separate areas of classical coordination chemistry and organometallic chemistry.

The Language of Coordination Chemistry

Compounds **4** through **10** can all be referred to as organometallic compounds, as they have direct metal-carbon bonds. In most of these examples the metal is in a low formal oxidation state (*e.g.*, 1+ in **10**, if the C_5H_5 group is taken as anionic†). Compounds **1, 2, 3,** and **11** are classical coordination compounds, compounds wherein the metal is

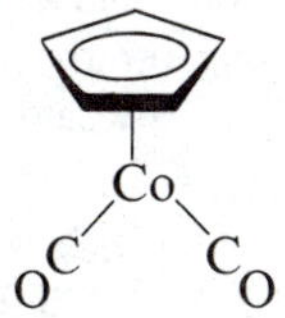

dicarbonyl*pentahapto*-
cyclopentadienylcobalt

10

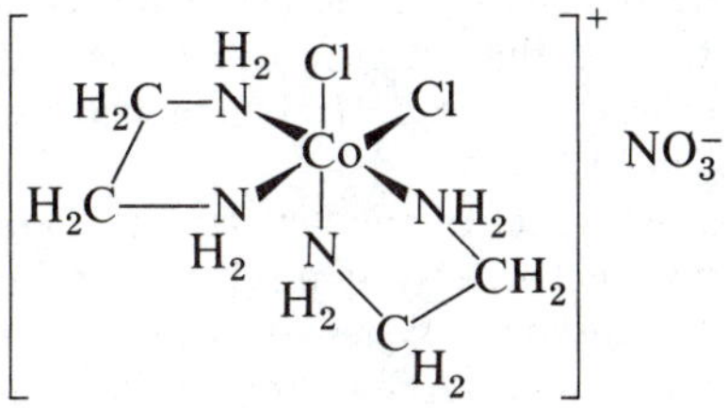

NO_3^-

cis-dichlorobis(ethylenediamine)
cobalt (+) nitrate

11

---

†As either a π radical (C_5H_5) or an anion ($C_5H_5^-$), this species presents another example of a π HOMO donor.

in a "normal" oxidation state (2+ or 3+) and there are no metal bonds to carbon. In all cases, the metal atom is referred to as the **central metal** or **coordinated metal** atom or ion. All of the groups attached directly to the central metal—whether ions or molecules—are the coordinating groups or **ligands**. A ligand attached directly through only one coordinating atom (or using only one coordination site on the metal) is called a **monodentate** ("one-toothed") ligand; carbon monoxide and chloride ion in the complexes above are monodentate ligands. A ligand that may be attached through more than one atom is **multidentate**, the number of actual coordinating sites being indicated by the terms bidentate, tridentate, tetradentate, and so forth. Multidentate ligands attached to a central metal by more than one coordinating atom are called **chelating ligands**. Their complexes are called **chelates**. The ethylenediamine molecule in compound **11** is a bidentate chelating ligand. Ligands such as the *pentahapto*cyclopentadienyl group,† C_5H_5, in compound **10** are not usually referred to as multidentate, although, strictly speaking, they are. A multidentate ligand that is attached to more than one central metal is called **bridging**. Therefore, the acetate ion in compound **12** is a bidentate chelating ligand, whereas it is a bidentate bridging group in **13**.[6,7] The whole assembly of one or more central metal atoms with their attached ligands is referred to as a **coordination complex** or, more simply, a complex.

diacetatodiaquozinc
12

diaquo-μ-tetraacetatodicopper
13

The ligands attached to the central metal may be anionic, neutral, or (less frequently) cationic. In classical coordination chemistry, the number of donor atoms (with one Lewis base electron pair per donor atom) bound directly to the central metal defines the **coordination number**. The number is clearly six for **11** and **12**, but in organometallic compounds containing π Lewis base ligands the coordination number of the central metal is less clearly defined in terms of the number of nearest atoms, for not every such atom possesses a Lewis base electron pair. In such cases (cf. Chapter 15) we fall back to counting the number of donor electron pairs (rather than donor atoms from each ligand). The ligands directly bound to the metal are said to be in the **inner coordination sphere**, and the counter-ions that balance out the charge remaining on the complex after the coordination number of the central metal has been "satisfied" are said to be **outer sphere ions** (*e.g.*, NO_3^- in **11**).

An important aspect of the language of coordination chemistry is the application of a systematic nomenclature system. Given the huge variety of compounds known, it is inevitable that a complete systematization of nomenclature involves idiosyncracies that are difficult to remember. A few simple rules do handle most cases; familiarity with them will make it possible for you to read comfortably this text and the inorganic literature (check their application to the compounds already given):

1. Cations are named first, then anions.
2. Within the coordination complex, the ligands are named before the metal.

---

† The Greek prefix "hapto" means fasten, with an obvious connotation in the context of **10**.

3. The charge on the complex, if an ion, is given in parentheses following the name of the metal; if the charge is negative, the suffix "ate" is added to the name of the metal (this is the Ewing-Basset convention).
4. Ligands are given in alphabetical order, for the most part.
5. Anionic ligands are given an "o" suffix; organic ligands (carbon the donor atom) and other neutral ligands are given their customary names as neutral species.
6. To indicate the number of times a simple ligand appears, use a prefix such as di, tri, tetra, penta, or hexa; for a complicated ligand, the ligand name is set off in parentheses with a prefix such as bis, tris, or tetrakis.
7. Bridging ligands are given the prefix μ-.

Probably the most difficult part of transition metal complex nomenclature arises with the organic ligands. Rarely are the IUPAC names used; rather, the practice is to use common names, which seem best learned through example and usage. Some fairly common ligands can be singled out here:

Name	Charge (abbr.)	Formula
ethylenediaminetetraacetato	4− (edta)	$[(O_2CCH_2)_2N(C_2H_4)N(CH_2CO_2)_2]^{4-}$
ethylenediamine	0 (en)	$H_2N(C_2H_4)NH_2$
acetylacetonato	1− (acac)	$[CH_3C(O)CHC(O)CH_3]^-$
glyoximato	1− (glyox)	$[HON(C_2H_2)NO]^-$
ammine	0	NH_3
amine	0	RNH_2, R_2NH, R_3N
aquo	0	H_2O
carbonyl	0	CO
nitrosyl	1+	NO^+
2,2′-bipyridine	0 (bipy)	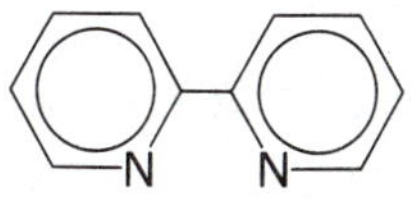
1,10-phenanthroline	0 (phen)	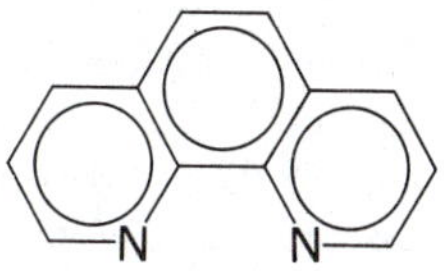

With reference to point 3 above, there is a long-standing alternate convention, called the Stock system, for indicating charge. The Stock system specifies the presumed oxidation state of the metal ion by a Roman numeral in parentheses following the name of the metal. For example, complex **11** appears as ··· cobalt(III) nitrate in the Stock system. Generally, there is no ambiguity in applying the charge count procedure to determine the oxidation number of the metal:

$$\text{metal charge} = \text{complex ion charge} - \text{ligand charges}$$

There are times, however, when considerable ambiguity is associated with this method (notably when an easily oxidized or reduced ligand is coordinated to the metal). On the other hand, the Ewing-Basset method of specifying the complex ion charge is experimentally rigorous, free of arbitrary electron counting conventions, and, therefore, to be preferred.

STUDY QUESTIONS

1. Match the formulas with the names.

   a) $[Cr(NH_3)_6]Cl_3$

   b) $K[Co(edta)]$

   c) $K_2[FeO_4]$

   d) $[Cr(NH_3)_2(H_2O)_3(OH)](NO_3)_2$

   e) $[Pt(py)_4][PtCl_4]$

   f) $\left[(H_3N)_4Co \begin{array}{c} H \\ O \\ \diagup \quad \diagdown \\ \quad\quad Co(en)_2 \\ \diagdown \quad \diagup \\ N \\ H_2 \end{array} \right] Cl_4$

   g) $[Ni(en)_2Cl_2]$

   h) $K[PtCl_3(C_2H_4)]$

   1) Dichlorobis(ethylenediamine)nickel
   2) Potassium trichloro(ethylene)platinate(1−)
   3) Potassium ethylenediaminetetraacetatocobaltate(1−)
   4) Tetrakis(pyridine)platinum(2+) tetrachloroplatinate(2−)
   5) Diamminetriaquohydroxochromium(2+) nitrate
   6) Tetraamminebis(ethylenediamine)-μ-hydroxo-μ-amidodicobalt(4+) chloride
   7) Potassium tetraoxoferrate(2−)
   8) Hexaamminechromium(3+) chloride

2. Give the Stock convention names for the complexes in question 1.

3. How are the names of H_2O and NH_3, in the above examples, exceptions to the rules? What is the difference in spelling for NH_3 and RNH_2?

THE TRANSITION METALS AND PERIODIC PROPERTIES

The transition elements display all of the properties common to metals: they have relatively low ionization potentials, they are good conductors of heat and electricity, and they are lustrous, malleable, and ductile. All of these properties are consistent with the band theory of bonding in elemental solids (Chapter 4). However, many of these properties are quite different in magnitude from those of the non-transition metals; also, we have not discussed their periodicity. It is well worth your time to investigate some of these physical properties for the insight they provide into the unique chemistry of the transition elements.

The most revealing properties of the elements are their binding energies, radii, and ionization potentials. The binding energy of an element can be expressed in terms of the enthalpy change for converting the element in its standard state at 298°K to an ideal gas at the same temperature. That is, the binding energy is equivalent to the heat of sublimation at 298°K. Binding energies[8] for the elements of the three transition metal periods are plotted in Figure 9-1. In all three periods, the binding energy at first increases, reaching a maximum value several times greater than those of the alkali metals, and then begins to fall off somewhere near the middle of the period. This behavior is not so strange as might first appear and may be explained reasonably well on the basis of the band theory. You encountered behavior similar to this in the discussion of homonuclear diatomics in Chapter 4. In proceeding across the second row of elements $(Li_2, \ldots, Ne_2)$, a maximum in atomization energy arises near the center of the series (N_2). Clearly, this behavior is characteristic of filling a set of bonding, non-bonding, and anti-bonding orbitals, with anti-bonding electron density introducing atom repulsion (in the directed orbital model this corresponds to lone pair repulsions). A similar situation is at hand here; important anomalies, however, prevent one from deriving more than a qualitative understanding of the trends.

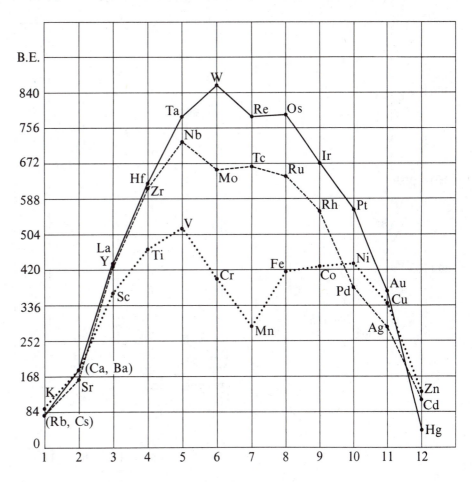

Figure 9-1. Binding energies (kJ/g-atom) of the transition elements as a function of the number of valence electrons.

First of all, the metals are macromolecules with a near *continuum* of orbital levels. Secondly, it is unclear to what extent the $(n + 1)s$ and $(n + 1)p$ ao's are important. Their separation from the nd ao's becomes smaller in the succeeding transition series ($n = 3$, 4, 5) so as to create additional bonding levels among the less strongly bonding and nonbonding d levels. This rationalizes the next important qualitative feature of Figure 9-1: the binding energies of the heavier elements tend to be greater than those of the corresponding lighter elements.

The radii[9] of the metals are plotted in Figure 9-2. The outstanding feature of this plot is obvious—there is a general decrease in radius for the metals of all three periods as the atomic number increases across a row, a minimum being reached with the Group VIII elements. This change is clearly related to d orbital penetration of core electrons; as the atomic number of an element in a period increases, there is a general contraction of the core and d orbitals, and a size decrease results. Anti-bonding orbital occupation in the later elements in each series tends to increase the M—M distance. Another observation that may be made is that there is a normal increase in size for the heavier elements in any given group; however, the increase is smaller between the fifth and sixth periods than between the fourth and fifth. This difference is again ascribed to the so-called "lanthanide contraction" (refer to Chapter 1†) and as a consequence of the latter may also be ascribed to greater involvement of $(n + 1)s$ and p ao's in bond formation.

---

†The principle here is $5d$ electron penetration of the $4f^{14}$ shell; see pp. 33–34.

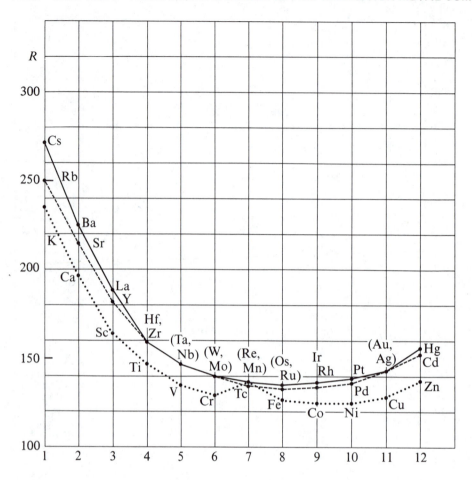

Figure 9-2. Metallic radii (pm) of the transition elements as a function of the number of valence electrons.

The ionization potential of a metal is an important characteristic of its chemistry; therefore, the variation in ionization potentials in a related series of elements can give you a feeling for the variation in chemical properties in that series. Plotted in Figure 9-3 are the first three ionization potentials[10] for each element in the transition metal series. (a) In each period, there is a general increase owing to the increase in effective nuclear charge across a period. (b) For the M^{2+} ions, all of which have d^n configurations (refer to page 33), a break in this trend occurs at the d^6 case, just as is found for the p^4 cases of the p block elements. (c) Another significant trend is that for elements of a given group, the normal tendency for decreasing ionization potential down the group is observed for the process $M^{2+} \rightarrow M^{3+}$ and for all ionization steps of Group III elements. The first and second ionization energies of the $5d$ elements following the lanthanide elements, however, are greater than those of the $3d$ congeners. Clearly, the effects of $5d$ penetration of the $4f^{14}$ shell are most pronounced early in the $5d$ period and for low ion charge. There are two rough generalizations to be made from Figure 9-3: (1) The elements early in the transition metal series have lower ionization potentials and should, therefore, reveal a more extensive chemistry of compounds in which the formal oxidation state of the metal is equal to the group number. The later elements have higher ionization potentials, however, and should reveal chemistries dominated by species containing the 2+ ions. (2) The second ionization is so easy that M^+ chemistries are likely to be less common than those of M^{2+}.

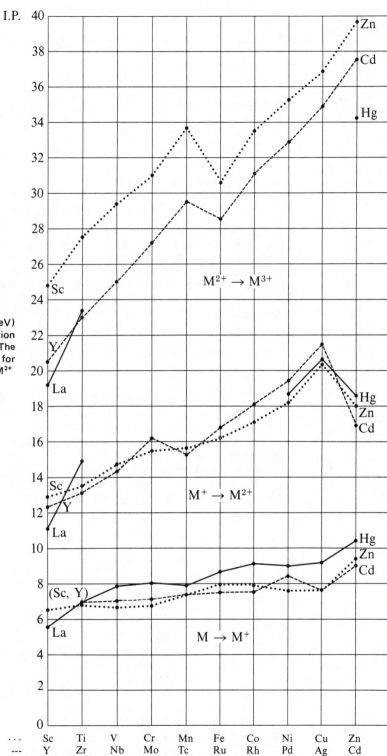

Figure 9-3. Ionization potentials (eV) of the transition elements as a function of the number of valence electrons. The abscissa values range from 3 to 12 for $M \rightarrow M^+$, from 2 to 11 for $M^+ \rightarrow M^{2+}$ from 1 to 10 for $M^{2+} \rightarrow M^{3+}$.

**STUDY
QUESTIONS**

4. Contrast the variation in E—E bond energy down a main group with that down a transition group. How might the comparison be interpreted in terms of (a) a difference in the number of valence ao types important to E—E bonding, or, equivalently, (b) the number of anti-bonding electrons in balance with the number of bonding electrons?

5. If the *d*-type bonding and anti-bonding mo's are completely filled for a crystal of Zn, how do you explain the fact that Zn is not a gas?

6. Why is there a "curve dislocation" at d^6 in Figure 9–3 for the $M^{2+} \rightarrow M^{3+}$ processes?

THE MOLECULAR ORBITAL MODEL

Ions of the metals just discussed form complexes that exhibit three properties which have long fascinated chemists: a rainbow of colors, magnetic behavior, and a great variety of structures. All of these can change dramatically by a seemingly simple change of a ligand. Taking color as an example: $[Co(NH_3)_6]Cl_3$ is yellow, while $[Co(NH_3)_5H_2O]Cl_3$ is rose. Regarding magnetism, $[Fe(H_2O)_6]Cl_2$ is attracted into a magnetic field, while $K_4[Fe(CN)_6]$ is repelled. Explanations of such phenomena have evolved over the past fifty years, beginning with the ideas of Pauling, Bethe, and van Vleck, and continuing to the present molecular orbital description of color, magnetism, and structure.

The Angular Overlap Model of *d* Electron Energies

TENETS

The **Angular Overlap Model**[11] (abbreviated AOM) is a simple first approximation to the full mo model and embodies all the characteristics of metal-ligand interactions important to an understanding of the principles of complex structure, magnetism, and color. Its primary value lies in its aid to estimating the ligand lone pair stabilization and central atom ao destabilization in MD_n structures. Therefore, it guides us to the key orbital interactions as a basis for molecular structure and reactivity.

To introduce the application of the model to *d* valence ao's, we will consider a ligand hybrid ao located along the +*z* axis and directed toward the metal ion centered at the origin:

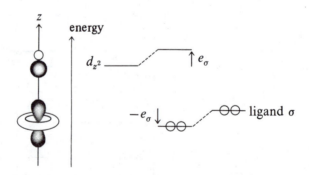

The overlap of the ligand σ ao and the metal d_{z^2} ao results in the creation of bonding and anti-bonding mo's, as you are well aware by now. The extent of metal-ligand orbital interaction (the downward shift in energy of the ligand ao and the upward shift in energy of the metal ao)† will be represented by the quantity e_σ (here defined as + signed).

---

†In actuality, the downward shift is slightly smaller than the upward shift.

The magnitude of e_σ depends on the magnitude of the metal-ligand overlap integral (the detailed characteristics of the metal and ligand ao radial and angular functions) and on the energy match between the ao's. *Therefore, e_σ is a property of both the metal and the ligand bound to it.*

It is of great interest to understand why some complexes adopt a square planar and others a tetrahedral geometry; it is important to know, then, how the metal-ligand interaction depends on the orientation of the ligand about the metal d_{z^2} and other d ao's. For example, moving the ligand down into the xy plane should reduce the d_{z^2}-

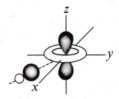

ligand interaction because it is now the less highly directed collar of the metal ao that overlaps the ligand ao. Accordingly, this *angular* change in the position of the ligand, without changing the metal-ligand distance, scales down the stabilization of the ligand ao, and the destabilization of the metal ao, to a fraction of e_σ. Even more dramatic is the effect of moving the ligand off the z axis by half the tetrahedral angle, for this is exactly

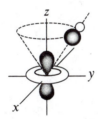

the angle that the d_{z^2} nodal cone makes with the z axis. In this event, the metal-ligand interaction goes to zero because the *angular* part of the overlap integral goes to zero.

The mathematical techniques actually used to compute the variation of the interaction between a metal ao and the ligand ao as a function of the angles θ and ϕ (Figure 1-1) about the metal ion are clearly beyond the purpose of the text. This in no way interferes with your use of the results (Table 9-1), which are evaluated for you for commonly encountered positions of ligands about a metal ion. Each ligand is considered to have one sigma ao and two pi type ao's (or mo's in some cases), and the table entries are the scaling factors for e_σ, the unit of σ interaction, and for e_π, the unit of π interaction, both defined by the standard orientations:

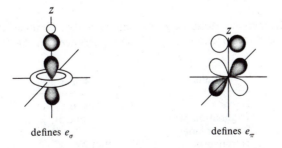

defines e_σ defines e_π

Recall the generalization from Chapter 4 that π interactions are normally less than the corresponding σ interactions; here $e_\pi < e_\sigma$. Additionally, e_σ is positive, whereas e_π may be positive (for π donors) or even negative (for good π acceptor ligands).

TABLE 9-1

**ANGULAR SCALING FACTORS FOR e_σ AND e_π
WHEN EACH LIGAND HAS 1σ AND 2π
ORBITALS.**

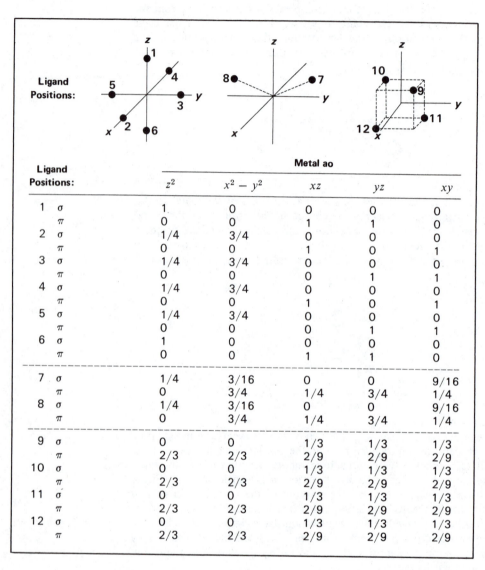

Ligand Positions:		Metal ao				
		z^2	$x^2 - y^2$	xz	yz	xy
1	σ	1	0	0	0	0
	π	0	0	1	1	0
2	σ	1/4	3/4	0	0	0
	π	0	0	1	0	1
3	σ	1/4	3/4	0	0	0
	π	0	0	0	1	1
4	σ	1/4	3/4	0	0	0
	π	0	0	1	0	1
5	σ	1/4	3/4	0	0	0
	π	0	0	0	1	1
6	σ	1	0	0	0	0
	π	0	0	1	1	0
7	σ	1/4	3/16	0	0	9/16
	π	0	3/4	1/4	3/4	1/4
8	σ	1/4	3/16	0	0	9/16
	π	0	3/4	1/4	3/4	1/4
9	σ	0	0	1/3	1/3	1/3
	π	2/3	2/3	2/9	2/9	2/9
10	σ	0	0	1/3	1/3	1/3
	π	2/3	2/3	2/9	2/9	2/9
11	σ	0	0	1/3	1/3	1/3
	π	2/3	2/3	2/9	2/9	2/9
12	σ	0	0	1/3	1/3	1/3
	π	2/3	2/3	2/9	2/9	2/9

For quick reference, the following list summarizes the common structures for two
through six ligands and the location of ligands with reference to Table 9-1.

Structure	Atoms
Linear	1 and 6
Trigonal planar	2, 7, and 8
Square planar	2–5
Tetrahedron	9–12
Trigonal bipyramid	1, 2, 7, 8, and 6
Square pyramid	1–5
Octahedron	1–6

d^* Orbital Energies and Occupation Numbers

To determine the first approximation values of the d ao energy shifts arising from overlap with the ligand orbitals, you simply sum over all ligands the coefficients of the appropriate type (σ, π) in each column of Table 9-1. For the octahedron, the upward shift in the $(z^2)^*$ mo energy is simply $1 + \frac{1}{4} + \frac{1}{4} + \frac{1}{4} + \frac{1}{4} + 1 = 3$ (the unit is e_σ). The same result is obtained for $(x^2 - y^2)^*$. Similarly, the d_{xz}, d_{yz}, and d_{xy} ao's shift as a degenerate set of three by an amount $4e_\pi$. The energy gap between the σ^* and π^* mo's is $3e_\sigma - 4e_\pi$, and this is the quantity defined as Δ_o.

$$\Delta_o \equiv 3e_\sigma - 4e_\pi$$

The d ao shifts for the D_{4h} and T_d structures are obtained similarly, and those shifts are summarized below in a form useful for future reference. *Note that the O_h and T_d sets are inverted with respect to one another.* The triply degenerate sets are called t and the doubly degenerate sets are called e, the symbols being codes for the symmetry properties of the orbitals.

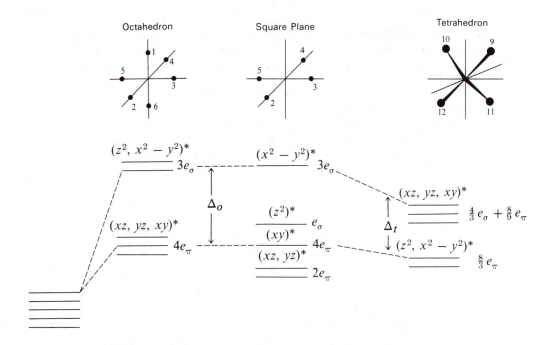

Several interesting features of these diagrams are immediately obvious. First of all, the ligand environment of the $(x^2 - y^2)^*$ and $(xy)^*$ mo's of the square planar geometry is the same as in the octahedral geometry, so the energy gap between them is simply Δ_o in both cases. For the square planar geometry, the position of the $(z^2)^*$ mo relative to the $(xy)^*$ orbital is dependent upon the relative magnitudes of e_σ and e_π, for if e_σ is smaller than $4e_\pi$ then the $(z^2)^*$ mo will lie below the $(xy)^*$ mo. Examination of the tetrahedral case shows the orbital splitting to be much smaller than in the octahedral case, and in fact this simple AOM approximation yields the important result (assuming no difference in M−L distance) that Δ_t is only about $\frac{1}{2}$ Δ_o:

$$\Delta_t = \frac{4}{3}e_\sigma + \frac{8}{9}e_\pi - \frac{8}{3}e_\pi = \frac{4}{3}e_\sigma - \frac{16}{9}e_\pi = \frac{4}{9}(3e_\sigma - 4e_\pi)$$

$$\Delta_t = \frac{4}{9}\Delta_o$$

So far, the issue of assigning valence electrons to the $t*$ and $e*$ mo's of the MD_6 case and to the $t*$ and $e*$ mo's of the $MD_4(T_d)$ case has been skirted. *First noting that all filled orbitals* (those of the metal as well as those of the ligands) *lie at lower energy than the $t*$ and $e*$ orbitals, it is only the valence d electrons of the metal ion that are left to occupy $t*$ and $e*$.* The critical question surrounding the assignment of a ground state configuration is whether or not Hund's rule of maximum unpaired electrons is still applicable to the full set of five $d*$ mo's. In other words, if Δ_o and Δ_t are fairly small, the spin correlation forces between electrons may determine that Hund's rule is to be followed. On the other hand, if the splitting of anti-bonding mo's is large enough, the lowest-energy configuration will result by half-filling, and then filling, the $t*$ ($e*$ in the T_d case) mo's before assigning electrons to the $e*$ ($t*$) mo's. The Hund's rule case, where electron repulsions (not the orbital energy splitting) dominate, is called the **high spin** case; the half-filling/filling situation is called the **low spin** case. For octahedral complexes, the concept of high and low spin cases is without meaning when there are fewer than four and greater than seven d electrons to be assigned to the anti-bonding mo's. Furthermore, in practice, it appears that the Δ_t *is usually too small for anything but the high spin case to result†* for MD_4 (although the distinction would be meaningless for metal ions with fewer than three and greater than six valence electrons).

Pursuing these ideas in a slightly more quantitative way, you want to determine whether a high spin configuration $\cdots (t*)^m(e*)^n$ is more or less stable than the low spin configuration $\cdots (t*)^{m+n}(e*)^0$. The energy difference between the two configurations may be approximately thought of as arising from two contributions; energy may be released on permitting the $ne*$ electrons to occupy the $t*$ mo's of lower energy, but at the expense of greater repulsions between the $mt*$ electrons and the n electrons in question (because the spin correlation energy for like-spin electrons is lost and because the charge correlation energy is increased when two electrons possess the same orbital function). Representing the (average) increase in repulsion energy for each of the n electrons as P (the pairing energy) and the difference in $t*$ and $e*$ orbital energies as Δ_o, the energy change on passing from high to low spin is $n(P - \Delta_o)$. Thus, if $P > \Delta_o$, the energy change will be positive and the high spin configuration is favored; whereas if $\Delta_o > P$, the energy change is negative and the low spin configuration is favored.

$$\Delta E = E_{LS} - E_{HS}$$
$$= (P - \Delta_o)$$

$$\Delta E = 2(P - \Delta_o)$$

The question now is, "Can one predict for a given complex whether a high or low spin configuration is appropriate?" To do so requires that you be able to estimate the relative magnitudes of P and Δ_o for any complex. Since molecular orbitals are delocalized over metal *and* ligand atoms, P quite obviously depends not only on the metal but also on the ligands in question. A similar dependency on the metal/ligand combination is expected for Δ_o. It is fortunate that in most cases, where the metal-ligand bonds are

---

†Only when extremely bulky, strong donor ligands are involved is this generalization excepted. $Cr(N(SiMe_3)_2)_3NO$ is a rare example: D. C. Bradley, *Chem. Britain*, **11**, 393 (1975).

highly polar, the anti-bonding mo's are mainly metal d ao's with some moderate to small contamination by the ligand orbitals, and the electron repulsions for the anti-bonding mo's are highly characteristic of the metal and the ion charge. For the estimation of Δ_O you must recognize that some metal/ligand combinations produce strong bonding and large Δ_O's while others result in small Δ_O's. *The terms "large" and "small" are used in relation to the size of P. At least one important generalization is possible: high metal ion charge leads to "large" Δ_O, and ligand properties of strong σ donor (e_σ large) and weak π donor (or strong π acceptor) character ($e_\pi \gtrsim 0$) lead to "large" Δ_O.*

To help clarify the competition between P and Δ_O in determining the low or high spin character of O_h complexes, the following tables present some values of P (these are nominal values calculated from spectra of *free* ions and so are probably up to 20 per cent higher than in complexes, where the repulsion energies are reduced by delocalization of the electrons onto the ligands) and some Δ_O values for H_2O as ligand (extracted from the electronic spectra of the complexes by methods to be given later). With the exception of Mn^{3+} and Co^{3+}, the Δ_O values fall in the range from 30 to 70 per cent of the *free ion P* value, and it is not surprising that the $M(OH_2)_6^{n+}$ are high spin cases. This is consistent with the expectation that H_2O does not form very strong adduct σ bonds and is a π donor (albeit with but a single $p\pi$ electron pair). The Mn^{3+} and Co^{3+} Δ_O's are 80 and 90 per cent of the respective *free ion P* values. Depending on how extensive the covalencies of the M–D bonds are, the *complex ion P* values could be reduced below the Δ_O's. In fact, $Mn(OH_2)_6^{3+}$ is a high spin case while $Co(OH_2)_6^{3+}$ is a low spin example. Thus, Mn^{3+} behaves as does Fe^{3+}, and it is not until Co^{3+} that the energy match between metal and H_2O orbitals is sufficiently good to produce a low spin complex.§

P VALUES[12]
(UNITS OF μm^{-1}†)

d^4	d^5	d^6	d^7
Cr^{2+} 2.04	Mn^{2+} 2.38	Fe^{2+} 1.92	Co^{2+} 2.08
Mn^{3+} 2.52	Fe^{3+} 2.99	Co^{3+} 2.36	

Δ_O VALUES[13] WITH H_2O
(UNITS OF μm^{-1}†)

d^3	d^4	d^5	d^6	d^7
V^{2+} 1.18	Cr^{2+} 1.40	Mn^{2+} 0.75	Fe^{2+} 1.00	Co^{2+} 1.00
Cr^{3+} 1.76	Mn^{3+} 2.10	Fe^{3+} 1.40	Co^{3+} ~2.1 (oxidizes water)	

At the other extreme of ligand behavior you find CN^-, a strong σ donor (e_σ large) also capable of acting as a π^* acceptor ($e_\pi < 0$). For example, Δ_O values with Fe^{2+} and Co^{3+} are large (3.3 and 3.5 μm^{-1}) and CN^- complexes are generally low spin. Now you can appreciate the facts that $Mn(H_2O)_6^{2+}$ has a ground electronic configuration with five unpaired electrons [Δ_O small $\Rightarrow$ a $(t^*)^3(e^*)^2$ configuration to follow Hund's rule] while $Mn(CN)_6^{4-}$ has only one unpaired electron [Δ_O large $\Rightarrow$ a $(t^*)^5(e^*)^0$ configuration]. NH_3 is an example of an intermediate ligand that gives low spin $Co(NH_3)_6^{3+}$ but high spin $Co(NH_3)_6^{2+}$ complexes. Here again [cf. $Co(OH_2)_6^{2+}$ and $Co(OH_2)_6^{3+}$] is the effect of increasing the metal ion charge. NH_3 is a fairly strong σ donor but a poor π donor ($e_\pi \approx$ 0, since the $NH_3 \pi$ mo's are actually N–H bonding orbitals).

Two interesting trends, which can be anticipated, are observed in P and Δ_O. For pairs of isoelectronic ions (Cr^{2+}, Mn^{3+}/Mn^{2+}, Fe^{3+}/Fe^{2+}, Co^{3+}), the ion of greater charge has greater effective nuclear charge for the valence shell electrons. As a result, there is an

---

†From the relationship $\lambda v = c$, we have $1/\lambda = v/c \equiv \tilde{v}$. Here, $\tilde{v}$ is the wave number of a photon of frequency v and is expressed in reciprocal μm. Its relation to photon energy comes from Einstein's equation $E = hv = hc\tilde{v}$.

§For the series Mn^{3+}, Fe^{3+}, Co^{3+} the effective nuclear charge steadily increases; the result is ion contraction and d electron stabilization to lower energy nearer that of the ligand lone pair.

increase in both P and Δ_o for a given ligand.† Similarly, for a series of ions of given charge (Cr^{2+}, Mn^{2+}, Fe^{2+}, Co^{2+}; Mn^{3+}, Fe^{3+}, Co^{3+}), the value of P is maximum for those ions with five $3d$ electrons. In other words, the pairing energy reaches a maximum for the half-filled shell, a trend you should recognize as having seen before, when you analyzed (in Chapter 1) the discontinuity in ionization potentials for the elements (B, C, N, O, F), (Al, Si, P, S, Cl), and the I.P. trend for $M^{2+} \rightarrow M^{3+}$ (p. 312). Note also the minimum in P for the d^6 configuration. Because of this, the low spin case arises more frequently for d^6 ions than for any other d^n case. Both the ionization potential trends for these elements and the trend in P for transition metal ions have a common origin in the spin and charge correlation energies of electrons in the same valence shell.

STUDY QUESTIONS

(These questions will take on added meaning in Chapter 12, Electron Transfer Reactions.)

7. Focusing your attention on the HOMO energies of $Co(NH_3)_6^{2+}$ and $Co(OH_2)_6^{2+}$, which half-cell potential should be more positive or less negative?

or

$$Co(NH_3)_6^{2+} \rightarrow Co(NH_3)_6^{3+} + e^-$$

$$Co(OH_2)_6^{2+} \rightarrow Co(OH_2)_6^{3+} + e^-$$

(Hint: e_σ is larger for NH_3 than H_2O.)

8. Which reaction is accompanied by the smaller change in metal-ligand distance?

or

$$Fe(CN)_6^{3-} \rightarrow Fe(CN)_6^{4-}$$

$$Co(NH_3)_6^{3+} \rightarrow Co(NH_3)_6^{2+}$$

(Answer on the basis of the change in numbers of t^* and e^* electrons.)

9. Using 3.4 μm^{-1} as a nominal value of Δ_o for CN^- with a transition metal ion, identify any metals on p. 319 for which $\Delta_t > P$.

COMPLEX STRUCTURES AND PREFERENCES

Now the tools are at hand for a discussion of the important consequences of d electron occupation of the σ^* and π^* mo's. To begin with, examine the d^0 case with an octahedral structure. Since the $(z^2)^*$ mo is destabilized by $3e_\sigma$, *there is a corresponding bonding orbital which is stabilized by the same amount*, and this orbital is doubly occupied with a pair of electrons; therefore, *the bonding orbital contributes $-6e_\sigma$ to the electronic energy* of the complex. Similarly, the bonding orbital associated with the $d_{x^2-y^2}$ ao is stabilized by $3e_\sigma$ and the associated electron pair also contributes $-6e_\sigma$ units to the molecular energy. The total energy of the bonding pairs is therefore 12 units of sigma interaction—a value equal to twice the number of ligands. This is a general result you can depend on. *The sum total σ bonding electron energies is equal to twice the number of ligands* (12 units for the octahedron, 8 for both tetrahedral and square planar geometries). By the same token, if each ligand possesses two π-type electron pairs, the stabilization of these electrons is *four times the number of ligands* (in units of e_π).

---

†P and Δ_o increase because of radial contraction of the d ao's, and Δ_o increases because of better energy match between the metal ($3d$, $4s$, $4p$) and ligand orbitals. In fact, because of better overlap, the $d\sigma^*$ mo's should be more sensitive to energy match improvement than the $d\pi^*$ mo's.

It is important to realize that the bonding electron stabilization of $8e_\sigma$ is *the same for both the square planar and tetrahedral geometries. That a given metal ion may favor one or the other structure derives from the way in which the d electrons are distributed among the anti-bonding (d*) mo's.* While it is *not necessary* to do so, the analysis may be simplified by assuming that the e_π factor is no more important than other minor corrections that a more thorough treatment makes to the e_σ parameter; π effects shall, therefore, be ignored. In this case the square planar and tetrahedral orbital diagrams reduce to

$$
\begin{array}{ll}
\underline{\qquad\qquad}\ (x^2 - y^2)^* \quad 3e_\sigma & \\[2ex]
\underline{\qquad\qquad}\ (z^2)^* \qquad e_\sigma & \underline{\qquad\qquad}\ (xz,\ yz,\ xy)^* \ \tfrac{4}{3}e_\sigma \\[3ex]
\underline{\underline{\qquad\qquad}}\ \text{zero} & \underline{\underline{\qquad\qquad}} \\[2ex]
\qquad D_{4h} & \qquad T_d
\end{array}
$$

Now for each d^n case, put in the appropriate number of electrons and add the cumulative energies of these anti-bonding electrons to the bonding electron energies of $-8e_\sigma$. Further, recognize the two extremes of electron configuration—high spin (complexes with weakly basic ligands) and low spin (complexes containing strongly basic ligands). As noted earlier, the orbital splitting of $\frac{4}{3}e_\sigma$ for the tetrahedral geometry is not likely to be greater than the d electron pairing energy, so you need only consider the high spin cases for this geometry—that is, each d^n configuration is achieved by first placing one electron in each orbital, starting from the bottom and working up, before assigning any orbital double occupancy. With the D_{4h} geometry you need to note that the 1.0 e_σ splitting of the $(z^2)^*$ mo from the lower triad should be less than the pairing energy, but that the splitting of the $(x^2 - y^2)^*$ mo from the $(z^2)^*$ mo by 2.0 e_σ may be greater than P.[†] Thus, low and high spin configurations must be distinguished for the D_{4h} geometry in the following way. In the low spin case, the lowest four orbitals are half-filled and then filled before occupancy is assigned to the $(x^2 - y^2)^*$ mo. For the high spin case, even $3e_\sigma$ is less than P and the filling of orbitals with electrons proceeds as in the tetrahedral case. General equations for the energies of the tetrahedral and square planar complexes are given below, as is an equation for the difference in energy between these two structures. We call this energy difference the **structure preference energy**, as it gives a measure of the preference for one structure over the other, based on the difference in d^* orbital occupations.

$$E(D_{4h}) = (n_{z^2} + 3n_{x^2-y^2} - 8)e_\sigma$$

$$E(T_d) = \left(\frac{4}{3}n_\sigma - 8\right)e_\sigma$$

$$E(D_{4h}) - E(T_d) = \left(n_{z^2} + 3n_{x^2-y^2} - \frac{4}{3}n_\sigma\right)e_\sigma$$

Of course, a tetrahedron is always favored over a square planar complex on the basis of VSEPR/steric forces, so the final structure is determined by the balance of these forces

---

[†]Assuming that $e_\pi \approx 0$, the generalization that $P > \frac{4}{9}\Delta_O = 1.3e_\sigma$ for all known cases establishes the rough guideline that orbitals split by less than $1.3e_\sigma$ are to be treated as "degenerate" by electron correlation forces.

and the d^* occupation differences. In practice, we find that the VSEPR/steric forces can be overcome if the structure preference energy is greater than $1\ e_\sigma$ (for observed values of e_σ).

The structure preference energies are plotted in Figure 9-4(A), from which it is apparent that the VSEPR/steric forces that favor the T_d geometry are unopposed for the d^n ions with $n = 0, 1, 2$, and 10 only. For all others, the magnitude of e_σ is highly important in determining the preference for D_{4h}. For sufficiently weak ligands (e_σ small, high spin case) it is clear that VSEPR forces are unopposed for the $n = 5, 6$, and 7 ions; for the $n = 3, 4, 8$, and 9 ions the observed structures are likely to be intermediate (i.e., a flattened tetrahedron, D_{2d}) between square planar and tetrahedral when e_σ is small. The geometries in Table 9-2 for Cl^- as a ligand are completely explained in this way. For the stronger ligands NH_3 and CN^- you find examples of a preference for a low spin, square planar geometry—as predicted by the low spin graph of Figure 9-4(A).

By way of conclusion, *the formation of a T_d or D_{2d} complex is likely for the $n > 2$ cases only by ligands that weakly interact with the metal ion because of intrinsically weak M—D bonds and/or because of unusual ligand steric requirements. More basic or less hindered ligands will prefer to bind the metal with a planar geometry.*

Some useful conclusions can also be drawn from a comparison of the six- and four-coordinate structures. This has been done graphically in Figure 9-4(B) and (C), where it is immediately seen that both high and low spin octahedral complexes are preferred over the high spin tetrahedral structures [Figure 9-4(B)]. This suggests that in addition to the basicity and steric factors just listed for the preparation of tetrahedral complexes, you must realize that stoichiometry control at a 4:1 ligand-to-metal-ion ratio is an important consideration for formation of tetrahedral complexes.

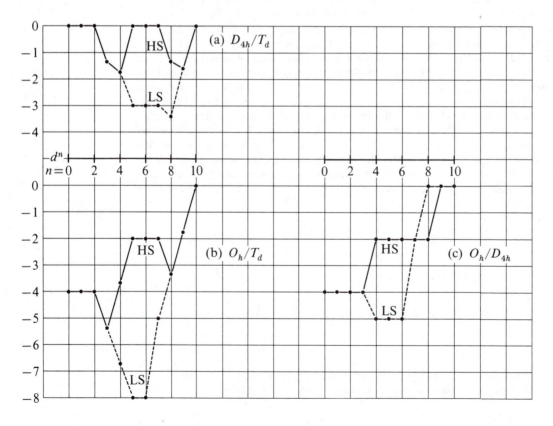

Figure 9-4. Structure preference energies. (A) $E(D_{4h}) - E(T_d)$; (B) $E(O_h) - E(T_d)$; (C) $E(O_h) - E(D_{4h})$; units = e_σ. A negative value means that the first named geometry is preferred.

TABLE 9–2

CORRELATION BETWEEN d^n AND SYMMETRY FOR FOUR-COORDINATE STRUCTURES

Complex	Structure	d^n	
$Ni(CO)_4$	T_d	10	
$Cu(NH_3)_4^{2+}$	D_{4h}	9	
$CuCl_4^{2-}$	D_{4h}	9	(Counter-ion = NH_4^+)
	D_{2d}	9	(Counter-ion = Cs^+)
$Ni(CN)_4^{2-}$	D_{4h}	8	
$NiCl_4^{2-}$	D_{2d}	8	
$CoCl_4^{2-}$	T_d	7	
$FeCl_4^-$	T_d	5	
$TiCl_4$	T_d	0	

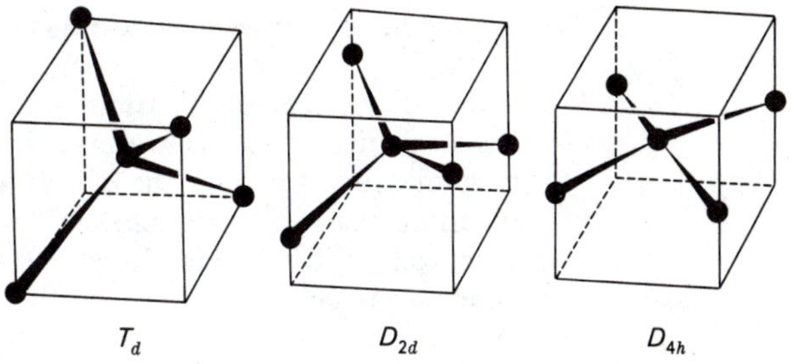

T_d $\qquad$ D_{2d} $\qquad$ D_{4h}

Experimental Evidence for a Structural Preference Energy and the Ligand Field Stabilization Energy

To further illustrate applicability of the structural preference energy, consider the semi-quantitative estimates[14] of the energies of the reactions

$$6H_2O_{(g)} + MCl_{4(g)}^{2-} \rightarrow 4Cl_{(g)}^- + M(OH_2)_{6(g)}^{2+}$$
$$\phantom{6H_2O_{(g)} + }T_d \phantom{MCl_{4(g)}^{2-} \rightarrow 4Cl_{(g)}^- + } O_h$$

for M = Mn through Zn. A plot of the reaction energies, taking that for M = Mn as the reference, is given in Figure 9–5. If you compare the deviations of the points in this graph

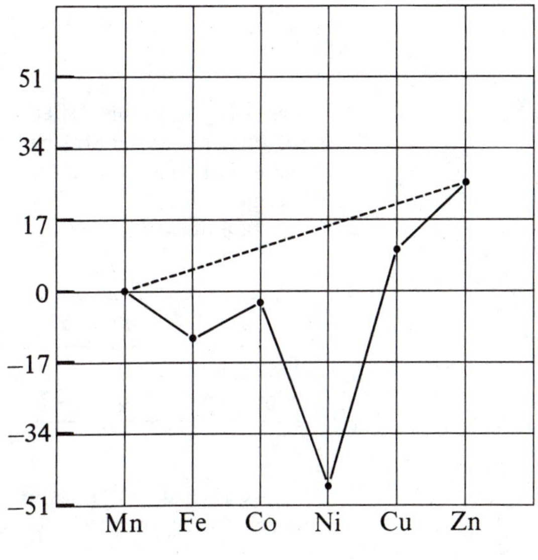

Figure 9–5. Enthalpies (kJ/mole) for the reaction $6H_2O + MCl_4^{2-} \rightarrow 4Cl^- + M(OH_2)_6^{2+}$, relative to that for M = Mn^{2+}.

from the smoothed data (dotted line), you see a variation with the number of d electrons that is remarkably similar to that on the right side of the high spin curve in Figure 9-4(B). Of particular importance is that both plots exhibit strong minima for Ni^{2+}, the d^8 case.

Further illustration of the utility of the AOM model arises in application to lattice and hydration energies and radii of the divalent transition metal ions.[15] Graphs of these quantities as a function of atomic number are shown in Figure 9-6, where you will notice (i) an underlying smooth trend in values, as evidenced by the d^0, d^5, and d^{10} cases, and (ii) rather characteristic deviations from this trend by the intervening cases. The direc-tions of the smooth trends are exactly what you should expect from the increase in ion effective nuclear charge across the series. The effect of increasing effective nuclear charge, in terms of the AOM model, is to increase progressively the magnitudes of e_σ and e_π by virtue of the fact that increasing effective nuclear charge lowers the energies of the d ao's so that they more effectively interact (by better energy match) with the ligand orbitals. Implicit in such an expectation, however, is an assumption that *each d electron has the same average anti-bonding role* to play in opposing the coordination of six ligands to the metal ion. A more specific statement of this background role for the d electrons is that they are evenly distributed among the five d orbitals so that the d^n charge cloud has a spherical shape. However, it is clear from preceding discussions that such is not, in fact, the case, for only the d^0, d^5 (high spin), and d^{10} configurations can have such a spherical charge distribution. In terms of the AOM, the idea that each d electron is spherically distributed in the d shell amounts to assigning each d electron an average anti-bonding energy

$$\frac{1}{5}[2 \cdot 3e_\sigma + 3 \cdot 4e_\pi] = \frac{6}{5}e_\sigma + \frac{12}{5}e_\pi$$

Regardless of the value of n in d^n for the metal ion, each MD_6^{2+} complex has an energy of $-12e_\sigma - 24e_\pi$ from ligand orbital electron stabilization. To this is added $n(\frac{6}{5}e_\sigma + \frac{12}{5}e_\pi)$ for the *reference* state, in which each of the n electrons plays an average anti-bonding role. For an *actual* high spin or low spin configuration, the correct energy is

$$-12e_\sigma - 24e_\pi + n_\pi[4e_\pi] + n_\sigma[3e_\sigma]$$

The difference between the energy of a given $t^{*m}e^{*n}$ configuration and the energy of the reference state is simply

$$E - E_r = n_\sigma\left[3e_\sigma - \frac{6}{5}e_\sigma - \frac{12}{5}e_\pi\right] + n_\pi\left[4e_\pi - \frac{6}{5}e_\sigma - \frac{12}{5}e_\pi\right]$$
$$= [3n_\sigma - 2n_\pi]\,\Delta_o/5$$

The following table reflects this relative electronic configuration energy brought on by the different destabilizing effects of the $d\sigma^*$ and $d\pi^*$ electrons. In the language developed along with the crystal field theory, these energies were called **ligand field stabilization energies**, LFSE.

In units of $(\Delta_o/5)$:

	d^0	d^1	d^2	d^3	d^4	d^5	d^6	d^7	d^8	d^9	d^{10}
Low spin	0	−2	−4	−6	−8	−10	−12	−9	−6	−3	0
High spin	0	−2	−4	−6	−3	0	−2	−4	−6	−3	0

A plot of LFSE against n of d^n has the appearance of Figure 9-7, where for simplicity Δ_o is presumed constant for all M^{2+} of a given charge (this is not a poor

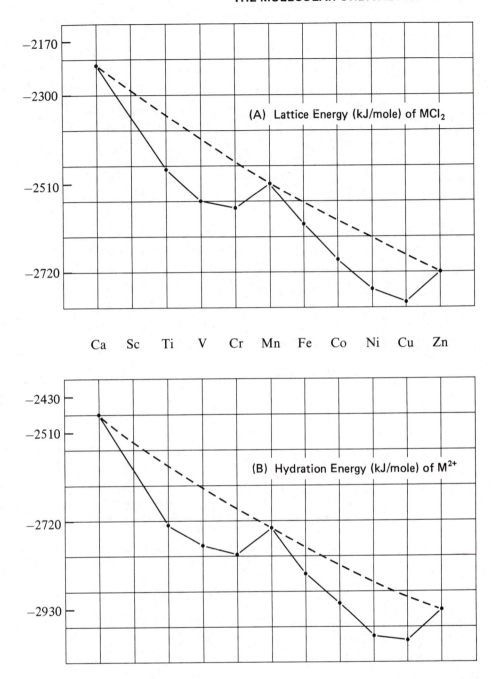

Figure 9–6. Some properties of M²⁺ and M³⁺. *Illustration continued on following page.*

approximation when all M come from the same period). Quite remarkably, the shape of the LFSE curve closely resembles those in Figure 9-6. While the LFSE diagram is an energy diagram, there is also a simple direct relationship between orbital occupation and metal-ligand bond distance (or metal ion radius). Each electron to occupy a $d\pi^*$ mo should lengthen the distance by an amount less than average. The deviations are cumulative as the $d\pi^*$ mo's are occupied, so that minima occur at the high spin d^3 and d^8 cases and at the low spin d^6 case [Figure 9-6(C) and (D)].

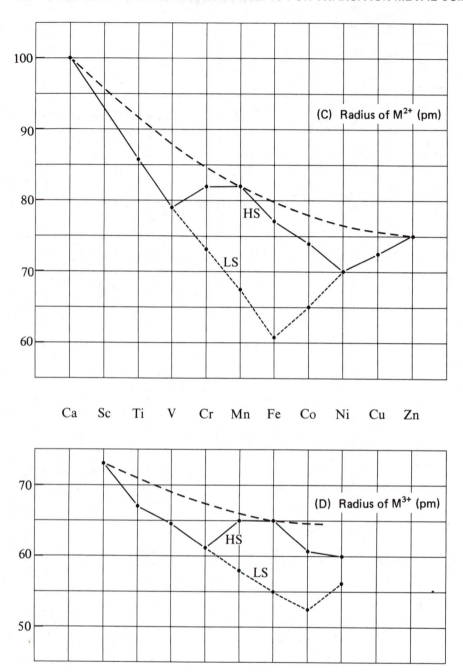

Ca   Sc   Ti   V   Cr   Mn   Fe   Co   Ni   Cu   Zn

Figure 9-6. *Continued.*

Jahn-Teller Distortions from O_h Geometry

Even though idealized MD_6 and MD_4 structures have been assumed for the comparisons of the preceding sections, comment is needed on the long established phenomenon of distorted MD_6 structures of d^4 (high spin), d^7 (low spin), and d^9 complexes. These complexes are all characterized by half-filled or filled t^* mo's but an odd number of electrons in the e^* mo's of $O_h MD_6$. The lowest energy configurations are, therefore, orbitally doubly degenerate. Generally, the distortions from O_h symmetry take the form of elongation of a pair of bonds *trans* to one another, with adoption of a D_{4h} symmetry.

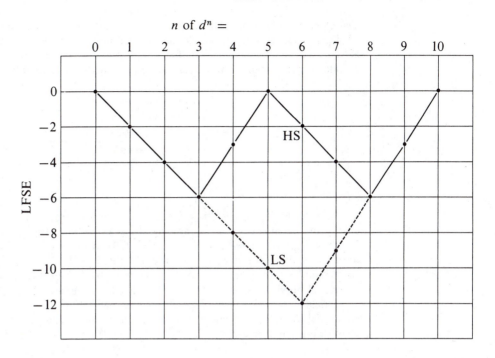

Figure 9–7. The relative electronic configuration energy of MD_6 complexes as a function of the number of d electrons.

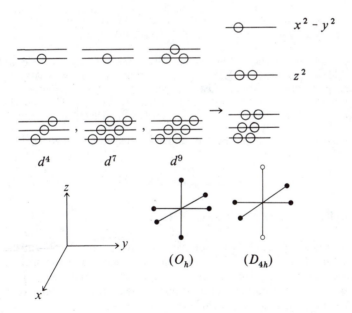

The distortions inherent to the d^4, d^7, and d^9 O_h cases are examples of a general theorem recognized and proved by Jahn and Teller,[16] to the effect that any structure producing an orbitally degenerate electronic configuration of a non-linear molecule is intrinsically unstable; in short, there exists another structure for the molecule that eliminates the orbital degeneracy and leads to lower molecular energy. Accordingly, the distortions are often referred to as **Jahn-Teller distortions**. The occurrence of an electronic degeneracy for a structure is identified by the possibility of a non-unique assignment of electrons to degenerate (equi-energetic) orbitals. In this vein, degeneracies also occur for $t*$ mo's, as in the d^2 case, and a perfect O_h structure is not possible. With $t*$ degeneracies, however, the driving force for the distortion arises from $M-D\pi$ interactions, which are much weaker than the σ interactions; these Jahn-Teller distortions in ground electronic states have so far escaped conclusive experimental detection.

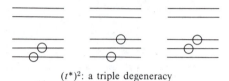

$(t*)^2$: a triple degeneracy

The driving force for this weakening of bonds along the z axis is a lowering of the energy of the molecule. By elongating the z axis bonds, the ligand/z^2 overlap (and e_σ for those ligands) decreases and the z^2 mo falls to lower energy. The three $\sigma*$ electrons, initially evenly distributed between the $\sigma*$ orbitals as $(z^2)^{1.5} (x^2 - y^2)^{1.5}$, now occupy the $\sigma*$ mo's as $(z^2)^2 (x^2 - y^2)^1$. The decrease by 0.5 electron in $(x^2 - y^2)$ means the equatorial bonds shorten, and e_σ for these ligands increases. The change in total energy in a particular complex will depend on the balance between axial bond weakening, equatorial bond strengthening, and change in ligand/ligand repulsion. A more complete analysis is given in reference 17.

STUDY QUESTIONS

10. Confirm the energies of the $d*$ mo's on page 317.

11. Determine which of the following might be expected to exhibit distorted structures (relative to O_h for MD_6 and T_d or D_{4h} for MD_4): $Cr(OH_2)_6^{3+}$, $Ti(OH_2)_6^{3+}$, $Fe(CN)_6^{4-}$, $CoCl_4^{2-}$, $Pt(CN)_4^{2-}$, $ZnCl_4^{2-}$, $Ni(en)_3^{2+}$, $FeCl_4^-$, $Mn(OH_2)_6^{2+}$.

12. Determine which of the following ligands is most likely to yield a tetrahedral complex MD_4 (rather than MD_6 or square planar MD_4) with Mn^{2+}: CN^-, NO_2^-, 2,6-dimethylpyridine, or NH_3. (Use the concepts of donor atom effective nuclear charge and hybridization to estimate an order based on e_σ. Then check to see that the relative e_π values do not alter the order. CN^- and NO_2^- are strong π acceptors.)

13. Give the structures and number of unpaired electrons for $Ni(en)_3^{2+}$, $NiCl_4^{2-}$, and $Ni(CN)_4^{2-}$.

ELECTRONIC STATES AND SPECTRA

With the foregoing concepts of complex structure and orbital sequencing well in hand, it is time to explore the link between those ideas and the striking color and magnetic properties of complexes. Specific examples of the challenge are found in Figure 9-8. How are the numbers and relative positions of the absorption bands to be explained? How does changing a metal ion cause shifts in these bands?

Ground and Excited States

The occurrence of absorption bands in the visible region of the electromagnetic spectrum is due to molecular transformation of photon energy into higher kinetic/potential energy of electrons in the molecule. The molecule is said to undergo a transition from one state to another. Our goal here is to qualitatively define these states.

To illustrate the concepts of **electronic configurations** and of **electronic states**, we will inspect the case of two electrons in the five d orbitals. As you have seen for the O_h and

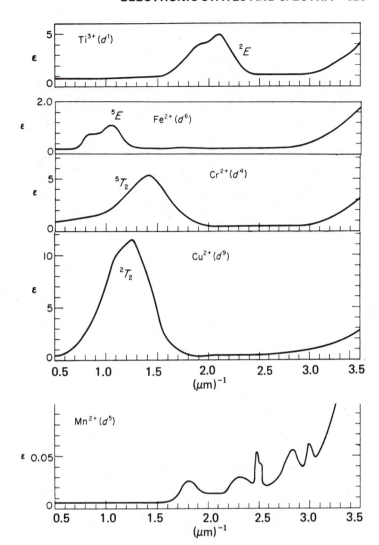

Illustration continued on next page.

Figure 9-8. Electronic spectra of $M(OH_2)_6{}^{n+}$ in the region from 0.5 to 3.5 μm^{-1}. Note that the ϵ scale is two orders of magnitude smaller for $Mn(OH_2)_6{}^{n+}$ than for the others. [From B. N. Figgis, "Introduction to Ligand Fields," Interscience, New York (1966).]

T_d structures, the degenerate set of five d orbitals of a transition metal ion is split into two sets: one set of three orbitals of equal energy (triply degenerate) that we shall label t, and the other set of two orbitals (doubly degenerate) that are labelled e. (The symbols t and e are simply agreed-upon codes for the symmetry properties of the orbitals under the operations of the O_h and T_d point groups.)

For two electrons assigned to these d orbitals, there are three general possibilities if we ignore the details of the assignments within the sets:

1. $(t)^2$
2. $(t)^1(e)^1$ } electron configurations
3. $(t)^0(e)^2$

For an O_h complex the e set is at higher energy than the t set, so the order of configuration energies is

$$(t)^2 < (t)^1(e)^1 < (e)^2$$

[Recall that the order of orbital and configuration energies is reversed (inverted) for a T_d complex.] It is from these *configurations* that we define the *states*. To understand the

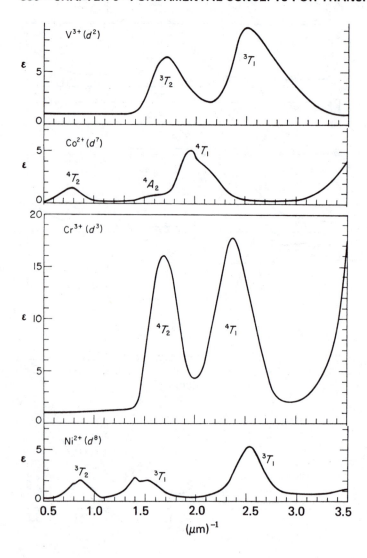

Figure 9-8. *Continued.*

idea of electronic states you must deal with exactly how the two electrons of the $(t)^2$ configuration are to be assigned to the t orbitals. The diagram

xz	yz	xy
↑↓		

↑	↓	

↑	↑	

microstates

shows just three of the 15 possibilities, each called a microstate. The energies of these microstates are not all the same because the repulsions between the electrons are different, depending upon whether they are of opposite spin in the same orbital, of opposite spin in different orbitals, or of the same spin in different orbitals. On the basis of experiment and theory, these microstates are subdivided into sets of common energy called states. *States, then, are groups of microstates of the same energy.* Because of electron repulsions, the microstate set symbolized by

↑	↑	

above should, and does, belong to the lowest state of the $(t)^2$ configuration. This lowest state is characterized by having half-filled degenerate orbitals with electrons of common spin. The number of unpaired electrons (n) is 2 in this case; one speaks of a **spin multiplicity** of $n + 1 = 3$, and the state is called a spin triplet.

So much for the spin characteristics of the *ground* state. What about the orbital characteristics? Of particular importance is the fact that there are three ways of assigning two electrons to two different orbitals out of a set of three (an equivalent view is that there are three ways of assigning the empty orbital):

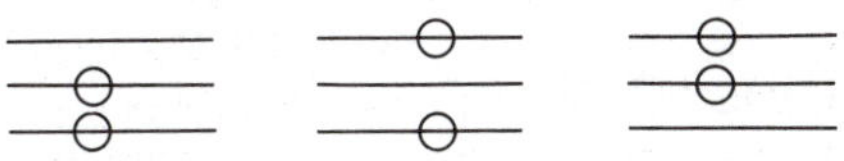

To convey the idea of triple orbital degeneracy of this state, we use the symbol T. Note the use of capital T for a state, and lower case t for three degenerate orbitals.

To convey both the spin and orbital characteristics of the $(t)^2$ ground state, we write ^3T. Table 9-3 summarizes the results of an analysis of the ground states for all d^n cases in O_h and T_d complexes.

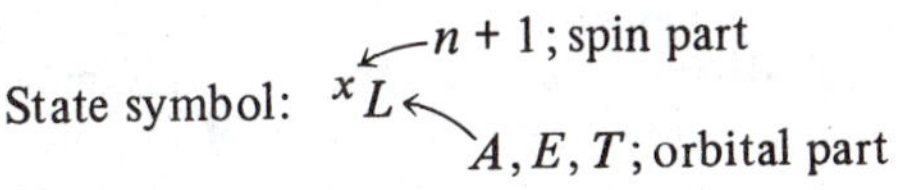

$$\text{State symbol:} \quad {}^xL \overset{\nwarrow\ n+1\,;\,\text{spin part}}{\underset{\searrow\ A, E, T\,;\,\text{orbital part}}{}}$$

The meanings to be associated with the labels A, E, and T are simply that the *orbital* wavefunctions for the states are, respectively, non-degenerate [$A \Rightarrow$ only one way (one

TABLE 9-3

GROUND CONFIGURATIONS AND STATES FOR MD$_6$ (O_h) AND MD$_4$ (T_d) COMPLEXES

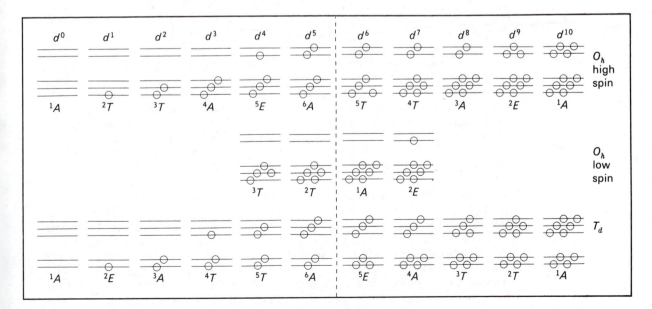

wavefunction) of assigning the electrons within the t^* and e^* sets], two-fold degenerate [$E \Rightarrow$ two ways (two equivalent wavefunctions) of assigning the t^* and e^* electrons], and three-fold degenerate [$T \Rightarrow$ three ways (three equivalent wavefunctions) of assigning the t^* and e^* electrons]. For example, considering the d^1 octahedral case, it is obvious that there are three ways of assigning the one electron among the triply degenerate orbitals of the metal (to xy^*, xz^*, or yz^*)—hence the ^2T symbolism. For the d^2 octahedral case there are also just three ways of assigning *like-spin* electrons among the three low-energy orbitals (^3T). For the d^4 case the ground state symbol depends on whether the configuration is high or low spin. For high spin there is only one way to fill in the first three electrons (like the preceding d^3 case) but two ways for assigning the fourth electron— hence the label is ^5E. For the low spin case there are three ways to assign the fourth electron, and the ground state symbol is accordingly ^3T. For the d^5 case, the high and low spin situations must be distinguished. The high spin case is orbitally non-degenerate and is called ^6A, while the low spin ground state is ^2T. As an exercise we suggest that you verify the A, E, and T orbital labels for the remaining cases.

Now you come to the first of the redundancies nature has so thoughtfully provided. The orbital labels for the d^6 through d^{10} sequence in the high spin part of the table are repetitions of the d^1 through d^5 section. The second part of the redundancy concerns the ground state labels for the tetrahedral geometry—the sequence is exactly *inverted* from that for the octahedral case. This is quite obviously due to the inversion relation between the orbital splitting of the O_h and T_d geometries. Thus, not only is it easy to just write down from inspection the ground state symbols for any d^n case, but you have as a check the fact that the $d^n(O_h)$ and $d^{10-n}(T_d)$ labels are identical.

Having established the ground state symbols, the next step is to inquire into the number of electronic excited states to which the molecule may go upon absorption of a photon. Were you to take a course in the theory of electronic absorption processes, one of the most important concepts you would learn is that transitions between states (ground and excited) of the same spin multiplicity are by far the most probable (will possess the greatest extinction coefficient). Therefore, let us consider for each of the high spin $O_h d^n$ cases what excited states of the *same multiplicity as the ground state* are possible for each excited configuration. The possibilities are given in Table 9-4. The d^1 and d^9 cases are very straightforward. For both there is only one excited configuration possible and that arises from excitation of an electron from a t orbital to an e orbital. For the d^1 case there are two ways that the single electron can be assigned to the e orbitals, and the excited state is ^2E. For the d^9 case the excitation produces a configuration with filled e orbitals and one-less-than-filled t orbitals. This "hole" in the t orbitals may be placed in any of three ways, so the configuration is triply degenerate (^2T). The customary notation for the single electronic transition in the d^1 case is $^2E \leftarrow {}^2T$ and that for the d^9 case is $^2T \leftarrow {}^2E$. *Note that these two cases of d^1 and $d^{10-1} = d^9$ are inversely related!* As you will see, nature has been kind to us again, for *all high spin d^n and d^{10-n} cases are inversely related.*

For d^2 and d^8 you encounter for the first time the possibility of two excited configurations achieved by excitation of one and then two electrons from the t to the e orbitals. Here also is encountered for the first time the need to consider microstate resolution by electron/electron repulsion. The first excited configuration of d^2 (i.e., t^1e^1) leads to a six-fold orbital degeneracy. For each of three ways of assigning the t electron, there are two ways of assigning the e electron *of the same spin*. Electron/ electron repulsions separate these six microstates into two groups—each of which is orbitally triply degenerate. Both are therefore ^3T states. For the doubly excited configuration (e^2) there is only one way to assign the two electrons *and* maintain the triplet spin multiplicity of the ground state. The symbol is, therefore, ^3A.

For the d^8 case, the first excitation produces a configuration t^5e^3. The t orbitals possess a "hole" that may be assigned in three ways, and for each of these assignments the "hole" in the e orbitals may be assigned in two ways. Again, the six-fold degeneracy of this configuration is resolved by electron/electron repulsions into two triply degenerate sets—both ^3T. The double excitation (the configuration is t^4e^4) produces a triple degeneracy in the t orbitals (the two "holes" may be assigned in only three ways) and the state for this configuration is also ^3T. To summarize the possible transitions, we write

TABLE 9-4

THE EXCITED CONFIGURATIONS AND STATES OF THE SAME SPIN MULTIPLICITY AS THE GROUND STATE OF MD_6 (O_h) COMPLEXES

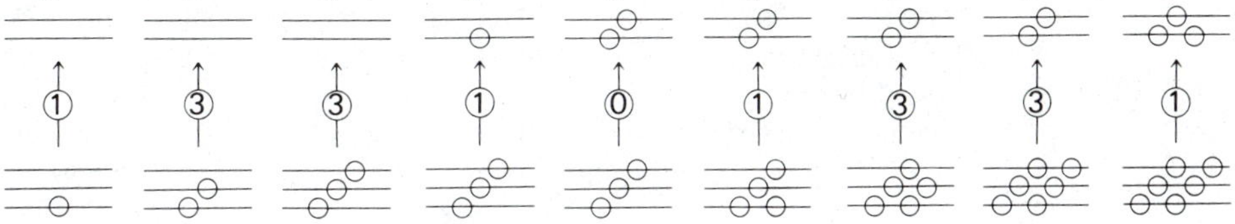

	d^1	d^2	d^3	d^4	d^5	d^6	d^7	d^8	d^9
Number of Transitions	1	3	3	1	0	1	3	3	1
Second Excited Configuration		^3A	^4T				^4A	^3T	
First Excited Configuration		^3T	^4T				^4T	^3T	
First Excited Configuration	^2E	$^3T'$	$^4T'$	^5T		^5E	$^4T'$	$^3T'$	^2T
Ground Configuration	^2T	^3T	^4A	^5E	^6A	^5T	^4T	^3A	^2E

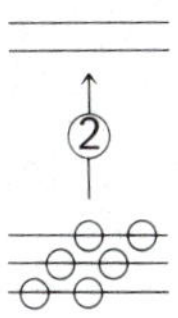

First Excited Configuration	$^1T'$
	^1T
Ground Configuration	^1A

d^2	d^8
$^3T' \leftarrow {}^3T$	$^3T' \leftarrow {}^3A$
$^3T \leftarrow$	$^3T \leftarrow$
$^3A \leftarrow$	$^3T \leftarrow$

Be sure to note the inversion relation between the lowest and highest states of these two cases: for d^2, in order of increasing energy,

$$^3T < (^3T' < {}^3T) < {}^3A$$

and for d^8 the order is

$$^3A < (^3T' < {}^3T) < {}^3T$$

For d^3 and d^7 a situation entirely analogous to that for d^2 and d^8 arises. The ground state of d^3 is ^4A. The first excited configuration (t^2e^1) gives two states—both ^4T. The second excited configuration (t^1e^2) gives only one quartet state, and that is orbitally triply degenerate so the symbol is ^4T. The inverse arrangement of high and low quartet states is encountered for the d^7 case: $^4T < (^4T' < {}^4T) < {}^4A$.

The d^4 and d^6 cases are as easy as the d^1 and d^9 cases. Only one *quintet* excited configuration is possible (t^2e^2 for d^4 and t^3e^3 for d^6). The excited state symbols are, logically enough, ^5T for d^4 and ^5E for d^6. Again the ground and excited states are inversely ordered.

For the d^5 high spin case, no excited configurations of the same spin multiplicity (6) as the ground state are possible! This is a critical prediction of electronic state theory, which is well supported by experiment—*high spin d^5 complexes exhibit no strong transitions in the visible region* and are "colorless." [The spectrum of $Mn(OH_2)_6^{2+}$ in Figure 9-8 exhibits only *weak* transitions to states of different spin multiplicity than the ^6A ground term.]

All this information about the ground and excited electron states of O_h complexes can now be conveniently summarized with the following diagrams:

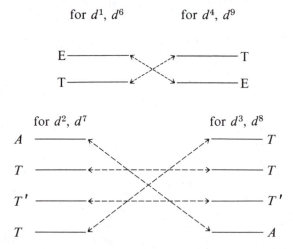

From these diagrams it is apparent that the $d^{1,6,4,9}$ octahedral complexes are characterized by the occurrence of only one absorption band in the visible region. The $d^{2,7,3,8}$ octahedral complexes should exhibit three bands in this region of the spectrum. Actual spectra of $M(OH_2)_6^{n+}$ complexes bear out these predictions; see Figure 9-8.

Even a cursory examination of these spectra should suggest to you, however, that the foregoing derivation of accessible excited states is rudimentary. All the bands show evidence of asymmetry or even resolved splitting. This means, of course, that where a single transition has been predicted there are actually several transitions in close proximity. One source of state splitting derives from the Jahn-Teller distortion of molecular structure. Just as each ground state has an associated lowest energy structure, each excited state has a distinct molecular architecture, which may be different from those of the ground states and other excited states. (Why, in terms of the change in t^*e^* occupancies, should this be the case?) In the event of an orbitally degenerate state, Jahn-Teller distortion will cause there to be at least two excited states, one of which will be slightly lower in energy than the other. Transitions to these molecular states could account for the splitting of a single band predicted to arise only by ignoring the occurrence of Jahn-Teller effects (cf. $Ti(OH_2)_6^{3+}$ in Figure 9-8).

Additionally, you notice that only two, not three, bands are found in the $n = 2$ and 3 spectra. The third transition occurs at such a high energy that it appears in the ultraviolet region, where it is lost among more intense transitions (above $3.5\ \mu m^{-1}$) of the charge transfer (CT) type. These CT transitions are very much like those discussed in Chapter 5 for I_2 adducts; in both cases the CT excitation is associated† with the promotion of an electron in a filled metal-ligand bonding orbital (σ or π) to a metal (acceptor) orbital (less than full d^* mo).

---

†With ligands of the π acceptor type, the ligand π^* mo's lie low enough in energy to also afford metal → ligand CT bands of the $d^* \to \pi^*$ type.

To close this preliminary examination of high spin complex spectra, we will simply state that the inversion relation already noted for the ground electronic states of octahedral and tetrahedral structures carries over into the excited states as well (you will have an opportunity to verify the statement by working study question 15). Thus, the energy level diagram for d^1 and d^6 octahedral geometry given above is appropriate for the $d^{4,9}$ tetrahedral cases as well. Similarly, the $d^{2,7}$ octahedral diagram is the same as the $d^{3,8}$ tetrahedral diagram.

Obtaining Δ from Spectra

Now that you have seen how to account for the number of d-d bands as a function of structure and d^{*n} configuration, the extraction of chemical information from the positions of those bands is the next important step. Following, you will find the rudiments of spectral analysis to extract the value of Δ_t or Δ_o, which reflect the metal-ligand σ and π bonding and thereby provide chemical information.

(i) For d^1, d^4, d^6, and d^9 cases (O_h or T_d), only a single $d \leftarrow d$ band is observed, and it has an energy Δ_o or Δ_t.

(ii) For d^2, d^3, d^7, and d^8 cases,† three bands are expected. For the d^3, d^8 (O_h) case, the lowest-energy transition gives Δ_o. For d^2, d^7 (O_h), it is the separation between the first and third bands (usually) that gives Δ_o. (Unfortunately, the third band cannot always be located, as noted in the last section; in such events more complete analysis is required.) The basis for these generalizations can be found by inspecting the state energy level diagrams in the Appendix.

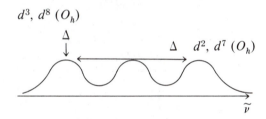

To illustrate the chemical interpretation process, consider Δ_o for $Cr(OH_2)_6^{3+}$ and $Ni(OH_2)_6^{2+}$ (d^3 and d^8 cases).

	Δ_o
$Cr(OH_2)_6^{3+}$	1.7 μm^{-1}
$Ni(OH_2)_6^{2+}$	0.9 μm^{-1}

A striking difference occurs in Δ_o, which is about half as large for Ni^{2+} as for Cr^{3+}. The ligand is the same in both complexes, so the difference must lie with the metal ions. Clearly the chromium complex has a greater charge, and this should lead to stronger metal-ligand bonds (greater e_σ, e_π, and Δ_o), but this is not the whole story. The Cr^{3+} complex has only three d electrons in t^* orbitals, so they play a π^* role in the complex; Ni^{2+}, however, has eight d electrons, and these are distributed so that six play the π^* role (located in t^*) and two play a σ^* role in e^* orbitals. Again you conclude that the Ni^{2+} complex should exhibit longer, weaker bonds and thus a smaller Δ_o.

---

†Remember that $d^{2,3}$ (T_d) ~ $d^{3,8}$ (O_h) and $d^{3,8}$ (T_d) ~ $d^{2,7}$ (O_h).

STUDY QUESTIONS

14. If CoF_6^{3-} is a high spin complex, what do you predict for $CoBr_6^{3-}$? Base your estimate on the relative σ bond strengths of HF and HBr. In fact, CoF_6^{3-} is the only known binary Co^{3+}/halogen complex because CoX_6^{3-}, X = Cl, Br, I, are unstable to internal redox to form $CoX_4^{2-} + X_2$.

15. Show by arguments like those in the text for the high spin octahedral cases that for $d^{4,9}$ tetrahedral complexes the ground and excited states are simply T and E, while those for $d^{3,8}$ tetrahedral complexes are the same as the states for $d^{2,7}$ octahedral ones (exclusive of spin multiplicity).

16. Refer to Figure 9–8 and notice that the band for $Fe(OH_2)_6^{2+}$ falls at lower energy than that of $Ti(OH_2)_6^{3+}$. Estimate from the spectra the values of Δ_o for these complexes and explain why one is greater than the other (consider both cation charge and d^* mo occupation numbers).

17. Referring to Figure 9–8, compare estimated Δ_o's for $Cr(OH_2)_6^{2+}$ and $Cu(OH_2)_6^{2+}$ and account for the fact that one is greater than the other.

18. Explain why FeF_6^{3-} is colorless while CoF_6^{3-} exhibits only a single absorption band in the visible region.

19. Give the number of t^* and e^* electrons and spin multiplicity for the ground configuration of each of the following; then predict the number of electronic absorption bands deriving from $t^* \leftrightarrow e^*$ electron promotion.

$$Fe^{3+} \ (O_h, HS)$$
$$Cu^{2+} \ (O_h)$$
$$Zn^{2+} \ (T_d)$$
$$Cr^{3+} \ (O_h)$$
$$Fe^{2+} \ (T_d)$$
$$Co^{2+} \ (T_d)$$

20. Which ion(s) in the preceding question shows the greatest preference for six-coordination?

The Spectrochemical Series

Through analysis of complex spectra, transition metal chemists have compiled considerable data on the parameter Δ_o. Inspection of the results has made possible several generalizations of use to the practicing chemist. For example, it is observed that if the ligands of MD_6 complexes are arranged in order of increasing Δ_o for each metal ion, the order is generally independent of the metal ion. It would appear that a rough† single-scale ordering of metal-ligand interactions is possible. A partial series, including members of commonly occurring ligand types, is (in order of increasing Δ_o):

$$I^- < Br^- < Cl^- < {}^*SCN^- < F^- \sim urea < OH^- < acetate < oxalate$$
$$< H_2O < {}^*NCS^- < glycine < pyridine \sim NH_3 < en < SO_3^{*2-}$$
$$< dipyridine \sim o\text{-}phen < {}^*NO_2^- < CN^-$$

(* labels the donor atom)

---

†Remember that even in the simple AOM model, Δ_o is a function of two parameters, e_σ and e_π.

From analysis of the magnitude of Δ_o in terms of e_σ and e_π, you expect to find that combinations of weak sigma and strong pi donor properties produce small Δ_o, while strong sigma donor and weak pi donor or strong pi acceptor character of a ligand should produce a large Δ_o. It comes as no surprise, then, that the halides appear early in the series and NO_2^- and CN^- appear late. Other comparisons, which might not make any sense at all if only relative sigma donor properties needed to be considered, make good sense upon including metal-ligand pi interactions. Recall that π donors make Δ_o smaller; for example, the higher position of H_2O than of OH^- is probably due to the fact that OH^- has two pi donor electron pairs while H_2O has only one.

Similarly, the higher positions of dipyridine and o-phenanthroline than of pyridine are believed to be due to the weaker pi donor character and stronger pi acceptor nature of the more delocalized amines (delocalization over an extended carbon frame tends to lower the energies of both π HOMO and π^* LUMO).

Quite generally, you see a relationship among nitrogen donors between Δ_o and p character of the donor lone pairs, although the certainty of this correlation is confused by the variation in π interactions (for example, dipyridine and o-phenanthroline appear higher than NH_3 and en). In addition to the sp character of the nitrogen lone pair, a factor in the low position of $*NCS^-$ relative to the other nitrogen donors may be its pi donor nature. Conversely, NO_2^- appears higher than the other "sp^2" hybridized nitrogen donors because of an appreciable π^* acceptor nature. It is reassuring that the overall order

$$(\text{halogen}) < (\text{oxygen}) < (\text{nitrogen}) < (\text{carbon})$$

is in keeping with our expectations of sigma donor strength.

Of course, the metal ion influences the magnitude of Δ_o through the overlap and energy match criteria; the following metal ion spectrochemical series generally follows expectations:

$$(\text{Mn}^{2+} < \text{Ni}^{2+} < \text{Co}^{2+} < \text{Fe}^{2+} < \text{V}^{2+}) < (\text{Fe}^{3+} < \text{Cr}^{3+} < \text{V}^{3+} < \text{Co}^{3+})$$
$$< \text{Mn}^{4+} < (\text{Mo}^{3+} < \text{Rh}^{3+} < \text{Ru}^{3+}) < \text{Pd}^{4+} < \text{Ir}^{3+} < \text{Re}^{4+} < \text{Pt}^{4+}$$

Here you find increasing Δ_o with increasing metal charge; within a series of like charged ions, the effect of improved metal-ligand orbital energy match (which improves as the d orbitals are stabilized by increasing *nuclear* charge) is in competition with the weakening effect of increasing occupation of the π^* and σ^* mo's; the heavier elements of a group generate a larger Δ_o than the lightest element of the same charge for two reasons. *Generally, the pairing energies of the heavier elements are smaller, meaning that low spin complexes are the rule.* This enhances Δ_o because low spin complexes minimize the number of electrons in the σ^* mo's. Additionally, a "transition metal contraction,"[†] symptomatic of a disproportionately high effective nuclear charge acting on the ligand electrons and greater involvement of the $(n + 1)s$ and p ao's, facilitates a stronger metal-ligand interaction, with e_σ being enhanced to a greater extent than e_π.

STUDY QUESTIONS

21. The Δ_O values for $Ru(OH_2)_6^{2+}$ and $RuCl_6^{3-}$ are nearly the same. Is this consistent with the positions of H_2O and Cl^- in the spectrochemical series? If not, how do you explain the similarity in Δ_O values?

22. Following are spectral data for some NiD_6^{2+} complexes. From the given band centroid energies (in μm^{-1}) for each complex, obtain Δ_O. Then arrange a ligand spectrochemical series and interpret trends in Δ_O. (DMSO = dimethylsulfoxide; DMA = N,N-dimethylacetamide.)

$Ni(DMSO)_6^{2+}$	0.77	1.30	2.40
$Ni(DMA)_6^{2+}$	0.76	1.27	2.38
$Ni(OH_2)_6^{2+}$	0.87	1.45	2.53
$Ni(NH_3)_6^{2+}$	1.08	1.75	2.82

Paramagnetism of Complexes

In addition to the visible-region spectral properties of transition metal complexes, there is another unique, valuable property of transition metal compounds. Their paramagnetism arises from the presence of unpaired electrons, and you have seen how this property is derived from n of d^{*n} and structure and, in the d^{4-7} cases, depends on the strength of metal-ligand bonding.

There is, of course, a characteristic magnetic field associated with the motion of a charged particle. The operation of solenoids is an example from everyday experience. Current passing through a coil of wire induces a magnetic field along the axis of the coil, and this field may be thought of as arising from a bar magnet at the coil center. For the electron there are two types of motion that may give rise to an associated magnetic field. The first of these is a quantum relativistic phenomenon for which no classical analogy is strictly possible. The intrinsic *spin angular momentum* of the electron is loosely thought of as arising from motion of the electron about its own axis and has associated with it a magnetic field represented by a magnetic dipole or moment, μ_s. It is not surprising that there is a relation between the magnitude of the angular momentum and the magnitude of the magnetic moment,

$$\mu_s = g_s \beta_e \sqrt{s(s + 1)} = g_s \beta_e \sqrt{3/4} \quad \text{since } s = 1/2 \text{ for a single electron}$$

where g_s has the magnitude 2.0023 (the difference from 2.00 is of little practical significance) and $\sqrt{s(s + 1)}$ gives the magnitude of the spin angular momentum. The Bohr magneton, β_e, is the only dimensioned quantity on the right side; its value is 9.27×10^{-24} joules/tesla. Therefore, μ_s has these units. The proper name given to the g factor is **magnetogyric ratio**,[§] a name to remind us that g_s is defined as the ratio of the magnetic moment to the angular momentum.

For an ion with n unpaired electrons, you can make use of the fact that the total spin angular momentum of the ion is $n/2$ to derive the general spin-only magnetic moment:

$$\mu_s = g_s \beta_e \sqrt{n(n + 2)/4} = \beta_e \sqrt{n(n + 2)}$$

[†] See p. 33.
[§] Also called the Landé g factor.

The experimental determination of the magnetic moment of an electron or electrons *in a molecule* yields a result consisting of contributions from both the spin and orbital moments, the *orbital motion* being the second way an electron acquires magnetism. For purposes of generality and simplicity in reporting results, the magnetogyric ratio in such cases can still be defined in terms of the ratio of total, observed moment to the spin (*not* total) angular momentum.

$$g = \frac{\mu_{obs}}{\beta_e \sqrt{s(s+1)}} = \frac{2\,\mu_{eff}}{\sqrt{n(n+2)}}$$

The last equality introduces the definition of the **Bohr magneton number**, $\mu_{eff} \equiv \mu_{obs}/\beta_e$. This is a dimensionless number frequently used in the research literature. Through this equation, the orbital contribution to the magnetic moment appears as an alteration of the spin-only factor, g_s. Thus, g becomes an important property of the molecule, in contrast to g_s, which is one of the universal constants and, therefore, is unrelated to molecular structure and the orbital motion of electrons.

Table 9-5 presents some typical g value ranges for the first row transition ion complexes so that you may see how significant are the orbital contributions. Note that if you subtract 1 from μ_{eff} and round to the nearest integer, you arrive at the number of unpaired electrons. This works well for both spin-only and observed Bohr magneton numbers (there are a very few exceptions for which the orbital contribution is large; see study question 25).

The theory for explaining these deviations has been satisfactorily developed but is sufficiently involved as to be beyond the level of this text. Fortunately, the deviations of g from the spin-only value are usually sufficiently small that the measurement of the magnetic moment of a complex reveals the number of unpaired electrons present. Magnetic measurements can, for example, help one determine whether a given complex is to be considered high- or low-spin octahedral. In at least one case (Ni^{2+}) the range of

TABLE 9-5

RANGES OF g VALUES FOR O_h AND T_d COMPLEXES

Ion	Configuration	g	$\mu_{eff} = \mu_s/\beta_e$ (spin only)
O_h			
Ti^{3+}, V^{4+}	d^1	1.85–2.08 ⎫	
Cu^{2+}	d^9	1.96–2.54 ⎭	1.7
V^{3+}, Cr^{4+}	d^2	1.91–2.05 ⎫	
Ni^{2+}	d^8	2.05–2.33 ⎭	2.8
Cr^{3+}, Mn^{4+}	d^3	1.91–2.06 ⎫	
Co^{2+}	d^7	2.22–2.68 ⎭	3.9
Co^{2+}, Ni^{3+}	d^7 (LS)	2.08–2.31	1.7
Cr^{2+}, Mn^{3+}	d^4	1.92–2.04 ⎫	
Fe^{2+}	d^6	2.08–2.33 ⎭	4.9
Cr^{2+}, Mn^{3+}	d^4 (LS)	2.26–2.33	2.8
Mn^{2+}, Fe^{3+}	d^5	1.89–2.06	5.9
Mn^{2+}, Fe^{3+}	d^5 (LS)	2.08–2.89	1.7
T_d			
Cr^{5+}, Mn^{6+}	d^1	1.96–2.08	1.7
Cr^{4+}, Mn^{5+}	d^2	1.84–1.98 ⎫	
Ni^{2+}	d^8	2.61–2.83 ⎭	2.8
Fe^{5+}	d^3	1.86–1.91 ⎫	
Co^{2+}	d^7	2.16–2.47 ⎭	3.9
Fe^{2+}	d^6	2.16–2.24	4.9
Mn^{2+}	d^5	1.99–2.09	5.9

g values helps to distinguish O_h from T_d complexes, and to some extent this is true of Co^{2+} complexes also.

Let us now sharpen somewhat our understanding of the measurement of magnetic moments, and in the process achieve some understanding of the origin of $g \neq g_s$. The concept of a molecular magnetic moment implicitly assumes that each molecule possesses a tiny bar magnet whose magnetic field strength is derived from the orbital and spin motions of the electrons in the molecule. Only unpaired electrons can contribute spin magnetism because paired electrons contribute offsetting spin magnets. Similarly, completely filled orbital shells contribute no orbital magnetism because for every electron revolving about the nucleus in one direction ($m_\ell = +n$) there is a partner revolving in the opposite sense ($m_\ell = -n$). For transition metal complexes it is the unfilled d^* mo's that are responsible for the presence of the molecular magnet.

For a sample of the complex in free space, thermal agitation causes complete disordering of these magnets; but in a magnetic field this chaos is partially resolved as the molecular magnets prefer to align their lines of force with those of the laboratory magnet.

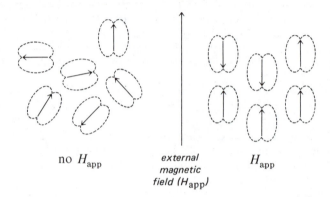

no H_{app} *external magnetic field (H_{app})* H_{app}

The extent to which this alignment occurs depends on the energy of the coupling between the laboratory and molecular magnets, relative to the thermal energy available. Consequently, the sample achieves a net magnetic polarization, in the field of the laboratory magnet, in proportion to the strength of the latter and to the strength of the molecular magnet, and inversely related to the temperature. The fractional increase in field strength through the sample itself is called the **magnetic susceptibility**, χ, a dimensionless quantity. The greater this susceptibility, the greater the interaction between the molecular and laboratory magnets. This provides a convenient experimental method for determining the susceptibility. If the sample were uniformly within a region of constant field strength, there would be no net force acting on the sample. On the other hand, if one end of the sample were in a region of high field strength and the other end in a region of low field strength, there would be a net force acting, which draws the sample into (paramagnetism) or out of (diamagnetism) the region of greater strength. An experimental arrangement such as that shown in Figure 9-9 illustrates the so-called Gouy technique.

Quite clearly, the susceptibility is concentration dependent; a low concentration of molecular magnets in the sample will result in a low susceptibility, all other factors being equal. Thus, the susceptibility χ is identified as a **volume susceptibility**. A **gram susceptibility**, χ_g, is obtained from χ by dividing the latter by the density of the sample. Finally, a **molar susceptibility**, χ_M, is defined by multiplying χ by the molecular weight.

$$\chi_M = \chi \left(\frac{\text{MW}}{d} \right) \text{m}^3 \text{ mole}^{-1}$$

where **MW** has units of kg mole^{-1} and d has units of kg m^{-3}, χ_M is a macroscopic property of a bulk sample, and some connection between it and the molecular magnetic moment, μ, is needed.

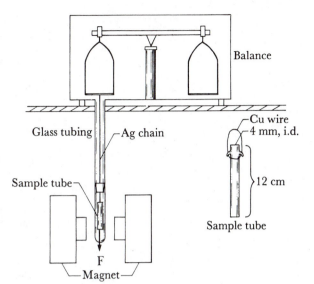

Figure 9-9. The Gouy apparatus and the force acting on a net paramagnetic sample. [From R. J. Angelici, "Synthesis and Technique in Inorganic Chemistry," 2nd ed., W. B. Saunders Company, Philadelphia, 1977.]

Before looking into this relationship between χ_M and μ, you must realize that every molecule contains a great number of paired electrons that do not, by arguments given earlier, contribute to μ but do affect the magnitude of χ_M from which μ is to be determined. This contribution from paired electrons *opposes* that from the unpaired electrons and so distinguishes the *diamagnetic* behavior of the *paired electrons* in the molecule from the *paramagnetic* property of *unpaired electrons*. The application of a magnetic field induces a circulation of all the electrons about the field direction, setting up, in loose parlance, a weak intramolecular electric current (note the analogy with an electric current generator). This induced current itself generates a magnetic field in opposition to the applied field (recall the right- and left-hand rules from your physics course). Thus, the field applied to a sample *induces* a molecular diamagnetism in opposition to the paramagnetism of the unpaired electrons. This, in effect, reduces the susceptibility of the sample below that due to the paramagnetism alone. Fortunately, the molecular diamagnetism is usually very much smaller than the paramagnetism, but pesky errors accrue in large molecules if its presence is neglected. Generally, it is satisfactory to assume the additivity of nominal atomic diamagnetic susceptibilities, and tables of these have been published for convenience. After correcting the measured χ_M for the molecular diamagnetic susceptibility, χ_M^P may be used to find g.

$$\chi_M^P = \chi_M + \left| \chi_M^D \right|$$

A complete, general, theoretical treatment[18] of the relation between χ_M and μ is not appropriate at this level. In the ideal case of non-coupled molecular magnets, the relationship is

$$\mu_{\text{eff}} = 798 \sqrt{\chi_M^P \cdot T}$$

where T is absolute temperature. The form of this equation ($\chi = C/T$) is that determined empirically by Curie and is known as the **Curie law.** For most paramagnetic materials, a plot of χ *vs.* T^{-1} yields a straight line, in agreement with the theory.

23. The magnetic susceptibility of a sample of $Ni(PPh_3)_2Cl_2$ was found to be $0.0756 \times 10^{-9} \ m^3 \ mole^{-1}$. Is the complex square planar or "tetrahedral"? The diamagnetic correction factors (in units of $10^{-9} \ m^3 \ mole^{-1}$) for the atoms are $Ni(-0.161)$, $P(-0.330)$, $Cl^-(-0.294)$, $H(-0.0368)$, and $C(-0.0754)$. What is the value of χ_M^P in unrationalized cgs units (see inside front cover)?

STUDY QUESTIONS

24. The complex (solid) $Hg[Co(SCN)_4]$ is frequently used to calibrate a Gouy susceptibility apparatus because its magnetic moment is high (4.5 Bohr magnetons). Can you eliminate any of the following as possibilities for the anion: T_d(high spin), T_d(low spin), D_{4h}(high spin), D_{4h}(low spin)?

25. Characterize the following complexes as high or low spin:

	μ_{eff}		μ_{eff}
$Co(NH_3)_6^{2+}$	5.0	$Pt(NH_3)_2Cl_2$	0.0
$Co(NO_2)_6^{4-}$	1.9	$Fe(CN)_6^{3-}$	2.3
CoF_6^{3-}	5.3	$Fe(OH_2)_6^{2+}$	5.3

Grouping these complexes into two categories based on whether their moments deviate from the spin-only value by more or less than 0.3 B.M., examine the $t*$ occupations and draw a conclusion.

26. For each complex below, decide whether it is high or low spin and calculate the spin-only magnetic moments:

RuF_6^{3-} (O_h) $Cu(NH_3)_4(OH_2)_2^{2+}$ (D_{4h}) $Fe(NO)(CN)_5^{2-}$ $(C_{4v})*$

$Pd(CN)_4^{2-}$ (D_{4h}) $Ru(CO)_5$ $(D_{3h})*$ $AuCl_4^-$ (D_{4h})

$NiCl_4^{2-}$ (D_{2d}) $Co(NH_3)_3(OH_2)_3^{3+}$ (C_{3v}) AlF_6^{3-} (O_h)

*CO and NO are strong π acceptor ligands.

27. Because of their close similarity to the naturally occurring vitamin B_{12} series of compounds, the "cobaloximes" have received considerable attention in recent years. Cyanobis(dimethylgloximato)(pyridine)cobalt is an example of such model complexes.

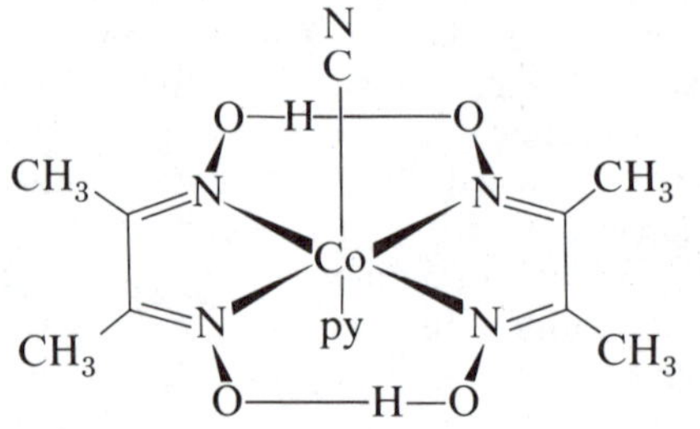

glyoxime =

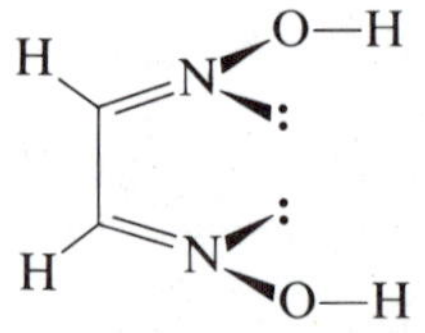

a) Predict a magnetic moment for this "cobaloxime" (the cobalt oxidation state is determined from the fact that the complex bears no net charge).
b) Of greater similarity to the natural cobalamins are the cobaloximes with CH_3^- in place of CN^-. Related to the role of the cobalamins as biological electron transfer reagents, methylcobaloximes experience irreversible chemical and polarographic reduction with loss of the CH_3 group. It is believed that the precursor to the electron transfer is the square pyramidal $Co(DMG)_2CH_3$. Identify the LUMO for this complex and, from its metal-ligand characteristics, explain how the loss of CH_3 is facilitated by the reduction of Co(III).

28. The methyl cobaloximes are among the most stable organo-transition metal compounds known, but they are photochemically reactive. Exposure to light leads to reactions via formation of CH_3 radicals. Suggest an excited electronic configuration

that could explain this property of the methyl(pyridine)cobaloximes (though formally octahedral, these complexes do present different ligand environments along the (x, y) and z axes).

CHECKPOINTS

1. The most common transition metal complex structures are
   a) MD_6, octahedral
   b) MD_4, tetrahedral
   c) MD_4, square planar
2. The stoichoimetry is determined by
   a) ligand steric bulk/VSEPR forces (favors low coordination number)
   b) the strength of the metal-ligand bond (favors high coordination number)
3. The stereochemistry is determined by the balance between ligand bulk/VSEPR forces and the destabilizing influence of the valence d electrons.
4. The angular overlap model is a simplified molecular orbital model for examining the antibonding nature of the d electrons and their influence on the properties of molecular
   a) structure (structure preference energy)
   b) color (electronic states from electronic configuration and e^-/e^- repulsion energy)
   c) paramagnetism (high spin/low spin)
5. The removal of the degeneracy of the five valence d orbitals leads to the concept of high and low spin complexes.
6. The energy splitting of the d orbitals reflects the sigma and pi bonding of a metal to its surrounding ligands and is intimately connected to the colors of complexes.

REFERENCES

[1] T. J. Kealy and P. L. Pauson, *Nature*, **168**, 1039 (1951); S. A. Miller, J. A. Tebboth, and J. F. Tremaine, *J. Chem. Soc.*, 632 (1952).
[2] L. Vaska and J. W. DiLuzio, *J. Amer. Chem. Soc.*, **83**, 2784 (1961).
[3] C. A. Tolman, *Science,* **181**, 501 (1973); *Chem. Soc. Revs.*, **1**, 337 (1972); R. Heck, "Organotransition Metal Chemistry; A Mechanistic Approach," Academic Press, New York, 1974.
[4] M. C. Baird, *Prog. Inorg. Chem.*, **9**, 1–159 (1968); F. A. Cotton, *Acc. Chem. Res.*, **2**, 240 (1969).
[5] J. M. Pratt, "Inorganic Chemistry of Vitamin B_{12}," Academic Press, New York, 1971; H. A. O. Hill, J. M. Pratt, and R. J. P. Williams, *Chemistry in Britain*, **5**, 156 (1969); G. N. Schrauzer, *Acc. Chem. Res.*, **1**, 97 (1968).
[6] J. N. Van Niekirk, F. R. L. Schoening, and J. F. de Wett, *Acta Cryst.*, **6**, 720 (1953).
[7] F. A. Cotton, B. G. DeBoer, M. D. LaPrade, J. R. Pipal, and D. A. Ucko, *Acta Cryst.,* **Sect. B, 27**, 1664 (1971).
[8] L. Brewer, *Science*, **161**, 115 (1968).
[9] A. F. Wells, "Structural Inorganic Chemistry," Third Edition, Oxford University Press, New York, 1962, p. 983. Remember that problems were encountered in Chapter 6 when we discussed the definition of ionic radii. Much the same difficulties arise here.
[10] C. E. Moore, "Ionization Potentials and Ionization Limits Derived From the Analyses of Optical Spectra," NSRDS-NBS 34, National Bureau of Standards, Washington, D.C. (1970).
[11] The references which can be cited for the AOM all presuppose at least some familiarity on the part of the reader with the mathematical machinery of LCAO-MO theory. Probably the least demanding in this regard is E. Larsen and G. N. LaMar, *J. Chem. Educ.*, **51**, 633 (1974), which may be consulted for references to C. K. Jørgensen's and C. E. Schäffer's writings in the research literature. Unfortunately, the *J. Chem. Educ.* reference has typographical errors on page 635. *Caveat emptor!*
[12] L. E. Orgel, *J. Chem. Phys.*, **23**, 1819 (1955); J. S. Griffith, *J. Inorg. Nucl. Chem.*, **2**, 1, 229 (1956).
[13] T. M. Dunn, *et al.*, "Some Aspects of Crystal Field Theory," Harper and Row, New York, 1965.
[14] A. B. Blake and F. A. Cotton, *Inorg. Chem.*, **3**, 9 (1964).
[15] Lattice and hydration energies from P. George and D. S. McClure, *Prog. Inorg. Chem.* (F. A. Cotton, Ed.), **1**, 381 (1959); radii from R. D. Shannon and C. T. Prewitt, *Acta Cryst.*, **B25**, 925 (1969), **B26**, 1076 (1970).
[16] H. A. Jahn and E. Teller, *Proc. Roy. Soc.*, **A161**, 220 (1937); H. A. Jahn, *ibid.*, **A164**, 117 (1938).
[17] K. F. Purcell and J. C. Kotz, "Inorganic Chemistry," W. B. Saunders Co., Philadelphia, 1977, p. 556.
[18] See, for example, S. F. A. Kettle, "Coordination Compounds," Nelson, London, 1969; B. N. Figgis, "Introduction to Ligand Fields," Interscience, New York, 1966.

Appendix to Chapter 9

State Energy Diagrams for MD$_6$ (O$_h$)

Where appropriate, low spin states are distinguished by dotted lines. $E' \equiv E/B$; $\Delta_o' \equiv \Delta_o/B$; B is a unit of electron repulsion energy which appears in the equations for the state energies.

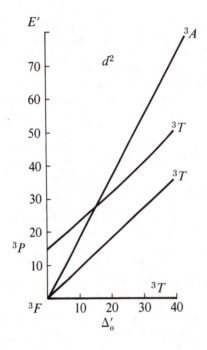

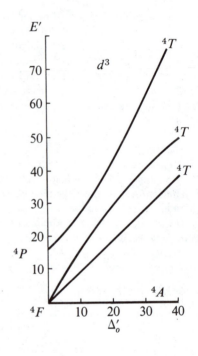

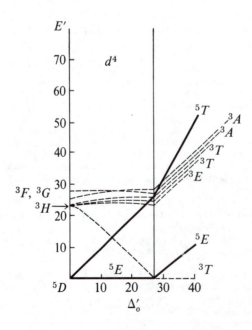

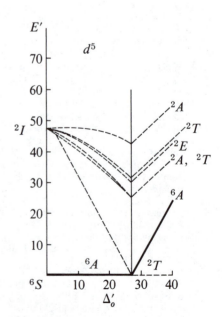

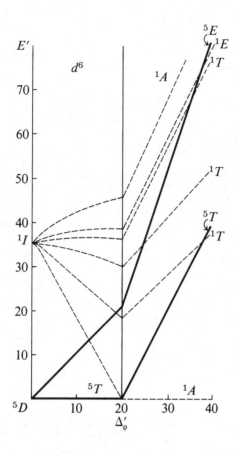

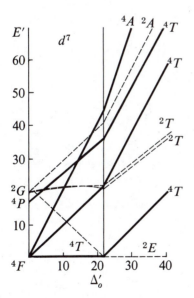

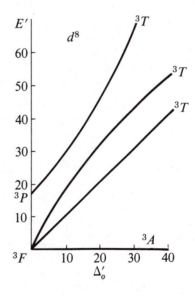

10. COORDINATION CHEMISTRY: STRUCTURAL ASPECTS

The transitions metals may form complexes with coordination numbers
from two through nine, and even higher coordination numbers are accessible if the metal ion involved is a lanthanide or actinide. In the preceding
chapter you explored the conditions under which a metal complex may be
six-coordinate and octahedral or four-coordinate and either square planar
or tetrahedral. In this chapter we examine in a more general way the
several stereochemical arrangements found for coordination numbers
from two through six and a few of the topologies found for higher coordination numbers. The application of the angular overlap model should
allow you to systematize to a great extent the structural chemistry of the
transition metals in their higher valence states.

GENERAL CONSIDERATIONS

Since metal compounds are characterized by a rich variety of stereochemical arrangements, the lack of a simple, consistent theory for the rationalization and prediction of
structures has heretofore been a serious handicap. However, in Chapter 9 we showed that
the stereochemistry of transition metal compounds is mainly determined by two factors:
(i) the number of $d\sigma^*$ electrons and (ii) ligand-ligand repulsion forces.† That is, if several
structures are possible for a given complex, the one adopted is that which minimizes the
role of the $d\sigma^*$ anti-bonding electrons; if this does not lead to a clear-cut structural
preference, the structure adopted is the one that minimizes repulsions between ligands.

A further structural constraint sometimes observed is that of ligand rigidity, wherein
the intrinsic structure of a ligand may be such that the donor atoms have definite positions and so define a "pocket" into which the metal atom or ion must fit. Such characteristics are peculiar to each ligand and will be explored as their complexes come into our
discussions. For the present we shall examine only the constraints that metal ions place
on the stereochemistry of complex ions.

Figure 10–1 is a series of plots of structural preference energies for the limiting two-,
three-, four-, five-, and six-coordinate geometries to be discussed. As demonstrated in
Chapter 9, these plots are constructed on the basis of the angular overlap model, and they
consider metal-ligand σ interactions only, since these generally dominate the selection of
one structure in preference to another. What we wish to do in this chapter is to compare
these theoretical predictions with real molecules to see how good our predictions are and
to attempt to lay down a set of general guidelines so that you are able to predict the gross
structure of any transition metal complex.

---

†For the sake of convenience, we shall refer to ligand-ligand and bond pair-bond pair repulsion
forces collectively as VSEPR forces. You should be aware that they are not strictly the same, however.
As used in earlier sections of the book, VSEPR forces arise from repulsions between stereochemically
active electron pairs, not atoms or groups of atoms.

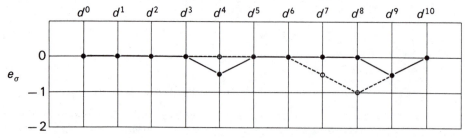

(A) Two-coordination. E(linear) − E(bent). Linear geometry favored over bent.

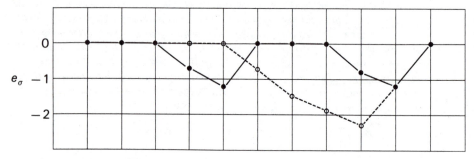

(B) Three-coordination. $E(\top) − E(\perp)$. $\top (C_{2v})$ favored over $\perp (D_{3h})$.

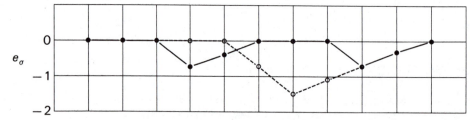

(C) Three-coordination. $E(\perp) − E(\perp)$. $\perp (C_{3v})$ favored over $\perp (D_{3h})$.

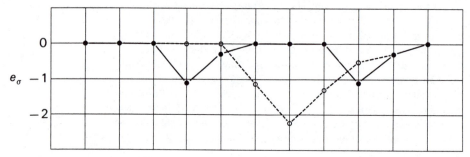

(D) Five-coordination. $E(\bigstar) − E(\divideontimes)$. $\bigstar (C_{4v})$ favored over $\divideontimes (D_{3h})$.

Figure 10-1. Structure preference energy plots for coordination numbers two through six. See Figure 9–4 for the corresponding plots for four-coordinate complexes. Plot (F) is the energy of the geometry favored by the angular overlap model relative to the octahedron. All are plotted in units of e_σ. In all cases, high spin is shown by the solid line and low spin by the dashed line.

Illustration continued.

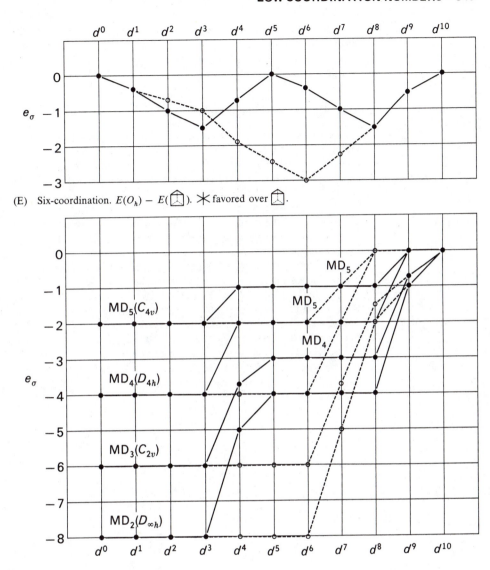

(E) Six-coordination. $E(O_h) - E(\text{⬠})$. ✳ favored over ⬠.

(F) $E(MD_6) - E(MD_x)$, where MD_x has the symmetry preferred on the basis of the angular overlap model. Note that MD_6 is favored over MD_x, $x < 6$.

Figure 10–1. *Continued*

LOW COORDINATION NUMBERS

Two-Coordinate Complexes

When the atoms of a given complex can describe different polyhedra or polygons, isomers can result; these are given the special name **polytopal isomers.**[1] For two-coordinate complexes the appropriate polytopal isomers (that is, the limiting geometries) are linear and 90° bent. VSEPR and steric forces favor the linear geometry in all cases, and the $d\sigma^*$ occupation forces never favor the bent geometry (Figure 10-1(A)). Therefore, you would predict that MD_2 complexes will generally be found to be linear (but ligand structural features requiring a bent geometry will be largely unopposed by the metal). Further, you would predict from Figure 10-1(F) that two coordinate complexes should be very rare for all but d^9 and d^{10} metal ions.

These predictions with regard to two-coordinate complexes have been experimentally confirmed. With perhaps only one or two exceptions, all known two-coordinate complexes are linear and are based on d^{10} species; for example, $[Ag(NH_3)_2]^+$, the Au(I) complex $[Au(PPh_3)_2]I$, and the Pd(0) complex $Pd[PPh(t\text{-}Bu)_2]_2$.[2]

Three-Coordinate Complexes[3]

Three different geometries have been considered for three-coordinate complexes: trigonal planar (D_{3h}), T-shaped (C_{2v}), and 90°-pyramidal (C_{3v}) [Figures 10–1(B) and (C)]. (The last is a simple variant of the D_{3h} structure wherein the metal is out of the plane of the donor atoms.) Generally speaking, the D_{3h} geometry is *not* favored by $d\sigma^*$ occupation forces, but it is favored by VSEPR/steric forces; thus, competition between these two opposing factors will be important. On the basis of $d\sigma^*$ forces, the most pronounced discrimination between the trigonal planar geometry on the one hand and the pyramidal or T-shaped geometries on the other hand is found for low-spin d^6 to d^9 ions. Pyramidal and T-shaped complexes are equally likely for d^6; otherwise, the T shape is preferred. Among high-spin complexes, the T structure preference is not large, but it reaches a maximum for d^4 and d^9 configurations.

Again as predicted by Figure 10–1(F), it is the metal ions with larger numbers of d electrons that have the best chance of forming three-coordinate complexes. The AOM predicts that a T-shaped complex should be most likely formed by a low-spin d^8 ion, and the first such complex was recently isolated: $[Rh(PPh_3)_3]^+$.[4] However, as suggested by Figure 10–1(B), most three-coordinate complexes are trigonal and planar, indicating that VSEPR/steric forces compete favorably with $d\sigma^*$ forces in most cases. A particularly rich source of trigonal planar complexes has been d^{10} Cu(I), and compound **1** is a typical example.

1 $[Cu(SPMe_3)_3]^+$

The ligand $[N(SiMe_3)_2]^-$ has been used to synthesize three-coordinate complexes of nearly all of the first row transition metals.[5] For example, $CrCl_3$ and $Li[N(SiMe_3)_2]$ give $Cr[N(SiMe_3)_2]_3$. Even though e_σ for an amide anion should be quite large, the fact that only three amide ligands bind the metal is a clue to important steric interactions between the ligands. That the ML_3 complexes are high spin suggests a lower e_σ than expected from ligand basicity alone. For the d^3 case of Cr^{3+}, the T structure preference is weak ($<e_\sigma$) and a trigonal structure is expected and found.

The three-coordinate complexes of the $[N(SiMe_3)_2]^-$ ligand are almost all colored. For example, $Ti[N(SiMe_3)_2]_3$ is bright blue and $Cr[N(SiMe_3)_2]_3$ is bright green. Their absorption spectra can be interpreted with the angular overlap model, but this is left to you as a study question.

Four-Coordinate Complexes

Tetrahedral and square planar compounds are polytopal isomers (the limiting geometries for four-coordinate complexes). In Chapter 9 (page 322; see Figure 9-4) we concluded that the formation of tetrahedral (T_d) or distorted tetrahedral (D_{2d}) complexes was likely for metal ions or atoms with more than two d electrons only when the ligands interact weakly with the metal; this weak interaction may be due to intrinsically weak M-D bonds and/or unusual ligand steric requirements. More basic or less hindered ligands will prefer to bind the metal with a planar geometry. Since the latter are the types of ligands perhaps most often used, square planar complexes are more commonly observed than tetrahedral or distorted tetrahedral ones.

TETRAHEDRAL COMPLEXES

Generally speaking, only neutral or anionic complexes have tetrahedral or distorted tetrahedral (D_{2d}) geometries. Most of those which are known are formed with weakly basic ligands interacting with elements of the first row of the transition metal series, particularly those of Group VIII. [Indeed, more tetrahedral complexes are known for Co(II) than for any other ion.] The combination of anion charge (ligand repulsion) and low basicity, of course, produces low e_σ values, a situation favorable to the tetrahedron. As noted on page 323, halide complexes of Fe(III) and Cu(II) (e.g., $[FeCl_4]^-$ and $[CuBr_4]^{2-}$, 2^6) are good examples of such behavior.

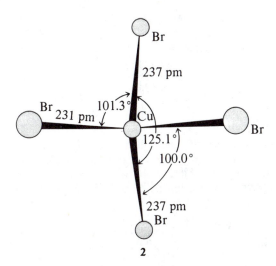

2

The D_{2d} geometry of $[CuX_4]^{2-}$ is expected on the basis of the angular overlap model (Chapter 9, page 322). In fact, more complete molecular orbital calculations for $[CuCl_4]^{2-}$ show that the flattened tetrahedron (D_{2d}) is about 8 kJ more stable than the T_d geometry (which is in turn about 75 kJ more stable than the planar D_{4h} configuration).[7] As a consequence, the distorted tetrahedron is often observed for four-coordinate Cu(II).

Just as you saw with three-coordinate complexes, ligand steric effects play a very important role in determining the geometry of four-coordinate complexes. For example, salicylaldimine complexes of Ni(II) and Cu(II) of the type shown as structure 3 are planar when the substituent R is an n-propyl group.[8]

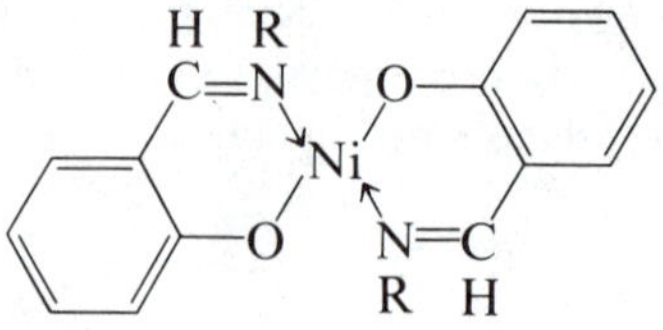

3

However, when R is an isopropyl group, there is apparently an equilibrium in solution between the *trans*-planar form and tetrahedral geometry. The magnetic moments of the Ni(II) complexes vary from about 1.8 to 2.3 B.M. depending on the ring substituent, corresponding to approximately 30 to 50 per cent tetrahedral form (if it is assumed that the tetrahedral molecule would have a magnetic moment of 3.3 B.M.). When the *t*-butyl group is used, however, steric hindrance forces the ligands considerably more out of plane and results in magnetic moments of at least 3.2 B.M., indicating ~95 per cent tetrahedral form.

SQUARE PLANAR COMPLEXES

Planar complexes have been very important to the development of coordination chemistry. Alfred Werner[9] first postulated their existence in 1893 to explain the existence of isomers of $Pt(NH_3)_2Cl_2$, and the deep red dimethylglyoxime complex of Ni(II) (**4**) was discovered by Chugaev in 1905; the latter compound is still commonly used in the gravimetric analysis of nickel.

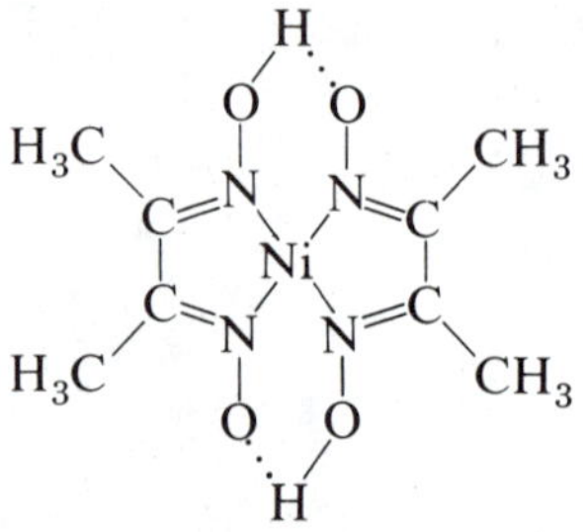

4

The overwhelming majority of square planar complexes are those having a d^8 metal ion, that is, Rh(I), Ir(I), Ni(II), Pd(II), Pt(II), and Au(III). Examples of d^8 square planar complexes include **4** and **5** as well as two complexes of great importance in organometallic chemistry, **6** (Wilkinson's catalyst)[10] and **7** (Vaska's compound).[11]

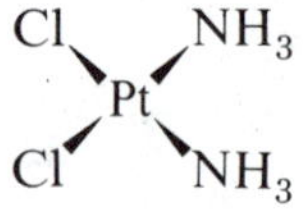

5
cis-diamminedichloro-
platinum

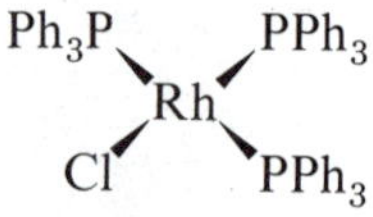

6
chlorotris (tri-
phenylphosphine)-
rhodium

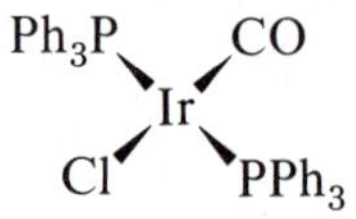

7
trans-carbonylchlorobis (triphenylphosphine) iridium

Five-Coordinate Complexes[12]

One of the most active areas of research in structural chemistry in recent years has concerned five-coordination.[13-15] We have already discussed examples of this in non-metal chemistry (*e.g.*, Chapter 7)—such compounds being most prevalent for Groups VA and VIIA—and now turn to those in transition metal chemistry.

Five-coordinate complexes are found for all of the first row transition metals from titanium to zinc, and occur about as frequently as square planar or tetrahedral four-coordinate complexes. Figure 10-1(F) illustrates the fact that six-coordinate complexes are almost always favored relative to lower coordinate ones (if only σ effects are considered), simply on the basis of thermodynamics. However, five-coordinate complexes are only about one to two e_σ units less favored than six-coordinate octahedral complexes, so there is often a very real competition between five- and six-coordination. We hasten to add, though, that this statement really applies only to the first row metals; five-coordination for second and third row metals is much less common. In part, this may be due to the fact that the heavier metals are characterized by larger e_σ values for ligand-metal interaction, and there is, therefore, a greater difference in energy between five- and six-coordination. Yet another reason for the preference of heavier metals for higher coordination may be simply their size relative to the first row metals; the heavier metals are larger, and six ligands can be accommodated with lower ligand-ligand repulsion.

Two limiting structures for five-coordinate complexes are considered in Figure 10-1 (D): the trigonal bipyramid (D_{3h}) and the square-based pyramid (C_{4v}). The latter is rarely found in non-transition metal compounds, whereas both geometries exist with about equal frequency for transition metal compounds. Indeed, the angular overlap model predicts that the square-based pyramid is significantly favored over the trigonal bipyramid for d^3 and d^8 high spin complexes or d^5 to d^7 low spin complexes. Otherwise, VSEPR and/or steric effects will generally favor the trigonal bipyramid.

In practice, the geometry actually observed for five-coordinate complexes is often intermediate between the two limiting structures. The reason for this is readily seen on viewing the polyhedra down their C_2 and C_4 axes as shown in Figure 10-2; a small pivoting of the basal bonds converts the square pyramid into a trigonal bipyramid. Indeed, the square pyramid is simply an intermediate state in the interchange of axial and equatorial groups in the trigonal bipyramid. This rearrangement—the Berry pseudo-rotation (Figure 10-3)—gives rise to the stereochemical non-rigidity of five-coordinate complexes, a phenomenon found important in the chemistry of silicon and phosphorus compounds (Chapter 7).

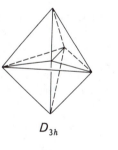
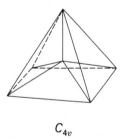

D_{3h} C_{4v}

(A)

Figure 10-2. Views of the two idealized five-coordination polyhedra; (A) conventional perspective framework diagrams. (B) comparison of the square pyramid viewed along the four-fold axis and the trigonal bipyramid viewed along one of its twofold axes.

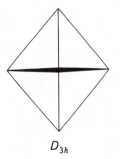
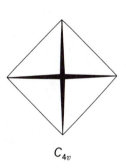

D_{3h} C_{4v}

(B)

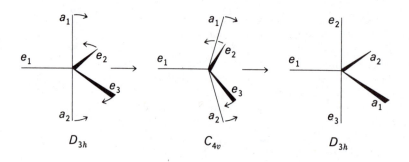

Figure 10-3. Schematic representation of the Berry intramolecular rearrangement for the trigonal bipyramid (also known as the pseudorotation process) involving a square pyramidal intermediate.

An interesting illustration of the fact that the trigonal bipyramid and square pyramid are energetically close to one another is found in the $[Ni(CN)_5]^{3-}$ anion. In fact, this complex is also an example of the importance of the size and charge of the counter-ion in the isolation of complexes. It has been observed that $[Ni(CN)_5]^{3-}$ cannot be isolated from aqueous solution at room temperature with K^+ as a counter-ion, but addition of $[Cr(en)_3]^{3+}$ does lead to precipitation of the anion as a stable salt, $[Cr(en)_3][Ni(CN)_5] \cdot 1.5 H_2O$.[16,17]

$$[Cr(en)_3][Ni(CN)_5] \text{ stable solid}$$

$$\Big\uparrow {\scriptstyle [Cr(en)_3]^{3+}}$$

$$[Ni(CN)_4]^{2-} + CN^- \rightleftharpoons [Ni(CN)_5]^{3-} \xrightarrow[K^+,\ conc.]{} K_2Ni(CN)_4 + KCN$$

$$-15° \Big\downarrow K^+$$

$$K_3Ni(CN)_5 \xrightarrow[room\ temp.]{}$$

The interesting feature of this salt is that there are two independent $[Ni(CN)_5]^{3-}$ ions per unit cell, one of which is a distorted trigonal bipyramid (**8**) and the other a distorted square pyramid (**9**). This should not be too surprising, however, as the angular overlap model predicts that the square-based pyramid is favored over the trigonal bipyramid by only about 0.5 e_σ unit. Experience shows that 0.5 e_σ is not decisive when in competition with VSEPR forces, and the structure may in fact be determined by crystal lattice forces.

8 $[Ni(CN)_5]^{3-}$, trigonal bipyramid

9 $[Ni(CN)_5]^{3-}$, square pyramid

Six-Coordinate Complexes

There are several possible geometries for six-coordinate complexes: planar hexagonal, trigonal prismatic, and octahedral, among others. One of Alfred Werner's greatest contributions to inorganic stereochemistry was his finding that octahedral geometry was adopted by the six-coordinate complexes known at that time.[9] Proof was provided by the observation that, when three bidentate ligands are coordinated to a metal center, two optically active enantiomers may be isolated (10) (see the discussion of stereoisomerism in Chapter 11). Had the geometry been planar hexagonal, no isomers would have been found (11) (there is an improper rotation-reflection axis, S_1, the existence of which precludes generation of enantiomers), while only two diastereomers (optically inactive) would be observed for the trigonal prism (12).

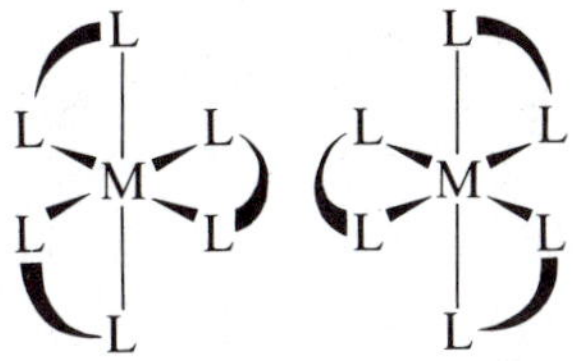

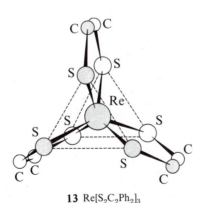

10    11 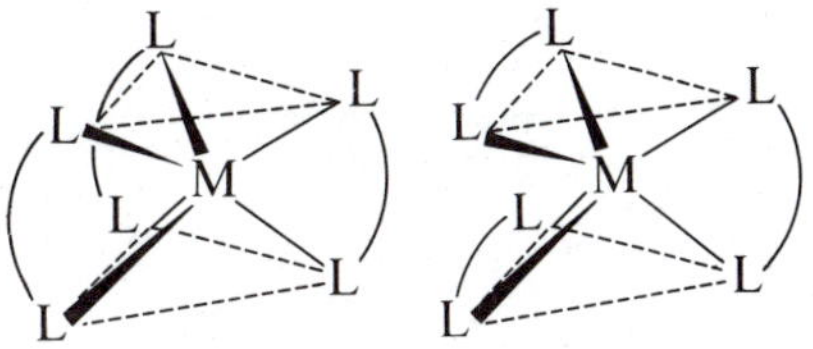   12

Therefore, six-coordinate complexes have always been assumed to be octahedral since Werner's work. In 1965 this long-held truism was invalidated: a trigonal prismatic rhenium compound, $Re[S_2C_2Ph_2]_3$ (13), was reported.[18] The quadrilateral sides of the

13 $Re[S_2C_2Ph_2]_3$

trigonal prism are almost perfectly square, there being little difference between intra- and interligand S—S distances (303 and 305 pm, respectively); further, the top and bottom of the prism are nearly eclipsed.

A trigonal prism may, of course, be formed from an octahedron by twisting one face relative to the other (Figure 10-4). Indeed, Bailar has suggested this as a mechanism for the racemization of tris-bidentate complexes (this phenomenon is discussed in Chapter 14).[19] Therefore, it is not surprising that, with continuing work on trigonal prismatic

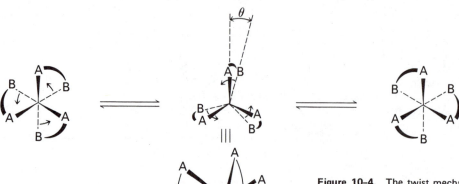

Figure 10–4. The twist mechanism for the racemization of tris-bidentate octahedral complexes, and illustration of the twist angle θ.

trigonal prismatic
intermediate

geometry, many additional examples have been found with the twist angle, θ, varying from the pure trigonal prism (TP, $\theta = 0°$) to the trigonal antiprism or octahedron (TAP or O_h, $\theta = 60°$).[20-22]

The geometry of the ligand is apparently one of the most critical factors in the formation of trigonal prismatic complexes. According to Wentworth, "a ligand is required which will maximize all bonding interactions and minimize all non-bonded contacts when the donor atoms are in TP array."[21] This has been achieved by rigid hexadentate ligands such as that (abbreviated PccBF) illustrated in compound **14**.

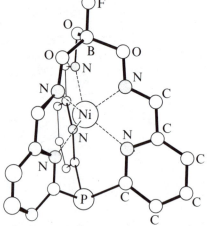

14 trigonal prismatic coordination in [fluoroborotris(2-aldoximo-6-pyridyl)phosphine]nickel(1+)

As with other coordination numbers and their polytopes, the nature of the metal ion is also clearly important in determining whether a complex will be octahedral or trigonal prismatic, and the angular overlap model can be used to analyze structural preference.

Figure 10-1(E) suggests that the octahedron is the preferred geometry for all configurations except d^0, high-spin d^5, and d^{10}; moreover, the preference is most decided for low-spin d^6 complexes. This is an especially interesting result in terms of complexes of the type $M(PccBF)^+$ (e.g., **14**) noted above. The Fe(II) complex is of particular interest, because the Fe(II) in this case is d^6 and low-spin, a configuration predicted to show maximum favoritism toward the octahedron relative to the trigonal prism. The Fe(II) complex is the only one involving the PccBF ligand that has a twist angle ($\theta = 22°$) significantly greater than zero; that is, the arrangement of ligating atoms about the metal ion is about one third of the way toward the octahedron. The other complexes of PccBF [Co(II), d^7 (high spin); Ni(II), d^8; Zn(II), d^{10}] all have twist angles of about 0°, and it is these complexes that the angular overlap model predicts will show much less preference for the octahedron.

1. Calculate structure preference energies for the three possible geometries for three-coordinate complexes, and verify Figures 10-1(B) and 10-1(C).

2. Based on the angular overlap model and other considerations such as those outlined in this chapter and in Chapter 9, predict the structure of each of the following molecules or molecular fragments (all are low-spin):

    a) $Fe(CO)_5$
    b) $Re(CO)_5$
    c) $Ni(CO)_3$
    d) $Cr(CO)_3$
    e) $Fe(CO)_3$
    f) $Cr(CO)_4$
    g) $Fe(CO)_4$
    h) $Co[N(SiMe_3)_2]_2$ (magnetic moment = 4.83 B.M.)

3. It was noted that complexes of the type $M[N(SiMe_3)_2]_3$ (M = Ti, V, Cr, and Fe) are all colored and that their spectra could be interpreted using the angular overlap method.

    a) Calculate the energies (in units of e_σ and e_π) of the $d\sigma*$ and $d\pi*$ orbitals using the methods of Chapter 9 (assume the trigonal structure).
    b) Give the symbols for the ground and first excited states for the complexes where M = Ti, V, Cr.
    c) In the case of the titanium complex, how many absorption bands are expected altogether? Give the symbol for any excited state beyond the first ($A \sim$ nondegenerate, $E \sim$ doubly degenerate).
    d) The iron complex has a magnetic moment of 5.94 B.M. Show the electron configuration that can account for this.

4. The addition of $PEtPh_2$ to $NiBr_2$ at $-78°C$ in CS_2 gives a red complex with the formula $(PEtPh_2)_2NiBr_2$, which is converted to a green complex of the same formula on standing at room temperature. The red complex is diamagnetic, but the green complex has a magnetic moment of 3.2 B.M.

    a) Which of these complexes is square planar and which is tetrahedral? State reasons for your choice.
    b) Rationalize the colors of these complexes in view of your structural choice. (Remember that the color seen is complementary to the color absorbed, and that photon energies decrease: blue > green > yellow > red.)

5. $NiCl_2$ and NiI_2 also form complexes with the ligand in problem 4. These complexes have the following characteristics: (i) $(EtPh_2P)_2NiCl_2$ is red and is diamagnetic at all temperatures. (ii) $(EtPh_2P)NiI_2$ is red-brown and is paramagnetic at all temperatures.

a) Which of these complexes is square planar and which is tetrahedral?
b) Attempt to rationalize why the $(EtPh_2P)_2NiX_2$ complexes assume different structures depending on the halide ion, X.

POLYHEDRA OF HIGH COORDINATION NUMBER[23,24]

Complexes of higher coordination number—seven, eight, nine, or more— are increasingly being studied. One reason for the interest in such complexes is that numerous geometries or polytopes are often possible. In the cases of five- and six-coordinate complexes you saw that relatively small motions could convert one geometry into another. It should not be surprising that, as the number of ligands increases, the motions necessary for interconversion of polytopes become even smaller, and the energies of the polytopes come closer together. Thus, for seven- and eight-coordinate complexes, several limiting geometries are accessible, with real molecules usually adopting geometries only approximating the limiting cases.

Before discussing some specific cases of higher coordinate complexes, it is useful to discuss briefly the conditions leading to the formation of such materials. With a given ligand, the ability of a metal ion to achieve a high coordination number appears to depend in part on a subtle balance of charge and ion size. That is, in order to attain optimum bonding conditions, the charge-to-radius ratio must be one that maximizes metal-ligand attraction and minimizes ligand-ligand repulsions. Eight-coordinate complexes, for example, are found predominantly for fifth and sixth period transition metal ions in Groups III to VI in oxidation states +3 to +5; furthermore, eight-coordinate complexes are very common for lanthanide ions in +3 oxidation states. It is, of course, these ions that are large enough to accommodate seven or more ions or molecules in a coordination sphere without severe ligand-ligand repulsions. Furthermore, these relatively highly charged ions exercise an attraction on the ligands that is sufficient to overcome the force of ligand-ligand repulsions.

The 3+, 4+, and 5+ ions of Groups III to VIB have electron configurations of d^0, d^1, or d^2 (a 3+ ion of a Group VI element would of course be d^3, but high-coordinate complexes of such ions are quite uncommon). This implies that d orbital occupation has some influence on the formation of high-coordinate complexes, and that arguments similar to those used with low-coordinate complexes also apply here. That is, low d orbital populations will minimize the σ anti-bonding role of the d electrons. In any complex of coordination number greater than six, there will be at least one d-like molecular orbital that is not strongly anti-bonding; therefore, population by one or two electrons will not seriously affect the σ bonding in the complex.

The nature of the ligand is another important factor to be considered. If the polarizability of an ionic ligand is high, a large amount of charge density is easily transferred to the metal ion, and fewer ligands will be necessary to arrive at electroneutrality. As a consequence, complexes of higher coordination number are formed readily by ligands that are small and/or of low polarizability (e.g., H^-, F^-, OH^-). However, most high-coordinate complexes are formed with chelating ligands having low steric requirements and donor atoms such as nitrogen and oxygen. The β-diketones, oxalate ion, EDTA, tropolonate (15), and nitrilotriacetate (16, NTA) are particularly good examples.

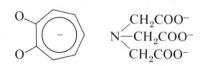

15   16

Seven-Coordinate Complexes[25]

Seven-coordinate complexes present a very complicated situation, and the literature concerning such compounds is often confusing. The three idealized polyhedra usually considered for seven-coordination are the C_{3v} capped octahedron (17), the D_{5h} pentagonal bipyramid (18), and the C_{2v} monocapped trigonal prism (19).[26]

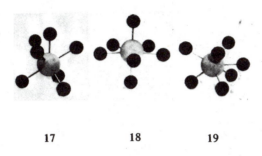

17	**18**	**19**

[From E. L. Muetterties and C. M. Wright, *Quart. Rev.,* **21**, 109 (1967).]

As may be seen in the models above, these polyhedra are very similar to one another, and the difference in energy between them is small. Therefore, which polyhedron a complex adopts will depend on a subtle interplay of metal ion size, metal ion electron configuration, and ligand basicity and geometry.

Complexes that closely resemble each of the idealized polyhedra are known. The heptacyanovanadate(4−) ion, for example, is a slightly distorted pentagonal bipyramid,[27] a geometry that also appears to be the preferred configuration for complexes of the type M(chelate)$_3$X when X is a ligand which forms a relatively strong, covalent M−X bond. Chlorotris(*N,N*-dimethyldithiocarbamato)titanium (20) is an example of this type of complex.[28]

20

Eight-Coordinate Complexes[29]

Eight-coordinate complexes have been observed rather commonly for heavier metal ions of Groups IV, V, and VI in oxidation states of +4 and +5. That is, ions such as Zr(IV), Mo(IV, V) and Re(V, VI) having d^0, d^1, and d^2 electron configurations are often

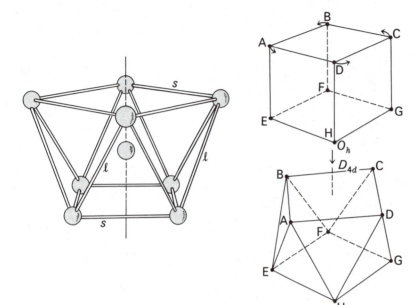

Figure 10-5. The square antiprismatic polyhedron and its mode of formation from the cube. [From S. J. Lippard, *Prog. Inorg. Chem.,* **8**, 109 (1967).]

involved. An explanation for this can, of course, be found in the angular overlap model of metal-ligand binding: the heavier metals are involved because they generally give rise to larger e_σ values owing to stronger M—L bonding, and the metals are from the earlier groups because such metal ions have low $d\sigma^*$ occupancy.

Several idealized geometries are possible for eight-coordinate complexes, the cube being the simplest and most symmetrical of these. However, although cubic geometry is observed in ionic crystal lattices, it has never been observed for molecular polyhedra. One reason for this, at least in the case of transition metals, is that only seven (of the nine) d, s, and p valence orbitals of the metal possess the proper symmetry for bonding. The reason for this can be seen by noting that the cube is equivalent to two interlocked tetrahedra. From Table 9-1 (p. 316) you see that the d_{z^2} and $d_{x^2-y^2}$ orbitals (which would be located at the cube faces) would have zero overlap with ligands at the cube corners. Therefore, only seven metal atomic orbitals [one ns, three np, and three $(n-1)$ d orbitals: d_{xy}, d_{xz}, and d_{yz}] are able to stabilize the eight electron pairs in the ligand σ orbitals. This, of course, does not preclude the formation of cubic complexes by the lanthanides or actinides, as these elements have f orbitals available. In any event, even if the proper orbitals were available, the cube maximizes interligand repulsions relative to the other possible eight-coordinate polyhedra.

The only two polyhedra actually observed for eight-coordinate complexes of the transition metals are simple distortions of the cube, both of which lead to a lessening of interligand repulsions in comparison to those found in the cubic geometry. The D_{4d} square antiprism may be obtained by simply rotating one face of a cube by 45° relative to the opposite face (Figure 10-5), while the D_{2d} dodecahedron may be formed by closing the cube along opposite faces (Figure 10-6); the net result of the latter is that the D_{2d} dodecahedron may be visualized as two interlocking trapezoids.† In both cases, in contrast to the cube, the valence ns, np, and $(n-1)$ d orbitals are all used for σ bonding.[23]

---

†The polyhedra in Figures 10-5 and 10-6 are labeled according to the notation of Hoard and Silverton (ref. 23). In the D_{4d} polyhedron, the sixteen edges are divided equally between two symmetry types, ℓ and s ($\ell = s$ for the idealized polyhedron). In the dodecahedron, on the other hand, there are eighteen edges divided into four classes: two a edges, four b edges, four m edges, and eight g edges ($g = m = a$ and $b = 1.250a$ for the idealized polyhedron).

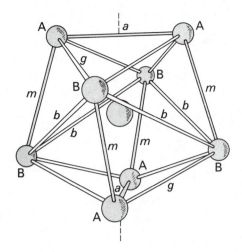

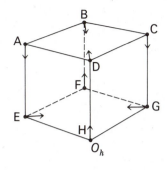

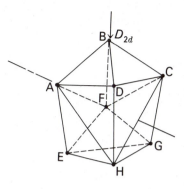

Figure 10-6. The dodecahedron and its mode of formation from the cube. [From S. J. Lippard, *Prog. Inorg. Chem.,* **8**, 109 (1967).]

Hoard has been largely responsible for opening up the field of high coordination number stereochemistry,[23] and it was he who first established D_{2d} dodecahedral geometry for eight-coordinate complexes in his study of $[Mo(CN)_8]^{4-}$. Numerous other complexes have since been found to be dodecahedral [*e.g.,* the tetrakis(oxalato)zirconate-(4−) ion **(21)**],[30] and it has begun to appear that the D_{2d} dodecahedron will be more common than the square antiprism. Among the best confirmed examples of the latter geometry are $[TaF_8]^{3-}$ and $[ReF_8]^{3-}$. Indeed, these octafluorometallates, and others such as $[MoF_8]^{2-}$, are the only square antiprismatic complexes with monodentate ligands.

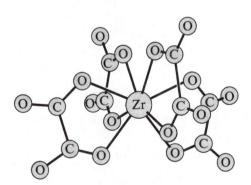

21 tetrakis(oxalato)zirconate(4−)

Complexes Having Coordination Numbers of Nine or Higher[31]

Transition metal ions surrounded by nine ligands or donor atoms are rare; perhaps the best examples of the anions are $[ReH_9]^{2-}$ and $[TcH_9]^{2-}$,[32,33] the structure of the former having been established by x-ray and neutron diffraction techniques. As shown in **22.** the hydride ligands clearly define a symmetrically tricapped trigonal prism. Although there are a number of possible idealized geometries for nine-coordinate complexes,

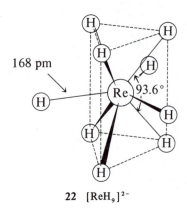

22 $[ReH_9]^{2-}$

it is interesting that only the symmetrically tricapped trigonal prism has been reported for such complexes. Further, it is important to note that, if a transition metal is to be nine-coordinate and all ligand σ orbitals are to be stabilized, then full utilization of the *d*, *s*, and *p* orbitals of the metal valence shell is required; Re(VII) and Tc(VII), with no *d* electrons, meet this criterion and avoid populating the *d*σ* molecular orbitals.

There are a few complexes known with coordination numbers of ten or higher, but they are all based on lanthanide or actinide elements. This is not surprising, since, from the point of view of ligand σ orbital stabilization, it is only these elements that have a sufficient number of valence shell orbitals.

CHECKPOINTS

1. Transition metals may form complexes with coordination numbers of two through nine.

2. Polytopal isomers are possible for any given coordination number. Such isomers are compounds of the same constitution, but in which the ligands arranged about the central metal describe different coordination polygons or polyhedra.

3. The two factors that govern the geometry of coordination complexes are (1) the number of *d*σ* electrons, and (2) ligand-ligand repulsions.

REFERENCES

[1] E. L. Muetterties, *J. Amer. Chem. Soc.,* **91**, 1636 (1969); see correction in **91**, 4943 (1969).

[2] M. Matsumoto, H. Yoshioka, K. Nakatsu, T. Yoshida, and S. Otsuka, *J. Amer. Chem. Soc.,* **96**, 3322 (1974).

[3] P. G. Eller, D. C. Bradley, M. B. Hursthouse, and D. W. Meek, *Coord. Chem. Rev.,* **24**, 1 (1977).

[4] Y. W. Yared, S. L. Miles, R. Bau, and C. A. Reed, *J. Amer. Chem. Soc.,* **99**, 7076 (1977).

[5] E. C. Alyea, D. C. Bradley, and R. G. Copperthwaite, *J. C. S. Dalton,* 1580 (1972); see also D. C. Bradley, *Chem. in Britain,* **11**, 393 (1975); E. C. Alyea, D. C. Bradley, R. G. Copperthwaite, and K. D. Sales, *J. C. S. Dalton,* 185 (1973).

[6] T. I. Li and G. D. Stucky, *Inorg. Chem.,* **12**, 441 (1973).

[7] J. Demuynck, A. Veillard, and U. Wahlgren, *J. Amer. Chem. Soc.,* **95**, 5563 (1973).

[8] R. H. Holm, G. W. Everett, and A. Chakravorty, *Prog. Inorg. Chem.,* **7**, 83 (1966).

[9] A. Werner, *Z. anorg. Chem.,* **3**, 267–330 (1893); this paper and others of Werner's papers of great importance to coordination chemistry have been published in translated and annotated form: G. B. Kauffman, *Classics in Coordination Chemistry,* Dover Publications, Inc., New York, 1968.

[10] D. Evans, G. Yagupsky, and G. Wilkinson, *J. Chem. Soc.* (A), 2660, 2665 (1968).

[11] L. Vaska, *Acc. Chem. Res.,* **1**, 335 (1968).

[12] A. R. Rossi and R. Hoffmann, *Inorg. Chem.,* **14**, 365 (1975).

[13] J. S. Wood, *Prog. Inorg. Chem.,* **16**, 227 (1972).

[14] B. F. Hoskins and F. D. Williams, *Coord. Chem. Rev.,* **9**, 365 (1972–1973).

[15] C. Furlani, *Coord. Chem. Rev.,* **3**, 141 (1968).

[16] K. N. Raymond, P. W. R. Corfield, and J. A. Ibers, *Inorg. Chem.,* **7**, 1362 (1968).

[17] F. Basolo, *Coord. Chem. Rev.,* **3**, 213–223 (1968).

[18] R. Eisenberg and J. A. Ibers, *Inorg. Chem.,* **5**, 411 (1966).

[19] J. C. Bailar, *J. Inorg. Nucl. Chem.,* **8**, 165 (1958).

[20] E. Larsen, G. N. LaMar, B. E. Wagner, J. E. Parks, and R. H. Holm, *Inorg. Chem.,* **11**, 2652 (1972).

[21] R. A. D. Wentworth, *Coord. Chem. Rev.,* **9**, 171 (1972–73); W. O. Gillum, R. A. D. Wentworth, and R. F. Childers, *Inorg. Chem.,* **9**, 1825 (1970).

[22] E. B. Fleischer, A. E. Gebala, D. R. Swift, and P. A. Tasker, *Inorg. Chem.,* **11**, 2775 (1972).

[23] Much of the pioneering work in this area has been done by J. L. Hoard, and his definitive paper on eight-coordination should be consulted [J. L. Hoard and J. V. Silverton, *Inorg. Chem.,* **2**, 235 (1963)]. In addition, there are several excellent reviews of the subject: R. V. Parish, *Coord. Chem. Rev.,* **1**, 439 (1966); S. J. Lippard, *Prog. Inorg. Chem.,* **8**, 109 (1967); E. L. Muetterties and C. W. Wright, *Quart. Revs.,* **21**, 109 (1967).

[24] M. G. B. Drew, *Coord. Chem. Rev.,* **24**, 179 (1977).

[25] R. Hoffmann, B. F. Beier, E. L. Muetterties, and A. R. Rossi, *Inorg. Chem.,* **16**, 511 (1977).

[26] E. L. Muetterties and C. M. Wright, *Quart. Rev.,* **21**, 109 (1967).

[27] R. A. Levenson, R. J. G. Dominguez, M. A. Willis, and F. R. Young III, *Inorg. Chem.,* **13**, 2761 (1974).

[28] D. F. Lewis and R. C. Fay, *J. Amer. Chem. Soc.,* **96**, 3843 (1974); see also A. N. Bhat, R. C. Fay, D. F. Lewis, A. F. Lindmark, and S. H. Strauss, *Inorg. Chem.,* **13**, 886 (1974).

[29] J. K. Burdett, R. Hoffmann, and R. C. Fay, *Inorg. Chem.,* **17**, 2553 (1978).

[30] G. L. Glenn, J. V. Silverton, and J. L. Hoard, *Inorg. Chem.,* **2**, 250 (1963).

[31] B. E. Robertson, *Inorg. Chem.,* **16**, 2735 (1977).

[32] K. Knox and A. P. Ginsberg, *Inorg. Chem.,* **1**, 945 (1962) **3**, 555 (1964).

[33] S. C. Abrahams, A. P. Ginsberg, and K. Knox, *Inorg. Chem.,* **3**, 558 (1964).

11. COORDINATION CHEMISTRY: ISOMERISM

In the previous chapter you learned of polytopal isomers: the trigonal prism and the octahedron (or trigonal antiprism) for coordination number six, and the tetrahedron and square plane for coordination number four. In this chapter we turn to other types of isomerism: constitutional isomerism and stereoisomerism. In the former, the empirical formulas of the isomers are identical, but the atom-to-atom bonding sequences differ. For example, the thiocyanate ion, NCS^-, may be bound either through N or through S in a complex ion such as $[Co(CN)_5(SCN)]^{3-}$. Stereoisomers, on the other hand, have the same empirical formulas and the same atom-to-atom bonding sequences, and describe identical coordination polyhedra; the difference between stereoisomers comes in the spatial arrangement of the atoms.

Chemical isomerism is a pervasive phenomenon. Since most of you were exposed to a rigorous study of organic chemistry before encountering this course, your ideas of isomerism were developed in that context. You know, therefore, that there are two basic forms of isomerism: **constitutional isomerism** and **stereoisomerism**.[1-3]

Constitutional isomerism, which has often been called structural or position isomerism, results when two or more molecules have the same empirical formula but the constituents of the molecules are arranged differently; that is, there is a difference in the atom-to-atom bonding sequence. For example, 1-butene (**1**) and 2-butene (**2**) differ in the placement of the double bond, while acetone (**3**) and propionaldehyde (**4**) differ in the position of the carbonyl group.†

$$H_3C-CH_2-CH=CH_2 \qquad H_3C-CH=CH-CH_3$$

1   **2**

$$H_3C-\overset{\overset{O}{\|}}{C}-CH_3 \qquad H_3C-CH_2-\overset{\overset{O}{\|}}{C}-H$$

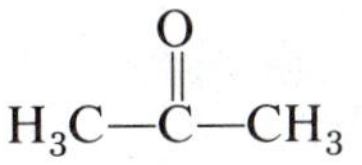

3 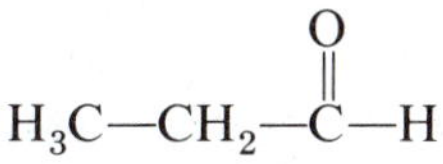   **4**

In contrast, *stereoisomerism* arises when two or more compounds have the same empirical formula and the same atom-to-atom bonding sequence but the atoms differ in their arrangement in space. Examples from organic chemistry would include the pair of enantiomers **R**- and **S**-3-bromo-2-methylpentane (**5** and **6**, respectively) or the

---

†The type of isomerism exhibited by acetone and propionaldehyde has sometimes been called "functional isomerism," since movement of the carbonyl group changes the chemical function of the molecule.

pair of diastereomers *cis*- and *trans*-2-butene (**7** and **8**). In addition, there are conformational isomers such as diequatorial (**9**) and diaxial (**10**) *trans*-1,2-dimethylcyclohexane.†

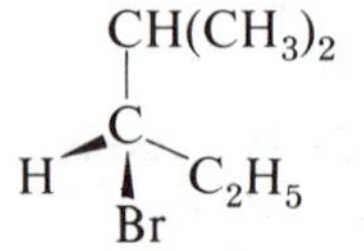

$$CH(CH_3)_2$$

5 = R 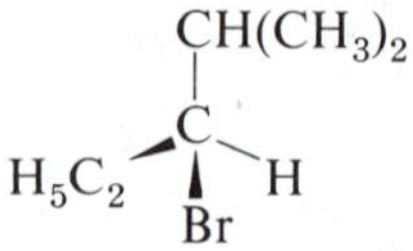   6 = S

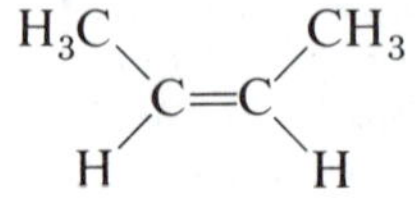

7 = *cis* 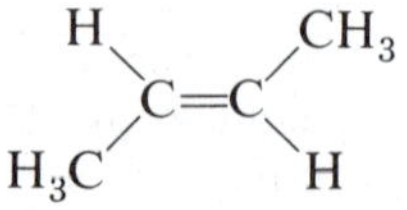   8 = *trans*

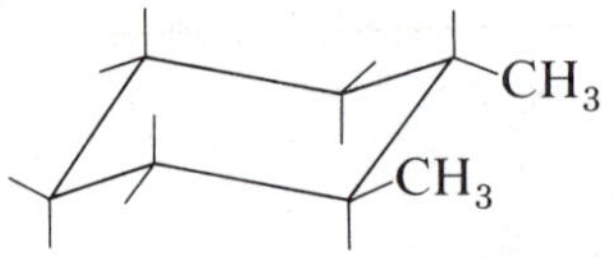

9 = diequatorial 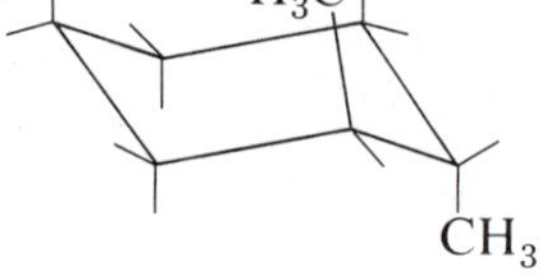   10 = diaxial

There are numerous examples of isomerism in inorganic chemistry, as well. For instance, there are many forms of constitutional isomerism, such as the so-called linkage isomers **11a** and **11b**, and a great many stereoisomers have been isolated and examined.

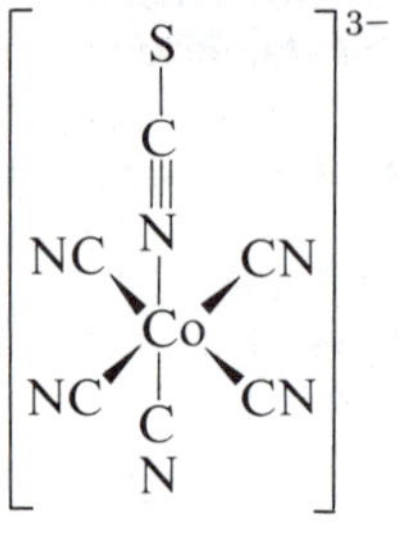

11a 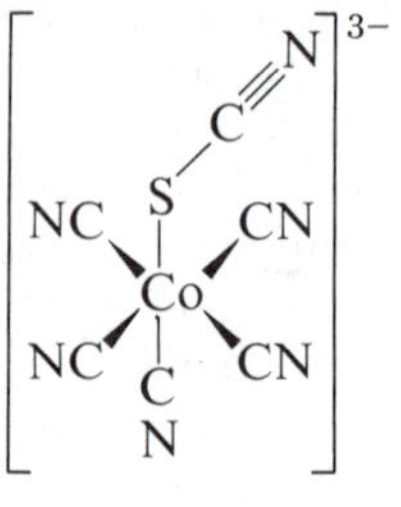   11b

One recent, unique example of the isolation of an enantiomer of a coordination compound is the report that the bacterium *Pseudomonas stutzeri* selectively removes one enantiomer, **12a**, from a racemic mixture of [Co(en)$_2$(phen)]$^{3+}$; after 60 hours under aerobic conditions the remaining enantiomer, **12b**, has an optical purity of 82%.[4]

In this chapter we shall examine some of the various forms of isomerism displayed by coordination compounds and the nomenclature by which they can be described. Before beginning our discussion, we might point out that the organization of this chapter according to isomer type follows, in large part, a classification scheme laid down by Alfred Werner in his work early in this century.[5-7]

†Stereoisomerism is important in literature as well. Lewis Carroll's story "Through the Looking Glass and What Alice Found There" is a journey into a mirror image world. The characters that best exemplify this world, in fact, are Tweedledum and Tweedledee, two inhabitants of the Looking Glass world that Carroll meant to be enantiomers. See "The Annotated Alice," Martin Gardner, ed., Clarkson N. Potter Publisher, New York, 1960.

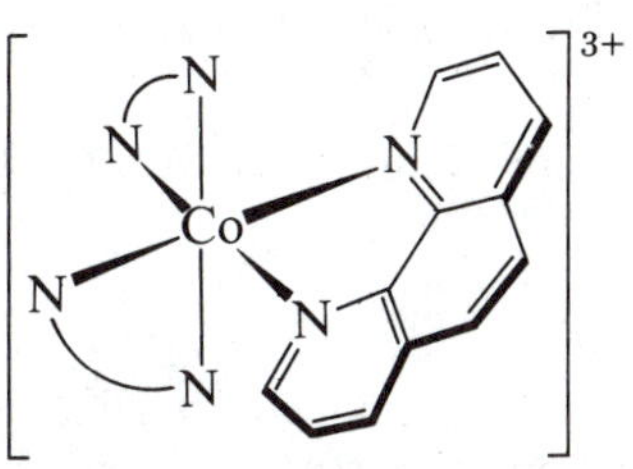

12a 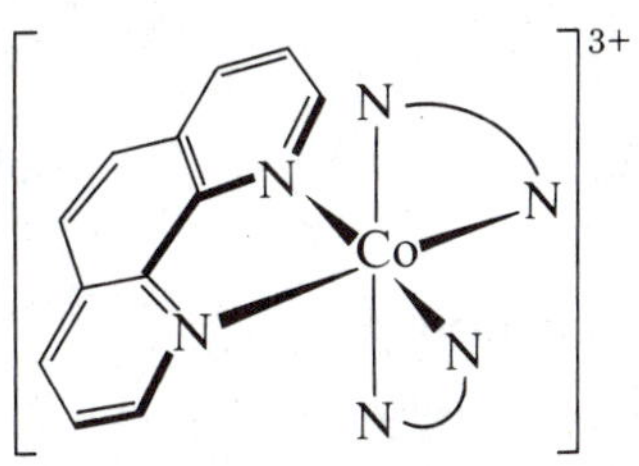   12b

CONSTITUTIONAL ISOMERISM

Constitutional isomers have identical empirical formulas, but the atom-to-atom connections differ. There are numerous forms of such isomerism, but only five of the most common are described below.

Hydrate Isomerism

The classical example of hydrate isomerism is the series of three hydrated chromium chloride complexes having the empirical formula $CrCl_3 \cdot 6H_2O$. The green commercial form of this compound (called Recoura's Green Chloride) is obtained from concentrated hydrochloric acid solutions and is assigned the molecular formula $[Cr(H_2O)_4Cl_2]Cl \cdot 2H_2O$. On dissolving this complex in water, however, the metal-bound Cl^- ions are successively replaced by water molecules, giving first light green $[Cr(H_2O)_5Cl]Cl_2 \cdot H_2O$ and finally violet $[Cr(H_2O)_6]Cl_3$.

Ionization Isomerism

This type of isomerism is somewhat similar to hydrate isomerism. However, instead of the exchange of coordinated ions for water, ionization isomers arise from the interchange of two ions between inner and outer coordination spheres. For instance, the complex formed from a cobalt(III) ion, five ammonia molecules, a bromide ion, and a sulfate ion exists in two forms, one dark violet (**A**) and the other violet-red (**B**). The dark violet form (**A**) gives a precipitate with barium chloride but none with silver nitrate. Form (**B**) behaves in an opposite manner. Therefore, the two complexes may be formulated as $A = [Co(NH_3)_5Br]SO_4$, and $B = [Co(NH_3)_5SO_4]Br$. Another example of ionization isomerism is the pair of complexes $trans$-$[Co(en)_2Cl_2]NO_2$ (green) and $trans$-$[Co(en)_2(NO_2)Cl]Cl$ (red).

Coordination Isomerism

This form of isomerism is possible *only for salts* wherein both the anion and cation contain a metal ion that can function as a coordination center. Isomerism arises from a different distribution of ligands between the two metal ions. For example, if two different metal ions (**M** and **M′**) and two different ligands (*a* and *b*) are involved, two of the possible coordination isomers would be $[Ma_x][M'b_x]$ and $[Mb_x][M'a_x]$.

One real example of coordination isomerism is found in two of the salts that have the empirical formula $CoCr(NH_3)_6(CN)_6$. They may be prepared, and their constitutions determined, by the following reactions:

$$[Co(NH_3)_6]Cl_3 + K_3[Cr(CN)_6] \rightarrow [Co(NH_3)_6][Cr(CN)_6] + 3\,KCl$$
$$\mathbf{13}$$

Compound **13** + $AgNO_3 \rightarrow$ insoluble salt identified as $Ag_3[Cr(CN)_6]$

$$[Cr(NH_3)_6]Cl_3 + K_3[Co(CN)_6] \rightarrow [Cr(NH_3)_6][Co(CN)_6] + 3\ KCl$$
14

Compound **14** + AgNO$_3$ → insoluble salt identified as Ag$_3$[Co(CN)$_6$]

Polymerization Isomerism

In organic chemistry the term polymer implies the linking of small radicals or molecules into chains, layers, or lattices. In coordination chemistry, however, the term polymerization isomerism has a slightly different meaning. Each member of a series of polymerization isomers has the same empirical formula, and the molecular formula of each is some multiple of the simplest formula. However, unlike organic polymers, inorganic polymerization isomers are not formed by the direct chemical bonding of small units to give larger units. To illustrate this, a series of polymerization isomers is listed in Table 11-1. (Notice also that the second and third and the fourth and fifth compounds are pairs of coordination isomers.)

Linkage Isomerism[8-10]

In organic chemistry it is well established that substituent groups such as NO$_2$ can form nitro $\left(R-N\begin{smallmatrix}O\\O\end{smallmatrix}\right)$ compounds or nitrites (R−O−N−O), CN can form nitriles (R−C≡N) or isonitriles (R−N≡C), SCN can form thiocyanates (R−S−C≡N) or isothiocyanates (R−N=C=S), and so on. In coordination chemistry these same groups can behave as **ambidentate** ligands, using either end of the ion in binding to a metal ion. So, in the case of the NO$_2^-$ ion, it can form nitrito (**15**) or nitro (**16**) complexes.

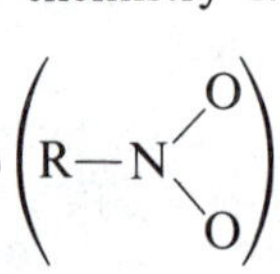

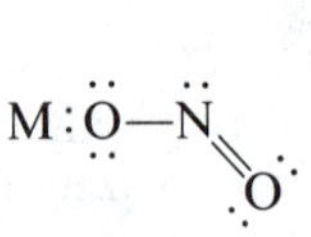

M:O̤−N̈⟨O̤   M:N⟨Ö: O̤
15 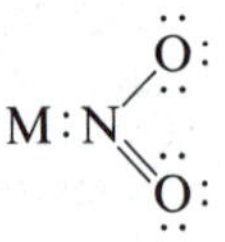   16

TABLE 11-1

POLYMERIZATION ISOMERS OF A PLATINUM(II) COMPLEX

Formula	Multiple of Simplest Formula Weight	Comments
[Pt(NH$_3$)$_2$Cl$_2$]	1	Yellow; *cis* isomer known as Peyrone's chloride, and the *trans* isomer often called Reiset's chloride.
[Pt(NH$_3$)$_4$][PtCl$_4$]	2	Known as Magnus's Green Salt. First Pt(II)-ammine discovered.
[Pt(NH$_3$)$_3$Cl][Pt(NH$_3$)Cl$_3$]	2	—
[Pt(NH$_3$)$_4$][Pt(NH$_3$)Cl$_3$]$_2$	3	Orange-yellow.
[Pt(NH$_3$)$_3$Cl]$_2$[PtCl$_4$]	3	—

Isomers that result from alternative modes of ligand coordination are called **linkage isomers**. For example, two cobalt(III)-NO_2 linkage isomers may be prepared according to the reactions:[11]

$$[Co(NH_3)_5Cl]^{2+} + H_2O \rightarrow [Co(NH_3)_5H_2O]^{3+} + Cl^-$$

$$[Co(NH_3)_5H_2O]^{3+} + NO_2^- \rightarrow [Co(NH_3)_5ONO]^{2+} + H_2O$$

light red
pentaamminenitritocobalt(2+) ion

$$[Co(NH_3)_5ONO]^{3+} \xrightarrow{H^+} [Co(NH_3)_5NO_2]^{2+}$$

yellow
pentaamminenitrocobalt(2+) ion

The yellow N-bonded or nitro isomer is stable for months if stored out of direct sunlight, but irradiation with ultraviolet light leads to isomerism to the light red O-bonded or nitrito form. Upon standing, the nitrito complex reverts completely to the nitro form again. The mechanism of interconversion, even in the solid state, is believed to be intra-molecular, proceeding through the three-center intermediate **17**.

$$[Co(NH_3)_5NO_2]^{2+} \underset{}{\overset{h\nu}{\rightleftharpoons}} \left[(NH_3)_5Co \cdots \begin{matrix} O \\ \| \\ N \\ \| \\ O \end{matrix} \right]^{2+} \rightleftharpoons [Co(NH_3)_5CoONO]^{2+}$$

17

In 1958 Ahrland, Chatt, and Davies pointed out that metal ions in their common oxidation states could be divided roughly into two classes, depending on their abilities to coordinate with certain donor atoms.[12] Those metals in class *a* (see Figure 11-1) generally bind most strongly with ligands whose donor atoms are second row elements (N, O, or F), and those in class *b* form their most stable complexes with elements of subsequent rows of the periodic table (P, S, Cl, etc.).† Since, with the exception of NO_2^- and

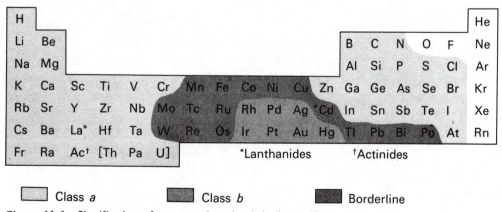

| | Class *a* | | | Class *b* | | | Borderline |

Figure 11-1. Classification of acceptor ions in their "normal" valence states into *a*, *b*, and *borderline* types.

---

†We shall explore in more detail the question of the thermodynamic stability of coordination complexes in Chapter 13.

CN⁻, all of the ligands exhibiting ambidentate behavior possess one donor atom from the second row of the Periodic Table and one from the third or higher row, the nature of the metal ion acceptor is clearly important. For example, the most stable complex formed between SCN⁻ and Pd(II) is the S-bonded linkage isomer $[Pd(SCN)_4]^{2-}$, while the analogous complex formed by Cd(II), an ion classified as being on the border between a and b, has both N- and S-bonded SCN⁻ ligands. On the other hand, the class a metal ion Co(III) forms the N-bonded complex ion $[Co(NH_3)_5NCS]^{2+}$.

The steric and electronic requirements of the other ligands in the complex also play a role in linkage isomerism, and their effects have been most thoroughly studied using the thiocyanate ligand. As the electron dot pictures below suggest,

$$M \leftarrow :\ddot{S} \diagdown C \equiv N. \qquad\qquad M \leftarrow :N \equiv C - \ddot{S}:$$

the anion will be linear if N-bonded and bent if S-bonded. In Figure 11-2 you see a complex wherein the SCN⁻ group adopts both modes of coordination. It is N-bonded when *cis* to the short NMe₂-Pd bond, and S-bonded when *cis* to the longer PPh₂-Pd bond. Although this is an appealing explanation for the differing modes of coordination of SCN⁻ in this compound, the coordination mode could also be influenced by the electronic effects of the *trans* donor atom. In most cases where this question has been examined, it seems apparent that both steric and electronic effects are important, and it is difficult to predict the dominance of one or the other of their influences.

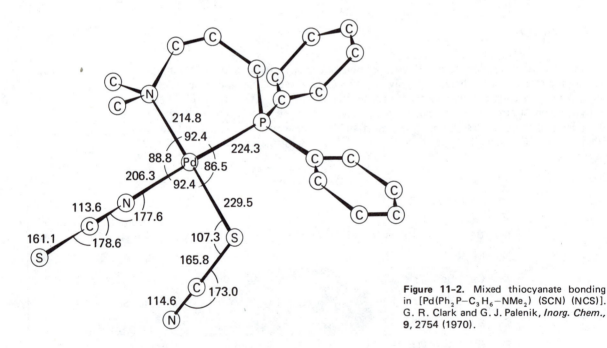

Figure 11-2. Mixed thiocyanate bonding in $[Pd(Ph_2P-C_3H_6-NMe_2) (SCN) (NCS)]$. G. R. Clark and G. J. Palenik, *Inorg. Chem.,* **9,** 2754 (1970).

STUDY QUESTIONS

1. Give an example of a coordination isomer of each of the following compounds:

    a) $[Co(en)_3][Cr(C_2O_4)_3]$

    b) $[Co(NH_3)_6][Co(NO_2)_6]$

    c) $[Pt^{II}(NH_3)_4][Pt^{IV}Cl_6]$

2. How many coordination isomers could be formed starting with $[Cu(NH_3)_4][PtCl_4]$? Write the formula for each isomer.

3. Alfred Werner first cited the nine known compounds of empirical formula $Co(NH_3)_3$ $(NO_2)_3$ as examples of polymerization isomerism.

   a) Write the formulas for at least four polymerization isomers based on $Co(NH_3)_3$ $(NO_2)_3$.
   b) Give the systematic name for each of these four isomers.

4. The relative yields of the various linkage isomers for Pt(II) complexes such as $[Pt(NCS)_3L]^-$ (L = NMe_3, PMe_3, $AsMe_3$, for example) have been studied [*J. C. S. Dalton*, 1683 (1977)]. Draw the structure for each of the possible linkage isomers when L = NMe_3.

STEREOISOMERISM[13]

Stereoisomerism has played a decisive role in coordination chemistry, since it was largely on the basis of the isolation of stereoisomers of four- and six-coordinate complexes that Werner was able to prove his revolutionizing concepts of inorganic stereochemistry.[8,14]

A GLOSSARY OF STEREOCHEMICAL TERMS[1-3, 15-17]

Stereoisomers: Two or more molecules that have the same empirical formula, the same atom-to-atom bonding sequence, and identical coordination polyhedra, but in which the atoms differ in their arrangement in space.

Enantiomer: A stereoisomer that has a non-superimposable mirror image.

Diastereoisomers: Stereoisomers that are not enantiomers; diastereoisomerism includes *cis-trans* isomerism as a sub-class.

Asymmetric: Applied to a molecule totally lacking in symmetry.

Dissymmetric: Applied to a molecule that lacks an S_n axis, that is, a rotation-reflection axis. Because it lacks an S_n axis, a dissymmetric molecule has neither a plane of symmetry (S_1) nor a center of symmetry (S_2). In order for a molecule to have an enantiomer, it must *not* have an S_n axis. However, a dissymmetric molecule may have a proper axis of rotation C_n (with $n > 1$).

Chiral compound: A molecule that is asymmetric or dissymmetric is chiral, and it is therefore not superimposable on its mirror image. Chirality denotes "handedness," and molecules of opposite chirality (enantiomers) are related to one another as the left hand is to the right.

Optical activity: The ability of chiral molecules to rotate the plane of plane polarized light.

Stereoisomers are compounds having the same molecular formula, the same atomic sequences, and identical coordination polyhedra (contrast polytopal isomers); however, their atoms differ in their spatial arrangement. They generally fall into two classes: *enantiomers* and *diastereomers*.

Enantiomers are molecules that are non-superimposable mirror images of one another; that is, they have opposite chirality. The term chirality (allegedly coined by Lord Kelvin in 1893) denotes "handedness," and molecules of opposite chirality are related to one another as the left hand is to the right. Chirality can arise from the presence of an "asymmetric" or chiral center such as a carbon atom (**5, 6,** and **18**) or a metal atom (**19**). To be a center of chirality, a carbon atom must be surrounded by four

Chirality

Many things natural and unnatural are chiral. The former is illustrated by the words of a song from the delightful English review, "At the Drop of a Hat."

"MISALLIANCE" FROM "AT THE DROP OF A HAT"
by Michael Flanders and Donald Swan

The fragrant honeysuckle spirals clockwise to the sun,
And many other creepers do the same.
But some climb anti-clockwise,
The bindweed does for one (or convolvulus to give her proper name).

Rooted on either side a door, one of each species grew,
And raced toward the window ledge above.
Each corkscrewed to the lintel in the only way it knew,
Where they stopped, touched tendrils, smiled, and fell in love.

Said the right-handed honeysuckle to the left-handed bindweed,
"Oh, let us get married if our parents don't mind.
We'd be loving and inseparable, inextricably entwined.
We'd live happily ever after," said the honeysuckle to the bindweed.

To the honeysuckle's parents it came as a great shock.
"The bindweeds," they cried, "are inferior stock.
They are uncultivated, of breeding bereft.
We twine to the right and they twine to the left."

Said the anti-clockwise bindweed to the clockwise honeysuckle,
"We'd better start saving, many a mickle mak's a muckle.
Then run away for a honeymoon and hope that our luck'll
Take a turn for the better," said the bindweed to the honeysuckle.

A bee who was passing remarked to them then,
"I said it before and I'll say it again,
Consider your offshoots if offshoots there be,
They'll never receive any blessing from me."

Poor little sucker, how will it learn,
When it is climbing which way to turn,
Right — left — what a disgrace.
Or it may go straight up and fall flat on its face.

Said the right-hand thread honeysuckle,
To the left-hand thread bindweed,
"It seems that against us all fate has combined,
Oh my darling, oh my darling, oh my darling columbine,
Thou art lost and gone for ever,
We shall never intertwine."

Together they found them the very next day,
They had pulled up their roots and just shriveled away,
Deprived of that freedom for which we must fight,
To veer to the left or to veer to the right!

different groups, but a chiral metal atom in an octahedral complex may be produced by the appropriate arrangement of at least three different types of monodentate ligands; for example, the Pt(II) compound $[Pt(NH_3)_2(NO_2)_2Cl_2]$ has been resolved into its enantiomers (**20a** and **20a'**). However, just as there are other ways for carbon compounds to be chiral [for example, an appropriately tetra-substituted biphenyl (**21**) does not have a chiral carbon, yet such compounds are chiral], there are six ways for chirality to be introduced into an octahedral complex; some of these are discussed in more detail in a later section of this chapter.

A necessary and sufficient condition for chirality in a molecule is that it lacks an improper axis of rotation—that is, a rotation-reflection axis, S_n. However, a less rigorous criterion often applied to decide whether or not a molecule is chiral is the lack of a center or plane of symmetry. The two enantiomers **20a** and **20a'** certainly possess neither of these. However, a center of symmetry is equivalent to an S_2 axis and a plane to an S_1 axis, so the main criterion is fulfilled. The less rigorous criterion should be applied with caution, however, since it is a well-recognized fact that a molecule may lack both a center and a plane of symmetry but still have a superimposable mirror image; compound **22** is an example of such a case.

The fact that chiral molecules lack the all-important S_n axis does not mean that they must also be *asymmetric*, that is, totally lacking in symmetry. Rather, in coordination chemistry, many of the compounds that have been resolved into their enantiomers have proper axes of rotation as in the tris(bidentate)metal complex **23**. In summary, the most important criterion for the existence of an enantiomer of a molecule is that it be *dissymmetric;* that is, that, while it may possess other elements of symmetry, it must lack an S_n axis.

All stereoisomers that are not enantiomers are *diastereoisomers* (sometimes referred to as *diastereomers*). Diastereoisomerism includes as a sub-class the well-known phenomenon of *cis-trans* or geometrical isomerism. For example, compounds **20a-e** are diastereoisomers; **20a'** is the enantiomer of **20a**. Diastereoisomers may be achiral (**20b-e**) or they may be chiral (**20a**).

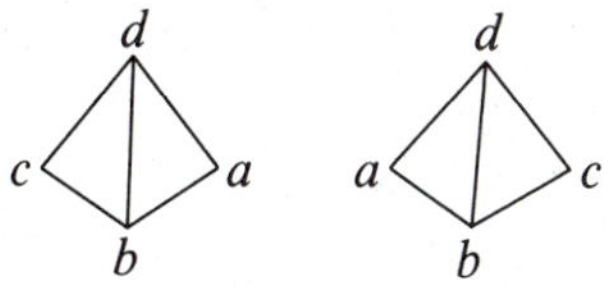

18 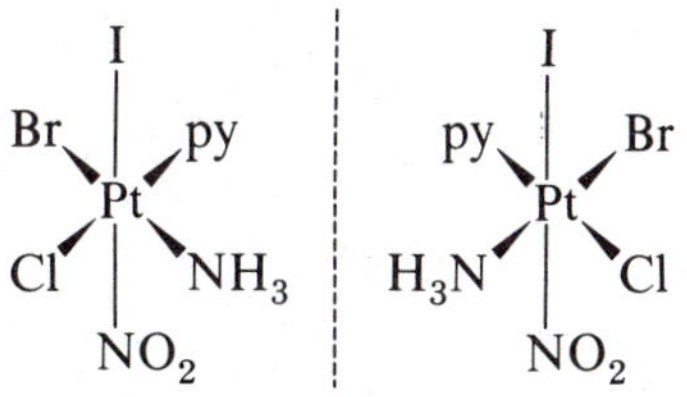   19

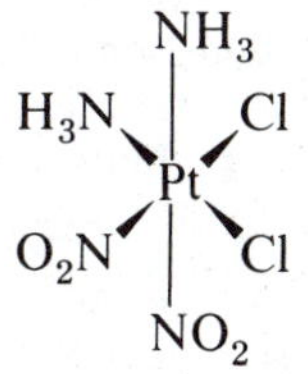

20a 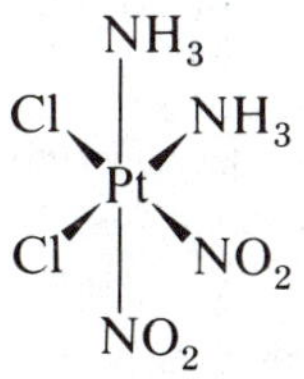   20a' 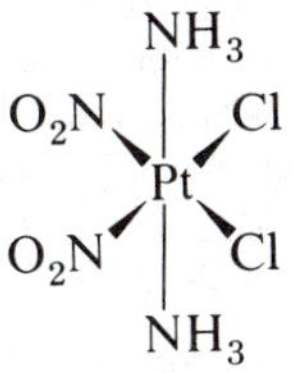   20b 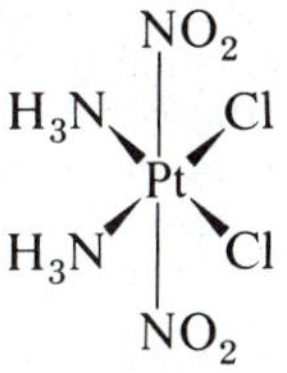   20c

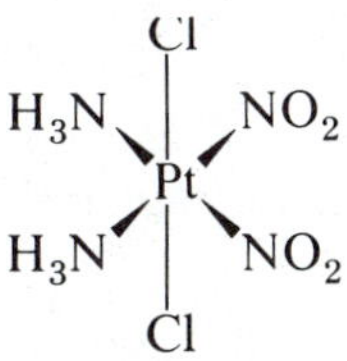

20d 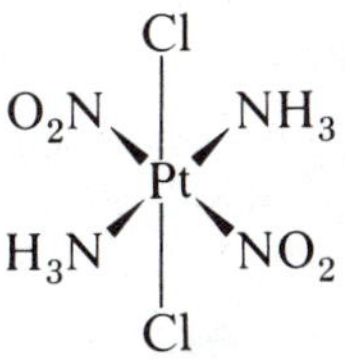   20e

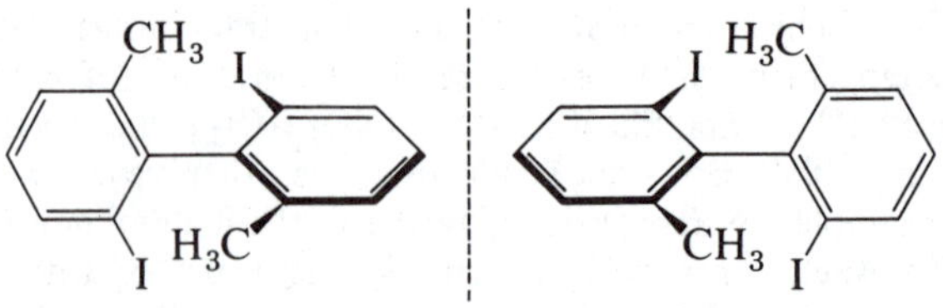

21

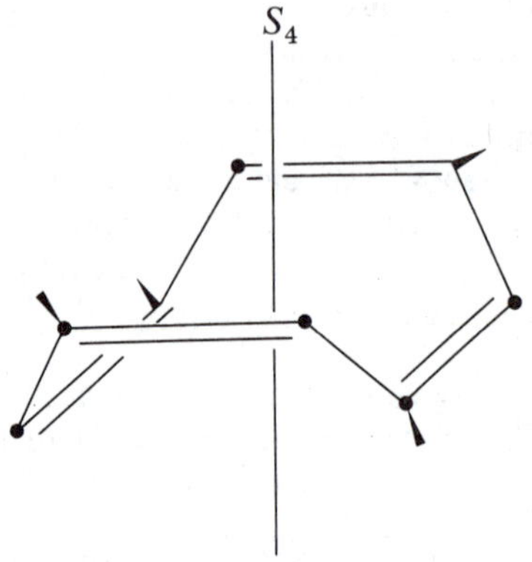

22

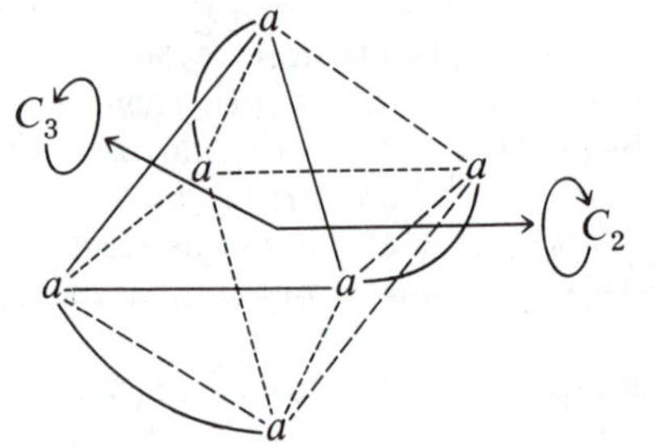

23

A very important difference between enantiomers and diastereomers is that the latter usually differ appreciably in their chemical and physical properties (in fact, in any property depending on molecular shape). However, two enantiomers differ only in their abilities to rotate plane polarized light.† Therefore, in order to separate two enantiomers (recall that with certain exceptions[18] a racemic mixture will arise in any normal chemical reaction producing chiral products), they must be first converted to diastereomers. For example, the enantiomers of *cis*-[Co(en)$_2$(NO$_2$)$_2$]$^+$ (24) can be separated into optically pure components by adding the chiral salt

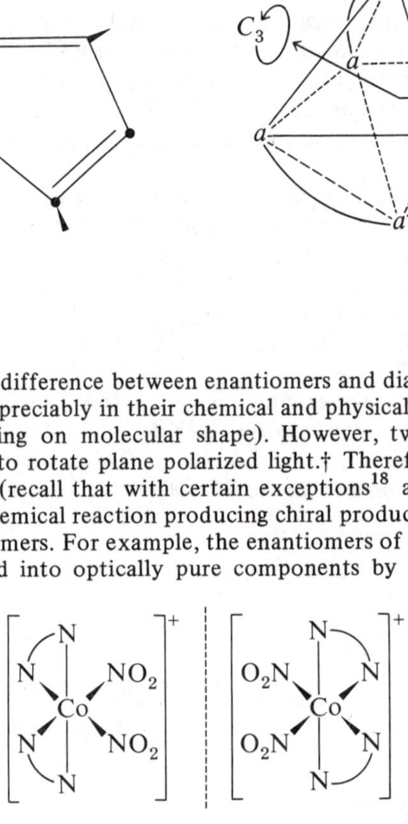

24

potassium antimonyl-*d*-tartrate to an aqueous solution of the racemic mixture.[19] The chiral tartrate forms salts with both the left-handed and right-handed isomers of the cobalt complex, these salts being diastereomeric because only one enantiomer of the anion is used. As diastereomers, they have different solubilities, and the salt *cis*-(−)$_D$-[Co(en)$_2$(NO$_2$)$_2$][(+)$_D$-SbOC$_4$H$_4$O$_6$] precipitates as yellow needles before the other diastereomer, *cis*-(+)$_D$-[Co(en)$_2$(NO$_2$)$_2$][(+)$_D$-SbOC$_4$H$_4$O$_6$].[20]

In the sections that follow, we shall examine briefly the types of stereoisomerism observed for the most important coordination numbers—four and six—and the consequences of that isomerism.

Four-Coordinate Complexes

An understanding of the stereochemistry of four-coordinate complexes was extremely important to the development of coordination chemistry. Alfred Werner's

---

†When such rotation is observed experimentally, the system is said to be optically active. The origin and measurement of optical activity are explored in more detail in K. F. Purcell and J. C. Kotz, "Inorganic Chemistry," W. B. Saunders, Philadelphia, 1977, pp. 644–652.

revolutionary, original paper in coordination chemistry (published in 1893)[5] was a reply to Blomstrand (1826-1897) and Jørgensen (1837-1914), whose formulation of four-coordinate complexes was based on ideas of organic structure prevalent in the middle of the nineteenth century. For example, they recognized that there were two complexes having the empirical formula $Pt(NH_3)_2Cl_2$, to which they assigned structures **25** and **26**.

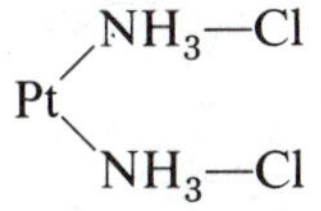

25

"platosammine chloride"

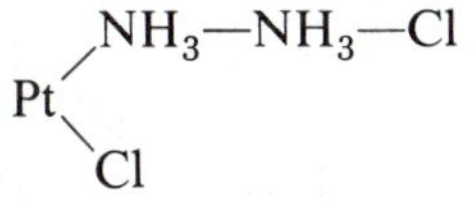

26

"platosemidiammine"

However, Werner rejected these structures for many reasons, but largely because they could not be used to rationalize the chemistry of the compounds. Instead, he proposed that the "platosammine chloride" was in fact the square planar complex *trans*-diammine-dichloroplatinum, and that the "platosemidiammine" was the square planar *cis* isomer.

$25 =$ $26 =$

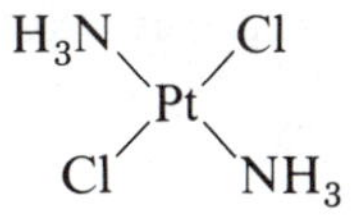

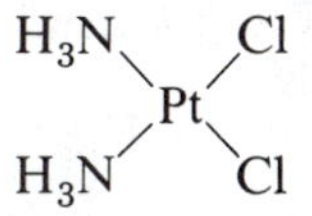

Proof of the correctness of Werner's proposal came in work by Grinberg in 1931.[21,22] Grinberg made the very reasonable assumption that when a chelating agent such as the oxalate ion coordinates to two positions of a square planar complex, then these two positions must be *cis*. Therefore, *cis*- and *trans*-diamminedichloroplatinum are predicted to react with oxalic acid as shown below, a prediction borne out by experiment.

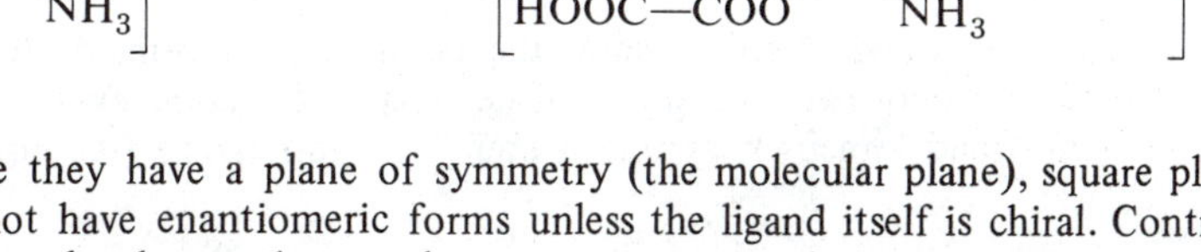

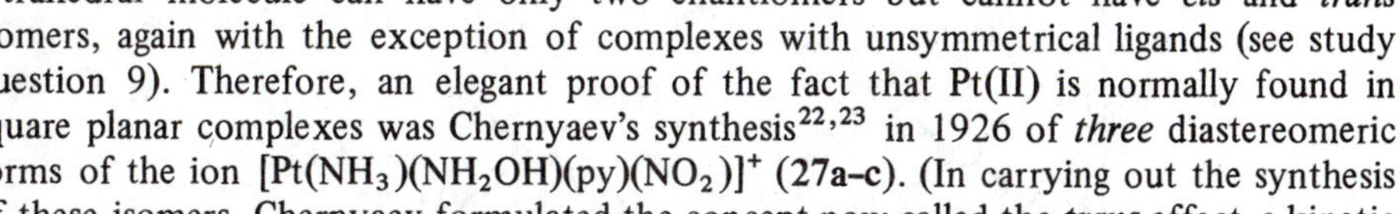

Because they have a plane of symmetry (the molecular plane), square planar complexes cannot have enantiomeric forms unless the ligand itself is chiral. Contrariwise, a tetrahedral molecule can have only two enantiomers but cannot have *cis* and *trans* isomers, again with the exception of complexes with unsymmetrical ligands (see study question 9). Therefore, an elegant proof of the fact that Pt(II) is normally found in square planar complexes was Chernyaev's synthesis[22,23] in 1926 of *three* diastereomeric forms of the ion $[Pt(NH_3)(NH_2OH)(py)(NO_2)]^+$ (**27a-c**). (In carrying out the synthesis of these isomers, Chernyaev formulated the concept now called the *trans* effect, a kinetic effect outlined in Chapter 13.)

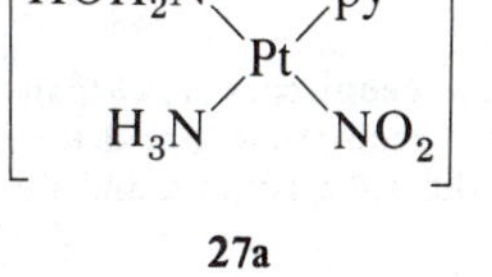

27a 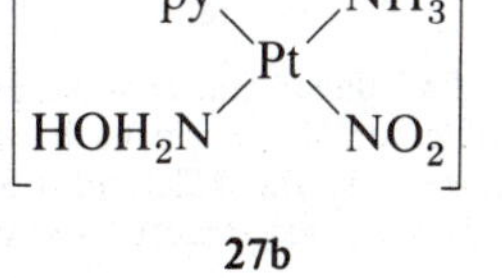   **27b** 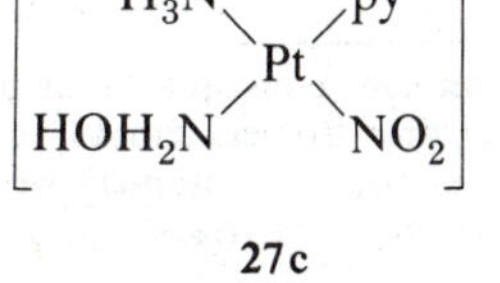   **27c**

In contrast to carbon-based compounds, there are no examples of tetrahedral transition metal complexes with four different ligands. Therefore, the only way to observe chirality for tetrahedral complexes is to use unsymmetrical, bidentate ligands, and for this reason the syntheses and stereochemistries of salicylaldimine, unsymmetrical β-diketone, and β-ketoamine (*e.g.,* **28**) complexes have been investigated extensively.[24]

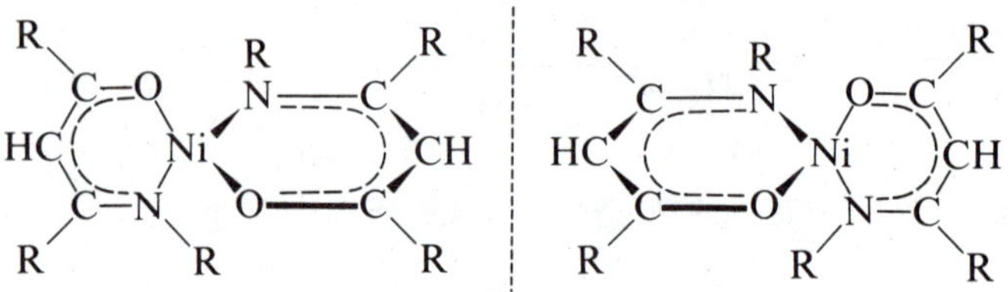

28

Only a very few tetrahedral bis-chelate complexes have been even partially resolved into their enantiomers, and most of them involve non-transition elements such as beryllium, boron, and zinc.[25] The few transition metal complexes that have been partially resolved racemize quite rapidly.

Six-Coordinate Complexes

The most elegant proof that many complexes are octahedral came from Werner's stereochemical studies, primarily on cobalt compounds. At the time Werner began work in this area, several isomeric series of salts of the type Coa_4b_2 were known. The Blomstrand-Jørgensen explanation for the existence of isomers of $[Co(NH_3)_4Cl_2]Cl$, for example, was that cobalt possessed three types of valence—α, β, and γ—and that the two isomers of the compound could be formulated as **29a** and **29b**. However, Werner

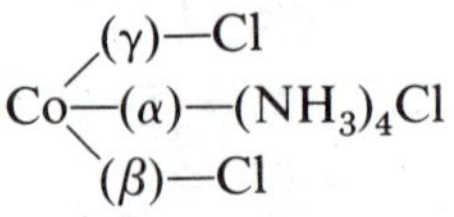

$$Co{<}\begin{array}{l}(\gamma)-Cl\\(\alpha)-(NH_3)_4Cl\\(\beta)-Cl\end{array} \qquad Co{<}\begin{array}{l}(\gamma)-Cl\\(\alpha)-Cl\\(\beta)-(NH_3)_4Cl\end{array}$$

29a 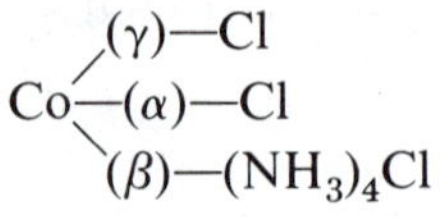   **29b**

proved that such compounds were actually the *cis* and *trans* isomers **30a** and **30b**. Further, he found that only two isomers could be isolated for complexes of this type, a fact that the Blomstrand-Jørgensen approach could not rationalize (see also p. 355).†

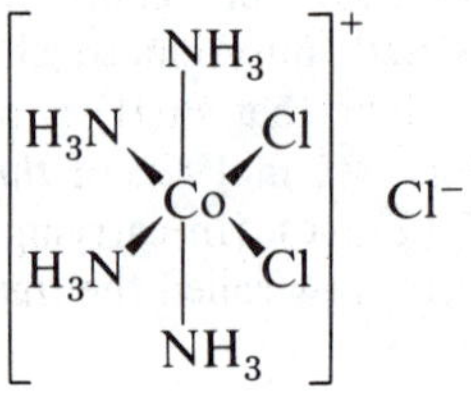

30a 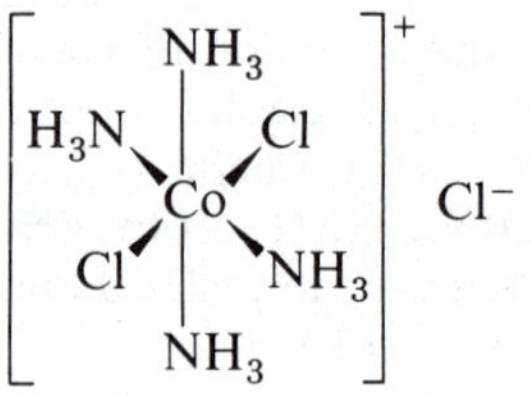   **30b**

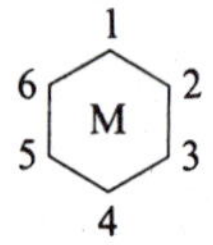

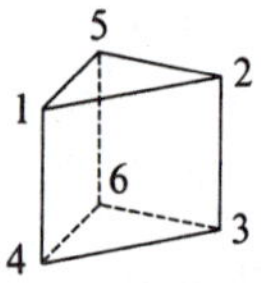

---

†Another consequence of this is that planar and trigonal prismatic geometries are virtually ruled out, as three isomers should be possible for Ma_4b_2 in each of these geometries. For a hexagonal planar geometry the isomers would have *b* in the 1,2, and 1,3, or the 1,4 positions, and the same would be true for a trigonal prismatic complex (see structures at left).

Perhaps the best proof of the overwhelming importance of the octahedron came with the resolution of *cis*-amminechlorobis(ethylenediamine)cobalt(2+) salts into their enantiomers (**31a** and **31b**), a feat accomplished in 1911 by an American, King, working in Werner's laboratory.[26]

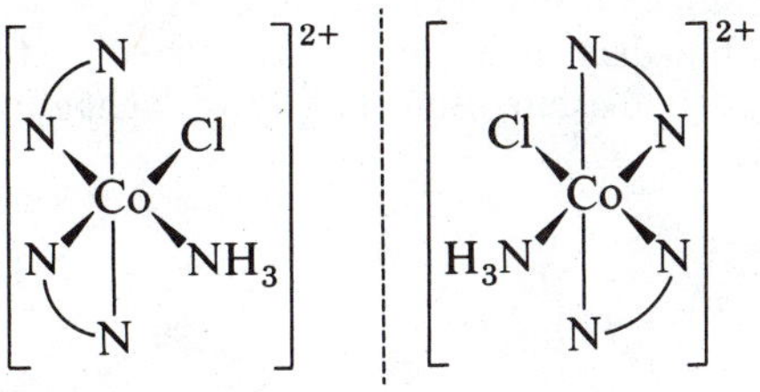

31a   **31b**

Some detractors of Werner's work, however, argued that the optical activity observed for these enantiomers arose in some mysterious way from the organic ligand. To answer these critics, therefore, Werner resolved into its enantiomers the completely inorganic complex ion tris[tetraammine-μ-dihydroxocobalt]cobalt(6+) (**32**); the inorganic ligand in this case clearly cannot be the center of chirality.

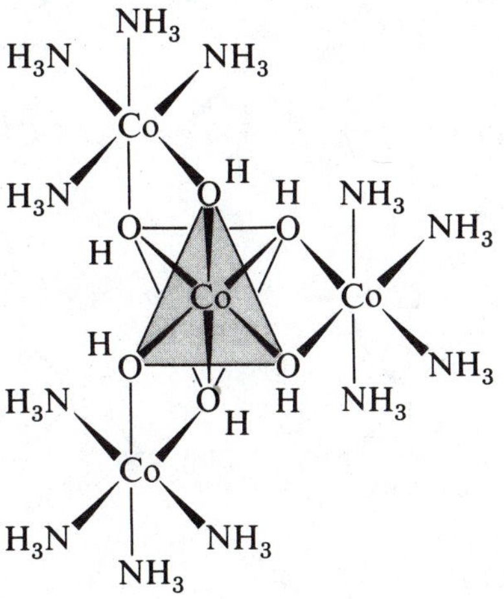

32

As you have seen from the brief discussion thus far, both "geometrical" and "optical" isomerism are possible for octahedral complexes, and there are several different ways of introducing chirality into six-coordinate complexes just as there are for four-coordinate ones. Indeed, there are six ways for octahedral complexes to be dissymmetric:[13c]

a. The distribution of unidentate ligands about the central metal.
b. The distribution of chelate rings about the central metal.
c. The coordination of unsymmetrical multidentate ligands.
d. The conformation of chelate rings.
e. The coordination of a chiral ligand.
f. The coordination of a donor atom that is asymmetric.

Our plan is to discuss only briefly types *a* through *c*. You will have to consult the specialist literature for more detailed information.[13]

Isomerism from Ligand Distribution and Unsymmetrical Ligands

As discussed further in Chapter 17, alkyl halides undergo so-called "oxidative addition" reactions with square planar metal complexes such as $IrCl(CO)(PPh_3)_2$ to give octahedral products. Although a single isomer is isolated in the reaction below, eight stereoisomers are possible: six diastereoisomers, two with enantiomers.

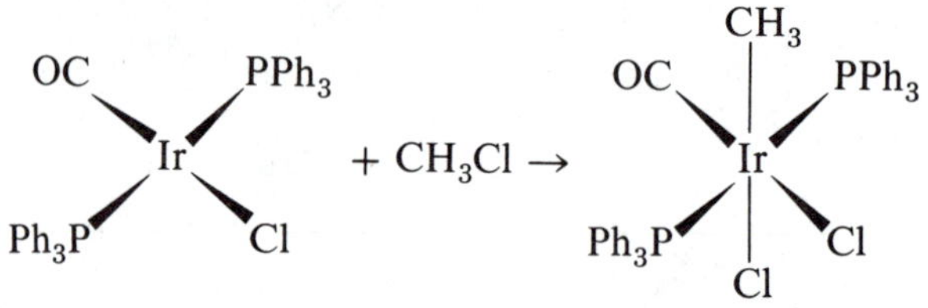

isolated product

33

Other Possible Isomers from Reaction of Ir(Cl)(CO)(PPh₃)₂ and CH₃Cl

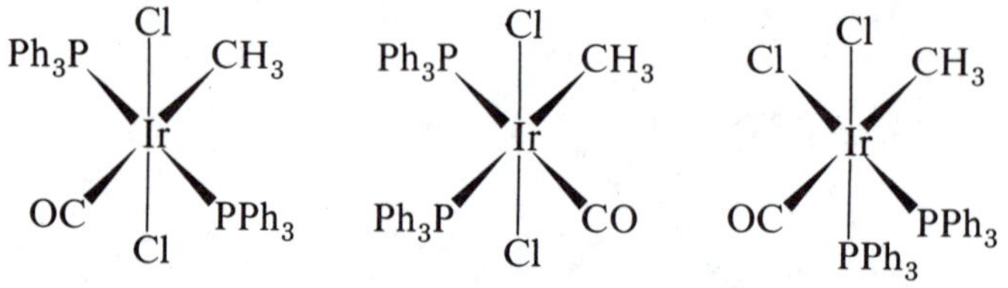

34   **35**   **36**

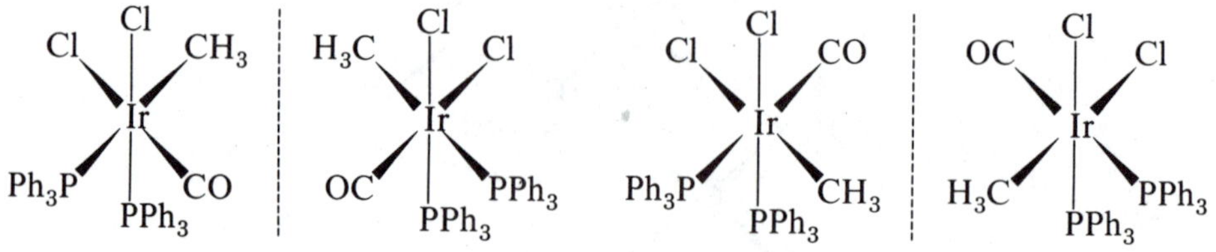

37   **38**

The derivation of the possible isomers for a given compound can usually be accomplished by the systematic application of common sense. For example, in compounds **33**, **34**, and **35** identical groups are placed *trans*, and the remaining ligands are permuted among the other four positions. This leads of necessity to a plane of symmetry in each, and thus to the non-existence of enantiomers. In the remaining diastereoisomers, like groups are placed *cis*. If both sets of identical groups are in the same plane, there is a plane of symmetry and no enantiomer (**36**). On the other hand, if both sets of identical *cis* ligands are in different planes, enantiomeric pairs result (**37** and **38**).

The derivation of the group of isomers above was relatively straightforward. However, it is easy to miss one or more isomers in more complicated cases, so many people have tried to describe more "foolproof" methods of isomer enumeration. Perhaps the most successful of these was developed by Bailar, whose paper should be consulted for the details.[27] A similar method is outlined in the appendix to this chapter.

Many coordination compounds involve chelating ligands, and isomerism may again arise as you saw with Werner's compounds, **31a** and **31b**. For example, four diastereoisomers are possible for the ion $[Co(en)(NH_3)_2(Cl)(H_2O)]^{2+}$. At least one plane of symmetry exists if both NH_3's (39) or both the Cl and H_2O (40) are in the same plane as the ethylenediamine. On the other hand, no plane or center of symmetry (*i.e.*, no S_n) exists if NH_3 and Cl or NH_3 and H_2O are in the same plane as the ethylenediamine (41 and 42).

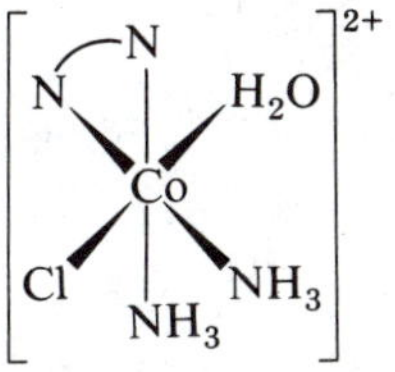

39

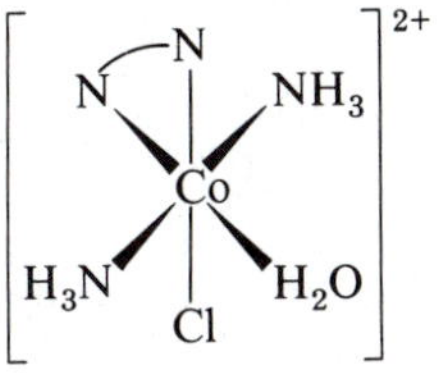

40

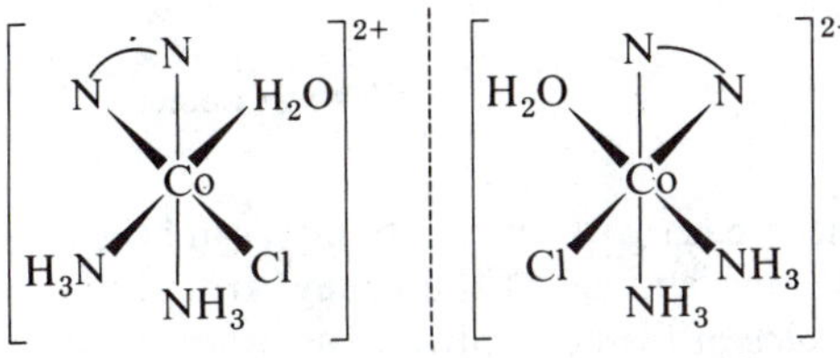

41

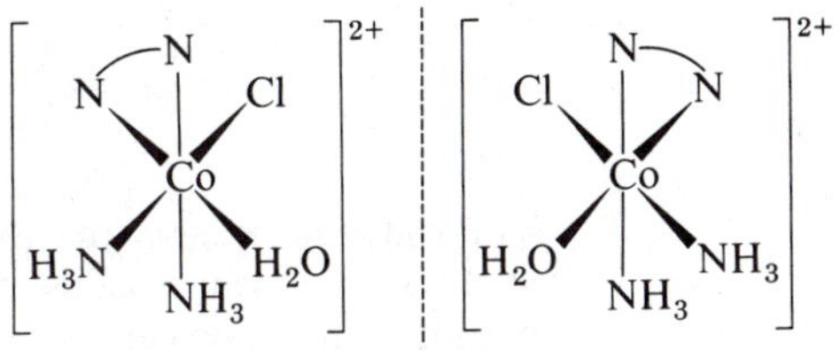

42

The third way to obtain dissymmetric complexes is to use unsymmetrical multidentate ligands. For example, in a complex of the glycinate ion, $Cu(H_2NCH_2COO)_2(H_2O)_2$, you must take into account the fact that, while the nitrogen and oxygen of one chelate must be *cis* to one another, they can be *trans* to the oxygen and nitrogen, respectively, of the other ligand. Thus, in contrast to a complex of the type M(symmetrical bidentate chelate)$_2$X$_2$ which would have only two diastereoisomers, there are five possible for $Cu(glycinate)_2(H_2O)_2$. The glycinate ligands can both be in the same plane with the N's *cis* (43) or *trans* (44). Both of these molecules have at least one plane of symmetry and neither is chiral. However, if the glycinate ligands are in different planes, three pairs of enantiomers are possible: glycinate N's *trans* (45), glycinate O's *trans* (46), or N's and O's *cis* (47).

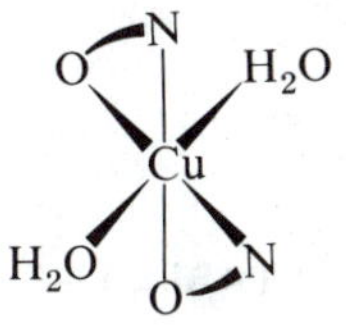

43

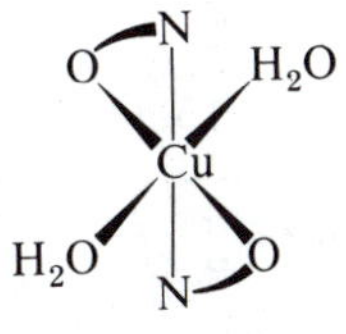

44

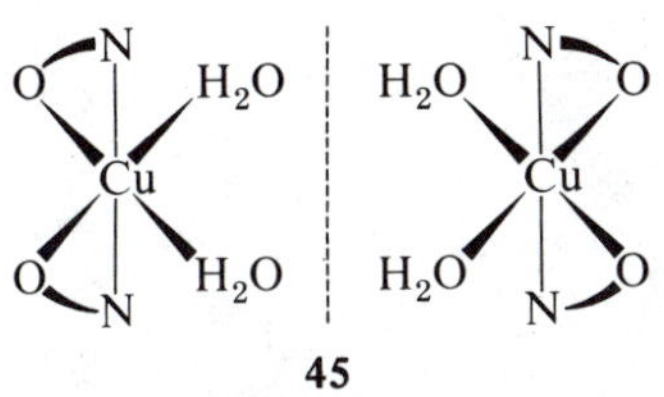

45

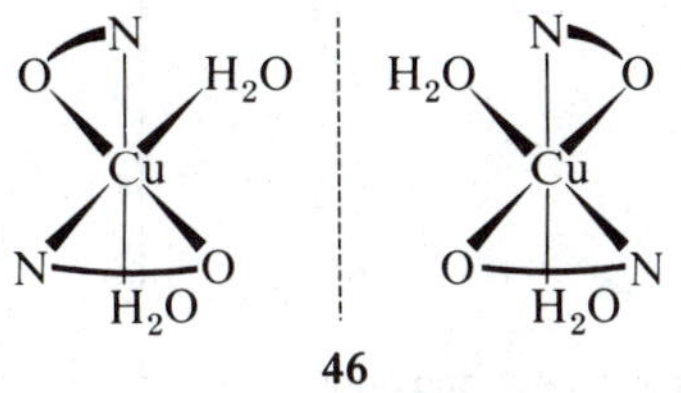

46

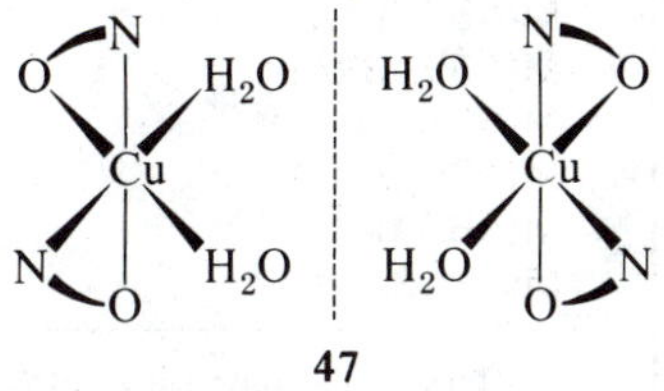

47

Before leaving this section, it is useful to add one last example in order to introduce some often-used nomenclature. Consider a complex formed from three unsymmetrical β-diketone ligands as in the following reaction.

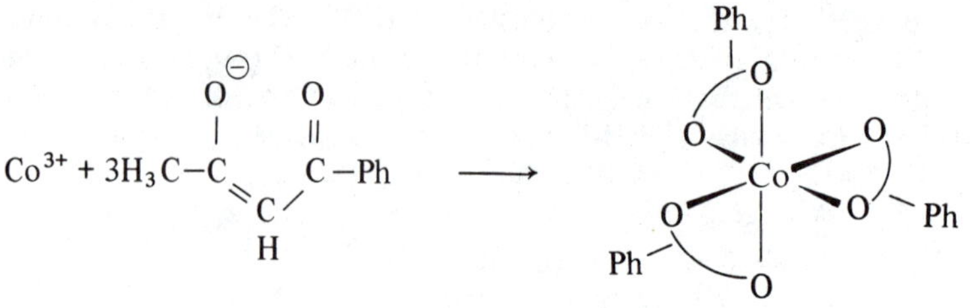

$$Co^{3+} + 3H_3C-C\underset{O^{\ominus}}{\overset{O^{\ominus}}{|}}=\underset{H}{\overset{C}{|}}\overset{O}{\overset{\|}{C}}-Ph \longrightarrow$$

48 = *fac* isomer

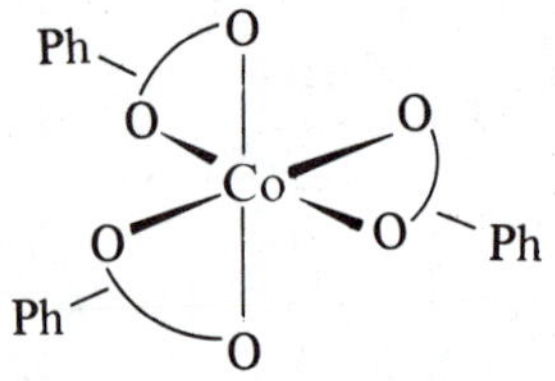

49 = *mer* isomer

You can readily show that two diastereoisomers, each with an enantiomer, can be generated. One of them is called the *fac* isomer (**48**), because all of the phenyl groups of the ligands lie near oxygen atoms located on one *face* of the octahedron. The other complex is called the *mer* isomer (**49**), because the phenyl groups lie near oxygen atoms located on the *mer*idian of the complex.

The method for isomer enumeration given in reference 27 can be used for all of the examples above and can be extended to others, some results for which are summarized in Table 11-2. A highly efficient "ordered pair" system for isomer identification is given in the appendix to this chapter (p. 386).

General Formula	Total Number of Stereoisomers	Pairs of Enantiomers
Ma_6	1	0
Ma_5f	1	0
Ma_4e_2	2	0
Ma_3d_3	2	0
Ma_4ef	2	0
Ma_3def	5	1
Ma_2cdef	15	6
$Mabcdef$	30	15
$Ma_2c_2e_2$	6	1
Ma_2c_2ef	8	2
Ma_3d_2f	3	0
$M(AA)(BC)ef$	10	5
$M(AB)(AB)ef$	11	5
$M(AB)(CD)ef$	20	10
$M(AB)(AB)(AB)$	4	2
$M(ABA)def$	9	3
$M(ABC)(ABC)$	11	5
$M(ABBA)ef$	7	3
$M(ABCBA)f$	7	3

TABLE 11-2

ISOMERS OF OCTAHEDRAL COMPLEXES[a,b]

[a] Lower case letters indicate monodentate ligands and upper case letters represent the donor atoms of chelating ligands.
[b] Table compiled from the following sources: W.E. Bennett, *Inorg. Chem.*, **8**, 1325 (1969); B.A. Kennedy, D.A. McQuarrie, and C.H. Brubaker, *Inorg. Chem.*, **3**, 265 (1964).

5. Verify that there are only two diastereoisomers for a compound such as $[Co(en)_2Cl_2]^+$.

6. Table 11–2 lists the number of stereoisomers possible for a number of different types of octahedral complexes. Verify the data for each of the following and draw the structure of each isomer generated. [If you need guidance, see appendix.]

   a) Ma_3def
   b) Ma_2cdef
   c) $M(AB)(AB)ef$

7. For each of the complexes listed below, derive the number of stereoisomers possible and indicate which are chiral. Draw the structure for each isomer. [See appendix for guidance.]

   a) $[Co(ethylenediamine)(NH_3)_2Cl_2]^+$
   b) M(ABA)(CDC) (two non-identical, tridentate ligands are attached to the metal)
   c) $M(ABCA)e_2$ (a tetradentate ligand and two monodentate ligands are attached to the metal ion)

8. The ligand 1,8-diamino-3,6-dithiaoctane, $H_2N-C_2H_4-S-C_2H_4-S-C_2H_4-NH_2$, forms numerous complexes with the Co(III) ion. Draw the isomers possible for a complex such as $[Co(NSSN)Cl_2]^+$, carefully showing each enantiomer and diastereomer. [See appendix for guidance.]

9. *Cis* and *trans* isomerism is possible for a four-coordinate, tetrahedral complex if an unsymmetrical ligand such as **R**-1,2-diaminopropane is used. Make a sketch or build a model of a complex of the type $(H_2N-CH_2CH(NH_2)CH_3)_2M$ to show how this is possible.

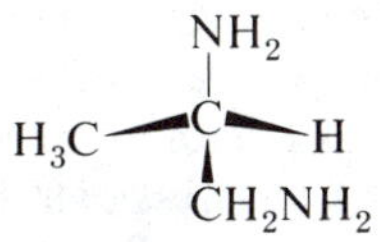

$$H_3C-\overset{\overset{\displaystyle NH_2}{|}}{\underset{\underset{\displaystyle CH_2NH_2}{|}}{C}}-H$$

R-1,2-diaminopropane

10. The square antiprism is often formed by eight-coordinate complexes. As discussed in Chapter 10, and as illustrated below, the square antiprism has two types of edges, ℓ and s. How many stereoisomers will a square antiprismatic complex of the type $M(AA)_4$ have (AA is a symmetrical bidentate ligand)? Indicate which are enantiomeric.

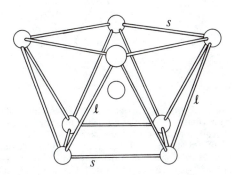

The Special Nomenclature of Chiral Coordination Compounds

At this point it is useful to introduce a special piece of nomenclature to deal with coordination compounds that are chiral because of the presence of multidentate ligands.[28] For example, the two $[Co(en)_3]^{3+}$ ions below are enantiomers. The absolute configuration of the one on the left is given the symbol Λ, and that of the one on the right is given the symbol Δ. The reason for this assignment can be seen by reference to a perhaps more familiar model, airplane or boat propellors.

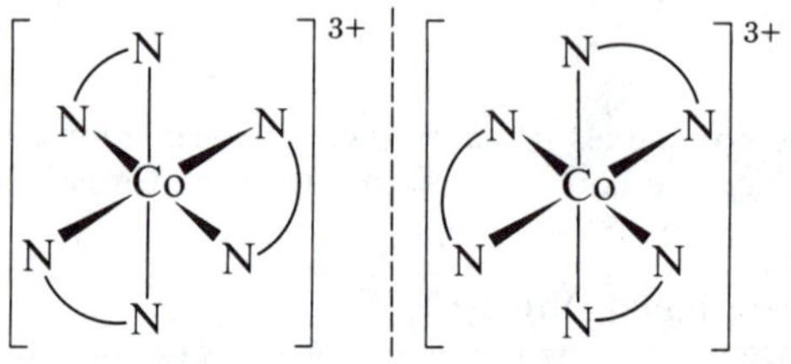

$$\Lambda \qquad\qquad\qquad \Delta$$

The three-bladed propellors at the top of Figure 11–3 are enantiomeric. The propellor on the left can describe a left-handed helix on rotating and moving through space, while the propellor on the right describes a right-handed helix. For those not familiar with such motion, this may require some explanation. Imagine that the propellor in Figure 11–3 is a dime-store pinwheel and that the "blades" of the pinwheel curl toward you; that is, the heavy edge in the figure is toward you when you look down the axis of rotation. If you move the pinwheel quickly through space, the pinwheel will turn in the direction indicated. If you fix your eye on one point on one blade, that point will describe a helix as the propellor rotates while moving through space. However, the two propellors in Figure 11–3 are different in that they have different "handedness." One propellor (A) describes a left-handed helix on moving through space. That is, if you use the thumb of your left hand to indicate the direction of motion, the fingers of your left hand curl in the same direction as the motion of the propellor blades. By the same test, the propellor at the bottom of Figure 11–3 is right-handed.†

In exactly the same way that a propellor or pinwheel can have "handedness" and describe left- or right-handed helices, a molecule such as a tris(bidentate)-metal complex can trace out a left- or right-handed helix on rotating and moving through space (Figure 11–4). Molecules having right-handed helicity are labeled Δ and those with left-handed helicity are labeled Λ.

An equivalent view of all of this, and one that will be directly applicable in the next section, is the following. Consider **50**, the usual drawing of an octahedron. The dashed

50

---

†If you still have difficulty telling the chirality or handedness of a propellor or a propellor-like molecule, try this. Imagine that you cup the propellor in your hand so that your hand is around the perimeter of the object, and your thumb is then parallel with the axis of rotation. Now imagine that you "walk" the fingers of this hand up the edge of one blade (from the back of the blade to the front), starting with your little finger and moving toward your first finger. If you can accomplish this operation with your left hand, the propellor will describe a left-handed helix or will have a left-handed screw axis; its chirality is given the symbol Λ. If, on the other hand, your right hand is required to carry out the "cupping and walking" exercise, the propellor is right-handed, and the object is given the chirality symbol Δ.

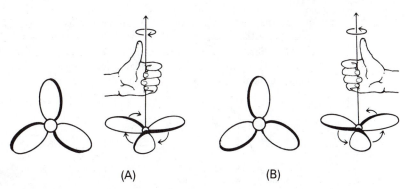

(A)   (B)

Illustration of the left- (Λ) and right-handedness (Δ) of enantiomeric three-bladed propellors.

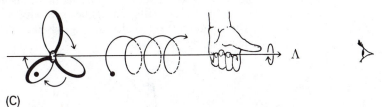

(C)

A left-handed propellor traces out a left-handed helix on traveling through space.

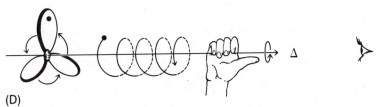

(D)

A right-handed propellor traces out a right-handed helix on traveling through space.

Figure 11-3. Illustration of the handedness of enantiomeric three-bladed propellors.

Figure 11-4. Illustration of the right (Δ) and left-handed (Λ) helicity of a tris(bidentate) complex. The pseudo-threefold axis of the octahedron is used as the defining axis.

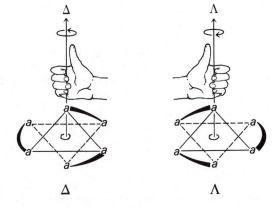

line at the back edge provides a reference edge or axis. Now concentrate on the heavy wedges to the upper-left front and to the upper-right front. A complex requires two chelating ligands to be chiral. The Λ or Δ configuration is immediately apparent if the complex is oriented so that one ligand lies along the back (reference) edge and the second chelating ligand lies along the upper-front-*left* edge (Λ) or upper-front-*right* edge (Δ).

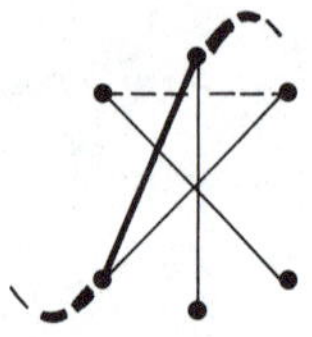

51   52

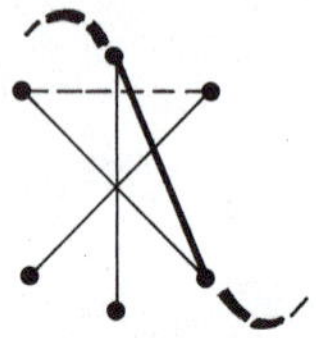

Λ   Δ

The front ligands appear as sections of the helices in Figures 11–3(C) and (D), respectively. The back-edge ligand defines the helix axis. This scheme works well for the chiral diastereomers of a tetradentate ligand like $H_2NC_2H_4NHC_2H_4NHC_2H_4NH_2$ (triethylenetetraamine ∿ trien).

Λ   Δ

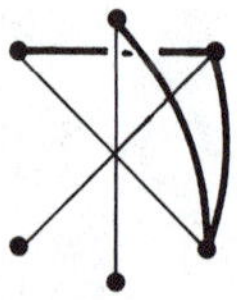

The key is to view the molecule with one ligand (or end of a multidentate ligand) along the back edge and the other ligand (or end of the same multidentate ligand) along the upper-left or upper-right front edges.

STUDY QUESTIONS

11. Give the correct chirality symbol—Λ or Δ—to each enantiomer generated by compounds **45–47**.

12. Give the correct chirality symbol for each enantiomer found for the compound in study question 8.

13. What are the chirality symbols for the two-bladed propellors below when their rotation axis is perpendicular to the page?

(A)   (B)

14. If a Co(III) complex of **S**-alanine $[Co(H_2NCH(CH_3)COO)_3]$ is prepared, four diastereoisomers can be isolated, while only two would be formed if glycine is used [glycinate ion = $H_2NCH_2COO^-$]. Why are there four in the case of the **S**-alanine complex?

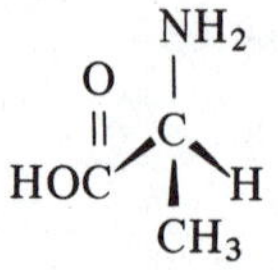

$$HOC \overset{\overset{NH_2}{|}}{\underset{\underset{CH_3}{|}}{\overset{O}{\overset{||}{\diagdown}} \overset{C}{\underset{}{}} \diagup H}} = \text{ S-alanine}$$

CHECKPOINTS

1. There are two classes of isomers of coordination compounds: constitutional isomers and stereoisomers.

2. Constitutional isomers have identical empirical formulas but different atom-to-atom connections. In coordination chemistry there are five common forms of such isomers: hydrate, ionization, coordination, polymerization, and linkage.

3. Stereoisomers have the same empirical formula and the same atom-to-atom bonding sequence, and describe identical coordination polyhedra, but the atoms differ in their arrangement in space.

4. Square planar, four-coordinate complexes can form only *cis* and *trans* diastereoisomers; enantiomers are possible only if the ligands are chiral. There are no known tetrahedral complexes with four different ligands, so chiral tetrahedral complexes arise only if unsymmetrical, bidentate ligands are used.

5. Both "optical" and "geometrical" types of isomerism are possible for octahedral complexes, and commonly arise from the distribution of monodentate or multidentate ligands about the central metal.

REFERENCES

[1] K. Mislow, "Introduction to Stereochemistry," W. A. Benjamin, Inc., New York, 1965.

[2] R. T. Morrison and R. N. Boyd, "Organic Chemistry," 3rd edition, Allyn and Bacon, Inc., Boston, 1975.

[3] A. L. Ternay, Jr., "Contemporary Organic Chemistry," 2nd edition, W. B. Saunders Company, Philadelphia, 1979.

[4] L. S. Dollimore, R. D. Gillard, and I. H. Mather, *J. C. S. Dalton*, 518 (1974).

[5] Alfred Werner's papers of greatest importance to coordination chemistry have been translated and annotated by G. B. Kauffman, "Classics in Coordination Chemistry," Dover Publications, Inc., New York, 1968.

[6] G. B. Kauffman, *Coord. Chem. Rev.*, **11**, 161 (1973).

[7] G. B. Kauffman, *Coord. Chem. Rev.*, **12**, 105 (1974).

[8] R. J. Balahura and N. A. Lewis, *Coord. Chem. Rev.*, **20**, 109 (1976).

[9] J. L. Burmeister, *Coord. Chem. Rev.*, **3**, 225 (1968); **1**, 205 (1966).

[10] A. H. Norbury and A. I. P. Sinha, *Quart. Revs.*, **24**, 69 (1970).

[11] W. L. Jolly, "The Synthesis and Characterization of Inorganic Compounds," Prentice-Hall, Inc., Englewood Cliffs, N. J., 1970, pp. 462–463.

[12] S. Ahrland, J. Chatt, and N. R. Davies, *Quart. Revs.*, **11**, 265 (1958).

[13] It is difficult to recommend any one or several books on the stereochemistry of coordination complexes, since so many have been written on this topic. However, the three listed below provide a good introduction.
   a. This book and the next provide a rather dated but excellent account of the early work in the field: "The Chemistry of Coordination Compounds," J. C. Bailar, Jr., ed., Reinhold Publishing Corp., New York, 1956.
   b. J. Lewis and R. G. Wilkins, "Modern Coordination Chemistry," Interscience, New York, 1960.
   c. An excellent survey of more recent work is given by C. J. Hawkins, "Absolute Configuration of Metal Complexes," Interscience, New York, 1971.

[14] A. Werner, *Zeitschrift für anorganische Chemie*, **3**, 267–330 (1893); this paper is included in ref. 5.

[15] E. L. Eliel, *J. Chem. Educ.*, **48**, 163 (1971).

[16] J. A. Hirsch, "Concepts in Theoretical Organic Chemistry," Allyn and Bacon, Boston, 1974.

[17] N. L. Allinger and E. L. Eliel, eds., "Topics in Stereochemistry," Vol. 1, Interscience, New York, 1967.

[18] G. R. Brubaker, *J. Chem. Educ.*, **51**, 608 (1974).

[19] S. Kirschner, "Optically Active Coordination Compounds," in "Preparative Inorganic Reactions," W. L. Jolly, ed., Vol. 1, Interscience, New York, 1964, p. 29.

[20] G. B. Kauffman and E. V. Lindley, Jr., *J. Chem. Educ.*, **51**, 424 (1974). This paper describes the resolution of $[Co(en)_2(NH_3)Br]^{2+}$, an experiment suitable for the undergraduate laboratory.

[21] A. A. Grinberg, *Helv. Chim. Acta*, **14**, 455 (1931).

[22] M. M. Jones, "Elementary Coordination Chemistry," Prentice-Hall, Inc., Englewood Cliffs, N. J., 1964, p. 59.

[23] F. Basolo and R. G. Pearson, *Prog. Inorg. Chem.*, **4**, 381 (1962).

[24] R. H. Holm, G. W. Everett, Jr., and A. Chakravorty, *Prog. Inorg. Chem.*, **7**, 83 (1966).

[25] F. Basolo and R. G. Pearson, "Mechanisms of Inorganic Reactions," 1st edition, Wiley, New York, 1958, p. 284.

[26] A. Werner, *Berichte der Deutschen Chemischen Gesellschaft*, **44**, 1887 (1911); see page 159 of ref. 5.

[27] J. C. Bailar, Jr., *J. Chem. Educ.*, **34**, 334 (1957); see also S. A. Meyper, *ibid.*, **34**, 623 (1957).

[28] *Inorg. Chem.*, **9**, 1 (1970). K. F. Purcell and J. C. Kotz, "Inorganic Chemistry," W. B. Saunders Company, Philadelphia, 1977, pp. 630–635.

Appendix to Chapter 11

Here we illustrate an "ordered pair" approach to finding all diastereomers of an octahedral complex.

A. All monodentate ligands.
   1. M$abcdef$
      a. The ligands are *arbitrarily* coded with letters of the alphabet. Since *trans* positions are stereochemically unique, we may completely define each diastereomer by denoting the three *trans* ligand pairs; for instance, (ab) (cd) (ef) symbolizes the structure

      b. To keep from counting the same diastereomer twice, we require that the first-written ligand in each parenthetical pair precede its partner in the master list ($a > b > c > d > e > f$):

$$(a\ c)\ (b\ d)\ (e\ f),$$

but not
$$(a\ c)\ (d\ b)\ (e\ f),$$

satisfies this "ordered pair" rule.

      c. Also to prevent duplication, we require the first ligand in each of the three pairs to be ordered. Thus

$$(a\ c)\ (b\ e)\ (d\ f)$$

but not
$$(a\ c)\ (d\ f)\ (b\ e)$$

satisfies the rule.

      d. These rules mean that we always have ligand a first in the first pair. Now consider *all possible* remaining ligands to be *trans* to a. This will define column headings as below.

      e. In each column you list the remaining possible *trans* pairings, obeying the "ordered pair" rule. Having picked a partner for a in the column head pair, only the highest member of the remaining four ligands can lead in the second pair. Also notice that once you have picked a pair for the second axis, no choice remains for the last *trans* pair.

from d.	($a\,b$)	($a\,c$)	($a\,d$)	($a\,e$)	($a\,f$)
from e.	(cd) ( ) (ce) ( ) (cf) ( )	(bd) ( ) (be) ( ) (bf) ( )	(bc) ( ) (be) ( ) (bf) ( )	(bc) ( ) (bd) ( ) (bf) ( )	(bc) ( ) (bd) ( ) (be) ( )

In this way you can write down the 15 diastereomers of M$abcdef$. It is now a simple matter to draw structures for these pairs. Those for the first two columns are

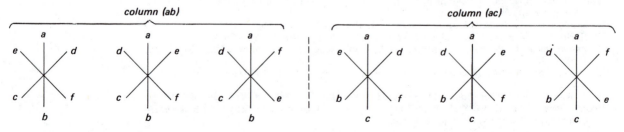

If two or more ligands are the same, the same procedure is followed, taking care not to duplicate in one column a pairing already written in an earlier column.

2. Ma_3def

from d. { (a a)   (a d)   (a e)   (a f)

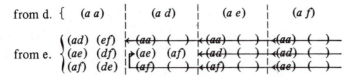

from e. { (ad) (ef)   (ae) (df)   (af) (de)

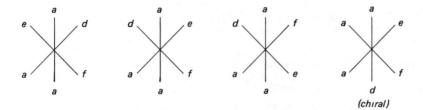

Lines are drawn through column entries that have been written down in a preceding column. For Ma_3def there are only four diastereomers:

Lines are drawn through column entries that have been written down in a preceding column. For Ma_3def there are only four diastereomers:

To determine the number of enantiomer pairs, inspect each structure for chirality (lack of an S_n axis). Only the last structure drawn lacks a reflection plane. There are, therefore, four diastereomers, one of which has an enantiomer for a total of five stereoisomers.

B. Multidentate ligands. As multidentate ligands (usually) form five- and six-membered rings with the metal, consecutive donor atoms in the ligand cannot appear *trans* to each other, that is, not in the same ordered pair. For example, in $[M(ABB'A')ef]$, the pairs (AB), (BB'), and $(B'A')$ are not allowed. An example of an $ABB'A'$ ligand is $H_2N(C_2H_4)NH(C_2H_4)NH(C_2H_4)NH_2$.

from d. {   (AB')   (AA')   (Ae)   (Af)

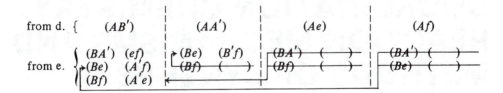

from e. { (BA') (ef)   (Be)   (Bf) (A'e)

Since A and A', and B and B', are equivalent donor atoms, ignore the primes when you check for redundancies in the column entries. In conclusion, $[M(ABB'A')ef]$ has four diastereomers, three of which have non-superimposable mirror images.

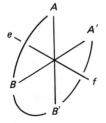

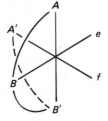
(chiral)

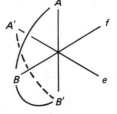
(chiral)

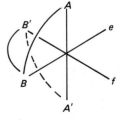
(chiral)

12. COORDINATION CHEMISTRY: REACTION MECHANISMS AND METHODS OF SYNTHESIS; ELECTRON TRANSFER REACTIONS

This is the first of three chapters that examine the main forms of re-activity displayed by coordination compounds and the usefulness of this reactivity in the synthesis of other coordination compounds.[1-4] The approach adopted here is to begin with changes occurring within the metal ion itself; that is, changes in oxidation state brought about by electron transfer from the metal ion of one compound to the metal ion of another.[5-10] Subsequent chapters deal with reactions involving changes in the inner coordination sphere of a complex; that is, the substitution of one ligand for another, rearrangement of ligands, and chemical reactions of the ligands themselves.

*In this chapter you shall see that electron transfer reactions may occur by either or both of two mechanisms: (1) **Outer sphere mechanism**—the transfer of an electron between two complexes whose inner or first coordination spheres remain intact. (2) **Inner sphere mechanism**—the transfer of an electron through a ligand that is simultaneously a member of the inner or first coordination spheres of both oxidant and reductant. Atom or group transfer may or may not accompany this electron transfer. However, it is often the case that, as an electron is transferred from reductant to oxidant through a bridging ligand, that ligand is transferred in the opposite direction.*

One of the most important topics in any introductory chemistry course is a discussion of oxidation-reduction reactions, reactions wherein there is a change in the formal oxidation states of the substances involved. However, it is often not pointed out that there are formally two types of reactions falling within the scope of redox processes: (i) reactions involving simple *electron transfer*, and (ii) reactions that can be considered as *atom transfer* reactions, both with and without electron transfer.

An example of a reaction that truly occurs by *electron transfer* is that between two different oxidation states of osmium.

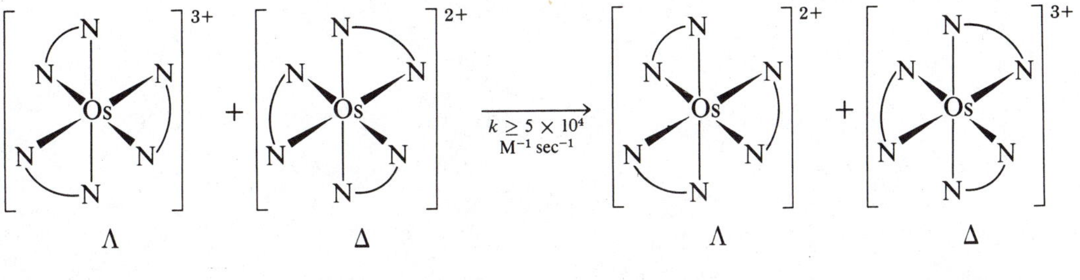

$$(12\text{-}1)$$

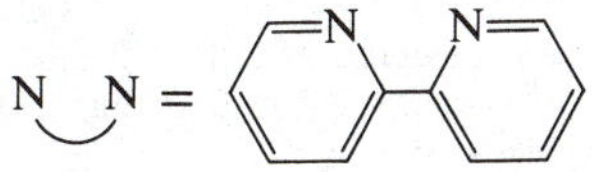

The two chiral reactants do not themselves racemize, but, when they are mixed, electron transfer leads to a rapid loss of optical activity since the Δ and Λ forms are equally distributed between the two oxidation states at equilibrium.[11] A great many other reactions also occur by direct transfer of electrons, and we shall turn to them momentarily.

In an *atom transfer reaction*, the fact that oxidation-reduction has occurred is a matter of the way in which one formally counts electrons. For example, in the oxidation of NO_2^- by OCl^-, the nitrogen oxidation state changes from +3 to +5, while chlorine changes from +1 to −1.

(12-2)
$$NO_2^- + OCl^- \rightarrow NO_3^- + Cl^-$$

By labeling the oxygen in OCl^-, the mechanism of this reaction has been shown to involve oxygen atom transfer (many such reactions were discussed in Chapter 7).[12]

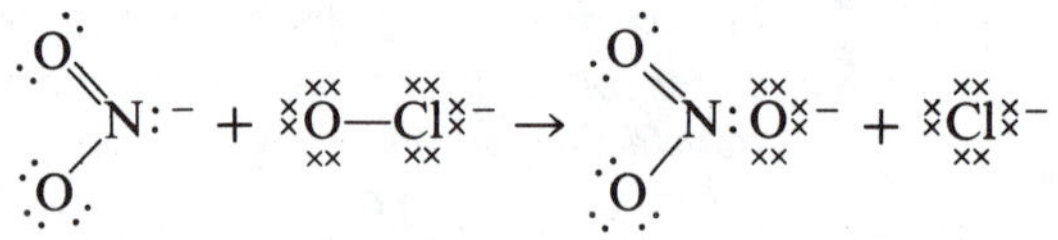

The change in formal oxidation states comes about because electrons that were formally non-bonding (the N lone pair) now become bonding (the N−O σ bond pair), and those that were previously counted as bonding (the O−Cl bond pair) are now non-bonding (the Cl^- lone pair).

Oxidation-reduction reactions in transition metal chemistry can also occur by atom transfer. For example, in acid solution

(12-3)

$$[Co^{III}(NH_3)_5Cl]^{2+} + [Cr^{II}(H_2O)_6]^{2+} \xrightarrow[-H_2O]{} [(NH_3)_5Co^{III}-Cl: \rightarrow Cr^{II}(H_2O)_5]$$

$$[Co(H_2O)_6]^{2+} + 5NH_4^+ + [Cr(H_2O)_5Cl]^{2+} \xleftarrow[+H_2O]{H^+} [(NH_3)_5Co^{II} \leftarrow\!-\; :Cl-Cr^{III}(H_2O)_5]$$

This reaction is similar to the NO_2^-/OCl^- reaction in that anion transfer has occurred. However, there has also been electron transfer in a direction opposite to the direction of ion transfer. In reaction (12-3), water on the Cr(II) center is substituted by a chloride [already attached to Co(III)] utilizing a lone pair of electrons; as the chloride moves from being cobalt-bound to being chromium-bound, there is a net flow of one electron in the opposite direction.†

$$Co^{+++}-Cl: + Cr^{++} \rightarrow [Co^{+++}\overset{e^-}{\underset{\curvearrowright}{-Cl:}} \rightarrow Cr^{++}] \rightarrow Co^{++} + :Cl-Cr^{+++}$$

---

†This reaction may be considered as either Cl atom transfer or Cl^- ion transfer + transfer of an electron. The two points of view are equivalent, but the latter is generally more consistent with the donor role of Cl^- ion complexes.

Reactions such as this occur by what is commonly known as an "inner sphere mechanism," because electron flow occurs by virtue of an atom or group held in common in the inner or first coordination spheres of the two metal centers; we shall discuss such reactions as a separate topic later.

Another type of atom transfer redox reaction is illustrated by the addition of molecular chlorine [equation (12-4)] or oxygen to an iridium(I) compound. In such

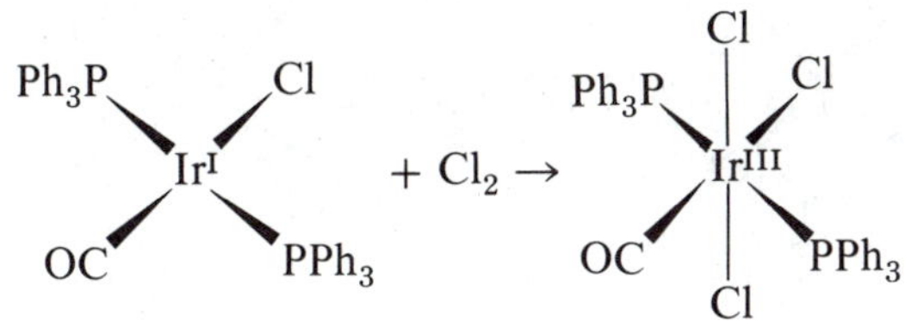

$$
\begin{array}{c}
Ph_3P \diagdown \quad \diagup Cl \\
Ir^I \\
OC \diagup \quad \diagdown PPh_3
\end{array}
\; + \; Cl_2 \; \rightarrow \;
\begin{array}{c}
\qquad Cl \\
Ph_3P \diagdown \; | \; \diagup Cl \\
Ir^{III} \\
OC \diagup \; | \; \diagdown PPh_3 \\
\qquad Cl
\end{array}
\qquad (12\text{-}4)
$$

an addition reaction there is no necessity to postulate the formal transfer of an electron or electrons, but, because of the way in which one formally counts electrons, the iridium is said to be oxidized to iridium(III) and chlorine is reduced to Cl^-. Reactions of this type have come to be known as "oxidative additions," and they are now recognized as an important class of reactions. They were first introduced in Chapter 5 (formation of O_2 adducts) and discussed again in Chapter 7 (phosphorus chemistry); one further example will be described in the section on Synthesis of Coordination Compounds later in this chapter, and we shall describe them in detail at a point where they are most pertinent, that is, in Chapter 17.

Study of redox reactions, especially those with electron transfer, has generally accelerated over the past decade; though far from complete, the scientific community's understanding has increased to the point where a reasonable interpretive framework is now in place. Indeed, it is now possible to begin an attack on redox reactions of ultimate importance—i.e., biological electron transfer.[13] The functions of hemoglobin (O_2 transport in plasma), myoglobin (oxygen storage in muscle), transferrin (iron storage in plasma), ferritin (iron storage in cells), and cytochrome c (electron transport) all involve in some manner reversible changes in the oxidation states of one or more iron atoms. For example, the transfer of iron from the gut to plasma is thought to involve the series of steps shown in Figure 12-1. Because of the obvious importance of these biological redox reactions, the case for a thorough study of electron transfer reactions in general hardly needs further justification.

MECHANISMS OF ELECTRON TRANSFER REACTIONS

Electron transfer reactions may occur by either or both of two mechanisms: outer sphere and inner sphere. In principle, an *outer sphere mechanism involves electron transfer from reductant to oxidant, with the coordination shells or spheres of each*

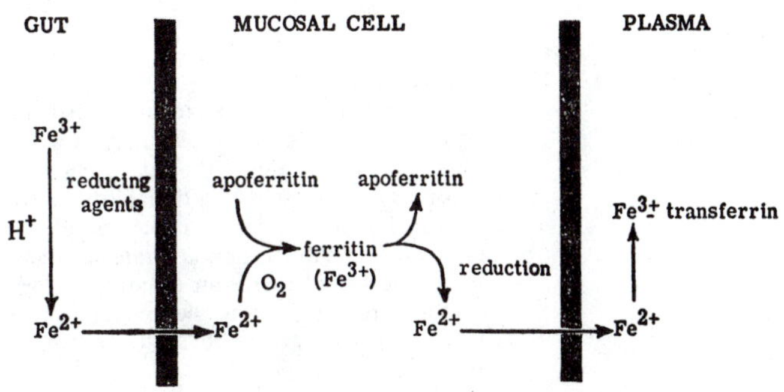

Figure 12-1. A possible mechanism for the absorption of iron from the gut into plasma. Following reduction from the +3 state, iron is taken up by the mucosal cell and is converted, with oxidation, into ferritin. Ferritin is an iron storage protein that consists of a shell of protein subunits surrounding a core of ferric hydroxyphosphate. Iron is then released and passed through the cell wall in the reduced form into the plasma, where it is stored in the +3 form in transferrin. [From R. R. Crichton, *Structure and Bonding*, **17**, 67 (1973).]

staying intact. That is, one reactant becomes involved in the outer or second coordination sphere of the other reactant, and an electron flows from reductant to oxidant. For example,

(12-5)

$$\left[\begin{array}{c}NC \quad \underset{C}{\overset{N}{C}} \quad CN \\ Fe \\ NC \quad \underset{N}{\overset{C}{C}} \quad CN\end{array}\right]^{4-} + \left[\begin{array}{c}Cl \quad \overset{Cl}{} \quad Cl \\ Ir \\ Cl \quad \underset{Cl}{} \quad Cl\end{array}\right]^{2-} \xrightarrow[\text{M}^{-1}\text{sec}^{-1}]{k = 4.1 \times 10^5} \left[\begin{array}{c}NC \quad \underset{C}{\overset{N}{C}} \quad CN \\ Fe \\ NC \quad \underset{N}{\overset{C}{C}} \quad CN\end{array}\right]^{3-} + \left[\begin{array}{c}Cl \quad \overset{Cl}{} \quad Cl \\ Ir \\ Cl \quad \underset{Cl}{} \quad Cl\end{array}\right]^{3-}$$

Such a mechanism is established when rapid electron transfer occurs between two substitution-inert complexes.† *An **inner sphere mechanism**, on the other hand, is one in which the reductant and oxidant share a ligand in their inner or primary coordination spheres, the electron being transferred across a bridging group* as in reaction (12-6).§

(12-6)

$$[Co(NH_3)_5NCS]^{2+}$$
$$+$$
$$[Cr(H_2O)_6]^{2+}$$

$$\xrightarrow{-H_2O}$$

$$\left[\begin{array}{c}H_2O \quad \overset{H_2O}{} \quad OH_2 \\ Cr \\ H_3N \quad \overset{H_3N}{} \quad NCS \quad H_2O \quad OH_2 \\ Co \\ H_3N \quad \underset{NH_3}{} \quad NH_3\end{array}\right]^{4+}$$

$$\xrightarrow{+H_2O/H_3O^+}$$

$$[Co(H_2O)_6]^{2+} + 5NH_4^+$$
$$+$$
$$[Cr(H_2O)_5SCN]^{2+}$$

There are clearly two prerequisites for the operation of an *inner sphere mechanism.* The *first requirement* is that *one reactant* (usually the oxidant) *possess at least one ligand capable of binding simultaneously to two metal ions,* however transiently. (As you shall see later, this requirement suggests a way to synthesize linkage isomers.) Although this bridging ligand is frequently transferred from oxidant to reductant in the course of electron transfer, this need not be the case: ligand transfer is not a requirement of an inner sphere mechanism. The *second requirement* for the inner sphere mechanism is that *one ligand of one reactant* (usually the reductant) *be substitutionally labile;* that is, *one ligand must be capable of being replaced by a bridging ligand in a facile substitution process.*†

There are numerous examples of both mechanistic types. It is useful to keep in mind the fact that the outer sphere mechanism is always possible, but, in certain cases, the inner sphere mechanism becomes competitive or even dominant.

---

†Coordination complexes may also undergo substitution reactions; that is, one or more of the ligands on the original complex may be substituted by other ligands. Generally, a complex is classified as inert to substitution if the rate of substitution is low, that is, a half-life greater than 1 minute. The opposite of inertness is kinetic lability; a complex is labile if its substitution reactions have half-lives less than 1 minute. In this chapter on electron transfer reactions, a substitution-inert complex is one that undergoes substitution at a rate substantially less than the rate of electron transfer.

§Reaction (12-6) is an example of "remote attack." That is, the reductant, Cr^{2+}, is attached to an atom "remote" from the oxidant instead of the atom directly bound to, or adjacent to, the oxidant, the nitrogen. You shall see other examples of both "adjacent" and "remote" attack (see page 407, for example).

Key Ideas Concerning Electron Transfer

Before beginning a discussion of the two main mechanistic pathways for electron transfer reactions, you should recognize that there are several key ideas that apply to both of these paths, inner- and outer-sphere.

1. In order for electron transfer to occur, one requirement seems to be that the molecular orbital in which the donated electron originates in the reducing agent, and the molecular orbital into which it is transferred in the oxidizing agent, must be of the same type. Efficient outer sphere electron transfer requires that both the donor and receptor orbitals be π^* orbitals, whereas inner sphere reactions require donor and receptor mo's to both be π^* or both be σ^*. (Ordinarily, one would expect the HOMO of the reducing agent to be the donor orbital and the LUMO of the oxidizing agent to be the receptor orbital.)

2. In most chemical reactions there is a chemical activation process that gets either or both of the reactants into the proper configuration for reaction. If either of the reagents in an electron transfer reaction does not meet the symmetry requirements noted above, even greater chemical activation—encompassing both structural deformation and electron configuration change—will be required. Such reactions will, in general, be slower than those requiring no electron configuration change.

3. There is a thermodynamic influence on the activation energy barrier. Reactions that require a HOMO and/or LUMO change can have lower barriers than might otherwise be predicted, if the reaction has a negative free energy change.

Outer Sphere Electron Transfer Reactions

The elementary steps involved in the outer sphere mechanism are:

Formation of a precursor (cage) complex	$Ox + Red \rightleftharpoons Ox\|Red$
Chemical activation of the precursor, electron transfer, and relaxation to the successor complex	$Ox\|Red \rightleftharpoons {}^-Ox\|Red^+$
Dissociation to the separated products	${}^-Ox\|Red^+ \rightleftharpoons Ox^- + Red^+$

(12-7)

The first step of the mechanism is the formation of the so-called precursor complex, wherein the distance between the reactant centers (the metal ions) is approximately that required for electron transfer, but their relative orientations and internal structures do not yet permit transfer.

The second step then involves both solvent cage and precursor structural changes to accommodate electron transfer. Within the precursor there must be both a reorientation of the oxidant and reductant complexes and, within those complexes, structural changes that define the chemical activation process for electron transfer. As the transition state is passed, there follows the completion of the electron transfer and final relaxation of the oxidant and reductant structure.

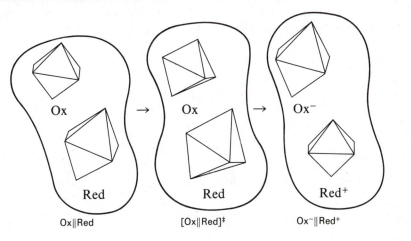

(12-8)

Ox‖Red   [Ox‖Red]‡   Ox⁻‖Red⁺

In the simplified schematic above, the oxidant is shown to enlarge while the reductant contracts in size. You know that these general changes are required because metal d electrons usually play an anti-bonding role. To accommodate the incoming electron, the oxidant must increase its metal-ligand distances, while the metal-ligand distances of the reductant (which loses a d^* electron) shorten. (In specific cases, there may be other, angular distortions of the ligands about the metal ions as well.) In short, partial bond-breaking occurs in the oxidant, while partial bond-making follows in the reductant. The schematic is also drawn to remind you that solvent reorganization to accommodate these precursor structural changes can be an important contributor to the activation free energy, $\Delta G^{\ddagger}$.

The final step in the outer sphere process, as in any reaction, is the separation of the product ions or molecules.

CHEMICAL ACTIVATION AND ELECTRON TRANSFER

Probably the most important determinant of the rate of electron transfer processes between complexes, whether outer- or inner sphere, is the π^* or σ^* nature of the electron donor mo of the reductant (the orbital from which the electron is transferred) and the receptor mo of the oxidant (the orbital into which the electron moves). Quite generally, you could expect more facile electron transfer when both donor and receptor mo are of the π^* type. One reason for this is that reductant/oxidant activation (changing M—L distance) is usually less for a change in $d\pi^*$ than for a change in $d\sigma^*$ electron density.

Another general principle is that the better the donor-receptor mo overlap and mixing are, the easier is the electron transfer. Since $d\pi^*$ electrons in O_h complexes are exposed everywhere but at the vertices of the octahedron (*e.g.*, the d_{xy} orbital, **1**), while $d\sigma^*$ electrons at the vertices (*e.g.*, the $d_{x^2-y^2}$ orbital, **2**) are more shielded from the surroundings, $\pi^* \rightarrow \pi^*$ electron transfer should be faster than $\sigma^* \rightarrow \sigma^*$ transfer.

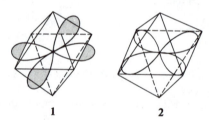

1   **2**

The representative data in Table 12-1 for outer sphere electron transfer reactions largely bear out such expectations. The first three examples require little chemical activation for M(II) → M(III) electron exchange because there is no change in the σ^* orbital configuration in either complex. For example, the change in metal-ligand distance in the Ru(III)-Ru(II) exchange illustrated below is only 4 pm.[14] For this reason, reactions such as these will generally be relatively rapid.

(12-9)

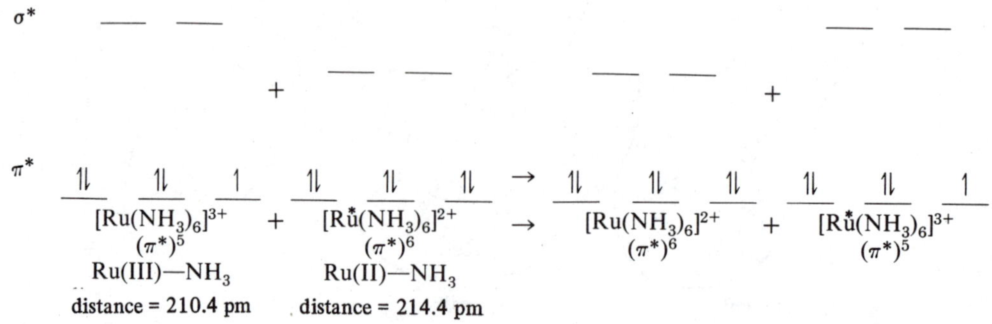

σ^* ___ ___   ___ ___

+ ___ ___   ___ ___ +

π^* $\uparrow\downarrow$ $\uparrow\downarrow$ $\uparrow$ $\uparrow\downarrow$ $\uparrow\downarrow$ $\uparrow\downarrow$ → $\uparrow\downarrow$ $\uparrow\downarrow$ $\uparrow\downarrow$ $\uparrow\downarrow$ $\uparrow\downarrow$ $\uparrow$

$[Ru(NH_3)_6]^{3+}$ + $[\overset{*}{Ru}(NH_3)_6]^{2+}$ → $[Ru(NH_3)_6]^{2+}$ + $[\overset{*}{Ru}(NH_3)_6]^{3+}$
$(\pi^*)^5$ $(\pi^*)^6$ $(\pi^*)^6$ $(\pi^*)^5$
Ru(III)—NH$_3$ Ru(II)—NH$_3$
distance = 210.4 pm   distance = 214.4 pm

TABLE 12-1

**SECOND ORDER RATE CONSTANTS FOR
SOME OUTER SPHERE ELECTRON
TRANSFER REACTIONS**

Reaction			Rate Constant k (M^{-1} sec^{-1})
		NET $\pi^* \to \pi^*$	
$Fe(phen)_3^{2+}$ $(\pi^*)^6$	+	$Fe(phen)_3^{3+}$ $(\pi^*)^5$	$\geq 3 \times 10^7$
$Ru(NH_3)_6^{2+}$ $(\pi^*)^6$	+	$Ru(ND_3)_6^{3+}$ $(\pi^*)^5$	8.2×10^2
$Ru(phen)_3^{2+}$ $(\pi^*)^6$	+	$Ru(phen)_3^{3+}$ $(\pi^*)^5$	$\geq 10^7$
		NET $\sigma^* \to \sigma^*$	
$Co(H_2O)_6^{2+}$ $(\pi^*)^5(\sigma^*)^2$	+	$Co(H_2O)_6^{3+}$ $(\pi^*)^6$	~ 5
$Co(NH_3)_6^{2+}$ $(\pi^*)^5(\sigma^*)^2$	+	$Co(NH_3)_6^{3+}$ $(\pi^*)^6$	$\leq 10^{-9}$
$Co(en)_3^{2+}$ $(\pi^*)^5(\sigma^*)^2$	+	$Co(en)_3^{3+}$ $(\pi^*)^6$	1.4×10^{-4}
$Co(phen)_3^{2+}$ $(\pi^*)^5(\sigma^*)^2$	+	$Co(phen)_3^{3+}$ $(\pi^*)^6$	1.1

One of the most striking features in Table 12-1 is the slow rate of the $[Co(NH_3)_6]^{2+}$/$[Co(NH_3)_6]^{3+}$ exchange in relation to the Fe(II)/Fe(III) and Ru(II)/Ru(III) systems. (Notice, however, the rate enhancement when phen is the ligand rather than NH$_3$ and ethylenediamine.) The most obvious distinction between the Co(II)/Co(III) systems and those of Fe(II)/Fe(III) and Ru(II)/Ru(III) in Table 12-1 is that the former requires change in the high-spin/low-spin conditions of the reductant and oxidant. This is just another way of saying that a change in the number of σ^* electrons in both oxidant and reductant must accompany electron transfer. And we should point out that, because of this, the change in Co—NH$_3$ distance is 17.8 pm, a change more than four times that in the analogous Ru(II)/Ru(III) exchange.[14] Presumably, major changes in Co—L bond distances must accompany electron transfer. As the associated energy changes (to some

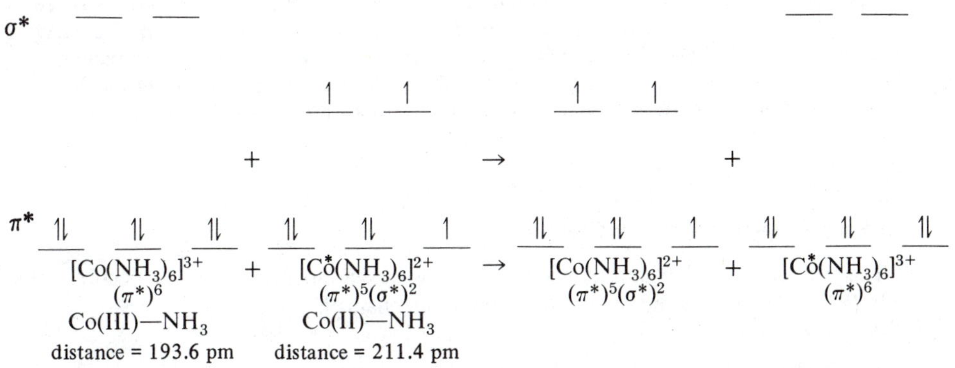

σ^* ⎯ ⎯ ⎯ ⎯

↑ ↑ ↑ ↑

$+$ → $+$ (12-10)

π^* ⇅ ⇅ ⇅ ⇅ ⇅ ↑ ⇅ ⇅ ↑ ⇅ ⇅ ⇅

$[Co(NH_3)_6]^{3+}$ $+$ $[C\overset{*}{o}(NH_3)_6]^{2+}$ → $[Co(NH_3)_6]^{2+}$ $+$ $[C\overset{*}{o}(NH_3)_6]^{3+}$

$(\pi^*)^6$ $(\pi^*)^5(\sigma^*)^2$ $(\pi^*)^5(\sigma^*)^2$ $(\pi^*)^6$

Co(III)—NH$_3$ Co(II)—NH$_3$

distance = 193.6 pm distance = 211.4 pm

degree accompanying or preceding electron transfer) are also large, we find a logical explanation for the rather large difference in velocities of high spin/high spin and low spin/low spin electron exchange on the one hand, and high spin/low spin exchange on the other. Furthermore, the remarkable effect of delocalized π acceptor ligands (*e.g.*, phen) on electron transfer rates is basically understood and confirms the idea that the outer sphere mechanism involves $\pi^* \to \pi^*$ electron transfer.

As noted above and in the section on Key Ideas, outer sphere reactions proceed by π^* electron transfer. If the ions involved do not have suitable electronic configurations in the ground state, then electronic excitation is required in addition to bond length changes. For Co(II)/Co(III) exchange reactions, the details of the chemical activation and electron transfer steps are difficult to conceptualize. However, one pathway, among several that can be illustrated, postulates the occurrence of Co(II) and Co(III) electronically excited *intermediates* prior to and following electron transfer.

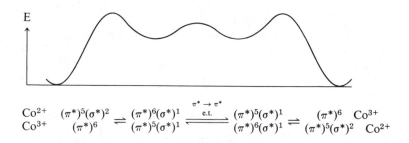

$$Co^{2+} \quad (\pi^*)^5(\sigma^*)^2 \rightleftharpoons (\pi^*)^6(\sigma^*)^1 \xrightleftharpoons[\text{e.t.}]{\pi^* \to \pi^*} (\pi^*)^5(\sigma^*)^1 \rightleftharpoons (\pi^*)^6 \quad Co^{3+}$$
$$Co^{3+} \quad (\pi^*)^6 \rightleftharpoons (\pi^*)^5(\sigma^*)^1 \xrightarrow{} (\pi^*)^6(\sigma^*)^1 \rightleftharpoons (\pi^*)^5(\sigma^*)^2 \quad Co^{2+}$$

CROSS REACTIONS AND THERMODYNAMICS

The reactions we have discussed thus far are *exchange reactions* occurring between two different oxidation states of the same compound. But what about *cross reactions*, reactions between two completely different ions or molecules? It is clear that, in order to achieve the maximum rate in an exchange reaction, one should have systems that effect $\pi^* \to \pi^*$ electron transfer. Cross reactions, however, are often rapid in spite of an otherwise large chemical activation (transfer involving at least one σ^* electron) because the large $\Delta E^{\ddagger}$ is partly "neutralized" by an attendant negative free energy change. Some examples of cross reactions are given in Table 12-2 along with standard electrochemical cell potentials; you will notice that there is a rough correlation between the rate of reaction and its free energy change (as given by $\Delta G^\circ = -nFE^\circ$).[15]

The outer sphere reductions of $[Ru(NH_3)_6]^{3+}$ and $[Co(NH_3)_6]^{3+}$ by Cr(II) and V(II) (see Table 12-2) provide some insight into the thermodynamics of the activation process. The Cr(II) outer sphere reductions should require greater $\Delta E^{\ddagger}$ than the V(II) reductions because the high spin ion $[Cr(H_2O)_6]^{2+}$ with a $(\pi^*)^3(\sigma^*)^1$ configuration requires activation toward the low spin $(\pi^*)^4$ configuration. On the other hand, little activation of $[V(H_2O)_6]^{2+}$ $[(\pi^*)^3]$ should be required, as there is no change in σ^* occupation on electron transfer. This expectation is clearly realized in Cr(II)/Cr(III) exchange ($k \le 10^{-5}$) compared with V(II)/V(III) exchange ($k \approx 10^{-2}$), where $\Delta G^\circ = 0$, and is generally seen in outer sphere Cr(II) and V(II) cross reactions even though ΔG° for Cr(II) reductions is more negative than for V(II) reductions.[16]

Inner Sphere Electron Transfer Reactions

The reduction of hexaamminecobalt(3+) by hexaaquochromium(2+) occurs rather slowly ($k = 10^{-3}$ M^{-1} sec^{-1}) by an outer sphere mechanism.

(12-11)

$$[Co(NH_3)_6]^{3+} + [Cr(H_2O)_6]^{2+} \xrightarrow{H^+} [Co(H_2O)_6]^{2+} + 6NH_4^+ + [Cr(H_2O)_6]^{3+}$$
$$(\pi^*)^6 \qquad\qquad (\pi^*)^3(\sigma^*)^1 \qquad\qquad (\pi^*)^5(\sigma^*)^2 \qquad\qquad\qquad (\pi^*)^3$$

TABLE 12-2

KINETIC PARAMETERS FOR SOME OUTER SPHERE REACTIONS

Reducing Agent	$\Delta E^{\ddagger}$ Required for Red. Ag.?	Oxidizing Agent	$\Delta E^{\ddagger}$ Required for Oxid. Ag.?	ΔE° (volts)	k_{obs} (M^{-1} sec^{-1})
Net $\sigma^{*} \rightarrow \sigma^{*}$					
Cr^{2+}	yes	$[Co(NH_3)_6]^{3+}$	yes	0.51	1.0×10^{-3}
Cr^{2+}	yes	$[Co(en)_3]^{3+}$	yes	0.17	3.4×10^{-4}
Cr^{2+}	yes	$[Co(phen)_3]^{3+}$	yes	0.83	3.0×10^{1}
Net $\sigma^{*} \rightarrow \pi^{*}$					
Cr^{2+}	yes	$[Ru(NH_3)_6]^{3+}$	no	0.51	2×10^{2}
Net $\pi^{*} \rightarrow \sigma^{*}$					
V^{2+}	no	$[Co(NH_3)_6]^{3+}$	yes	0.355	1.0×10^{-2}
V^{2+}	no	$[Co(en)_3]^{3+}$	yes	0.015	7.2×10^{-4}
V^{2+}	no	$[Co(phen)_3]^{3+}$	yes	0.675	3.8×10^{3}
Net $\pi^{*} \rightarrow \pi^{*}$					
V^{2+}	no	$[Ru(NH_3)_6]^{3+}$	no	0.355	80

However, if one ammonia ligand on Co(III) is substituted by Cl^{-}, reaction (12-12) now occurs.

$$[Co(NH_3)_5Cl]^{2+} + [Cr(H_2O)_6]^{2+} \xrightarrow{\text{H}^+} [Co(H_2O)_6]^{2+} + [Cr(H_2O)_5Cl]^{2+} + 5NH_4^+ \qquad (12\text{-}12)$$
$$(\pi^*)^6 \qquad\qquad (\pi^*)^3(\sigma^*)^1 \qquad\qquad (\pi^*)^5(\sigma^*)^2 \qquad\qquad (\pi^*)^3$$

and with a substantially greater rate ($k = 6 \times 10^5$ M^{-1} sec^{-1}). It is clear that different mechanistic pathways are probably utilized by these two reactions, a fact that Taube has elegantly demonstrated.[5,6,17] Indeed, it is likely that an inner sphere mechanism is followed in the reduction of [Co(NH$_3$)$_5$Cl]$^{2+}$; that is, the Cl^{-} ligand, while still attached to the Co(III), replaces an H$_2$O at Cr(II) to give an intermediate such as **3**, and electron transfer follows.

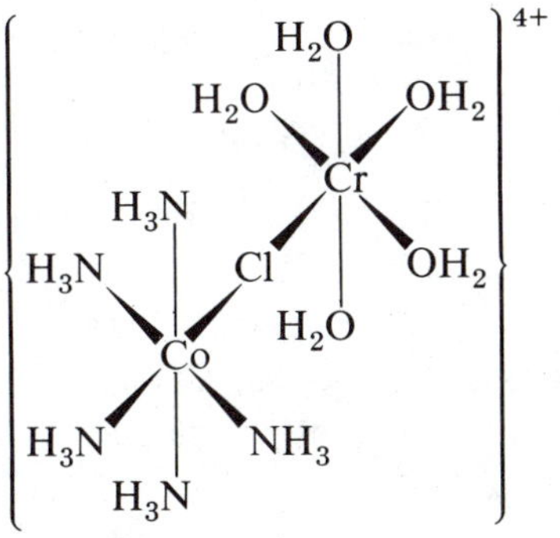

3

The paragraph below describes the experimental proof of this pathway.

The reductant, $[Cr(H_2O)_6]^{2+}$, is substitution labile,† as is the product $[Co(H_2O)_6]^{2+}$ (the specific rate constant for the exchange of chromium(II)-coordinated water with solvent water is greater than 10^9 sec^{-1}, while that for exchange with $[Co(H_2O)_6]^{2+}$ is 2×10^5 sec^{-1}). On the other hand, the two chloro complexes are substitution inert. Therefore, the only way in which chloride could be transferred from Co(III) to Cr(II) in a reaction where $k \approx 10^5$ M^{-1} sec^{-1} would be through some complex resulting from attack of $CoCl^{2+}$ on Cr^{2+}. The Cl^- must have been passed *directly* from one ion to the other. An alternative mechanism might have been outer sphere electron transfer followed by release of Cl^- by Co(II) to the solution and substitution of the free chloride onto Cr(III). However, this is quite unlikely since the specific rate constant for the replacement of water in $[Cr(H_2O)_6]^{3+}$ by Cl^- is 3×10^{-8} M^{-1} sec^{-1}. Just as convincing, though, was another experiment; upon adding free, radioactive Cl^- to the $[Co(NH_3)_5Cl]^{2+}/$ $[Cr(H_2O)_6]^{2+}$ solution, no radio-chloride was found in the product, $[Cr(H_2O)_5Cl]^{2+}$.

The general conclusion that can be drawn from Taube's experiment is that *an inner sphere mechanism can be unequivocally assigned when both the oxidant and the oxidized reducing agent are substitution inert and when ligand transfer from oxidant to reductant has accompanied electron transfer.* This criterion has been successfully applied to numerous reactions of $[Cr(H_2O)_6]^{2+}$ with $[Co(NH_3)_5X]^{n+}$ and $[Cr(H_2O)_5X]^{n+}$. However, we must emphasize that *ligand transfer is not a requirement of the inner sphere mechanism.*

The elementary steps in a generalized inner sphere mechanism (in aqueous solution) can be represented by:

(12-13)

Formation of precursor complex	$Ox-X + Red(H_2O) \rightleftharpoons Ox-X\cdots Red + H_2O$
Activation of precursor complex and electron transfer	$Ox-X\cdots Red \rightleftharpoons \bar{Ox}\cdots X-\overset{+}{Red}$
Dissociation to separated products	$\bar{Ox}\cdots X-\overset{+}{Red} + H_2O \rightleftharpoons Ox(H_2O)^- + RedX^+$

Such reactions generally show second order kinetic behavior. In some cases, the rate determining step is formation of the precursor complex. However, in most instances the rate determining step is rearrangement and electron transfer within the intermediate, or fission of the successor complex. Precursor formation and subsequent electron transfer are the most interesting aspects of these reactions and so are discussed briefly.

FORMATION OF PRECURSOR COMPLEXES

As you shall see in the next chapter on substitution reactions, octahedral complexes with a d^3 metal ion are relatively inert to substitution, while high spin d^4 and d^5 complexes are labile to substitution. For example, compare the following reactions for V^{2+} [a $(\pi^*)^3$ ion], Cr^{2+} [a $(\pi^*)^3(\sigma^*)^1$ ion], and Fe^{2+} [a $(\pi^*)^4(\sigma^*)^2$ ion]. Such differences

(12-14) $[V(H_2O)_6]^{2+} + NCS^- \rightarrow [V(H_2O)_5NCS]^+ + H_2O$ $k = 28$ $M^{-1}sec^{-1}$

(12-15) $[Cr(H_2O)_6]^{2+} + H_2O^* \rightarrow [Cr(H_2O)_5(H_2O^*)]^{2+} + H_2O$ $k = 10^9$ sec^{-1}

(12-16) $[Fe(H_2O)_6]^{2+} + H_2O^* \rightarrow [Fe(H_2O)_5(H_2O^*)]^{2+} + H_2O$ $k = 10^6$ sec^{-1}

in substitution rates will clearly be important in determining relative inner sphere electron transfer rates for V(II) and Cr(II). In fact, the data for V(II) reductions in Table 12-3

†See footnote on page 392.

suggest that when this ion functions as an inner sphere reductant, the reaction rate is controlled by the rate at which water is lost from the coordination sphere of $[V(H_2O)_6]^{2+}$.

The conclusions regarding V(II) reductions in Table 12-3 are in direct contrast with those concerning Cr(II) and Fe(II) reductions. The rates of their electron transfer reactions are considerably less than their rates of water substitution.

ELECTRONIC STRUCTURE OF THE OXIDANT AND REDUCTANT

The outer sphere reduction of $[Co(NH_3)_6]^{3+}$ by $[Cr(H_2O)_6]^{2+}$ is quite slow ($k = 1.6 \times 10^{-3}$ M^{-1} sec^{-1}). However, when one ammonia ligand is replaced by a halide, the following reaction, cited earlier, occurs:

$$[Co(NH_3)_5Cl]^{2+} + [Cr(H_2O)_6]^{2+} \xrightarrow{H^+} [Co(H_2O)_6]^{2+} + [Cr(H_2O)_5Cl]^{2+} + 5NH_4^+ \qquad (12\text{-}12')$$
$$(k = 6 \times 10^5 \text{ M}^{-1} \text{ sec}^{-1})$$

As you have already seen, it is presumed that the mechanism is inner sphere since the halide ion has been transferred to the oxidized reducing agent. However, what is more important now is that the reaction has been tremendously accelerated relative to the outer sphere mechanism for the similar reaction. Since a similar effect is observed for numeous other acidopentaamminecobalt(2+) complexes, some general explanation is called for.

The answer is found in a consideration of the symmetries of the reductant orbital from which the electron is lost and the oxidant orbital into which the electron moves.

TABLE 12-3

SECOND ORDER RATE CONSTANTS (k, M^{-1} sec^{-1}) FOR THE INNER SPHERE REDUCTION OF VARIOUS Co(III) COMPLEXES AND SOME Ru(III) COMPLEXES AT 25°. NOTE THAT R = Co(NH$_3$)$_5$

Oxidant	Reductant [HOMO]		
	$V^{2+}[(\pi^*)^3(\sigma^*)^0]$ net π^* to σ^*	$Cr^{2+}[(\pi^*)^3(\sigma^*)^1]$ net σ^* to σ^*	$Fe^{2+}[(\pi^*)^4(\sigma^*)^2]$ net σ^* to σ^*
RF^{2+}	—	2.5×10^5	6.6×10^{-3}
RCl^{2+}	7.6	6×10^5	1.4×10^{-3}
RBr^{2+}	25	1.4×10^6	7.3×10^{-4}
RI^{2+}	13	3.4×10^6	—
RN$_3^{2+}$	13	3.0×10^5	8.7×10^{-3}
RNCS^{2+}	0.3	1.9×10^1	3.0×10^{-3}
RSCN^{2+}	—	see text	1.2×10^{-1}
RC$_2$O$_4$H^{2+}	12.5	4.0×10^2	4.3×10^{-1}
Ru(NH$_3$)$_5$Cl^{2+}	—	3.5×10^4	—
Ru(NH$_3$)$_5$Br^{2+}	—	2.2×10^3	—
Ru(NH$_3$)$_5$I^{2+}	—	$<5 \times 10^2$	—

Data taken from: H. Taube, "Electron Transfer Reactions of Complex Ions in Solution," Academic Press, New York (1970), page 51; J. A. Stritar and H. Taube, *Inorg. Chem.*, **8**, 2284 (1969); J. Espenson, *Inorg. Chem.*, **4**, 121 (1965).

Examination of the data in Tables 12-2 and 12-3 for outer sphere and inner sphere reactions, respectively, gives the following general observations:

HOMO	LUMO	Example	Approximate acceleration on going from outer sphere to inner sphere mechanism for similar reaction
σ^*	σ^*	Cr^{2+}/Co^{3+}	10^{10}
σ^*	π^*	Cr^{2+}/Ru^{3+}	10^{2}
π^*	σ^*	V^{2+}/Co^{3+}	10^{4}
π^*	π^*	V^{2+}/Ru^{3+}	no acceleration

The obvious conclusion is that, while some rate acceleration has generally occurred on going from an outer sphere to an inner sphere mechanism in comparable reactions, the effect is most marked when both HOMO and LUMO are σ^*. This is somewhat surprising, since, as you saw in the discussion of outer sphere mechanisms, there is a requirement of considerable chemical activation because of bond distance changes in transfers to or from σ^* orbitals. Furthermore, effective mixing of σ^* donor and receptor mo's should not be as readily achieved as if both were of the more "exposed" π^* type. These impediments to rapid reaction are apparently largely overcome when the bridging ligand is intimately involved with donor and receptor mo's and when bridging group transfer accompanies electron transfer, as it does in reactions such as (12-12). Orbital following arguments are particularly useful in explaining this phenomenon. Such arguments were developed by Woodward and Hoffmann to explain certain types of organic reactions, and you may well have studied them in your organic course.[18] Unfortunately, a full exposition of the orbital following argument applied to electron transfer is beyond the scope of this text. However, if you imagine the simple fragment $M^{ox}-X \cdots M'^{red}$ [$M^{ox} = (d\sigma^*)^0$ and $M'^{red} = (d\sigma^*)^1$] as a model for the precursor complex for inner sphere $\sigma^* \rightarrow \sigma^*$ electron transfer, the essence of the orbital following argument is that, as X^- transfers from M to M', the orbital containing the odd electron becomes concentrated on M.[19]

THE ROLE OF THE BRIDGING LIGAND[9,10]

A glance at Table 12-3 shows that the nature of the bridge ligand is important, and extensive research has been done to understand its role in inner sphere transfer. As Haim has recently pointed out, "The role of the bridging ligand is . . . dual. It brings the metal ions together (thermodynamic contribution) and mediates the transfer of the electron (kinetic contribution)."[9] The thermodynamic contribution arises from factors important to the stability of the intermediate complex, and the kinetic contribution arises from factors such as oxidant-reductant reorganization and matching of donor and receptor molecular orbital types. Both inorganic and organic bridges have been studied. The former will be mentioned in the "Synthesis" section later in this chapter, but some aspects of organic bridges are discussed briefly here.

Studies of inner sphere reactions with *organic bridging ligands* are extremely interesting and informative, in that they show that reaction rates can be controlled by bridge steric effects, the point of attack by the reductant on the bridge, and the electronic structure of the bridge, including its reducibility.[9,10,20] Most such studies have centered on carboxylate complexes (such as **4**) and amine complexes (such as **5**). Second order kinetics are still observed for these reactions, and the proposed mechanisms all involve attack at a remote site rather than at the atom attached directly to the oxidant metal center.

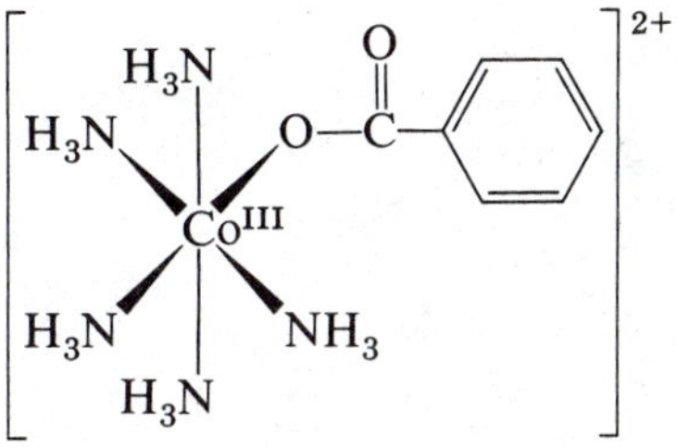

4

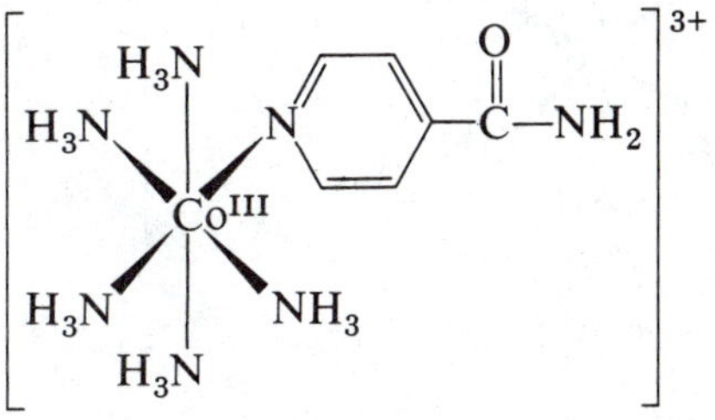

5

Pictured below are a few of the many acidopentaamminecobalt(III) complexes that have been observed to react with Cr(II). Attack by the Cr(II) ion is at the carbonyl oxygen of the carboxylate group, and it is clear that the rate constants for reduction by $[Cr(H_2O)_6]^{2+}$ fall with increasing steric bulk of the organic group (compounds **6** to **8**). Further, if another basic group is added, chelation of the reducing agent [Cr(II) in this case] is possible, and the rate constants increase on going from **7** to **10**, for example.

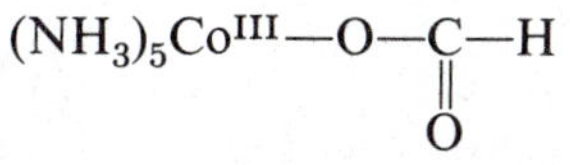

$(NH_3)_5Co^{III}-O-C-H$

$k_f = 7.2$ M^{-1}sec^{-1} for Cr(II)

6

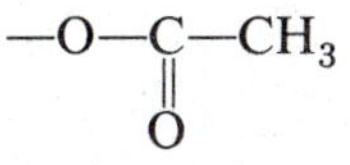

$-O-C-CH_3$

$k_f = 0.35$

7

$-O-C-C-CH_3$ with CH$_3$ groups

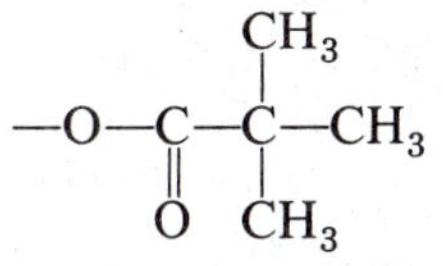

$k_f = 9.6 \times 10^{-3}$

8

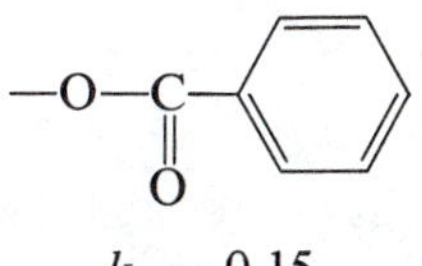

$k_f = 0.15$

9

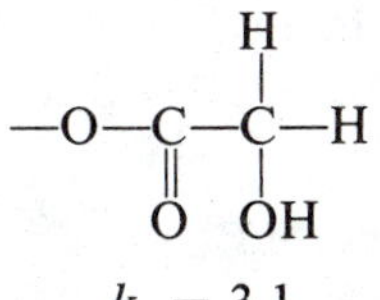

$k_f = 3.1$

10

Electron transfer *via* attack at a site even more remote than the carboxyl carbonyl oxygen is possible and has clearly been observed in the reactions of various Co(III) and Ru(III) complexes of substituted pyridines.[21] For example, pentaamminepyridinecobalt (3+) reacts relatively slowly with Cr(II) ($k_f = 4 \times 10^{-3}$ M^{-1} sec^{-1}) to give $[Cr(H_2O)_6]^{3+}$ *by an outer sphere pathway;* however, the isonicotinamide complex (**11**) reacts much more rapidly ($k_f = 17.4$ M^{-1} sec^{-1}), and the first Cr(III) product observed is **12**. These latter observations suggest that the attack of Cr(II) is at the remote carbonyl group in **11**. This attack site is expected from the Lewis base properties of amides (as discussed in Chapter 5), where the carbonyl oxygen is usually the preferred donor site when compared with the −NR$_2$ group.

(12-17)

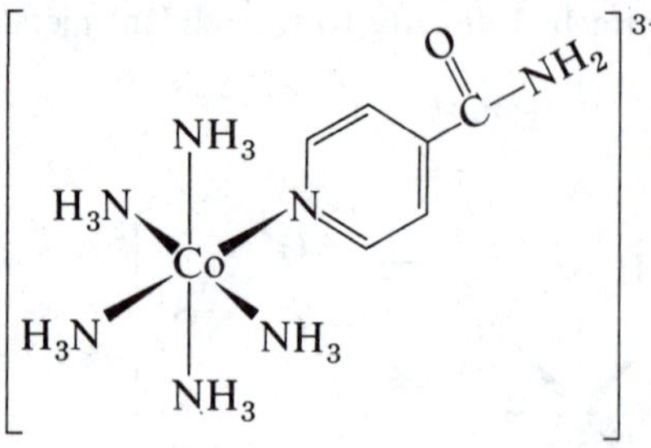

$$\left[\begin{array}{c} \text{(Co complex with NH}_3\text{ and pyridine carboxamide ligand)} \end{array} \right]^{3+} + [\text{Cr}(H_2O)_6]^{2+} \rightarrow$$

11

$$[\text{Co}(H_2O)_6]^{2+} + 5\,NH_3 + \left[\begin{array}{c} \text{(Cr complex with H}_2O\text{ and pyridine carboxamide ligand)} \end{array} \right]^{3+}$$

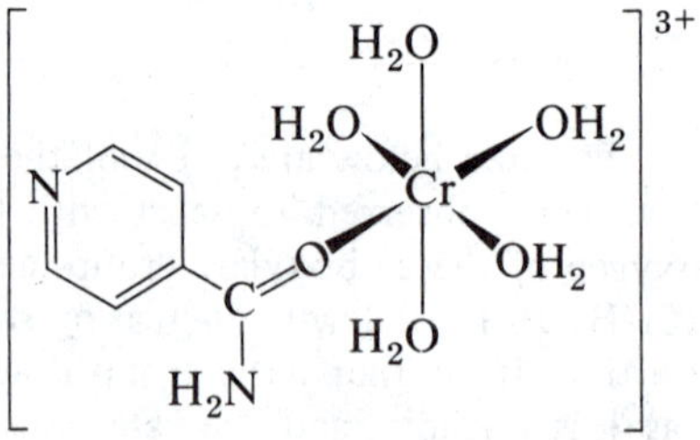

12

Two-Electron Transfers

To this point we have limited the discussion to reactions in which the oxidant and reductant change by only one unit in their formal oxidation state. However, two-electron changes have certainly been observed for the transition metals, and it may be shown that such reactions fit within the general framework presented thus far.

One of the best understood two-electron transfer reactions in transition metal chemistry is the Pt(II)-catalyzed exchange of free Cl$^-$ for chloride bound to a Pt(IV) compound.[22-24] That is,

(12-18)

$$trans\text{-}[\text{Pt(en)}_2\text{Cl}_2]^{2+} + {}^*\text{Cl}^- \xrightarrow[{[\text{Pt(en)}_2]^{2+}}]{} trans\text{-}[\text{Pt(en)}_2{}^*\text{ClCl}]^{2+} + \text{Cl}^-$$

The rate law for this reaction is

$$\text{Rate}_f = k[\text{Pt(II)}][\text{Pt(IV)}][\text{Cl}^-]$$

and the mechanism to account for this involves rapid addition of free chloride to the Pt(II) compound to form a five-coordinate, monopositive ion that then forms a six-coordinate, inner sphere complex with the Pt(IV) reactant. The two platinum atoms are now in very similar environments, and the transfer of two σ^* electrons, with accompanying anion transfer in the opposite direction, can readily occur.

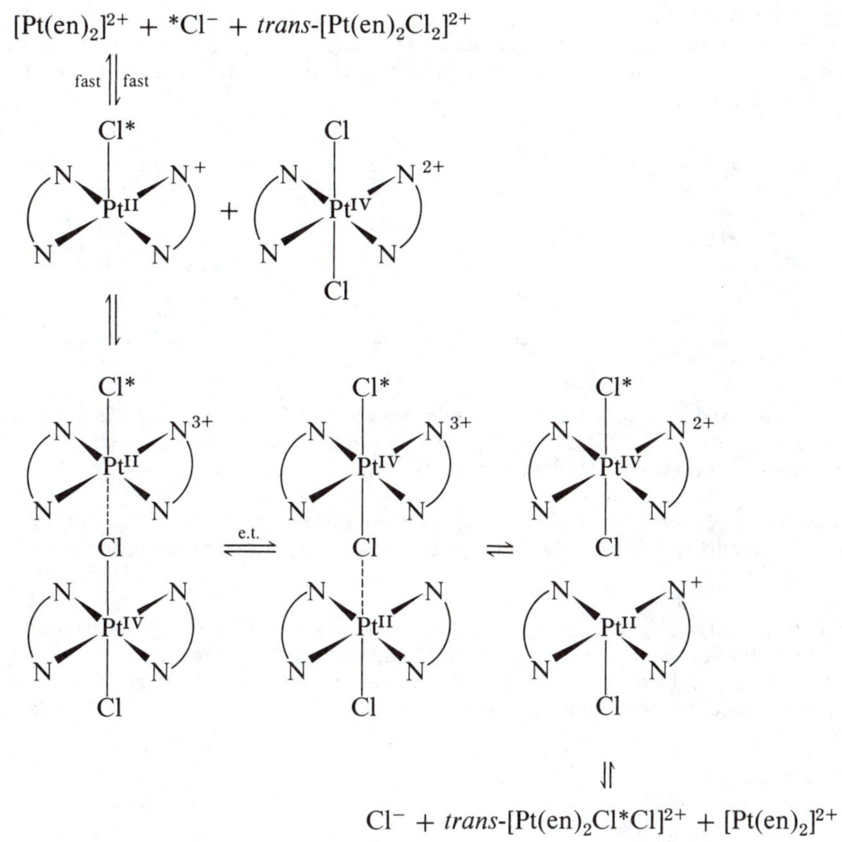

$$[Pt(en)_2]^{2+} + {}^*Cl^- + trans\text{-}[Pt(en)_2Cl_2]^{2+}$$

fast ⇅ fast

(structures: Pt^{II} with Cl^*, charge $+$) $\quad+\quad$ (structure: Pt^{IV} with two Cl, charge $2+$)

⇅

(12–19)

(three Pt^{II}/Pt^{IV} bridged structures) e.t. ⇌ ... ⇌

⇅

$$Cl^- + trans\text{-}[Pt(en)_2Cl^*Cl]^{2+} + [Pt(en)_2]^{2+}$$

The mechanism of this reaction was in large part established through the use of a radioactive tracer and a chiral Pt(IV) complex. That is, the exchange of radiochloride was found to occur at the same rate as the change in *optical rotation* in the reaction of trans-$[Pt(\ell\text{-pn})_2Cl_2]^{2+}$ with $[Pt(en)_2]^{2+}$.† Clearly, the rate determining step in the reaction

$$[Pt(en)_2]^{2+} + trans\text{-}[Pt(\ell\text{-pn})_2Cl_2]^{2+} \rightarrow [Pt(\ell\text{-pn})_2]^{2+} + trans\text{-}[Pt(en)_2Cl_2]^{2+} \qquad (12\text{--}20)$$

involves both the initially free chloride and the chiral, six-coordinate Pt(IV) compound.

---

1. The outer sphere electron exchange between $[IrCl_6]^{2-}$ and $[IrCl_6]^{3-}$ has been examined. Would you expect the reaction to be relatively fast or relatively slow? That is, in comparison with the rate constants in Table 12–1, will the iridium system have a rate constant close to that for the ruthenium compounds in Table 12–1 or closer to the cobalt compounds in that table?

STUDY QUESTIONS

---

†ℓ-pn is the ℓ-form of $H_2N-CHMe-CH_2-NH_2$. The idea behind the experiment described in the text is that the optical rotations of the reactant, *trans*-$[Pt(\ell\text{-pn})_2Cl_2]^{2+}$, and of the product, $[Pt(\ell\text{-pn})_2]^{2+}$, are different. Therefore, the rate of change of optical rotation from that characteristic of reactant to that appropriate for the product should give the rate of the reaction.

2. In the table below are given the rate constants for the exchange of an electron between the hexaaquo metal ions listed. All occur by outer sphere transfer. Rationalize these rate constants in terms of the general rules for relative rates of outer sphere transfers.

Metal Ion Pair	Rate Constant (M^{-1} sec^{-1})
Fe^{2+}/Fe^{3+}	4
Cr^{2+}/Fe^{3+}	2.3×10^3
V^{2+}/Fe^{3+}	1.8×10^4
Cr^{2+}/Cr^{3+}	2×10^{-5}
V^{2+}/V^{3+}	1×10^{-2}

You will need to know some reduction potentials of the metal ions involved in this question. For V^{3+}, Cr^{3+}, and Fe^{3+}, the reduction potentials are, respectively, -0.255 v, -0.41 v, and $+0.771$ v (all *vs.* the normal hydrogen electrode).

3. Reference 16 says that if V(II) reductions are faster than Cr(II) reductions of a common oxidant, this is considered indicative of an outer sphere reduction. Explain why this is the case.

4. Three complexes of dicarboxylic acids are illustrated below. Each reacts with Cr(II) with reduction of the bound Co(III). Complexes **A** and **B** react by an inner sphere mechanism, and complex **C** reacts by an outer sphere mechanism. What does this imply about the point of attack of Cr(II) on **A** and **B**? Why does **A** react more slowly than **B**?

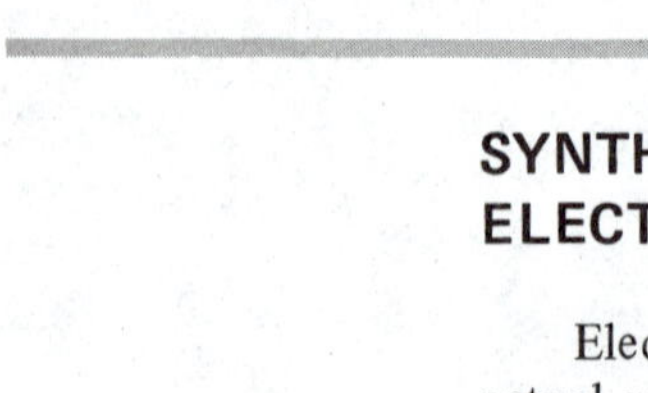

A $k_f = 0.075$ M^{-1}sec^{-1}    **B** $k_f = 0.2$ M^{-1}sec^{-1} 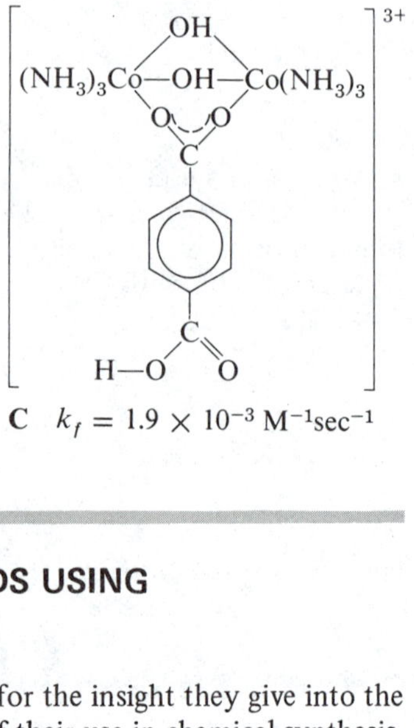   **C** $k_f = 1.9 \times 10^{-3}$ M^{-1}sec^{-1}

SYNTHESIS OF COORDINATION COMPOUNDS USING ELECTRON TRANSFER REACTIONS[25]

Electron transfer reactions have been studied largely for the insight they give into the actual process of transfer. However, there are examples of their use in chemical synthesis, and we shall discuss a few of the situations in which they have been used or observed.

Enantiomer Synthesis

As described in the previous chapter, the resolution of chiral compounds into their optically active enantiomers has been enormously important to the development of coordination chemistry. The resolution of $[Co(en)_3]^{3+}$, for example, was first described by Alfred Werner in 1912.[26] This was achieved by adding one enantiomer of a chiral anion [say, the (+) form of the chiral tartrate ion] to the solution of the racemic complex. The pair of enantiomeric complex ions will then form a pair of salts, (+)(+)

and (−)(+). These salt pairs are diastereomeric, and differ in their solubility as a result. In this particular case, the (+)(+) salt precipitates from solution, leaving the (−)(+) form in the mother liquor. A similar process was described in Chapter 11 (page 374).

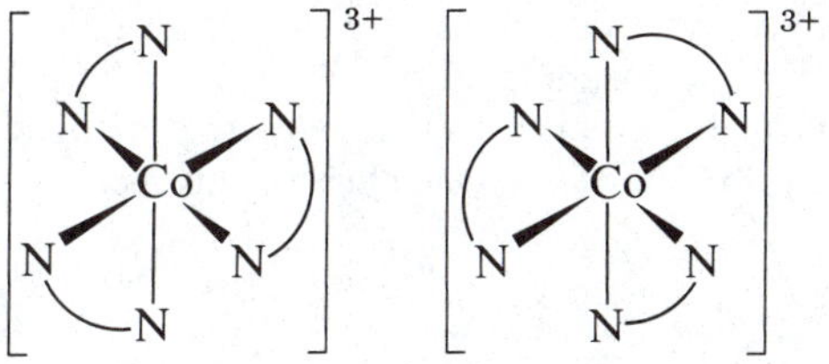

$$\Lambda\text{-}(+)\text{-}[Co(en)_3][(+)\text{-tartrate}]\,Cl \qquad\qquad \Delta\text{-}(-)\text{-}[Co(en)_3]^{3+}$$

precipitate

(12-21)

Assuming that the (+) and (−) enantiomers were originally formed in equal amounts, half of the original material can be isolated in one form or the other. However, by clever use of an electron transfer reaction you can convert *all* of the original material—both (+) and (−) enantiomers—to your choice of one *or* the other of the enantiomers. That is, a 100% yield can be obtained of one or the other of the enantiomers. The tris-ethylenediamine complexes of Co(II) and Co(III) undergo electron transfer by an outer sphere mechanism (see Table 12-1). More important, though, for our present purposes is the fact that $[Co(en)_3]^{2+}$ is configurationally labile; that is, if it could be obtained optically pure in one enantiomer or the other, it would racemize rapidly. Therefore, combination of the lability of the 2+ oxidation state and electron transfer between the oxidation states can be used in the isolation of (+)- or (−)-$[Co(en)_3]^{3+}$ in yields approaching 100%.[27] The essence of the procedure is as follows:

After precipitation of $[(+)\text{-}Co(en)_3][(+)\text{-tartrate}]Cl$ by the procedure outlined in reaction sequence (12–21), the mother liquor contains a preponderance of the more soluble diastereomer $[(-)\text{-}Co(en)_3][(+)\text{-tartrate}]Cl$. At this point a small amount of ethylenediamine and $CoCl_2 \cdot 6H_2O$, which presumably form $Co(en)_3Cl_2$, are added to the mother liquor. There then ensues a reaction between $[Co(en)_3]^{2+}$, formed with equal amounts of the (+) and (−) enantiomers, and (−)-$[Co(en)_3]^{3+}$.

$$\frac{1}{2}(+)\text{-}[Co(en)_3]^{2+} + \frac{1}{2}(-)\text{-}[Co(en)_3]^{2+} + (-)\text{-}[Co(en)_3]^{3+} \rightarrow$$

$$\frac{1}{2}(+)\text{-}[Co(en)_3]^{3+} + \frac{1}{2}(-)\text{-}[Co(en)_3]^{3+} + (-)\text{-}[Co(en)_3]^{2+}$$

(12-22)

The latter forms (−)-$[Co(en)_3]^{2+}$, which then rapidly racemizes to (+)- and (−)-$[Co(en)_3]^{2+}$, in effect replenishing the stock of $[Co(en)_3]^{2+}$ introduced by adding ethylenediamine and $CoCl_2 \cdot 6H_2O$. The other product of the electron transfer reaction is a racemic mixture of the enantiomers of the Co(III) complex. The (+) isomer of the newly formed 3+ complex is precipitated from solution to leave the (−) enantiomer, which can again undergo electron transfer with the Co(II) complex to again produce a racemic mixture of $[Co(en)_3]^{3+}$, the (+) enantiomer of which again precipitates from solution. Continuation of this cycle can obviously be used to convert virtually all of the (−) enantiomer originally present to the (+) enantiomer; alternatively, by starting with a tartrate anion of opposite chirality, all of the (+) enantiomer can be converted to (−) enantiomer. That is, the cycle of reactions can in principle lead ultimately to complete conversion of one enantiomer to another. In practice, however, the yield is only somewhat in excess of 75%.

The resolution of $[Co(en)_3]^{3+}$ as described above is a process catalyzed by a lower valent state of the same complex. Although this at first glance may seem to be an isolated case, reactions catalyzed in this manner are not at all uncommon, and some have synthetic implications.[28] One example is the Pt(II)-Pt(IV) two-electron transfer reaction described previously (p. 402). This has obvious synthetic utility, but the subject is left for you to explore in a Study Question at the close of this section.

Linkage Isomer Synthesis

One of the best ways to prove that an inner sphere electron transfer reaction has occurred is to be able to show that atom or group transfer has accompanied oxidation state change. As you have previously seen, the group that is transferred (which must be at least potentially ambidentate) is usually attached to the oxidizing agent, and in the course of the reaction it also becomes attached to the reducing agent. This means that one may be able to begin the reaction with an unsymmetrical ligand such as SCN⁻, attached to the oxidizing agent through either N or S, and finish the reaction with the ligand attached to the now-oxidized reducing agent through either S or N, respectively [reaction (12-23)].

(12-23)
$$\begin{Bmatrix} M^{III}-S\diagdown{}_{\displaystyle C}\\ {}_{\textstyle \diagup}\!\!\!\equiv N \\ M^{III}-N\equiv C-S \end{Bmatrix} + M'(II) \rightarrow \begin{Bmatrix} S-C\equiv N-M'^{III} \\ {}_{\textstyle N}\!\!\equiv C \diagdown S-M'^{III} \end{Bmatrix} + M(II)$$

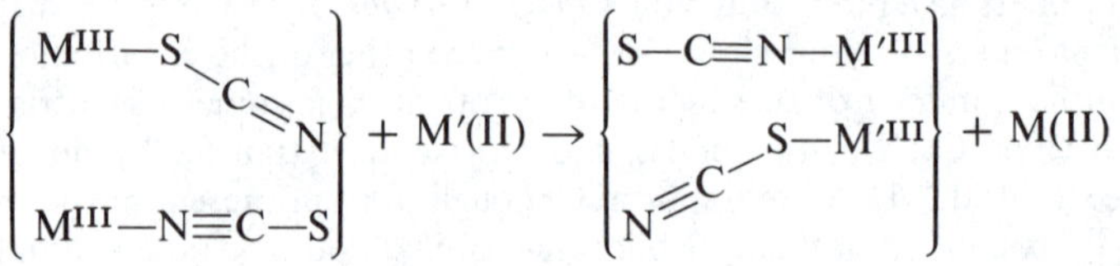

In principle, it is possible to run the reaction both ways, and, given that the product ions or molecules are inert to substitution, linkage isomers could be produced. No one has yet discovered a pair of linkage isomers that will produce another pair of linkage isomers by electron/group transfer. However, one example of the synthesis of a linkage isomer by inner sphere electron transfer is the following reaction, where the product is the S-bonded isomer.[29,30]

(12-24)
$$[Co(NH_3)_5NCS]^{2+} + [Co(CN)_5]^{3-} \xrightarrow[k = 1.1 \times 10^6 \, M^{-1}sec^{-1}]{}$$
$$[Co(CN)_5SCN]^{3-} + Co^{2+}(aq) + 5NH_3$$

The fact that it is indeed the S-bonded isomer that comes out of the electron transfer reaction is substantiated by the observation that it is quite different from the N-bonded isomer previously isolated from a substitution reaction (12-25).[31]

(12-25)
$$[Co(CN)_5(OH_2)]^{2-} + NCS^- \rightarrow [Co(CN)_5NCS]^{3-} + H_2O$$
$$\downarrow \text{ heat at } 150°/6 \text{ hr}$$
$$[Co(CN)_5SCN]^{3-}$$

One interesting aspect of S-bonded SCN⁻ ligands is that they can be attacked either at the coordinated sulfur end ("adjacent" attack) or at the free nitrogen end ("remote" attack). Indeed, using $[Cr(H_2O)_6]^{2+}$ as the reducing agent, the rates of attack at the

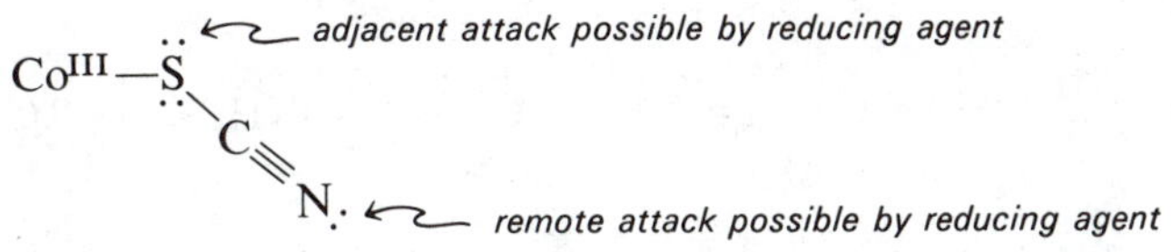

$$\text{Co}^{\text{III}} \text{—} \overset{\displaystyle ..}{\underset{\displaystyle \diagdown}{\text{S}}} \qquad \longleftarrow \text{ adjacent attack possible by reducing agent}$$

$$\underset{\displaystyle \ddot{\text{N}} \cdot \quad \longleftarrow \text{ remote attack possible by reducing agent}}{\overset{\displaystyle \|}{\text{C}}}$$

remote and adjacent sites on the SCN^- ligand of $[Co(NH_3)_5SCN]^{2+}$ are nearly equal.[30] Thus, starting with S-bonded thiocyanate on the oxidizing agent does not necessarily mean that one will achieve the synthesis of the N-bonded product.

$[Co(CN)_5]^{3-}$: A Redox and Catalytic Reagent

We have briefly described the use of $[Co(CN)_5]^{3-}$ as an electron transfer agent, but it has a much broader chemical behavior.[32] For example, it has been widely studied as a reagent in reactions of the general type

$$2[Co(CN)_5]^{3-} + X\text{—}Y \rightarrow [Co(CN)_5X]^{3-} + [Co(CN)_5Y]^{3-} \qquad (12\text{-}26)$$

where $X\text{—}Y$ can be H_2, H_2O, Br_2, I_2, H_2O_2, CH_3I, or RX.[33-36] Although not necessarily correct in a mechanistic sense, in formal terms each Co(II) center has been oxidized to Co(III) by electron transfer to X and Y, and the reduced species X^- and Y^- have added to the Co(III) centers. Such reactions are often called "oxidative additions." Although it is apropos to discuss the chemistry of $[Co(CN)_5]^{3-}$ at this point, the *general* discussion of "oxidative addition" reactions is delayed to Chapter 17.

Pentacyanocobaltate(3–) is generally prepared and used *in situ* by dissolving a Co(II) salt, a soluble source of cyanide, and potassium or sodium hydroxide in water. Although the formula is usually written as $[Co(CN)_5]^{3-}$, there is experimental evidence that the species existing in aqueous solution is $[Co(CN)_5(H_2O)]^{3-}$;[37] however, since the water molecule is apparently lost readily prior to reaction (see Chapter 13), the five-coordinate representation is not entirely inaccurate. This is especially so because the five-coordinate complex can be isolated from non-aqueous solvents under the following conditions:[38]

$$CoCl_2 + 5Et_4N^+CN^- \xrightarrow[\text{dry, deoxygenated DMF}]{} [Et_4N]_3[Co(CN)_5] \qquad (12\text{-}27)$$

yellow, translucent needles
$\mu_{\text{eff}} = 1.77 \text{ B.M.}$

The structure of the pentacyanocobaltate(3–) ion in the salt $[Et_4N]_3[Co(CN)_5]$ is square pyramidal,[39] as predicted by the structure preference energy plots in Figure 10-1.

The reactions of $[Co(CN)_5]^{3-}$ that are of greatest interest in terms of synthesis are those with organic compounds.[35,36] For example, if benzyl bromide is added to a water-methanol solution of $[Co(CN)_5]^{3-}$ prepared *in situ*, the observed reaction is

$$2[Co(CN)_5]^{3-} + PhCH_2Br \xrightarrow[\text{inert atmosphere}]{} [PhCH_2Co(CN)_5]^{3-} + [Co(CN)_5Br]^{3-} \qquad (12\text{-}28)$$

The yellow crystalline salt $Na_3[PhCH_2Co(CN)_5] \cdot 2H_2O$ is stable to alkali in the absence of air, but it reacts with acid to give an unidentified intermediate which, upon addition of base, releases phenylacetonitrile.[40] The following scheme was devised to account for this reaction, which is also observed when R is *n*-alkyl, benzyl, or phenyl.

$$\underset{\text{Co}-\text{C}\equiv\text{N}}{\overset{R}{|}} \xrightarrow{\text{H}^+} \underset{\text{Co}-\overset{+}{\text{C}}=\text{NH}}{\overset{R}{|}} \rightarrow \overset{+}{\text{Co}}-\text{CR}=\text{NH} \xrightarrow{\text{OH}^-} \text{RC}\equiv\text{N}$$

When R is an olefinic residue, however, reaction with acid leads directly to the olefin; for example,

$$2[\text{Co(CN)}_5]^{3-} + \text{CH}_2{=}\text{CHCH}_2\text{I} \rightarrow [\text{Co(CN)}_5(\text{CH}_2\text{CH}{=}\text{CH}_2)]^{3-} + [\text{Co(CN)}_5\text{I}]^{3-}$$

$$\downarrow \text{H}^+$$

$$\text{H}_3\text{C}-\text{CH}{=}\text{CH}_2$$

(12-29)

Examination of the kinetics of reactions such as (12–28) indicates that they are first order in both the cobalt compound and the organic halide. Further, the rate is dependent on the strength of the carbon-halogen bond; that is, the rate constants increase by roughly a factor of 10^3 to 10^4 on going from RCl to RBr to RI. The mechanism proposed to account for these facts is an atom transfer redox reaction involving free radical intermediates.

(12-30)

$$[\text{Co}^{\text{II}}(\text{CN})_5]^{3-} + \text{RX} \rightarrow [\text{Co}^{\text{III}}(\text{CN})_5\text{X}]^{3-} + \text{R} \cdot \quad \textit{(rate determining)}$$

$$[\text{Co}^{\text{II}}(\text{CN})_5]^{3-} + \text{R} \cdot \rightarrow [\text{RCo}^{\text{III}}(\text{CN})_5]^{3-} \quad \textit{(fast)}$$

The pentacyanocobaltate(3−) ion also reacts with H_2, but in a concerted process and not a free radical one such as those discussed above.[33] The reaction proceeds according to equation (12–31) to give hydridopentacyanocobaltate(3−),

(12-31)

$$2[\text{Co}^{\text{II}}(\text{CN})_5]^{3-} + \text{H}_2 \rightleftharpoons 2[\text{Co}^{\text{III}}(\text{CN})_5\text{H}]^{3-}$$

The equilibrium constant is approximately 10^5 M^{-1} over a wide range of H_2 partial pressures. This has important implications, because $[\text{Co(CN)}_5]^{3-}$ has been used as a catalyst for homogeneous hydrogenations† such as reaction (12–32).[41-44]

(12-32) $$\text{H}_2\text{C}{=}\text{CH}-\text{CH}{=}\text{CH}_2 + \text{H}_2 \xrightarrow[\text{[Co(CN)}_5]^{3-}]{} \text{H}_2\text{C}{=}\text{CH}-\text{CH}_2-\text{CH}_3 + \text{H}_3\text{C}-\text{CH}{=}\text{CH}-\text{CH}_3$$

Indeed, Pratt and Craig state that "there is probably no other catalyst which can hydrogenate such a wide variety of substrates under such mild conditions."[45] The following conversions, among others, may be effected:

conjugated olefins → olefins

$\text{PhCH}{=}\text{CH}_2 \qquad \rightarrow \text{PhCH}_2\text{CH}_3$

$\text{PhCHO} \qquad \rightarrow \text{PhCH}_2\text{OH}$

$\text{PhNO}_2 \qquad \rightarrow \text{PhN}{=}\text{NPh} + \text{PhNH}-\text{NHPh} + \text{PhNH}_2$

$\text{C}_6\text{H}_4\text{O}_2$ (quinone) $\rightarrow \text{C}_6\text{H}_6\text{O}_2$ (hydroquinone)

---

†By "homogeneous hydrogenation" we mean to imply that the reaction is carried out with all reactants in the same phase, that is, all in solution.

Two mechanisms have been proposed to account for such hydrogenations [reactions (12-33) and (12-34)], and both involve the preliminary formation of $[Co^{III}(CN)_5H]^{3-}$.

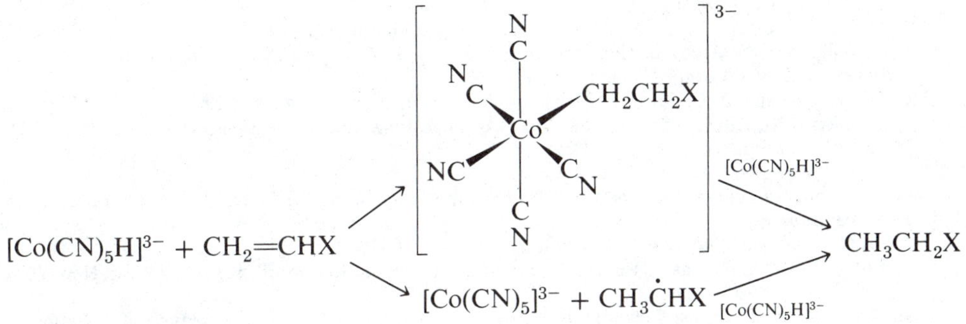

$$[Co(CN)_5H]^{3-} + CH_2{=}CHX$$

(12-33)

(12-34)

In one mechanism the elements of Co—H add across the C=C double bond to give an organocobalt complex, and the now-reduced olefin is released by further reaction with the cobalt hydride (12-33). The other mechanism is a free radical process involving two successive hydrogen transfers from $[Co^{III}(CN)_5H]^{3-}$ (12-34).

---

5. Refer back to the discussion of two-electron transfer, and suggest a method for the synthesis of trans-$[Pt(en)_2Br_2]^{2+}$ from trans-$[Pt(en)_2Cl_2]^{2+}$. [See R. Johnson and F. Basolo, *J. Inorg. Nucl. Chem.*, **13**, 36 (1960).]

6. The anion $[Co(CN)_5(SCN)]^{3-}$ was prepared by an electron/atom transfer reaction (p. 406). When the preparation of the other linkage isomer, $[Co(CN)_5(NCS)]^{3-}$, was attempted by the inverse of reaction (12-24) (that is, $[Co(NH_3)_5(SCN)]^{2+}$ + $[CO(CN)_5]^{3-}$), only the S-bonded isomer was obtained [C. J. Shea and A. Haim, *Inorg. Chem.*, **12**, 3013 (1973)]. Does this imply adjacent or remote attack by Co(II) on the SCN⁻ ligand? Explain fully.

7. $[Co(CN)_5]^{3-}$ reacts with hydrogen peroxide, H_2O_2, to give only $[Co(CN)_5OH]^{3-}$. Write a free radical mechanism for this reaction.

CHECKPOINTS

1. Electron transfer reactions may occur by either or both of two mechanisms:
   a. Outer sphere—the coordination spheres of the oxidant and reductant stay intact, and electron transfer occurs by an interaction of these coordination spheres.
   b. Inner sphere—transfer of electrons is mediated by a ligand held in common by the oxidant and reductant.

2. Outer sphere electron transfer is most effective if both the donor and receptor orbitals are π^*.

3. Inner sphere reactions are most efficient when donor and receptor orbitals are both σ^* or both π^*.

4. When the orbital symmetry requirement is not met, then greater activation is required, and, in general, such reactions will be slower than those requiring no electron configuration change.

REFERENCES

The following books and review article represent a selection of the literature available on mechanistic inorganic chemistry in general.

[1] F. Basolo and R. G. Pearson, "Mechanisms of Inorganic Reactions," 2nd Edition, John Wiley and Sons, New York (1967).

[2] M. L. Tobe, "Inorganic Reactions Mechanisms," Nelson, London (1972).

[3] R. G. Wilkins, "The Study of the Kinetics and Mechanism of Reactions of Transition Metal Complexes," Allyn and Bacon, Inc., Boston (1974).

[4] R. G. Pearson and P. C. Ellgen, "Mechanisms of Inorganic Reactions in Solution," in "Physical Chemistry, An Advanced Treatise," Vol. VII, H. Eyring, ed., Academic Press, New York (1975). An excellent, up-to-date, but brief review.

The next group of books and review articles are representative of the material available on electron transfer reactions.

[5] H. Taube, *J. Chem. Educ.*, **45**, 452 (1968).

[6] H. Taube, "Electron Transfer Reactions of Complex Ions in Solution," Academic Press, New York (1970).

[7] N. Sutin, "Free Energies, Barriers, and Reactivity Patterns in Oxidation-Reduction Reactions," *Acc. Chem. Res.*, **1**, 225 (1968).

[8] H. Taube and E. S. Gould, "Organic Molecules as Bridging Ligands in Electron Transfer Reactions," *Acc. Chem. Res.*, **2**, 321 (1969).

[9] A. Haim, "Role of the Bridging Ligand in Inner Sphere Electron Transfer Reactions," *Acc. Chem. Res.*, **8**, 265 (1975).

[10] J. K. Burdett, *Inorg. Chem.*, **17**, 2537 (1978); M. J. Weaver, *ibid.*, **18**, 402 (1979).

[11] M. W. Dietrich and A. C. Wahl, *J. Chem. Phys.*, **38**, 1591 (1963).

[12] H. Taube, *Record Chem. Progr. Kresge-Hooker Sci. Lib.*, **17**, 25 (1956).

[13] L. E. Bennett, *Prog. Inorg. Chem.*, **18**, 1 (1973).

[14] H. C. Stynes and J. A. Ibers, *Inorg. Chem.*, **10**, 2304 (1971).

[15] Indeed, excellent correlations have been found between $\log k$ and $\Delta G°$. D. P. Rillema, J. F. Endicott, and R. C. Paul, *J. Amer. Chem. Soc.*, **94**, 394 (1972).

[16] If V(II) reductions are faster than Cr(II) reductions of a common oxidant, this is considered indicative of an outer sphere mechanism. See, for example, M. Hery and K. Wieghardt, *Inorg. Chem.*, **15**, 2315 (1976).

[17] H. Taube, H. Myers, and R. L. Rich, *J. Amer. Chem. Soc.*, **75**, 4118 (1953); H. Taube and H. Myers, *J. Amer. Chem. Soc.*, **78**, 2103 (1954).

[18] R. T. Morrison and R. N. Boyd, "Organic Chemistry," 3rd Edition, Allyn and Bacon, Inc., Boston (1973), Chapter 29.

[19] Orbital following arguments applied to electron transfer reactions are explained in K. F. Purcell and J. C. Kotz, "Inorganic Chemistry," W. B. Saunders, Philadelphia, 1977, pp. 673–675.

[20] C. A. Radlowski and E. S. Gould, *Inorg. Chem.*, **18**, 1289 (1979). This paper is one of the latest in a series dealing with electron transfer through organic structural units.

[21] F. Nordmeyer and H. Taube, *J. Amer. Chem. Soc.*, **90**, 1162 (1968); R. G. Gaunder and H. Taube, *Inorg. Chem.*, **9**, 2627 (1970).

[22] F. Basolo, P. H. Wilks, R. G. Pearson, and R. G. Wilkins, *J. Inorg. Nucl. Chem.*, **6**, 161 (1958).

[23] R. C. Johnson and F. Basolo, *J. Inorg. Nucl. Chem.*, **13**, 36 (1960).

[24] F. Basolo, M. L. Morris, and R. G. Pearson, *Disc. Faraday Soc.*, **29**, 80 (1960).

[25] J. L. Burmeister and F. Basolo, "The Application of Reaction Mechanisms to Synthesis of Coordination Compounds," in "Preparative Inorganic Reactions," Vol. 5, W. L. Jolly, ed., Interscience Publishers, New York (1968).

[26] A. Werner, *Chemische Berichte*, **45**, 121, 3061 (1912).

[27] D. H. Busch, *J. Amer. Chem. Soc.*, **77**, 2747 (1955); C. S. Lee, E. M. Gorton, H. M. Neumann, and H. R. Hunt, Jr., *Inorg. Chem.*, **5**, 1397 (1966).

[28] For some other examples see the following references:
   a. J. V. Rund, F. Basolo, and R. G. Pearson, *Inorg. Chem.*, **3**, 658 (1964).
   b. R. D. Gillard, J. A. Osborn, and G. Wilkinson, *J. Chem. Soc.*, 1951 (1965).
   c. M. C. Hughes and D. J. Macero, *Inorg. Chem.*, **13**, 2739 (1974); D. M. Soignet and L. G. Hargis, *ibid.*, **11**, 2921 (1972); B. R. Baker and B. Dev Mehta, *ibid.*, **4**, 849 (1965).

[29] I. Stotz, W. K. Wilmarth, and A. Haim, *Inorg. Chem.*, **7**, 1250 (1968).

[30] C. J. Shea and A. Haim, *Inorg. Chem.*, **12**, 3013 (1973).

[31] J. L. Burmeister, *Inorg. Chem.*, **3**, 919 (1964).

[32] B. M. Chadwick and A. G. Sharpe, *Adv. Inorg. Chem. Radiochem.*, **8**, 84 (1966).

[33] J. Halpern and M. Pribanic, *Inorg. Chem.*, **9**, 2616 (1970).

[34] P. B. Chock, R. B. K. Dewar, J. Halpern, and L-Y. Wong, *J. Amer. Chem. Soc.*, **91**, 82, (1969).

[35] J. Halpern and J. P. Maher, *J. Amer. Chem. Soc.*, **86**, 2311 (1964); 87, 5361 (1965).

[36] P. B. Chock and J. Halpern, *J. Amer. Chem. Soc.*, **91**, 582 (1969).

[37] J. M. Pratt and R. J. P. Williams, *J. Chem. Soc.*, A, 1291 (1967).

[38] D. A. White, A. J. Solodar, and M. M. Baizer, *Inorg. Chem.*, **11**, 2160 (1972).

[39] L. D. Brown and K. N. Raymond, *Inorg. Chem.*, **14**, 2590 (1975).

[40] J. Kwiatek and J. K. Seyler, *J. Organometal. Chem.*, **3**, 433 (1965).
[41] J. Kwiatek, I. L. Mador, and J. Seyler, *J. Amer. Chem. Soc.*, **84**, 304 (1962).
[42] J. Kwiatek and J. K. Seyler, *J. Organometal. Chem.*, **3**, 421 (1965).
[43] J. M. Pratt and P. J. Craig, *Adv. Organometal. Chem.*, **11**, 331 (1973).
[44] M. G. Burnett, P. J. Connolly, and C. Kemball, *J. Chem. Soc.*, **A**, 991 (1968).
[45] See page 434 of ref. 43.

13. COORDINATION CHEMISTRY: REACTION MECHANISMS AND METHODS OF SYNTHESIS; SUBSTITUTION REACTIONS

In the first of the three chapters devoted to the mechanisms of the reactions of coordination compounds, we focused on reactions that occur by transfer of electrons to or from coordinated metal ions. In electron transfer reactions that proceed by the inner sphere mechanism, one or more ligands of one reactant (usually the reductant) are substituted by a ligand from the other reactant (usually the oxidant), and the substituting ligand forms a bridge between the two metal ions. The details of this substitution step were not stressed in the previous chapter, although it was clearly important to the successful functioning of the mechanism. Moreover, it should be evident that substitution reactions in general are of utmost importance, since many of the reactions of transition metal complexes involve ligand substitutions. It is for this reason that we turn now to a discussion of the principles of substitution reactions and their use in the synthesis of coordination compounds.

The focus in this chapter is on substitution reactions—reactions wherein ligand-metal bonds are broken and new ones formed in their stead.[1-4] We can reach back more than three-quarters of a century, to Alfred Werner's second paper on coordination compounds, to illustrate substitution reactions.[5]

"If dichrocobalt chloride is dissolved in water at room temperature, a solution is obtained which is colored green for a moment, but *only for a moment;* the color changes very quickly to blue and then to violet, a reliable indication that the dichro salt does not remain unchanged in solution. According to the observations which we have made with praseo salts, it seems unquestionable to us that the change in color is caused by the transformation of the triaminepraseo salt into the triamine-purpureo salt and the triamineroseo salt."

That is, the following substitution reactions occur:

$$[Co(NH_3)_3(H_2O)Cl_2]Cl + H_2O \rightarrow [Co(NH_3)_3(H_2O)_2Cl]Cl_2 \qquad (13\text{-}1)$$

dichrocobalt chloride
or
triaminepraseocobalt chloride
　　　　　　　　　　　　　　　　　　　triaminepurpureocobalt chloride

$$[Co(NH_3)_3(H_2O)_2Cl]Cl_2 + H_2O \rightarrow [Co(NH_3)_3(H_2O)_3]Cl_3 \qquad (13\text{-}2)$$

　　　　　　　　　　　　　triamineroseocobalt chloride

To introduce the principles of substitution reactions as they apply to coordination compounds, we shall turn first to those typical of square planar complexes (reaction 13–3) and then to those of octahedral complexes.

(13-3)

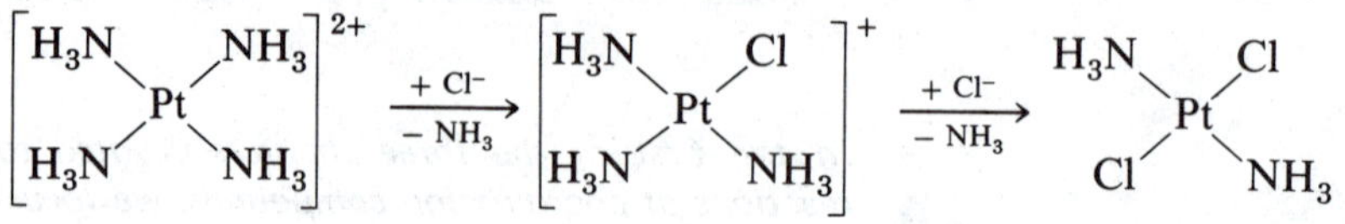

$$\left[\begin{array}{c}H_3N \quad NH_3 \\ Pt \\ H_3N \quad NH_3\end{array}\right]^{2+} \xrightarrow[-NH_3]{+\,Cl^-} \left[\begin{array}{c}H_3N \quad Cl \\ Pt \\ H_3N \quad NH_3\end{array}\right]^{+} \xrightarrow[-NH_3]{+\,Cl^-} \begin{array}{c}H_3N \quad Cl \\ Pt \\ Cl \quad NH_3\end{array}$$

Reiset's salt

Far less is known about the substitution reactions of coordination compounds of other geometries or of other coordination numbers, so we shall introduce material on such compounds only peripherally (see Study Question 10).

REPLACEMENT REACTIONS AT FOUR-COORDINATE PLANAR REACTION CENTERS[6,7]

As was pointed out in Chapter 9, by far the greatest number of square planar complexes known are based on transition metal ions with d^8 electron configurations, especially those metal ions listed below.

Group VIII			Group IB
—	—	Ni(II)	
—	Rh(I)	Pd(II)	
—	Ir(I)	Pt(II)	Au(III)

Of these, Pt(II) offers the most useful combination of properties for the study of substitution reactions: (i) Pt(II) is more stable to oxidation than Rh(I) or Ir(I). (ii) Pt(II) complexes are always square planar, unlike Ni(II) complexes which can often be tetrahedral. (iii) Pt(II) chemistry has been thoroughly studied, and its substitution reactions proceed at rates convenient for laboratory study. This latter property is in contrast with Ni(II) complexes, for example, which can undergo substitution 10^6 more rapidly than Pt(II) complexes. For these reasons and others, the remainder of this discussion is largely concentrated on complexes of Pt(II).[8,9]

The General Mechanism of Square Planar Substitution Reactions

In almost all of the substitution reactions studied with square planar complexes, the observed rate law has the form

$$-\frac{d[ML_3X]}{dt} = (k_s + k_Y[Y])[ML_3X]$$

for the general reaction

$$ML_3X + Y \rightarrow ML_3Y + X$$

This rate law has been rationalized in terms of two parallel pathways, both involving an associative or A mechanism (Figure 13-1). In the k_Y pathway the nucleophile Y attacks the metal complex and the reaction passes through a five-coordinate transition state and

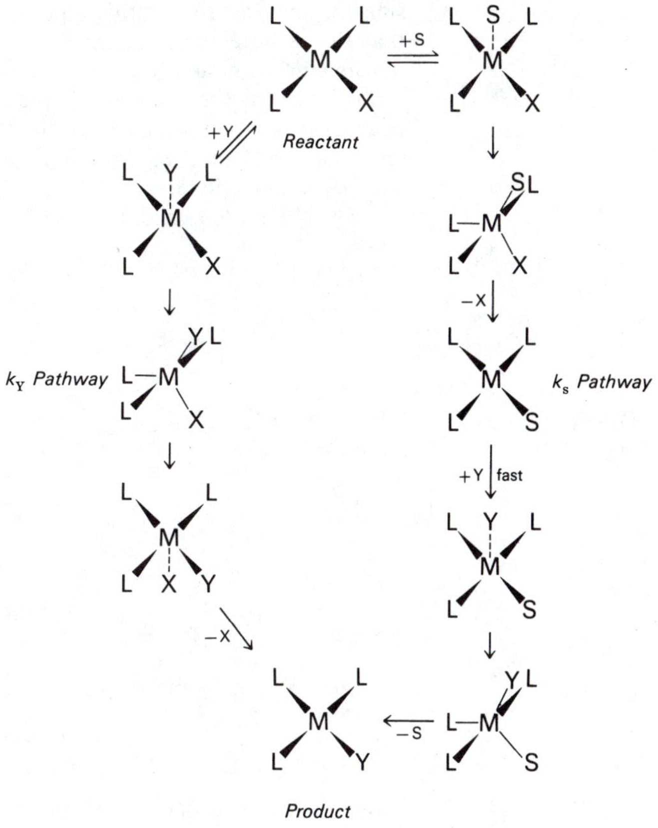

Figure 13-1. The associative pathways utilized in the substitution of one ligand for another at a square planar reaction center.

intermediate, the former presumably having a regular trigonal bipyramidal structure. Since *the geometry of the original complex is invariably maintained*—that is, the ligands *cis* and *trans* to the substituted ligand remain in that arrangement—the entering and leaving groups as well as the ligand originally *trans* to the leaving group are in the trigonal plane. The k_s pathway also involves the formation of a trigonal bipyramidal transition state, except that the solvent is the entering group. Therefore, the latter pathway is also associative, and k_s is actually a pseudo first-order constant ($= k$[solvent]).

Factors Affecting the Reactivity of Square Planar Complexes of Pt(II) and Other d^8 Metal Ions

As in the case of any chemical reaction, there is a variety of factors that affect the reactivity of square planar complexes, among them the following:

a. the nature of the group entering the complex
b. the effect of other groups in the complex
  i. ligands *trans* to the leaving group
  ii. ligands *cis* to the leaving group
c. the nature of the leaving group
d. the nature of the central metal ion

At least for Pt(II) complexes, the first three factors have been arranged in order of decreasing importance, although this is not always the case for other d^8 ions.

INFLUENCE OF THE ENTERING GROUP

Rates of reactions proceeding by an associative mechanism must depend in some manner on the nature of the entering group. The usual parameter to which one refers in substitution reactions of this type is the **nucleophilicity** of the reactant. You will recall that this is different from the **basicity** of a compound, the latter term being thermodynamic in origin and being defined, for example, by the pK_a of the conjugate acid of a Lewis base. On the other hand, the nucleophilicity is a parallel, but not necessarily an equivalent, concept which refers to the ability of the Lewis base to act as the entering group and influence the reaction rate in a nucleophilic substitution; that is, nucleophilicity is a kinetic term.

Relative nucleophilicities have been measured for various entering groups in reaction 13-4,[10]

(13-4)

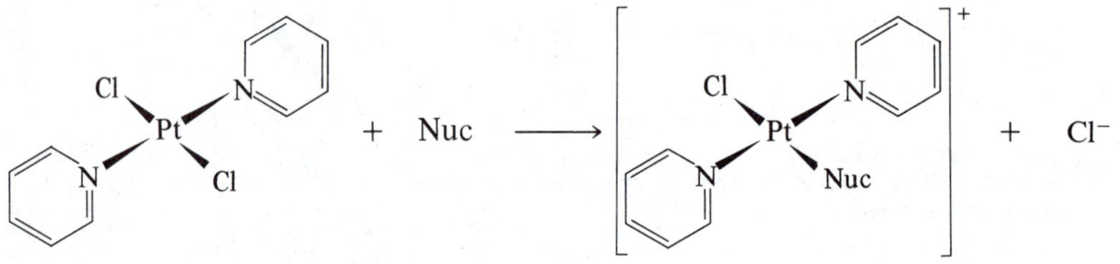

and the order of these is as follows:

$$CH_3OH < F^- < Cl^- < NH_3 < piperidine < pyridine < NO_2^- < Br^- < (CH_2)_4S <$$
$$I^- < Ph_3Sb < Ph_3As < CN^- < Ph_3P$$

Within the overall order, you should notice the following trends:

a. The nucleophilicity of the halides and pseudohalides increases in the order

$$F^- < Cl^- < Br^- < I^- < CN^-$$

b. Of the Group VA bases, amines are considerably less nucleophilic than those based on the heavier elements of the group. For the latter, their nucleophilicity decreases on going down the group, a trend opposite to that noted for the halides. That is,

$$Ph_3Sb < Ph_3As < Ph_3P$$

c. Sulfur nucleophiles are better than those based on oxygen.
d. The trends in the nucleophilicity and basicity of a series of entering groups are not the same. For example, the basicities (as given by pK_a values) of the amines listed above increases in the order

$$pyridine < ammonia < piperidine$$

Such trends as these can obviously be quite important when considering the synthesis of a particular complex.

INFLUENCE OF LIGANDS *trans* TO THE ENTERING GROUP[2,4,11,12]

An understanding of the stereochemistry of coordination compounds played an important role in the development of chemical theory. For example, one important stereochemical question, which was not settled to everyone's satisfaction until the 1920's, was whether the complexes of platinum(II) were square planar or tetrahedral. Since some understanding of kinetics—the effect of ligands on the ease of substitution of groups to which they are *trans*—played an important role in the answer to the basic stereochemical question, it is worthwhile to discuss briefly the experiments involved and the conclusions that may be drawn.

One approach to solving the problem of the correct structure of platinum(II) complexes was to synthesize deliberately the isomers of a complex of the general formula Pt*abcd*.[13] If the complex were square planar, three diastereoisomers (1) should be isolated, while only two enantiomers (2) should arise if the complex were tetrahedral. In the

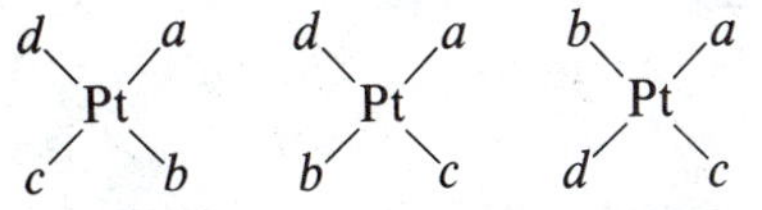

1920's the Russian chemist Chernyaev believed, as did most chemists, that Pt(II) complexes were square planar, and he recognized that the stereospecific synthesis of the three diastereoisomers of Pt*abcd* should be possible[13a] using an effect first mentioned by Alfred Werner—"*trans* elimination."[14] That is, assuming Pt(II) complexes are square planar, certain groups can more readily than others cause the elimination and substitution of groups *trans* to themselves. If it is assumed that the *trans* labilizing effect of the following groups decreases in the order given—$NO_2^- > Cl^- > NH_3 \sim$ pyridine $\sim NH_2OH$—one can design synthetic paths to each of the three isomers, **3a–c**.

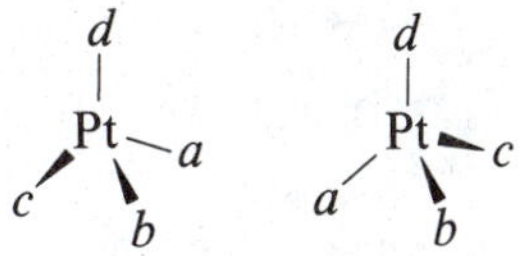

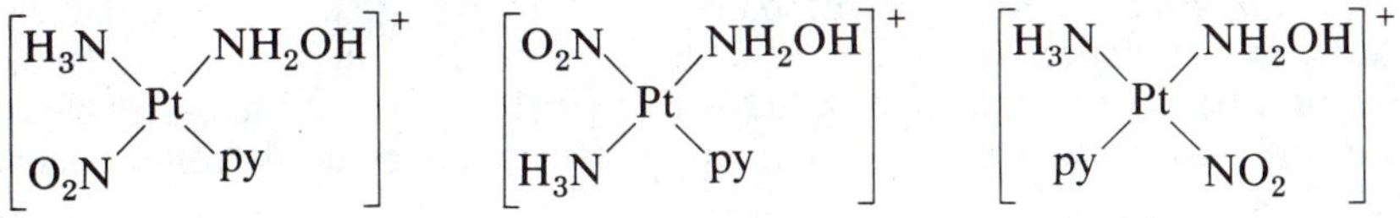

For example,

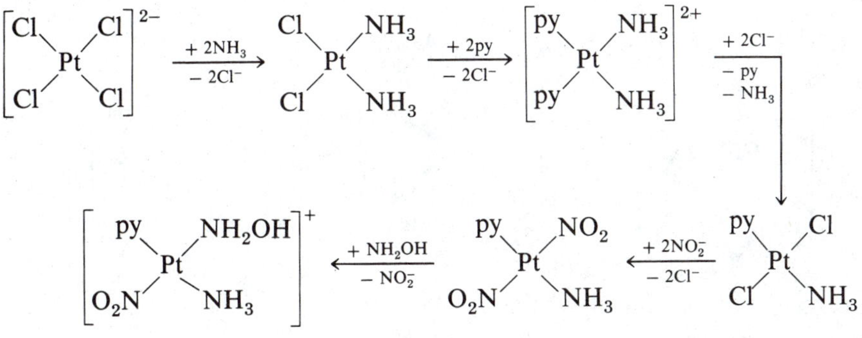

(13-5)

All of the steps in this synthesis follow the assumed order of *trans* labilizing effect— except step 2. However, if you accept the order of *trans* effects for what it is—an empirical order—you can accept another empirical guide: other things being equal, a metal-halogen bond is generally more labile than a metal-nitrogen bond.[15]

Since Werner's and Chernyaev's original work, considerable effort has gone into probing the effects wrought by *trans* ligands in square planar (and octahedral) complexes and into understanding the causes of such effects. It has come to the point where the inclusion of the "*trans* effect" in an inorganic chemistry course is virtually mandatory, even if it is, as Tobe has said,[16] "among the legendary beasts of modern inorganic chemistry." We shall therefore proceed to discuss the more "theoretical" ideas surrounding the beast and return later in the chapter to its use in synthesis.

The *trans effect* is perhaps best defined as *the effect of a coordinated ligand upon the rate of substitution of ligands opposite to it.*[12] For the particular case of Pt(II) complexes, the labilizing effect of typical ligands is generally in the order

$$H_2O \sim OH^- \sim NH_3 \sim \text{amines} < Cl^- \sim Br^- < SCN^- \sim I^- \sim NO_2^- \sim C_6H_5^-$$
$$< CH_3^- \sim S{=}C(NH_2)_2 < \text{phosphines} \sim \text{arsines} \sim H^- < \text{olefins} \sim CO \sim CN^-$$

As an example of the kinetic *trans* effect, consider the following reaction.[17]

(13-6)

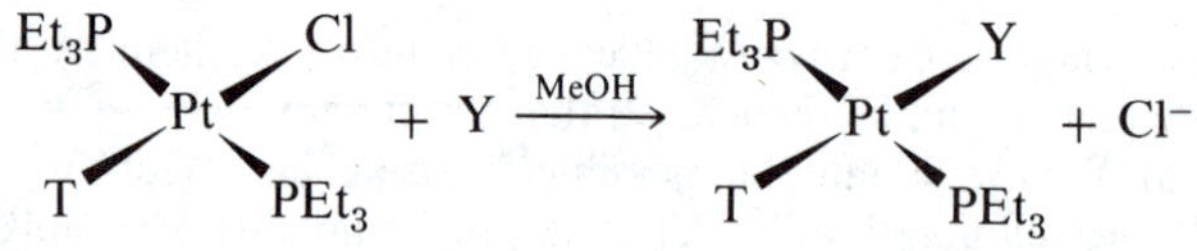

$$T = Cl^-, CH_3^-, \text{ and } C_6H_5^-$$

When Y is Br^-, N_3^-, NO_2^-, or pyridine (all relatively weak nucleophiles), the *trans* effect is in the order given above: $CH_3^- > C_6H_5^- > Cl^-$. However, when the better nucleophile I^- is used, the *trans* effect order now becomes $CH_3^- > Cl^- > C_6H_5^-$. Although the reason for this inversion in *trans* effects is not at all clear, two other observations can be made with assurance: (i) the *trans* effect depends in part on the nature of the entering group, and (ii) the effects of *trans* ligands and entering groups cannot be neatly compartmentalized.

The language used above was chosen carefully to differentiate the **trans effect**—a kinetic phenomenon—from the **trans influence**—a thermodynamic phenomenon.[11] That is, ligands can *influence* the ground state properties of groups to which they are *trans* such as the *trans* metal-to-ligand bond distance, the vibrational frequency, or a host of other parameters. In the series below, for example, the *trans* Pt—Cl bond length decreases on going from **4** to **6**.[18] If the usual assumption can be made concerning the

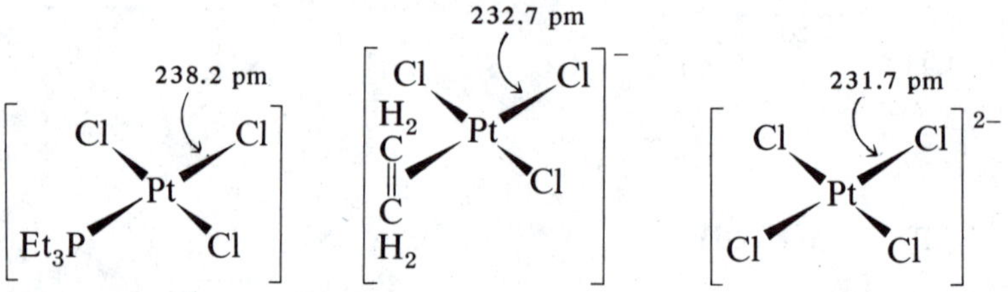

relation between bond length and bond strength, the Pt—Cl bond increases in strength in the order: *trans* ligand = phosphine < olefin < chloride. Based on more extensive data, the structural *trans* influence order has been given as

$$\sigma{-}R \sim H^- \geq \text{carbenes} \sim PR_3 \geq AsR_3 > CO \sim RNC \sim C{=}C \sim Cl^{\cdot-} \sim NH_3$$

Perhaps the explanation of the *trans* effect that has best stood the test of time is that of Langford and Gray,[3] an idea examined more recently by others through molecular orbital calculations.[19,20] For σ bonding ligands, these calculations show that the strength of the Pt–N bond in *trans*-Pt(Cl$_2$(NH$_3$)T (**7**) decreases as T is changed from H$_2$O to CH$_3^-$ in the order H$_2$O ≥ H$_2$S ≥ Cl$^-$ ≫ PH$_3$ ≥ H$^-$ ≥ CH$_3^-$.

$$\underset{\mathbf{7}}{\overset{\displaystyle \text{Cl} \diagdown \quad \diagup \text{NH}_3}{\underset{\displaystyle \text{T} \diagup \quad \diagdown \text{Cl}}{\text{Pt}}}}$$

In general, the *trans*-Pt–N bond is weakened as T becomes a better labilizer of the *trans*-ammonia. Therefore, at least for σ bonding ligands, there must be a connection between the ability of T to act as a labilizer of the *trans* position and its effectiveness in the sense of the *trans* influence. However, Drago argues that the weakening on going from the poor *trans* labilizer H$_2$O to the good *trans* labilizer H$^-$ is only a relatively small fraction of the total bond strength, so bond weakening cannot be the total argument.[19] Rather, stabilization of the transition state by T must also be an important contribution to the *trans* effect.

The effectiveness of ligands such as olefins in labilizing *trans* ligands can also be explained by transition state stabilization. An olefin binds to a metal in two ways: donation of π electron density from the olefin into a metal orbital of σ symmetry (structure **8**) and back-donation of metal electron density from a metal orbital of π symmetry (say d_{yz}) into an olefin orbital of the same symmetry, the olefin π* orbital (**9**). As the entering group in a substitution reaction approaches, the electron density on the metal is increased, and any metal-T atomic orbital interaction that can remove this excess density should stabilize the transition state and lead to rate enhancement. This is, of course, precisely the effect of metal-to-ligand dπ-π* bonding, and ligands capable of this—CO, CN$^-$, and olefins—will have large *trans* effects but not necessarily a large *trans* influence.

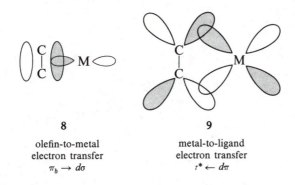

8	**9**
olefin-to-metal electron transfer	metal-to-ligand electron transfer
$\pi_b \rightarrow d\sigma$	$\pi^* \leftarrow d\pi$

OTHER INFLUENCES ON THE REACTIVITY OF SQUARE PLANAR COMPLEXES

As noted in the introduction to this section, *cis* ligands, as well as those in the *trans* position, have some influence on the rates of the reactions of square planar complexes. However, the effect of a *cis* ligand is often several hundred times less than that of the same ligand located in a *trans* position.[21]

The leaving group must also have some effect on reactivity, as it is the bond between this group and the metal that is broken. Thus, it is not surprising that good correlations have been found between the pK_a of the leaving group (*i.e.*, its basicity) and log k_s, for example.[22]

Finally, there is the effect of the metal ion. In general, the tendency to undergo substitution is in the order Ni(II) > Pd(II) ≫ Pt(II).[21] This order of reactivity is in the same order as the tendency to form five-coordinate complexes. More ready formation of a five-coordinate intermediate complex leads to stabilization of the transition state and to rate enhancement.

STUDY QUESTIONS

1. Outline the preparation of the *cis* and *trans* isomers of $[Pt(NH_3)(NO_2)Cl_2]^-$, given that the substituent *trans* effects are in the order $NO_2^- > Cl^- > NH_3$.

2. a. The square planar complex $[Pt(amine)(DMSO)Cl_2]$ can be synthesized from $[PtCl_4]^{2-}$, DMSO, and the appropriate amine (DMSO = dimethylsulfoxide, Me_2SO, a ligand that can bind to a metal *via* either the S or the O). Given that the *trans* effects of the ligands are in the order DMSO (S-bonded) $> Cl^- >$ amine, describe a preparation of the *cis* and *trans* isomers of the desired compound.
   b. If *cis*-$[Pt(amine)_2(DMSO)Cl]^+$ is heated with one equivalent of HCl, one ligand is substituted with chloride. What is the composition of the final product and what is its structure?

3. The reaction below (where Pr = *n*-propyl) can be carried out in either hexane or methanol,

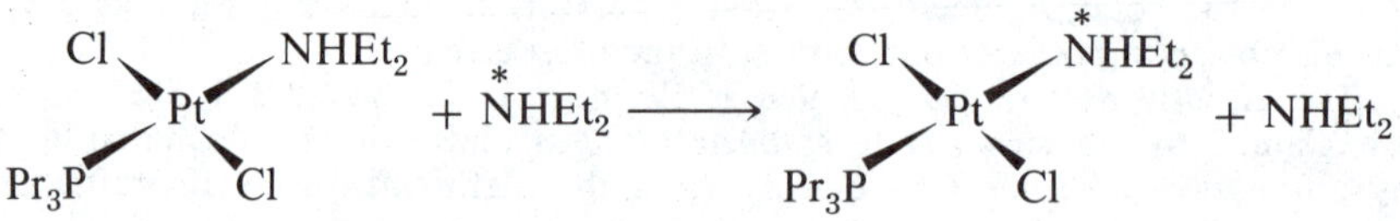

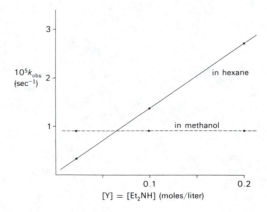

and observed rate constants obtained in these solvents are plotted against the concentration of $\overset{*}{N}HEt_2$ in the following figure. Note that the slope of the methanol line is zero while that of the hexane line is positive, but the hexane line has an intercept of zero. What are the mechanistic implications of these observations?

[Data from T. P. Cheeseman, A. L. Odell, and H. A. Raethel, *Chem. Commun.*, 1496 (1968).]

4. Why do phosphines have a high *trans* effect? (Hint: Consider the way in which they are bonded to a transition metal ion.)

SUBSTITUTION REACTIONS OF OCTAHEDRAL COMPLEXES[1-4],[23],[24]

The vast majority of the complexes formed by the transition metals are octahedral, so an examination of their substitution reactions is especially important. Before beginning our discussion, however, we might point out several general features of octahedral substitutions and some similarities and differences with the reactions of square planar complexes.

(1) Most of the research on the rates and mechanisms of substitution reactions of square planar complexes has been done with Pt(II) complexes. Compounds of this ion combine several factors favorable to kinetic studies: they may be readily prepared (as you shall see later in the chapter), and their reactions are slow enough to be followed by conventional means. Similarly, in kinetic studies of octahedral complexes, the majority of the work has been done with Co(III) complexes, that is, until the advent of methods for the examination of fast reactions (half-lives less than one minute).[25] Further, based on Alfred Werner's extensive work, numerous complexes of Co(III) are known,[26] and many undergo reaction at convenient rates.

(2) Reactions at square planar centers occur by associative activation; only a few examples of dissociative activation are known. In contrast, substitution reactions of the complexes of divalent, first row transition metals and of Co(III) occur by dissociative activation. This change from *a* to *d* activation with change in coordination number is not surprising because additional coordination positions are not as available in octahedral complexes as they are in square planar complexes. However, evidence has been accumulating in recent years that substitution reactions of complexes of trivalent ions—except for Co(III)—may occur by associative activation (in particular, I_a).[27] This apparent change in mechanism with metal ion charge is not unreasonable, since bond-making could occur more readily than bond-breaking with a trivalent ion than with a divalent ion.†

(3) Studies on octahedral complexes have been largely limited to two types of reactions:

(i) *Replacement of coordinated solvent (water).* Perhaps the most thoroughly studied replacement reaction of this type is the formation of a complex ion from the hydrated metal ion in solution. When the entering group is an anion, as in reaction 13-7.

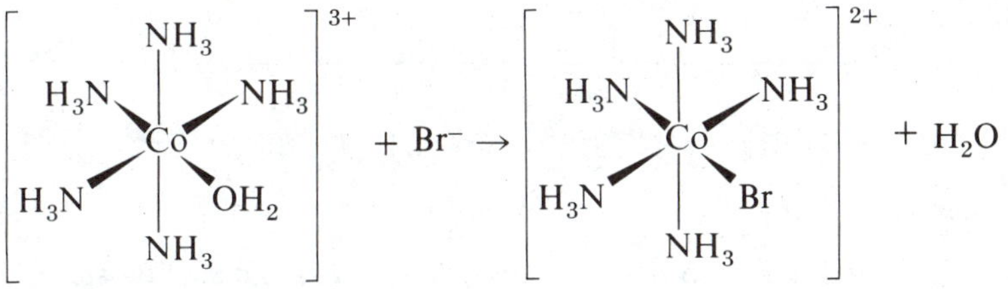

$$\tag{13-7}$$

the reaction is often called **anation**.[28]

(ii) *Solvolysis.* Since the majority of such reactions have been carried out in aqueous solution, **hydrolysis** is a more appropriate term. Hydrolysis reactions have been done under both acidic (13-8) and basic (13-9) conditions.[29,30]

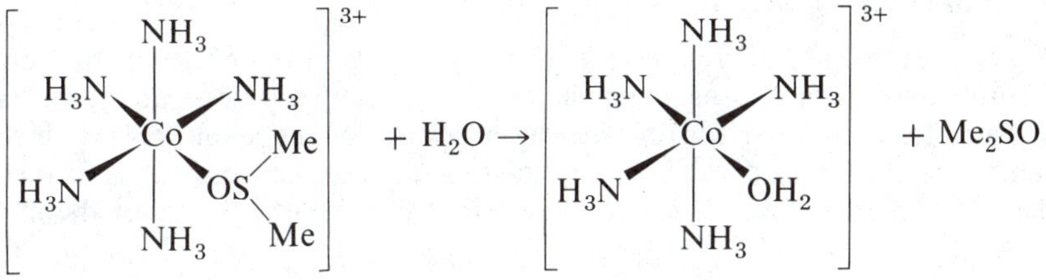

$$\tag{13-8}$$

---

†Reactions of Co(III) have long been considered as typical of other transition metal ions. However, it would now appear that complexes of this ion may in fact be anomalous in their behavior; reasons for this are uncertain. See ref. 27.

(13-9)

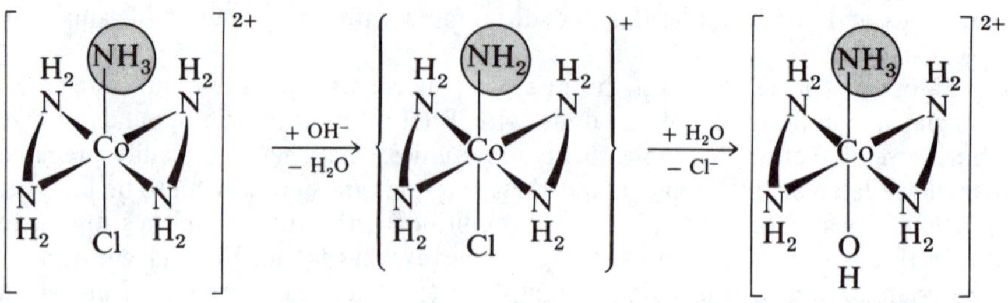

(the *Conjugate Base* or *CB* mechanism)

There is a very interesting and fundamental difference in the mechanisms under these two pH ranges, a fact that we shall explore later in the chapter.

You will notice that no mention is made of reactions involving the *direct* interchange of two anions, a reaction type commonly observed with four-coordinate Pt(II) complexes. Instead, an octahedral complex must first lose a coordinated anion by hydrolysis and then replace the newly coordinated solvent by the other anion.[31] (This sequence is, of course, akin to the k_s path for square planar Pt(II) complexes.) This observation

(13-10)

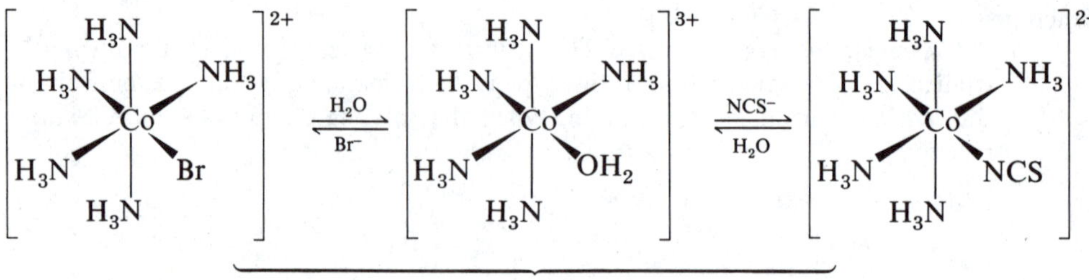

net reaction

$$[Co(NH_3)_5Br]^{2+} + NCS^- \rightleftharpoons [Co(NH_3)_5NCS]^{2-} + Br^-$$

is an important one for you to ponder. Firstly, hydrolysis and anation are the reverse of one another, so the two reactions follow the same reaction coordinate in opposite directions. Secondly, both reactions apparently occur by dissociative activation. That is, since one anion does not directly replace another, but is instead replaced by water, this means either that water is always the superior nucleophile *or* that bond formation plays an insignificant role in the activation process. The superior nucleophilicity of water may generally be safely discounted.

Replacement of Coordinated Water

A great deal of information regarding substitution reactions of octahedral complexes has been obtained by a thorough examination of the simple process of the exchange of bound water for bulk water† or for other ligands when the metal ion is placed in aqueous solution.[24] We shall first discuss the mechanism of water replacement on divalent ions and then turn to an analysis of the dependence of rates on the nature of the metal ion.

---

†The number of coordinated waters for almost all metal ions has been determined. Although almost all of these ions—whether from the main groups, the transition metals, or the lanthanides—have coordination numbers of six, some alkaline earth ions (*e.g.*, Be^{2+} and Ca^{2+}) have coordination numbers of four, and some lanthanide ions have higher coordination numbers. See ref. 25.

THE MECHANISM OF WATER REPLACEMENT

The alternative mechanisms for the replacement of coordinated water by another ligand are illustrated in Figure 13-2. In one of these—the dissociative or D mechanism— the original metal complex dissociates a water molecule to produce a five-coordinate, solvent caged intermediate in the rate-determining step.

The other two mechanisms both involve, as a first step, the formation of an ion- or reactant-pair. Manfred Eigen, who shared the Nobel Prize in 1967 for his work on fast reactions, found that the following stoichiometric mechanism generally best describes the replacement of water by another ligand:

$$[M(H_2O)_6]^{n+} + X \rightleftharpoons [M(H_2O)_6]^{n+} \cdots X \tag{13-11}$$

$$[M(H_2O)_6]^{n+} \cdots X \rightarrow [M(H_2O)_5X]^{m+} + H_2O \tag{13-12}$$

The first step in this process is the formation of the reactant-pair. Since the incoming ligand is held in the solvent layer immediately surrounding the metal ion, this is often called an **outer sphere complex** to convey the fact that the new ligand is contained in the outer coordination sphere of the metal ion. [We referred to this as the cage complex in Chapter 7 (pp. 219–221) and there the equilibrium constant for its formation was labeled K_{dif}.] Following the formation of the reactant-pair, a seven-coordinate intermediate may be formed in the rate-determining step. This pathway is associative and is labeled A.

If the experimental data clearly indicate the presence of an intermediate, the reaction can be labeled D or A as appropriate. However, if there is no apparent intermediate, the

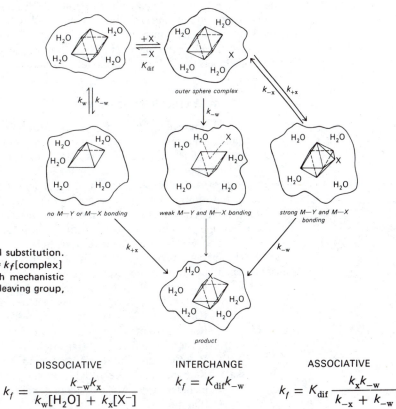

Figure 13-2. Mechanisms of octahedral substitution. For each mechanism the rate law Rate = k_f[complex] [X] applies, where k_f is given for each mechanistic type. The group Y in the figure is the leaving group, H_2O.

DISSOCIATIVE

$$k_f = \frac{k_{-w}k_x}{k_w[H_2O] + k_x[X^-]}$$

INTERCHANGE

$$k_f = K_{dif}k_{-w}$$

ASSOCIATIVE

$$k_f = K_{dif}\frac{k_x k_{-w}}{k_{-x} + k_{-w}}$$

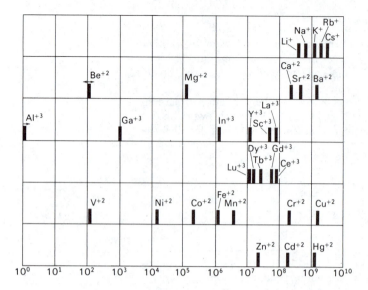

Figure 13-3. Characteristic rate constants (sec^{-1}) for the substitution of inner sphere water molecules in a number of metal ions. [From C. M. Frey and J. Stuehr, "Kinetics of Metal Ion Interactions with Nucleotides and Base Free Phosphates," in "Metal Ions in Biological Systems," ed., H. Sigel, Marcel Dekker, Inc., New York, vol. 1, 1974, p. 69.]

reaction is assigned an interchange mechanism ($= I$) in which there is a smooth interchange of entering and leaving groups (the pathway in the middle of Figure 13-2). As you have already seen in this chapter and in Chapter 7, there is the additional possibility that the interchange activation may be predominantly associative or dissociative. That is, the mechanism is labeled I_a if bond making is apparently more important than bond breaking in the transition state; the opposite situation, when bond breaking is more important than bond making, is labeled I_d.

The rate law observed for the substitution of an aquo complex is first order in both metal ion and entering ligand; that is, Rate = k_{obs}[complex] [X]. Unfortunately, all of the rate laws given in Figure 13-2 for the various possible mechanisms can reduce to the simple, observed law under easily attained conditions. The accumulated evidence strongly favors I_d as the mechanism of water substitution on an aquated divalent transition metal ion, especially for later members of the series. However, recent evidence suggests that earlier members may show associative behavior, the changeover from I_a to I_d occurring around Fe(II).[32]

RATES OF WATER REPLACEMENT[4,24,25]

First order rate constants for the substitution of inner sphere water molecules by other ligands on both transition metal and non-transition metal ions fall over a very wide range of values. Figure 13-3 is a tabulation of these constants by periodic group for a large number of ions, and, in Figure 13-4, the constants have been plotted against the number of d electrons for the di- and trivalent ions of the first row transition metals. Several observations can be made regarding these data.

1. The rate constants for the substitution reactions of a given ion are approximately the same no matter what the nature of the entering group. This may be taken as evidence for a mechanism involving dissociative activation.

2. The metal ions of Figures 13-3 and 13-4 can be placed in four reasonably distinct classes depending on their substitution rates.

Class I.   Exchange of water is very fast and is essentially diffusion controlled ($k \gtrsim 10^8$ sec^{-1}). The class encompasses ions of Periodic Groups IA, IIA (except Be^{2+} and Mg^{2+}), and IIB (except Zn^{2+}), plus Cr^{2+} and Cu^{2+}.

Class II.   Rate constants are in the range from 10^4 to 10^8 sec^{-1}. This class includes most of the first row transition metal divalent ions (the exceptions are V^{2+}, Cr^{2+}, and Cu^{2+}) and Mg^{2+} and the lanthanide 3+ ions.

Class III.   Rate constants are roughly in the range from 1 to 10^4 sec^{-1}. This class includes Be^{2+}, Al^{3+}, V^{2+}, and some of the 3+ ions of the first row transition metals.

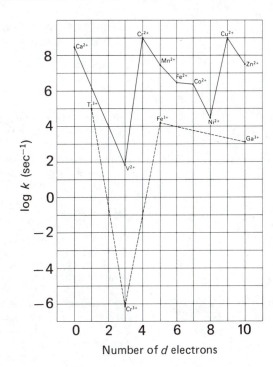

Figure 13-4. Rate constants for the exchange of water molecules in the first coordination sphere of $[M(H_2O)_6]^{n+}$ in aqueous solution. Data largely taken from R. G. Pearson and P. C. Ellgen in "Physical Chemistry, An Advanced Treatise," Volume VII, H. Eyring, ed., Academic Press, New York, 1975, Chapter 5.

Class IV. Rate constants are roughly in the range from 10^{-3} to 10^{-6} sec^{-1}. The following ions fall within the class: Cr^{3+}, Co^{3+}, Rh^{3+}, Ir^{3+}, and Pt^{2+}.

Like so many other thermodynamic and kinetic properties, there are numerous factors, originating with the metal ion, that affect the rate of exchange of bound water. In the case of Class I ions, for instance, ionic size is obviously of considerable importance. As the ions become smaller, the substitution rate slows, suggesting that the rates are controlled by the effective charge on the ion or some other related property (*e.g.*, orbital overlap between the metal ion and the departing ligand). The same general trend is observed in Group IIA and Group IIB, although the rate for Be^{2+} is considerably slower than might be expected. This latter effect is apparently due to hydrolysis reactions, which complicate the kinetics of simple substitution. Notice the large difference in Mg^{2+} and Ca^{2+} rates, a difference that is thought to play a role in their different behavior as enzymatic activators.

There is obviously less correlation between rate and ion size for ions outside of Groups I and II. For example, Cr^{2+}, Ni^{2+}, and Cu^{2+} have nearly identical radii. However, Cr^{2+} and Cu^{2+} are in Class I whereas Ni^{2+} is in Class II. Fortunately, there is a rather simple explanation for this observation. That is, Cu^{2+} (d^9) and Cr^{2+} (d^4) complexes are often structurally distorted, with bonds to axial groups longer and weaker than bonds to equatorial groups (see the discussion of the Jahn-Teller effect in Chapter 9). Therefore, the ground state structures for Cr^{2+} and Cu^{2+} are not far removed from the transition state structures, and the axial water molecules, which are held less tightly, can exchange more rapidly.

To explain the behavior of the other 2+ ions of the first row transition metals, you can pursue a more careful analysis of the correlation between rate and d electron configuration in the section which follows. That such a correlation may exist is seen in Figure 13-4; these plots resemble some of the plots of physico-chemical data *vs.* electron configuration that were discussed in Chapter 9 and thus suggest a rate–electron configuration correlation.

Trivalent metal ions react somewhat more slowly than divalent ions, so the trivalent ions largely constitute Classes III and IV. In the case of transition metal ions, however, there must again be an orbital occupation effect, as the pattern of rate constants *vs.* d electron configuration is approximately the same as that for the divalent ions of the same configuration.

ORBITAL OCCUPATION EFFECTS ON SUBSTITUTION REACTIONS OF OCTAHEDRAL COMPLEXES

Basolo and Pearson introduced the idea that there is a connection between d orbital energies and the relative kinetic inertness or lability of transition metal complexes.[2] The basic notion behind this approach is that a significant contribution to the activation energy in a substitution reaction is the change in d orbital energy on going from the ground state of the complex ion to the transition state. Any loss in energy is presumed to be proportional to the activation energy. Using this approach one can, in principle, answer the following questions: (i) Will the complex be kinetically labile or inert? If the structure preference energy, SPE, of the octahedron (*i.e.*, the difference in d electron energies between the octahedron and a transition state structure) is less than some arbitrary value, the complex is predicted to be labile; if the SPE is more than that value, the complex will be inert. (ii) What is the most probable transition state structure? For a given d orbital occupation, the difference in energy between the octahedron and one of several transition state structures may be less than that for the others; this will be the favored pathway for reaction.

According to Figure 13–4, the high spin d^3 through d^{10} hexaaquo complexes have relative labilities in the order: V^{2+} (inert) $\ll Cr^{2+} > Mn^{2+} > Fe^{2+} > Co^{2+} > Ni^{2+} \ll Cu^{2+} > Zn^{2+}$. As an example of the use of the angular overlap model in rationalizing this order, consider ions with configurations d^3 through d^7.[33] Since such complexes are thought to react generally by dissociative activation in octahedral substitution, assume, for the same of simplification, a D mechanism and trigonal bipyramidal or square pyramidal structures for the transition state. Further assume that a difference in energy of one unit of e_σ or less between the octahedron and the five-coordinate transition state means that the activation energy barrier is low and that the complex can be kinetically labile; conversely, a difference of two or more units of e_σ means that the complex is kinetically inert.

You can show, for example, that the difference in energy between a d^3 octahedral and a square pyramidal transition state is $-2e_\sigma$ (see Figure 10–1), and the difference between the octahedron and a trigonal bipyramid is $-3e_\sigma$. Thus, in either pathway there is a considerable loss in energy, so the activation energy is relatively large and d^3 complexes are predicted to be inert by a D mechanism proceeding through either of two possible five-coordinate intermediates. In contrast, you will find there is only one unit of e_σ loss in energy on going from the octahedron to either five-coordinate geometry for high spin d^5-d^7 ions, and these complexes are predicted to be relatively labile. Notice, however, that Figure 13–4 shows there is actually a decline in lability proceeding across this series. The reason for this decline is clearly the increase in Z^* on going from Mn^{2+} to Co^{2+}, an increase leading to greater values of e_σ and greater $E^{\ddagger}$.

Finally, we note that similar calculations for low spin d^6 ions show great differences in energy between the octahedron and the transition state, and these ions are predicted to be kinetically inert. As you have seen thus far in this chapter, Co(III) complexes are indeed substitutionally inert.

Solvolysis or Hydrolysis

The hydrolysis of metal complexes under acidic conditions has been extensively studied, as have base- and metal-ion-catalyzed hydrolyses. The first two types will be discussed in the following sections, while metal-ion-catalyzed reactions are mentioned in the next chapter.

HYDROLYSIS UNDER ACIDIC CONDITIONS

As mentioned earlier, hydrolysis reactions are the reverse of anation reactions, and the forward and reverse reactions traverse the same reaction coordinate in opposite directions. The hydrolysis of acidopentaamminecobalt(3+) salts has been studied extensively, with results that bear further on the thesis that substitution reactions of octahedral metal complexes generally occur by an I_d mechanism.

(13-13)
$$[Co(NH_3)_5X]^{2+} + H_2O \underset{k_{anation}}{\overset{k_{hydrolysis}}{\rightleftharpoons}} [Co(NH_3)_5H_2O]^{3+} + X^-$$

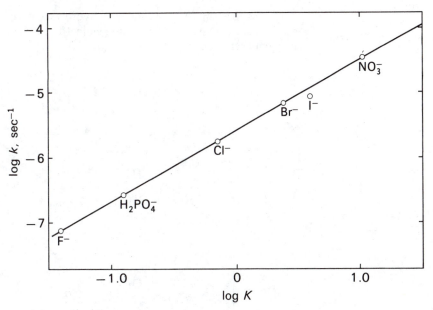

Figure 13-5. Correlation of the acid hydrolysis rates with the equilibrium constants in the reaction $Co(NH_3)_5 X^{2+} + H_2O \rightarrow Co(NH_3)_5 H_2O^{3+} + X^-$. [Reproduced from "Ligand Substitution Processes," by Cooper H. Langford and Harry B. Gray, with permission of the publishers, Addison–Wesley/W. A. Benjamin Inc., Advanced Book Program, Reading, Mass.]

Figure 13-5 is a plot of log $k_{\text{hydrolysis}}$ against log K, where K is the equilibrium constant for the hydrolysis reaction and is equal to $k_{\text{hydrolysis}}/k_{\text{anation}}$. An important conclusion can be drawn from this plot. Earlier in this chapter (p. 424) we pointed out that the rate of the anation reaction does not depend on the nature of X^-. This and other information were used to support an I_d mechanism for anation. On the other hand, Figure 13-5 shows that the rate of hydrolysis is dependent on the nature of X^-. This dependence on the basicity of the leaving group is taken as evidence for dissociative activation. As there is no experimentally apparent intermediate, the mechanism is assigned an I_d label.[34]

BASE-CATALYZED HYDROLYSIS: THE CONJUGATE BASE OR CB MECHANISM†[35]

It is fair to say that octahedral substitution reactions are not sensitive to the nature of the entering group—with one exception. In basic media, Co(III) complexes having ligands of the type NH_3, RNH_2, or R_2NH are sensitive to the nature of the entering group. In this case OH^- is the apparent entering ligand, and *base catalyzed reactions are generally much more rapid than anations or hydrolyses in acid solution.* Furthermore, the rate law for the general reaction

$$R_5MX + OH^- \rightarrow R_5MOH + X^-$$

is

$$\frac{-d[\text{complex}]}{dt} = k_{\text{obs}}[\text{complex}][OH^-]$$

(13-14)

The clear-cut second order kinetics of the reaction early prompted many people to suggest that hydrolyses in basic media occurred by an A or I_a mechanism, a suggestion clearly at odds with the previously observed generality of D or I_d mechanisms. Such seeming contradictions need explanation.

---

†The CB mechanism has already been discussed in connection with nitrogen chemistry; see page 239.

The mechanism of base-catalyzed hydrolyses is controversial. However, it is now generally agreed that the correct stoichiometric mechanism, illustrated for the case of pentaamminechlorocobalt(2+), is

(13-15)

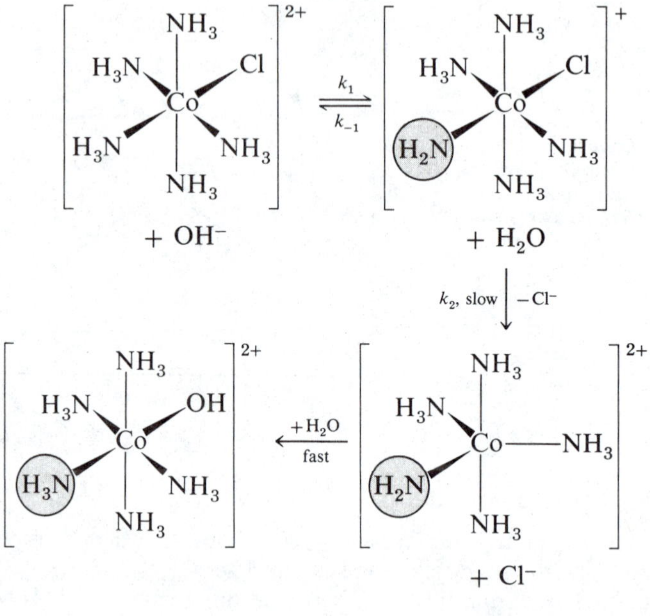

You will notice that the chief feature of this mechanism is the reversible removal of a proton from an ammonia ligand in the first step. This same mechanism was discussed for substitution at amine nitrogen in Chapter 7 (page 239). Here, as there, this creates the conjugate base of the original complex, hence the name "conjugate base" or CB mechanism. This step is generally very rapid, perhaps 10^5 times faster than the release of Cl^- in the second step.

STUDY QUESTIONS

5. Cd^{2+} and Hg^{2+} exchange coordinated water very rapidly and are in Class I. Why is Zn^{2+} in Class II and not Class I?

6. Why does Ga^{3+} exhibit a greater rate of water exchange than Al^{3+}? What would the rate of exchange be for Tl^{3+} with respect to Al^{3+} and Ga^{3+}? What about Tl^+?

7. With one exception, octahedral complexes based on low spin d^3-d^7 ions are inert to substitution by pathways involving either a trigonal bipyramidal or square pyramidal intermediate. The exception to this statement is a d^7 complex proceeding through a trigonal bipyramidal intermediate. Prove these predictions by making appropriate calculations based on the angular overlap model.

8. Cr^{2+} and Cu^{2+} are expected to show pronounced Jahn-Teller distortions (see page 326). Explain how this can lead to the observation of great kinetic lability.

If you have studied Chapter 7 on "Reaction Pathways," the next two questions will be of interest.

9. The rate laws in Figure 13-2 can all reduce to the simple form Rate = k_{obs} [Complex][X] under certain circumstances. Show how this is possible in each case.

10. Perhaps the chief conclusion to be reached from the discussion of reaction mechanisms in this chapter is that four-coordinate planar complexes react predominantly by an associative mechanism, whereas octahedral complexes react predominantly by dissociative pathways. The question that is now of interest is the mechanism of substitution in five-coordinate complexes. Clearly, both associative and dissociative mechanisms are possible, since a five-coordinate complex can potentially either add or lose a ligand. Some research has recently been published on such reactions, and this question concerns one of those publications.

The replacement of a phosphine ligand on an iron- or cobalt-dithiolene complex proceeds according to the stoichiometry:

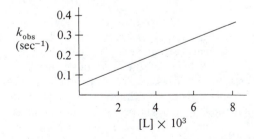

X and L = phosphines

a. For a stoichiometrically dissociative mechanism, the reactions involved would be

$$M(S_2C_2Ph_2)_2X \underset{k_2}{\overset{k_1}{\rightleftharpoons}} M(S_2C_2Ph_2)_2 + X$$

$$M(S_2C_2Ph_2)_2 + L \xrightarrow{k_3} M(S_2C_2Ph_2)_2L$$

Write the rate law for this mechanism, assuming that the four-coordinate intermediate is in a steady state. What is the expression for k_{obs} when excess L is used (that is, you operate under pseudo-first order conditions)? To what does this last expression reduce when $k_3 \gg k_2$?

b. For a stoichiometrically associative mechanism, the reactions involved would be

$$M(S_2C_2Ph_2)_2X + L \underset{k_2}{\overset{k_1}{\rightleftharpoons}} M(S_2C_2Ph_2)_2XL \xrightarrow{k_3} M(S_2C_2Ph_2)_2L + X$$

Assuming steady state conditions for the six-coordinate intermediate, write a rate law for this two-step mechanism. Again assume pseudo-first order conditions in L and write the expression for k_{obs}.

c. Given below is a plot of k_{obs} vs. [L] for the reaction of $Co(S_2C_2Ph_2)_2PPh_3$ + L. When L = P(OEt)$_3$, $k_{obs} = 0.05 + 42.5[L]$.

From this plot, and the fact that added X (the leaving group) has a small effect on the reactions of the cobalt complex, what can you conclude about the mechanism of the reaction? That is, is the reaction strictly dissociative or strictly associative, or are both pathways used? If both pathways are used, which predominates? [If you wish further information on reactions of five-coordinate complexes, consult the following, more recent references: D. A. Sweigart and P. Heidtmann, *J. C. S. Chem. Commun.*, 556 (1973); *J. C. S. Dalton,* 1686 (1975).]

SYNTHESIS OF COORDINATION COMPOUNDS BY SUBSTITUTION REACTIONS[36-42]

In this chapter and the previous ones on coordination chemistry, we have discussed a great variety of square planar and octahedral coordination complexes as well as some of the other geometries and coordination numbers. However, you must realize that you cannot normally purchase these compounds from Fisher Scientific Company, Alfa Products, or some other supplier of inorganic chemicals. Rather, most compounds must be synthesized from starting materials such as $CrCl_3$, $CoCl_2 \cdot 6H_2O$, $NiCl_2 \cdot 6H_2O$, K_2PtCl_4, and so on, and most of these syntheses involve the substitution of water and/or halide ion in the primary coordination sphere of the starting material. As an example, consider the following preparation:

Preparation of hexaamminechromium(3+) nitrate.[39] This preparation is interesting because it illustrates the application of the CB mechanism to chemical synthesis. If anhydrous $CrCl_3$ is allowed to react with liquid ammonia, $[Cr(NH_3)_5Cl]Cl_2$ will form as the major product. The remaining chromium-bound chloride ion is displaced very slowly. However, you might recall that base catalyzed hydrolysis reactions occur very rapidly compared with those in neutral solution. Therefore, it is possible to replace the remaining Cl^- by a base catalyzed solvolysis in liquid ammonia where the amide ion, NH_2^-, is the base and the entering ligand is NH_3. Mechanistically, the reaction probably follows a path similar to that shown on page 427.

(13-16)

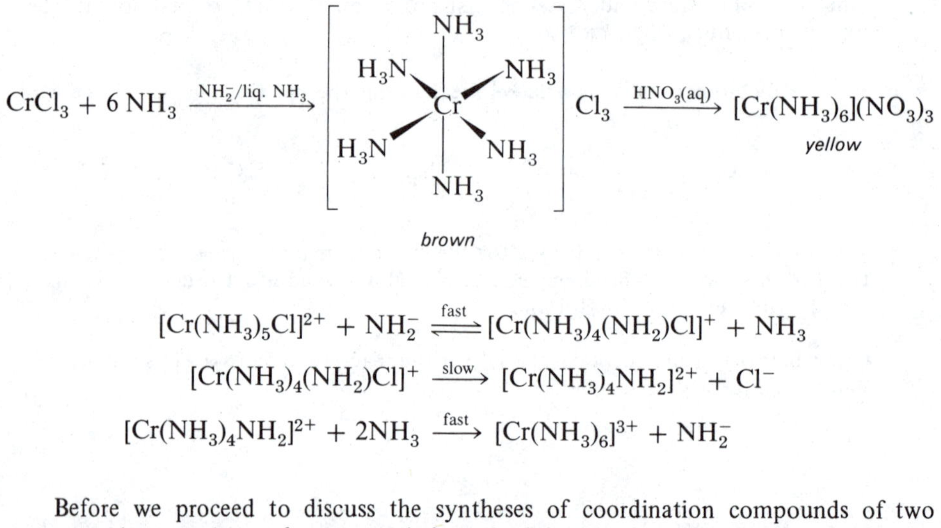

(13-17)

$$[Cr(NH_3)_5Cl]^{2+} + NH_2^- \xrightleftharpoons{fast} [Cr(NH_3)_4(NH_2)Cl]^+ + NH_3$$

$$[Cr(NH_3)_4(NH_2)Cl]^+ \xrightarrow{slow} [Cr(NH_3)_4NH_2]^{2+} + Cl^-$$

$$[Cr(NH_3)_4NH_2]^{2+} + 2NH_3 \xrightarrow{fast} [Cr(NH_3)_6]^{3+} + NH_2^-$$

Before we proceed to discuss the syntheses of coordination compounds of two representative transition elements—cobalt and platinum—there is one additional topic that we should take up, a discussion of the thermodynamic stability of coordination compounds. We have already seen that complexes of some transition metal ions, Co(III) for example, are notably kinetically inert with respect to substitution. However, this does not necessarily mean that the complex will persist in solution under ordinary conditions for an indefinite length of time. Such would be the case only for a complex of high thermodynamic stability. For example, $[Co(NH_3)_6]^{3+}$ may be inert to substitution, but it is completely unstable thermodynamically in acid solution with respect to formation of NH_4^+ and $[Co(H_2O)_6]^{3+}$. On the other hand, although the d^8 ion $[Ni(CN)_4]^{2-}$ has a dissociation constant for replacement of CN^- by H_2O of 10^{-30} (indicating great thermodynamic stability), bound CN^- is exchanged very rapidly for free CN^- in solution (indicating substitutional lability).

Thermodynamic Stability of Coordination Compounds[43-45]

The thermodynamic stability of coordination compounds is usually defined in terms of the equilibrium constant for the reaction

$$M_{(aq)}^{m+} + L^{n-} \rightleftharpoons ML_{(aq)}^{(m-n)+}$$

$$K = \frac{[ML^{(m-n)+}]}{[M^{m+}][L^{n-}]}$$

(13-18)

For example, for the reaction of aqueous Cu(II) with NH_3:

$$[Cu(H_2O)_4]^{2+} + NH_3 \rightleftharpoons [Cu(H_2O)_3NH_3]^{2+} + H_2O$$

$$K = 1.66 \times 10^4$$

(13-19)

The constant K is usually referred to as the **stoichiometric stability constant**.

More often than not, more than one water molecule on the original metal ion may be substituted by another ligand, and it is assumed that this may occur in a sequence of steps, each step characterized by its own stoichiometric equilibrium constant, K_n. For example, the reaction of tetraaquocopper(2+) with ammonia (13-19) is just the first of four possible steps, and the stoichiometric equilibrium constant for the first step would be K_1. The subsequent steps would then be as follows:

Step 2 $\quad [Cu(H_2O)_3NH_3]^{2+} + NH_3 \rightleftharpoons [Cu(H_2O)_2(NH_3)_2]^{2+} + H_2O$

$$K_2 = \frac{[Cu(H_2O)_2(NH_3)_2^{2+}]}{[Cu(H_2O)_3NH_3^{2+}][NH_3]} = 3.16 \times 10^3$$

(13-20)

Step 3 $\quad [Cu(H_2O)_2(NH_3)_2]^{2+} + NH_3 \rightleftharpoons [Cu(H_2O)(NH_3)_3]^{2+} + H_2O$

$$K_3 = \frac{[Cu(H_2O)(NH_3)_3^{2+}]}{[Cu(H_2O)_2(NH_3)_2^{2+}][NH_3]} = 8.31 \times 10^2$$

(13-21)

Step 4 $\quad [Cu(H_2O)(NH_3)_3]^{2+} + NH_3 \rightleftharpoons [Cu(NH_3)_4]^{2+} + H_2O$

$$K_4 = \frac{[Cu(NH_3)_4^{2+}]}{[Cu(H_2O)(NH_3)_3^{2+}][NH_3]} = 1.51 \times 10^2$$

(13-22)

Summing up these four steps, the overall reaction that occurs in this case is:

$$[Cu(H_2O)_4]^{2+} + 4NH_3 \rightleftharpoons [Cu(NH_3)_4]^{2+} + 4H_2O$$

(13-23)

and an equilibrium constant for the overall process is:

$$K_{total} = \frac{[Cu(NH_3)_4^{2+}]}{[Cu(H_2O)_4^{2+}][NH_3]^4} = 6.58 \times 10^{12}$$

The overall equilibrium constant or K_{total} is, of course, just the product of the stepwise constants; that is, $K_{total} = K_1 \cdot K_2 \cdot K_3 \cdot K_4$. Usually, K_{total} is given the symbol β and is equal to $K_1 \cdot K_2 \cdot K_3 \cdots \cdot K_n$.

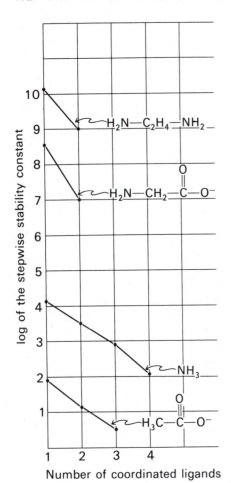

Figure 13-6. Log K values for the reactions of $Cu(H_2O)_4^{2+}$ with various ligands to illustrate the effect of the addition of successive ligands on complex stability. Ethylenediamine and glycine are bidentate and ammonia and acetate are monodentate.

In general, it is to be expected that the addition of each successive ligand to a metal ion in place of water will be less favored than the previous addition. This is illustrated by the constants for $Cu(H_2O)_4^{2+}$ and by Figure 13-6. There are clearly many reasons—statistical, steric, and electrostatic—for this observed decrease.

The main factors affecting the thermodynamic stability of transition metal complexes are outlined below, and each will be discussed in turn.

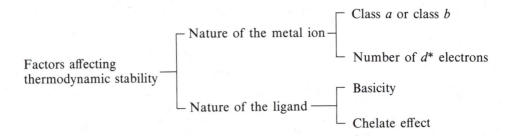

In Chapter 11 we discussed the phenomenon of linkage isomerism. In order to rationalize the preference of a metal ion for one end or another of a ligand that is potentially ambidentate, we used the Ahrland-Chatt-Davies *empirical* classification of metal ions into three groups: class *a*, class *b*, and borderline (see Figure 11-1 and Table 13-1).[46] As mentioned previously, this classification was made on the basis of the relative stabilities of complexes with ligands having donor atoms from Group VA, VIA, or VIIA.

TABLE 13-1

THE THREE CLASSES OF METAL IONS

Class a	Borderline	Class b
H^+, Li^+, Na^+, K^+	Fe^{2+}, Co^{2+}, Ni^{2+}	Cu^+, Ag^+, Au^+, Tl^+
Be^{2+}, Mg^{2+}, Ca^{2+}, Sr^{2+}	Cu^{2+}, Zn^{2+}, Pb^{2+}	Hg_2^{2+}, Hg^{2+}, Pd^{2+}, Pt^{2+}
Mn^{2+}, Al^{3+}, Sc^{3+}, Ga^{3+}		Pt^{4+}, Tl^{3+}
In^{3+}, La^{3+}		
Cr^{3+}, Co^{3+}, Fe^{3+}		
Ti^{4+}, Sn^{4+}		

If the stability of the complexes is greatest with the lightest element of each of these Groups as the donor atom (see Table 13-2), the ions are placed in class a. Conversely, class b ions form the least stable complexes with the lightest element of each Group as donor atom. Furthermore, class b ions form stable complexes with CO and olefins, while those of class a do not form such complexes.

Some ions frequently form complexes whose stabilities cannot be predicted on the basis of the order of generally observed stabilities in Table 13-2, and these ions are placed in the borderline class. The 2+ transition metal ions in this class (Table 13-1) are, of course, just the ones in which we are often most interested. Fortunately, it has been known for some time that the stability of complexes of the borderline ions with a given ligand is almost invariably in the order

$$Mn^{2+} < Fe^{2+} < Co^{2+} < Ni^{2+} < Cu^{2+} > Zn^{2+}$$

This order is known as the **Irving-Williams series**, and it is illustrated in Figure 13-7 for several different ligands. Although the figure shows only the trend for K_1, the Irving-Williams series usually holds for K_2 and K_3 (at least) as well.

The stability constant for a metal ion and its ligands is related to $\Delta G°$ for the reaction ($\Delta G° = -2.303\ RT \log K$). Since $\Delta G° = \Delta H° - T\Delta S°$, the next obvious question is which of the two thermodynamic functions, $\Delta H°$ or $\Delta S°$, is the controlling factor in the Irving-Williams series. The plots in Figure 13-8 provide an answer. For the two ligands having an amino donor group, ethylenediamine and glycine, the plot of $\Delta H°$ against metal ion shows the same trend as the Irving-Williams series. Therefore, at least for ligands having the amino donor group, the reaction is enthalpically controlled. On the other hand, the malonate ion reactions are all endothermic, so the stability of its complexes must arise from a large, positive $\Delta S°$ of reaction. (The more detailed data in Table 13-3 back up this conclusion.) The positive $\Delta S°$ apparently arises from the extensive desolvation of the carboxylate group upon complexation of that group with the metal ion; an anion in aqueous solution is a good "organizer" of solvent, and this organization is lost when the carboxylate binds to the metal ion. The positive $\Delta H°$ of the

TABLE 13-2

METAL ION CLASSIFICATION AND LIGAND DONOR ATOM TRENDS

The trend in complex stability with ligand donor atom type for class a metal ions.	The trend in complex stability with ligand donor atom type for class b metal ions.
$F > Cl > Br > I$	$F < Cl < Br < I$
$O \gg S > Se > Te$	$O \ll S \sim Se \sim Te$
$N \gg P > As > Sb$	$N \ll P > As > Sb$

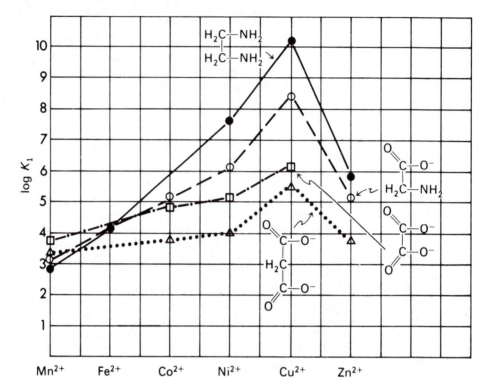

Figure 13-7. Illustration of the Irving-Williams series, the trend in log K_1 values as M^{2+} is changed.

carboxylate ion complexation reaction also has its origin in ligand desolvation. That is, water is bound more strongly to the anionic carboxylate than to the dipolar amine group. Apparently, more energy is required to desolvate the carboxylate than the amine, and the net reaction is endothermic.†

The plots of log K or $\Delta H°$ against metal ion that were discussed above should look familiar to you. They have the same general form as other plots of thermodynamic functions against number of $d*$ electrons that were discussed in Chapter 9 [see Figure 9-6(A), for example]. Therefore, it is evident that the Irving-Williams series arises from many of the same factors (*e.g.*, d orbital occupation, $Z*$ of the metal ion, d/TSAO overlap and energy match) that control the previously discussed trends.

The basicity of a ligand will most certainly be related to complex stability, but this effect is not always as predictable as one might wish.§ At first glance it would seem reasonable to presume that complex stability should, in general, be directly related to ligand basicity, more stable complexes being formed from more basic ligands. Indeed, there are reasonable correlations between ligand basicity (as measured by the pK_a of the ligand conjugate acid) and log K—but good correlations exist only for series of closely related ligands. Little or no correlation of any predictive value exists even for a series of

---

†This endothermicity arises in spite of the fact that, in the reaction products, the ion-ion M^{2+}-carboxylate interaction should be more exothermic than the ion-dipole M^{2+}-amine interaction.

§By "basicity" here we mean the free energy change or equilibrium constant for $H^+ + L = HL^+$.

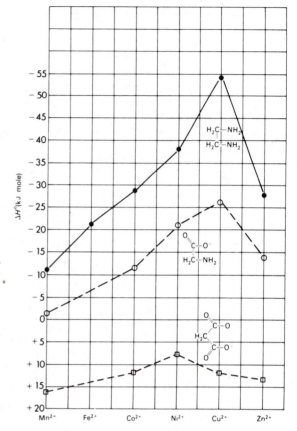

Figure 13–8. The trend in ΔH° values (for the reaction $M^{2+} +$ $L \rightleftharpoons ML^{n+}$) as a function of M^{2+}.

TABLE 13–3

ΔH° **AND** ΔS° **FOR THE REACTION** $M^{2+} + L \rightleftharpoons ML^{n+}$ **WHERE L IS ETHYLENEDIAMINE, GLYCINATE, OR MALONATE**[a]
($\Delta H =$ kJ/mol and $\Delta S =$ J/deg-mol)

		Metal Ion, M^{2+}					
		Mn^{2+}	Fe^{2+}	Co^{2+}	Ni^{2+}	Cu^{2+}	Zn^{2+}
H_2C-NH_2	ΔH°	−11.7	−21.3	−28.9	−37.2	−54.4	−28.0
H_2C-NH_2	ΔS°	12.6	12.6	16.7	23.0	22.6	16.7
$C-O^-$ H_2C-NH_2	ΔH°	−1.26	−	−11.7	−20.5	−25.9	−13.8
	ΔS°	56.5	−	57.3	49.8	77.0	53.1
$C-O^-$ H_2C $C-O^-$	ΔH°	15.40	−	12.1	7.9	11.9	13.1
	ΔS°	114.6	−	113.0	104.6	148.1	117.2

[a] All data taken from S. J. Ashcroft and C. T. Mortimer, "Thermochemistry of Transition Metal Complexes," Academic Press, New York, 1970.

ligands such as NH_3, pyridine, and imidazole, where the ligating atom is only nitrogen;† and the situation is even worse when considering ligands having different donor atoms (*e.g.*, RNH_2, RCO_2^-, and RS^-).

In many of the complexes we have discussed in this and previous chapters, there are one or more ligands that span two or more coordination positions; that is, they are chelating ligands. It is fair to ask, finally, why such complexes are so numerous and have been studied so thoroughly. The answer lies in thermodynamics—*complexes of chelating ligands are in general more stable thermodynamically than those with an equivalent number of monodentate ligands.* This special effect, which is strikingly illustrated in Figure 13-6, is called the *chelate effect*, an effect which can be explained as arising from a large positive entropy change when two or more ligands are replaced by one.[46a]

The chelate effect is illustrated by the thermodynamically favorable exchange of two ammonia molecules for an ethylenediamine on Cu(II).

(13-24)

$$[Cu(NH_3)_2(H_2O)_2]^{2+} + en \rightleftharpoons [Cu(en)(H_2O)_2]^{2+} + 2NH_3$$

$$\Delta G° = -17.2 \text{ kJ/mol} \qquad \Delta H° = -7.9 \text{ kJ/mol} \qquad \Delta S° = +31 \text{ J/deg-mol}$$

$$\log K = 3.0$$

As the enthalpy and entropy changes for the NH_3/en exchange show, these factors both favor the reaction and provide some insight into the chelate effect. $\Delta S°$ is evidently positive because two particles (the copper complex and en) have been converted into three particles (the new copper complex and two NH_3 molecules). This, of course, should generally be the case when a chelating ligand displaces monodentate ligands, and this provides a driving force for the displacement. The enthalpy change also favors the NH_3/en exchange. However, $\Delta H°$ is small because the donor groups of the two ligands are somewhat similar (pK_a for NH_4^+ is 9.3 and pK_a for $H_2NC_2H_4NH_3^+$ is 9.6), and the Cu–L bonds are almost isoenergetic. The entropy change doubles the thermodynamic driving force in this ligand exchange at room temperature.

The *size of the chelating ring* is also important. Generally, five-membered chelate rings are more stable than six-membered rings, which are in turn more stable than seven-membered rings. The importance of ring size is illustrated in Figure 13-9 for a series of dicarboxylate ligands.

Ring size effects are also evident in Figure 13-7, which we introduced earlier to illustrate the Irving-Williams series. That is, for most metal ions, five-membered ring chelates formed by ethylenediamine, oxalate, and glycinate are more stable than the six-membered ring chelates of the malonate ion. However, Figures 13-7 and 13-8 and Table 13-3 illustrate another important point. That is, for chelate rings of the same size with the same metal ion, the log K values will usually be in the order $(O-O)^{2-} < (O-N)^- < N-N$. *In general*, diamine complexes will be more stable than amino acid complexes, which will in turn be more stable than dicarboxylate complexes, at least for the first row 2+ transition metal ions from Co^{2+} to Zn^{2+}. The explanation for this is found in the data in Table 13-3. As mentioned earlier (page 433), carboxylate complexes derive stability from large changes in entropy, an effect attributable, in part, to desolvation of the anion

---

† As an example, consider the following data for Cu^{2+} complexes of the type $[CuL_4]^{2+}$:

	Pyridine	Imidazole	Ammonia
pK_a	5.25	6.95	9.3
$\log \beta_4$	6.63	12.5	12.8

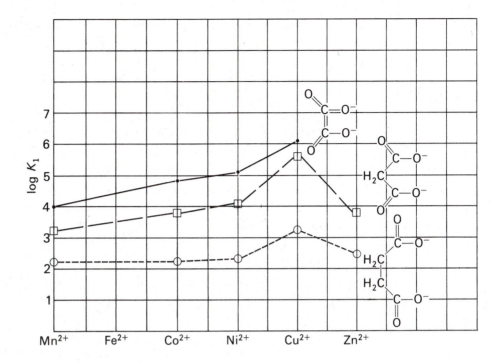

Figure 13-9. Dependence of complex stability (log K_1) on the size of the chelate ring.

upon complexation. On the other hand, diamines form quite stable complexes because of a very favorable $\Delta H°$.

STUDY
QUESTIONS

11. Given below are thermodynamic data for the reactions of halides and pseudo-halides with Fe(III) and Hg(II) in water.

Metal Ion	Halide	$\Delta H°$ (kJ/mol)	$\Delta G°$ (kJ/mol)	$\Delta S°$ (J/deg-mol)
Hg^{2+}	F^-	3.6	-5.9	33.5
	Cl^-	-24.7	-38.5	46.4
	Br^-	-40.2	-53.6	45.6
	I^-	-75.3	-73.2	-8.4
	CN^-	-96.2	-97.1	2.9
Fe^{3+}	F^-	31.4	-27.9	200.8
	Cl^-	35.6	8.5	146.4
	Br^-	25.5	-3.4	96.2
	SCN^-	-6.3	-17.1	36.4
	N_3^-	-6.7	-28.5	72.4

Plot $\Delta G°$, $\Delta H°$, and $T\Delta S°$ against the halide in the order given, and notice the great difference in the plots for the two different metal ions. Comment on these differences. For example, which factor—$\Delta H°$ or $\Delta S°$—controls the reactions of Fe(III) and Hg(II)? Does complex stability reflect the factors that went into the classification of metal ions into a, b, and borderline groups? (See Tables 13-1 and 13-2.)

12. Figure 13-8 shows the trend in $\Delta H°$ for diamine, amino acid, and dicarboxylate ligands. Supportive data are presented in Table 13-3. Speculate on the reason that

$\Delta H°$ for dicarboxylate ligands is greater than zero, whereas that for diamine ligands is less than zero.

13. One of the most important ligands in inorganic chemistry is the acetylacetonate ion, $[H_3C-CO-CH-CO-CH_3]^-$, a bidentate ligand that forms complexes with virtually all metal ions. The thermodynamic data for the formation of 1:1 complexes (reaction below) are given in tabular form below. Assuming room temperature, calculate $\Delta G°$ or log K. Plot $\Delta H°$ and $\Delta G°$ or log K vs. metal ion as illustrated in Figures 13-7 and 13-8, and compare the resulting plots with those for diamine and di-

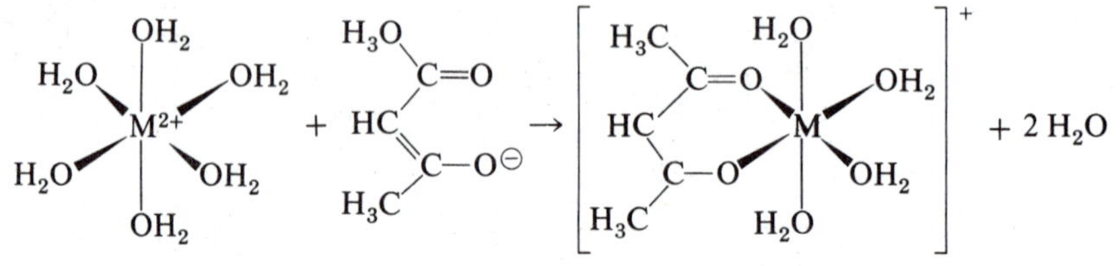

carboxylate ligands in Figures 13-7 and 13-8. Also compare the data below with those in Table 13-3. Comment on similarities and differences.

	Mn^{2+}	Fe^{2+}	Co^{2+}	Ni^{2+}	Cu^{2+}	Zn^{2+}	
$\Delta H°$	-10.5	—	-5.0	-28.0	-19.7	-7.9	(kJ/mol)
$\Delta S°$	46	—	87.8	50.2	92	71.2	(J/deg-mol)

The Synthesis and Chemistry of Some Cobalt Compounds[47]

To discuss the chemistry of cobalt is to discuss the history of chemistry in modern times. The element has, of course, been known for centuries in the form of "cobalt glass," the beautiful blue color coming from the presence of Co(II) in a tetrahedral oxide environment. A systematic development of cobalt chemistry, however, has occurred within the past two centuries. In 1798, B. M. Tassaert observed that on placing a cobalt(II) salt in aqueous ammonia, the solution gradually turned brown in the presence of air and finally wine red on boiling. From this solution at least two compounds, **13** and **14**, can be isolated, and others such as **15** can be obtained if excess chloride ion is added.

$CoCl_3 \cdot 6NH_3$

"luteocobaltic chloride"
yellow

10

13

$CoCl_3 \cdot 5NH_3 \cdot H_2O$

"roseocobaltic chloride"
red

11

14

$CoCl_3 \cdot 5NH_3$

"purpureocobaltic chloride"
purple

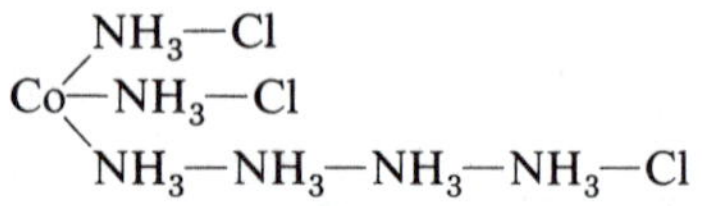

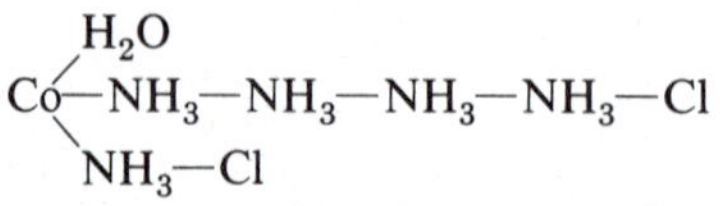

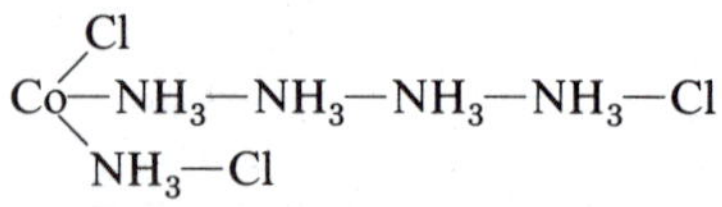

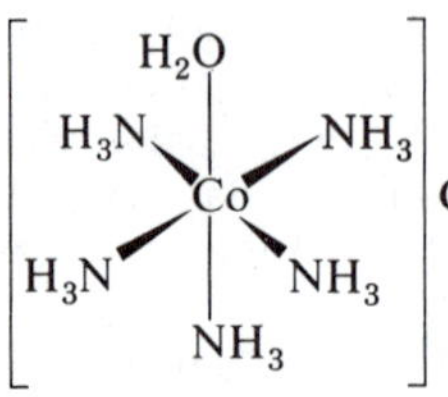

12

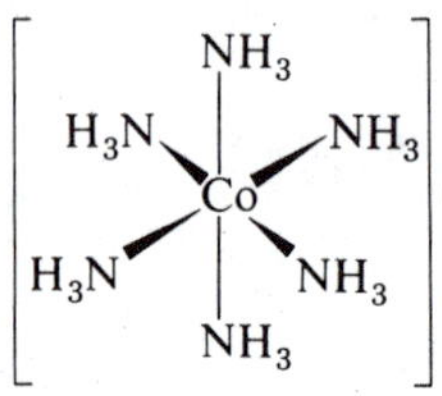

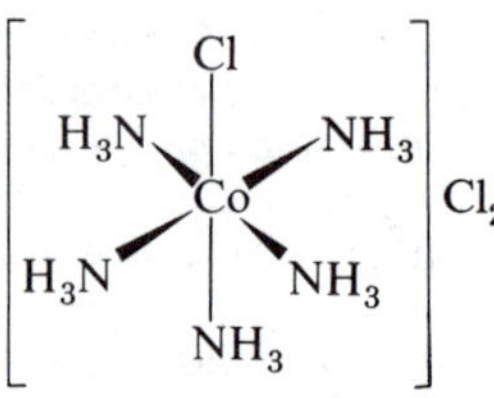

15

Although chemists in the earlier part of the 19th century formulated them as **10, 11,** and **12,** Alfred Werner recognized in 1893 that their structures were probably based on the octahedron. From that point on, Co(III) compounds played an important role in the development of Werner's ideas on coordination chemistry, and he and his students synthesized about 700 cobalt compounds and published approximately 127 papers on these compounds before his career was prematurely ended with his death in 1919 at the age of 52.[14,26]

There are probably two reasons for the usefulness of cobalt (III) compounds in the development of chemical theory: (i) As you have already seen in this chapter, Co(II) compounds are labile to substitution, whereas those of Co(III) are inert. (ii) Cobalt complexes of great variety are readily prepared. We now turn to that aspect of transition metal chemistry.

As is often the case in synthetic chemistry in general (see Chapter 8), much of the preparative work in cobalt chemistry can begin with the halides. Reaction of cobalt(II) oxide or carbonate with aqueous HCl gives the pink compound $CoCl_2 \cdot 6H_2O$, a salt that presumably contains the aquated, octahedral Co^{2+} ion. Dehydration at 150°C *in vacuo*

$$CoO_{(s)} + 2HCl + 5H_2O \rightarrow \left[\begin{array}{c} H_2O \\ H_2O \quad | \quad OH_2 \\ \diagdown \quad | \quad \diagup \\ Co \\ \diagup \quad | \quad \diagdown \\ H_2O \quad \quad OH_2 \\ | \\ H_2O \end{array} \right] Cl_2$$

(13-25)

$$\Big\downarrow \begin{array}{l} \text{heat } in\ vacuo \\ -H_2O \quad \text{or react with } SOCl_2 \\ (H_2O + SOCl_2 \rightarrow 2HCl + SO_2) \end{array}$$

$$Co_{(s)} + Cl_2 \rightarrow CoCl_2$$
blue

or by reaction with $SOCl_2$ gives anhydrous $CoCl_2$, the solid state structure of which is of the $CdCl_2$ type (see page 64). Cobalt(II) chloride, which is blue, can also be prepared by the direct reaction of cobalt metal with chlorine. It is this anhydrous salt that is the indicator in silica gel desiccants; as the desiccant absorbs water and loses its effectiveness, the cobalt(II) compound hydrates to the blue-violet monohydrate, the violet dihydrate, and finally the pink hexahydrate.

Aside from the 0 and +1 oxidation states of cobalt, which are more typical of organometallic compounds of the element, the common oxidation states of cobalt in coordination compounds are 2+ and 3+. Of the two possible hexaaquo complexes of cobalt in these latter oxidation states, the 2+ ion is thermodynamically the more stable form. Hexaaquocobalt(3+) is a powerful oxidant capable of oxidizing water to O_2, and

$$4Co^{3+}_{(aq)} + 2H_2O \rightarrow 4Co^{2+}_{(aq)} + 4H^+ + O_2$$

(13-26)

this capability is presumably the reason that simple salts of Co(III) can be isolated only with the anions that are not readily oxidized. For example, Co(III) fluoride, which is stable in the absence of water, can be prepared from CoF_2 and F_2 at 250°C (the reaction is reversible above 350°C). On the other hand, the chloride, bromide, and iodide of Co(III) are not known. The most common form of cobalt(III) in a simple salt is the sulfate, $Co_2(SO_4)_3 \cdot 18H_2O$; the compound can be prepared by oxidation of the corresponding Co(II) compound with O_3 or F_2 in 8 NH_2SO_4. The salt is stable when dry, but it decomposes in water with evolution of O_2.

In the presence of coordinating ligands other than water, Co(III) is stabilized relative to Co(II); that is, the higher valent species is no longer a strong oxidizing agent.

$$[Co(H_2O)_6]^{3+} + e^- \rightleftharpoons [Co(H_2O)_6]^{2+} \qquad E° = +1.84 \text{ volts } vs. \text{ S.H.E.}$$
$$[Co(NH_3)_6]^{3+} + e^- \rightleftharpoons [Co(NH_3)_6]^{2+} \qquad E° = +0.108 \text{ volts}$$
$$[Co(CN)_6]^{3-} + e^- \rightleftharpoons [Co(CN)_5]^{3-} + CN^- \qquad E° = -0.8 \text{ volts}$$

This change in reduction potential, a common phenomenon in coordination chemistry, has been related to complex stability.[48]

The hexaammine complexes of Co(II) and Co(III) are both more stable than the corresponding hexaaquo complexes; $\Delta H°$ for formation of $[Co(NH_3)_6]^{2+}$ from the hexaaquo ion is -54.4 kJ/mol, whereas the corresponding value for $[Co(NH_3)_6]^{3+}$ is -237.7 kJ/mol. This indicates that the ammonia molecules are more tightly bound to Co(III) or Co(II) than are the water molecules in the corresponding aquo complexes. As a result, the effective positive charge on the metal ion in the oxidizing agent $[Co(NH_3)_6]^{3+}$ is lower than that on Co(III) in $[Co(H_2O)_6]^{3+}$. In terms of the electrochemical properties of these complexes, this means that the oxidizing ability of the Co(III)-ammine complex is less than that of the Co(III)-aquo complex. Of course, by the same argument, the reducing ability of the Co(II)-ammine complex should be greater than that of the Co(II)-aquo complex. Thus, the stronger the M−ligand bond, the more unstable are the σ^* electrons (i.e., the metal d electrons), because e_σ is larger. Accordingly, it becomes more favorable to remove the σ^* electron(s), and the Co(II) complex becomes more reducing while Co(III) becomes less oxidizing.

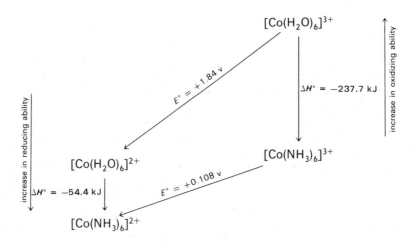

Because of these changes in electrochemical properties, it is clear that coordination of a transition metal ion by ligands other than water can usually change the oxidizing and reducing ability of the metal ion, a property important to consider in synthetic chemistry. For example, if this idea is combined with your knowledge of kinetic behavior, the synthesis of an ion such as $[Co(NH_3)_6]^{3+}$ is rationalized. That is, you now realize that you cannot simply place a Co(III) salt in aqueous ammonia and expect the desired hexaamminecobalt(III) ion; much of the Co(III) would be converted to Co(II). For this reason one must begin with a Co(II) salt in water. Because high spin d^7 ions are labile to substitution, water coordinated to Co(II) can be displaced in favor of ammonia in an ammoniacal solution to give $[Co(NH_3)_6]^{2+}$, an ion thermodynamically more stable than the starting material, $[Co(H_2O)_6]^{2+}$. Oxidation of hexaamminecobalt(II) ion is then easily accomplished by any of a host of common laboratory oxidants. The usual laboratory preparation of $[Co(NH_3)_6]^{3+}$, of course, combines all of these steps into a single operation. This compound is commonly prepared by air oxidation of $CoCl_2 \cdot 6H_2O$ in a strongly ammoniacal solution in the presence of activated charcoal as a catalyst.[37,49,50] After several hours the initially red solution has turned yellow-brown and the orange product, $[Co(NH_3)_6]Cl_3$, can be isolated by filtration. Separation of the charcoal catalyst from the product is achieved by re-dissolving the complex in hot, dilute HCl and filtering to remove the charcoal; the product precipitates from solution on cooling. Other syntheses of Co(III) complexes can be done in very similar ways.

Although $[Co(NH_3)_6]Cl_3$ is the ultimate product from the oxidation of ammoniacal solutions of $CoCl_2 \cdot 6H_2O$, it has been known for more than 100 years that a brown intermediate product can be isolated.[51,52] This brown material has two cobalt ions in the same molecule and has the formula $[(NH_3)_5Co \cdot O_2 \cdot Co(NH_3)_5]X_4$. Although Alfred Werner and his students did extensive research on this compound and others like it,[53] the true nature of these compounds was not understood until recently.

We mentioned above that water molecules coordinated to Co(II) can be displaced by ammonia. One of the possible products of this displacement is $[Co(NH_3)_5H_2O]^{2+}$, and it is now known that the last remaining water can be displaced by molecular oxygen according to the equation:[54,55]

$$[Co(NH_3)_5H_2O]^{2+} + O_2 \underset{k_{-1}}{\overset{k_1}{\rightleftharpoons}} [Co(NH_3)_5O_2]^{2+}$$

$$k_1 = 2.5 \times 10^4 \text{ M}^{-1} \text{ sec}^{-1}$$

(13-27)

The mononuclear oxygen complex then reacts with the remaining pentaammineaquo complex to give the final binuclear cobalt compound.

$$[Co(NH_3)_5O_2]^{2+} + [Co(NH_3)_5H_2O]^{2+} \underset{k_{-2}}{\overset{k_2}{\rightleftharpoons}} [(NH_3)_5Co \cdot O_2 \cdot Co(NH_3)_5]^{4+}$$

(13-28)

$$k_{-2} = 56 \text{ sec}^{-1}$$

brown

16

The overall process, with an equilibrium constant of 6.3×10^6 M^{-2} and a $\Delta H°$ of about -125 kJ/mol, is reversible over short periods of time, say ten minutes. After a longer period of time, however, decomposition to give mononuclear Co(III) complexes such as $[Co(NH_3)_6]^{3+}$ is observed. The binuclear complex, **16**, is reasonably stable as a solid and in strongly ammoniacal solutions, but it is decomposed in acid solution according to the reaction

$$[(NH_3)_5Co \cdot O_2 \cdot Co(NH_3)_5]^{4+} + 10H^+ \rightarrow 2Co^{2+} + 10NH_4^+ + O_2$$

(13-29)

This decomposition to return molecular oxygen, and the reversibility of the process over short periods of time, are of obvious interest, and such complexes have been thoroughly studied as synthetic carriers of molecular oxygen.

The structure of the binuclear complex has only recently been determined (**17**),[56] and it is now clear that the cobalt ions are bridged by an O—O linkage. Of greatest interest is the length, 147 pm, of the O—O bond. On comparing this with O—O distances in some diatomic oxygen species in Table 13-4, it is obvious that the bridging O—O is very similar to a peroxide ion, O_2^{2-}. (Therefore, although the bonding in the Co—O—O—Co bridge is polar covalent, one way to count electrons is to assume that two Co^{3+} ions are bound to a peroxide ion, O_2^{2-}.) Indeed, the structure of the Co—O—O—Co

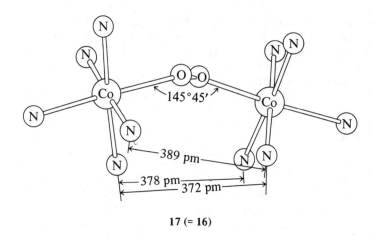

17 (= 16)

unit somewhat resembles that of H_2O_2. Because of this clear similarity to peroxides, the systematic name of the binuclear complex is decaammine-μ-peroxo-dicobalt(4+) ion.

The peroxo complex can be oxidized with, for example, peroxydisulfate ($S_2O_8^{2-}$) to a green paramagnetic compound, $[(NH_3)_5Co \cdot O_2 \cdot Co(NH_3)_5]^{5+}$, where the O—O bridge bond length is now 131 pm.

(13-30)
$$2[(NH_3)_5Co \cdot O_2 \cdot Co(NH_3)_5]^{4+} + S_2O_8^{2-} \rightarrow 2[(NH_3)_5Co \cdot O_2 \cdot Co(NH_3)_5]^{5+} + 2SO_4^{2-}$$

brown, diamagnetic   *green, paramagnetic* $(S = \frac{1}{2})$

16

Comparison with the data in Table 13–4 indicates that the bridge now resembles the paramagnetic superoxide ion, O_2^-, and this green complex may be thought of as having a $Co^{3+} \cdot O_2^- \cdot Co^{3+}$ unit where the paramagnetism formally arises from the superoxide bridge; however, recent electron spin resonance studies show that the odd electron is delocalized onto two equivalent Co^{3+} ions.[56a]

In addition to the extensive series of cobalt-ammine complexes, cobalt also forms an important group of complexes with the bidentate ligand ethylenediamine. The discussion of chiral compounds in Chapter 11 used $[Co(en)_3]^{3+}$ and *cis*-$[Co(en)_2(NH_3)Br]^{2+}$ as examples, because these compounds played such an important part in Alfred Werner's proof of the octahedron as the appropriate coordination geometry for many complex

Species	O—O Bond Distance, pm
O_2^+	112
O_2	120
O_2^- (superoxide)	132
O_2^{2-} (peroxide)	149

TABLE 13-4

BOND DISTANCES FOR SOME DIATOMIC OXYGEN SPECIES[a]

[a]A. G. Sykes and J. A. Weil, *Prog. Inorg. Chem.*, **13**, 1 (1970).

compounds. Ethylenediamine complexes can often be prepared by starting with a Co(III)-NH_3 complex and using the thermodynamic driving force of the "chelate effect" to displace NH_3 in favor of the bidentate, chelating diamine. Generally, however, the tris(ethylenediamine)cobalt(3+) ion is prepared by bubbling air through a solution of $CoCl_2 \cdot 6H_2O$ and ethylenediamine in dilute HCl. By adjusting the amount of HCl and ethylenediamine used, *trans*-dichlorobis(ethylenediamine)cobalt(1+) is obtained instead.[57,58]

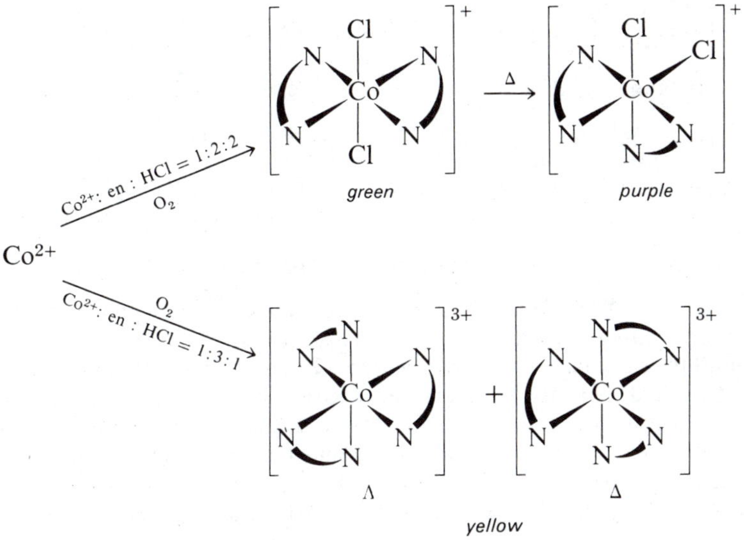

(13-31)

In the discussion of mechanisms of hydrolysis of Co(III) complexes, we pointed out that hydrolysis in basic media is considerably faster (by perhaps 10^6) than hydrolysis in acidic or neutral solution. For example, the half-life of the hydrolysis of *trans*-$[Co(en)_2Cl_2]^+$ in neutral or acidic solution is a matter of many minutes to hours at room temperature or slightly above, whereas the hydrolysis of the same ion is virtually instantaneous in base.

$$\textit{trans}\text{-}[Co(en)_2Cl_2]^+ \text{ (green)} \xrightarrow[t_{1/2} = \text{minutes to hours}]{H_2O,\ pH = 7} [Co(en)_2(H_2O)Cl]^{2+} \text{ (pink)}$$

$$\textit{trans}\text{-}[Co(en)_2Cl_2]^+ \text{ (green)} \xrightarrow[\text{instantaneous}]{H_2O,\ pH > 7} [Co(en)_2(OH)(Cl]^+ \text{ (purple)}$$

(13-32)

The latter reaction is presumed to occur by the CB mechanism which may be outlined as follows:

$$[Co(en)_2Cl_2]^+ + OH^- \xrightarrow{\text{fast}} [Co(en)(en{-}H)Cl_2] + H_2O \qquad (13\text{-}33a)\dagger$$

$$[Co(en)(en{-}H)Cl_2] \xrightarrow{\text{slow}} [Co(en)(en{-}H)Cl]^+ + Cl^- \qquad (13\text{-}33b)$$

$$[Co(en)(en{-}H)Cl]^+ + H_2O \xrightarrow{\text{fast}} [Co(en)_2(OH)Cl]^+ \qquad (13\text{-}33c)$$

†The symbol en–H stands for the $H_2NC_2H_4NH^-$ ion.

This mechanism suggests that the base catalyzed reaction may be turned to preparative uses *if* the reaction is run in a non-aqueous solvent. That is, if the five-coordinate intermediate can be generated as in reaction 13–33b by base catalysis, an added anion can compete effectively for the open site in a non-aqueous solvent. Such a reaction has actually been carried out. If *trans*-$[Co(en)_2Cl(NO_2)]^+$ is dissolved in DMSO (dimethylsulfoxide) and a small amount of OH^- is added, the Cl^- ligand is rapidly replaced by added anions such as NO_2^-.[59] The first two steps of this reaction (13–34) are presumed to be similar to those above (13–33a and b).

(13-34)

$$[Co(en)_2(NO_2)Cl]^+ + Base \xrightarrow{fast} [Co(en)(en-H)(NO_2)Cl] + BaseH^+$$

$$[Co(en)(en-H)(NO_2)Cl] \xrightarrow{slow} [Co(en)(en-H)(NO_2)]^+ + Cl^-$$

$$[Co(en)(en-H)(NO_2)]^+ + NO_2^- \xrightarrow{fast} [Co(en)(en-H)(NO_2)_2]$$

$$[Co(en)(en-H)(NO_2)_2] + BaseH^+ \xrightarrow{fast} [Co(en)_2(NO_2)_2]^+ + Base$$

The Synthesis and Chemistry of Some Platinum Compounds[8,9,60,61]

Although there is evidence that platinum was known to the ancients, it was apparently confused with silver, so there is no historical record of a separate metal called platinum until the early part of the 18th century. Originally discovered in South America, platinum was found in relative abundance in Russia in 1819. As a result, much of the early work on platinum chemistry was done by the Russian school. Currently, however, platinum is produced chiefly in South Africa and Canada as well as the Soviet Union.

The chemistry of platinum encompasses three oxidation states—0, II, and IV—and some representative compounds in these oxidation states are **18**, **19**, and **20**. Platinum(0)

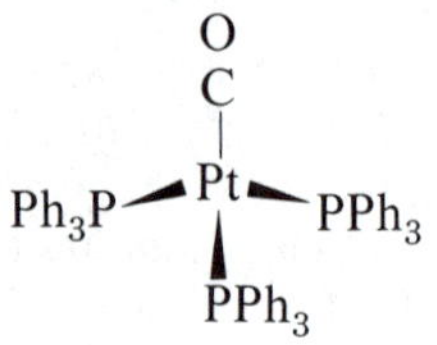

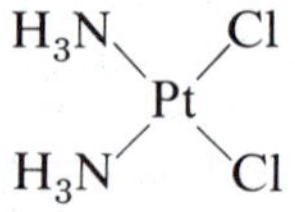

Pt–Cl = 233 pm
Pt–N = 201 pm

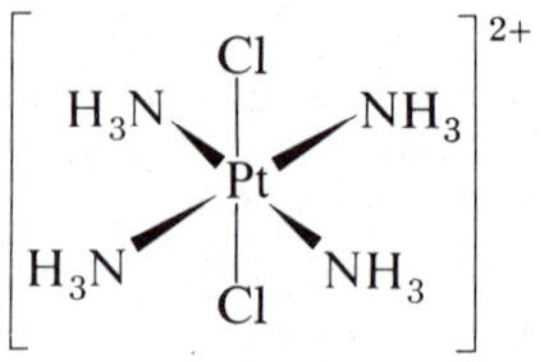

Pt–Cl = 225 pm
Pt–N = 203 pm

18   **19**   **20**

compounds are a very recent development, having been first synthesized about 1957.[62] They are representative of the tremendous development in the chemistry of low valent compounds in general. Many platinum(0) compounds have catalytic activity, and some are clusters of the metal. Both of these topics are among the more important recent developments in inorganic chemistry and their discussion is best postponed to the chapters on organometallic chemistry, where the chemistry of low-valent metals in general is discussed.

Platinum(II) compounds are especially abundant, and, as seen from previous discussions of their importance in the development of chemical theory (see Chapter 11, p. 375, and this chapter, p. 417), they have an interesting and rather systematic chemistry. Entrance into platinum(II) chemistry is often gained through the halides as starting materials. The following reactions are typical of metallic platinum and lead to just the starting materials needed.

(13-35)

$$Pt\ sponge \xrightarrow[\text{2. add KCl}]{\substack{\text{1. dissolve in}\\ \text{aqua regia}}} K_2PtCl_6 \xrightarrow{(N_2H_6)(SO_4)_2} K_2PtCl_4$$

yellow
potassium
hexachloroplatinate(2−)

red (86% yield)
potassium
tetrachloroplatinate(2−)

$$Pt + xCl_2 \rightarrow \begin{array}{c} \xrightarrow{\text{250-300°C}} PtCl_4 \text{ (red-brown; sol. } H_2O, \text{ acetone)} \\ \downarrow \Delta \\ \xrightarrow{\text{500°C}} PtCl_2 \text{ (olive green; insol. } H_2O) \end{array}$$

(13-36)

With M_2PtX_4 and PtX_2 as starting materials, a wealth of other Pt(II) compounds may then be synthesized; just a few of these are outlined below.

a. Preparation of $[PtL_4]^{2+}$. There are several routes to ions of this type; two of them are

$$PtL_2X_2 + 2L + 2AgNO_3 \rightarrow [PtL_4](NO_3)_2 + 2AgX$$

(13-37)

$$2[PtCl_4]^{2-} + 4L \rightarrow [PtL_4][PtCl_4] + 4Cl^-$$

(13-38)

The second reaction gives $[PtL_4][PtCl_4]$, a class of compounds generally known as Magnus' salts, and it should be apparent that these can also be prepared by direct reaction of the cation with the anion. For example, the product of reaction 13-39 is "Magnus' green salt."

$$[Pt(NH_3)_4]Cl_2 + K_2PtCl_4 \rightarrow 2KCl + [Pt(NH_3)_4][PtCl_4]$$

(13-39)

One of the starting materials for this reaction, $[Pt(NH_3)_4]Cl_2$, was synthesized by Magnus in 1828 and is presumably the first metal-ammine complex ever prepared.

b. Preparation of $PtLL'X_2$. The following reaction is useful when L = L'.

$$M_2PtX_4 + 2L \rightarrow 2MX + PtL_2X_2$$

(13-40)

A specific example is the preparation of the bright yellow complex $Pt(en)Cl_2$.[63]

$$K_2PtCl_4 + H_2N-C_2H_4-NH_2 \rightarrow 2KCl + \begin{array}{c} H_2 \\ N \\ H_2C \diagdown \diagup Cl \\ | \quad Pt \\ H_2C \diagup \diagdown Cl \\ N \\ H_2 \end{array}$$

(13-41)

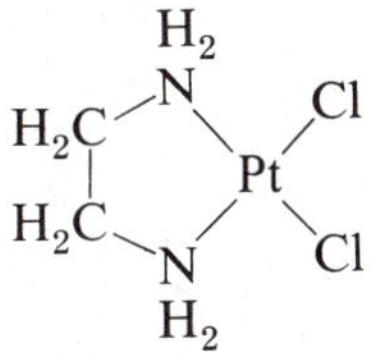

21

As might be anticipated from reaction 13-38, however, a by-product is the violet complex $[Pt(en)_2][PtCl_4]$. Fortunately, the by-product and the desired compound have different solubilities in liquid ammonia (the salt is not soluble), and this fact allows their separation.

Two of the most important complexes in platinum chemistry from an historical point of view are *cis*- and *trans*-$Pt(NH_3)_2Cl_2$.[64] The *cis* isomer is prepared by the direct reaction of K_2PtCl_4 with aqueous ammonia. Notice that this reaction is apparently controlled by the *trans* effect. Once one NH_3 has been attached to the Pt(II) ion, the ligand most readily replaced is one of the Cl^- ions *trans* to each other.

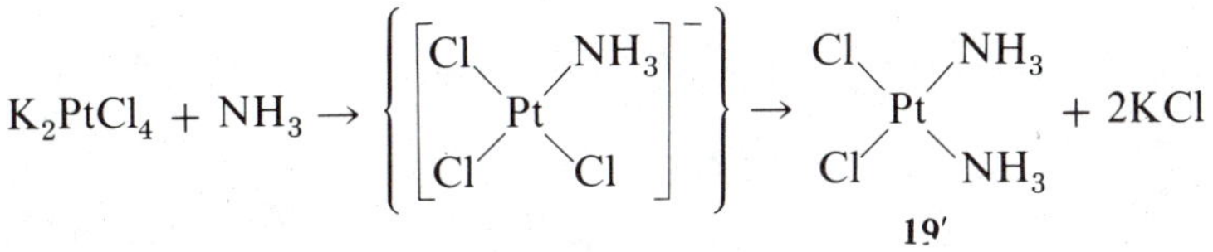

$$K_2PtCl_4 + NH_3 \rightarrow \left\{ \begin{bmatrix} Cl & NH_3 \\ Pt \\ Cl & Cl \end{bmatrix}^- \right\} \rightarrow \begin{array}{c} Cl \diagdown \diagup NH_3 \\ Pt \\ Cl \diagup \diagdown NH_3 \end{array} + 2KCl$$

(13-42)

19'

In the reaction above, one must be careful to avoid the formation of Magnus' green salt as in reaction 13-38. The synthesis of the *trans* isomer, on the other hand, takes advantage of the fact that $[Pt(NH_3)_4]^{2+}$ is formed in high concentrations of ammonia. This tetraammine complex is first prepared from $[PtCl_4]^{2-}$,

(13-43)
$$K_2PtCl_4 + 4NH_3 \rightarrow [Pt(NH_3)_4]Cl_2 + 2KCl$$

and it is then heated in the presence of excess Cl^- (13-44).

(13-44)

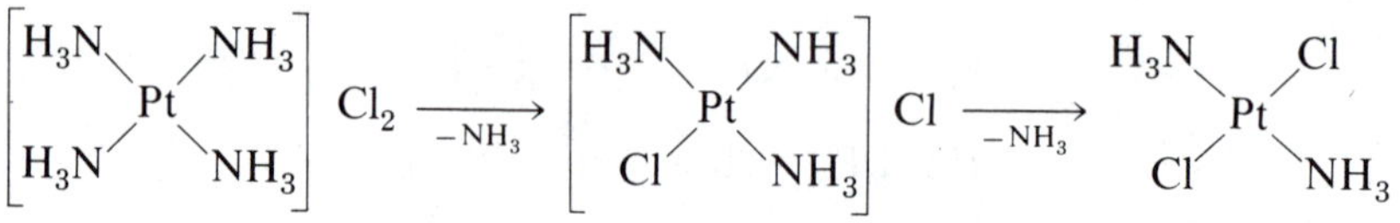

This is, of course, yet another illustration of the *trans* effect. Once the first NH_3 has been replaced by Cl^-, the *trans* effect leads to the desired *trans* isomer.

The preparation of $PtLL'X_2$ where $L \neq L'$ is more difficult and generally follows the sequence of reactions

(13-45)
$$PtL_2X_2 + HX \rightarrow [PtLX_3]^- + HL^+$$
$$\xrightarrow{L'} PtLL'X_2 + X^-$$

Compounds of the type PtL_2X_2 are clearly important, both historically and as starting materials for other platinum(II) compounds. However, they have recently taken on an added importance, because Rosenberg found that *cis*-$Pt(NH_3)_2Cl_2$ and *cis*-$Pt(en)Cl_2$ are effective anti-tumor agents.[65] This has led to a veritable avalanche of research on platinum chemistry in the past few years,[66-71] and new facets of its chemistry are being uncovered. In particular, it is now recognized that only the *cis* isomer has any biological activity, apparently because the complex PtL_2X_2 must lose both of the X ligands and form a complex with some biochemically important polymer that can function as a bidentate ligand. Although the "bidentate ligand" is currently thought to be a portion of the DNA molecule, the exact binding site is not known, nor is the reason for the effectiveness of *cis*-PtL_2X_2 compounds in general known. Nonetheless, testing of various platinum(II) complexes continues in order to find the most effective anti-cancer drug.[69,70] Since biochemical functioning can often be controlled by seemingly minor changes in a drug, it is important to be able to tailor-make a variety of new platinum compounds in order to more carefully probe their functioning. It is on this synthetic problem that we can bring to bear much of what has been discussed thus far in this chapter on substitution reactions.

If we presume that the platinum compound functions by forming a complex with some site in a protein that can act as a bidentate ligand, structure **22**, then we can discuss some requirements for the type of platinum-containing drug that will be most effective.

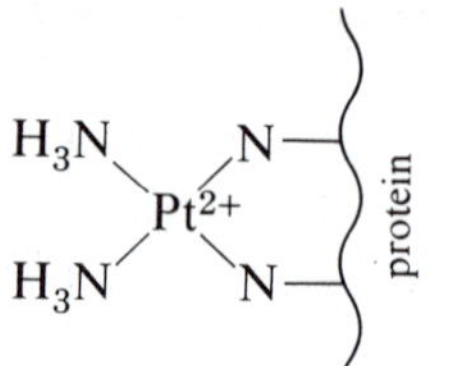

22

(i) A *cis* compound of the type PtL_2X_2 is required with two L ligands that are good *trans* labilizing groups.

(ii) The two ligands L that remain after loss of X should be reasonably strongly bound.

(iii) If it is assumed that the final Pt-protein complex must remain undissociated once formed, the Pt-protein bonds must be thermodynamically stable. Further, the remaining L ligands must not be such good *trans* labilizers that the Pt-protein bonds are easily broken; that is, the final complex should be kinetically stable.

Against the requirements for a Pt-containing drug laid out above, consider the following facts:

(i) Pt(II) is a class *b* metal ion, and the following order of metal-ligand bond strength is generally observed in the laboratory:

$$CN^- > OH^- > NH_3 > SCN^- > I^- > Br^- > Cl^- > F^- \geq H_2O$$

and

$$RS^- > R_2S \geq NH_3$$

A ligand in this series cannot be replaced by one to the right of it in a thermodynamically favorable reaction, provided that the two ligand concentrations are approximately equal. Therefore, a good X ligand in *cis*-PtL_2X_2 drug would be Cl^-, since it can be replaced in a thermodynamically favorable reaction by biologically important groups such as RS^-, R_2S, and RNH_2.

(ii) The *trans* effect generally decreases down the series

$$CN^-, \text{olefin} > SR_2, PR_3 > NO_2^- > I^- > Br^- > Cl^- > NH_3, \text{py}, RNH_2 > OH^- > H_2O$$

At first glance one would say that L should be a group well to the left of X in *trans* effect series. In fact, however, this is not desirable, since an L ligand with a large *trans* effect will also labilize the Pt-protein bond to dissociation. Because of this factor, a compound such as *cis*-$Pt(R_3P)_2Cl_2$ would be a poor choice as an anti-tumor drug.

Two of the compounds found most effective thus far as anti-tumor agents are two of the most common, **19** and **21**. Both generally meet the criteria spelled out above. The

19'   21'

Cl^- ligand is a good leaving group because the thermodynamic and kinetic stability of the Pt—Cl bond is low. Both ethylenediamine and NH_3 have a relatively low *trans* effect and so do not labilize the Pt-protein bond, once formed. Finally, ethylenediamine is a chelating ligand and effectively blocks the two remaining positions on the drug to further substitution by monodentate ligands; the "chelate effect" helps render its displacement thermodynamically unfavorable.

Platinum(IV) is a d^6 ion that forms neutral, cationic, and anionic octahedral complexes. The complexes are always diamagnetic because heavier transition metals, especially those in higher formal oxidation states, have large e_σ values. The preparation of Pt(IV) complexes is usually done by so-called oxidative addition reactions (such as 13–46) or by ligand exchange. The most notable feature of Pt(IV) complexes is their

(13-46)

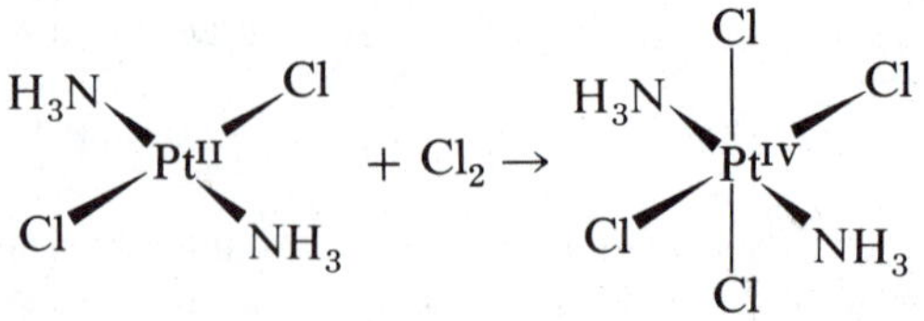

$$H_3N \diagdown \underset{Cl}{\overset{Cl}{Pt^{II}}} \diagup {}^{Cl}_{NH_3} + Cl_2 \rightarrow H_3N \diagdown \underset{Cl}{\overset{Cl}{\underset{Cl}{Pt^{IV}}}} \diagup {}^{Cl}_{NH_3}$$

kinetic stability. For example, the ion $[Pt(en)_3]^{4+}$ was resolved into its enantiomers (page 382) by Alfred Werner in 1917. Once separated, the enantiomers are extremely stable to racemization, a feature that you will appreciate more after the discussion of mechanisms of rearrangement in Chapter 14. Yet another example of the kinetic stability of Pt(IV) complexes is illustrated by the isolation of two of the fifteen possible diastereomers of the compound $Pt(NH_3)(NO_2)pyClBrI$.[60]

STUDY QUESTIONS

14. In the course of experiments on the biological functioning of platinum compounds, *cis-* and *trans-*$[Pt(NH_3)_2Cl_2]$ were allowed to react with methionine, with the results outlined below. What can these reactions tell about the ability of Cl^- to act as a leaving group and about the *trans* effect of the thioether sulfur?

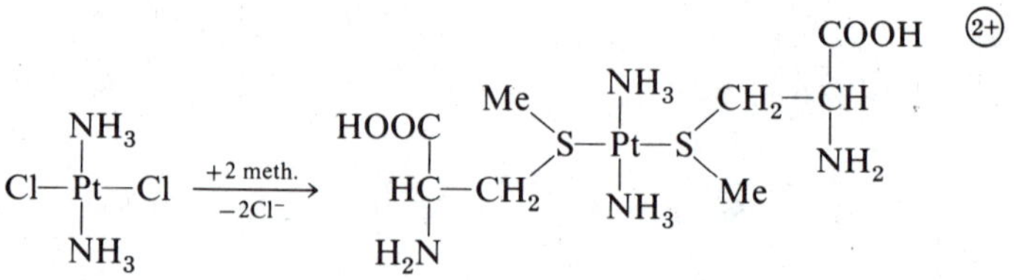

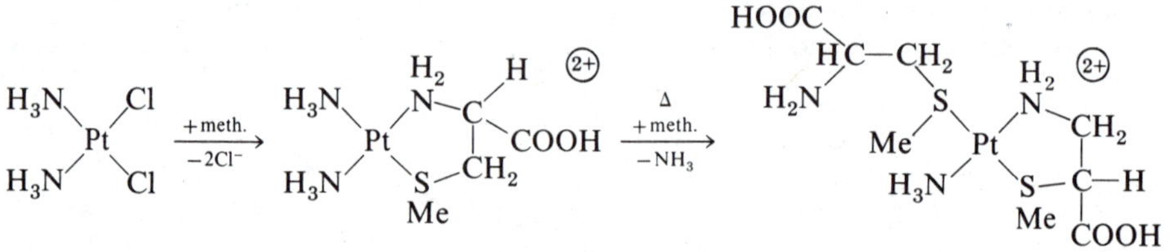

15. What spectral transition(s) gives rise to the blue color of cobalt glass? Why are octahedral Co(III) complexes usually not blue, in contrast with tetrahedral Co(II) complexes? (Hint: Compare the energies of absorbed photons.)

16. Outline the synthesis of *trans-*$Pt(py)(PEt_3)Cl_2$ starting with $[PtCl_4]^{2-}$.

17. The compound below has been found to be effective against certain types of cancer. Devise a synthetic route to the compound starting with $[PtCl_4]^{2-}$.

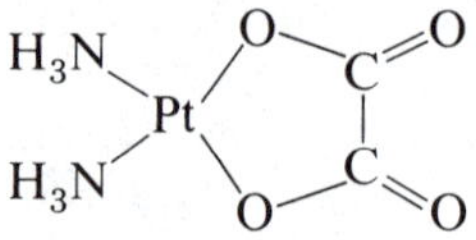

CHECKPOINTS

1. Ligand substitutions on square planar complexes generally occur by I_a or, less often, by A mechanisms. The first group entering may be either the substituting ligand or the solvent.

2. Octahedral complexes, especially those of the divalent ions of the later transition metals, react generally by I_d mechanisms. Recent evidence suggests that complexes of the divalent ions of the early metals and those of the trivalent ions, except for Co(III), *may* react by I_a mechanisms.

3. In contrast with square planar complexes, direct replacement of one ligand for another (except when ligand = solvent) does not occur on an octahedral complex reacting by a dissociative pathway. Rather, the leaving group is first substituted by solvent, which is subsequently replaced by the new ligand.

REFERENCES

[1] M. L. Tobe, "Inorganic Reaction Mechanisms," Thomas Nelson and Sons, London, 1972.

[2] F. Basolo and R. G. Pearson, "Mechanisms of Inorganic Reactions," 2nd edition, John Wiley and Sons, New York, 1967.

[3] C. H. Langford and H. B. Gray, "Ligand Substitution Processes," W. A. Benjamin Inc., 1965.

[4] R. G. Pearson and P. C. Ellgen, "Mechanisms of Inorganic Reactions in Solution" in "Physical Chemistry, An Advanced Treatise," volume VII, H. Eyring, ed., Academic Press, New York, 1975.

[4a] R. G. Wilkins, "The Study of Kinetics and Mechanism of Reactions of Transition Metal Complexes," Allyn and Bacon, Inc., Boston, 1974.

[5] A. Werner and A. Miolati, *Z. phys. Chem.*, 14, 506–21 (1894). This paper is available in translation in G. B. Kauffman, "Classics in Coordination Chemistry," Dover Publications, Inc., New York, 1968.

[6] L. Cattalini, *Prog. Inorg. Chem.*, 13, 263 (1970).

[7] A. Peloso, *Coord. Chem. Rev.,* 10, 123–181 (1973).

[8] U. Belluco, "Organometallic and Coordination Chemistry of Platinum," Academic Press, New York, 1974.

[9] F. R. Hartley, "The Chemistry of Platinum and Palladium," John Wiley and Sons, New York, 1973.

[10] R. G. Pearson, H. Sobel, and J. Songstad, *J. Amer. Chem. Soc.*, 90, 319 (1968).

[11] T. G. Appleton, H. C. Clark, and L. E. Manzer, *Coord. Chem. Rev.*, 10, 335–422 (1973).

[12] F. Basolo and R. G. Pearson, *Prog. Inorg. Chem.*, 4, 381 (1962).

[13] See also our previous discussion of this problem on p. 375 of Chapter 11.

[13a] G. B. Kauffman, *J. Chem. Educ.*, 54, 86 (1977).

[14] A. Werner, *Z. anorg. Chem.*, 3, 267–330 (1893). This important paper in inorganic chemistry has been translated and annotated by G. B. Kauffman, "Classics in Coordination Chemistry, Part I: The Selected Papers of Alfred Werner," Dover Publications, Inc., New York, 1968.

[15] See the section on "The Thermodynamic Stability of Coordination Complexes" beginning on page 431.

[16] See ref. 1 (page 54).

[17] U. Belluco, M. Graziana, and P. Rigo, *Inorg. Chem.*, 5, 1123 (1966).

[18] G. Bushnell, A. Pidcock, and M. A. R. Smith, *J. C. S. Dalton,* 572 (1975).

[19] S. S. Zumdahl and R. S. Drago, *J. Amer. Chem. Soc.*, 90, 6669 (1968).

[20] D. R. Armstrong, R. Fortune, and P. G. Perkins, *Inorg. Chim. Acta*, 9, 9 (1974). See also J. K. Burdett, *Inorg. Chem.*, 16, 3013 (1977).

[21] F. Basolo, J. Chatt, H. B. Gray, R. G. Pearson, and B. L. Shaw, *J. Chem. Soc.*, 2207 (1961).

[22] R. Romeo and M. L. Tobe, *Inorg. Chem.*, 13, 1991 (1974).

[23] N. Sutin, *Ann. Rev. Phys. Chem.*, 17, 119 (1966).

[24] A. McAuley and J. Hill, *Quart. Rev.*, 23, 18 (1969).

[25] K. Kustin and J. Swinehart, *Prog. Inorg. Chem.*, 13, 107–158 (1970).

[26] F. Morral, *Adv. Chem. Ser.*, 62, 70 (1967).

[27] T. W. Swaddle, *Coord. Chem. Rev.*, 14, 217 (1974); see also M. Maestri, F. Bolletta, N. Serpone, L. Moggi, and V. Balzani, *Inorg. Chem.*, 15, 2048 (1976). Although both of these papers conclude that *a* activation is consistent with experiment in many cases, especially for Cr(III) complexes, the matter of associative activation substitution reactions of octahedral complexes in general is still a matter of conjecture and discussion.

[28] W. L. Reynolds, I. Murati, and S. Ašperger, *J. C. S. Dalton,* 719 (1974).

[29] W. L. Reynolds, M. Biruš, and S. Ašperger, *J. C. S. Dalton,* 716 (1974).

[30] D. A. Buckingham, I. I. Olsen, and A. M. Sargeson, *J. Amer. Chem. Soc.*, 90, 6654 (1968).

[31] R. G. Pearson and J. W. Moore, *Inorg. Chem.*, **3**, 1334 (1964).

[32] F. K. Meyer, K. E. Newman, and A. E. Merbach, *J. Amer. Chem. Soc.*, **101**, 5588 (1979).

[33] A similar approach has recently been developed: J. K. Burdett, *J. C. S. Dalton*, 1725 (1976).

[34] C. H. Langford, *Inorg. Chem.*, **4**, 265 (1965).

[35] M. L. Tobe, *Acc. Chem. Res.*, **3**, 377 (1970).

[36] The best source of information on the synthesis of coordination compounds is *Inorganic Syntheses*, published by the McGraw-Hill Book Company.

[37] Another excellent source of information on synthesis is the "Handbook of Preparative Inorganic Chemistry," G. Brauer, ed., Vols. 1 and 2, Academic Press, New York, 1965.

[38] The following is a review of coordination compound synthesis in general: J. L. Burmeister and F. Basolo, "Synthesis of Coordination Compounds," in *Preparative Inorganic Chemistry*, **5**, 1 (1968).

[39] This reference and the next three are primarily textbooks of preparative inorganic chemistry and contain much useful information on coordination compound synthesis. R. J. Angelici, "Synthesis and Technique in Inorganic Chemistry," 2nd Ed., W. B. Saunders Company, Philadelphia, 1977.

[40] W. L. Jolly, "The Synthesis and Characterization of Inorganic Compounds," Prentice-Hall, Inc., Englewood Cliffs, N. J., 1970.

[41] G. Pass and H. Sutcliffe, "Practical Inorganic Chemistry," 2nd Ed., Chapman and Hall Ltd., London, 1974.

[42] G. C. Schlessinger, "Inorganic Laboratory Preparations," Chemical Publishing Co., Inc., New York, 1962. This is a particularly good source of the syntheses of many "common" coordination compounds.

[43] S. J. Ashcroft and C. T. Mortimer, "Thermochemistry of Transition Metal Complexes," Academic Press, New York, 1970. All of the thermodynamic data quoted in this section, and all of the figures and tables in this section, are derived from data in this source.

[44] R. J. Angelici in G. L. Eichhorn, "Inorganic Biochemistry," Elsevier Scientific Publishing Company, New York, 1973, p. 63.

[45] F. J. C. Rossotti in J. Lewis and R. G. Wilkins, "Modern Coordination Chemistry," Interscience Publishers, New York, 1960, p. 1.

[46] S. Ahrland, J. Chatt, and N. R. Davies, *Quart. Rev.*, **11**, 265-276 (1958); S. Ahrland, *Structure and Bonding*, **5**, 118 (1968).

[46a] C. F. Bell, "Metal Chelation," Clarendon Press, Oxford, 1977; H. K. J. Powell, *Chem. Brit.*, **14**, 220 (1978) and references therein.

[47] D. Nicholls in "Comprehensive Inorganic Chemistry," J. C. Bailar, Jr., H. J. Emeleus, R. Nyholm, and A. F. Trotman-Dickenson, eds., Pergamon Press, New York, 1973, vol. 3, p. 1053.

[48] V. Gutmann, *Structure and Bonding*, **15**, 141 (1973).

[49] J. Bjerrum and J. P. McReynolds, *Inorg. Syn.*, **2**, 216 (1946).

[50] The catalytic activity of charcoal may be a general phenomenon for reactions involving $Co-N$ bonds. For example, the racemization of $[Co(en)_3]^{3+}$, which we noted in Chapter 12 as being catalyzed by electron transfer, can also be catalyzed by charcoal. See J. A. Broomhead, F. P. Dwyer, and J. W. Hogarth, *Inorg. Syn.*, **6**, 186 (1960).

[51] A. G. Sykes and J. A. Weil, *Prog. Inorg. Chem.*, **13**, 1 (1970).

[52] The syntheses of a wide range of binuclear cobalt compounds are given by R. Davies, M. Mori, A. G. Sykes, and J. A. Weil, *Inorg. Syn.*, **12**, 197 (1970).

[53] A. W. Chester, *Adv. Chem. Ser.*, **62**, 78 (1967).

[54] M. Mori, J. A. Weil, and M. Ishiguro, *J. Amer. Chem. Soc.*, **90**, 615 (1968).

[55] J. Simplicio and R. G. Wilkins, *J. Amer. Chem. Soc.*, **91**, 1325 (1969).

[56] W. P. Schaefer, *Inorg. Chem.*, **7**, 725 (1968).

[56a] D. A. Summerville, R. D. Jones, B. M. Hoffman, and F. Basolo, *J. Chem. Educ.*, **56**, 157 (1979).

[57] *Cis*- and *trans*-dichlorobis(ethylenediamine)cobalt chloride: J. C. Bailar, Jr., *Inorg. Syn.*, **2**, 222 (1946).

[58] Tris(ethylenediamine)cobalt(3+) ion: J. A. Broomhead, F. P. Dwyer, and J. W. Hogarth, *Inorg. Syn.*, **6**, 183 (1960).

[59] R. G. Pearson, H. H. Schmidtke, and F. Basolo, *J. Amer. Chem. Soc.*, **82**, 4434 (1960).

[60] S. E. Livingston in "Comprehensive Inorganic Chemistry," J. C. Bailar, Jr., H. J. Emeleus, R. Nyholm, and A. F. Trotman-Dickenson, eds., Pergamon Press, New York, 1973, vol. 3, p. 1330.

[61] So many of the fundamental ideas of chemistry were developed in Europe in the 19th and early 20th centuries that we tend to forget that American chemists made significant contributions as well. One example is James Lewis Howe (1859-1955), who did especially important laboratory and bibliographic research in the chemistry of the Group VIII metals, particularly on ruthenium. In view of Howe's contributions, it is significant that he spent much of his career at a small college in the South, Washington and Lee University, where for a number of years he was the chemistry department. One of the authors had the privilege of attending that college as an undergraduate shortly after Howe's death. He has fond memories of seeing Howe's laboratory equipment and papers in a dusty corner of the attic of the chemistry building, Howe Hall. Another memory is of Howe's daughter, Guendolen, who carried on the family's association with the University after her father's death; although not a chemist herself, she frequently attended weekly seminars, bringing cookies and cakes in a wicker basket. For a short biography of Howe and a discussion of his work see G. B. Kauffman, *J. Chem. Educ.*, **45**, 804 (1968).

[62] L. Malatesta and M. Angoletta, *J. Chem. Soc.,* 1186 (1957); R. Ugo, *Coord. Chem. Rev.,* 3, 319 (1968).

[63] G. L. Johnson, *Inorg. Syn.,* 8, 242 (1966).

[64] G. B. Kauffman and D. O. Cowan, *Inorg. Syn.,* 7, 239 (1963).

[65] B. Rosenberg, L. Van Camp, J. E. Trosko, and V. H. Mansour, *Nature,* 222, 385 (1969).

[66] B. Rosenberg, *Naturwissenschaften,* 60, 399–406 (1973).

[67] B. Rosenberg, *Platinum Metals Review,* 15, 42–51 (1971).

[68] A. J. Thomson, R. J. P. Williams, and S. Reslova, *Structure and Bonding,* 11, 1 (1972).

[69] M. J. Cleare and J. D. Hoeschele, *Platinum Metals Review,* 17, 2–13 (1973).

[70] M. J. Cleare, *Coord. Chem. Rev.,* 12, 349–405 (1974).

[71] S. J. Lippard, *Acc. Chem. Res.,* 11, 211 (1978).

14. COORDINATION CHEMISTRY: REACTION MECHANISMS AND METHODS OF SYNTHESIS; MOLECULAR REARRANGEMENTS AND REACTIONS OF COORDINATED LIGANDS

In the previous two chapters, we first examined reactions that intimately involved the metal ion—electron transfer reactions—and then reactions wherein ligands of the first coordination sphere were replaced by others—substitution reactions. In this, the final chapter on reactions of coordination complexes, we turn to reactions in which the ligands remain attached to the central metal ion but in which these ligands rearrange within the coordination sphere or undergo addition, substitution, or oxidation-reduction.

MOLECULAR REARRANGEMENTS[1-5]

The first part of this chapter is devoted to reactions in which ligands in the first coordination sphere remain attached to the central metal ion but rearrangement of the ligands occurs. As you shall see, this may lead to *cis-trans* isomerism or to a change in chirality, or both, or to polytopal isomerism.

A study of molecular rearrangements can be important, as they have some bearing on the synthesis of coordination compounds. Although you may set out to prepare one isomer of a particular complex, another may in fact result. Because of its importance, the examination of molecular rearrangements has been a very active area of research lately. However, in spite of all the work, you should keep in mind the fact that the mechanistic aspects of many rearrangements are still simply a matter of conjecture.

Four-Coordinate Complexes

When ethyldiphenylphosphine is added to nickel(II) bromide, a green complex, **1a**, is precipitated.[6] On dissolution of this product in chloroform or nitromethane, a green solution with a hint of red is obtained, while dissolution in benzene or carbon disulfide gives a red solution with no green color. On cooling to $-78°C$, the red solution will give dark red crystals that slowly turn green again on standing at room temperature.

The color changes observed in the reaction above are due to a molecular rearrangement, one that we shall explore as an illustration of the polytopal rearrangements possible for four-coordinate complexes and of the experimental methods used to attack the structural and mechanistic problem.

The thermochromic behavior of compound **1** is understood upon examination of the effective magnetic moments and UV-visible spectra of the two solids and the magnetic moment of the green compound in solution. The green solid (**1a**) has a magnetic moment of 3.20 B.M., and, as shown in Figure 14–1, clearly absorbs at longer wavelengths than does the red solid (**1b**). On the other hand, the magnetic moment of the red solid is 0.0 B.M. The obvious conclusion is that the green solid is the tetrahedral isomer, whereas the red solid is the square planar isomer. The lowest energy *d-d* transition for tetrahedral complexes always occurs at lower energies than the corresponding band for square planar complexes, since $\Delta_t < \Delta_{sp}$ (page 317). Further, a tetrahedral d^8 complex (such as **1a**) would be expected to have two unpaired electrons, whereas a square planar d^8 complex (such as **1b**) should be diamagnetic.

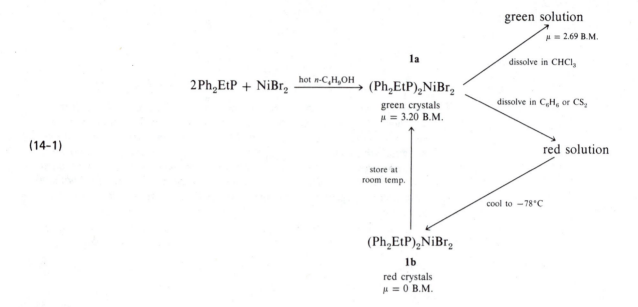

(14-1)

In solution in CH_2Cl_2, however, both isomers exist together, as indicated by the fact that the effective magnetic moment of the compound under these conditions is only 2.69 B.M., less than the expected moment for two unpaired electrons.

The mechanism of this polytopal isomerism has been examined by nuclear magnetic resonance spectroscopy,[7] and it is found that two mechanistic pathways are apparently followed: intra- and intermolecular. As you can see in the diagram below, only a small

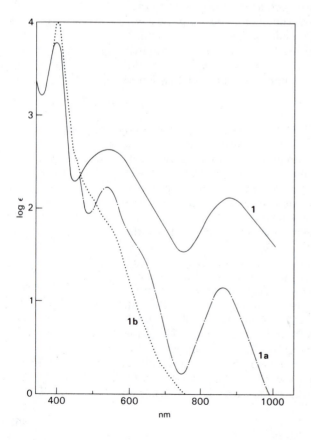

Figure 14-1. Absorption spectra of $NiBr_2(PPhEt_2)_2$: **1**, benzene solution of either isomer; **1a**, solid tetrahedral isomer—green; **1b**, solid square planar isomer—red. Units of absorption for spectra **1a** and **1b** are arbitrary. [Reprinted with permission from R.G. Hayter and F.S. Humiec, *Inorg. Chem.*, **4**, 1701 (1965). Copyright by the American Chemical Society.]

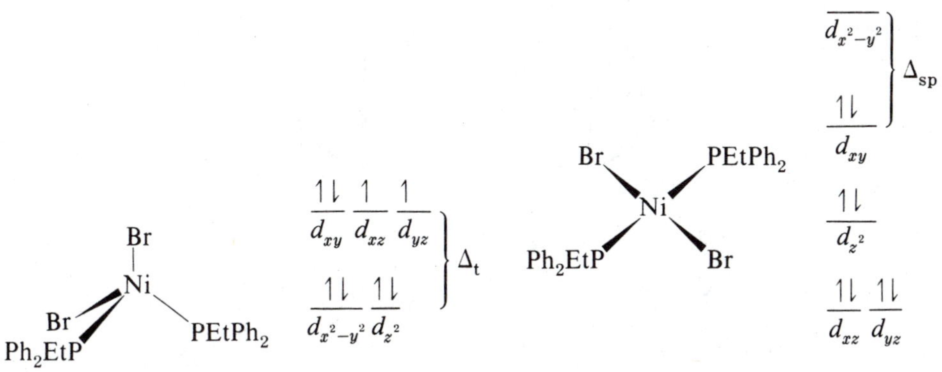

flattening motion along the pseudo-S_4 axis is necessary to convert one isomer into the other, and this intramolecular route is the one normally followed. However, in the presence of excess ligand, a bimolecular path, presumably similar to that used in substitution reactions of square planar complexes (Chapter 13), becomes more important than the intramolecular path.

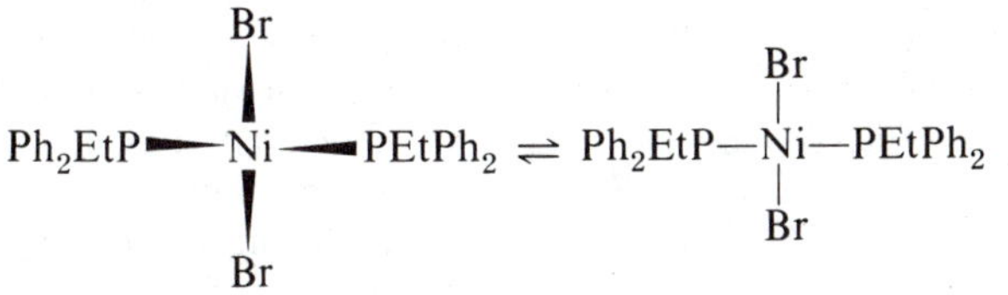

$$ca. \text{ 70\% tetrahedral at 40°C}$$

$$k^{298} = 1.6 \times 10^5 \text{ sec}^{-1}$$
$$\Delta H^{\ddagger} = 9.0 \text{ kcal/mol}$$
$$\Delta S^{\ddagger} = -4.7 \text{ e.u.}$$

(14-2)

Several general conclusions have been drawn regarding the thermodynamics of the tetrahedral ⇌ square planar equilibrium for complexes of the type L_2NiX_2. First, the fraction of tetrahedral isomer increases in the order X = Cl < Br < I and L = R_3P < ArR_2P < Ar_2RP < Ar_3P (where Ar is an aryl group such as phenyl and R is an alkyl group such as methyl or ethyl). Both of these trends can be understood readily in terms of steric effects. As discussed in Chapters 9 and 10 (see Figure 9-4), the square planar geometry is generally favored by d^8 complexes, whether high spin or low spin (if only σ bonding is considered). The fact that a tetrahedral ⇌ square planar interconversion exists at all means that these isomers are close to one another in energy, and the steric bulk of the ligands and their mutual repulsion can strongly influence the geometry of the complex. This is apparently the case for the phosphine complexes in question. As the steric bulk of the ligand increases, the fraction of the tetrahedral isomer increases. Because the tetrahedron has greater bond angles, it is better able to accommodate bulky ligands.

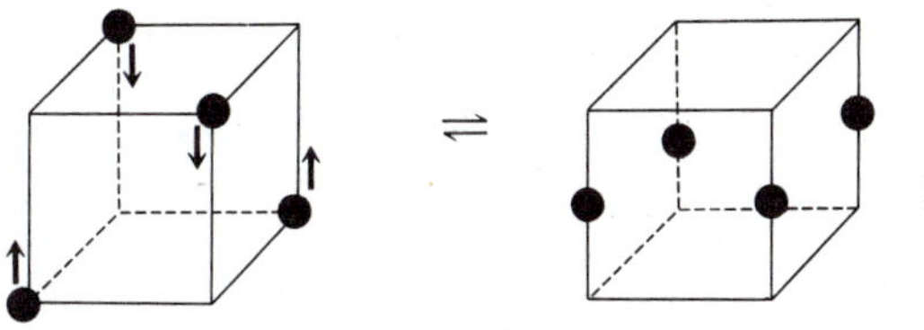

Changes in molecular chirality represent another form of molecular rearrangement. Consider for example a tetrahedral complex formed from unsymmetrical ligands [e.g., $M(A-B)_2$]; such a molecule is chiral, and a pair of enantiomers with Δ and Λ configurations will be generated (the C_2 axis is the helix axis). An example is the pair of complexes formed from a β-thioketoamine, 2a and 2b. Three pathways for the interconversion

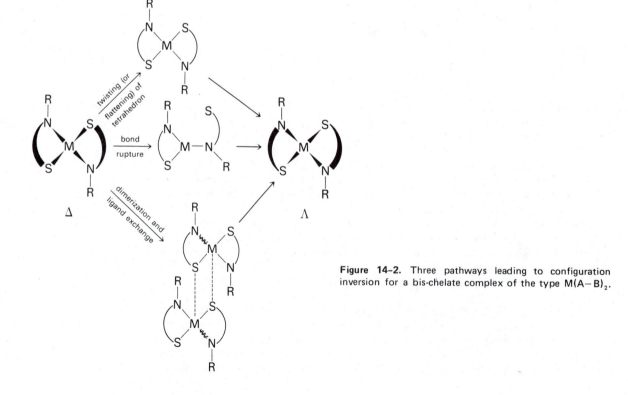

of Δ and Λ enantiomers are illustrated in Figure 14-2.[8] The first pathway involves a twisting or flattening of the tetrahedral complex to form a square planar complex and then further twisting to re-form the tetrahedron. The second path involves breaking a metal-to-ligand bond to give a three-coordinate intermediate; reformation of the bond may occur to give a complex having either the original or the opposite chirality. Finally, a third path arises when one of the donor atoms (like S) is capable of bridging; this path involves the formation of a complex between two square planar units *via* intermolecular M-S interactions. Breaking a M-N bond and re-formation of a M-N bond with the other

Figure 14-2. Three pathways leading to configuration inversion for a bis-chelate complex of the type M(A—B)₂.

metal, followed by dissociation of the dimer, could lead to inversion with concomitant ligand exchange. In general, tetrahedral twisting or flattening seems to be the pathway followed, although there is evidence in many cases that the other paths compete, often quite favorably.

Photochemistry is an area of great current interest to inorganic chemists; a photo-induced reaction germane to the material in this chapter is the photochemical isomerization of square planar *cis*-platinum(II) complexes.[9] For example,

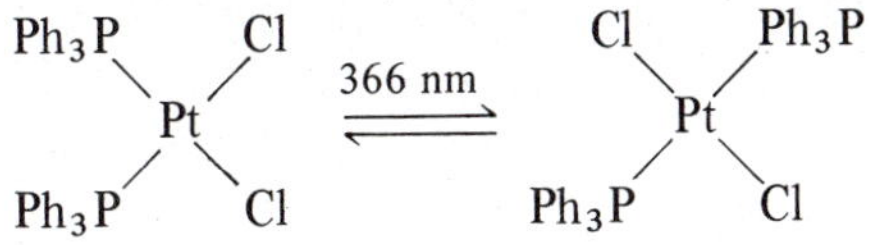

(14-3)

Such transformations of PtL_2X_2 can also be chemically catalyzed by X^- or L, and the mechanism has been the subject of considerable debate.[10] At least in the case where X^- or L has not been added, the following pathway is reasonable:

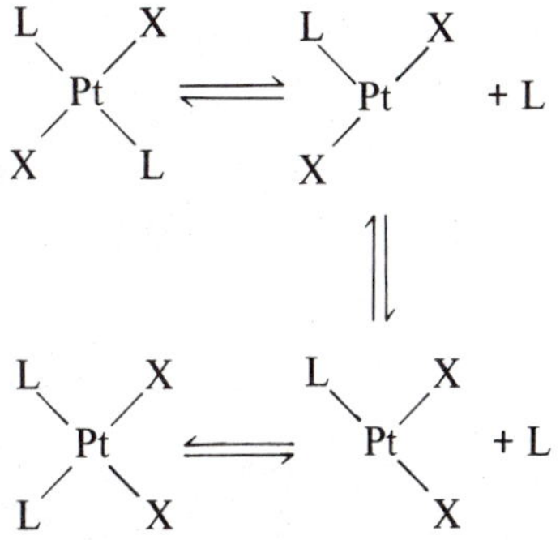

The second step is plausible from the information in Figure 10-1, from which you see that a low-spin d^8 complex exhibits a weak structural preference for the T shape; the trigonal transition state is not so high in energy as to effectively block the geometrical isomerization in the second step. Other mechanisms involving a coordinated solvent molecule are also plausible.[11]

When added ligands catalyze the process, a fluxional five-coordinate intermediate may form (*cf.* the associative substitution mechanism for square planar complexes), and the ligands interchange positions *via* a Berry pseudorotation (see p. 354); loss of the catalyzing ligand gives the now-isomerized complex.

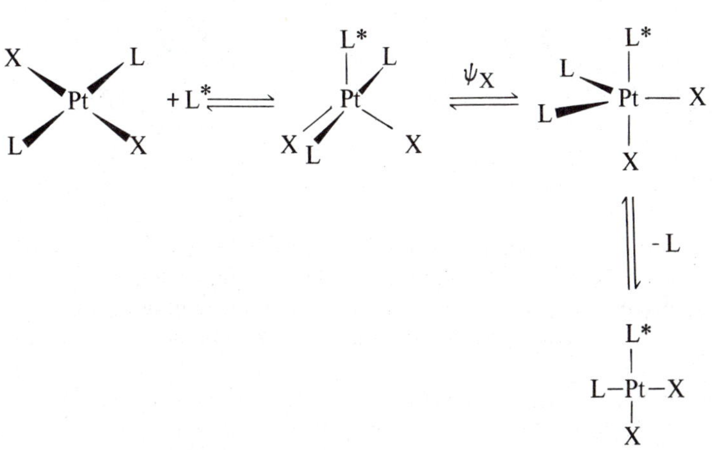

Six-Coordinate Octahedral Complexes[1,2]

Intramolecular rearrangements in six-coordinate complexes are more complicated than those of four-coordinate ones, and great experimental ingenuity is required to obtain useful information. Much of the work that has given definition to the problem of deciphering these rearrangement mechanisms has been done with complexes such as tris(*o*-phenanthroline)chromium(3+) (**3**) and with complexes having unsymmetrical chelating ligands attached to a trivalent metal ion such as Cr(III) or Co(III) (*e.g.,* **4**).[12-14] It is the latter type of complex which we shall discuss as a model for rearrangements in octahedral complexes.

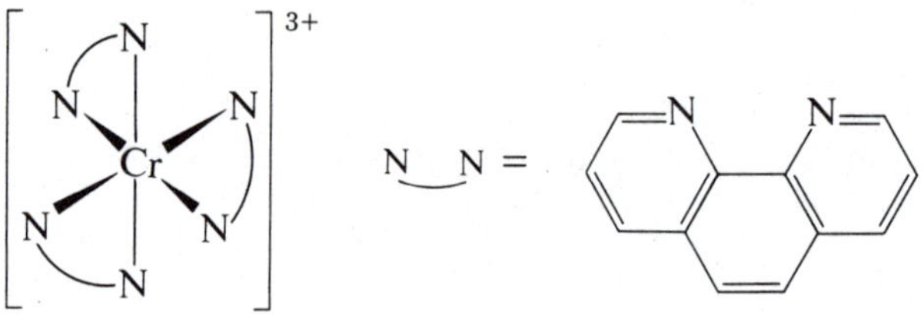

3

(14-4)

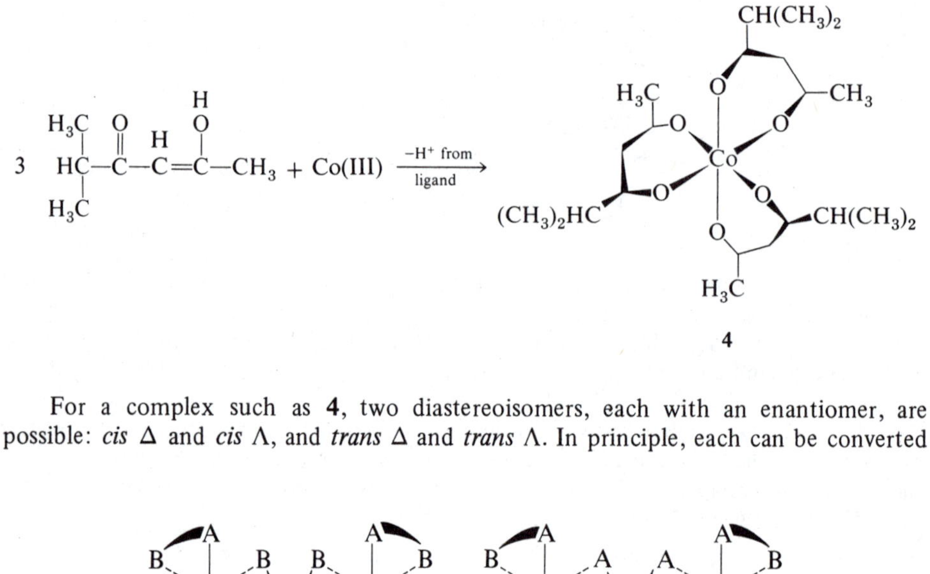

$$3 \quad \underset{H_3C}{\overset{H_3C}{}}HC - \overset{O}{\overset{\|}{C}} - \overset{H}{\overset{|}{C}} = \overset{O}{\overset{|}{C}} - CH_3 + Co(III) \xrightarrow[\text{ligand}]{-H^+ \text{ from}}$$

4

For a complex such as **4**, two diastereoisomers, each with an enantiomer, are possible: *cis* Δ and *cis* Λ, and *trans* Δ and *trans* Λ. In principle, each can be converted

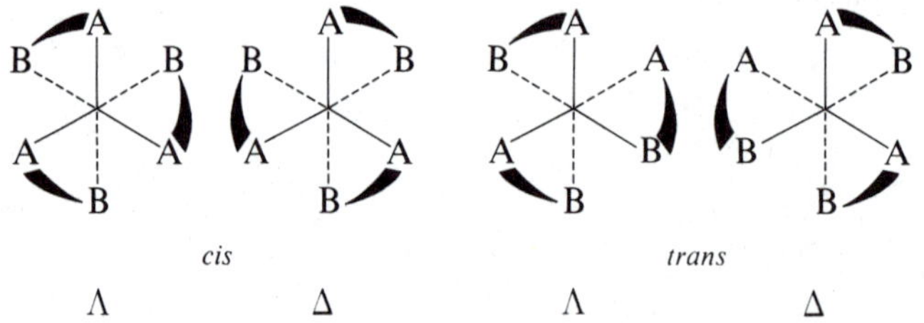

cis   Λ   Δ   *trans*   Λ   Δ

into any or all of the others by an intramolecular process such as a twisting motion and/or several different bond breaking/bond making modes. What follows is a very elementary discussion of these pathways. For those interested in inorganic stereochemistry, you will find this a fascinating topic, and you are referred to more advanced treatments.[1,2]

Before looking at the intramolecular twist, the first of several possible rearrangement mechanisms, you should be made aware again of the symmetry properties of a M(A—B)₃ complex such as **4**. As seen in Figure 14-3, a *cis* Δ (or a M(A—A)₃ complex such as **3**)

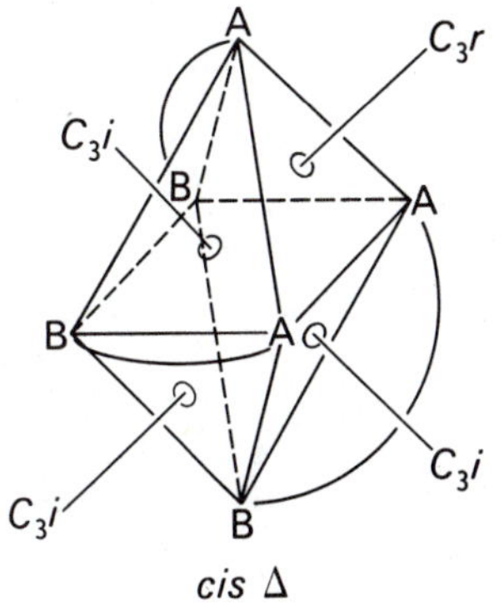

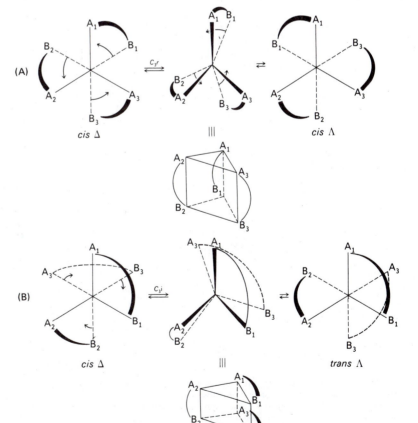

Figure 14-3. The *cis* Δ enantiomer of a complex of the type (unsym-chel)$_3$ M = (A—B)$_3$ M showing the real ($C_3 r$) and imaginary ($C_3 i$) axes of the *cis* complex.

will have two types of C_3 symmetry axes: (i) a real C_3 axis (= $C_3 r$) passing through the parallel octahedral faces with the three A's and the three B's at the vertices, and (ii) three "imaginary" C_3 axes (= $C_3 i$) passing through the other three sets of parallel faces for which one vertex atom is different from the other two.

In the twist mechanism for rearrangement, one trigonal face of the octahedron is rotated with respect to the opposite face about one of the C_3 axes. The transition state structure is the other six-coordinate polytope, the trigonal prism, and further rotation about the same axis leads to a complex having the opposite chirality. Figure 14-4 depicts this process for rotation about the real and "imaginary" C_3 axes of the *cis* Δ enantiomer.

Figure 14-4. The twist motion leading to configuration inversion in a complex having three identical, unsymmetrical ligands. The axes about which the twisting motion occurs are illustrated in Figure 14-3. Example (A) is often called a ((trigonal twist" because the symmetry of the transition state structure (assuming A = B) is D_{3h}. Example (B) is called a "rhombic twist" because the symmetry of the transition state (assuming A = B) is C_{2v}.

Rotation about the C_3r axis proceeds through a trigonal prism with the ligands spanning the square edges only, and is called the *trigonal twist*. Rotation about C_3i gives eclipsed ligands along parallel triangle edges and is called the *rhombic twist*. These processes lead to configuration inversion only, and to inversion with geometrical isomerization, respectively. These results, together with those of the *trans* enantiomer (see study question 2), lead to an important conclusion: *The intramolecular twist always leads to inversion of chirality but not necessarily to cis-trans isomerization. Geometrical isomerization without racemization is not possible by the twist mechanism;* to accomplish this, another process, bond rupture, must be postulated.

Bond rupture can lead to two different intermediate structures: the trigonal bipyramid (TBP) and the square pyramid (SP). In the case of bond rupture leading to a TBP intermediate, the five-coordinate complex may have either a "dangling axial ligand" or a "dangling equatorial ligand." These choices are illustrated in Figure 14-5 for the *cis* Λ enantiomer. Because of the real C_3 symmetry of the *cis*-M(A—B)$_3$ complex, there are two types of bonds that may be broken (a and b), and, as a result, there are four possible intermediates; only those generated by breaking bond b are illustrated (see study question 4 for an analysis of this process for a *trans* isomer). Complete analysis of the path proceeding through a TBP intermediate leads to two important conclusions: *(i) Each of the enantiomers may directly convert by a TBP rupture to any of the others, except for cis Δ to cis Λ. (ii) "Equatorial" intermediates give only geometrical isomerization; no inversion can occur.*

Intramolecular rearrangement by the path involving square pyramid intermediates is illustrated in Figure 14-6, also for the *cis* Λ enantiomer. Complete analysis of this pathway leads to yet another important result: *Square pyramidal intermediates can lead to interconversion of all isomers.* Notice that this is in direct contrast with the other paths.

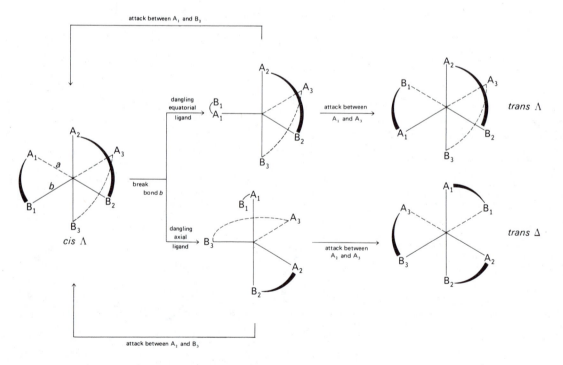

Figure 14–5. Illustration of the rearrangement of the *cis* Λ enantiomer by the bond rupture mechanism involving a trigonal bipyramidal intermediate. There are three, symmetry equivalent M—A bonds (= a) and three, symmetry equivalent M—B bonds (= b) in the *cis* isomer. Only the intermediates from breaking bond b are shown; two more arise from breaking bond a.

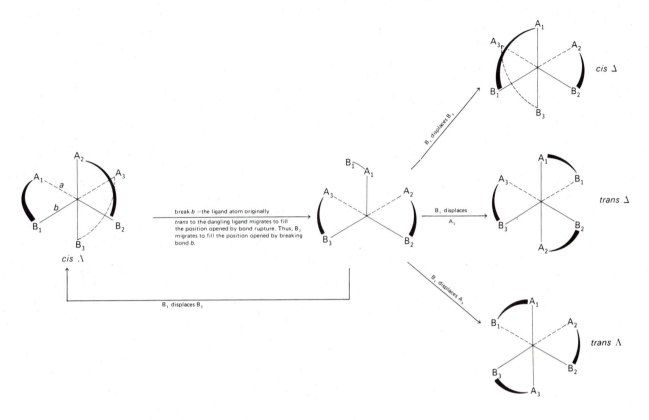

break b —the ligand atom originally
trans to the dangling ligand migrates to fill
the position opened by bond rupture. Thus, B₂
migrates to fill the position opened by breaking
bond b.

Figure 14-6. Illustration of the rearrangement of the *cis* Λ enantiomer by the bond rupture mechanism involving a square pyramidal intermediate. As in Figure 14–5, only an intermediate from breaking bond *b* is shown.

Nuclear magnetic resonance methods can be used to obtain the rates of isomerization of compounds such as **4**, and polarimetry can be applied to measure racemization rates. Using such methods, it has been found that for $[Cr(o\text{-phen})_3]^{3+}$ racemization apparently occurs by a twisting mode,[15] but, for complexes such as $[Cr(C_2O_4)_3]^{3-}$ and **4**, rearrangement occurs largely by a bond rupture pathway with TBP-axial intermediates.†[12,13,15a]

1. It was noted on page 457 that PtL_2X_2 complexes may be isomerized if catalytic amounts of L or X^- are added and that the mechanism may involve a fluxional five-coordinate intermediate. Write out a mechanistic pathway to show how this may happen for added X^-.

2. The twisting mechanism for rearrangements of octahedral complexes was shown only for a *cis* isomer. Taking the *trans* Δ isomer on the following page, show the results for the trigonal and rhombic twisting mechanisms of a *trans*-M(A−B)₃ complex.

---

†Recent work indicates that octahedral low spin d^6 complexes must become high spin in order to achieve a trigonal prismatic transition state, a requirement not easily met with Co(III). See reference 15a.

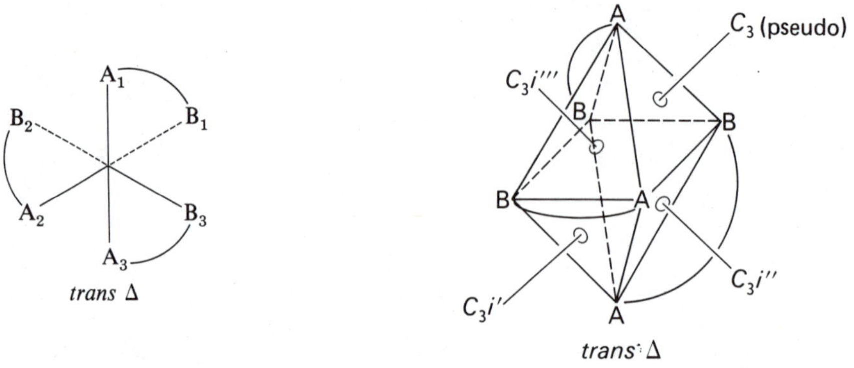

trans Δ

trans Δ

3. Figure 14–5 illustrates the rearrangement of the *cis* Λ enantiomer by the bond rupture mechanism with TBP intermediates. What products arise if bond *a* is broken in this enantiomer and a TBP-equatorial intermediate forms?

4. Taking the *trans* Λ enantiomer below, and considering the bond rupture mechanism with TBP intermediates, show that only axial intermediates will lead to inversion of chirality and that equatorial intermediates will result in isomerization.

trans Λ

5. The rates of *cis-trans* isomerization of trivalent metal complexes of trifluoroacetyl-acetonate (see below) have been studied extensively, and these rates decrease in the order (Fe,In) > Mn > Ga > Al > V > Co > Ru > Rh. Assuming that the isomerization occurs through the twist mechanism (*i.e.*, it passes through a trigonal prismatic transition state), relate the relative rates of isomerization to the structure preference energies for the two possible six-coordinate polytopes (see Chapters 9 and 10 and ref. 15a). [*Hint:* First decide on spin state.]

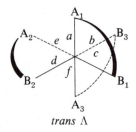

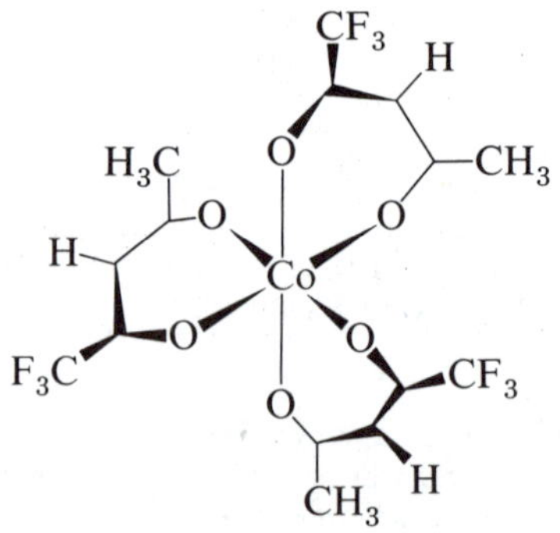

REACTIONS AT COORDINATED LIGANDS[16-20]

We come finally to the periphery of the coordination complex and to reactions that occur at the coordinated ligands themselves. It is obvious that a metal ion should be able to influence strongly the reactions at coordinated ligands, as compared with the uncoordinated ligand, by causing changes in intra-ligand charge distribution (both by σ withdrawal and by π donation or withdrawal). This may lead to enhancement of a particular reaction of the coordinated ligand, force a given reaction to occur by a different mechanism, or give a product not attainable without prior metal coordination. In addition, metal coordination may have a "template effect"; that is, the coordination of ligands may place them in the correct geometry for their later union, thereby overcoming, by means of a favorable ΔH for reaction, an unfavorable ΔS for reaction in the absence of coordination. Alternatively, metal coordination may stabilize a particular tautomer, a form not stable or of little consequence without metal coordination.

In the discussion that follows, we intend to outline only a few examples of the ways in which reaction can occur at coordinated ligands. These have been grouped into two broad classifications: (i) reactions occurring because of metal ion polarization of the coordinated ligand and (ii) those in which a "template effect" is clearly operative. The one class of reactions not really covered below consists of those involving true catalysis, that is, reactions involving the activation and transformation of small molecules as in the Wacker process (14–5), for example.

$$[PdCl_4]^{2-} + C_2H_4 + H_2O \rightarrow CH_3CHO + Pd + 4Cl^- + 2H^+ \qquad (14\text{-}5)$$

Instead, these are covered in more detail in Chapter 18.

Reactions Due to Metal Ion Polarization of Coordinated Ligands

HYDROLYSIS OF AMINO ACID ESTERS AND AMIDES AND OF PEPTIDES

Carboxypeptidase A is a zinc-containing digestive enzyme that hydrolyzes the carboxyl-terminal peptide bond in polypeptide chains.[21]

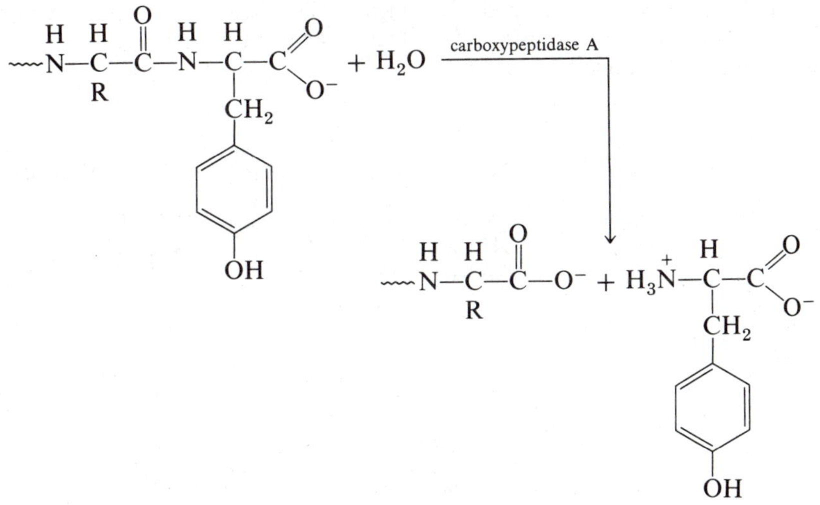

$$(14\text{-}6)$$

Its structure has recently been determined and its functioning has been probed.[22] On the basis of this work, one suggested mechanism involves metal ion polarization of a peptide carbonyl group, labilizing the carbon to attack by a neighboring carboxyl oxygen (from glutamine, residue number 270).

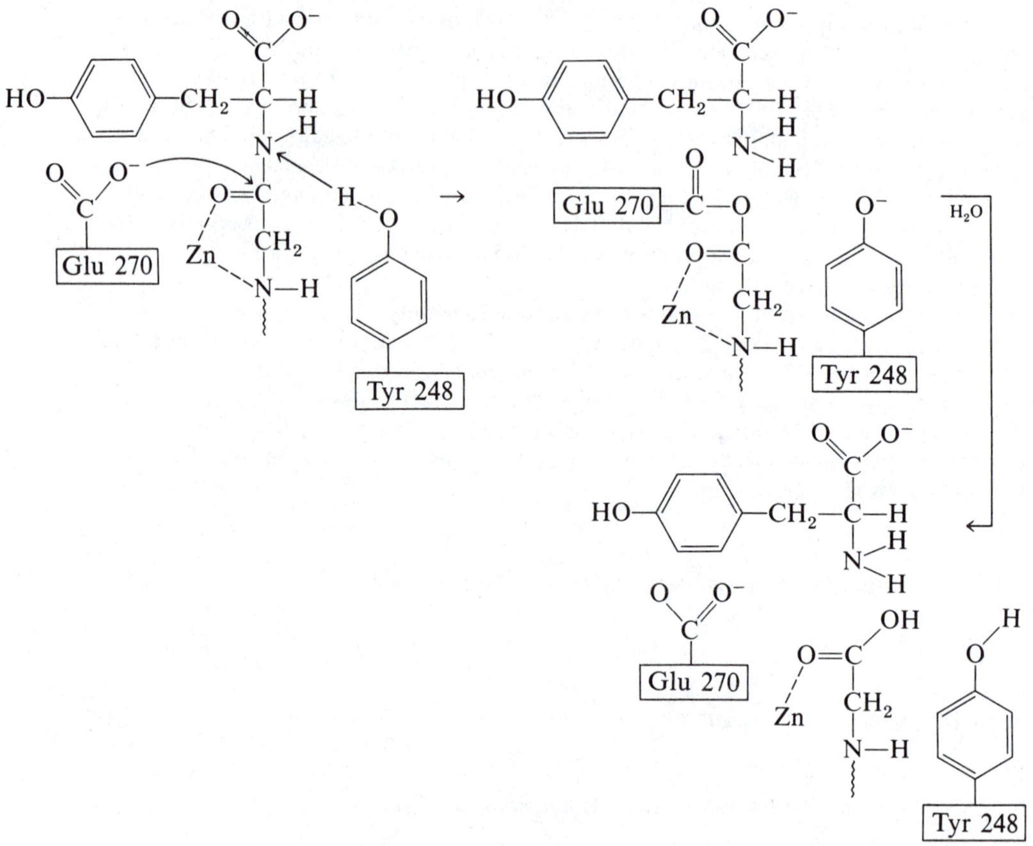

Glu 270

Tyr 248

Glu 270

Tyr 248

H_2O

Glu 270

Tyr 248

This is by no means yet the proved mechanism by which the enzyme operates, so a great deal of work is continuing in enzyme studies; the discussion that follows relates to inorganic model systems.

Amino acid esters are hydrolyzed very rapidly in the presence of ions such as Cu^{2+}, Co^{2+}, and Mg^{2+}, even at approximately neutral pH's where the amino acid esters are normally quite stable to hydrolysis.[18,23-27] As an example, consider reaction 14-7 for the hydrolysis of phenylalanine ethyl ester.

(14-7)

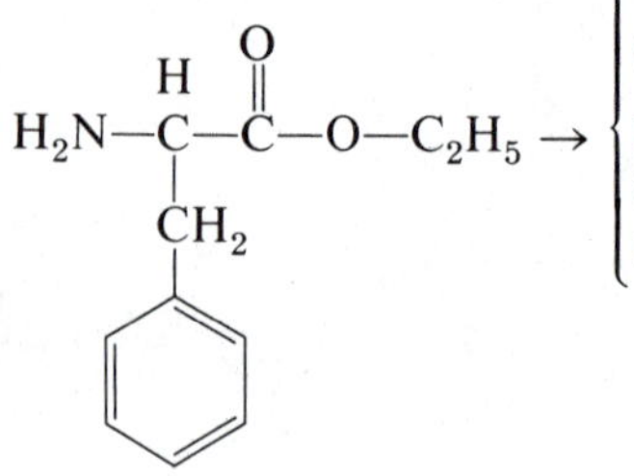

Catalyst	k (sec^{-1}) (when [ester] = 0.0138 M)
H$^+$ (pH = 2.3)	1.46×10^{-11}
OH$^-$ (pH = 7.3)	5.8×10^{-9}
Cu^{2+} (0.0775 M)	2.7×10^{-3}

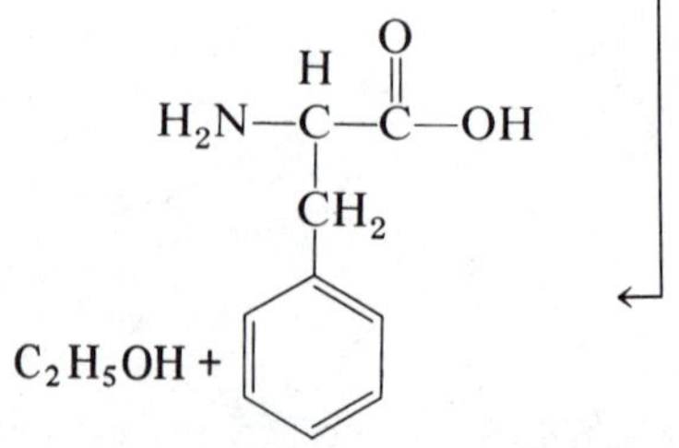

$C_2H_5OH +$

Such hydrolysis reactions have been studied extensively, and the evidence is overwhelming that a metal ion-amino acid complex such as **5** is the kinetically important species and that the rate-determining step in the hydrolysis is the attack of water or hydroxide on the 1:1 complex.

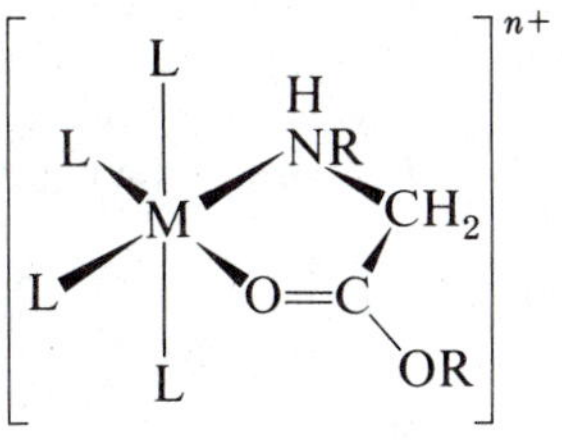

5

For a variety of metal ion catalyzed hydrolyses of amino acid esters in the pH range from 7 to 9, the overall rate is given by

$$\text{Rate} = k_{OH}[ME^{2+}][OH^-] + k_{H_2O}[ME^{2+}][H_2O]$$

where ME^{2+} is the 1:1 metal-ester complex. The specific rate constants, k_{OH} and k_{H_2O}, have been determined for ethyl glycinate hydrolysis with the metal ions Co^{2+}, Ni^{2+}, Cu^{2+}, and Zn^{2+}. These are listed, along with the stability constants for the 1:1 metal-ester complexes, in Table 14–1. There are two main points of interest concerning these data: (i) the values of k_{OH} are about 10^9 greater than k_{H_2O} (because OH^- is a better nucleophile than H_2O), and (ii) the values of k_{OH} more or less increase with complex stability. It is the latter point that enables you to propose an explanation for the fact that great rate acceleration for hydrolysis is observed when an amino acid ester *chelates* with a metal ion. When chelation with a metal ion occurs, resonance structures **6** and **7** are important. The C−O−M angle in **6** is in theory (*i.e.*, VSEPR) close to $120°$, an angle leading to considerable ring strain. On the other hand, the C−O−M angle in **7** is about $109°$, an angle much closer to the theoretical angle ($108°$) for a symmetric five-membered chelate ring. Thus, ring strain is reduced in **7** relative to **6** and we presume that **7** becomes the more important resonance structure as complex stability increases. It seems logical to conclude that, as **7** becomes more important, the electrophilicity of the carbonyl carbon increases, and the rate of OH^- attack is increased.

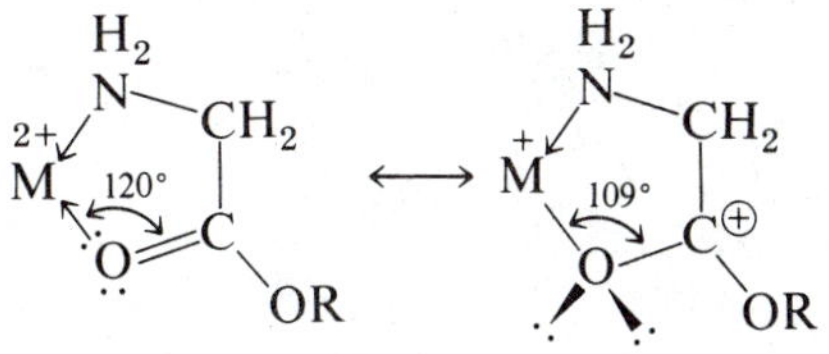

6   **7**

TABLE 14–1

SPECIFIC RATE CONSTANTS FOR THE Co²⁺, Ni²⁺, Cu²⁺, AND Zn²⁺ CATALYZED HYDROLYSIS OF ETHYL GLYCINATE AND THE STABILITY CONSTANTS FOR THE M²⁺:ETHYL GLYCINATE COMPLEXES[a]

Metal Ion	$10^{-4} \, k_{OH}$ $M^{-1} \, sec^{-1}$	$10^6 \, k_{H_2O}$ $M^{-1} \, sec^{-1}$	$K_{stability}$ (for 1:1 complex)
Co^{2+}	0.99	8.3	26.8
Ni^{2+}	0.398	3.5	198
Cu^{2+}	7.64	0.77	7400
Zn^{2+}	2.33	7.07	62

[a] J. E. Hix, Jr., and M. M. Jones, *Inorg. Chem.*, **5**, 1863 (1966).

More recent work on the hydrolysis of amino acid esters and amides and of peptides has been done with Co(III) complexes.[28,29] The hydrolysis of 8 (see Figure 14–7) is an important example, because it is closely related to real biological systems.[29] Under basic conditions (pH = 9), coordinated halide is first removed in a base-assisted or CB mechanism (see page 427). This produces a five-coordinate intermediate (9) that is relatively long-lived, and two nucleophiles — water and the amide carbonyl group — compete for the sixth coordination position.† This competition means that amide hydrolysis may

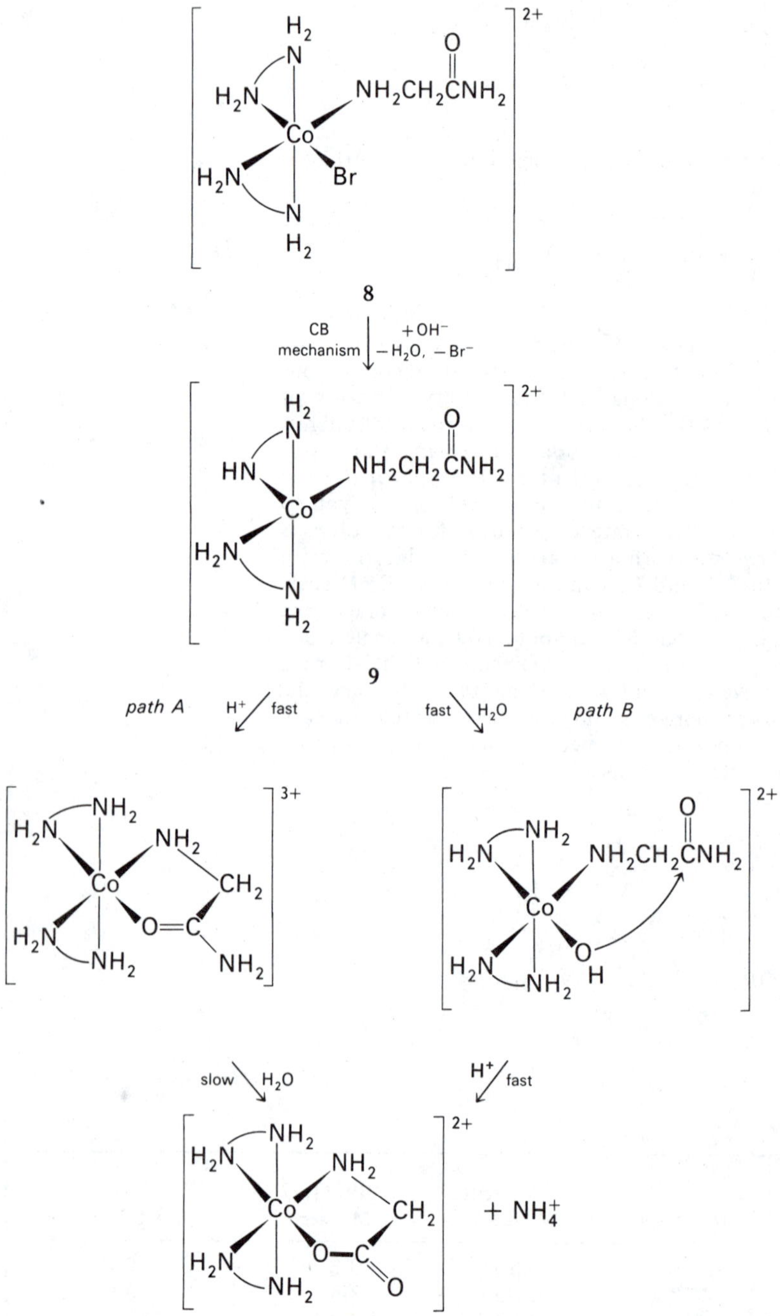

Figure 14–7. The two mechanistic pathways for the base-catalyzed hydrolysis of a Co(III) complex of an amino acid amide.

†If anions are added in quantity, they can compete for the open coordination position in 9. Further reaction is then blocked. This observation, of course, indicates the existence of a five-coordinate intermediate.

occur by two paths, A and B in Figure 14-7. Indeed, ^{18}O tracer studies indicate (i) that path A accounts for 46% of the product and path B for 54%, and (ii) that the product distribution is pH independent. Perhaps of greater importance, however, is the observation that paths A and B have decidedly different efficiencies. *Inter*molecular hydrolysis by path A is approximately 10^4 times more rapid than the rate of hydrolysis of the un-coordinated amino acid amide; in contrast, *intra*molecular hydrolysis by path B is *at least* 10^7 more rapid than hydrolysis of the uncoordinated molecule.

There are a large number of other nucleophilic reactions that occur because the ligand is polarized by a coordinated metal ion. Two important ones are mentioned below.

ALDOL CONDENSATION

When an amino acid is coordinated to a metal ion, the α-C—H bond is polarized and therefore susceptible to attack by a nucleophile. For example, OH⁻ can apparently lead to formation of an enolate ion, because reaction of bis(glycinato)copper (**10**) with acetaldehyde in base ultimately gives a complex of threonine (**11**). The reaction can be generalized to other amino acids.

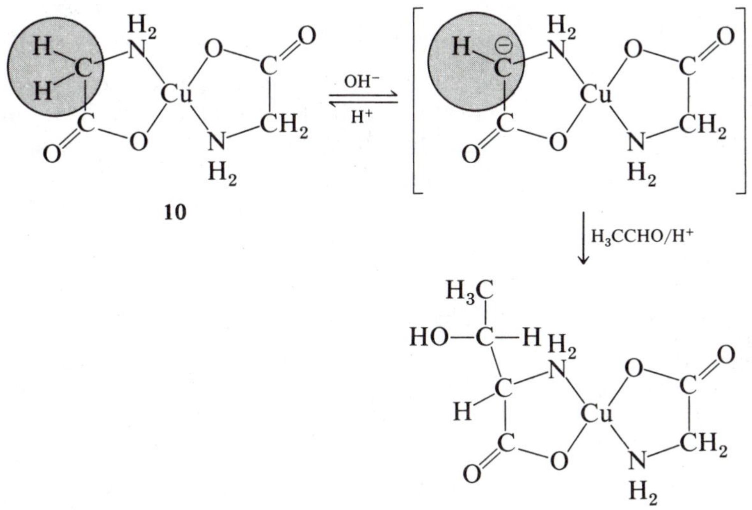

10

$$H_3CCHO/H^+$$

11

(14-8)

IMINE FORMATION, HYDROLYSIS, AND SUBSTITUENT EXCHANGE[30]

You should recall from your courses in organic chemistry that aldehydes and ketones may be converted to **Schiff bases** (imines) by an acid-catalyzed reaction with a primary amine:

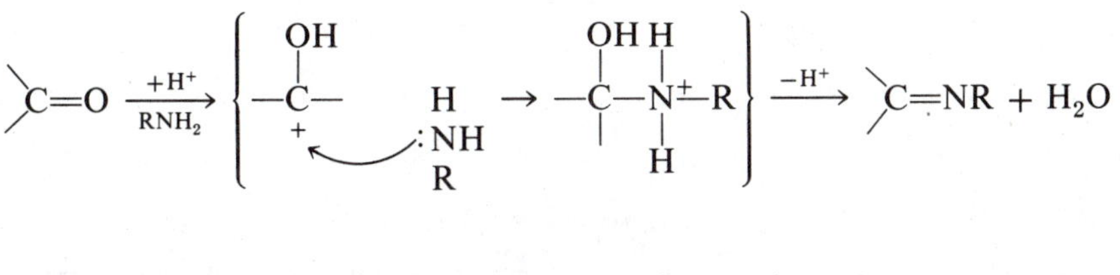

Since a metal ion can sometimes function as a catalyst in place of a proton, it is possible that an imino complex (**13**) may be formed by reaction of an amine with an aldehyde complex, **12**.

(14-9)

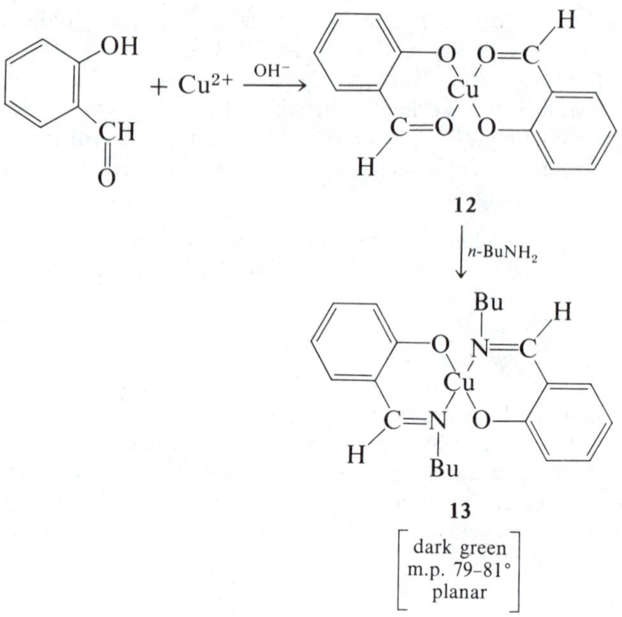

12

13

⎡ dark green
m.p. 79–81°
planar ⎤

Aldimines (the ligand in **13**; it would be called a ketimine if formed from a ketone) constitute an important class of ligands, particularly as it is possible to interchange nitrogen substituents by a **transamination reaction**, the process connecting **13** to **14**.

(14-10)

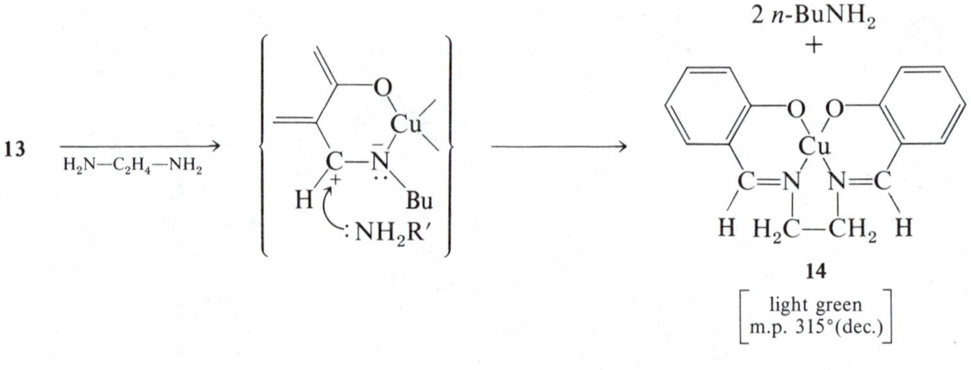

13 $\xrightarrow{H_2N-C_2H_4-NH_2}$

14

⎡ light green
m.p. 315°(dec.) ⎤

The Template Effect and Macrocyclic Ligands[19,20,31]

Imine-forming reactions are extremely important in the synthesis of multidentate ligands such as the one in compound **14** and in the formation of macrocyclic ligands, that is, ligands that are at least quadridentate and that can completely encircle the metal ion. In fact, the formation of macrocycles, as in reaction 14-11 (or in the synthesis of the ligand in compound **14** in Chapter 10), serves to illustrate another important feature of the reactions of coordinated ligands—the **coordination template effect**.

(14-11)

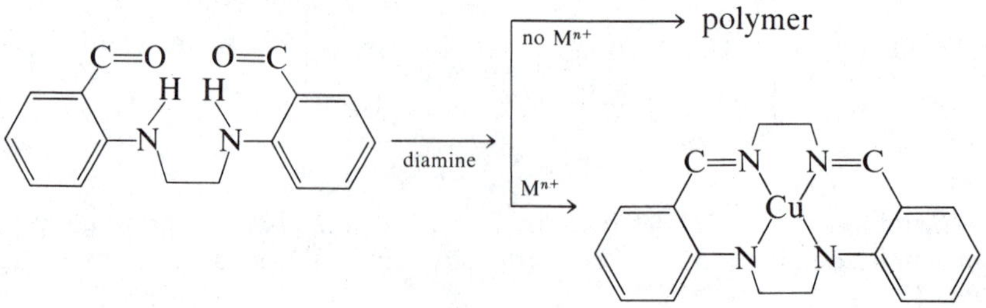

One of the most systematically studied syntheses[32] of a macrocyclic ligand illustrates the template effect as well as the points we have discussed thus far in this chapter. Busch wished to synthesize the potentially quadridentate ligand 16, but direct reaction of an α-diketone with 2-aminoethanethiol gave a thiazoline (15) instead.[33]

$$(14\text{-}12)$$

15   16

However, chemical tests for mercaptan indicated that a small amount of the desired compound, 16, existed in equilibrium with 15. Therefore, the diketone and the amino-thiol were allowed to react in the presence of nickel(II) acetate, and the Ni(II) complex of the desired ligand (17) was produced in 70% yield. It was theorized that 16 was

17

actually an intermediate in the formation of the thermodynamically more stable thiazo-line, 15. Ni(II) apparently stabilizes this intermediate by chelation, an effect sometimes referred to as a **thermodynamic template effect**.[34]

In earlier work Busch had found that coordinated RS⁻ could function as a nucleo-phile and would readily undergo alkylation. Therefore, in order to utilize 16 as the core of a macrocyclic ligand, 17 was allowed to react with an organic dihalide (18); complexes of the type illustrated by 19 were produced in good yield.[32] Busch has pointed out that such reactions illustrate the **kinetic template effect**; that is, a metal ion has been utilized to hold reactive groups in the proper geometry so that "a stereochemically selective multistep reaction" may occur.

$$(14\text{-}13)$$

17 +   →

18   19

As just demonstrated, template reactions can lead to macrocyclic ligands, a class of molecules or ions that has been increasingly studied for its possible relevance to bio-chemically important ligands such as the porphyrins (20).[20]

Aside from that, macrocyclic ligands are of interest because they often give complexes having unusual metal-donor atom bond angles and distances, and because they are capable of stabilizing unusual oxidation states of metals such as Cu(III). Further, metal complexes of macrocyclic ligands often possess considerably greater thermodynamic and kinetic stability (with respect to ligand dissociation) than their open chain analogs (these effects

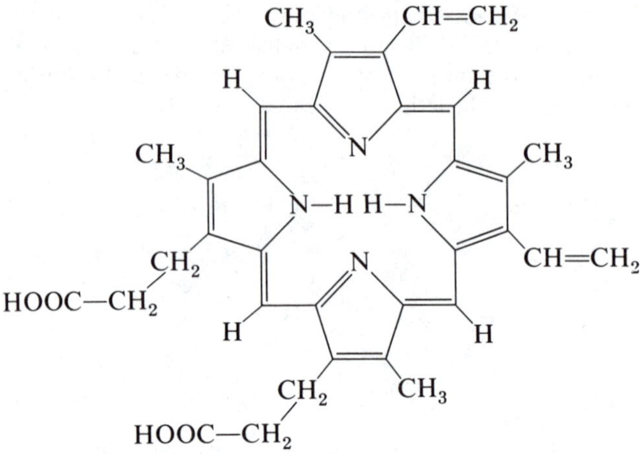

20

protoporphyrin IX

have been referred to collectively as the "macrocyclic effect"; see study question 10).[35] In fact, macrocyclic ligands can impart kinetic stability even when the central metal ion is thought to lead normally to labile complexes. This property makes macrocyclic ligands ideal for the study of ligand reactions as influenced by metal ion, since the possibility of metal-ligand dissociation during reaction is much reduced.

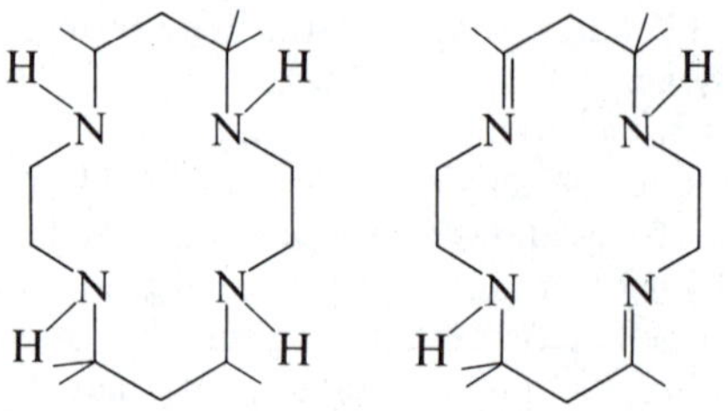

Me$_6$[14]ane N$_4$ Me$_6$[14]4,11-diene N$_4$

21   **22**

Because of their relationship to the porphyrins, perhaps the most thoroughly studied macrocyclic ligands have been those containing four nitrogen atoms capable of coordination. Of the ligands of this type, those having 14- and 16-membered rings are most common (compounds **21** to **27**†).[36,37]

---

†Each compound is designated by the abbreviation commonly used in the literature. Taking compound **22** as an example, Me$_6$ indicates the number of substituent methyl groups, [14] is the ring size, and 4,11 is the location of the C=N double bonds. The full, systematic name of this ligand would be 5,7,7,12,14,14-hexamethyl-1,4,8,11-tetraazacyclotetradeca-4,11-diene. See V. L. Goedken, P. H. Merrell, and D. H. Busch, *J. Amer. Chem. Soc.*, **94**, 3397 (1972), for the details of the naming system used.

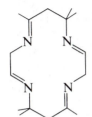

Me$_6$[14]1,4,8,11-tetraene N$_4$

23

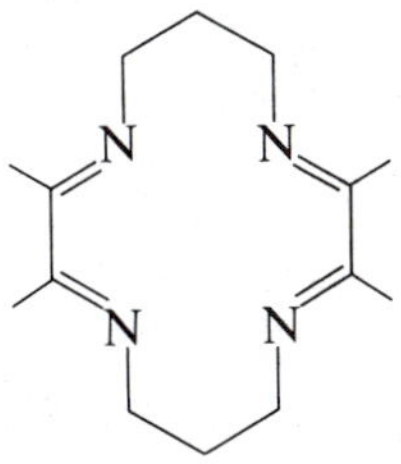

Me$_4$[14]1,3,8,10-tetraene N$_4$

24

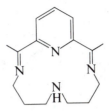

25

14-membered ring macrocyclic ligands

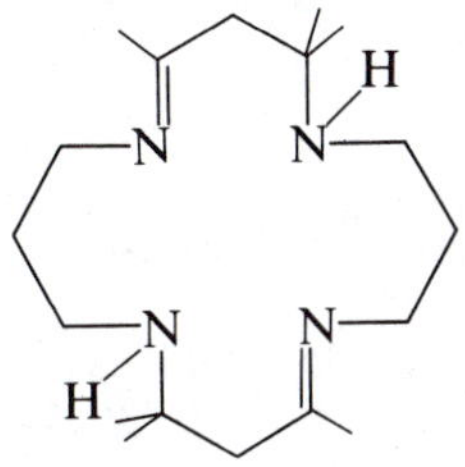

Me$_6$[16]4,12-diene N$_4$

26

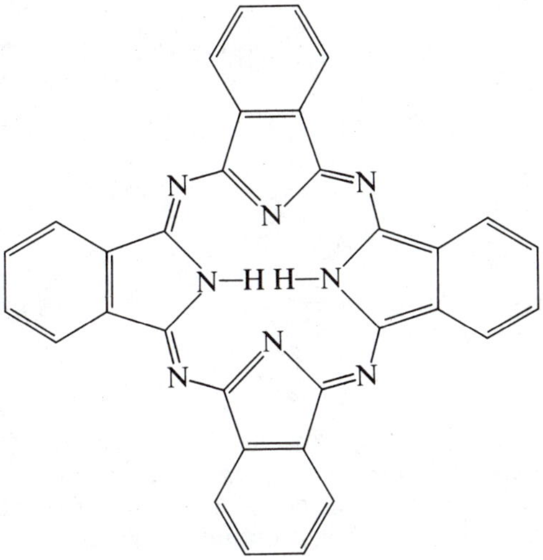

phthalocyanine

27

16-membered ring macrocyclcic ligands

STUDY QUESTIONS

6. A possible mechanism for the hydrolysis of a peptide by carboxypeptidase is given on page 464. Why does the interaction of tyrosine, residue 248, play such an important role in this reaction?

7. The intermediate in path A in Figure 14–7 uses the carbonyl oxygen and not the amide nitrogen to bind to the cobalt ion. On the basis of the discussion in Chapter 5, p. 149, is it reasonable to expect this binding mode?

8. It was noted in the text that water and the amide carbonyl group are competitors for the open coordination site in the five-coordinate intermediate **9** in Figure 14–7. What evidence is there that OH$^-$ is not a competitor? [*Hint:* Is the reaction pH dependent?]

9. Suggest a mechanism for the formation of the imine complex in the following reaction. [*Hint:* Consult the section on hydrolysis under basic conditions in Chapter 13.]

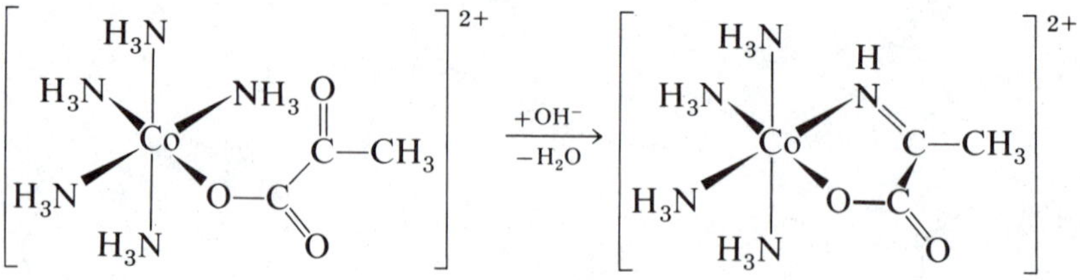

10. On page 469 the special thermodynamic and kinetic stability of complexes of macro-cyclic ligands is noted. Considering the following reaction for the formation and dissociation of such a complex, explain the thermodynamic and kinetic factors behind the "macrocyclic effect."

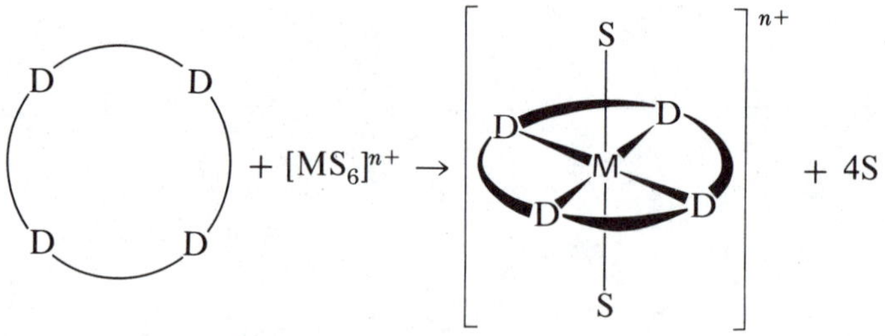

CHECKPOINTS

1. Coordination complexes may undergo *cis-trans* isomerization or changes in chirality (or both of these), or polytopal isomerization.

2. These changes in molecular geometry may occur by intramolecular twisting motions and/or by various bond rupture pathways.

3. Various reactions — among them, ligand hydrolysis, aldol condensation, and imine formation — are possible at coordinated ligands, in part owing to metal ion polarization of the ligands.

4. When a metal ion holds several ligands in close proximity, inter-ligand condensation may occur to form macrocyclic ligands.

REFERENCES

[1] N. Serpone and D. G. Bickley, *Prog. Inorg. Chem.*, **17**, 391 (1972).

[2] J. J. Fortman and R. E. Sievers, *Coord. Chem. Rev.*, **6**, 331–375 (1971).

[3] R. H. Holm and M. J. O'Connor, *Prog. Inorg. Chem.*, **14**, 241 (1971).

[4] E. L. Muetterties, *Acc. Chem. Res.*, **3**, 266 (1970).

[5] J. P. Jesson, "Stereochemistry and Stereochemical Nonrigidity in Transition Metal Hydrides," in "Transition Metal Hydrides," E. L. Muetterties, ed., Marcel Dekker, New York (1971).

[6] R. G. Hayter and F. S. Humiec, *Inorg. Chem.*, **4**, 1701 (1965).

[7] G. N. LaMar and E. O. Sherman, *J. Amer. Chem. Soc.*, **92**, 2691 (1970).

[8] R. H. Holm, *Acc. Chem. Res.*, **2**, 307 (1969); S. S. Eaton and R. H. Holm, *Inorg. Chem.*, **10**, 1446 (1971).

[9] S. H. Mastin and P. Haake, *Chem. Commun.*, 202 (1970).

[10] W. J. Louw, *Inorg. Chem.*, **16**, 2147 (1977); R. Romeo, *ibid.*, **17**, 2040 (1978).

[11] J. H. Price, J. P. Birk, and B. B. Wayland, *Inorg. Chem.*, **17**, 2245 (1978).

[12] J. G. Gordon, II, and R. H. Holm, *J. Amer. Chem. Soc.*, **92**, 5319 (1970).

[13] R. C. Fay and T. S. Piper, *J. Amer. Chem. Soc.*, **85**, 500 (1963); A. Y. Girgis and R. C. Fay, *ibid.*, **92**, 7061 (1970).

[14] An experiment involving rearrangements of six-coordinate complexes has been devised for the advanced undergraduate: D. E. Kranbuehl, P. M. Metzger, D. W. Thompson, and R. C. Fay, *J. Chem. Educ.*, **54**, 119 (1977).

[15] G. A. Lawrence and D. R. Stranks, *Inorg. Chem.*, **16**, 929 (1977); E. L. Blinn and R. G. Wilkins, *ibid.*, **15**, 2952 (1976).

[15a] K. F. Purcell, *J. Amer. Chem. Soc.*, **101**, 5147 (1979). This paper shows that octahedral low spin d^6 complexes must become high spin for rearrangement to occur through a trigonal prismatic transition state. However, recall from Figure 10-1(e) that it is just such complexes that have the largest energy difference between the octahedron and the trigonal prism. Thus, for low spin d^6 complexes with large e_σ values [Co(III), Rh(III), and Ru(II)] there is little tendency to racemize by this pathway. But, owing to smaller e_σ, Fe(II) complexes are able to use this mechanism.

[16] J. P. Collman, "Reaction of Ligands Coordinated with Transition Metals," in "Transition Metal Chemistry," Vol. 2, R. L. Carlin, ed., Marcel Dekker, New York (1966).

[17] Q. Fernando, *Adv. Inorg. Chem. Radiochem.*, 7, 185–261 (1965).

[18] M. M. Jones, "Ligand Reactivity and Catalysis," Academic Press, New York (1968).

[19] L. F. Lindoy and D. H. Busch, "Complexes of Macrocyclic Ligands," in "Preparative Inorganic Reactions," Vol. 6, W. L. Jolly, ed., Wiley-Interscience, New York (1971).

[20] L. F. Lindoy, *Chem. Soc. Rev.*, **4**, 421 (1975).

[21] L. Stryer, "Biochemistry," W. H. Freeman and Company, San Francisco (1975).

[22] W. N. Lipscomb, *Acc. Chem. Res.*, **3**, 81 (1970).

[23] M. L. Bender and B. W. Turnquest, *J. Amer. Chem. Soc.*, **79**, 1889 (1957).

[24] H. L. Conley, Jr., and R. B. Martin, *J. Phys. Chem.*, **69**, 2914, 2923 (1965).

[25] J. E. Hix and M. M. Jones, *Inorg. Chem.*, **5**, 1863 (1966).

[26] M. L. Bender, "Mechanisms of Homogeneous Catalysis from Protons to Proteins," Wiley-Interscience, New York (1971).

[27] In contrast with the catalysis of amino acid ester hydrolysis by metal ions, the hydrolysis of simple esters is not affected by metal ions. For example, Cu(II) neither interacts with ethyl acetate nor effects its hydrolysis. Apparently a simple ester cannot displace H_2O from the M^{2+} coordination sphere as effectively as a chelating amino acid ester (*cf.* Chapter 13). M. L. Bender and B. W. Turnquest, *J. Amer. Chem. Soc.*, **79**, 1656 (1957).

[28] M. D. Alexander and D. H. Busch, *J. Amer. Chem. Soc.*, **88**, 1130 (1966).

[29] D. A. Buckingham, D. M. Foster, and A. M. Sargeson, *J. Amer. Chem. Soc.*, **92**, 6151 (1970); D. A. Buckingham, C. E. Davis, D. M. Foster, and A. M. Sargeson, *ibid.*, **92**, 5571 (1970).

[30] D. F. Martin, "Metal Complexes of Ketimine and Aldimine Compounds," in "Preparative Inorganic Reactions," Vol. 1, W. L. Jolly, ed., Wiley-Interscience, New York (1964).

[31] D. H. Busch, *Acc. Chem. Res.*, **11**, 392 (1978).

[32] M. C. Thompson and D. H. Busch, *J. Amer. Chem. Soc.*, **86**, 3651 (1964).

[33] M. C. Thompson and D. H. Busch, *J. Amer. Chem. Soc.*, **86**, 213 (1964).

[34] For a recent example of the "thermodynamic template effect," see L. F. Lindoy and D. H. Busch, *Inorg. Chem.*, **13**, 2494 (1974).

[35] F. P. Hinz and D. W. Margerum, *J. Amer. Chem. Soc.*, **96**, 4993 (1974); *Inorg. Chem.*, **13**, 2941 (1974).

[36] A. B. P. Lever, *Adv. Inorg. Chem. Radiochem.*, 7, 28 (1965).

[37] N. F. Curtis, *Coord. Chem. Rev.*, 3, 3–47 (1968).

THE SIXTEEN AND EIGHTEEN
ELECTRON RULE

15. FROM CLASSICAL TO ORGANOMETALLIC TRANSITION METAL COMPLEXES AND THE SIXTEEN AND EIGHTEEN ELECTRON RULE

This chapter is a bridge between the previous six chapters on the chemistry of transition metal complexes involving the metals in their higher oxidation states—with ligands such as amines, halides, and oxygen donors—and chapters treating those complexes wherein the metal is in a lower formal oxidation state and is bound through carbon to ligands such as CO, CH_3, C_6H_6, and C_2H_4. That is, our next topic is organometallic chemistry. Although some of the compounds we shall discuss are based on main group metals, a large part of organometallic chemistry is concentrated on transition metal complexes. We present here an empirical rule, the "sixteen and eighteen electron rule" or the "effective atomic number rule" (= EAN), a device that will be an invaluable aid in systematizing the structural and reaction chemistry of transition metal complexes.

THE SIXTEEN AND EIGHTEEN ELECTRON RULE[1]

Before beginning the chapters on organometallic chemistry, we should discuss the "sixteen and eighteen electron rule" or "effective atomic number rule" (EAN). This rule allows the systematization of much of the structural chemistry of transition metal organometallic compounds (Chapter 16) and is a useful guide to their reactions (Chapter 17).

The transition metal complexes dealt with thus far in this book are often referred to as "classical" because their discovery and characterization reach back into the history of modern chemistry. Generally speaking, such complexes contain the metal in an oxidation state of 2+ or higher and involve ligands (such as amines and other nitrogen donors, oxygen donors, and halides) which bind the metal primarily by σ bonding. However, in the subsequent three chapters, we deal with low oxidation state metal compounds having ligands that are capable of, and even require, **synergic** ligand $\rightarrow$ metal σ/metal $\rightarrow$ ligand π electron flow.

First introduced in Chapter 5, the synergic interaction arises when the σ donor has a low energy π acceptor orbital (LUMO) while the σ acceptor in the adduct configuration has a π donor orbital (HOMO). An important consequence of this two-way donor-acceptor bonding (**1** and **2**) is that a bonding role can be given to otherwise non-bonding

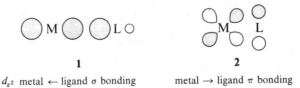

1

d_{z^2} metal $\leftarrow$ ligand σ bonding

2

metal $\rightarrow$ ligand π bonding

or even σ^* electrons on the metal. It is an interesting, empirical fact that, as compared with classical complexes, M-L adducts of this type more closely follow rules relating structure to number of valence electrons. Consider, as an example, $[Cr(NH_3)_6]^{3+}$ and $Cr(CO)_6$. The total number of valence electrons about Cr^{3+} in the hexaamminechromium (3+) ion is 15 (three from Cr^{3+} and 12 from the six NH_3's), whereas the total for hexacarbonylchromium is 18 (six from Cr(0) and 12 from the six CO's). There are a total of nine d, s, and p valence orbitals available to a transition metal, and all of these orbitals can be utilized fully in the carbonyl complex. In molecular orbital terms, this means that, for $Cr(CO)_6$, six bonding mo's arise from overlap of *all six* ligand lone pairs with the two d orbitals of σ symmetry, the s orbital, and the three p orbitals. In addition, and most importantly, the three metal $d\pi$ orbitals are full, and it is these orbitals that participate in metal $\rightarrow$ ligand π electronic flow and further stabilize the complex. Otherwise, $Cr(CO)_6$, a complex formed from a zero-valent metal, would suffer too great a build-up of negative charge on the metal.

In general, organic complexes of the transition metals are formed with those metals in low oxidation states (−1, 0, and +1), since it is only in that condition that the metal will have populated orbitals of π symmetry and a low Z^* to allow metal $\rightarrow$ ligand π electronic flow. With certain well-defined exceptions, *stable organometallic compounds of the transition metals will have a total of 18 valence electrons about the metal;* in other words, *they will have the "effective atomic number" (EAN) of the next higher inert gas.* The *exception* mentioned is the fact that *molecules having only 16 valence electrons can often be just as stable* as, or even more stable than, 18-electron molecules of the same metal. This is especially true of metals at the bottom right corner of the transition metal series, that is, Rh, Pd, Ir, and Pt. Recall from Chapters 9 and 10 (*cf.* Figure 10–1) that it is just these elements that have the least tendency to be six-coordinate. Thus, in oxidation states such as Ir(I) and Pt(II), such metals can form stable 16-electron molecules.

The application of the EAN rule is quite simple. If a metal is to accommodate six adduct bond pairs and follow the EAN rule, the metal must have three electron pairs of its own. That is, the metal atom or ion involved will have a d^6 configuration (**3**), and the structures of such complexes will be expected to be roughly octahedral. Metals with four valence electron pairs of their own require the presence of five adduct bond pairs and should exhibit structures based on the trigonal bipyramid or square pyramid (**4**). Metals with five valence electron pairs should adopt tetrahedral or square planar structures by acquisition of four ligand electron pairs (**5**).

3

d^6 + 6 ligand bond pairs = MD_6
(V^-, Cr^0, Mn^+, Fe^{2+})

4

d^8 + 5 ligand bond pairs = MD_5
(Mn^-, Fe^0, Co^+)

5

d^{10} + 4 ligand bond pairs = MD_4
(Co^-, Ni^0, Cu^+)

Some compounds that illustrate the 18 electron rule are given on the following page. In counting electrons, ligands such as CO, PPh_3, halides, H^-, and alkyl and aryl groups (*e.g.*, CH_3^- and $C_6H_5^-$) are considered two electron donors. NO and the π bonded organic ligands deserve special comment, and we return to that momentarily.

MD$_6$: M $= d^6 =$ V$^-$, Cr^0, Mn$^+$, Fe^{2+}

$V(CO)_6^-$ $Cr(CO)_6$ $Mn(CO)_6^+$ $Fe(CO)_4H_2$

$H_3CMn(CO)_5$

MD$_5$: M $= d^8 =$ Mn$^-$, Fe^0, Co$^+$

$Mn(CO)_5^-$ $Fe(CO)_5$ $Fe(CO)_3(PPh_3)_2$ $Co(CO)_4H$

$Mn(CO)_4NO$

MD$_4$: M $= d^{10} =$ Fe^{2-}, Co$^-$, Ni^0, Cu$^+$

$Fe(CO)_4^{2-}$ $Co(CO)_4^-$ $Ni(CO)_4$ $Cu(PPh_3)_3Cl$

$Co(CO)_3NO$

18 ELECTRON MOLECULES

As mentioned before, there are numerous stable organometallic compounds that have only 16 valence electrons about the metal. Such complexes are invariably based on d^8 metal ions and are square planar (four ligand bond pairs), or they are based on d^{10} metal atoms with three ligands (three ligand bond pairs) and are presumably trigonal. Examples include those shown below.

MD$_4$: $d^8 =$ Rh$^+$ or Ir$^+$; Pd^{2+} or Pt^{2+}

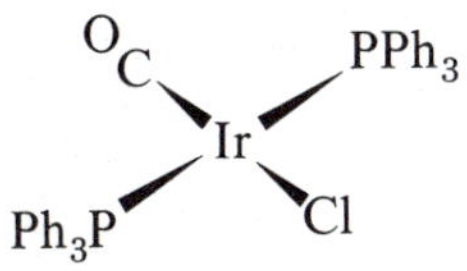

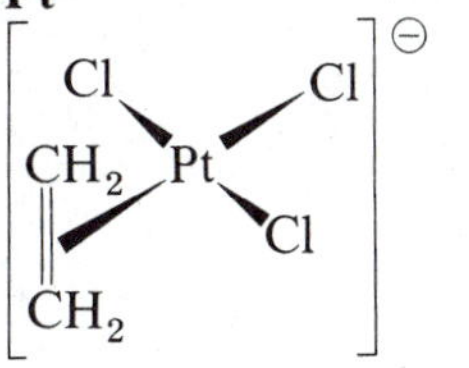

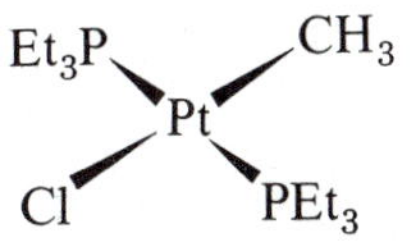

MD$_3$: $d^{10} =$ Pd^0 or Pt^0

$Pt(PPh_3)_3$

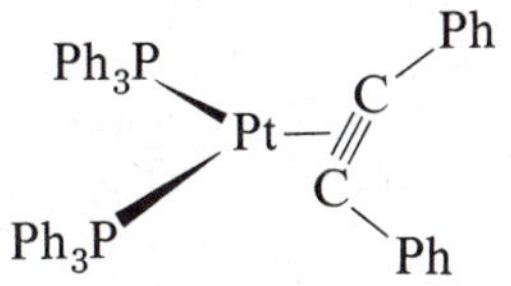

16 ELECTRON MOLECULES

We turn now to the convention used to determine the number of electrons contributed by a ligand to the valence shell or the metal. As mentioned previously, this is no problem for ligands of the classical type (amines, phosphines, halides, etc.), but some of the π donor ligands encountered in organometallic chemistry are less familiar to you at this point and deserve special comment. Our point of view for electron counting—and it is but one of several that we could take—is simply to view the complexes as donor/acceptor adducts wherein each ligand is thought of as an electron pair donor and the metal as an electron pair acceptor. Little needs to be added at this point about the donors of one electron pair: CO, H$^-$, PF$_3$, PPh$_3$, and C$_2$H$_4$. (C$_2$H$_4$ donates a pair of π electrons in binding to Pt^{2+} in [Pt(C$_2$H$_4$)Cl$_3$]$^-$.) Ligands such as NO$^+$, C$_5$H$_5^-$, C$_6$H$_6$, and H$_2$C=CH−CH=CH$_2$ require comment.

1 PAIR DONORS

As described in Chapter 4, **nitric oxide** is an odd-electron molecule possessing not only a nitrogen donor lone pair but also one electron in the π^* mo's. Our convention is to treat NO as coordinated NO$^+$ with the extra electron counted among the metal or other ligand electrons. Other conventions may treat NO as a three-electron donor or as NO$^-$, the latter formed by taking an electron from the metal and assigning it to the ligand.

However, the convention used here is based on the analogy of NO^+ to CO, CN^-, and N_2, which constitute an isoelectronic series. The assignment of the π^* electron to the metal for electron counting purposes facilitates this comparison but generally has little or no physical significance (but neither do the other two conventions). The NO^+ ligand *in the complex* will involve its π^* mo's with the metal $d\pi$ ao's to a considerable extent, and there will be, in the end, some net electron density distributed between the metal and the NO^+ π^* mo's that is independent of electron counting devices. Electron counting in NO complexes is illustrated by the following example.

$$Mn(CO)_4NO$$

tetracarbonylnitrosylmanganese

$$
\begin{aligned}
NO^+ &= 1\ \sigma\ \text{pair} \\
4CO &= 4\ \sigma\ \text{pairs} \\
Mn^- &= \underline{4\ \text{pairs}} \\
&\quad 9\ \text{pairs or} \\
&\quad 18\ \text{electrons}
\end{aligned}
$$

Before proceeding further with this discussion of electron counting conventions, we must insert a word concerning **nomenclature**. The systematic name of compound **6** is tricarbonyl(η^5-cyclopentadienyl)manganese and that of **7** is (η^6-benzene)tricarbonyl-

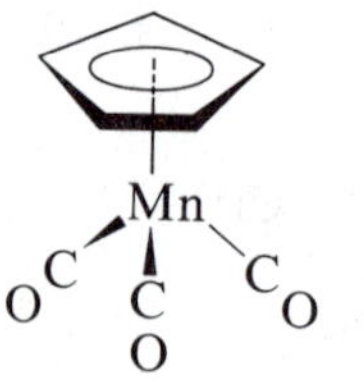

6 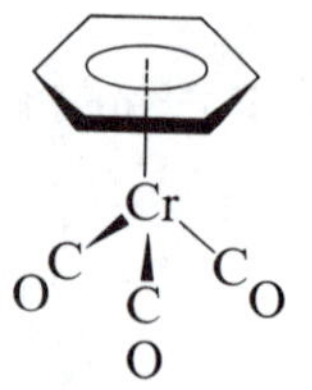   **7**

chromium. The notation η^n is a shorthand designating the number of ligand atoms ($= n$) formally bound to the metal. The letter η (Greek *eta*) stands for *hapto*, from the Greek word *haptein* meaning "to fasten." Thus, the C_5H_5 ligand in **6** is said to be a *pentahapto*-cyclopentadienyl group, whereas C_6H_6 in **7** is *hexahapto*benzene.[2] The utility of this system will become more obvious as you study the examples in this chapter and subsequent ones. Finally, we call your attention to the fact that the nomenclature of transition organometallic compounds is otherwise no different from that for classical compounds as laid out in Chapter 9 (p. 308).

Now to return to electron counting conventions. The ligand designated η^1-C_5H_5 is the cyclopentadiene molecule with a metal in place of a hydrogen. The structure of **monohapto**cyclopentadienyl places it in the category of σ bonded alkyl and aryl groups,

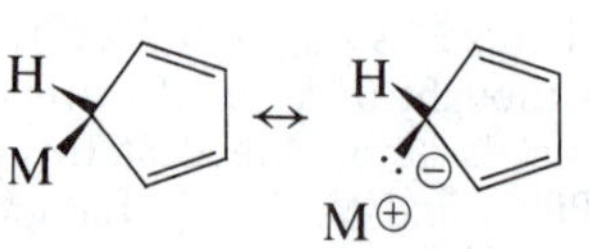

which must be thought of as carbanions (R^-) in order to conform to the electron pair scheme.

3 PAIR DONORS

*Pentahapto*cyclopentadienyl is one of the most important ligands in organometallic chemistry. In the η^5 configuration the cyclopentadienyl group is positioned so as to have all five carbons equidistant from the metal.

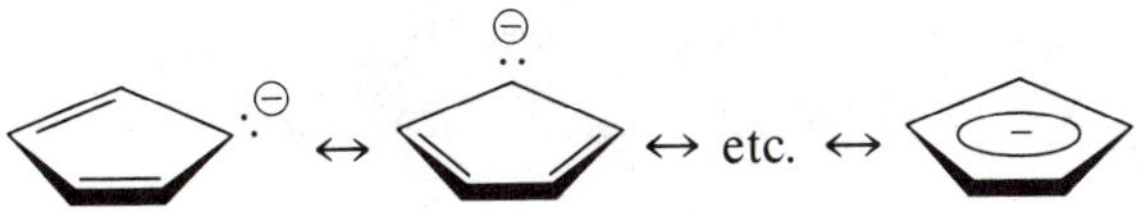

Here you are to think of the ligand as a carbanion and a three-pair π donor. $Fe(CO)_2(\eta^5\text{-}C_5H_5)(\eta^1\text{-}C_5H_5)$ has two kinds of $C_5H_5^-$ ligands, and the electrons are counted as follows.

$$Fe(CO)_2(\eta^5\text{-}C_5H_5)(\eta^1\text{-}C_5H_5)$$

dicarbonyl(*monohapto*cyclopentadienyl)(*pentahapto*cyclopentadienyl)iron

$$
\begin{array}{l}
2CO = 2\ \sigma\ \text{pairs} \\
\eta^5\text{-}C_5H_5^- = 3\ \pi\ \text{pairs} \\
\eta^1\text{-}C_5H_5^- = 1\ \sigma\ \text{pair} \\
Fe^{2+} = 3\ \text{pairs} \\
\hline
 9\ \text{pairs or} \\
 18\ \text{electrons}
\end{array}
$$

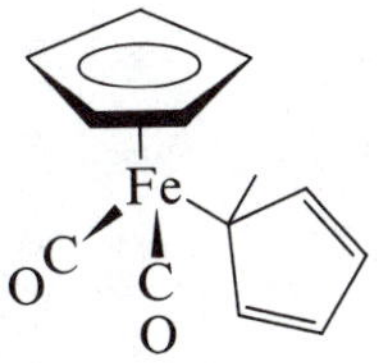

8

Another three-pair π donor is **hexahaptobenzene**, as illustrated by bis(η^6-benzene) chromium.

$$
\begin{array}{l}
2C_6H_6 = 6\ \pi\ \text{pairs} \\
Cr = 3\ \text{pairs} \\
\hline
 9\ \text{pairs}
\end{array}
$$

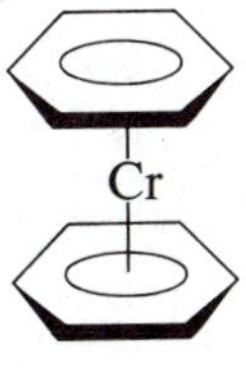

9

2 PAIR DONORS

The simplest two-pair π donor is derived from **trihaptoallyl**, $H_2C{=}CH{-}CH_2$. Like cyclopentadienyl, C_3H_5 contains an odd number of electrons when considered as a neutral hydrocarbon. However, our convention is to treat it as an anion and therefore as a two-pair π donor. In this configuration all three carbons are bound to the metal, and it is designated as $\eta^3\text{-}C_3H_5^-$.

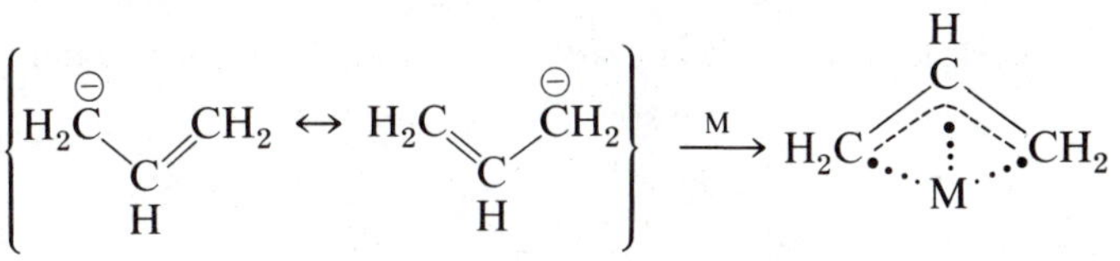

For example, valence electrons in $Mn(CO)_4(\eta^3\text{-}C_3H_5)$ are counted as follows.

$$Mn(CO)_4(\eta^3\text{-}C_3H_5)$$

(*trihapto*allyl)tetracarbonylmanganese

$$\eta^3\text{-}C_3H_5^- = 2 \ \pi \ \text{pairs}$$
$$4CO = 4 \ \sigma \ \text{pairs}$$
$$Mn^+ = \underline{3 \ \text{pairs}}$$
$$9 \ \text{pairs}$$

Among the neutral two-pair donors are

$\eta^4\text{-}C_4H_4$	$\eta^4\text{-}C_4H_6$	$\eta^4\text{-}C_7H_8$
*tetrahapto*cyclobutadiene 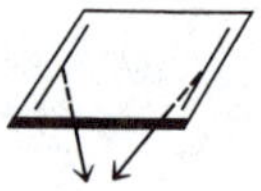	*tetrahapto*butadiene 	*tetrahapto*norbornadiene 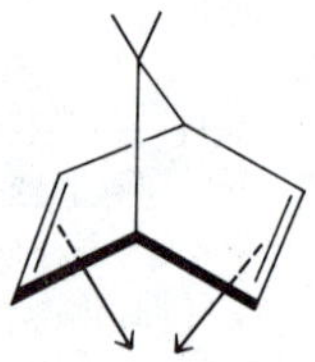

The interesting hydrocarbon **cycloheptatriene**, C_7H_8, is an isomer of norbornadiene, and has an intriguing flexibility in the number of donor pairs it can provide.

$$C_7H_8 = \quad = \text{cycloheptatriene}$$

Most obviously it could act as a neutral, three-pair π donor as in $(\eta^6\text{-}C_7H_8)Cr(CO)_3$ (**10**).

10

tricarbonyl(*hexahapto*cycloheptatriene)chromium

In $Fe(CO)_3(\eta^4\text{-}C_7H_8)$, however, it appears as a neutral, two-pair π donor like the dienes mentioned above.

11

Loss of H^+ or H^- from C_7H_8 leads to ions that can also function as ligands in a most interesting manner. For example, H^- abstraction (using Ph_3C^+, for example) from $(\eta^6\text{-}C_7H_8)Mo(CO)_3$ gives a complex of the **tropylium** ion, $C_7H_7^+$, **12**. The *heptahapto*-$C_7H_7^+$ ion is considered to be a three-pair or six-electron donor, so the central molybdenum atom is surrounded by 18 valence electrons.

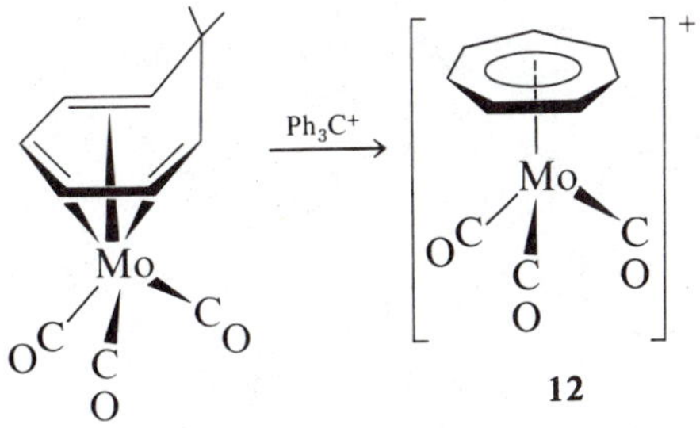

12

The final example of a hydrocarbon π donor is **cyclooctatetraene**, C_8H_8. Generally this ligand functions in a two-pair donor capacity as in $Fe(CO)_3(\eta^4\text{-}C_8H_8)$ and $Co(\eta^5\text{-}C_5H_5)(\eta^4\text{-}C_8H_8)$.

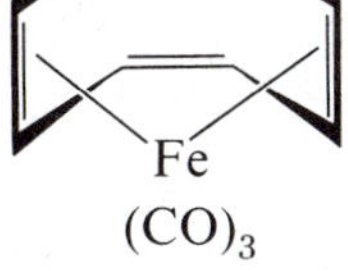

13

tricarbonyl(1,2,5,6-*tetrahapto*cyclo-octatetraene)iron

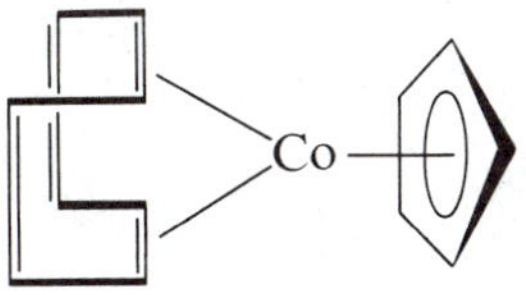

14

(1,2,5,6-*tetrahapto*cyclo-octatetraene)(*pentahapto*-cyclopentadienyl)cobalt

For your reference when the chemistries of these complexes are discussed in Chapters 16 and 17, Table 15–1 summarizes the donor and coordination characteristics of hydrocarbon ligands.

You may have noticed that we have made no mention of MD_x cases in which an odd number of electrons surround the metal, a situation of course not "permitted" by the 16 and 18 electron rule. However, there are cases wherein the empirical formula of the molecule would lead one to believe that there is an odd number of valence electrons. For example, there are formally 17 electrons about cobalt in $Co(CO)_4$. Such a species does not exist as a stable entity under normal conditions, though, as it "dimerizes" to form $(OC)_4Co-Co(CO)_4$ **(15)**, an even-electron molecule with a metal-metal bond. Each Co atom now has a share in 18 valence electrons. Another example of metal-metal bond formation to satisfy the 18 electron rule is **16**.

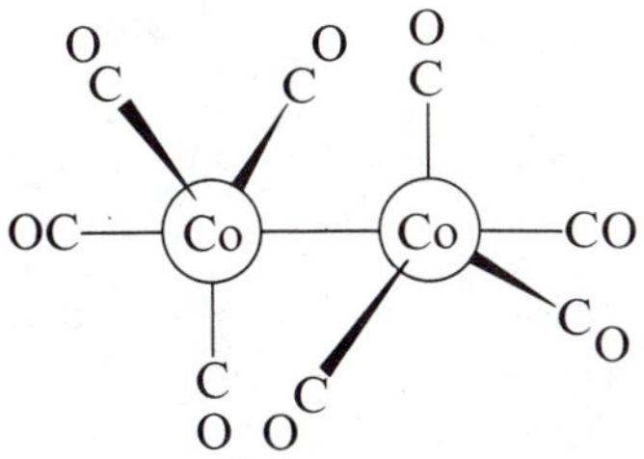

15

octacarbonyldicobalt

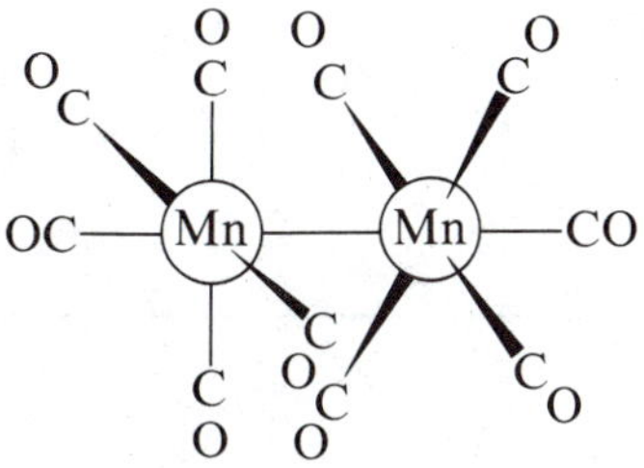

16

decacarbonyldimanganese

TABLE 15-1

**THE DONOR/COORDINATION CHARACTERISTICS
OF CERTAIN HYDROCARBON π DONORS**

Ligand	Donor Pairs	Coordination
η^2-C_2H_4	1	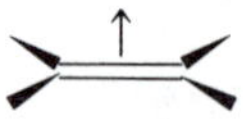
η^3-C_3H_5 (allyl)	2	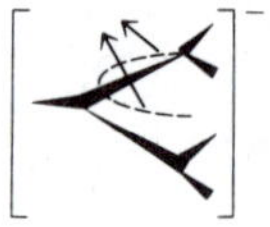
η^3-C_7H_7 (cycloheptatrienyl)		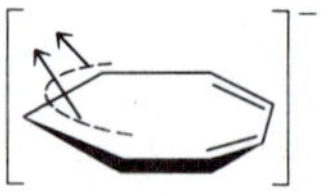
η^4-C_4H_6 (butadiene)		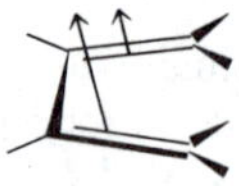
η^4-C_7H_8 (norbornadiene)		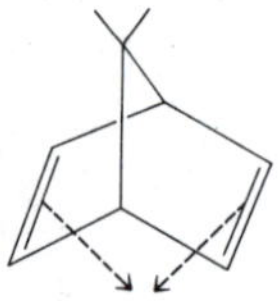
η^4-C_7H_8 (cycloheptatriene)		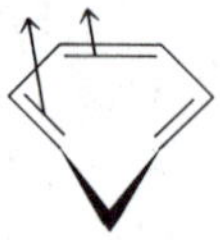
η^4-C_8H_8 (cyclooctatetraene)		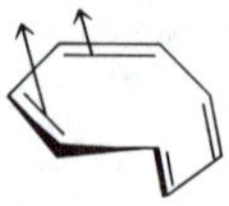 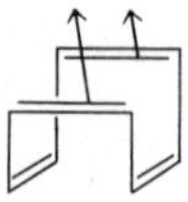
η^5-C_5H_5 (cyclopentadienyl)	3	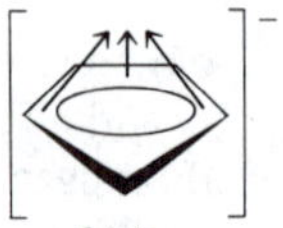
η^6-C_6H_6 (benzene)		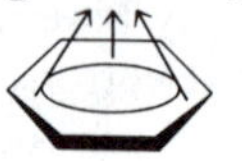
η^6-C_7H_8 (cycloheptatriene)		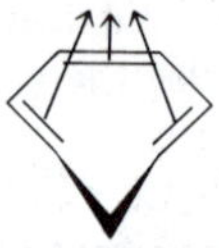
η^6-C_8H_8 (cyclooctatetraene)		
η^7-$C_7H_7^+$ (tropylium)		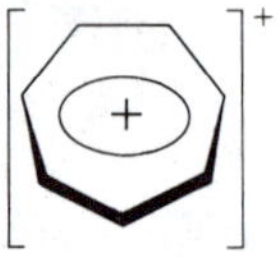

Note that dimerization of formally odd electron fragments is not the only way in which metal-metal bonded species can arise. Mononuclear metal carbonyls can form complexes of higher nuclearity by losing CO and forming M—M bonds, both with and without CO bridges between the metals. The best known examples are the iron carbonyls.

$$2Fe(CO)_5 \rightarrow Fe_2(CO)_9 + CO$$
$$3Fe(CO)_5 \rightarrow Fe_3(CO)_{12} + 3CO$$

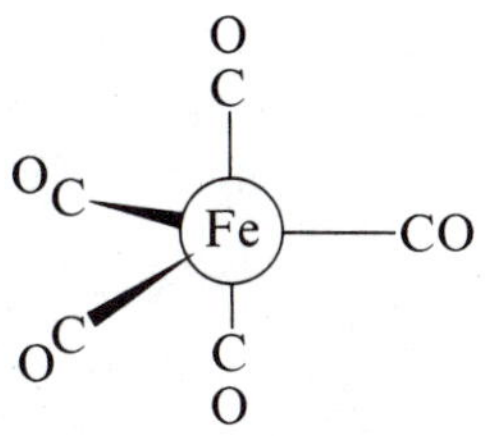

Fe(CO)$_5$
pentacarbonyliron
(D_{3h})

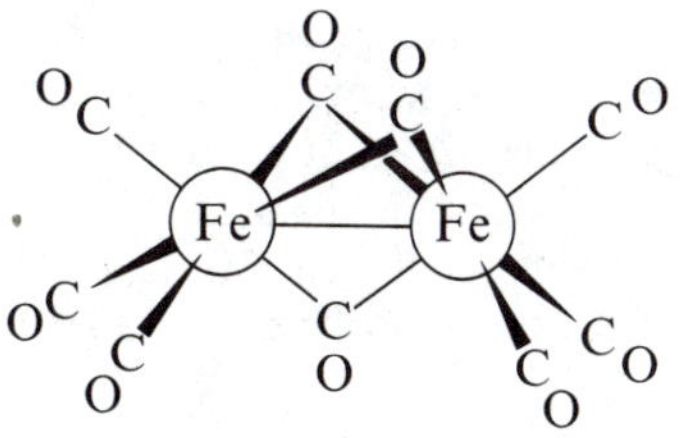

Fe$_2$(CO)$_9$
nonacarbonyldiiron
(D_{3h})

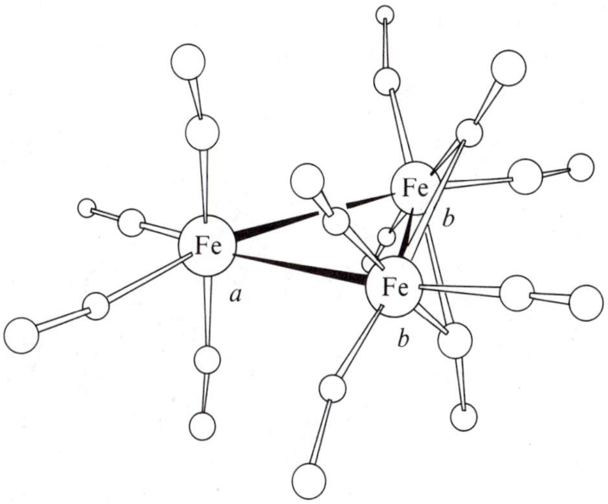

Fe$_3$(CO)$_{12}$
dodecacarbonyltriiron
(C_{2v})

The electron count for the bridging CO group derives from the description of the Fe—(CO)—Fe bridge bond as a three-center, two-electron case. The carbon lone pair of electrons is delocalized over both iron atoms and, therefore, makes a contribution of ½ pair to each iron valence shell. For Fe$_2$(CO)$_9$ the electron count for each iron is the same by symmetry and arises as follows.

$$3CO_t = 3 \text{ } \sigma \text{ pairs}$$ (the subscript t means terminal)
$$3CO_b = 1\tfrac{1}{2} \text{ } \sigma \text{ pairs}$$ (the subscript b means bridging)
$$1Fe\text{—}Fe = \tfrac{1}{2} \text{ pair}$$
$$Fe = \underline{4 \text{ pairs}}$$
$$9 \text{ pairs}$$

For Fe$_3$(CO)$_{12}$ there are two kinds of Fe atoms:

apex Fe ($= a$)	base Fe ($= b$)
$4CO_t = 4$ σ pairs	$3CO_t = 3$ σ pairs
$2Fe\text{—}Fe = 1$ pair	$2CO_b = 1$ pair
$Fe = \underline{4 \text{ pairs}}$	$2Fe\text{—}Fe = 1$ pair
9 pairs	$Fe = \underline{4 \text{ pairs}}$
	9 pairs

As a final note, it is sometimes found that structures with bridging rather than terminal CO's are favored where none are "required" by the EAN. The conversion takes the general forms

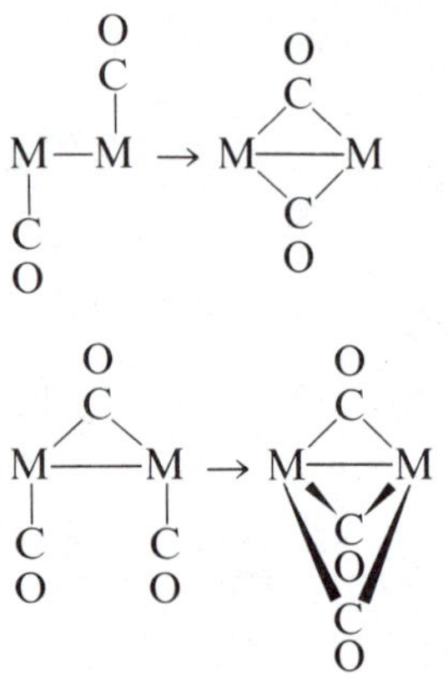

and it occurs by pairs of CO ligands, in accordance with the "rule of 18." Cases in point are $Fe_2(CO)_9$ and $Fe_3(CO)_{12}$. A more striking example is octacarbonyldicobalt. In hydrocarbon solvents this complex has all terminal CO groups, but in the solid state the structure exhibits two bridging CO groups. This example makes the important point that often the energy distinction between two bridging and two terminal CO groups is subtle and easily influenced by environmental factors such as solvation and lattice forces.

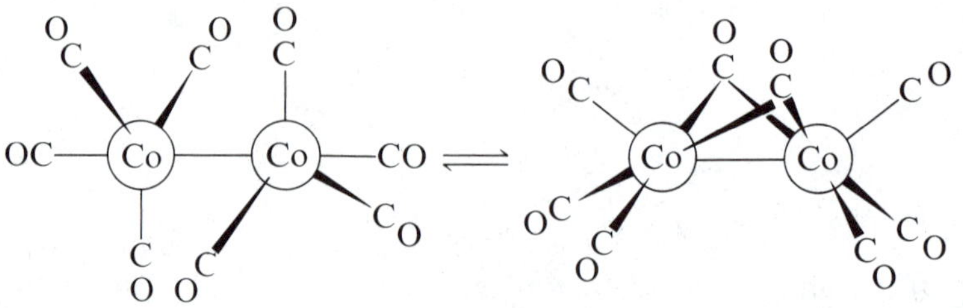

$Co_2(CO)_8$ in solution $Co_2(CO)_8$ in solid state

Occasionally, 14-electron molecules have been postulated as intermediates in reactions of organometallic compounds,[1] but such species were never adequately confirmed. Recently, however, a 14-electron bis(acetylene)platinum(0) complex, $Pt(PhC\equiv CPh)_2$, was reported.[3] Although more such molecules are certain to be isolated in the future, for the moment 14-electron molecules are the exception rather than the rule. Nonetheless, their existence as reaction intermediates may be more probable than was once thought.

STUDY QUESTIONS

1. Apply the 18 electron rule to the following combinations of metals and ligands to predict possible complex stoichiometries (give one example each of a neutral, cationic, anionic, and dimeric species).

   a) Fe, CO, NO
   b) Ni, PF_3, CO
   c) Co, $C_5H_5^-$, NO
   d) Fe, norbornadiene, CO

   e) Pd, cyclooctatetraene, Cl
   f) Mn, cycloheptatriene, PF_3
   g) Fe, PF_3, H

2. Show how the valence electrons are counted in each transition metal species in the following reactions:

   a) $Ni(CO)_4 \xrightarrow{-CO} [Ni(CO)_3] \xrightarrow{PPh_3} Ni(PPh_3)(CO)_3$

b)

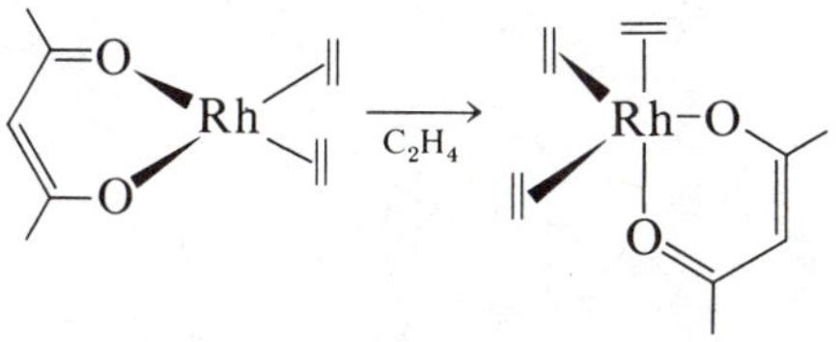

c)

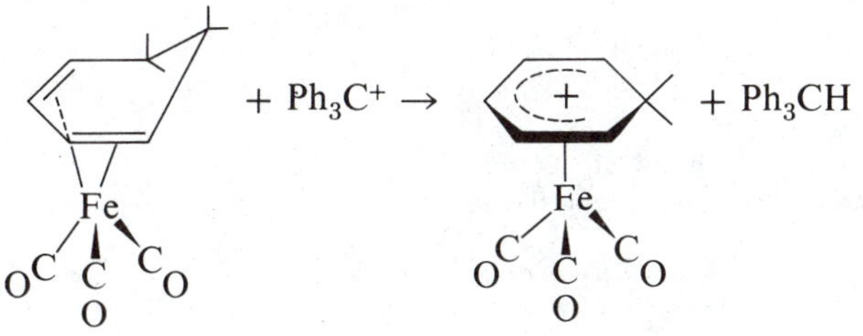

$+ Ph_3C^+ \rightarrow$

$+ Ph_3CH$

d)

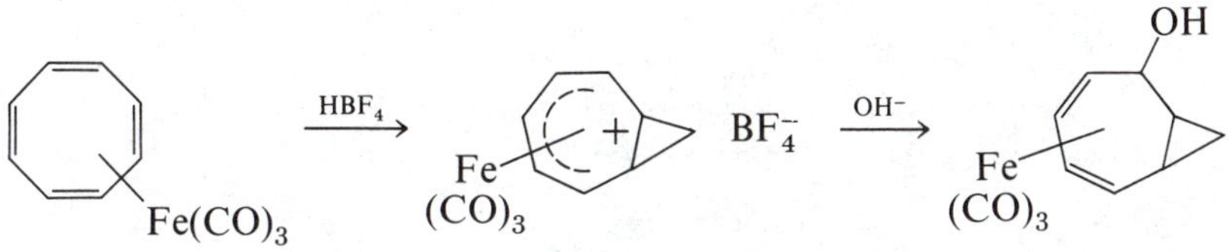

$\xrightarrow{HBF_4}$

BF_4^-

$\xrightarrow{OH^-}$

e) 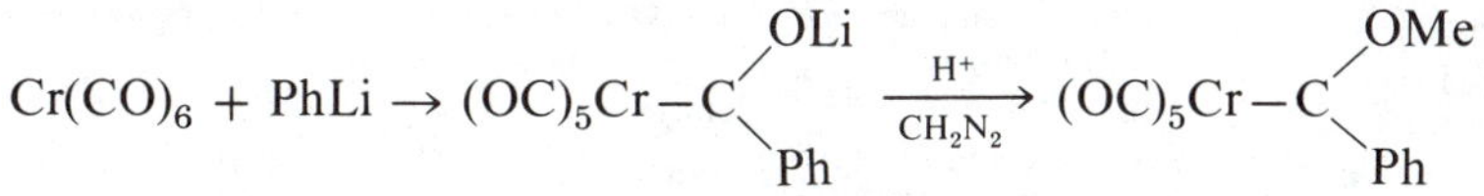 $Cr(CO)_6 + PhLi \rightarrow (OC)_5Cr-C\begin{smallmatrix}OLi\\ \\Ph\end{smallmatrix} \xrightarrow[CH_2N_2]{H^+} (OC)_5Cr-C\begin{smallmatrix}OMe\\ \\Ph\end{smallmatrix}$

3. Metals of the first transition series form an interesting series of compounds containing $C_5H_5^-$ and CO. For the metals Cr through Ni, write the formulas for the neutral compounds containing both $C_5H_5^-$ and CO, and draw the structure of each.

4. It is believed that the catalytic role of organometallic complexes in effecting organic reactions may be understood in terms of successive steps generating 16- and 18-electron complexes from 18- and 16-electron molecules, respectively. Thus, the "hydroformylation" of olefins

$$CH_2 = CHR + H_2 + CO \rightarrow RCH_2CH_2CHO$$

is catalyzed by $HCo(CO)_4$ according to the following scheme. Identify each cobalt-containing species as a 16- or 18-electron molecule.

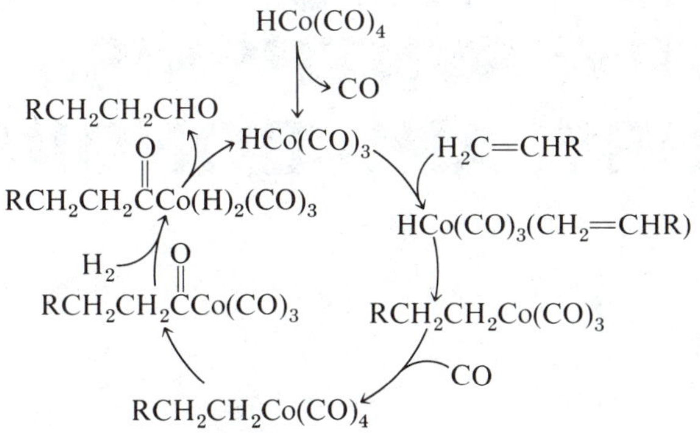

[1] C. A. Tolman, *Chem. Soc. Rev.*, **1**, 337 (1972).
[2] F. A. Cotton, *J. Organometal. Chem.*, **100**, 29 (1975).
[3] M. Greene, D. M. Grove, J. A. K. Howard, J. L. Spencer, and F. G. A. Stone, *J. C. S. Chem. Commun.*, 759 (1976).

REFERENCES

16. ORGANOMETALLIC CHEMISTRY: SYNTHESIS, STRUCTURE, AND BONDING

In the Prologue to this book, we surveyed the various areas of inorganic chemistry to arrive at some feeling for their current relative importance. The sub-area of organometallic chemistry was clearly prominent. To see why this should be the case, we turn now to a systematic look into this fascinating field. In this chapter we shall survey the various types of compounds with metal-carbon bonds, with particular emphasis being given to the methods by which they are synthesized, the peculiarities of the metal-carbon bond, and their structural chemistry. In organizing the field, we break the topic down into an examination of those compounds formed by σ donors—both anionic (**A**, ligand L = CH_3^-) and neutral donors (**B**, L = CO)—and compounds formed by chain (**C**, L = $H_2C=CH-CH=CH_2$) or cyclic (**D**, L = C_6H_6) π donors. In Chapter 17 we shall give more complete attention to the reactivity of the various types of compounds.

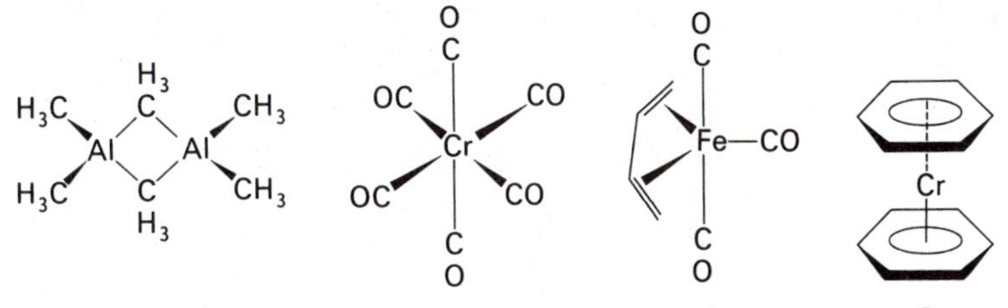

INTRODUCTION

We turn now to one of the most interesting and important areas in which modern inorganic chemists are involved: organometallic chemistry, a broad, interdisciplinary field whose sphere of interest includes all compounds wherein a metal, usually in a low valence state, is bonded through carbon to an organic molecule, radical, or ion.

The beginnings of the field can be traced to the discovery of Zeise's salt in 1831,[1] the synthesis of zinc alkyls by Frankland in 1848,[2] and Victor Grignard's Nobel Prize winning synthesis and exploitation of organomagnesium halides in 1900.[3]

$$C_2H_4 + K_2PtCl_{4(aq)} \rightarrow K^+[(C_2H_4)PtCl_3]^- + KCl \qquad (16\text{-}1)$$

$$Zn + C_2H_5I \rightarrow C_2H_5ZnI \rightarrow \frac{1}{2}(C_2H_5)_2Zn + \frac{1}{2}ZnI_2 \qquad (16\text{-}2)$$

$$Mg + C_2H_5Br \xrightarrow[\text{ether}]{} C_2H_5MgBr \qquad (16\text{-}3)$$

Of greater importance to the more recent development of the field, however, has been the synthesis of such compounds as ferrocene[4,5] and the extensive exploration of their chemistry that followed.[6,7,8]

(16-4)

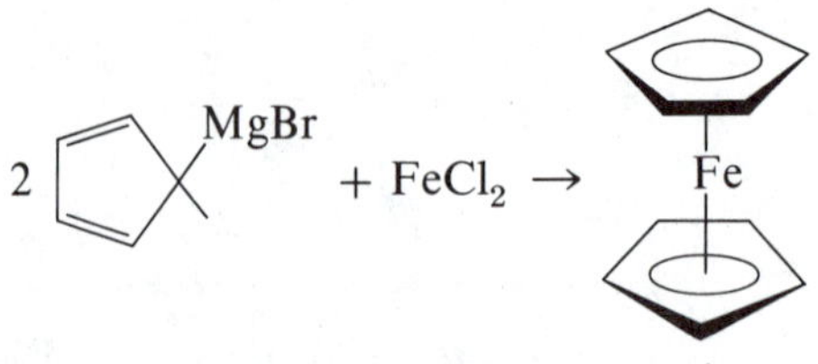

$$2 \overset{MgBr}{\underset{}{\bigtriangleup}} + FeCl_2 \rightarrow \underset{\mathbf{1}}{Fe}$$

The fascination of organometallic chemistry lies not only in its tremendously varied chemistry and structural and bonding forms, but also in the actual and potential practical uses of such materials. For example, alkylaluminum compounds form the basis of the Ziegler-Natta catalyst system, which is widely used in industry for the homogeneous polymerization of ethylene and propylene. Trialkyltin oxides and acetates have found wide application as fungicides, and dialkyltin compounds as stabilizers—anti-oxidants and ultraviolet radiation filters—for polyvinyl chloride and rubber. In addition, silicon compounds are extensively used in the form of silicones (organosilicon polymers) (reaction 16-5), which can be made into rubbery substances (elastomers); 14 million pounds were produced in the U.S. in 1969. And finally, transition metal organometallic compounds play important roles in industrial processes such as the OXO reaction, the hydroformyla-

(16-5)

$$R_2SiCl_2 + H_2O \longrightarrow 2\ HCl + \left(-O-\underset{\underset{R}{|}}{\overset{\overset{R}{|}}{Si}}-O-\underset{\underset{R}{|}}{\overset{\overset{R}{|}}{Si}}-O-\underset{\underset{R}{|}}{\overset{\overset{R}{|}}{Si}}-O-\right)_n$$

tion of olefins (16-6). Aldehydes obtained in the OXO process can be hydrogenated to give alcohols, many of which are used in plasticizers; for example, 100 million pounds

(16-6)

$$C_5H_{11}CH{=}CH_2 + CO + H_2 \xrightarrow[Co_2(CO)_8]{} C_5H_{11}\underset{\underset{H}{\overset{|}{C=O}}}{\overset{|}{CH}}{-}CH_3 + C_5H_{11}CH_2CH_2CHO$$

isooctaldehyde   *n-octaldehyde*

of isooctanol was produced in the U.S. in 1969. Other industrial processes are represented by metal-catalyzed olefin hydrogenation, the conversion of ethylene to acetaldehyde (the Wacker process, reaction 16-7),

(16-7)

$$H_2C{=}CH_2 + \tfrac{1}{2}O_2 \xrightarrow{PdCl_4^{2-}/H_2O} H_3C{-}CHO$$

and the "water gas shift" reaction (16-8); the latter is used to obtain hydrogen for the Haber process (NH_3 synthesis) or the Fisher-Tropsch reaction (hydrocarbon production from CO and H_2).[9]

(16-8)

$$C + H_2O \longrightarrow CO + H_2 \quad \textit{(water-gas reaction)}$$

$$CO + H_2O \longrightarrow CO_2 + H_2 \quad \textit{(shift reaction)}$$

In this chapter and the two that follow, we shall explore the syntheses and structures of major types of organometallic compounds and the reaction pathways that are characteristic of such compounds, especially those of the transition metals.

A NOTE ON THE ORGANIZATION OF ORGANOMETALLIC CHEMISTRY

The field of organometallic chemistry is relatively new; systematic study in the area by large numbers of people did not really begin until the 1950's. When this systematic work did get underway, chemists naturally worked on the chemistry of a particular element or group of elements, and, furthermore, divided themselves rather naturally into main group or transition metal chemists. Therefore, the literature of the field and texts in the area have often sub-divided organometallic chemistry by the part of the periodic table being studied. At least initially, this made some sense, because there is some periodicity in the types of metal-carbon bonds formed (see Figure 16-1). In the past five years or so, however, it has become increasingly obvious that such divisions are artificial.

Because there no longer seems to be any reason to organize organometallic chemistry by element type, we have chosen to organize these chapters by ligand and reaction type. As we mentioned in Chapter 5, there are in general two types of carbon donor ligands: σ and π. Therefore, in this chapter we shall first discuss molecules formed by σ donors—metal alkyls and aryls, metal carbonyls, and metal carbenes and carbynes—and then turn to complexes formed between metals and π donors (that is, molecules such as ferrocene, **1**).

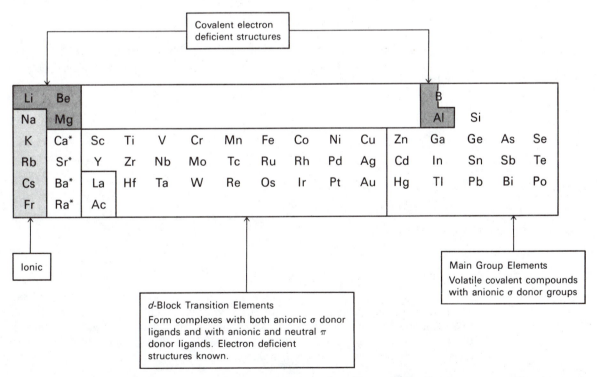

Figure 16-1. Types of bonding in organometallic compounds. (*The nature of metal-carbon bonding in alkaline earth compounds has not been clearly established.)

CARBON σ DONORS

Organometallic compounds formed from carbon σ donor ligands may be subdivided into two classes: (a) compounds in which the organic group can be considered an anionic σ donor, and (b) compounds in which the organic group is a neutral σ donor and π acceptor. Compounds of the first type include metal alkyls and aryls such as phenyl sodium (an ionic compound according to Figure 16-1); dimeric trimethylaluminum, **2** (a volatile "electron deficient" compound according to Figure 16-1);[10] volatile covalent compounds such as $(CH_3)_4 Sn$; and involatile transition metal compounds such as $Cr(CH_2 CPhMe_2)_4$ (**3**).

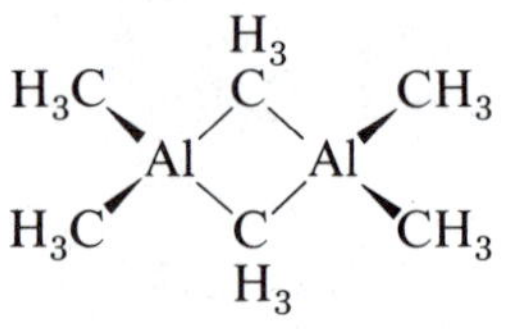

2

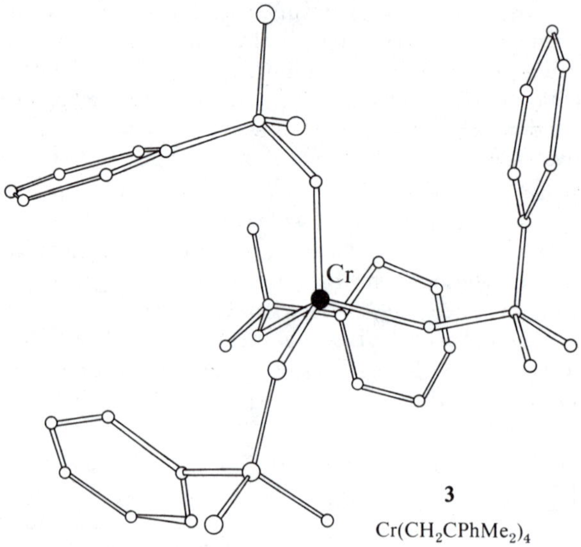

3

$Cr(CH_2CPhMe_2)_4$

Type (b) compounds are almost exclusively formed by transition metals in their low valence states, and they are exemplified by complexes with carbon monoxide and carbene (**4**) and with isonitriles (**5**).

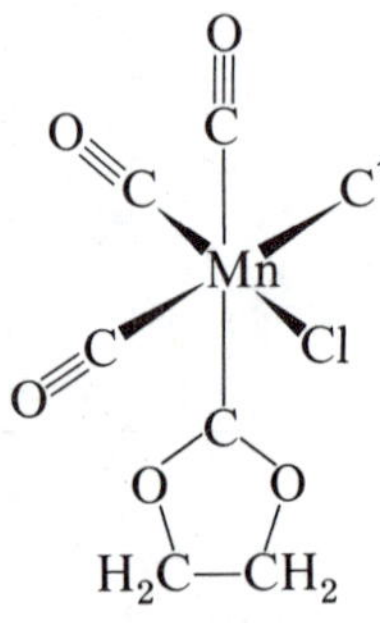

4 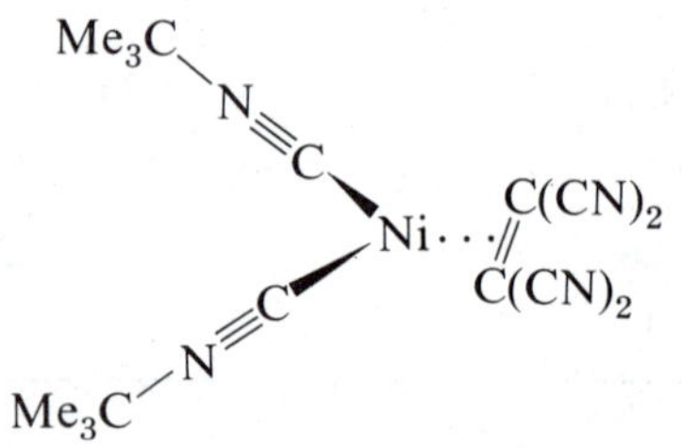   **5**

Because of the natural division of carbon σ donor ligands into two sub-types, this first section will follow this general organizational scheme; that is, we shall consider first the synthesis, structure, and bonding in compounds of type (a) and then those same aspects for complexes with carbon monoxide.

The Synthesis of Metal Alkyls and Aryls[11]

In the sections that follow, we attempt to introduce to you some of the ways currently available for the synthesis of metal-carbon σ bonds. As you read through these sections, note the relative advantages and disadvantages of the different methods. Notice also the similarities and differences between main group metals and transition metals.

DIRECT REACTION OF A METAL WITH AN ORGANIC HALIDE

The direct reaction of metals with alkyl and aryl halides may be used to produce numerous organometallic derivatives. Because of the simplicity of the reaction, the products are frequently of industrial importance. Furthermore, such reactions often produce many of the more reactive representatives of each class of compounds. For example, direct reaction of alkyl halides with Li, Mg, Al, or Zn (e.g., reactions 16-2 and 16-3) leads to some of the most reactive metal alkyls known; these compounds may then be used to produce alkyl derivatives of metals for which the direct reaction does not work. A current listing of metals that can react directly with organic halides to produce organometallic compounds is given in Figure 16-2. (Notice that this figure also includes, for future reference, those metals that can react directly with CO or with hydrocarbons such as olefins and arenes. Such reactions are discussed later in the chapter.[12-14]

Alkyl lithium compounds are usually prepared by the reaction of a lithium metal dispersion with an alkyl chloride or bromide (reaction 16-9) in a solvent such as petroleum ether, cyclohexane, benzene, or ether; the reaction must be carried out in an atmosphere of dry nitrogen because of the air-sensitivity of the organometallic product. Generally speaking, alkyl iodides cannot be used, because unreacted RI can quickly react with LiR to produce the coupling product as shown in reaction 16-11.

$$2\,Li + n\text{-}C_4H_9Cl \rightarrow n\text{-}C_4H_9Li + LiCl \qquad (16\text{-}9)$$

$$2\,Li + C_2H_5I \rightarrow C_2H_5Li + LiI \qquad (16\text{-}10)$$

$$C_2H_5Li + C_2H_5I \rightarrow C_4H_{10} + LiI \qquad (16\text{-}11)$$

1																	18
1 H																	2 He
3 Li ◇	4 Be ◇											5 B	6 C	7 N	8 O	9 F	10 Ne
□ 11 Na ◆	□ 12 Mg ◇											□ 13 Al ◆	14 Si ◆	15 P ◆	16 S	17 Cl	18 A
□ 19 K ◆	20 Ca ◇	21 Sc	□ 22 Ti	□ 23 V	□ 24 Cr	□ 25 Mn	□ 26 Fe ■	□ 27 Co ■	□ 28 Ni ■	29 Cu ◆	30 Zn ◆	31 Ga	32 Ge ◆	33 As ◆	34 Se ◇	35 Br	36 Kr
□ 37 Rb ◆	38 Sr ◇	39 Y	40 Zr	41 Nb	□ 42 Mo ■	43 Tc	44 Ru ■	45 Rh ■	□ 46 Pd ◆	47 Ag ◆	48 Cd	49 In	50 Sn ◆	51 Sb ◆	52 Te ◆	53 I	54 Xe
□ 55 Cs ◆	56 Ba ◆	57 La *	72 Hf	73 Ta	□ 74 W ■	75 Re	76 Os	77 Ir	□ 78 Pt ◆	79 Au ◆	80 Hg ◆	81 Tl ◆	82 Pb ◆	83 Bi ◆	84 Po	85 At	86 Rn
87 Fr	88 Ra	89 Ac *															

◆ react with RX with or without solvent
◇ react with RX in a solvent
■ react with CO
□ react with unsaturated hydrocarbon

Figure 16-2. The chemical elements that react directly with organic halides, carbon monoxide, or unsaturated hydrocarbons. [Adapted from E. G. Rochow, *J. Chem. Educ.*, **43**, 58 (1966).]

Sodium alkyls and aryls are not usually made by direct reaction of sodium with organic halides for the simple reason that the very reactive organosodium compound can react rapidly with any excess organic halide to produce a coupled product in a reaction sequence similar to that illustrated by reaction 16-11. This is the Wurtz-Fittig coupling reaction taught in introductory organic courses. In order to prepare sodium alkyls or aryls on a laboratory scale, the method of choice is the reaction of sodium with a mercury alkyl or aryl; this reaction is discussed in more detail on page 494.

The class of organometallic compounds perhaps most familiar to chemists in general is that of the Grignard reagents; *i.e.*, **alkyl- or arylmagnesium halides**. These compounds are also made directly from the metal and organic halides; again, the chloride and bromide are used in preference to the iodide in order to reduce coupling reactions. In contrast with organolithium reagents, however, Grignard reagents do not attack ethers to any appreciable extent; indeed, the solvent of choice in the preparation and use of organomagnesium halides is an ether.

Organic derivatives of mercury are extremely important as synthetic reagents. Alkylmercury halides and dialkylmercury compounds may be prepared by reaction of mercury with the organic halide, but only if the metal is in the form of a sodium amalgam. That this reaction requires the use of amalgamated mercury suggests the importance of thermodynamics in determining the course of some organometallic reactions. From the following equations, you see that there is little driving force for the direct reaction of mercury with methyl chloride. However, if the chloride is taken up in the form of its sodium salt, the very high heat of formation of NaCl (-411.0 kJ/mol) provides the driving force for the reaction.

(16-12)
$$2\ Hg_{(liq)} + 2\ CH_3Cl_{(g)} \rightarrow Hg(CH_3)_{2(g)} + HgCl_{2(s)} \qquad \Delta H°_{rxn} = +\ 30.5\ kJ$$

$$HgCl_{2(s)} + 2\ Na_{(s)} \rightarrow Hg_{(liq)} + 2\ NaCl_{(s)} \qquad \Delta H°_{rxn} = -597.9\ kJ$$

(16-13)
$$Hg_{(liq)} + 2\ Na_{(s)} + 2\ CH_3Cl_{(g)} \rightarrow$$
$$Hg(CH_3)_{2(g)} + 2\ NaCl_{(s)} \qquad \Delta H°_{rxn} = -567.4\ kJ$$

Since the preparation of an organometallic compound by the direct reaction of a metal with an organic halide is formally an oxidation-reduction reaction, it is evident that the reaction will work only when the metal is sufficiently electropositive. It is partly for this reason that organoboron compounds cannot be prepared by the direct method. However, the reaction is possible for the other members of group IIIA, especially aluminum. Aluminum metal—in the form of foil or turnings—reacts smoothly with alkyl halides in a hydrocarbon solvent to give **alkylaluminum sesquihalides**.

(16-14)
$$2\ Al + 3\ C_2H_5I \rightarrow (C_2H_5)_3Al_2I_3$$

The sesquihalides are, like all organoaluminum compounds, extremely air-sensitive, reacting readily with oxygen and water; however, they do not have sharp boiling points [for example, $(CH_3)_3Al_2Cl_3$ has a boiling range of 127° to 148°], as they are in reality an equilibrium mixture of organoaluminum halides.

(16-15)
$$2\ (CH_3)_3Al_2Cl_3 \rightleftharpoons (CH_3)_2Al_2Cl_4 + (CH_3)_4Al_2Cl_2$$

Group IVA elements also react directly with alkyl halides, but only when catalyzed by copper or some other suitable metal. The most important and widely studied system is the reaction of elemental **silicon** with methyl chloride in the presence of a copper catalyst at 285°C (16-16). This industrially important reaction was first described in the literature after World War II, although the problem of the efficient large-scale production of organosilanes was solved in 1940 by General Electric.[15]

The overall reaction of silicon and methyl chloride is simple,

$$2\ CH_3Cl + Si \xrightarrow[235°C]{Cu} (CH_3)_2SiCl_2 \tag{16-16}$$

but many by-products such as $SiCl_4$, CH_3SiCl_3, and $(CH_3)_3SiCl$ are obtained. Under appropriate conditions, the product consists of approximately 65% of $(CH_3)_2SiCl_2$, 25% of CH_3SiCl_3, and 5% of $(CH_3)_3SiCl$; the remainder is a complex mixture of products. After careful separation of dimethyldichlorosilane from the other products, the former is hydrolyzed under controlled conditions to yield commercially important polyorganosiloxanes or silicones (reaction 16-5). The function of the copper catalyst is not entirely clear, but the most recent view is that it brings about the formation of methyl radicals and transforms methyl chloride into a more reactive form by a reaction such as 16-17.

$$2\ Cu + CH_3Cl \rightarrow CuCl + CuCH_3 \tag{16-17}$$

Methylcopper, like many other simple transition metal alkyls, is quite unstable toward decomposition to the free metal and methyl radicals.

REACTIONS OF ANIONIC ALKYLATING AGENTS WITH METAL HALIDES OR OXIDES

Perhaps the most useful and generally applied method of synthesis of metal-carbon σ bonds is the reaction of a metal halide or oxide with an alkyl or aryl derivative of lithium, sodium, magnesium, or aluminum.

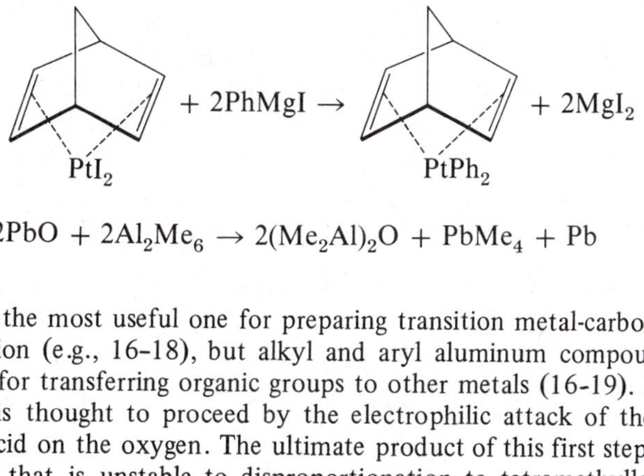

$$(16-18)$$

$$2PbO + 2Al_2Me_6 \rightarrow 2(Me_2Al)_2O + PbMe_4 + Pb \tag{16-19}$$

Of these methods, the most useful one for preparing transition metal-carbon σ bonds is the Grignard reaction (e.g., 16-18), but alkyl and aryl aluminum compounds are also excellent reagents for transferring organic groups to other metals (16-19). Reaction 16-19, for example, is thought to proceed by the electrophilic attack of the aluminum-containing Lewis acid on the oxygen. The ultimate product of this first step is dimethyllead, a compound that is unstable to disproportionation to tetramethyllead and lead metal (16-20).

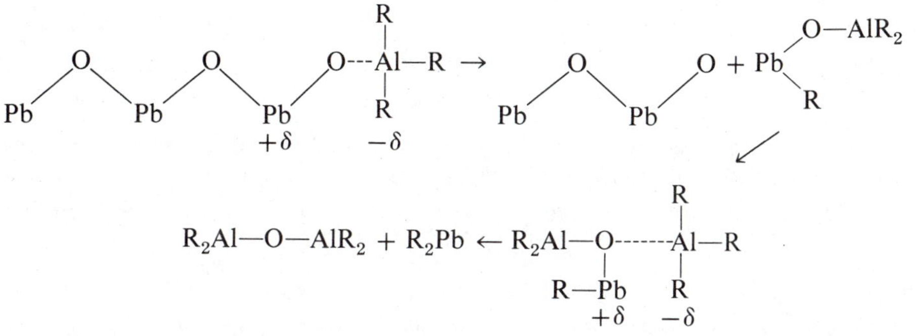

$$(16-19')$$

$$2R_2Pb \rightarrow Pb + PbR_4 \tag{16-20}$$

REACTION OF A METAL WITH A MERCURY ALKYL OR ARYL

Just as with the first two methods of metal-carbon bond synthesis, the efficacy of this method is due to very favorable thermodynamics. For example, ΔH_{rxn}° for the interaction of metallic aluminum with dimethylmercury is -477.0 kJ.

(16-21)
$$3\ HgMe_{2(liq)} + 2\ Al_{(s)} \rightarrow 3\ Hg_{(liq)} + Al_2Me_{6(liq)}$$

The $M + R_2Hg$ reaction is formally an oxidation-reduction reaction. Therefore, it should apply best to the more electropositive metals; indeed, only the elements of Groups IA, IIA, and IIIA (except boron) have been observed to participate, and experimentally it is found that all those metals that react directly with organic halides under relatively mild conditions will also react with organomercury compounds.†

One of the great advantages of the $M + R_2Hg$ reaction is that it can be run in non-ethereal solvents. Therefore, it can be used to prepare ether-free compounds of aluminum, gallium, or beryllium, compounds that are exceptionally good Lewis acids (see Chapter 5) and would bind ether in the product. In addition, it may be used for the preparation of organosodium compounds [recall that the direct reaction ($M + RI$) could not be used in this case because of radical coupling reactions] and organolithium compounds.

METALATION REACTIONS: METAL-HYDROGEN EXCHANGE

Reactions of this type have been studied for a number of years and are more generally used to prepare compounds with metal-carbon σ bonds than would appear at first glance.[16]

Lithiation reactions (reactions 16-22 and 16-23) are used to prepare many other alkyl and aryl lithium compounds, which are in turn synthetically useful intermediates. Such reactions are thought to involve nucleophilic attack of the hydrocarbon portion of the Li-containing reagent (*e.g.*, the butyl group) on a hydrogen atom of the compound undergoing metalation.[17] Thus, in order for metalation to occur, the H atom must be relatively acidic.

(16-22)

(16-23)

Intramolecular metalation reactions (*e.g.*, 16-24) involving transition metals have recently been discovered and hold great promise for providing numerous new, stable molecules containing metal-carbon σ bonds.[18] For example, palladium chloride reacts with azobenzene to give a compound having $Pd-C$ σ bonds.

---

†Alternatively, one would predict that the reaction will form any metal alkyl whose $M-C$ bond energy is greater than the $Hg-C$ bond energy; since mercury-carbon bonds are thermodynamically weak, it should be possible to transfer the mercury alkyl or aryl group to a great many other metals.

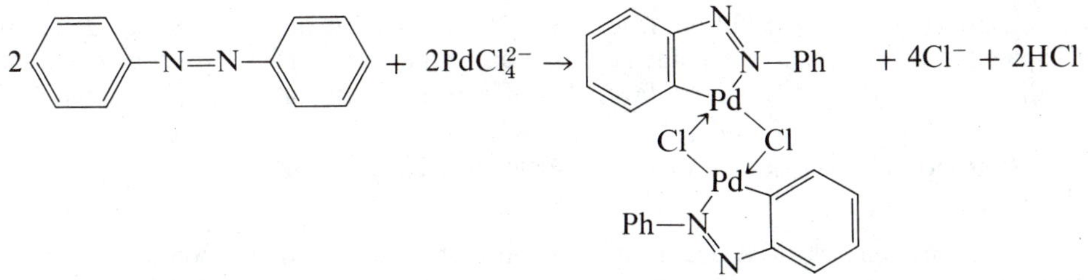

$$2 \text{ (PhN=NPh)} + 2PdCl_4^{2-} \rightarrow \text{[complex]} + 4Cl^- + 2HCl \qquad (16\text{-}24)$$

OXIDATIVE ADDITION REACTIONS†

Oxidative addition reactions have generally come to mean a class of reactions of transition metal complexes in which oxidation, *i.e.*, an increase in *formal* oxidation number of the central metal, is accompanied by an increase in the coordination number.[19-21] Such reactions are commonly observed for compounds having metals in low spin d^7, d^8, or d^{10} configurations. Examples of oxidative addition reactions that lead to the formation of metal-carbon σ bonds are outlined below. In each case, the number of d electrons of the central metal is given, as well as the number (in parentheses) of valence electrons surrounding that metal.

d^7 (17 valence shell electrons) $\rightarrow d^6$ (18 valence shell electrons)

$$ML_5 \xrightarrow[+L']{-1e^- \text{ from M}} ML_5L'$$

$$2 \, [Co(CN)_5]^{3-} + MeI \rightarrow [Co(CN)_5Me]^{3-} + [Co(CN)_5I]^{3-} \qquad (16\text{-}25)$$

d^8 (16 valence shell electrons) $\rightarrow d^6$ (18 valence shell electrons)

$$ML_4 \xrightarrow[+2L']{-2e^- \text{ from M}} ML_4(L')_2$$

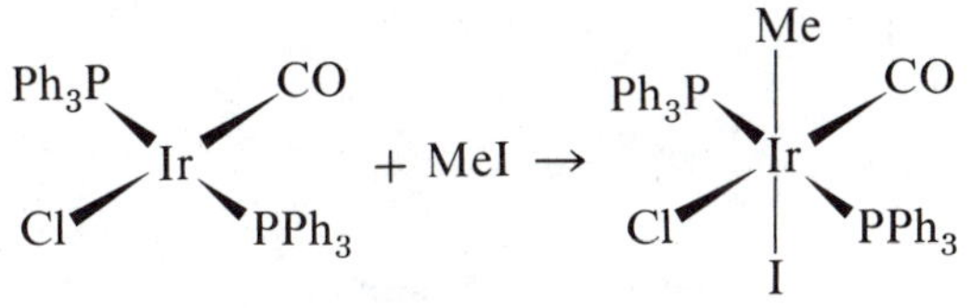

$$\text{[Ir complex]} + MeI \rightarrow \text{[Ir complex]} \qquad (16\text{-}26)$$

$$ML_5 \xrightarrow[-L + 2L']{-2e^- \text{ from M}} ML_4(L')_2$$

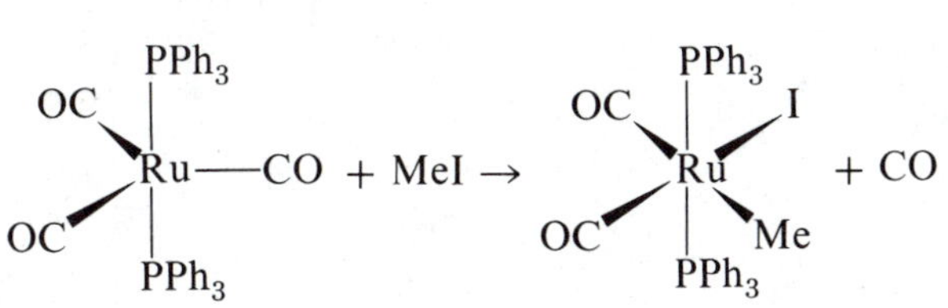

$$\text{[Ru—CO complex]} + MeI \rightarrow \text{[Ru complex]} + CO \qquad (16\text{-}27)$$

d^{10} (16 valence shell electrons) $\rightarrow d^8$ (16 valence shell electrons)

$$ML_3 \xrightarrow[-L + 2L']{-2e^- \text{ from M}} ML_2(L')_2$$

$$M(PEt_3)_4 \rightleftharpoons PEt_3 + M(PEt_3)_3$$

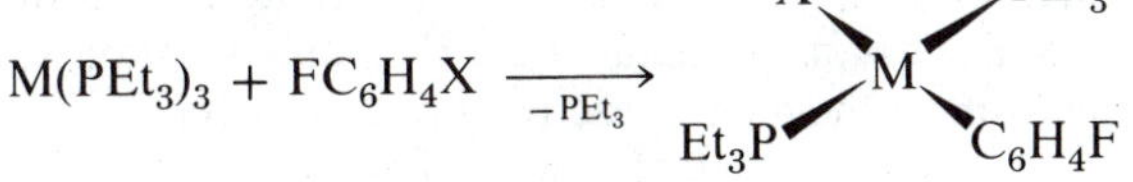

$$M(PEt_3)_3 + FC_6H_4X \xrightarrow[-PEt_3]{} \text{[M complex]} \qquad (M = Ni, Pd, Pt) \qquad (16\text{-}28)$$

---

†Oxidative addition reactions were first encountered in Chapters 5 and 7.

Although oxidative addition reactions will be discussed in much greater detail in Chapter 17, we can make the point here that an obvious driving force for such reactions is the attainment of the highly favored 18-electron configuration (see Chapter 15).

REACTIONS OF METAL-CONTAINING ANIONS WITH ORGANIC HALIDES

Reactions such as these are most commonly observed for anions containing Group IVA metals and those of the transition metal series, with those of metal carbonylates, or anions derived from metal carbonyls, being especially important.[22] Some reactions of the molybdenum-containing anion $[(\eta^5\text{-}C_5H_5)Mo(CO)_3]^-$ as a Lewis base are outlined in the next three equations. Each is a straightforward metathetical exchange of anions.

(16-29)

(16-30)

(16-31)

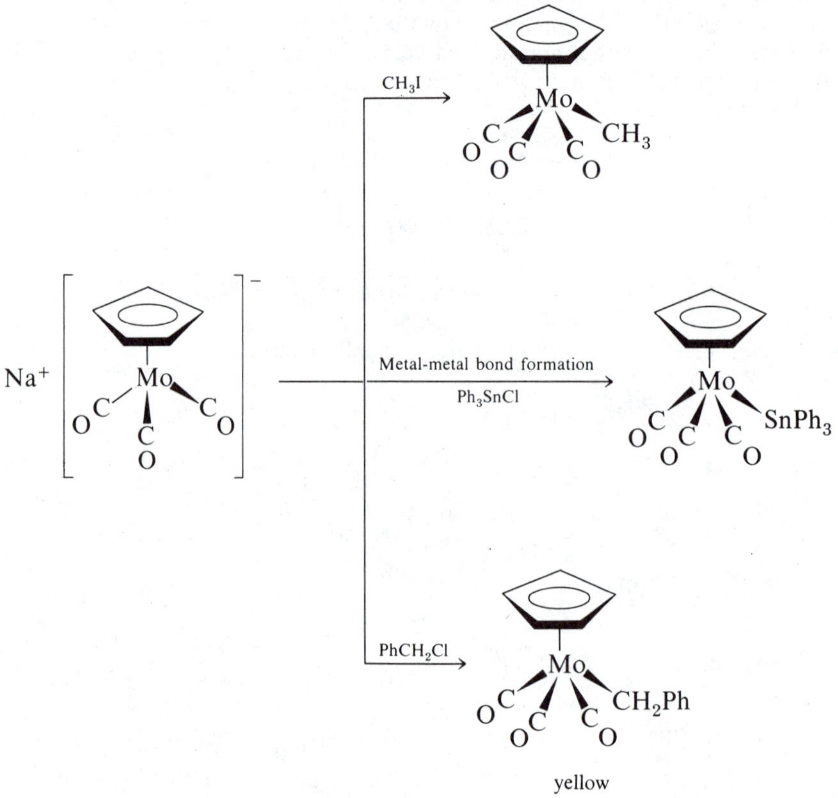

yellow

The Group IVA elements—Si, Ge, Sn, and Pb—also form strongly basic anions of the type R_3M^-. The organic groups R are generally aromatic rather than aliphatic, as the latter do not allow the formation of particularly stable anions, presumably because alkyl groups do not stabilize the negative charge as well as do aryl groups. These anions are synthesized by the cleavage of metal-metal bonds (16-32), or by the reaction of an alkali metal with a triorganometal halide (reactions 16-33 and 16-34). The alkali metal must be in a very reactive form. Therefore, the preparations are sometimes carried out in liquid ammonia (see page 156 for a discussion of metal-liquid ammonia solutions), or sodium and potassium may be in the form of the highly reactive liquid alloy NaK.

(16-32)
$$Ph_3Ge{-}GePh_3 + NaK \xrightarrow{Et_2O + THF} Ph_3GeK + Ph_3GeNa$$

$$Me_3SnBr + 2Na \xrightarrow{\text{liq. } NH_3} Me_3Sn\text{—}SnMe_3 + 2NaBr \qquad (16\text{-}33)$$

$$\xrightarrow{2Na} 2Me_3SnNa$$

$$Me_3SnCl + 2C_{10}H_8Na \rightarrow Me_3SnNa + NaCl + 2C_{10}H_8 \qquad (16\text{-}34)$$

A very convenient alternative is the use of the sodium-naphthalene ion pair, $Na^+C_{10}H_8^-$, or some other metal-arene ion pair.†

1,2-ADDITION OF METAL COMPLEXES TO UNSATURATED SUBSTRATES[23]

In this section we shall consider the general process

$$M\text{—}G + \diagup\hspace{-0.6em}C{=}C\hspace{-0.6em}\diagdown \rightarrow M\text{—}\overset{|}{C}\text{—}\overset{|}{C}\text{—}G \qquad (16\text{-}35)$$

where G is hydrogen, oxygen, a halogen, or an organic group.§

Hydrometalations. Reactions of this type are among the most widely studied in chemistry, which is why they appear in various contexts in Chapters 6, 7, and 8. Most important of these is the **hydroboration** reaction:[24] the addition of the elements of diborane to olefins and acetylenes.

$$2\ \overset{\overset{\displaystyle Me}{|}}{Me}C{=\!\!=}\overset{\overset{\displaystyle Me}{|}}{C}H + \tfrac{1}{2}B_2H_6 \rightarrow \left[\overset{\overset{\displaystyle Me}{|}}{Me}C\underset{\underset{\displaystyle H}{|}}{\overset{\overset{\displaystyle Me}{|}}{}}\overset{\overset{\displaystyle Me}{|}}{\underset{\underset{\displaystyle H}{|}}{C}}\text{—}BH \right]_2 \qquad (16\text{-}36)$$

disiamylborane

---

†Sodium dissolves readily in THF or 1,2-dimethoxyethane solutions of aromatic molecules containing two or more joined (biphenyl), fused (naphthalene or anthracene), or conjugated (1,4-diphenylbutadiene) rings. The sodium reduces the arene to the highly reactive radical anion and is

$$Na + C_{10}H_8 \xrightarrow[\text{1,2-dimethoxyethane}]{} \left[\begin{array}{c} Me \quad Me \\ O \quad\ \ O \\ \diagdown\!Na\!\diagup \\ O \quad\ \ O \\ Me \quad Me \end{array} \right]^+ \left[\begin{array}{c} \text{naphthalene} \end{array} \right]^{\overline{\cdot}}$$

itself taken into solution as the solvated cation.

§The 1,2-additions of metal-containing molecules to olefins and acetylenes are frequently called "insertion reactions" by organometallic chemists, since such reactions can be considered as the insertion of the unsaturated molecule into a metal-element bond. The use of this terminology emphasizes the role of the metal-containing compound and allows the inclusion of reactions such as (16-35) in a much larger class of reactions. In addition to the reactions discussed in this section, insertion reactions include those wherein the molecule being inserted may be CO_2, SO_2, CS_2, or organic compounds such as ketones, aldehydes, or isocyanates, among many others.

(16-37)

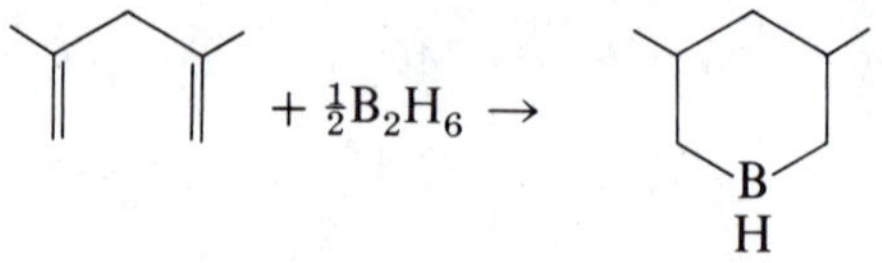

$$\text{(structure)} + \tfrac{1}{2}B_2H_6 \rightarrow \text{(structure)}$$

3,5-dimethylborinane

The boron-containing products of hydroborations need not be isolated; they can be converted *in situ* into a variety of organic derivatives. For example, oxidation of the intermediate alkylboranes with alkaline peroxide replaces the boron with an OH group with retention of configuration. Analysis of the oxidation products from reactions such as 16-38 through 16-40 indicates that the addition of the elements of diborane to olefins is *cis* and largely anti-Markownikoff; furthermore, attack is always at the less

(16-38)†

$$CH_3(CH_2)_3CH{=}CH_2 \xrightarrow[\text{2. [O]}]{\text{1. HB}} CH_3(CH_2)_3\underset{\underset{OH}{|}}{CH}{-}CH_3 + CH_3(CH_2)_3CH_2CH_2OH$$

6%   94%

(16-39)

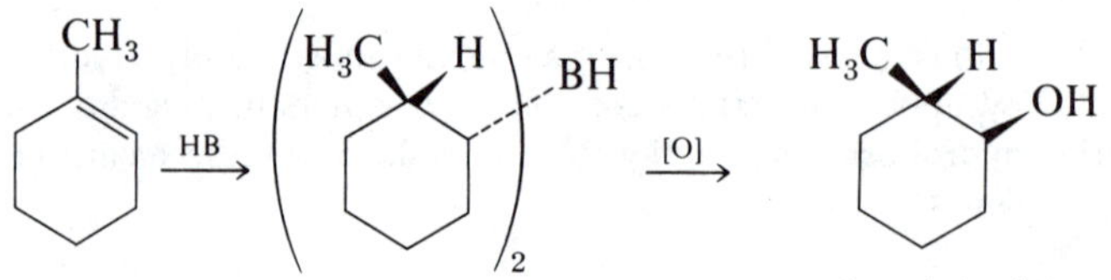

trans-2-methylcyclohexanol

(16-40)

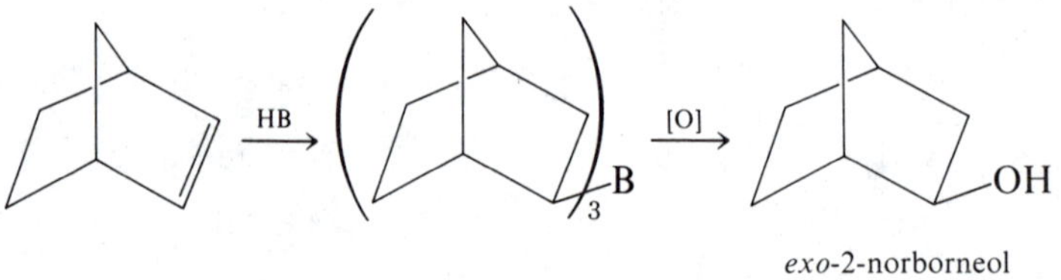

exo-2-norborneol

hindered face of a double bond (see reaction 16-40). Based on these results, and many others, the mechanism must involve a concerted 1,2-addition; the end of the double bond to which the boron adds is dictated by the carbon substituents.

(16-41)

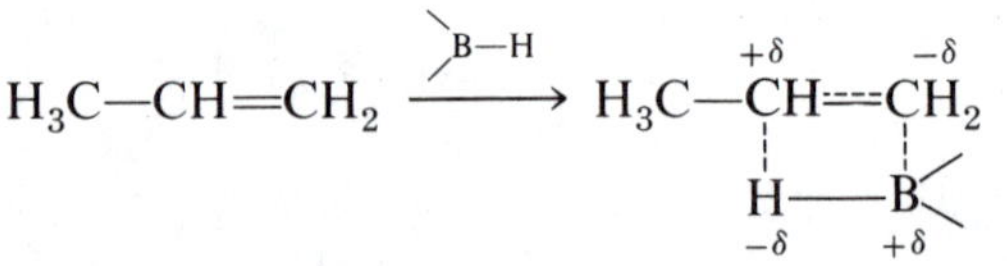

$$H_3C{-}CH{=}CH_2 \xrightarrow{\text{B}{-}\text{H}} H_3C{-}\overset{+\delta}{CH}{=\!=\!=}\overset{-\delta}{CH_2}$$
$$\underset{-\delta}{H}{-\!-\!-}\underset{+\delta}{B}{\big\langle}$$

Professor H. C. Brown, who developed the concept of hydroboration, was awarded the Nobel Prize in chemistry in 1979 for his work, the major thrust of which has been to devise methods whereby the intermediate alkylboranes can be converted easily and

---

†HB is a symbol for the addition of the elements B and H (from B_2H_6) to an olefin in an ethereal solution at temperatures of 0° to 25°. The symbol [O] designates *in situ* oxidation with H_2O_2 under basic conditions.

directly into organic compounds of all types;[24] such reactions were discussed briefly in Chapter 7 (p. 228).

Hydroaluminations (the addition of Al—H to unsaturated systems) are generally similar to hydroborations in that both are anti-Markownikoff and *cis*. However, there are some important differences.

$$(i\text{-Bu})_2\text{AlH} + \text{MeC}\!\equiv\!\text{CPh} \underset{100°}{\overset{50°}{\rightleftharpoons}} \begin{array}{c}\text{Me}\quad\quad\text{Ph}\\ \diagdown C\!=\!C \diagup \\ \text{H} \quad\quad \text{Al}(i\text{-Bu})_2\end{array} \qquad (16\text{-}42)$$

As is the case in reaction 16-42, hydroaluminations are most commonly carried out with diorganoaluminum hydrides such as $(i\text{-Bu})_2\text{AlH}$ or Et_2AlH. Such compounds, in fact, are an integral part of the industrial synthesis of trialkylaluminum compounds by the *direct synthesis* method developed by Ziegler.[25] In the first step of this process (16-43), activated aluminum and hydrogen react with preformed $R_3\text{Al}$ to form a dialkylaluminum hydride. This then reacts with an olefin in a hydroalumination reaction to give additional aluminum trialkyl.

$$\text{Al} + \frac{3}{2}\,\text{H}_2 + 2\,\text{Et}_3\text{Al} \rightleftharpoons 3\,\text{Et}_2\text{AlH} \qquad (16\text{-}43)$$

$$3\,\text{Et}_2\text{AlH} + 3\,\text{CH}_2\!=\!\text{CH}_2 \rightleftharpoons 3\,\text{Et}_3\text{Al} \qquad (16\text{-}44)$$

An important feature of the hydroalumination reaction is its ready reversibility. For example, heating $(i\text{-Bu})_3\text{Al}$ at 160-180° in a nitrogen atmosphere results in the smooth evolution of isobutene and formation of $(i\text{-Bu})_2\text{AlH}$. Reversibility is an important aspect of hydrometalations in general. Although dialkylboron hydrides cannot be isolated by the thermal disproportionation of trialkylboranes, the reaction must be reversible to some extent, as the olefin isomerization process (reaction 16-45) proves. On the other hand,

(16-45)

it is frequently difficult to isolate transition metal alkyls because of their very rapid decomposition to metal hydride and olefin.

The reason for this, as you shall see later in this chapter (p. 508) and in Chapter 17, is not that transition metal-carbon σ bonds are thermodynamically weak, but rather that one or more pathways for their decomposition are readily available.[26] One such pathway is the so-called β-**elimination** of olefin with concomitant formation of the metal hydride in a process entirely similar to the dehydroalumination reaction discussed above.

(16-46)

$$L_nM\text{---}CH_2CH_2R \rightleftharpoons \left[L_nM\text{-----}\begin{array}{c} H \quad CHR \\ \| \\ CH_2 \end{array} \right] \rightleftharpoons L_nMH + H_2C\text{=}CHR$$

Hydrometalations involving transition metal–hydrogen bonds are also important examples of this reaction type,[27-32] because the insertion of olefins into transition metal–hydrogen bonds is implicated in numerous industrially important (both actual and potential) processes that are catalyzed by transition metals. An example of such a process is the hydroformylation of olefins by $HCo(CO)_4$ (basically the addition of H_2 and CO to an olefin; *cf.* reaction 16-6). In general, important catalytic systems involve Co(I), Pt(II) (both d^8 systems), or Rh(III) (d^6) complexes. The subject of homogeneous catalysis by transition metal compounds is of such importance that Chapter 18 is devoted to the topic.

Organometalations. Reactions such as these involve the insertion of alkenes and alkynes into metal-carbon σ bonds. The most important of these are **carbalumination** reactions—the addition of the elements Al and C to unsaturated systems.

If triethylaluminum is heated to 90 to 120° with ethylene at a pressure of 100 atmospheres, ethylene is inserted into the Al−C bond at the rate of about one mole of ethylene added per mole of R_3Al in one hour. This so-called *growth reaction* (16–47), discovered by Karl Ziegler, is in competition with dehydroalumination, the β-elimination of $>$Al−H to form a 1-alkene (16–48).

(16-47)

$$Et_3Al + 3m\,C_2H_4 \rightarrow Al \begin{array}{c} \diagup(C_2H_4)_nC_2H_5 \\ \text{---}(C_2H_4)_pC_2H_5 \\ \diagdown(C_2H_4)_qC_2H_5 \end{array} \quad (n+p+q)/3 = m$$

(16-48)

$$R_2Al(C_2H_5) \rightarrow R_2AlH + C_2H_4$$

Even at relatively low temperatures, the number of addition reactions does not exceed eliminations by more than 100; *i.e.*, the maximum chain length that can be formed is only about C_{200}. In practice, the growth reaction is more suitable for the synthesis of straight chain aliphatic compounds between C_4 and about C_{30}.

(16-49)

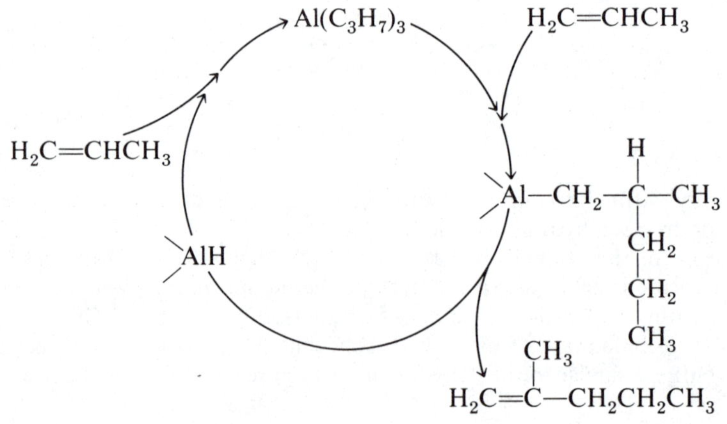

As we have already discussed in the hydrometalation section above, one of the most important features of organoaluminum chemistry is the reversibility of the hydroalumination reaction. In combination with carbalumination, several reactions of practical importance can be performed. For example, the net result of the reaction of propene with tri-*n*-propylaluminum is the catalytic dimerization of propene. The carbalumination step leads to a β-branched alkylaluminum compound that is unstable with respect to dehydroalumination. The product of this process, 2-methyl-1-pentene, is important, because it can be converted to the industrially useful material isoprene, $CH_2=CHC(CH_3)=CH_2$ (see bottom of p. 500).

STUDY QUESTIONS

1. Synthesis of metal alkyls:

   a) In order to carry out an infrared study of the structure of trimethylaluminum, suppose you wish to have the fully deuterated compound $Al_2(CD_3)_6$. Outline a synthesis of the ether-free material from appropriate starting materials. (The compound CD_3I can be purchased.)

   b) Synthesize Ph_4Sn from appropriate starting materials.

   c) Outline a synthesis of a catenated Group IVA compound such as Ph_6Ge_2 from appropriate materials.

   d) Synthesize Me_2Cd and Me_3Ga from appropriate starting materials.

2. A useful synthetic method for many organometallic compounds is to let a metal react directly with a dialkyl- or diarylmercury compound (see reaction 16–21). Can this method be applied to B? to Si?

3. Lithioferrocene is prepared as in reaction 16–23. How may this compound be used to prepare the two compounds illustrated below?

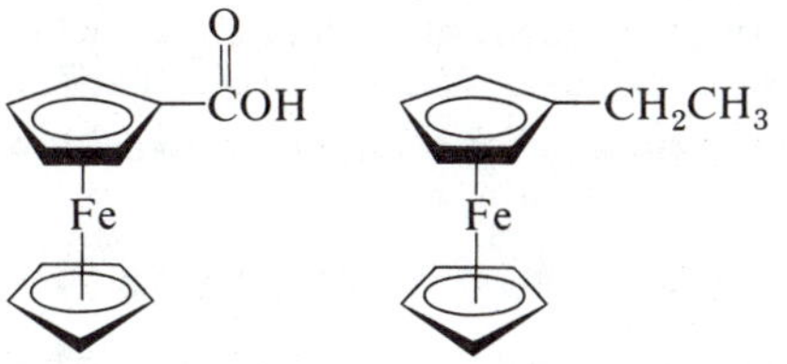

4. Group IIIA alkyls and aryls are considerably more air-sensitive than Group IVA alkyls and aryls. Suggest why this may be the case.

5. Use a hydrometalation reaction to prepare $HO-CH_2CH_2CH_2CH(CH_3)_2$ from $H_3C-CH=CH-CH(CH_3)_2$.

6. Karl Ziegler observed that the absorption rate of ethylene by dissolved triethylaluminum to form diethylbutylaluminum increases when more solvent is added. The rate of absorption of olefin is proportional to the square root of the total solution volume for a given amount of starting material. What does this observation suggest about the nature of the reacting species? Assume rate $\propto$ [reactive Al-containing species] [olefin].

7. Refer to the oxidative addition reactions on page 495. Suggest a method for the synthesis of the two compounds below.

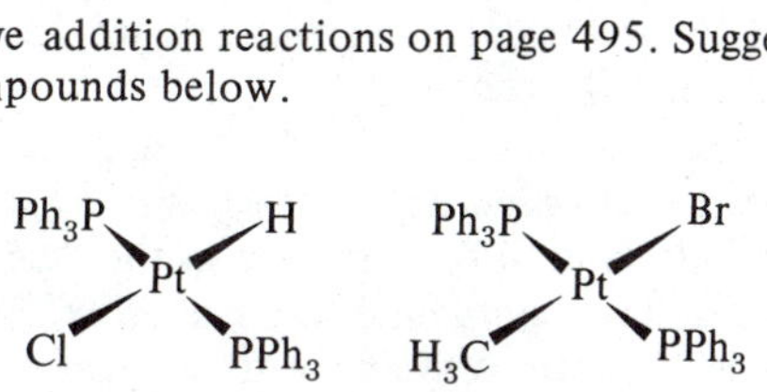

8. Osmium(0) and iridium(I) compounds most readily undergo oxidative addition reactions. However, reactivity falls off rapidly as one goes to the right in the Periodic Table; that is, Pt(II) and Au(III) compounds are much less susceptible to this type of reaction. Suggest why this may be the case.

9. Starting with Me_3SnX and other appropriate reagents, suggest two synthetic routes to the mixed alkyl compound Me_3SnCH_2Ph.

10. Will the lithium of BuLi substitute for H_a or H_b in the compound below?

11. Answer the questions below using the following thermochemical data (in kJ):

Element	ΔH_f° [MCl$_2$ (s)]	ΔH_f° [MMe$_2$ (g)]	Bond dissociation energy for the M–Me bond
Zn	−415.1	54.8	175.7
Cd	−391.6	109.6	139.3
Hg	−224.3	93.3	122.1

a) Calculate ΔH_{rxn}° for: $2\,M + 2\,CH_3Cl(g) \rightarrow M(CH_3)_2(g) + MCl_2(s)$ [ΔH_f° of $CH_3Cl(g) = -80.8$ kJ/mol].

b) Which factor most influences ΔH_{rxn}°, the heat of formation of the metal halide or that of the metal alkyl? A plot of ΔH_{rxn}° *vs.* metal atomic number is very informative in this regard.

c) Explain, using the thermochemical data above, why HgR_2 compounds could be quite useful in transferring R groups to other, more electropositive metals.

d) Using ΔH_{rxn}° as a criterion for reaction spontaneity, which of the reactions below might be synthetically useful?

$$2HgMe_2(g) + Pb(s) \rightarrow PbMe_4(liq)\ [\Delta H_f^\circ = 98.3\ kJ] + 2Hg$$

$$3ZnMe_2(g) + 2Al(s) \rightarrow Al_2Me_6\ [\Delta H_f^\circ = -301.2\ kJ] + 3Zn(s)$$

$$2HgMe_2(g) + Sn(s) \rightarrow SnMe_4\ (liq)\ [\Delta H_f^\circ = -52.3\ kJ] + 2Hg\ .$$

Structure and Bonding in Metal Alkyls and Aryls[33,34]

The majority of metal alkyls and aryls are constructed and bonded in entirely predictable ways; exceptions are the so-called "electron deficient" compounds of Groups IA, IIA, and IIIA and some transition metal derivatives.

Many of the simpler alkyls and aryls of Group IA, IIA, and IIIA metals are self-associated—they form dimers, trimers, and higher molecular weight species. Self-association of compounds through bridging atoms is a phenomenon that pervades inorganic chemistry (see Chapter 8 for non-metal examples), and we discuss here those aspects that apply specifically to organometallic chemistry.

Diborane, dimeric trimethylaluminum, and the dimeric aluminum halides can be considered the prototypes of self-associated species. We have already discussed the structure and three-center, two-electron bridge bonding in the "electron deficient" ion $B_2H_7^-$ in Chapter 5 (pp. 128–129). This compound was said to be "electron deficient," but not because there are insufficient electrons available to account for bonding; rather the term is used in the topological sense to indicate that there are not enough valence electron pairs to account for all of the nearest neighbor atom-atom connections on a one-pair-per-neighbor basis. The same approach used to account for bonding in $B_2H_7^-$ can be applied to "electron deficient" organometallic molecules.

As noted on page 129, the aluminum halides (when X = Cl, Br, I, but not F) have dimeric structures with conventional two-center, two-electron bridge bonds (6). **Organoaluminum chlorides**, **bromides**, and **iodides** are also dimeric, the halogen atoms continuing to be the preferred bridging elements (7).

6

7

Organoaluminum hydrides and **fluorides**, however, are unusual in that they form trimers or tetramers, the halide or H atom being the preferred bridging element again. Because H and F are small in relation to Al, steric congestion about aluminum is low and this facilitates formation of these highly associated species. Also, because of the nondirectional character of the hydrogen $1s$ orbital that is utilized in bridging, an Al—H—Al bridge can assume almost any angle and still result in good Al—H overlap. Therefore, it should not be surprising that diorganoaluminum hydrides (8) are found to be trimeric, with very obtuse Al—H—Al angles.

(R = Me)
Al—F = 180.8 pm
Al—C = 194.2 pm
Al—F—Al = 148°
F—Al—F = 94°

8

9

In the case of the fluorides, the halogen is also the bridging element, but the Al—F—Al angle is greater than 140° (9). It would appear that bridging fluoride characteristically adopts very obtuse angles.

Trialkyl- or triarylaluminum compounds are associated as well, but bridging is accomplished through an organic group. The majority of such compounds are dimeric, except in a few cases where steric hindrance prevents bridging (e.g., tri-isobutylaluminum). Trimethylaluminum is perhaps the best characterized organoaluminum compound. This compound, which has a melting point of 15.0° and a boiling point of 132°, is a dimer as a pure liquid, in the vapor phase, or in solution in non-basic solvents. Like all other organoaluminum compounds, it is extremely sensitive to both water and oxygen, and reacts readily with a variety of Lewis bases (see Chapter 5).

Since dimeric trimethylaluminum is the prototype of other dimeric organoaluminum compounds, its structure is shown again on p. 504. Its features are important to an understanding of bonding in other dimeric organoaluminum compounds and in other electron deficient organometallic main group compounds in general. The $C_{terminal}$—Al—$C_{terminal}$ angle implies that the aluminum atoms are sp^2 hybridized, and the short Al—Al distance suggests some degree of metal-metal bonding. This latter effect leads necessarily to the final important feature, the very acute Al—C_{bridge}—Al angle.

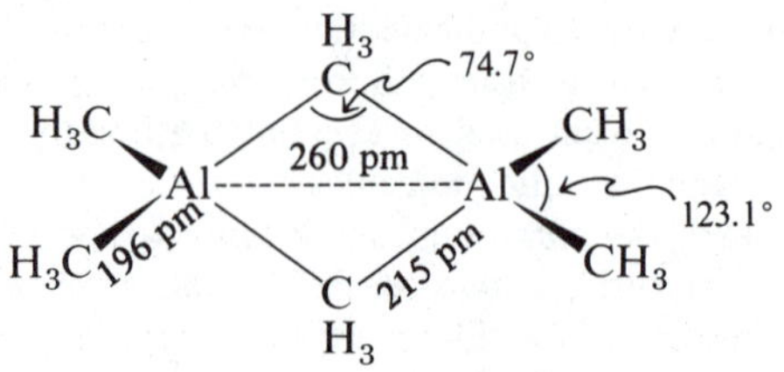

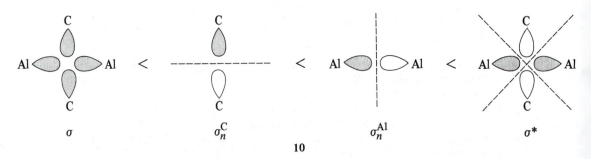

2'

The orbital structure for the four-atom bridge unit in Al_2Me_6 can be described by the four-orbital pattern in Figure 4-6, page 106. Using two Al sp^2 and two C sp^3 hybrid orbitals, we have, in increasing order of energy (only σ and σ_n^C are occupied):

10

To this pattern we add two Al $p\pi$ molecular orbitals:

11

Overlap of π_{Al} with σ_n^C stabilizes the latter and augments Al–C_{bridge} bonding. The two pairs of bridge electrons occupy the σ and the $(\sigma_n^C + \pi_{Al})$ molecular orbitals. Note the Al–Al bonding feature of σ when the Al–Al distance is short.

Molecular orbital calculations also show that **boron alkyls** may dimerize in the same manner if only the extent of B–B overlap is considered. However, if the boron atoms move close enough to form a B–B bond (the B–B bond length in B_2Cl_4 is 170 to 180 pm), serious steric interactions between bridge and terminal alkyl groups would destabilize the dimer.

None of the heavier metals of Group IIIA form associated organometallics. In order to achieve the appropriate bridge geometry, the metal atoms may have to move too close together and strong metal-metal inner shell repulsions would result.

Because they have approximately the same charge-to-radius ratio, Be^{2+} and Al^{3+} resemble each other chemically, and their compounds are frequently structurally similar. **Beryllium** has only two valence electrons, and so forms compounds of the type Me_2Be. The beryllium atom can satisfy its electron deficiency and coordinative unsaturation by self-association or by forming complexes with Lewis bases such as NMe_3. Thus, Me_2Be is a self-associated, polymeric substance in the solid phase, while other alkyl- and aryl-beryllium compounds are predominantly dimeric. The structure of $(Me_2Be)_x$ strongly suggests that bonding follows the Me_6Al_2 pattern. The Be-C_{bridge}-Be angle is very acute ($66°$) and the Be–C bond is significantly longer (193 pm) than the sum of the covalent radii (Be = 91 pm; C = 77 pm), thereby implying that the Be−C bridge bond is less than a full two-electron bond. Both of these features are shared with Me₆Al₂. However, the angles about the Be atoms are more nearly tetrahedral than was the case in Me₆Al₂.

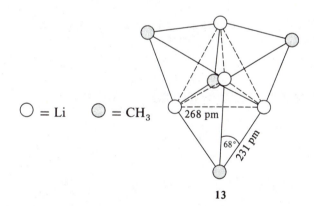

12

The other Group II metal alkyls such as Me₂Zn, Me₂Cd, and Me₂Hg are all monomeric. This lack of self-association can be explained by noting that, outside of the transition metals of Groups VI to VIII, metal-metal bond formation is a rare occurrence with heavier metals; even when it does occur (as it does for metals of Group IVA), the bond stability drops drastically on descending the Periodic Table. In part, the reason for this is that, as the heavier metals move within bonding distance, metal-metal inner shell repulsions become relatively more destabilizing, while the ao overlaps decrease from increasing radial diffuseness.

As you might have anticipated from the behavior of the organometallic derivatives of the lighter, main group metals, **alkyl-** and **aryllithium** compounds are also associated, although they form only tetramers, and hexamers in the solid or vapor phase or in nonbasic solvents.[35] Methyllithium, for example, is tetrameric in the solid and in ether. A crystal x-ray diffraction study suggests that the structure (13) is composed of a tetrahedron of lithium atoms with methyl groups located symmetrically in the faces of the tetrahedron. As in the other electron deficient molecules examined thus far, the Li-C-Li angle is very acute (68.30°), and the Li−C distances (231 pm) are larger than the sum of their respective covalent radii (Li = 134 pm; C = 77 pm).

$\bigcirc$ = Li $\bigcirc$ = CH₃   268 pm   68°   231 pm

13

Grignard reagents are one of the most important classes of organometallic compounds having M−C σ bonds.[36] Although they are commonly formulated as RMgX (*cf.* page 269), the nature of Grignard reagents in solution is considerably more complicated, and there are many unanswered questions surrounding their constitution.[37] All of the accumulated experimental evidence suggests that RMgX is the first species formed upon reaction of RX with Mg in a basic solvent.

$$yRX + yMg + xyS \rightleftharpoons (RMgXS_x)_y \qquad (16\text{-}50)$$

However, the extent to which RMgX is solvated, its degree of association (= observed molecular weight ÷ formula weight), and the possibility of disproportionation to R₂Mg and MgX₂ all depend on the nature of the solvent, the organic group, and X (see Chapter 5, page 152 ff). For example, ethylmagnesium bromide and chloride are highly associated in diethyl ether (Figure 16–3), whereas EtMgBr is a monomer in tetrahydrofuran (THF). The behavior of EtMgCl in THF is unique but important to our understanding of Grignard reagent structure; concentration of the EtMgCl/THF solution leads to essentially quantitative yields of Et_2Mg and the very interesting self-associated species $[EtMg_2Cl_3 \cdot 3THF]_2$ (**14**), which contains six- and five-coordinate Mg and three-coordinate Cl.

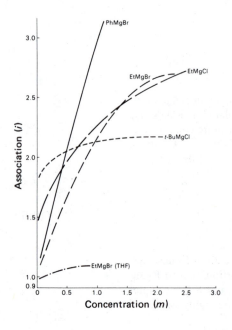

14

The existence of **14** implies that equilibria such as 16–51 and 16–52 must exist; reaction 16–51 is further substantiated by the isolation of MgI_2 when Grignard reagents are prepared from Mg and RI in THF.

(16-51)
$$2EtMgCl \rightleftharpoons Et-Mg\underset{Cl}{\overset{Et}{\diagdown\diagup}}Mg-Cl \rightleftharpoons Et_2Mg + MgCl_2$$

(16-52)
$$EtMgCl + MgCl_2 \rightleftharpoons Et-Mg\underset{Cl}{\overset{Cl}{\diagdown\diagup}}Mg-Cl$$

Figure 16–3. Degree of association of various organomagnesium halides in diethyl ether. [From E. C. Ashby, *Quart. Revs.*, **21**, 259 (1967).]

One of the most important transition metal organometallic compounds is the co-enzyme of **vitamin B$_{12}$ (15)**.[38] For our present purposes, the most interesting property of this molecule, and that of vitamin B$_{12}$ model complexes such as bis-(dimethylglyoxi-mato)methyl(pyridine)cobalt (often abbreviated methylpyridinecobaloxime) **(16)**,[39] is the striking stability of the Co—C σ bond. In neutral, aqueous solution, or as solids at room temperature, the coenzyme, other alkylcobalamins, and alkylcobaloximes are thermally and oxidatively stable.

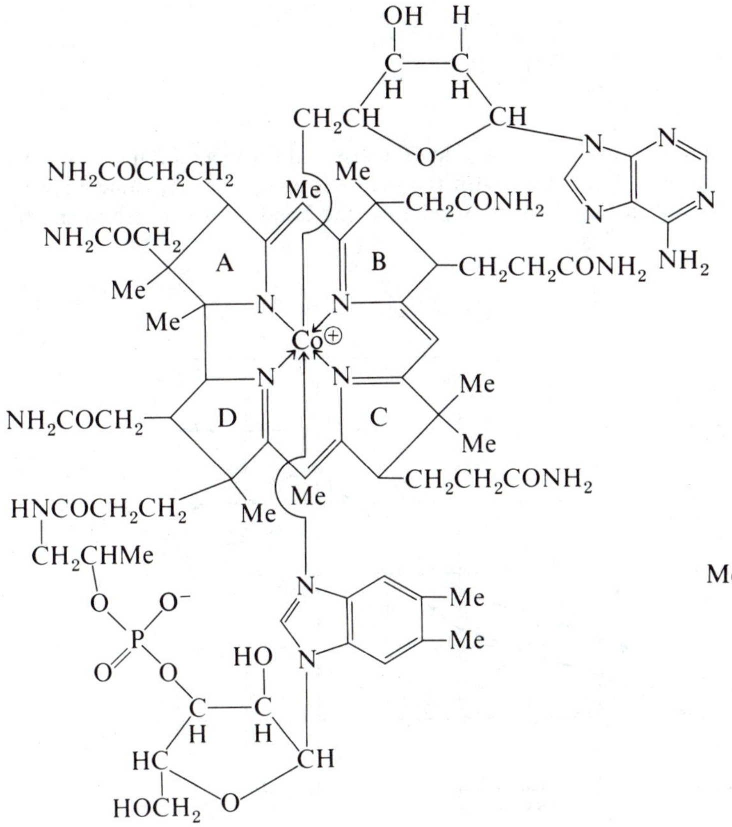

15

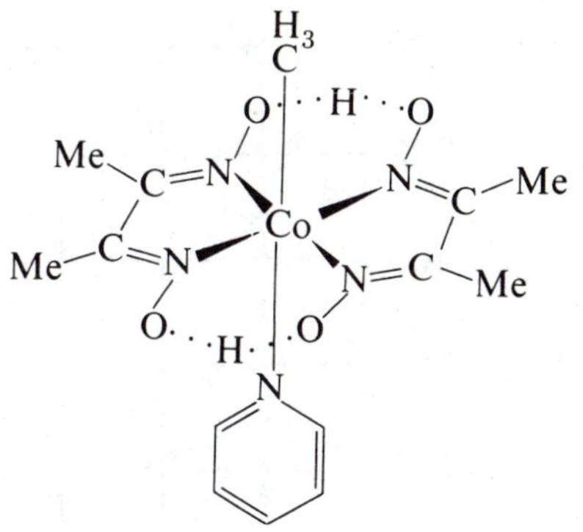

16

The very great stability of the Co—C bond in vitamin B$_{12}$ coenzyme and model systems is in decided contrast to numerous other **transition metal alkyl** compounds, as mentioned on page 499. Originally, it was thought that this instability was an intrinsic thermodynamic property, that the isolation of alkyls of the early transition metals would be especially difficult, and that ligands such as (η^5-C$_5$H$_5$), CO, and R$_3$P are required to achieve stability. However, Lappert and Wilkinson have recently isolated compounds such as Cr(CH$_2$CPhMe$_2$)$_4$ **(3)**, Ti(CH$_2$SiMe$_3$)$_4$, and Mo$_2$(CH$_2$SiMe$_3$)$_4$ **(17)**, which are quite stable to decomposition.†[33,40] Most importantly, the availability of such compounds has permitted the determination of average metal-carbon bond strengths, some of which are plotted in Figure 16-4 for Ti and Zr compounds of the type MX$_4$. When X is a σ-bonded alkyl group, the metal-carbon bond energies are about 160 to 335 kJ, a range similar to that observed for main group metal-carbon bonds. In addition, it should be noted that while the M—O, M—N, and M—Cl bond energies are greater than that of M—C, this is the

---

†Note that none of these compounds can suffer the β-elimination reaction described on page 499.

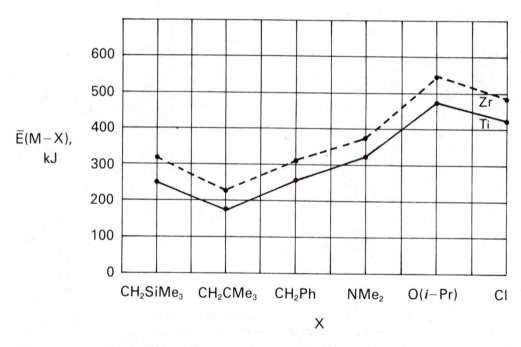

17[41]

$Mo_2(CH_2SiMe_3)_6$

same general trend (M—O > M—Cl > M—N > M—C) observed for main group metals such as Si, Ge, and Sn. However, in contrast with the main group metals, the bond between carbon and the heavier metal (Zr—C) is stronger than the bond between carbon and the lighter metal (Ti—C).

Figure 16-4. Average M—X bond energies in molecules of the type TiX_4 and ZrX_4. [After M. F. Lappert, D. A. Patil, and J. B. Pedley, *Chem. Commun.*, 830 (1975).]

**STUDY
QUESTIONS**

12. Knowing that you can purchase Me_3SiCH_2Cl, suggest a preparation for $Ti(CH_2SiMe_3)_4$ from chloromethyltrimethylsilane and other appropriate starting materials.

13. The structure of $Cr(CH_2CPhMe_2)_4$ is given on page 490). Does the observed structure agree with the prediction of the angular overlap model? What structure do you predict for $Ti(CH_2SiMe_3)_4$?

14. Dimethylberyllium is polymeric in the solid phase but monomeric in the gas phase. This behavior contrasts with that of trimethylaluminum, the latter compound being dimeric in both solid and gas phases. The reason for this contrasting behavior is that hyperconjugation in $BeMe_2$ (illustrated below) competes favorably with polymeriza-

$$H_3C-\overset{\ominus}{Be}=C\overset{\oplus}{=}H_3$$

tion, whereas the opposite is apparently true for $AlMe_3$. Considering the atomic orbitals involved in metal-carbon π bonding, why should Be so readily participate in hyperconjugation, but not Al?

Metal Carbonyls[42-46]

Transition metal carbonyls are frequent starting materials for building other compounds. Not only can the carbon monoxide be replaced by a wide variety of ligands, but any carbonyl groups that remain frequently stabilize the molecule with respect to oxidation and/or thermal decomposition. Furthermore, the CO ligands can be used as a probe of the electronic and molecular structure of a compound; such information can be obtained by examining the frequency and intensity of the C—O stretching modes in the infrared region. For these reasons and others, the study of the chemical and physical properties of metal carbonyls and their derivatives has been extensively pursued.

In this section we shall discuss the preparation and physical properties of simple binary carbonyls and their anions and cations. Subsequent sections of this chapter deal with derivatives of metal carbonyls.

THE SYNTHESIS OF METAL CARBONYLS

The known **neutral, binary metal carbonyls** are listed in Table 16-1. The shaded elements have not been reported to form simple, readily isolable binary carbonyls;[47] these metals form carbonyl complexes only when the metal is complexed with other ligands as well, or they form only anionic complexes such as $[Pt_6(CO)_{12}]^{2-}$. Almost all of the carbonyls listed in Table 16-1 are commercially available.

TABLE 16-1

THE KNOWN NEUTRAL BINARY METAL
CARBONYLS. SHADED ELEMENTS FORM
ONLY ANIONIC CARBONYL COMPLEXES OR
ONLY AFFORD M—CO BONDS WHEN THE
METAL IS COORDINATED TO OTHER LIGANDS.

III	IV	V	VI	VII	VIII			I	II
	Ti	$V(CO)_6$	$Cr(CO)_6$	$Mn_2(CO)_{10}$	$Fe(CO)_5$ $Fe_2(CO)_9$ $Fe_3(CO)_{12}$	$Co_2(CO)_8$ $Co_4(CO)_{12}$	$Ni(CO)_4$	Cu	
	Zr	Nb	$Mo(CO)_6$	$Tc_2(CO)_{10}$ $Tc_3(CO)_{12}$	$Ru(CO)_5$ $Ru_3(CO)_{12}$	$Rh_2(CO)_8$ $Rh_4(CO)_{12}$ $Rh_6(CO)_{16}$	Pd	Ag	
	Hf	Ta	$W(CO)_6$	$Re_2(CO)_{10}$	$Os(CO)_5$ $Os_3(CO)_{12}$	$Ir_2(CO)_8$ $Ir_4(CO)_{12}$	Pt	Au	

Two methods of preparation are used for the **simplest metal carbonyls**: (1) direct reaction of the metal with CO; and (2) reductive carbonylation. The only metals that can directly react with CO under mild conditions are iron and nickel (see Figure 16-2).†[47a] The remainder require the use of the second method, that is, reduction of a salt of the metal (using an active metal or H_2) in the presence of CO.[47b] For example,

(16-53)

$$VCl_3 + 4\ Na + 6\ CO(200\ atm) \xrightarrow[160°C]{diglyme} [Na(diglyme)_2][V(CO)_6] + 3\ NaCl$$

$$\downarrow \text{HCl-Et}_2O$$

$$V(CO)_6$$

or

(16-54)

$$2\ Co(H_2O)_4(OAc)_2 + 8\ (Ac)_2O + 8\ CO(160\ atm) + 2\ H_2(40\ atm) \rceil$$
$$Co_2(CO)_8 + 20\ HOAc \leftarrow \rule{1cm}{0.4pt}\rfloor$$

Methods for the preparation of **polynuclear carbonyls** depend on the particular metal. For example, photolysis of $Fe(CO)_5$ in glacial acetic acid gives an excellent yield of $Fe_2(CO)_9$.§ One way in which $Fe_3(CO)_{12}$ may be obtained is oxidation of a solution of the anion $[HFe(CO)_4]^-$, which is prepared in turn from the pentacarbonyl:

(16-55)

$$Fe(CO)_5 + 2\ OH^- \rightarrow [HFe(CO)_4]^- + HCO_3^-$$
$$3\ HFe(CO)_4^- + 3\ MnO_2 \rightarrow Fe_3(CO)_{12} + 3\ OH^- + 3\ MnO$$

Metal carbonyl anions and their derivatives are especially important as intermediates in organometallic chemistry, as you have already seen on page 496. However, they are also interesting in themselves. Not only are the simple metal carbonyl anions much more numerous than neutral carbonyls, but for some metals only anionic carbonyls (*e.g.*, $[Nb(CO)_6]^-$ and $[Ta(CO)_6]^-$) have been isolated. Methods for the preparation of metal carbonyl anions include:

1. Reaction of a metal carbonyl with a base such as an amine or OH^-.

(16-56)

$$Fe(CO)_5 \xrightarrow[Et_4NI]{NH_3/H_2O} (Et_4N)(HFe(CO)_4)$$
$$\xrightarrow{Et_3N/H_2O} (Et_3NH)(HFe_3(CO)_{11})$$

2. Reaction with an alkali metal.

(16-57)

$$2Na + Fe(CO)_5 \longrightarrow Na_2Fe(CO)_4$$

You might note that the $Fe(CO)_4^{2-}$ ion has found use as a synthetic reagent in organic chemistry,[48] an aspect of metal carbonyl chemistry we shall discuss in Chapter 17.

Metal carbonyl cations are not nearly so numerous as anions. The probable reason for this is that anions can disperse the excess negative charge over the entire molecule by back-bonding electron density from the metal to the CO ligand (see below; also Chapter 5, p. 133), an effect that strengthens the M—CO bond. On the other hand, removal of an

---

†Other metals also react directly with CO, but they require considerably more severe conditions than does Fe or Ni.

§See Chapter 8, page 266, for a description of the apparatus used in the experiment.

electron from a metal carbonyl to produce a cation leads to a decrease in M—CO $d\pi$-$p\pi^*$ bonding, a bond-weakening effect. Carbonyl cations can be produced in several ways, one of them being the disproportionation of the parent carbonyl:

$$Co_2(CO)_8 + 2\ PPh_3 \rightarrow [(Ph_3P)_2Co(CO)_3^+][Co(CO)_4^-] + CO \tag{16-58}$$

METAL CARBONYLS: PROPERTIES AND STRUCTURES

Some properties of the simplest carbonyl compounds, primarily of the first row elements, are collected in Table 16-2. You should note that, with the exception of the mononuclear carbonyls of the iron sub-group and $Ni(CO)_4$, all of the known metal carbonyls are solids; and that all of the mononuclear carbonyls are colorless, while the polynuclear compounds are colored, the color becoming darker as the number of metal atoms increases. For example, while $Fe(CO)_5$ is a colorless liquid, $Fe_2(CO)_9$ forms golden-yellow plates, and $Fe_3(CO)_{12}$ is a very dark green-black.

Investigation of the structures of metal carbonyls has occupied chemists for many years. The structure of $Ni(CO)_4$ was first established in 1935, but many others have only recently been defined unambiguously. The mononuclear carbonyls all have structures expected on the basis of their formulas; that is, O_h for $M(CO)_6$ where M = V, Cr, Mo, and W; D_{3h} for $M(CO)_5$ where M = Fe, Ru, and Os;† and T_d for $Ni(CO)_4$.

It is the polynuclear carbonyls that are of the greatest structural interest. You found in Chapter 15 that, in order to satisfy the EAN rule for a neutral metal carbonyl, metals with odd atomic numbers must form either paramagnetic compounds [e.g., $V(CO)_6$] or polynuclear compounds with metal-metal bonds [e.g., $Mn_2(CO)_{10}$], supplemented in some cases with bridging carbonyl groups [e.g., $Co_2(CO)_8$]. In the case of polynuclear carbonyls, the 18 electron rule suggests that two arrangements are equally likely for the carbonyl subunit $M_2(CO)_2$; the two metals, bonded to each other, may each have a terminal CO ligand, or the two metals may share two CO ligands as bridging groups. $Co_2(CO)_8$ adopts both of these configurations, the former (18a) in solution and the latter (18b) in the solid state. The reasons for the preference for one form or another are not

18a	18b
$Co_2(CO)_8$ in solution	$Co_2(CO)_8$ in the solid state

entirely clear. However, it would appear that the larger the metal radius, the less likely it is that CO bridging will occur. Therefore, owing to the decrease in metal radius on moving across the transition metal series, $Mn_2(CO)_{10}$ (19) does not have CO bridges, whereas $Co_2(CO)_8$ does, at least in the solid state. $Fe_2(CO)_9$ (20) can satisfy the 18 electron rule with two structures, one with three CO bridges and an Fe—Fe bond (the observed structure), or another with only one CO bridge and an Fe—Fe bond.

---

†Although a trigonal bipyramidal structure is observed for $M(CO)_5$, a square pyramid is in fact predicted on the basis of the angular overlap model (see Chapter 10). However, recall that if the difference in energy between two competing structures is less than $1e_\sigma$, as it is in this case, then VSEPR/steric forces can govern the final structure.

TABLE 16-2

PROPERTIES OF METAL CARBONYLS

Compound	Color	Melting Point (°C)	Structure	IR $\nu(CO)(cm^{-1})$	Comments
$V(CO)_6$	black-green	70(d)	O_h	1976^a	Paramagnetic by $1e^-$
$Cr(CO)_6$	white	130(d)	O_h	2000^b	Sublimes; Cr–C = 192 pm $\Delta_0 = 3.2\ \mu m^{-1\ c}$
$Mo(CO)_6$	white	—	O_h	2004^b	Sublimes; Mo–C = 206 pm $\Delta_0 = 3.22\ \mu m^{-1\ c}$
$W(CO)_6$	white	—	O_h	1998^b	Sublimes; W–C = 207 pm $\Delta_0 = 3.22\ \mu m^{-1\ c}$
$Mn_2(CO)_{10}$	golden yellow	154	D_{4d}	$2044(m)^d$ $2013(s)$ $1983(m)$	Mn–Mn = 293 pm $\Delta G_f^\circ = -1677$ kJ
$Tc_2(CO)_{10}$	white	160	D_{4d}	$2065(m)^d$ $2017(s)$ $1984(m)$	
$Re_2(CO)_{10}$	white	177	D_{4d}	$2070(m)^d$ $2014(s)$ $1976(m)$	
$Fe(CO)_5$	pale yellow	−20	D_{3h}	$2034(s)^e$ $2013(vs)$	b.p. 103°; Fe–C_{axial} = 181 pm Fe–C_{eq} = 183 pm $\Delta G_f^\circ = -746$ kJ
$Ru(CO)_5$	colorless	−22	D_{3h}	$2035(s)^f$ $1999(vs)$	Unstable with respect to light-catalyzed decomposition to $Ru_3(CO)_{12}$
$Os(CO)_5$	colorless	−15	D_{3h}	$2034(s)^f$ $1991(vs)$	Very unstable with respect to $Os_3(CO)_{12}$
$Fe_2(CO)_9$	golden yellow	d	D_{3h}	$2082(m)$ $2019(s)$ $1829(s)$ (bridging CO)	Fe–Fe = 246 pm
$Co_2(CO)_8$	orange-red	51(d)	C_{2v} (solid) D_{3d} (solution)	$C_{2v}{}^g$ 2112, 2071, 2059, 2044, 2031, 2001 } term. / D_{3d} 2107, 2069, 2042, 2031, 2023, 1991 / 1886, 1857 } bridge	Co–Co = 254 pm
$Ni(CO)_4$	colorless	−25	T_d	2057	b.p. 43°; Ni–C = 184 pm Very toxic due to ready decomp. to Ni + 4 CO

[a] In cyclohexane: W. Hicker, J. Peterhaus, and E. Winter, *Chem. Ber.*, **94**, 2572 (1961).

[b] Gas phase: L. H. Jones, *Spectrochim. Acta*, **19**, 329 (1963).

[c] N. A. Beach and H. B. Gray, *J. Amer. Chem. Soc.*, **90**, 5713 (1968).

[d] N. Flitcroft, D. K. Huggins, and H. D. Kaesz, *Inorg. Chem.*, **3**, 1123 (1964).

[e] Gas phase: L. H. Jones and R. S. McDowell, *Spectrochim. Acta*, **20**, 248 (1964).

[f] F. Calderazzo and F. L'Eplattenier, *Inorg. Chem.*, **6**, 1220 (1967).

[g] K. Noack, *Spectrochim. Acta*, **19**, 1925 (1963); *Helv. Chim. Acta*, **47**, 1555 (1964). This reference should be consulted for relative band intensities.

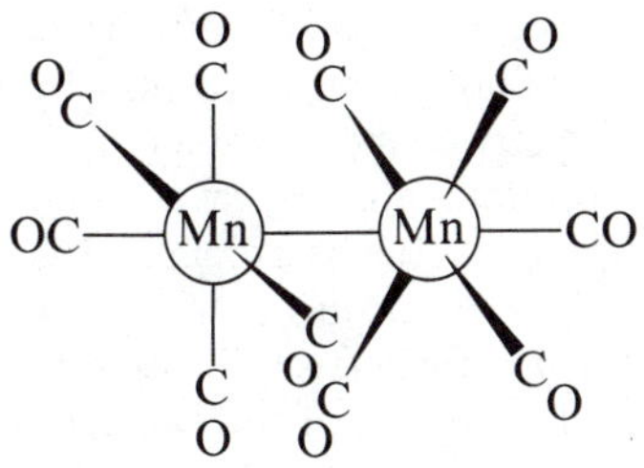

19

Mn₂(CO)₁₀

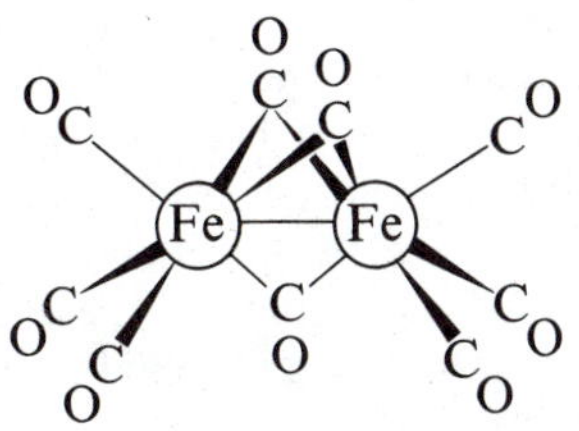

20

Fe₂(CO)₉

The more complex polynuclear carbonyls are metal clusters (*e.g.*, **21** and **22**). A comparison of **21** and **22** shows that the generalization that metals with larger radii do not allow CO bridging is true for the polynuclear clusters as well as for the bimetallic compounds.

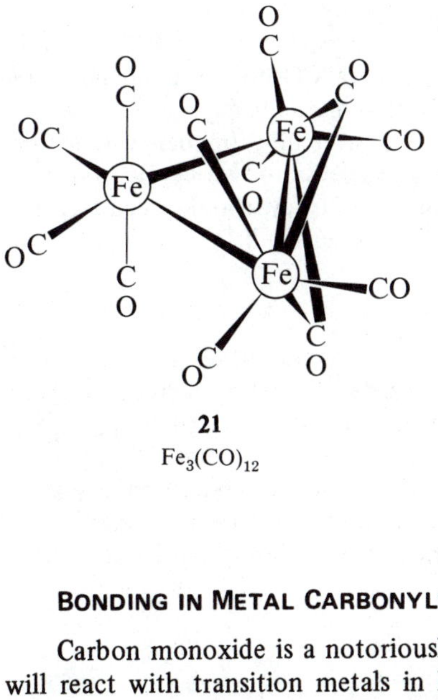

21

Fe₃(CO)₁₂

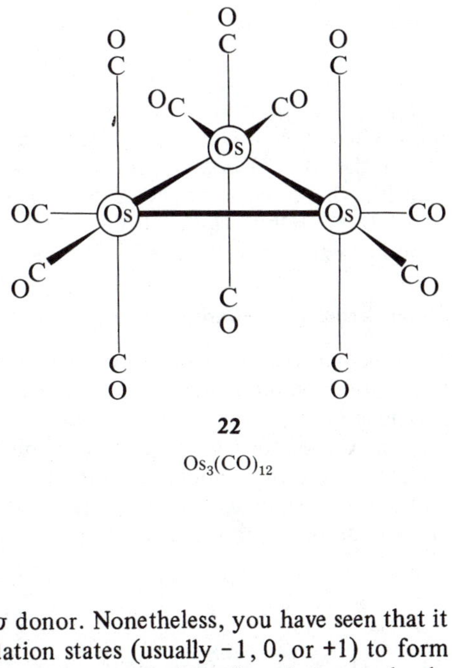

22

Os₃(CO)₁₂

BONDING IN METAL CARBONYLS

Carbon monoxide is a notoriously poor σ donor. Nonetheless, you have seen that it will react with transition metals in low oxidation states (usually −1, 0, or +1) to form complexes that are often quite stable with respect to oxidation, dissociation, and substitution. The feature of metal-CO bonding that is presumed to be responsible for this stability is the interaction of filled metal *d* orbitals of π symmetry with the empty π* or anti-bonding orbitals of the CO (Chapter 5, p. 133) (**23**).

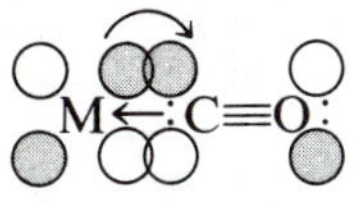

23

You have already seen in Chapter 9 (p. 315) that such an interaction causes an increase in Δ owing to a lowering of the energy of the *t* orbitals. In other terms, delocalization of electron density into ligand π* orbitals–a process called back-bonding–causes an increase in the metal effective nuclear charge, an increase that should eventually limit the extent of metal $t \to \pi^*$ electron flow. However, the increase in Z^* has the further effect of producing, by the simple expedient of increased attraction by the metal, a stronger ligand $\to$ metal σ bond than might otherwise be possible. As the amount of electron density delocalized from the CO σ orbital onto the metal increases, though, this diminishes Z^* and further metal $t \to \pi^*$ electron flow is possible. The result of this two-way electron flow is that the metal-ligand bond is stronger than the sum of isolated ligand-to-metal σ bonding and metal-to-ligand π bonding effects. This mutual strengthening of two effects to produce a result greater than the sum of the two acting individually is called *synergism*.

Some evidence for the back-bonding model of metal-CO bonding is the reduction in the stretching frequency of the C—O bond upon coordination with a metal, the variation in stretching frequency with metal oxidation state. The stretching frequency of free CO is 2140 cm^{-1}, whereas it is only 2000 cm^{-1} in Cr(CO)$_6$. The explanation for this is that back-bonding populates the CO π^* orbital, thereby resulting in C—O bonding weakening. Alternatively, you can explain this effect in terms of the canonical forms for M—CO bonding (**24**).

$$\overset{-}{\ddot{M}}-C\equiv\overset{+}{O}:\leftrightarrow M=C=\ddot{O}$$

24

It may be seen from these structures that increased metal-CO double bond character implies decreased C-O π bond order and a lower CO stretching frequency.

If the M—CO bonding model is correct, the C—O stretching frequency (and presumably the force constant) should increase in the isoelectronic series [V(CO)$_6$]$^-$ < Cr(CO)$_6$ < [Mn(CO)$_6$]$^+$, while the M—C stretching frequency should decrease. The experimental results in Table 16-3 support the model extremely well.[49]

Metal-Carbene and -Carbyne Complexes[50-54]

The synthesis, structure, and chemistry of metal-carbene complexes are of considerable current interest, owing to the possibility of their intermediacy in olefin metathesis reactions (a topic to be discussed in Chapter 18). Molecular orbital calculations suggest that the C atom of CO becomes more positive upon formation of a metal carbonyl complex, a feature that should make the CO carbon more susceptible to nucleophilic attack. Such attack has indeed been observed, especially with carbanions from lithium alkyls, and the ultimate product is a metal-carbene complex. Reaction 16-59, which represents

(16-59)

25   **26**

	Stretching Frequencies (cm^{-1})		
	[V(CO)$_6$]$^-$	Cr(CO)$_6$	[Mn(CO)$_6$]$^+$
M—C	460	441	416
C—O	1859	1981	2101

TABLE 16-3

INFRARED RESULTS FOR THE ISOELECTRONIC SERIES [V(CO)$_6$]$^-$, Cr(CO)$_6$, [Mn(CO)$_6$]$^+$

one of the earliest routes to metal-carbenes, has been extended to a number of other metals, and other synthetic routes now exist as well. Well over 300 metal-carbene complexes are known, and most of the later transition metals are represented. Metal electron configurations range from d^3 to d^{10}, with d^5 and d^9 as yet unrepresented. Metal oxidation states range from 4+ to 0 and coordination numbers from 2 to 7.

Previously it was thought that one of the groups attached to the carbene carbon should be capable of π bonding with that carbon. Recently, carbenes without such substituents have also been found to act as ligands, one of the most important examples being CH_2 in **27**.[54] Notice that the CH_2 plane is perpendicular to the C_{methyl}-Ta-$C_{carbene}$ plane and that the Ta-CH_2 bond is considerably shorter than a "normal" Ta-C bond (*i.e.*, Ta-CH_3). These structural features, and the fact that the CH_2 group does not rotate about the Ta-CH_2 axis, have been taken to indicate that there is nearly a full double bond between Ta and C.

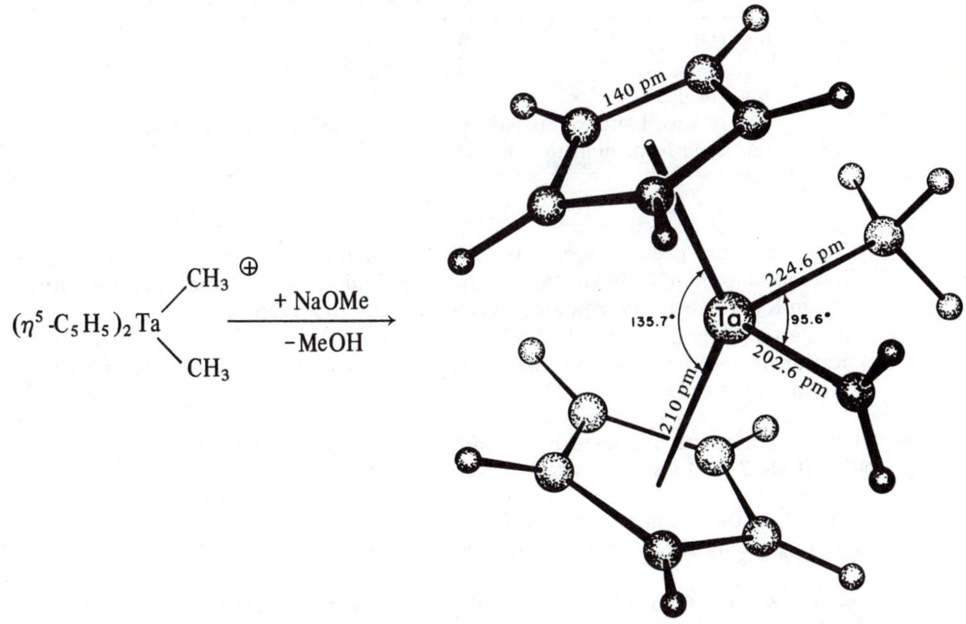

$$(\eta^5\text{-}C_5H_5)_2 Ta \overset{CH_3}{\underset{CH_3}{\big\langle}}{}^{\oplus} \xrightarrow[-MeOH]{+\,NaOMe}$$

(16-60)

27

[From R. R. Schrock, *Acc. Chem. Res.*, **12**, 98 (1979).]

Among the reactions of carbene complexes is that with boron trihalide, a reaction that has led to a most startling development: the synthesis of metal-carbyne complexes such as **28**. In this case, the metal-ligand bond is quite short (just as was the case with metal-carbenes), and the M-C-R angle is usually about 180°. Both features suggest

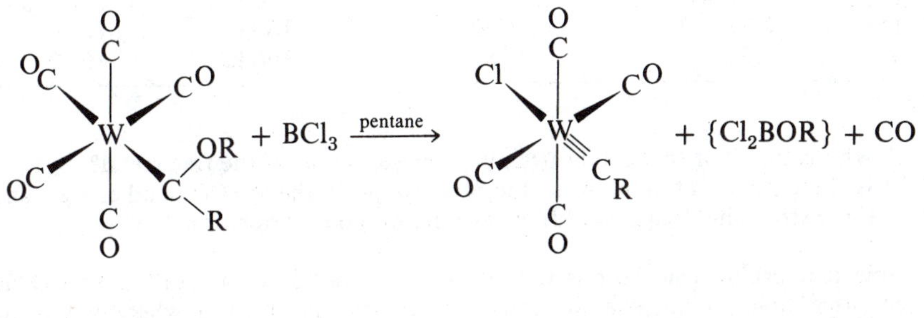

(16-61)

28

electronic structure **29**, a structure that may be derived by assuming that an *sp* hybridized carbyne carbon can donate a pair of electrons to the metal. The metal may in turn back-donate electrons from two orthogonal π-type orbitals (*e.g.*, d_{xz} and d_{yz}).

$$XL_4M\equiv C-R$$

29

STUDY QUESTIONS

15. Given below are three metal carbonyl derivatives, and $C-O$ stretching frequencies for each:

Compound	$C-O$ Stretching Frequency (cm^{-1})
$F_3Si-Co(CO)_4$	2128, 2073, 2049
$Cl_3Si-Co(CO)_4$	2125, 2071, 2049
$Me_3Si-Co(CO)_4$	2100, 2041, 2009

(the relative band intensities are always weak, medium, and strong on going from high to low energy)

a) There are two possible structures for the compounds above, both based on the trigonal bipyramid. Draw these structures and, assuming that the substituent group R_3Si is just a sphere, assign each structure to a point group.
b) Suggest a method for the synthesis of $Me_3Si-Co(CO)_4$.
c) Rationalize the changes in the $C-O$ stretching frequencies as the substituent group is changed.

16. What structures are predicted by the angular overlap model for $Fe(CO)_5$, $Co(CO)_4^-$, and $Ni(CO)_4$? Do these predictions agree with experiment?

17. Verify that the metal atoms in $Os_3(CO)_{12}$, $Fe_3(CO)_{12}$, and the two forms of $Co_2(CO)_8$ satisfy the "effective atomic number" rule.

18. You are given the infrared spectra of two different metal carbonyls of the type $LMn(CO)_5$. Both spectra exhibit three $C-O$ stretching frequencies. The bands in spectrum *A* appear at 2134, 2043, and 2019 cm^{-1} and the bands in spectrum *B* appear at 2111, 2012, and 1990 cm^{-1}. Which spectrum corresponds to $(CH_3)Mn(CO)_5$ and which to $(CF_3)Mn(CO)_5$? Explain your choice.

19. The data given below pertain to the hexacarbonyls of Group VIB metals.

	Ionization Potential of the Gaseous Metal	Ionization Potential of Gaseous $M(CO)_6$	Bond Energy per M$-$CO Bond	$C-O$ Stretching Frequency
Cr	6.76 eV	8.15 eV	110 kJ	2000 cm^{-1}
Mo	7.38	8.23	150 kJ	1984
W	7.98	8.56	180 kJ	1960

a) Why is the IP of the metal carbonyl so close to that of the free metal?
b) Is there a correlation between the $M(CO)_{6(g)}$ IP, the M$-$CO bond energy, and the $C-O$ stretching frequency? If so, rationalize your correlation.

20. Molecular orbital calculations indicate that the carbon of a coordinated CO should be susceptible to nucleophilic attack. Suggest a mechanism whereby OH^- could attack a metal carbonyl to result in a metal hydride (*e.g.*, reaction 16-55).

21. A metal carbene complex can be generated in the following reaction:

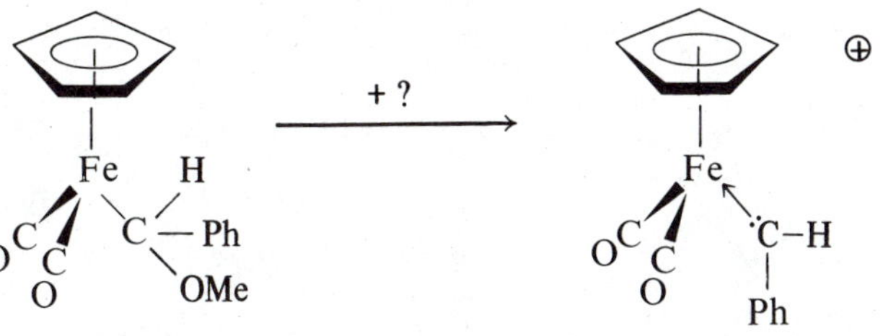

What reagent is needed to effect this transformation?

CARBON π DONORS

In this, the second main portion of the chapter, we shall consider compounds whose existence clearly depends on the formation of complexes between transition metals in low valence states and ligands that can act as π donors. Many complexes of this type are implicated in metal-catalyzed processes such as hydrogenation, polymerization, hydroformylation, and cyclization. Thus, metal π donor complexes constitute an important class of organometallic compounds that deserve intensive study.

Hydrocarbon π donors were discussed briefly in Chapter 15, and examples were listed in Table 15–1. Recall that there can be complexes formed by olefins, which are thought of as donating an even number of π electrons to a metal or metals (**30** and **31**),[55-57] and acetylenes, which may donate one pair of π electrons (**32**) or both pairs (**33**).[58] The allyl anion may be π bonded, and effectively donates four electrons in doing so (**34**).[59] In addition, there are truly aromatic π donors such as the *pentahapto*cyclopen-

tadienyl anion [that is, a $C_5H_5^-$ ligand in which all five carbon atoms are involved in bonding to the metal (**35**)], benzene and its derivatives (**36**),[60,61] and the η^7-$C_7H_7^+$ ion (tropylium or *heptahapto*cycloheptatrienyl ion) (**37**);[62] all are capable of donating six π electrons. (For nomenclature of these compounds, see p. 478.)

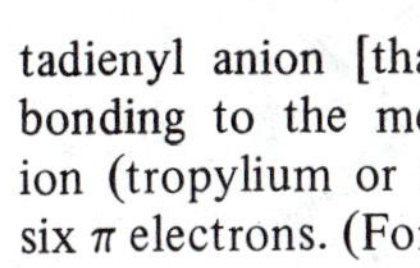

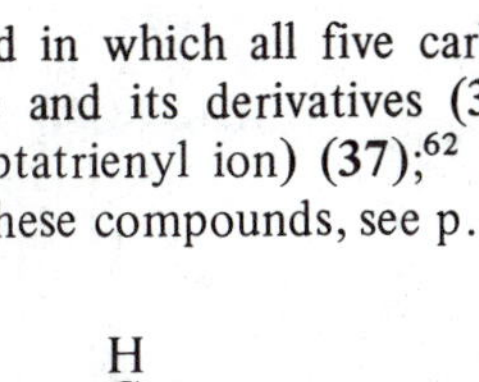

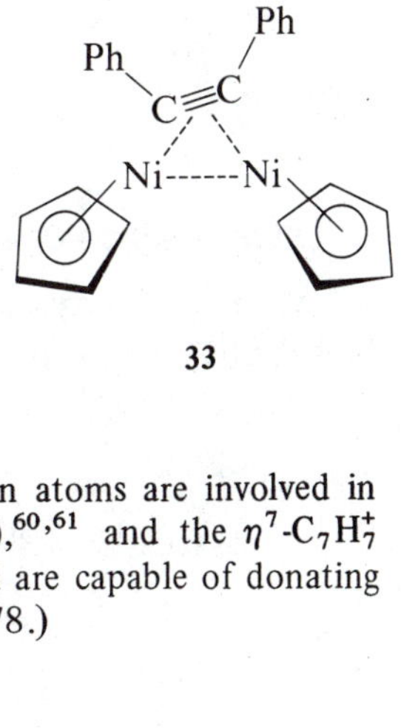

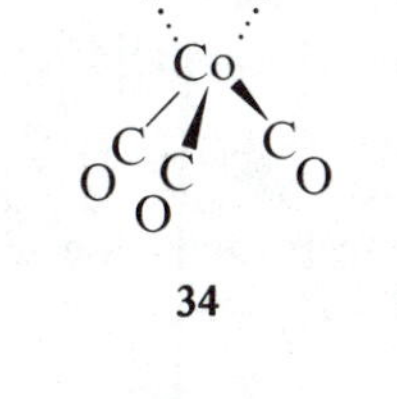

34

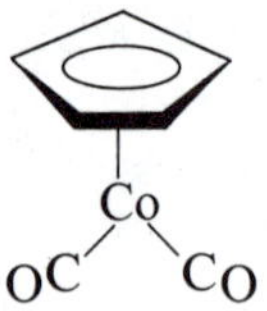

35

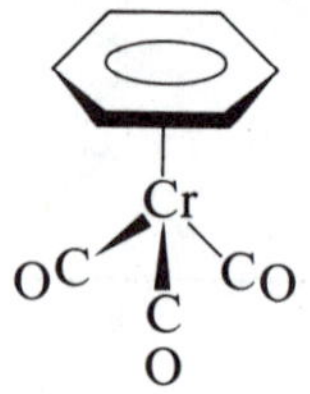

36

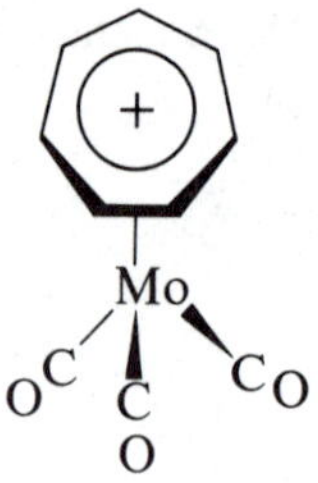

37

Chain π Donor Ligands (Olefins, Acetylenes, and π-Allyl)

Except for a few scattered reports, the vast majority of metal complexes with π donor ligands involve low valent transition metals, especially those of Groups VIB through VIII. The reason for the necessity of a low valent metal (*i.e.*, oxidation states −1 to +2) becomes clear on examining the currently accepted model for bonding between metals and olefins. This model is the Dewar-Chatt-Duncanson (or DCD) model.[63] Its essence is that the ligand donates π electron density to a metal orbital of σ symmetry directed to the center of the ligand π system (**38a**), and the metal in turn back-bonds electron density into a ligand π* orbital (**38b**). The result is a synergism that, as in the case of the metal carbonyls, leads to relatively strong bonding. Refinements to this simple model are described below. However, even in this simple form it is clear that such bonding can occur effectively only with a low valent (low Z^*) metal with populated π symmetry orbitals (*i.e.*, a metal late in the transition series).

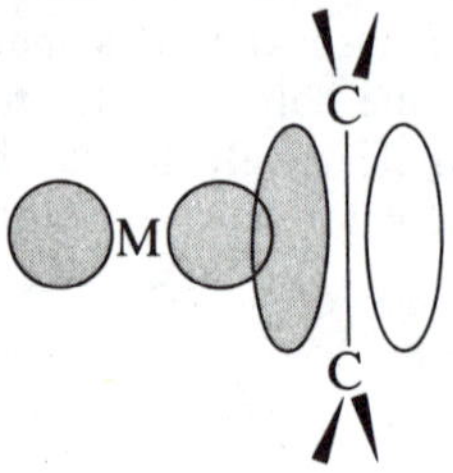

$$M \leftarrow L\pi \qquad\qquad M \rightarrow L\pi^*$$

38a   **38b**

SYNTHESIS OF OLEFIN, ACETYLENE, AND π-ALLYL COMPLEXES

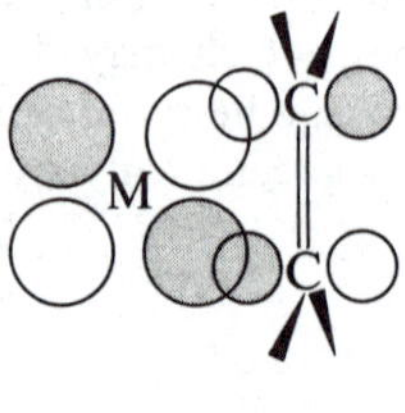

Olefin and acetylene complexes may be prepared by the direct addition of the ligand to a 16-electron metal-containing substrate (reaction 16–62),

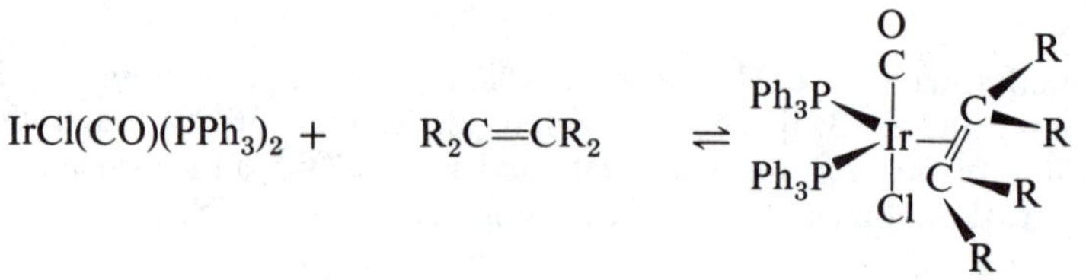

$$IrCl(CO)(PPh_3)_2 + R_2C{=}CR_2 \rightleftharpoons$$

(16–62)

Olefin	Relative Equilibrium Constant
$H_2C{=}CH_2$	1
$(NC)HC{=}CH(CN)$	1,500
$(NC)_2C{=}C(CN)_2$	140,000

but more generally their preparation involves the displacement of some weakly bound ligand (reactions 16–63 and 16–64).

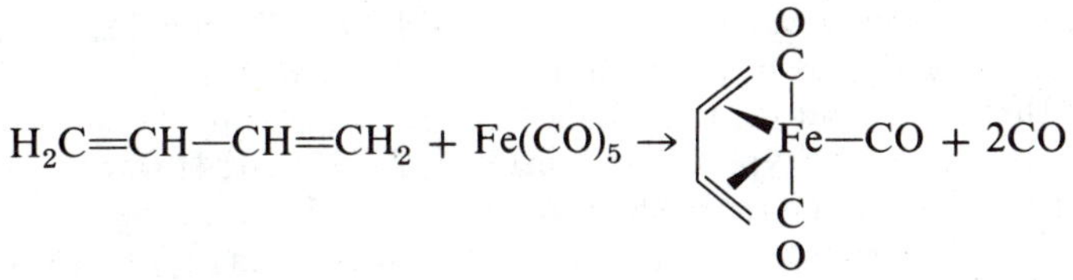

$$H_2C=CH-CH=CH_2 + Fe(CO)_5 \rightarrow \text{[complex]} Fe-CO + 2CO \qquad (16\text{-}63)$$

$$[(PhC\equiv N)_2PdCl_2] + \text{[norbornadiene]} \rightarrow \text{[Pd complex]} + 2PhC\equiv N \qquad (16\text{-}64)$$

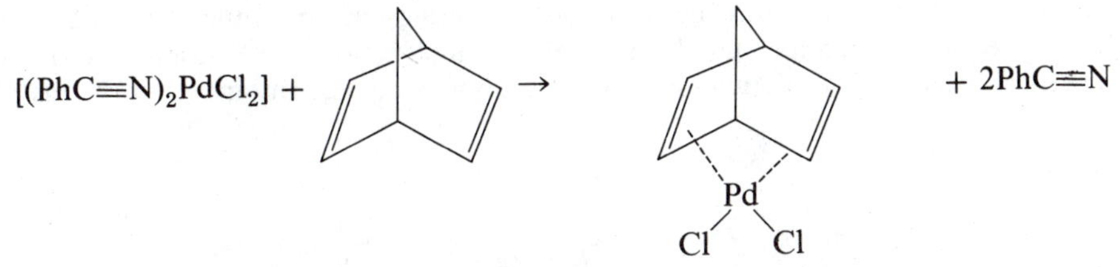

$$\Delta H_{rxn} = -55.6 \text{ kJ/mole}$$

You should be aware that in the case of acetylene complexes, ligand coupling is often observed as a subsequent reaction. For example, at least eight compounds have been isolated from the reaction of diphenylacetylene with $Fe(CO)_5$; three are shown below.

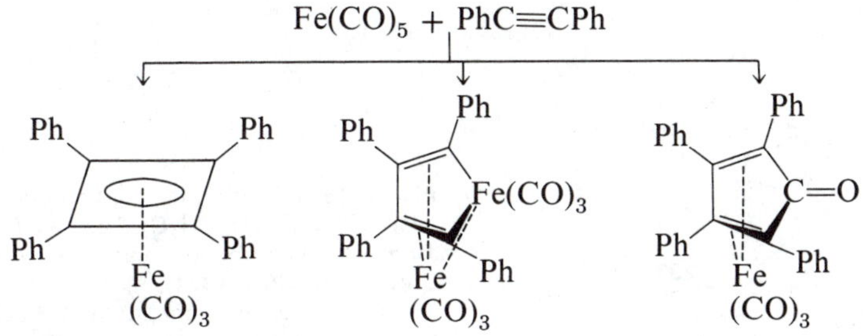

$$Fe(CO)_5 + PhC\equiv CPh \qquad (16\text{-}65)$$

Although a number of π-allyl complexes are known, their preparation invariably proceeds through a σ-bonded compound, even though the latter may only be a transient intermediate. For example, metal carbonyl anions can react with allyl chloride to give a σ-bonded allyl derivative (**39**), and heat or light then causes expulsion of a CO ligand and formation of a π-allyl or η^3-allyl complex (**40**).

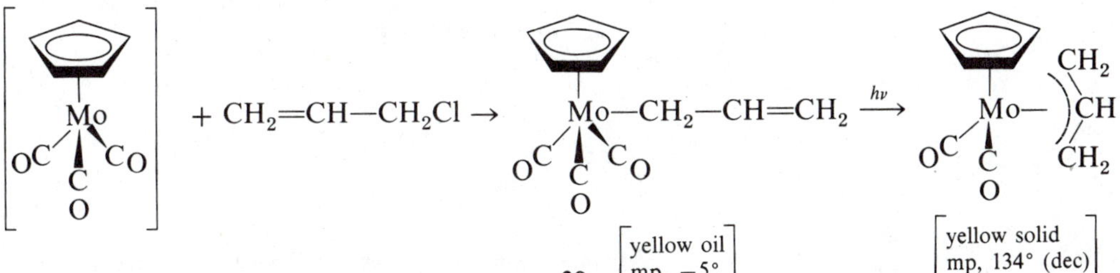

$$\text{[Mo anion]} + CH_2=CH-CH_2Cl \rightarrow Mo-CH_2-CH=CH_2 \xrightarrow{h\nu} \text{[Mo } \pi\text{-allyl]} \qquad (16\text{-}66)$$

$$\begin{bmatrix} \text{yellow oil} \\ \text{mp, } -5° \end{bmatrix} \qquad \mathbf{39}$$

$$\begin{bmatrix} \text{yellow solid} \\ \text{mp, } 134° \text{ (dec)} \end{bmatrix} \qquad \mathbf{40}$$

STRUCTURE AND BONDING IN OLEFIN, ACETYLENE, AND π-ALLYL COMPLEXES[64]

The molecular structures of numerous mono-olefin and -acetylene complexes have been determined by x-ray techniques. Representative of these are **Zeise's salt**, $K[PtCl_3(C_2H_4)]$ (Figure 16-5), and $(PPh_3)_2Pt(PhC\equiv CPh)$ (Figure 16-6). The important features to notice in these structures are: (i) The length of the coordinated olefinic or acetylenic C—C bond is greater than the C—C bond in the free hydrocarbon. The C=C bond in ethylene has gone from 134 pm in free C_2H_4 to 137 pm in Zeise's salt, whereas the $C\equiv C$ bond in Ph_2C_2 has increased from about 120 pm to 132 pm in the platinum complex. (ii) The $C=C$ bond (or $C\equiv C$ bond) is perpendicular to the molecular plane in 16-electron molecules of the type L_3M(olefin) (*e.g.*, Zeise's salt; **41a**), whereas it is in the molecular plane in 16-electron molecules such as L_2M(olefin) (*e.g.*, the Pt-acetylene complex in Figure 16–6; **41b**) and in 18-electron molecules of the type L_4M(olefin) (*e.g.*, $Ir(PPh_3)_2Br(CO)[(NC)_2C=C(CN)_2]$; **41c**). (iii) Upon complexation, the originally planar olefin becomes non-planar, with the substituents bending away from the metal and other ligands; acetylenes become non-linear upon complexation. (iv) Complexation is symmetrical. That is, the olefinic or acetylenic carbons are equidistant from the metal.

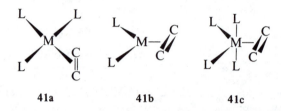

41a   **41b**   **41c**

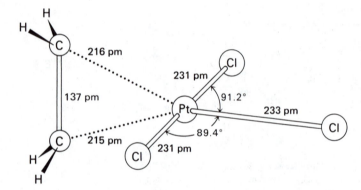

Figure 16–5. Molecular configuration of Zeise's salt, $K[PtCl_3(C_2H_4)]H_2O$. Adapted from J. A. J. Jarvis, B. T. Kilbourn, and P. G. Owston, *Acta Cryst.*, **B27**, 366 (1971); M. Black, R. H. B. Mais, and P. G. Owston, *ibid.*, **B25**, 1753 (1969). See also R. A. Love, *et al., Inorg. Chem.*, **14**, 2653 (1975) for a neutron diffraction structure.

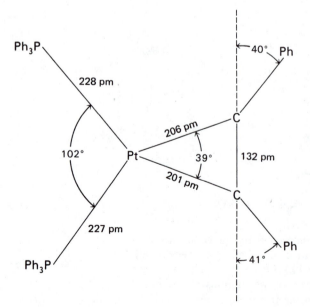

Figure 16–6. The molecular structure of bis(triphenyl-phosphine)diphenylacetyleneplatinum. [From J. O. Glanville, J. M. Stewart, and S. O. Grim, *J. Organometal Chem.*, **7**, P9 (1967).]

The DCD model suggests a synergic relation between ligand-to-metal $\pi \to \sigma$ bonding (38a) and metal-to-ligand $d\pi \to \pi^*$ electron flow (38b). Both donation of π electron density from the olefin to the metal and accumulation of electron density in the olefin π^* orbital would be expected to lower the π bond order of the coordinated olefin. As a consequence, the coordinated olefinic or acetylenic bond would be lengthened, an effect that has been observed in most cases. The lowered bond order is also manifested by a decrease in the C=C or C≡C stretching frequency; for example, $\nu_{C=C}$ drops from 1623 cm^{-1} for the free olefin to 1526 cm^{-1} in Zeise's salt, and $\nu_{C≡C}$ drops from 1657 cm^{-1} in the uncomplexed acetylene to 1230 cm^{-1} in $(Ph_3P)_2Pt(PhC≡CH)$. In general, $\Delta\nu_{C=C}$ and $\Delta\nu_{C≡C}$ are in the range from -50 to -250 cm^{-1} for divalent platinum and palladium complexes, whereas decreases of 400 to 500 cm^{-1} accompany complexation of acetylene to zero-valent metals; the latter has been taken as evidence of a much stronger metal-ligand interaction in the zero-valent complexes.

One of the more interesting properties of metal-olefin complexes, and one that gives us some further insight into metal-olefin bonding, is the fact that the olefin has in some cases been found to rotate about the metal-olefin bond axis (42).

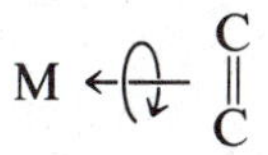

42

Reflection on the DCD model of metal-olefin bonding shows how this may happen. The ligand-to-metal σ bond, which accounts for about 75% of the bonding inter-action in Zeise's salt, can form whether the C=C bond is in the molecular plane or perpendicular to it (see 38a). Similarly, since out-of-plane $d\pi \to \pi^*$ back-bonding can take place through overlap of a d_{xz} orbital (43), the same interaction can, in principle, take place between the metal d_{yz} orbital and the C=C π^* orbital when the olefin is in the molecular plane (44).

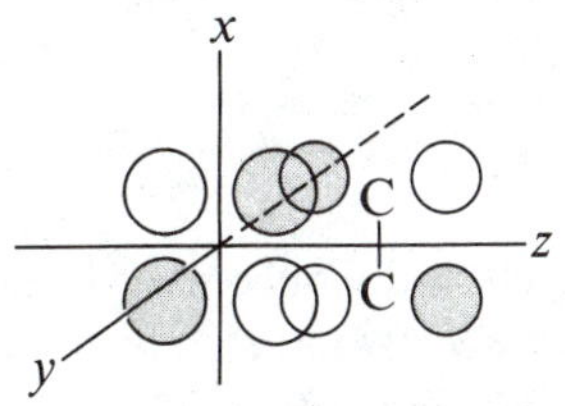

$d_{xz} \to \pi^*$

43

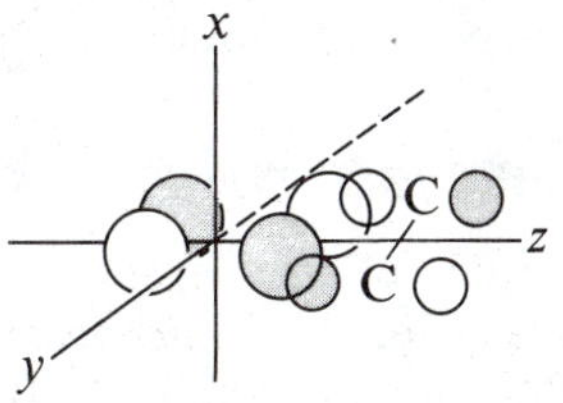

$d_{yz} \to \pi^*$

44

The η^3-allyl ligand is bound to a metal, as in bis(η^3-2-methylallyl)nickel (45),[59] in such a way that the three allylic carbon atoms are equidistant from the metal. Because of this structure, it should not be surprising that the bonding model applied to π-allyl complexes is similar to the metal-olefin DCD model.

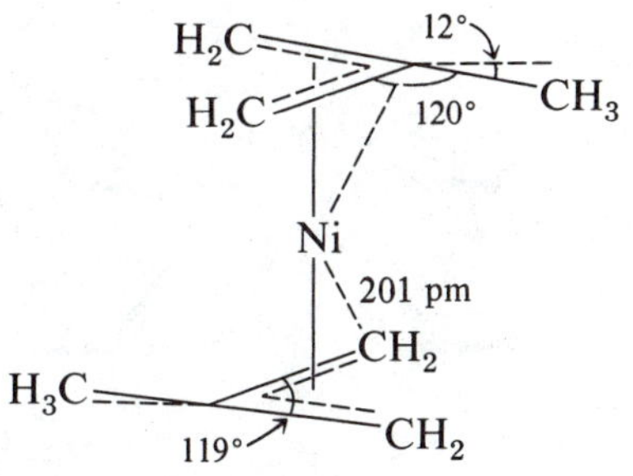

45

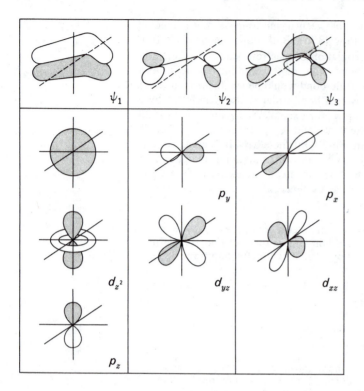

Figure 16–7. π-Allyl molecular orbitals and the metal orbitals with which they can most effectively overlap to produce substantial metal-ligand bonding. The metal lies below the C_3 plane, along the vertical axis. [Adapted from M. L. H. Green and P. L. I. Nagy, *Adv. Organometal. Chem.*, **2**, 325 (1964).]

The three $p\pi$ orbitals of the allyl group combine to form three mo's in the pattern familiar from Chapter 4 (see Figure 16–7 and Figure 4–1). Ligand-to-metal electron donation may then occur by overlap of ψ_1 with the metal s, d_{z^2}, and p_z orbitals and by overlap of ψ_2 with the metal p_y and d_{yz} orbitals.† The metal-to-ligand back-bond then derives from d_{xz} overlap with ψ_3.

Complexes with Cyclic π Donors

The discussion of complexes formed with cyclic π donors centers largely on those of the six π electron donors, the **pentahapto**cyclopentadienyl ion and the **hexahapto**benzene molecule. There are basically three types of complexes formed by η^5-$C_5H_5^-$, η^6-C_6H_6, and related molecules:

(i) $(\pi$-R$)_2$M. Symmetrical "sandwich" type complexes represented by ferrocene **(46)**, dibenzenechromium **(47)**, and $(\eta^5$-cyclopentadienyl)$(\eta^6$-benzene)manganese **(48)**.

(ii) $(\eta^5$-$C_5H_5)_2ML_x$. Bent metallocenes where L represents some other ligand such as H$^-$, R$^-$, halide, olefin, or NO; examples include compounds **49** and **50**.

(iii) $(\pi$-R$)ML_x$. Complexes in this class contain only one cyclic π donor ligand in addition to other ligands L; compounds **51** and **52** are examples.

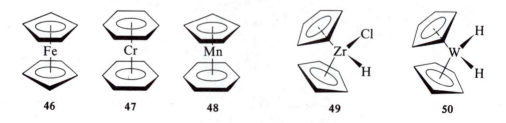

†Note that ψ_1 and ψ_2 are both occupied by electron pairs in the $C_3H_5^-$ ion.

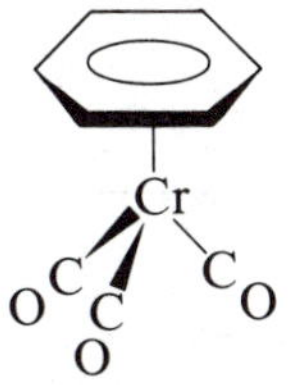

51 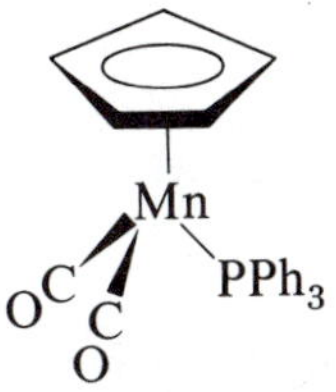   **52**

SYNTHESIS AND PROPERTIES

The preparation of symmetrical bis(η^5-cyclopentadienyl)metal compounds[65] —often called **metallocenes**† —begins with the preparation of the parent molecule C_5H_6, a diene that is, in turn, obtained by thermally cracking dicyclopentadiene (**53**).

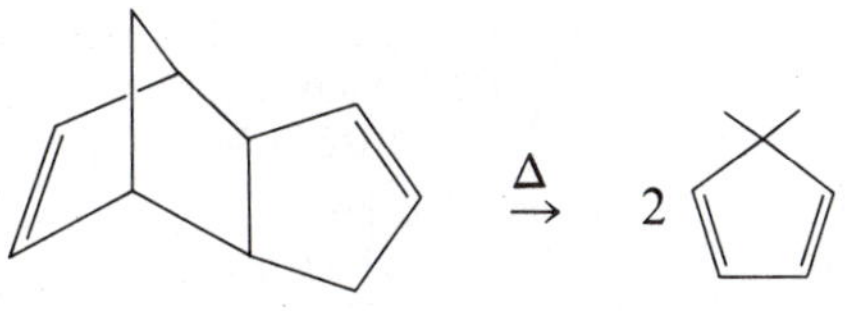

53

$$\qquad (16\text{-}67)$$

The basic problem, then, in any synthetic procedure is to remove H^+ from the diene, and several useful methods are available (16-68).

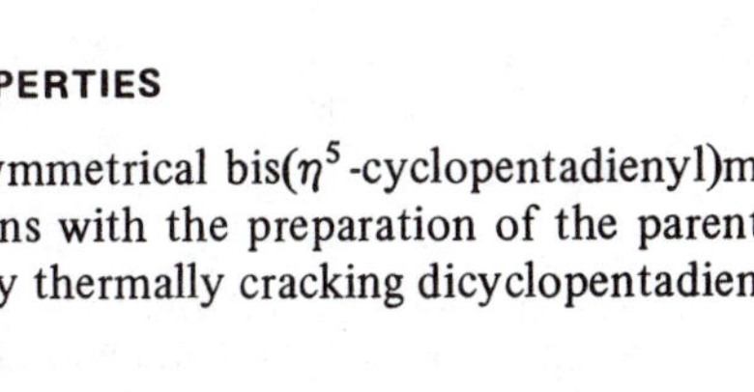

$$\qquad (16\text{-}68)$$

The hydrolytically unstable $C_5H_5^-$ salts are not usually isolated but are used directly in a reaction with an anhydrous salt of the metal to produce the desired metallocene. As an alternative, it is possible to use a hydrated metal salt if the cyclopentadienide ion can be generated *in situ*, and if H_2O is removed by a side-reaction, as in reaction 16-70. (See Table 16-4 for a summary of properties of the common metallocenes.)

$$2NaC_5H_5 + CoCl_2 \rightarrow (\eta^5\text{-}C_5H_5)_2Co + 2NaCl \qquad (16\text{-}69)$$

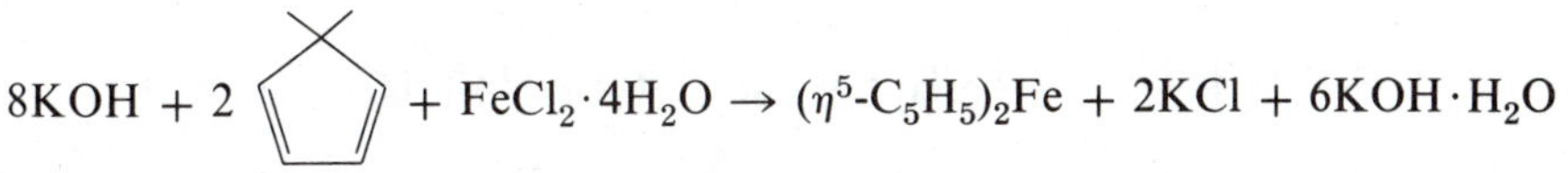

$$8KOH + 2 \; \text{[cyclopentadiene]} + FeCl_2 \cdot 4H_2O \rightarrow (\eta^5\text{-}C_5H_5)_2Fe + 2KCl + 6KOH \cdot H_2O \qquad (16\text{-}70)$$

---

†The name "metallocene" was applied to bis(η^5-cyclopentadienyl)metal compounds soon after the discovery of ferrocene. Hence, the congeners of ferrocene are called ruthenocene and osmocene, for example.

TABLE 16-4

PROPERTIES OF SOME METALLOCENES

Compound	Color	Melting Point (°C)	Remarks
''$(C_5H_5)_2Ti$''	dark green	>200(d)	Actually exists as $[(C_5H_5)(C_5H_4)TiH]_2$.[a]
$(C_5H_5)_2V$	purple	167	Very air-sensitive, $\mu = 3.84$ B.M.
$(C_5H_5)_2Cr$	dark red	173	Very air-sensitive, $\mu = 3.20$ B.M.
$[(C_5H_5)_2Mo]_x$[b]	red-brown solid		Diamagnetic.
$[(C_5H_5)_2W]_x$[b]	red solid		
$(C_5H_5)_2Mn$	dark brown	173	Air-sensitive. Hydrolyzed with characteristic crackling sound. Heating under N_2 to 158° turns the brown solid to pale pink solid, which melts at 173°. $\mu = 5.86$ B.M.
$(C_5H_5)_2Fe$	orange	173	Air-stable. Can be chemically oxidized to blue-green solid.
$(C_5H_5)_2Co$	purple-black	174	Air-sensitive. Oxidized to stable yellow $[(C_5H_5)_2Co]^+$ salts. $\mu = 1.73$ B.M.
$(C_5H_5)_2Ni$	dark green	173	Oxidizes slowly in air to give rather unstable yellow-orange $[(C_5H_5)_2Ni]^+$. $\mu = 2.86$ B.M.

[a] H. H. Brintzinger and J. E. Bercaw, *J. Amer. Chem. Soc.*, **92**, 6182 (1970); A. Davison and S. S. Wreford, *ibid.*, **96**, 3017 (1974).
[b] J. L. Thomas, *J. Amer. Chem. Soc.*, **95**, 1838 (1973).

The preparation of **dibenzenechromium** and similar molecules (Table 16-5) presents a somewhat different problem.[60] Since these are clearly complexes between neutral ligands and a metal in a zero (or perhaps +1) oxidation state, the problem is to bring a reduced metal in a reactive form in contact with the ligand. Therefore, the bis-arenes have usually been prepared by the aluminum reduction method developed by Fischer (*e.g.*, reaction 16-71).[67]

(16-71)

$$3\ CrCl_3 + 2\ Al + AlCl_3 + 6\ C_6H_6 \rightarrow 3\ [(\eta^6\text{-}C_6H_6)_2Cr][AlCl_4]$$

$$2\ [(\eta^6\text{-}C_6H_6)_2Cr]^+ + S_2O_4^{2-} + 4\ OH^- \rightarrow 2\ (\eta^6\text{-}C_6H_6)_2Cr + 2\ SO_3^{2-} + 2\ H_2O$$

TABLE 16-5

**PROPERTIES OF SOME BIS-π-ARENE
METAL COMPLEXES**[a]

Compound	Color	Melting Point (°C)	Remarks
$(C_6H_6)_2V$	black	227	Oxidized rapidly in air to the red-brown cation $[(C_6H_6)_2V]^+$.
$(\eta^6\text{-}C_6H_5F)_2V$	red	—	Sublimes, air-sensitive.
$(C_6H_6)_2Cr$	brown-black	284	Oxidized readily to the yellow cation $[(C_6H_6)_2Cr]^+$.
$(\eta^6\text{-}C_6H_5Cl)_2Cr$	olive green	89–90	Sublimes, air-stable.
$(C_6H_5F)_2Cr$	yellow	96–98	Sublimes, air-stable.
$[1,4\text{-}C_6H_4(CF_3)_2]_2Cr$	amber	150–152	Air-stable to 266°.
$(C_6H_6)_2Mo$	green	115	Very air-sensitive.
$(C_6H_6)_2W$	yellow-green	160(d)	Less air-sensitive than the Mo analog.
$[\eta^6\text{-}C_6(CH_3)_6]_2Mn^+$	pink-white	—	Diamagnetic.
$[\eta^6\text{-}C_6(CH_3)_6]_2Fe^{2+}$	orange	—	Can be reduced with dithionite to the deep violet Fe(I) complex and to the extremely air-sensitive black, paramagnetic ($2e^-$) Fe(0) complex.
$[\eta^6\text{-}C_6(CH_3)_6]_2Co^+$	yellow	—	Paramagnetic ($2e^-$).

[a] H. Zeiss, P. J. Wheatley, and H. J. S. Winkler, "Benzenoid-Metal Complexes," Ronald Press Company, New York, 1966; M. L. H. Green, "Organometallic Compounds, Volume Two: The Transition Elements," G. E. Coates, M. L. H. Green, and K. Wade, eds., Methuen, London, 3rd edition, 1968; K. J. Klabunde and H. F. Efner, *Inorg. Chem.*, **14**, 789 (1975); P. S. Skell, D. L. Williams-Smith, M. J. McGlinchey, *J. Amer. Chem. Soc.*, **95**, 3337 (1973).

A number of **mixed sandwich compounds** are also now known. For example, all of the members of the isoelectronic series **54-58** have been prepared,[68] and their syntheses illustrate the imaginative methods used in organometallic chemistry.

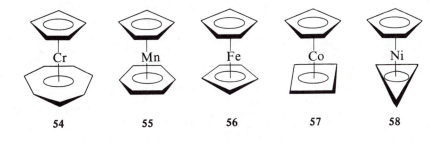

| 54 | 55 | 56 | 57 | 58 |

For example, $(\eta^5$-cyclopentadienyl)$(\eta^6$-benzene)manganese is prepared from $(\eta^5$-$C_5H_5)$ MnCl and PhMgBr in THF. Reaction 16–72 presumably proceeds through formation of a σ-phenyl derivative;† upon hydrolysis this intermediate is converted to the π-benzene complex, and the metal is reduced.

(16-72)

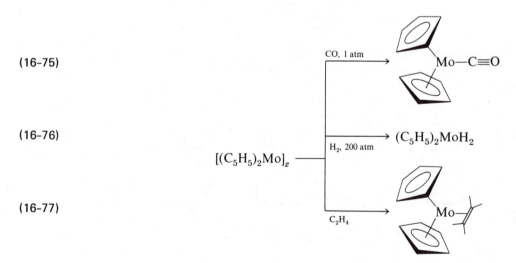

The preparation of **bent metallocenes** of the type $(\eta^5$-$C_5H_5)_2ML_x$ has not yet been systematized. However, considerable work has been done with the bis(η^5-cyclopentadienyl)metal complexes of Group VIB (chromocene, molybdenocene, and tungstenocene), and some of their reactions lead to compounds of the desired type.

Preparation of Molybdenocene.§

(16-73) $$MoCl_5 + 3NaC_5H_5 + 2NaBH_4 \rightarrow \quad Mo\langle{}^H_H \quad + 5NaCl + \tfrac{1}{2}C_{10}H_{10} + B_2H_6$$

(16-74) $$(C_5H_5)_2MoH_2 \xrightarrow{\Delta,\ CHCl_3} (C_5H_5)_2MoCl_2 \xrightarrow{Na/Hg,\ THF} [(C_5H_5)_2Mo]_x$$

red-brown solid; sparingly sol. in THF

Preparation of Bent Molybdenocene Derivatives.

(16-75) $\xrightarrow{CO,\ 1\ atm}$ Mo—C≡O

(16-76) $[(C_5H_5)_2Mo]_x$ ——$\xrightarrow{H_2,\ 200\ atm}$ $(C_5H_5)_2MoH_2$

(16-77) $\xrightarrow{C_2H_4}$ Mo—

†Note the relationship with π-allyl formation; page 519.

§Molybdenocene is a $16e^-$ molecule, a fact related to its "polymeric" structure and to its propensity for bent metallocene formation.

With the exception of ferrocene and bis-arenes with very electronegative substituents, most metallocenes and bis-arenes are oxidatively unstable. Therefore, much research has been done with the oxidatively more stable η^5-cyclopentadienylmetal and η^6-benzene-metal carbonyls. In general, π-bonded cyclopentadienylmetal carbonyls are prepared from metallocenes and CO or from the metal carbonyl and $C_5H_5^-$ (reaction 16-78) or dicyclopentadiene. Reaction 16-79 is a particularly good example of the latter. Although the intermediates in 16-79 have been isolated in separate reactions, the pathway depicted was not confirmed until recently, when the important $(\eta^5\text{-}C_5H_5)M(CO)_2H$ was isolated in the reaction of $Ru_3(CO)_{12}$ with cyclopentadiene.

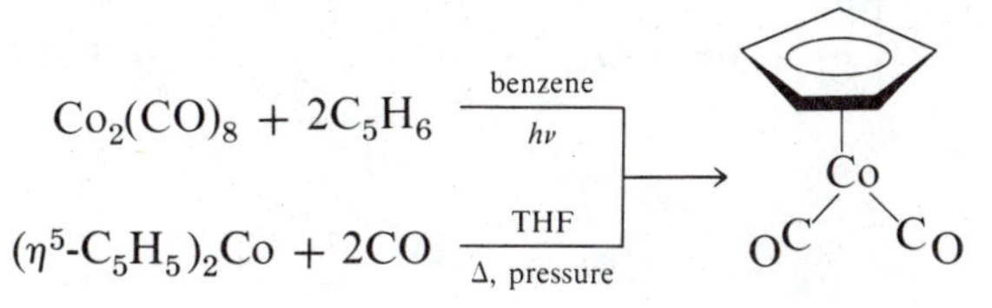

$$(16\text{-}78)$$

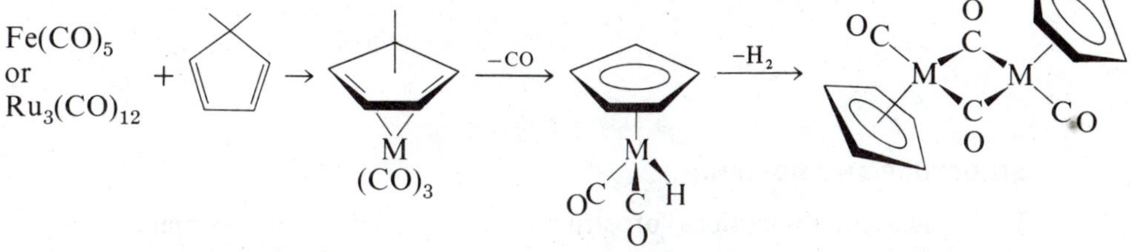

$$(16\text{-}79)$$

Arenemetal carbonyls, especially those of Group VIB, are readily synthesized from the appropriate metal carbonyl and the arene. If the potential ligand is a high boiling liquid (e.g., mesitylene), the carbonyl may be refluxed under N_2 in the liquid arene; heat brings about the dissociation of the carbonyl and its replacement by the arene.[69]

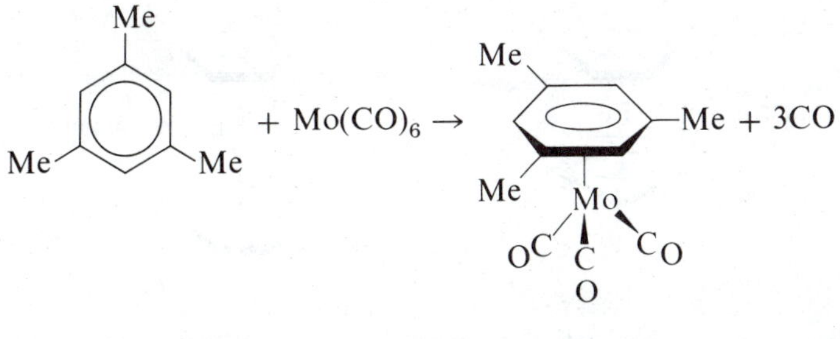

$$(16\text{-}80)$$

A more useful method, however, is the prior formation of a labile complex of the carbonyl with ligands such as diglyme or acetonitrile. Neither of these ligands is capable of forming strong metal-to-ligand $d\pi\text{-}p\pi$ bonds; therefore, the metal-ligand bond is weak, and the ligand is easily replaced. The intermediate may be isolated and then treated with arene, or the metal carbonyl may simply be refluxed in diglyme, for example, already containing the arene.

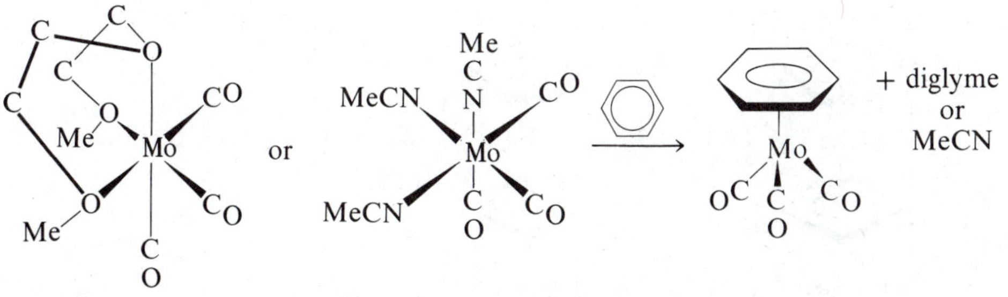

$$(16\text{-}81)$$

One of the most sought-after molecules in organic chemistry has been **cyclobutadiene**. Although it was recently isolated at low temperatures by photolytic methods, it was first prepared in 1965 in the form of its iron carbonyl complex **(59)**.[70] The reaction is important for two reasons: (i) It illustrates the fact that *gem* and *vicinal* dihalides may be dehalogenated with metal carbonyls.[71] (ii) It represents an increasingly important direction being taken by organic and organometallic chemists—the isolation and study of highly reactive organic molecules in the form of their metal complexes.

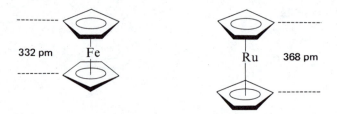

(16-82)

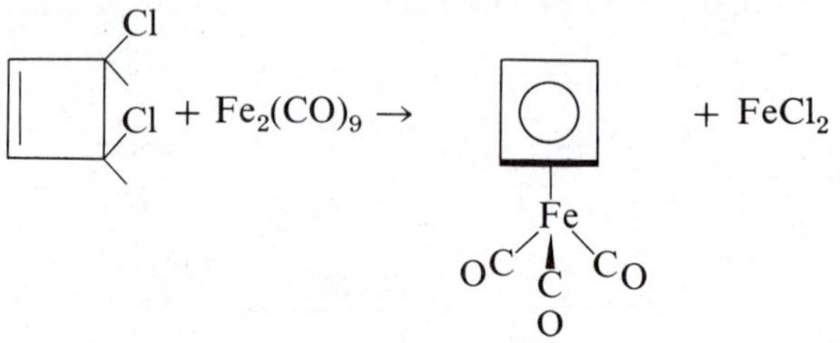

59

[pale yellow prisms]

STRUCTURE AND BONDING

The fundamental questions of structure and bonding can be approached by examining a few representative compounds: ferrocene, bis(η^5-cyclopentadienyl)rhenium hydride, bis(η^6-benzene)chromium, (η^6-benzene)tricarbonylchromium, and bis(η^8-cyclooctatetraene)uranium.

In the solid state the cyclopentadienyl rings of ferrocene are staggered (D_{5d} symmetry), whereas those of ruthenocene and osmocene are eclipsed (D_{5h} symmetry).

332 pm   Fe   Ru   368 pm

These arrangements are apparently the result of crystal packing forces, because the barrier to rotation of the ferrocene rings, measured in the gas phase, is quite small (~4 kJ). Furthermore, there is abundant evidence that the rings rotate freely when ferrocene and its derivatives are placed in solution.

The electrochemical properties of ferrocene can provide some insight into the bonding in the molecule. Ferrocene, for example, can lose an electron to give the deep blue-green, paramagnetic ferricenium ion. This redox reaction is chemically and electro-

(16-83)

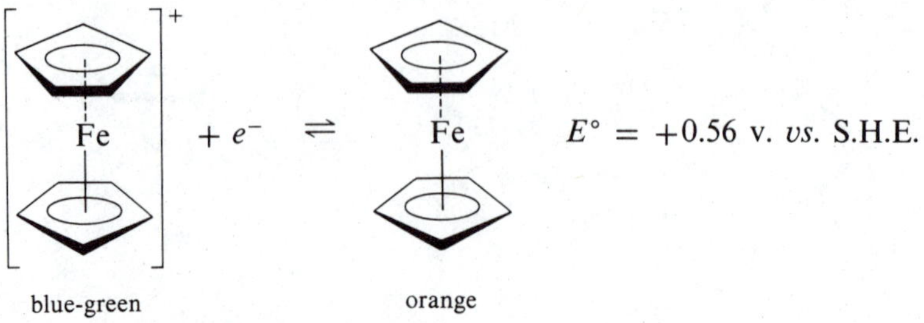

$$\left[\text{Fe} \right]^{+} + e^{-} \rightleftharpoons \text{Fe} \qquad E^{\circ} = +0.56 \text{ v. } vs. \text{ S.H.E.}$$

blue-green   orange

chemically reversible, and the potential shows a sensitivity to the nature of the substituents in that more electronegative groups make the ferricenium ion easier to reduce ($E°$ increases). The $E°$ for reduction of unsubstituted ferricenium ion (which is about the same as $E°$ for MnO_4^- functioning as an oxidant in basic solution) is somewhat less positive than that of the Fe(III)/Fe(II) couple in aqueous solution (+0.77 v). These facts imply that the electron involved in the FcH^+/FcH couple is in a molecular orbital with some ligand character but largely localized on the iron atom. Thus, any molecular orbital picture that is derived for ferrocene must take the redox reaction and similar observations into account.

Ferrocene serves as a prototype for metal binding by π donor ligands, and its molecular orbital diagram is shown in Figure 16-8.[72,75] The orbital levels in Figure 16-8 are of three types. The six lowest energy orbitals are occupied in all metallocenes by the six electron pairs from the $C_5H_5^-$ ligands and assist M—Cp bonding. The eight highest energy mo's are ligand-like and empty. The five orbitals at the center are most like the metal d orbitals and are occupied by electrons associated with the metal; it is from these orbitals that most chemical properties of ferrocene in particular, and metallocenes in general, derive. These chemical properties depend on how many electrons are located in the middle five mo's, the two lowest of which are strongly M—Cp bonding and the two highest of which are M—Cp anti-bonding. The ligand-metal combinations responsible for the center five mo's are depicted in the boxed area on page 531.

Based on the molecular orbital picture outlined above, it should not be surprising that ferrocene is the least reactive of the metallocenes yet discovered. The molecule has 18 valence electrons (six from each $C_5H_5^-$ and six from Fe^{2+}), and these nine pairs of

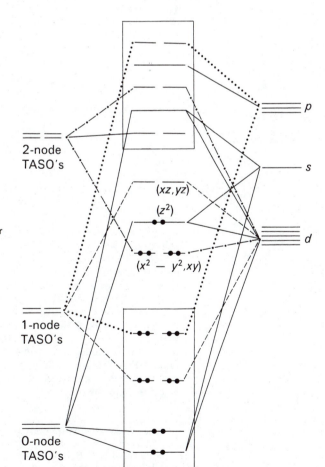

Figure 16-8. A qualitative molecular orbital diagram for ferrocene.

electrons are just accommodated in the bonding and non-bonding mo's. However, cobaltocene (a 19-electron molecule based on d^7 Co^{2+}) and nickelocene (a 20-electron molecule based on d^8 Ni^{2+}) have electrons in the high energy anti-bonding orbitals. This fact is reflected in their electrochemical properties as noted above; both the cobalt and nickel compounds are readily oxidized. Vanadocene (15 valence electrons, d^3 V^{2+}) and chromocene (16 valence electrons, d^4 Cr^{2+}), on the other hand, are electron deficient, and they will add additional ligands that can contribute more electrons. For example, $(C_5H_5)_2Mo$ reacts with H$_2$ (reaction 16–77) to give **60**. The cyclopentadienyl rings have bent back from their parallel position in order to accommodate the H atoms (with little or no loss in metal-C$_5$H$_5$ bond strength), and the Mo atom now has the favored 18-electron configuration: 6 electrons from each of the C$_5$H$_5^-$ ligands, 2 electrons from each of the two H$^-$ ligands, and 2 electrons from Mo^{4+}.

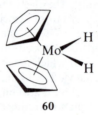

60

Besides ferrocene, the metallocenes that have provoked perhaps the greatest amount of discussion are titanocene and manganocene. Reduction of $(\eta^5\text{-}C_5H_5)_2$ TiCl$_2$ gives a green compound with the empirical formula $(C_5H_5)_2$Ti, and it was included as such in the earliest tabulations of metallocenes as titanocene. Later work, however, showed the molecule to be dimeric, and numerous proposals were advanced to account for this. Very recently the structure of the dimer was found to be **61** (cf. Table 16–4).†

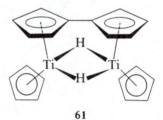

61

[titanocene = μ-(η^5:η^5-fulvalene)-di-μ-hydrido-
bis(cyclopentadienyltitanium)]

Manganocene is an amber solid that can be prepared from Na$^+$C$_5$H$_5^-$ and anhydrous MnCl$_2$ in THF. Early investigations of the chemistry of $(C_5H_5)_2$Mn led chemists to believe that it may be in a class apart from the covalent metallocenes such as ferrocene. That is, manganocene reacts rapidly with ferrous salts to give ferrocene, and its cyclopentadienyl groups exchange rapidly with C$_5$H$_5^-$ added to its solutions. These properties, coupled with the fact that the compound has a magnetic moment of 5.86 ± 0.05 B.M. (very close to the spin-only moment for five electrons) in dilute solid solution, indicated to early workers that the compound was ionic, *i.e.*, $(C_5H_5^-)_2$Mn^{2+}. More recently, however, this idea was put to rest as further, extensive studies of the magnetic behavior of manganocene strongly suggest that the magnetic moment simply arises from the fact that the energies of the five metal-like mo's are very similar, and Hund's rule is obeyed in filling electrons into the orbitals.§

---

†A molecule with the formula $(\eta^5\text{-}C_5H_5)_2$Ti would be a 14-electron species; dimerization to give **61**, however, gives rise to a more stable 16-electron molecule.

§Note that $(\eta^5\text{-}C_5H_5)_2$Mn is a 17-electron molecule; it appears that the high pairing energy of d^5 metals is at work here.

Ferrocene Molecular Orbitals

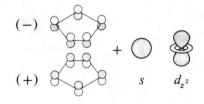

$(-)$ ... $(+)$ s d_{z^2}

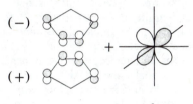

d_{yz}

d_{xz}

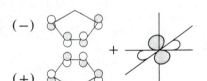

d_{xy}

$d_{x^2-y^2}$

Dibenzenechromium (**47**) is the best example of the other type of symmetrical metal sandwiches. One of the major questions that arose after the discovery of the compound was whether it could best be described as a triene complex (**47′**) or a complex of benzene (**47″**).

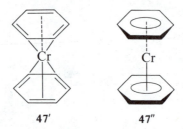

47′   47″

Some early x-ray studies, in fact, suggested the former; but more careful work recently has established reasonably well that dibenzenechromium is indeed a benzene complex of true D_{6h} symmetry. Because of this, one can build a bonding model for dibenzenechromium that is quite analogous to that of ferrocene. Just as with ferrocene, there are six occupied, low energy molecular orbitals and five mid-range molecular orbitals with considerable d character. (Ten high energy, empty mo's complete the scheme.) The valence electrons of the molecule (2×6 benzene electrons and 6 Cr^0 electrons) fill nine molecular orbitals, the HOMO being an orbital best described as essentially a Cr d_{z^2} orbital, just as in ferrocene. Thus, you would expect that the properties of $(C_6H_6)_2Cr$ which depend on electron loss will be essentially those of a perturbed chromium atom.

One of the chief differences between ferrocene and dibenzenechromium concerns their stabilities with respect to oxidation. As discussed previously, the iron compound (with Fe formally in the 2+ oxidation state) may be oxidized chemically or electrochemically to $[(\eta^5\text{-}C_5H_5)_2Fe^{III}]^+$, but it is nonetheless stable in air in the solid state or in solution under normal conditions. However, dibenzenechromium and many other bis-arenes of Cr, Mo, or W (wherein the metal is formally in the 0 oxidation state) oxidize very readily in air to give the paramagnetic species $[(arene)_2M]^+$.† The reason for this difference is that Z^* for Fe^{II} is greater than for Cr^0.

You can extrapolate from the conclusion above to the concept that any effect that enhances the charge on the central metal should render it less likely to lose electrons. Indeed, replacing one arene ligand in a bis-arene with three carbonyl groups leads to a great increase in oxidative stability. For example, the Group VIB arenemetal tricarbonyls (**51**) are usually oxidatively stable as solids, although they often oxidize slowly in solution.

51′

There is considerable evidence that carbonyl groups are more efficient π acceptors of electron density than are cyclopentadienyl or benzene ligands. Therefore, inclusion of

---

†Note that there are exceptions to this statement. For example, all of the air stable bis-arene compounds have very electronegative groups attached to the benzene ring; see Table 16–5.

CO ligands in place of C_6H_6 may indirectly lower the energy of all of the metal based mo's and lessen the reducing ability of the metal.

Lanthanide and actinide organometallic chemistry, where there is the possibility of f orbital utilization, is a rapidly growing area of interest.[73] One of the most important compounds to come out of this area thus far is "uranocene" (62).[74] The green crystalline, water stable, pyrophoric compound is prepared by the reaction of cyclooctatetraene dianion (COT^{2-}) with UCl_4 in THF at $0°C$. COT^{2-}, prepared in turn by the reaction of

$$\text{(structure)} + 2K \xrightarrow{\text{THF}} 2K^+ \text{(structure)} \qquad (16\text{-}84)$$

$$2C_8H_8^{2-} + UCl_4 \xrightarrow{\text{THF/0°C}} \text{(structure)} \qquad (16\text{-}85)$$

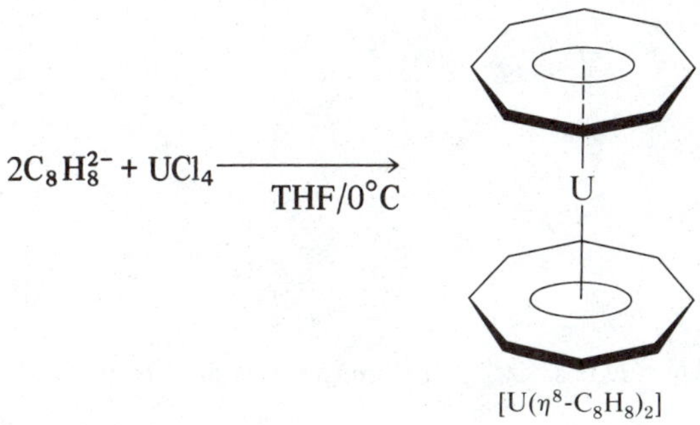

$$[U(\eta^8\text{-}C_8H_8)_2]$$

62

cyclooctatetraene with metallic potassium in THF, is a planar, aromatic ion having ten π electrons. Placing two such anions face to face generates orbitals that can combine to form molecular orbitals with uranium f orbitals such as $f_{z(x^2-y^2)}$ (cf. Table 1-1 and page 35) (63).

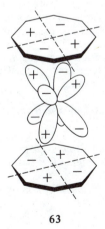

63

---

22. The first row transition metals from vanadium through nickel can form complexes of the type $[(\eta^5\text{-}C_5H_5)M(CO)_x]_n$. Using the EAN rule, predict the molecular formula for the compound formed by each first row transition metal. Draw the structure for each of the compounds.

STUDY QUESTIONS

23. Suggest syntheses for each of the following:

a) 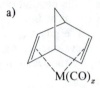   where M = Fe or Mo

b)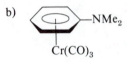

c) compound **31**

d)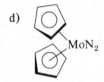

24. Verify that the EAN rule is satisfied for compounds **54–58** (p. 525).

25. Predict the products for the reaction of molybdenocene, $[(C_5H_5)_2Mo]_x$, with N_2 and RX (*e.g.*, CH_3I).

26. The bis(*pentahapto*cyclopentadienyl) compounds of V, Cr, Co, and Ni are all paramagnetic; the magnetic moment for each compound is listed in Table 16–4. Account for each of these compounds using the mo diagram suggested for ferrocene (Figure 16–8).

27. What is the function of the excess KOH in reaction 16–70?

28. Dibenzenechromium, $(C_6H_6)_2Cr$, is quite susceptible to air-oxidation. However, some other bis-arenes listed in Table 16–5 are apparently stable to air-oxidation. Why is a compound such as $(C_6H_5F)_2Cr$ more stable to oxidation than $(C_6H_6)_2Cr$?

29. In reaction 16–62 the equilibrium constant increases as the hydrogen atoms of ethylene are substituted by CN groups. Suggest a reason for this increase based on the mechanism of metal-olefin binding.

30. Why is it necessary to postulate an iron-iron interaction for the di-iron compound produced in reaction 16–65?

31. Devise an experiment that would allow you to prove that the olefin may rotate about the metal-olefin axis in such complexes (*cf.* structure **42**).

32. One of the compounds in the series **46–52** is only a 16 electron molecule. Which is it?

CHECKPOINTS

1. Organometallic compounds may be grouped into those involving σ donor ligands (Class A) and those involving π donor ligands (Class B).

2. Organometallic compounds in Class A generally fall into two sub-classes:
   (a) compounds in which the organic group can be considered an anionic σ donor (e.g., CH_3^-).
   (b) compounds wherein the organic group is a neutral σ donor and π acceptor (e.g., CO).

3. Compounds of Class A(a) may be synthesized by numerous methods: direct reaction of a metal with an organic halide, reaction of a metal halide with an anionic alkylating agent, reaction of a metal with an organomercury compound, oxidative addition, and hydro- or organometalation, among others.

4. Bonding in Class A(a) compounds is straightforward, except for so-called "electron deficient" compounds with bridging alkyl or aryl groups (e.g., aluminum and lithium compounds).

5. Class A(b) compounds are formed between low-valent transition metals and CO, isonitriles, carbenes, and carbynes. They may be synthesized by a variety of methods, including direct reaction and displacement of other, more weakly bonding ligands.

6. Class B compounds are formed between low-valent transition metals and π donor ligands such as olefins and arenes. Bonding is much the same as in Class A(b) compounds; that is, the ligand donates π bonding electrons to the metal, which then back-donates metal electron density into ligand π^* molecular orbitals.

7. The synthesis of Class B molecules likewise involves a variety of methods, including direct addition or ligand replacement for metal-olefin complexes and reaction of a metal halide with an anionic donor such as the cyclopentadienide ion.

REFERENCES

[1] W. C. Zeise, *Pogg. Ann.*, **21**, 497 (1831).

[2] E. Frankland, *J. Chem. Soc.*, **2**, 263 (1848).

[3] V. Grignard, *Compt. rend.*, **130**, 1322 (1900). Grignard received the Nobel Prize in 1912 for his work. See also E. C. Ashby, *Quart. Rev.*, **21**, 259 (1967).

[4] T. J. Kealy and P. L. Pauson, *Nature*, **168**, 1039 (1951).

[5] S. A. Miller, J. A. Tebboth, and J. F. Tremaine, *J. Chem. Soc.*, 632 (1952).

[6] For a very personal discussion of the activity in the chemical community immediately following the announcement of the discovery of ferrocene, see G. Wilkinson, *J. Organometal. Chem.*, **100**, 273 (1975). As mentioned in the Prologue to this book, Wilkinson (now at Imperial College, London) and Fischer at the University of Munich shared the Nobel Prize in 1973 for their work on the chemistry of metallocenes.

[7] M. Rosenblum, "Chemistry of the Iron Group Metallocenes," Interscience Publishers, New York, 1965.

[8] A very nice accounting of the history of organometallic chemistry is given by J. S. Thayer, "Organometallic Chemistry: A Historical Perspective," *Adv. Organometal. Chem.*, **13**, 1–49 (1975).

[9] Information on the chemical industry, the use of catalysts, and the production of inorganic and organic chemicals is found in a very interesting book: B. G. Reuben and M. L. Burstall, "The Chemical Economy: A Guide to the Technology and Economics of the Chemical Industry," Longmans, London, 1973.

[10] For the structure of trimethylaluminum, see R. G. Vranka and E. L. Amma, *J. Amer. Chem. Soc.*, **89**, 3121 (1967).

[11] The best general source of information on metal alkyl and aryl synthesis is the set of books by Coates, Green, and Wade.
   a. "Organometallic Compounds" (Volume I, G. E. Coates and K. Wade, "The Main Group Elements"), Methuen, London, 1967.
   b. "Organometallic Compounds" (Volume II, M. L. H. Green, "The Transition Elements"), Methuen, London, 1968.

[12] E. G. Rochow, *J. Chem. Educ.*, **43**, 58 (1966).

[13] P. L. Timms, *Adv. Inorg. Chem. Radiochem.*, **14**, 121 (1972).

[14] K. J. Klabunde, *Acc. Chem. Res.*, **8**, 393 (1975).

[15] E. G. Rochow and A. C. Smith, Jr., *J. Amer. Chem. Soc.*, **75**, 4103 (1953); J. J. Zuckerman, *Adv. Inorg. Chem. Radiochem.*, **6**, 383 (1964).

[16] H. Gilman, *Adv. Organometal. Chem.*, **7**, 1 (1968).

[17] H. Gilman and J. W. Morton, *Organic Reactions*, **8**, 258 (1954).

[18] G. W. Parshall, *Acc. Chem. Res.*, **3**, 139 (1970); J. Dehand and M. Pfeffer, *Coord. Chem. Rev.*, **18**, 327 (1976).

[19] J. Halpern, *Acc. Chem. Res.*, **3**, 386 (1970).

[20] J. P. Collman and W. R. Roper, *Adv. Organometal. Chem.*, **7**, 53 (1968).

[21] M. F. Lappert and P. W. Lednor, *Adv. Organometal. Chem.*, **14**, 345 (1976).

[22] R. B. King, *Acc. Chem. Res.*, **3**, 417 (1970).

[23] There are three general reviews of reactions of this type:
   a. M. F. Lappert and B. Prokai, *Adv. Organometal. Chem.*, **5**, 225 (1967).
   b. R. F. Heck, *Adv. Chem. Series*, **49**, 181 (1965).
   c. J. P. Candlin, K. A. Taylor, and D. T. Thompson. "Reactions of Transition Metal Complexes," Elsevier Publishing Company, New York, 1968, pp. 119–151.

[24] Unless otherwise noted, the information on hydroborations was taken from the following sources.
   a. H. C. Brown, "Hydroboration," W. A. Benjamin, Inc., New York, 1962.
   b. G. Zweifel and H. C. Brown, in "Organic Reactions," Volume 13, A. C. Cope, ed., John Wiley and Sons, New York, 1963, Chap. 1.
   c. H. C. Brown, *Acc. Chem. Res.*, **2**, 65 (1969).
   d. H. C. Brown, "Organic Syntheses via Boranes," Wiley-Interscience, New York, 1975.
   e. H. C. Brown, "Boranes in Organic Chemistry," Cornell University Press, Ithaca, New York, 1972.

[25] K. Ziegler, in "Organometallic Chemistry," ACS Monograph No. 147, H. Zeiss, ed., Reinhold, New York, 1960, pp. 194–269.

[26] P. S. Braterman and R. J. Cross, *J. C. S. Dalton*, 657 (1972); *Chem. Soc. Rev.*, **2**, 271 (1973).

[27] R. F. Heck, *Acc. Chem. Res.*, **2**, 10 (1969).

[28] R. F. Heck, *Adv. Chem. Ser.*, **49**, 181 (1965).

[29] C. A. Tolman, *Chem. Soc. Rev.*, **1**, 337 (1972).

[30] R. Cramer, *Acc. Chem. Res.*, **1**, 186 (1968).

[31] R. F. Heck, "Organotransition Metal Chemistry," Academic Press, New York, 1974.

[32] A. J. Chalk and J. F. Harrod, *Adv. Organometal. Chem.*, **6**, 119 (1968).

[33] P. J. Davidson, M. F. Lappert, and R. Pearce, *Chem. Rev.*, **76**, 219 (1976).

[34] J. P. Oliver, *Adv. Organometal. Chem.*, **15**, 235 (1977).

[35] The best source of general information on organolithium compounds is B. J. Wakefield, "The Chemistry of Organolithium Compounds," Pergamon Press, New York, 1974.

[36] For a general review of Grignard reagent structure and composition, see E. C. Ashby, *Quart. Rev.*, **21**, 259 (1967).

[37] J. Toney and G. D. Stucky, *J. Organometal. Chem.*, **28**, 5 (1971).

[38] J. M. Pratt, "Inorganic Chemistry of Vitamin B_{12}," Academic Press, New York, 1972.

[39] G. N. Schrauzer, *Acc. Chem. Res.*, **1**, 97 (1968).

[40] R. R. Schrock and G. W. Parshall, *Chem. Rev.*, **76**, 243 (1976).

[41] A most interesting feature of **17** is that there is a Mo–Mo triple bond.

[42] J. C. Hileman, "Metal Carbonyls," in "Preparative Inorganic Reactions," Vol. 1, W. L. Jolly, ed., Interscience, New York, 1964.

[43] E. W. Abel and F. G. A. Stone, *Quart. Rev.*, **23**, 325 (1969).

[44] E. W. Abel and F. G. A. Stone, *Quart. Rev.*, **24**, 498 (1970).

[45] P. S. Braterman, *Structure and Bonding*, **10**, 57 (1972).

[46] F. A. Cotton, *Prog. Inorg. Chem.*, **21**, 1 (1976).

[47] This statement should be qualified. Low temperature matrix isolation techniques have led to the observation of such otherwise unstable species as $Re(CO)_5$ and $Pt(CO)_2$: G. A. Ozin and A. Vander Voet, *Acc. Chem. Res.*, **6**, 313 (1973).

[47a] R. Job, *J. Chem. Educ.*, **56**, 556 (1979).

[47b] A. P. Hagen, T. S. Miller, D. L. Terrell, B. Hutchinson, and R. L. Hance, *Inorg. Chem.*, **17**, 1369 (1978).

[48] J. P. Collman, *Acc. Chem. Res.*, **8**, 342 (1975).

[49] D. M. Adams, "Metal-Ligand and Related Vibrations," St. Martin's Press, New York, 1968; L. M. Haines and M. H. B. Stiddard, *Adv. Inorg. Chem. Radiochem.*, **12**, 53 (1969); P. S. Braterman, *Structure and Bonding*, **26**, 1 (1976).

[50] D. J. Cardin, B. Cetinkaya, and M. F. Lappert, *Chem. Rev.*, **72**, 545 (1972).

[51] D. J. Cardin, B. Cetinkaya, M. J. Doyle, and M. F. Lappert, *Chem. Soc. Rev.*, **2**, 99 (1973).

[52] F. A. Cotton and C. M. Lukehart, *Prog. Inorg. Chem.*, **16**, 487 (1972).

[53] E. O. Fischer, *Adv. Organometal. Chem.*, **14**, 1 (1976).

[54] R. R. Schrock and P. R. Sharp, *J. Amer. Chem. Soc.*, **100**, 2389 (1978); R. R. Schrock, *Acc. Chem. Res.*, **12**, 98 (1979).

[55] H. W. Quinn and J. H. Tsai, *Adv. Inorg. Chem. Radiochem.*, **12**, 217 (1969).

[56] J. H. Nelson and H. B. Jonassen, *Coord. Chem. Rev.*, **6**, 27 (1971).

[57] F. R. Hartley, *Chem. Rev.*, **69**, 799 (1969).

[58] F. L. Bowden and A. B. P. Lever, *Organometal. Chem. Rev. A*, **3**, 227 (1968).

[59] M. L. H. Green and P. L. I. Nagy, *Adv. Organometal. Chem.*, **2**, 235 (1964).

[60] H. Zeiss, P. J. Wheatley, and H. J. S. Winkler, "Benzenoid-Metal Complexes," Ronald Press, New York, 1966.

[61] W. E. Silverthorn, *Adv. Organometal. Chem.*, **13**, 47 (1975).

[62] M. A. Bennett, *Adv. Organometal. Chem.*, **4**, 353 (1966).

[63] M. J. S. Dewar, *Bull. Soc. Chem. Fr.*, **18**, C79 (1951); J. Chatt and L. A. Duncanson, *J. Chem. Soc.*, 2939 (1953).

[64] S. D. Ittel and J. A. Ibers, *Adv. Organometal. Chem.*, **14**, 33 (1976).

[65] R. L. Pruett, "Cyclopentadienyl and Arene Metal Carbonyls," in "Preparative Inorganic Reactions," Volume 2, W. L. Jolly, ed., Interscience, New York, 1965.

[66] W. L. Jolly, *Inorg. Syn.,* **11**, 120 (1968).

[67] E. O. Fischer, *Inorg. Syn.,* 6, 132 (1960).

[68] M. D. Rausch, *Pure and Applied Chem.,* **30**, 523 (1972).

[69] R. J. Angelici, "Synthesis and Technique in Inorganic Chemistry," 2nd ed., W. B. Saunders Co., Philadelphia, 1977.

[70] R. Pettit and J. Henery, *Organic Reactions*, **50**, 21 (1970).

[71] M. Ryang, *Organometal. Chem. Rev., A,* **5**, 67 (1970).

[72] J. W. Lauher and R. Hoffmann, *J. Amer. Chem. Soc.,* **98**, 1729 (1976). This paper summarizes much of what is currently known about bonding in metallocenes of all types. It is a lucid and readable paper and is especially recommended as a starting place for study in this area.

[73] E. C. Baker, G. W. Halstead, and K. N. Raymond, *Structure and Bonding,* **25**, 23 (1976).

[74] A. Streitwieser, Jr., U. Muller-Westerhoff, G. Sonnichsen, F. Mares, D. G. Morrell, K. O. Hodgson, and C. A. Harmon, *J. Amer. Chem. Soc.,* **95**, 8644 (1973).

[75] A. Haaland, *Acc. Chem. Res.,* **12**, 415 (1979).

17. ORGANOMETALLIC COMPOUNDS: REACTION PATHWAYS

In this chapter the types of reactions usually encountered in organo-metallic chemistry are surveyed. Briefly, these are

i. *Association/dissociation reactions: primarily reactions involving the formation of Lewis acid-base type complexes or ligand protonation reactions.*

ii. *Electrophilic and nucleophilic substitution reactions: the substitution of one ligand, atom, or group for another.*

iii. *Addition/elimination reactions: including 1,2-additions of M—R to double bonds, 1,1-additions of M—R to the carbon of CO (carbonyla-tion), and the 1,1-addition of R—X to an unsaturated metal center (oxidative additions). Also discussed are the reverses of these reaction types—β-elimination of M—H from alkyls, decarbonylation, and reduc-tive elimination.*

iv. *Rearrangements: including intermolecular ligand redistribution and the intramolecular phenomenon of stereochemical non-rigidity.*

Our focus in this chapter will be on reactions of organometallic compounds, such as:

acid-base combination reactions

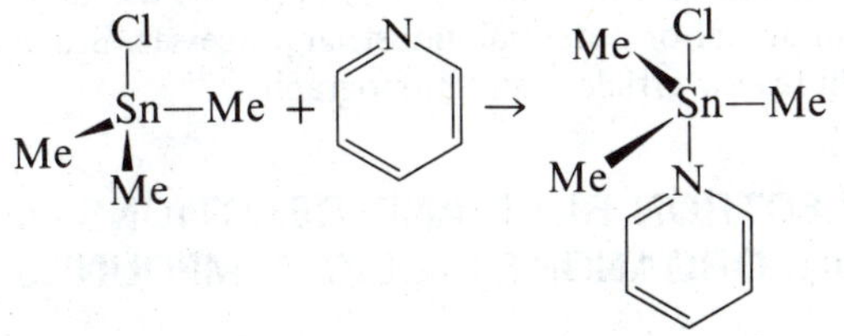

(17-1)

ligand substitution reactions

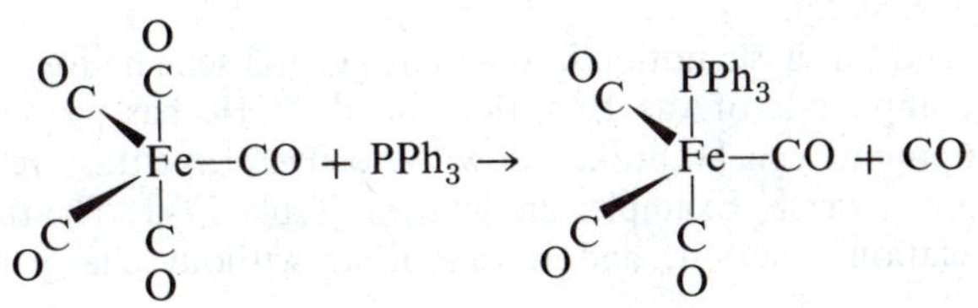

(17-2)

electrophilic attack on a coordinated ligand

(17-3)

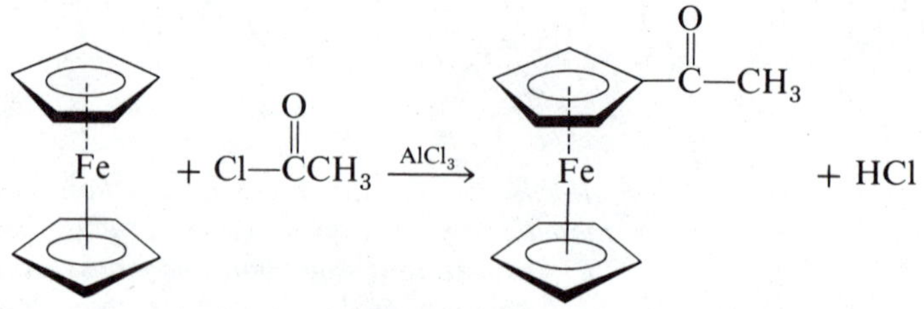

1,2-addition of an organometallic compound across a double bond (alternatively, insertion of an olefin into a metal–X bond)

(17-4)

$$Me_3SnH + H_2C{=}CHPh \rightarrow Me_3Sn{-}CH_2CH_2Ph$$

olefin or **acetylene coupling** with **CO insertion**

(17-5)

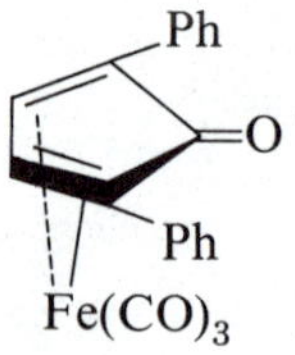

redistribution reactions

(17-6)

$$SnMe_4 + SnCl_4 \rightleftharpoons Me_3SnCl + MeSnCl_3$$

These types of reactions and others are described in this chapter. The emphasis is generally on reaction mechanisms and, where possible, on the usefulness of the reactions in organic synthesis or in actual or potential industrial processes. Broader and more complete coverage is found in review articles[1] and monographs.[2-15]

THE 16 AND 18 ELECTRON RULE AND REACTIONS OF TRANSITION METAL ORGANOMETALLIC COMPOUNDS

First, we must discuss again one of the more important organizing devices in inorganic chemistry: the 16 and 18 electron rule. As also discussed in Chapter 15, this rule states that thermodynamically or kinetically stable, diamagnetic organometallic compounds of the transition metals have a total of 16 or 18 valence electrons. In this chapter we add a corollary to this rule: *Reactions of transition metal organometallic compounds generally proceed by elementary steps that involve only 16 or 18 valence electron species as intermediates.*†

Tolman has expanded this basic notion into a more general scheme for organizing the reactions of organic compounds of the transition metals.[16] He has proposed that the reactions of such compounds can be broken down into five elementary reaction types, each with its microscopic reverse. Examples are given in Table 17-1. The first type is an acid association/dissociation reaction, and it can occur without change in the total

---

†14 valence electron species have been implicated in a few reactions; see page 563 and ref. 17.

TABLE 17-1

CLASSIFICATION OF REACTIONS OF TRANSITION METAL ORGANOMETALLIC COMPOUNDS[a]

Reaction	ΔNVE	ΔOS	ΔN	Example	Reverse Reaction	ΔNVE	ΔOS	ΔN
Lewis acid ligand dissociation	0	−2	−1	$HCo(CO)_4 \rightleftarrows H^+ + Co(CO)_4^-$	Lewis acid association	0	+2	+1
	0	0	−1	$Cp_2WH_2 \cdot BF_3 \rightleftarrows BF_3 + Cp_2WH_2$		0	0	+1
Lewis base dissociation	−2	0	−1	$Pt(PPh_3)_4 \rightleftarrows PPh_3 + Pt(PPh_3)_3$	Lewis base association	+2	0	+1
Reductive elimination	−2	−2	−2	$H_2IrCl(CO)L_2 \rightleftarrows H_2 + IrCl(CO)L_2$	Oxidative addition	+2	+2	+2
Insertion	−2	0	−1	$MeMn(CO)_5 \rightleftarrows Me-\overset{\overset{O}{\|}}{C}-Mn(CO)_4$	Deinsertion	+2	0	+1
Oxidative coupling	−2	+2	0	$\text{Fe(CO)}_3 \rightleftarrows \text{Fe(CO)}_3$	Reductive decoupling	+2	−2	0

[a] NVE = number of valence electrons, OS = oxidation state, and N = coordination number.

number of valence electrons surrounding the metal. Furthermore, since a Lewis acid does not bring or carry away a pair of electrons in a reaction, acid association or dissociation can occur with either 16 or 18 electron molecules. However, Lewis base dissociation, the second reaction type, is restricted to 18 electron molecules, because the base carries away two electrons to give a 16 electron species. Conversely, Lewis base association can occur only with a 16 electron molecule. Reductive elimination, insertion, and oxidative coupling are restricted to 18 electron molecules because all three reactions lead to a loss of two valence electrons by the metal.[17] The reverse of each of these processes—oxidative addition, deinsertion, and reductive coupling—can occur, therefore, only with 16 electron molecules, because all lead to an increase in the number of valence electrons.

Because this chapter is meant to cover the reactions of both main group and transition metal organometallic compounds, Tolman's scheme is not the most convenient for organizing the chapter. Rather, we have put the chemistry of organometallic compounds as a whole into a somewhat more general framework, which only broadly resembles Tolman's scheme. Thus, the first two sections, on association reactions and on substitution, contain examples of the first two reaction types in Table 17-1. The third section, on addition/elimination reactions, encompasses both insertion/deinsertion and oxidative addition/reductive elimination reactions. The next chapter deals with catalysis, a reaction type especially characteristic of transition metals and a type wherein the 16 and 18 electron rule will find its greatest usefulness.

ASSOCIATION REACTIONS

The Lewis Acidity and Basicity of Organometallic Compounds

The interaction of Lewis acids and bases pervades chemistry, so much so that the principles of such reactions were discussed in Chapter 5. In Chapter 8 (page 244) some chemistry of main group metal complexes was described. Thus, in this brief section, we discuss only some aspects of transition metal basicity.

That main group metal compounds function as acids is easily accepted, and you are accustomed to thinking of transition metals as acids. But then, one is used to considering only transition metal compounds in which the metal is in a formal oxidation state of 2+ or higher. In the low oxidation states characteristic of organometallic compounds, however, the metal may function as a site of Lewis basicity in a reaction that is the first of Tolman's five elementary reaction types.

There are currently many examples of metal basicity,[18,19] and such compounds are of great interest because of possible importance in metal-catalyzed reactions. In particular, **Vaska's compound**—carbonylchlorobis(triphenylphosphine)iridium—and its rhodium analog react with boron trihalides to form 1:1 complexes with a $M \rightarrow B$ donor-acceptor linkage.

(17-7)

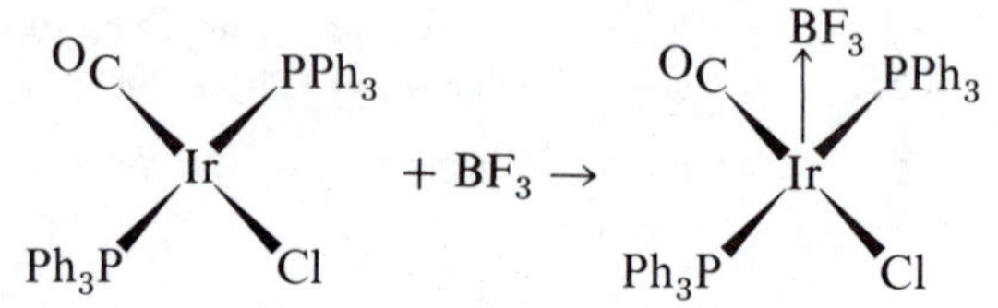

Most importantly, however, the rhodium compound reacts only with BCl_3 and BBr_3 but not with BF_3, whereas the iridium compound reacts with BF_3; since the order of acid strength is clearly $BF_3 < BCl_3 < BBr_3$ (see Chapter 5), this implies that the iridium compound is more basic than the rhodium derivative. Although there are exceptions, this same trend—an increase in basicity on descending a group—is observed for other metal sub-groups; the reason for the inversion in behavior when compared with main group bases is not yet clear.

Metal-containing anions also function as nucleophiles in displacement reactions of several types. We discussed in Chapter 16 the use of such reactions as a method of synthesis of metal-carbon bonds (p. 496).

(17-8)

$$CH_3\overset{O}{\overset{\|}{C}}{-}Cl + [Fe(CO)_2(\eta^5\text{-}C_5H_5)]^- \rightarrow Cl^- + CH_3\overset{O}{\overset{\|}{C}}{-}Fe(CO)_2(\eta^5\text{-}C_5H_5)$$

Indeed, the relative nucleophilicities of a number of metal-containing anions have been measured by determining their rates of reaction with an alkyl halide (Table 17-2). On the basis of these data and experiments with other acids, some general statements regarding metal basicity or nucleophilicity can be made:[19,20]

1. With the exception of some compounds of the iron sub-group, metal basicity and nucleophilicity generally increase on descending a given group. This is clearly opposite the usual trend noted in main group chemistry.

2. CO is one of the strongest known π acceptor ligands. Therefore, replacement of CO by less capable π acids (such as PPh_3 and η^5-C_5H_5) increases the relative nucleophilicity of the metal.

3. In contrast to main group bases, metal bases frequently undergo large geometrical changes upon reaction with an acid.

(17-9)

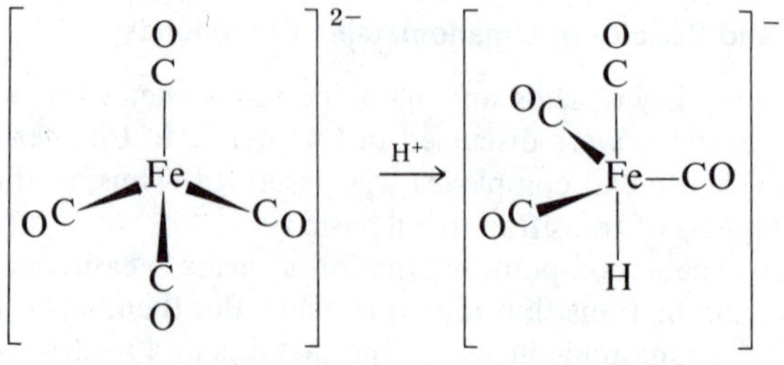

TABLE 17-2

RELATIVE NUCLEOPHILICITIES OF SOME METAL CARBONYL ANIONS[a,b]

Anion	Relative Nucleophilicity
$(C_5H_5)Fe(CO)_2^-$	70,000,000
$(C_5H_5)Ru(CO)_2^-$	7,500,000
$(C_5H_5)Ni(CO)^-$	5,500,000
$Re(CO)_5^-$	25,000
$(C_5H_5)W(CO)_3^-$	500
$Mn(CO)_5^-$	77
$(C_5H_5)Mo(CO)_3^-$	67
$(C_5H_5)Cr(CO)_3^-$	4
$Co(CO)_4^-$	1

[a] The data are taken from R. E. Dessy, R. L. Pohl, and R. B. King, *J. Amer. Chem. Soc.,* **88**, 5121 (1966).

[b] The nucleophilicity of $Co(CO)_4^-$ has arbitrarily been set at 1.

4. Main group anions of the type Ph_3M^- (M = Sn, Ge) are *more* nucleophilic than those containing transition metals. This is not unexpected, since the negative charge of the Group IVA anions is not stabilized as well as in a transition metal anion with π acceptor ligands.

In addition to reacting with BX_3 and H^+, $(\eta^5\text{-}C_5H_5)_2Fe_2(CO)_4$ and other organo-metal carbonyls form complexes with triorganoaluminum compounds.[21] However, in addition to attacking the metal, aluminum alkyls may also bind to the oxygen of the CO ligand as in $(\eta^5\text{-}C_5H_5)_2Fe_2(CO)_4 \cdot 2AlEt_3$ (**1**).

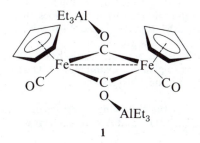

1

Although CO functions as a Lewis base toward BH_3 and transition metals through the carbon atom, the iron-containing Et_3Al adduct (**1**) represents one of the first observations that the carbon monoxide molecule could similarly utilize the oxygen lone pair, albeit only when the carbon is bound to a metal.

Ligand Protonation

Treatment of a transition metal organometallic compound with a strong protonic acid can result in ligand protonation as well as metal protonation. For example, buta-dienetricarbonyliron (**2**) eventually gives **5** upon reaction with anhydrous HCl, and reaction 17-11 shows that ligand rearrangement can occur subsequent to the ligand

(17-10)

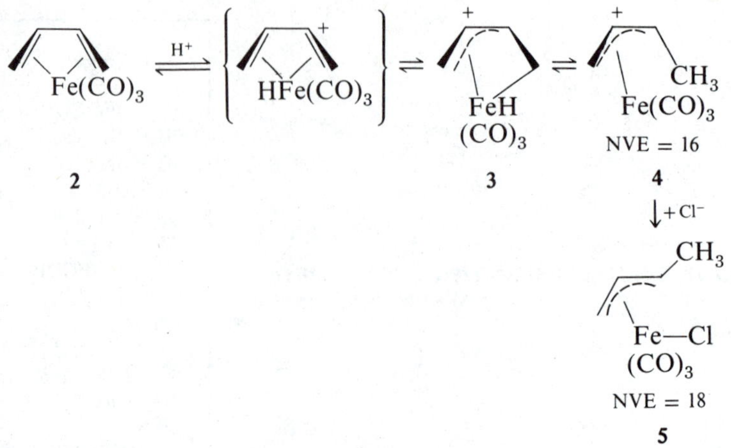
H⁺
Fe(CO)3
HFe(CO)3
FeH
(CO)3
CH3
Fe(CO)3
NVE = 16
2
3
4
+Cl⁻
CH3
Fe—Cl
(CO)3
NVE = 18
5

protonation.[22] The mechanism and stereochemistry of ligand protonation, which can be important to our understanding of electrophilic substitutions on ligands, has been the subject of several studies.[23]

(17-11)

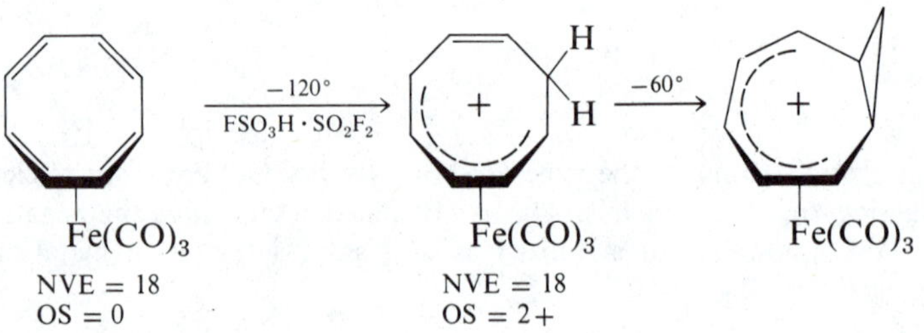
−120°
FSO3H · SO2F2
H
H
−60°
Fe(CO)3
NVE = 18
OS = 0
Fe(CO)3
NVE = 18
OS = 2+
Fe(CO)3

STUDY QUESTIONS

1. The Al−N bond in $Me_3N{\cdot}AlCl_3$ is shorter than in the $AlMe_3$ analog. Account for this in terms of Al hybrid orbital *s/p* character, as in Chapter 3.

2. In reaction 17-7, would the stretching frequency of the CO ligand be expected to increase or decrease upon adduct formation? Does this reaction fit Tolman's 16/18 electron rule? Calculate ΔNVE, ΔOS, and ΔN for this reaction.

3. The iron atom in $Fe(CO)_5$ is a weak Lewis base. However, replacement of a CO ligand with a phosphine to give, for example, $Fe(CO)_4PPh_3$ causes the metal basicity to be enhanced. Why should this be the case?

4. Place the steps of each of the following reactions into one of the categories in Table 17-1 and calculate ΔNVE, ΔOS, and ΔN for each step.

a) 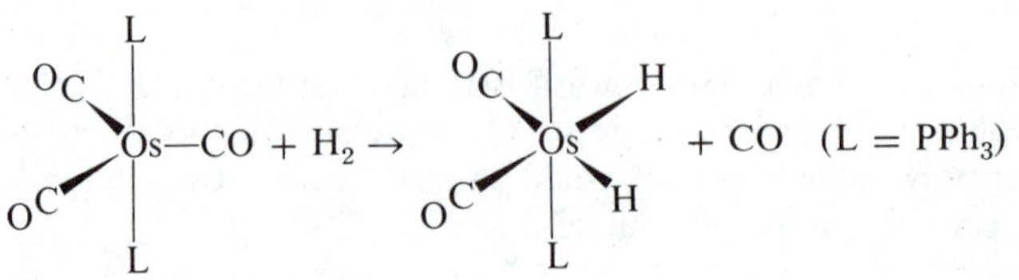
L
OC
Os—CO + H2 →
OC
L
L
OC
H
Os
+ CO (L = PPh3)
OC
H
L

b) 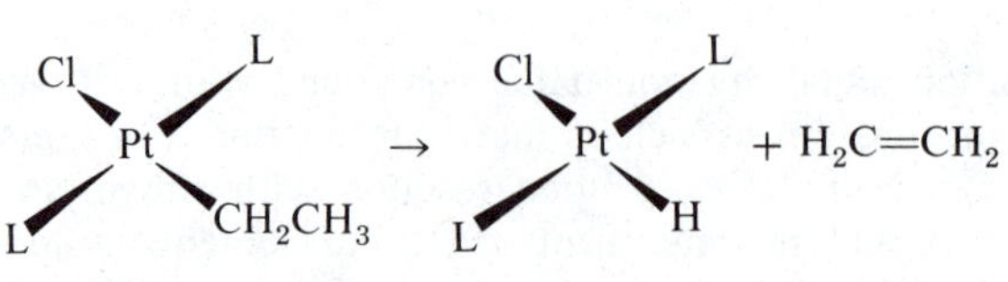
Cl
L
Pt
L
CH2CH3
→
Cl
L
Pt
L
H
+ H2C=CH2

c)

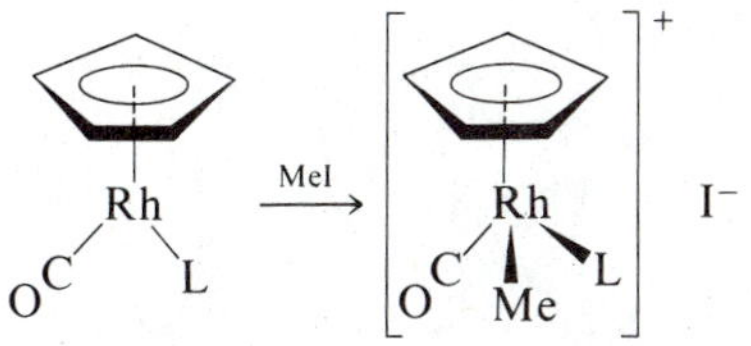

d)

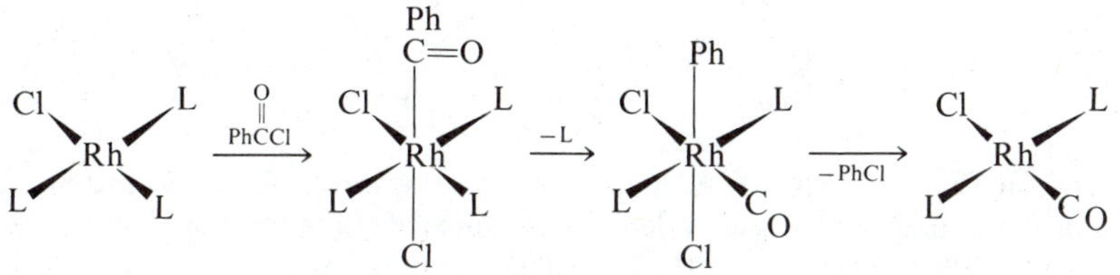

SUBSTITUTION REACTIONS[7,24]

Nucleophilic Ligand Substitutions[25,26]

The importance of metal carbonyls lies in the fact that they are the starting place for the syntheses of so many other organometallic compounds. Therefore, the mechanism and stereochemistry of the nucleophilic substitution of CO by another ligand or ligands have been studied extensively, and some general principles have evolved that help in understanding these reactions and in predicting methods for the syntheses of new compounds.

Many substitution reactions of metal carbonyls proceed according to a rate law that is first order in metal carbonyl and zero order in the substituting ligand; fewer substitutions are first order in both. This generalization was partly established by a very careful examination of the substitution of CO by isotopically labeled CO or phosphine in $Ni(CO)_4$ (17-12).[27]

$$2Ni(CO)_4 + 3L \; (= CO \text{ or } PPh_3) \rightarrow Ni(CO)_3L + Ni(CO)_2L_2 + 3CO$$

(17-12)

The following dissociative ($= D$) mechanism was proposed:

$$Ni(CO)_4 \xrightarrow{k_{sub}} Ni(CO)_3 + CO \quad (slow)$$

(17-13)

$$Ni(CO)_3 + L \rightarrow Ni(CO)_3L \qquad (fast)$$

(17-14)

The hypothesis that Ni—CO dissociation is rate determining is consistent with a $\Delta H^{\ddagger}$ of about 105 kJ/mole (the *mean* Ni—CO bond dissociation energy is 146 kJ/mol) and a positive $\Delta S^{\ddagger}$.

The kinetic behavior of ligand substitution reactions such as [28]

$$Cr(CO)_6 + (n\text{-Bu})_3P \rightarrow [(n\text{-Bu})_3P]Cr(CO)_5 + CO$$

(17-15)

(17-16)

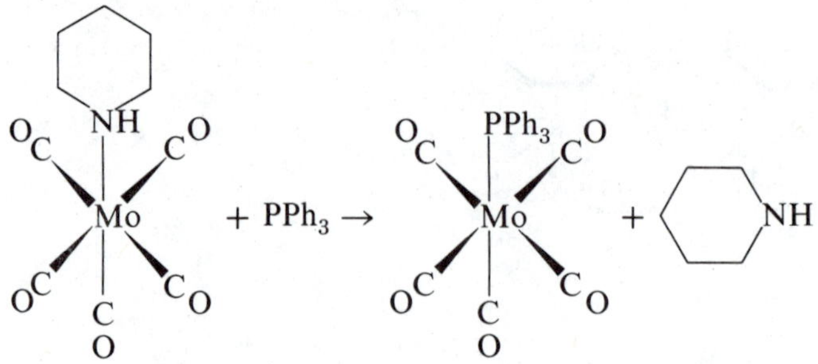

lies between that for $Ni(CO)_4/L$ and that usually observed for substitution reactions of octahedral complexes of higher valent metal ions (*cf.* Chapter 13). That is, the rate law for 17-15 and 17-16 is given by the general expression

$$\text{Rate of disappearance of starting material} = k_1[\text{substrate}] + k_2[\text{nucleophile}][\text{substrate}]$$

This rate law is reminiscent of those for substitution reactions of square planar complexes, but its interpretation is very different.[29] A very careful study of reaction 17-16 clearly shows that the rate law is consistent with the following overall mechanism.[30]

(17-17)

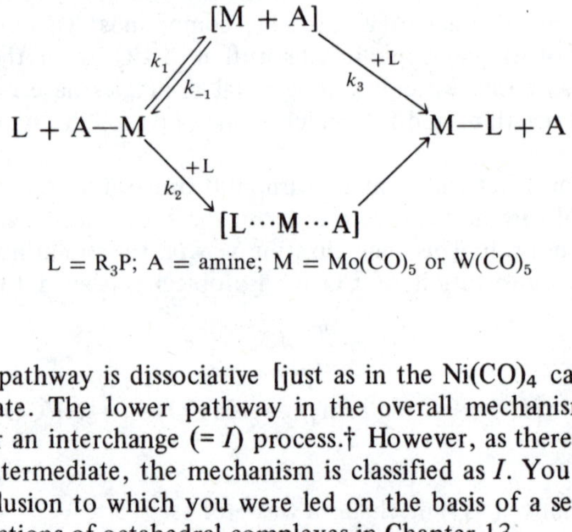

$L = R_3P; \ A = \text{amine}; \ M = Mo(CO)_5 \text{ or } W(CO)_5$

That is, the upper pathway is dissociative [just as in the $Ni(CO)_4$ case] and leads to an $M(CO)_5$ intermediate. The lower pathway in the overall mechanism can be either an associative ($= A$) or an interchange ($= I$) process.† However, as there is no evidence for a seven-coordinate intermediate, the mechanism is classified as I. You will recall that this is exactly the conclusion to which you were led on the basis of a second order rate law for substitution reactions of octahedral complexes in Chapter 13.

That both pathways in 17-17 are dissociatively activated (D and I_d) is in agreement with Tolman's rules (Table 17-1) for reactions of transition metal organometallic compounds. A dissociatively activated pathway will see the starting material change from an 18 electron molecule to a 16 electron species and then back to an 18 electron molecule. An associatively activated path would necessarily involve a 20 electron species, a most unlikely event based on the current state of understanding of organometallic reactions. On the basis of the study of reaction 17-16 and others like it, it may be concluded that *all instances of substitution at metal carbonyl centers in which there is a distribution of 18 electrons probably occur by dissociative pathways, either D or I_d.*

---

†Refer to Chapter 7 for a complete discussion of rate laws and the various mechanisms possible for a chemical reaction.

Electrophilic and Nucleophilic Attack on Coordinated Ligands[31,32]

In the previous section, the substitution of one ligand for another was discussed. In this part, we examine briefly substitutions on the ligands themselves, reactions that most frequently occur by electrophilic mechanisms.

The chemistry of ferrocene and its derivatives has been studied more extensively than that of perhaps any other organometallic compound.[15,33] The cyclopentadienyl rings are aromatic, and much of the chemistry of ferrocene or its derivatives may be predicted on this basis. The major reaction types observed with ferrocene (and to some extent rutheno-cene, osmocene, and tricarbonyl-η^5-cyclopentadienylmanganese, among others) are as follows:

1. **Acetylation.** This reaction is of some historical importance, because it established the aromatic character of ferrocene.

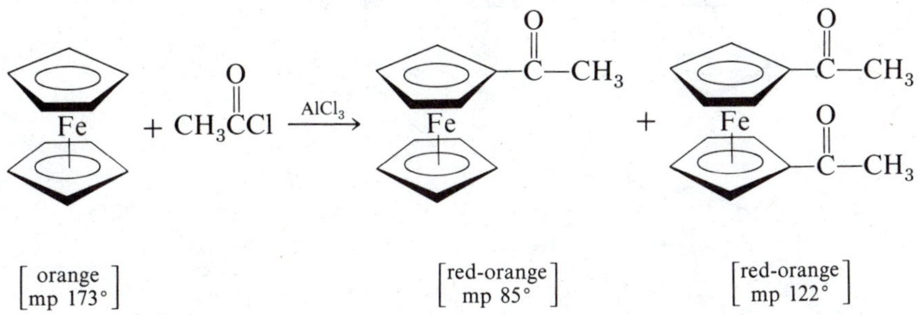

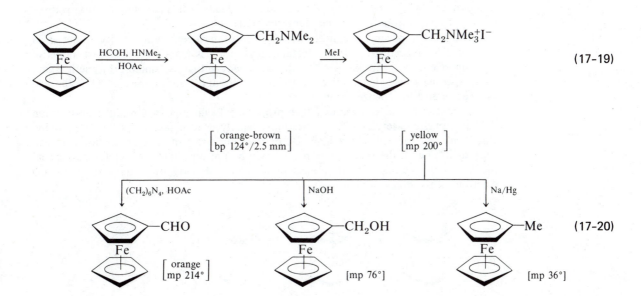

(17–18)

The reaction may be catalyzed by any Lewis acid (most often $AlCl_3$), and proceeds to give both mono- and 1,1'-disubstituted products, with the latter being formed in very small amounts. If disubstitution on one ring does occur, it is most frequently *ortho*.

2. **Aminomethylation.** Dimethylamine and formaldehyde undergo a Mannich reaction with ferrocene to give dimethylaminomethylferrocene, a compound useful in the preparation of many other derivatives.

(17–19)

(17–20)

3. Metalation. Lithioferrocene (Chapter 16, p. 494) and chloromercuriferrocene, two readily prepared organometallic compounds, are widely used as intermediates in the preparation of other ferrocene derivatives.

(17-21)

(17-22)

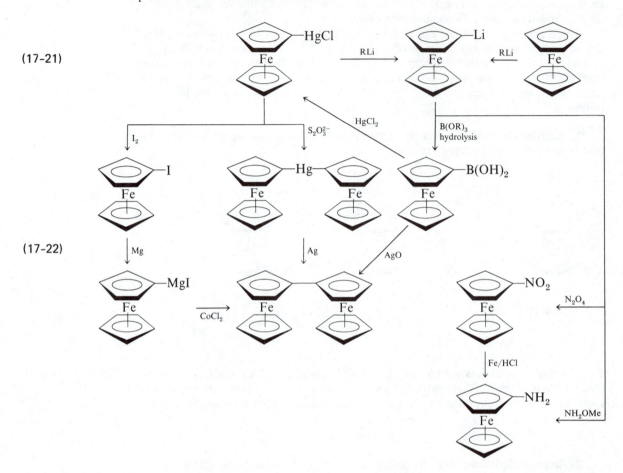

4. Nitration and Halogenation. The direct introduction of $-NO_2$ or $-X$ into ferrocene is not possible, as the metallocene is oxidized to ferricenium ion or other degradation products under the reaction conditions. Instead, both nitro and halo derivatives must be prepared from the metalated intermediates above.

The most important chemical characteristic of ferrocene is its ability to act as an electron releasing substituent; the metallocenyl group is among the strongest inductive electron releasing agents known.[34,35] In agreement with this, aminoferrocene is found to be a stronger base than aniline by a factor of about 20, and ferrocene carboxylic acid is a weaker acid than benzoic acid.

Ferrocene has an anomalously high propensity to undergo electrophilic substitution in that it may be acetylated 10^6 times more rapidly, and mercurated 10^9 more rapidly, than benzene. To explain the high rate of reaction, direct metal participation has been suggested.[36] Given the fact that ferrocene may be protonated at the metal in strong acid media, it has been argued that an electrophile could similarly attack the weakly bonding electrons of the iron (reaction 17-23);

(17-23)

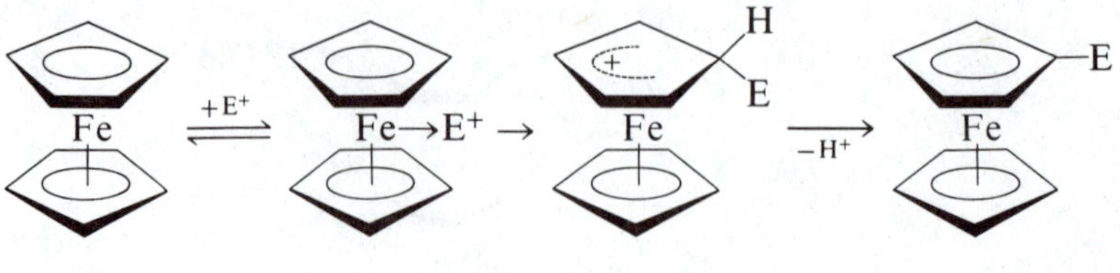

this would be followed by transfer of the electrophile to the C_5 ring in an *endo* manner and by proton elimination.

In contrast to the properties of $C_5H_5^-$ coordinated to $[Fe(C_5H_5)]^+$, a Group VIB metal tricarbonyl group lowers the electron releasing ability of C_6H_5X in $(\eta^6\text{-}C_6H_5X)$ $M(CO)_3$ (M = Cr, Mo, W). [For example, aniline is a stronger base $(pK_b = 11.70)$ than $(\eta^6\text{-}C_6H_5NH_2)Cr(CO)_3$ $(pK_b = 13.31)$, and benzoic acid $(pK_a = 5.68)$ is a weaker acid than $(\eta^6\text{-}C_6H_5COOH)Cr(CO)_3$ $(pK_a = 4.77)$.[37]] Most importantly, this suggests a new synthetic route to benzene derivatives. Most reactions occurring at an aromatic ring (*e.g.*, Friedel-Crafts acylation and alkylation) are electrophilic substitutions. To carry out a nucleophilic substitution requires the attachment of a strongly electron-withdrawing group to the ring prior to the desired substitution; the electron-withdrawing group must then be removed later, a process not often easily done. However, an electron-withdrawing group such as the π-bonded $-M(CO)_3$ group can be easily attached to benzene and then removed simply by mild oxidizing agents. For example, π-bonded benzene or chlorobenzene is converted in high yield to phenylisobutyronitrile after a few hours at room temperature (17–24).[38]

$$(17\text{-}24)$$

5. If reaction 17–15 is carried out using $Mo(CO)_6$ or $W(CO)_6$ instead of the chromium compound, comparison of k_1 and k_2 values (obtained at the same temperature) is quite informative. For example, while the k_2/k_1 ratio is 0.7 for $Cr(CO)_6$, it is 9.6 for $Mo(CO)_6$ and 34.8 for $W(CO)_6$ (all data having been obtained at $112°C$). [See J. R. Graham and R. J. Angelici, *Inorg. Chem.*, 6, 2082 (1967).] Considering the structure of the transition state in an interchange process, suggest a reason for the relatively greater importance of the interchange mechanism as the metal atomic weight increases.

6. Refer to reactions 17–18 through 17–22, and outline:

   a) three ways to prepare ferrocenyldiphenylphosphine, $(C_5H_5)Fe(C_5H_4PPh_2)$.
   b) a preparation of ferrocenyl carboxylic acid.

ADDITION AND ELIMINATION REACTIONS[8,39]

The 1,1- and 1,2-addition of metal-containing compounds to unsaturated substrates, or the 1,1-oxidative addition of organic halides and other molecules to low-valent metals in compounds of low coordination number, constitutes one of the most important forms of organometallic reactivity. Such reactions have been implicated in the homogeneous hydrogenation and polymerization of olefins, in the formation of metal-carbene complexes, in rearrangement reactions of metal carbonyls, and in many others. In Chapter 16 we discussed the usefulness of hydro- and organometalations in the synthesis of organometallic compounds of the main group elements. In this section we shall discuss briefly a few examples of addition reactions of organometallic compounds themselves, with a special emphasis on mechanism and stereochemistry.

1,2-Additions to Double Bonds

In this portion we are interested in reactions of the type

$$X_n'M\text{—}X + Z\text{=}Y \rightarrow X_n'M\text{—}Z\text{—}Y\text{—}X \qquad (17\text{-}25)$$

and such reactions are exemplified by additions to the following types of compounds.

a) Alkenes and Alkynes:

(17-26) $Et_6Al_2 + R-CH=CH_2 \rightarrow [R-CH(Et)-CH_2-AlEt_2]_2$

(17-27)

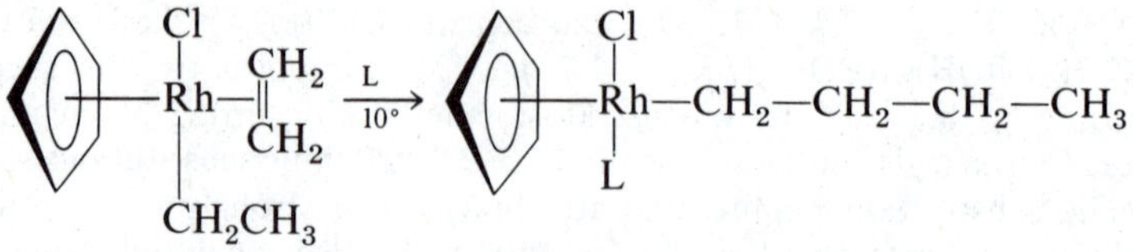

(17-28)

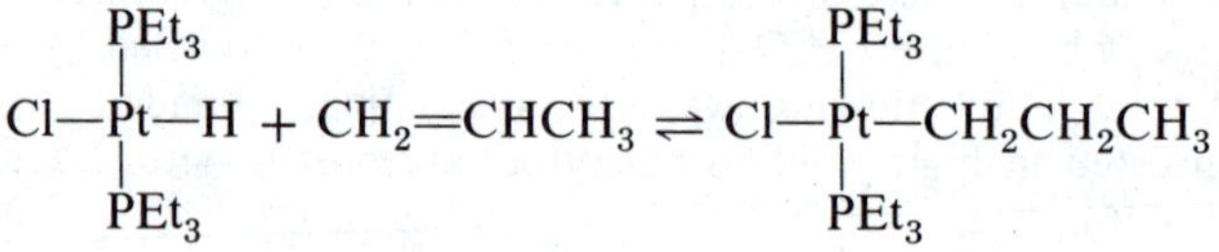

b) Carbon Dioxide and Ketones:

(17-29)

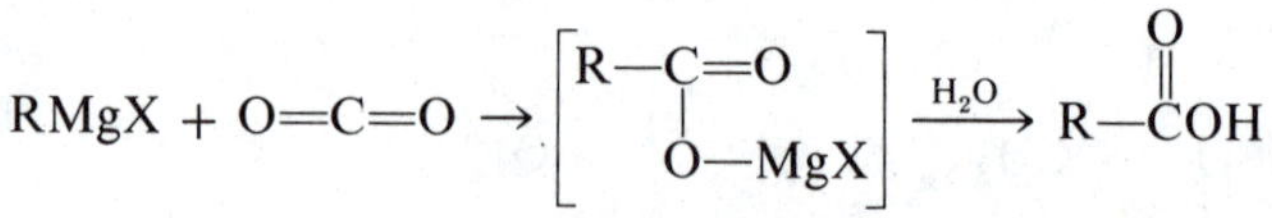

(17-30)

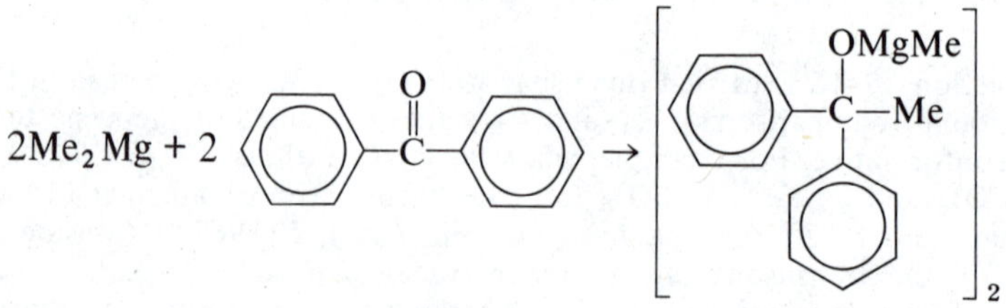

c) Cyanides and Isocyanates:

(17-31)

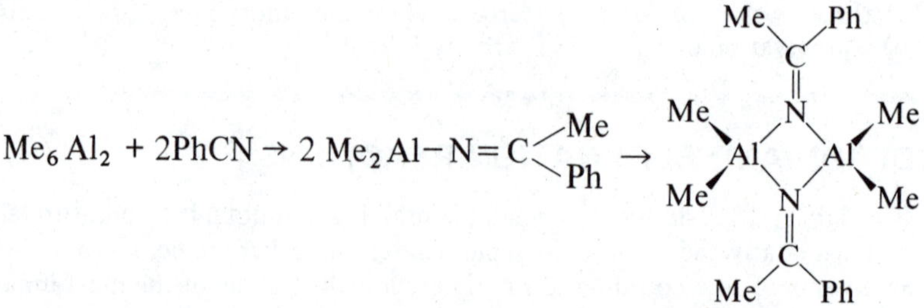

(17-32)

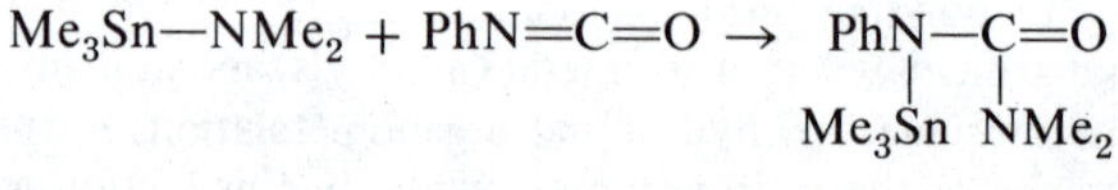

All of these reactions apparently occur by the same general mechanism: initial coordination of the organometallic $X_n'M-X$ to the substrate ZY followed by migration of X to the substrate.[39a]

(17-33)

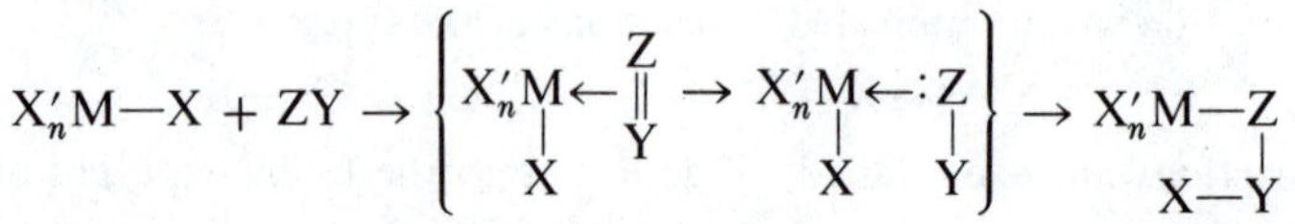

As illustrations of this general scheme, we shall examine a few mechanisms in detail. However, one last point is pertinent; addition reactions are often referred to by organo-metallic chemists as **insertion reactions**, because the substrate ZY is inserted into the M—X bond.

Because of the importance of the alkylaluminum-catalyzed "growth reaction," that is, the polymerization of olefins in the presence of aluminum alkyls (Chapter 16, p. 500), the kinetics and mechanism of the addition of organoaluminum compounds to C—C unsaturated systems has been extensively explored.[40] One important piece of mechanistic evidence, is that the rate of olefin or acetylene absorption by triethylaluminum is proportional to the square root of the Et_6Al_2 concentration. Similarly, the triphenyl-aluminum reaction with diphenylacetylene (17–34) is one-half order in the arylaluminum compound.[41] On the other hand, tri-*iso*-butylaluminum absorbs ethylene at a rate proportional to the concentration of $(i\text{-Bu})_3Al$. These observations make sense when you know that almost all organoaluminum compounds, except for tri-*iso*-butylaluminum, are largely dimeric in solution in non-basic solvents and in the gas phase at normal temperatures. Therefore, the reactions are really first order in organoaluminum monomer. This, of course, has a parallel in hydroborations (*cf.* Chapter 7, p. 228), where the reactive species is BH_3 and not B_2H_6.

Triphenylaluminum reacts with diphenylacetylene to give the addition compound $Ph_2Al—C(Ph)=CPh_2$.[41]

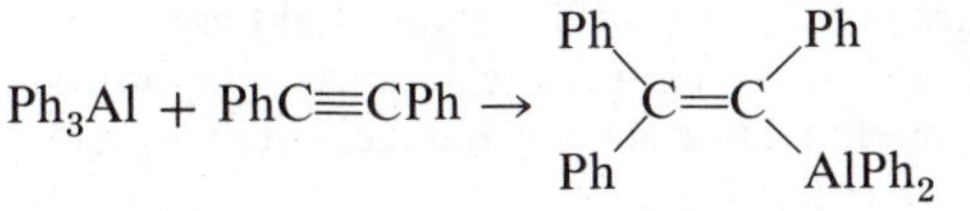

$$Ph_3Al + PhC{\equiv}CPh \rightarrow \quad (17\text{-}34)$$

If an unsymmetrical acetylene is used, three different products are possible. Compound **8** can arise from either *cis* or *trans* addition, but **6** and **7** can only be the result of *cis* or *trans* addition, respectively.

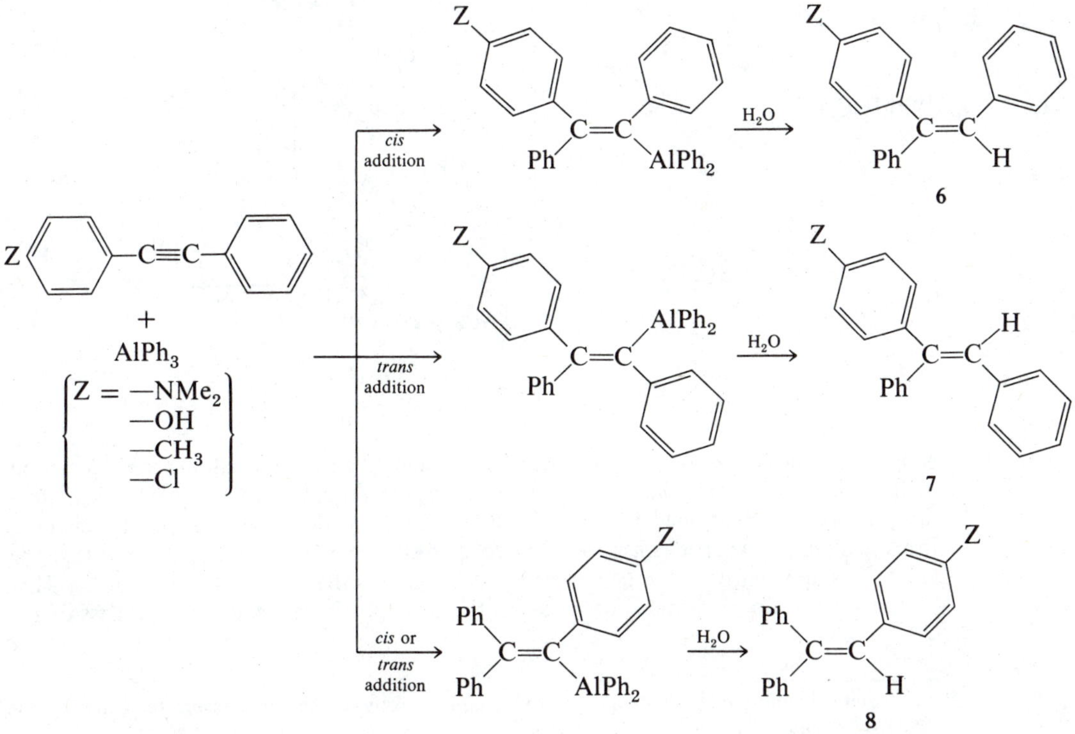

(17-35)

After hydrolysis of the addition product (which is known to proceed with retention of configuration), mixtures of **6** and **8** are isolated, thereby proving that addition proceeds only in a *cis* manner as in hydroborations (*cf.* Chapters 7 and 16). Furthermore, the observation that the relative amount of compound **6** increases as Z becomes more electron-releasing supports the idea of electrophilic attack at an acetylenic carbon and suggests a transition state such as **9**. In this case, attack of the more positively charged Al occurs preferentially at the carbon β to the ring, and the incipient carbonium center α to the ring is then stabilized by electron-releasing substituents, Z, in the transition state.

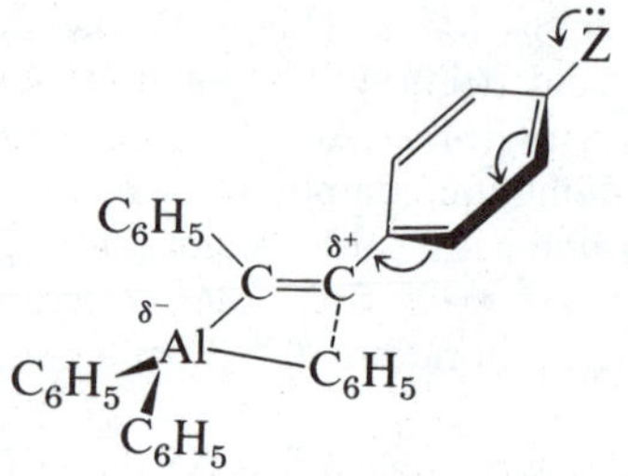

9

Olefin arylation has been accomplished under mild conditions through the addition of an "aryl palladium salt" to an olefin. The detailed composition of the arylating agent, which is made from phenylmercuric acetate and palladium acetate (17–36), is not known, but it is represented as "Ph—Pd—OAc."[42]

(17-36)
$$Ph—Hg—OAc + Pd(OAc)_2 \rightarrow \text{"Ph—Pd—OAc"} + Hg(OAc)_2$$

The olefin arylation reaction itself undoubtedly proceeds by addition of this "aryl palladium salt" to the olefin (17–37), followed by elimination of a palladium hydride.

(17-37)

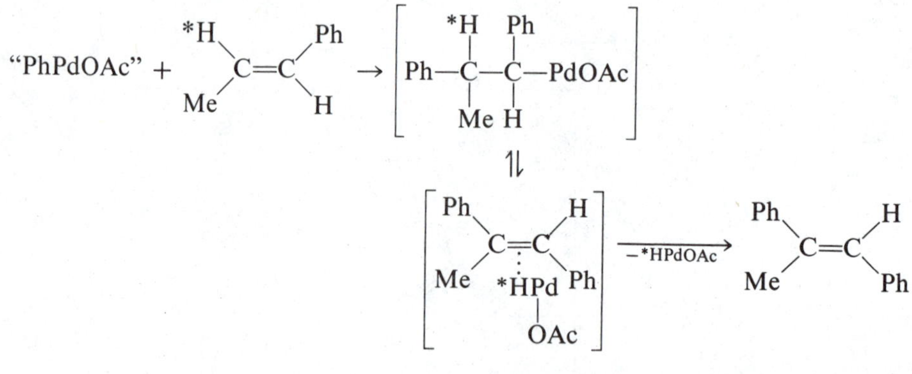

Notice that the palladium generally becomes bound to a hydrogen atom from a β carbon in the course of the elimination reaction; hence, the process is often called a *β-elimination*, a reaction mentioned in Chapter 16 (p. 499) and described later in this chapter. Investigation of the mechanism and stereochemistry of olefin arylation shows that the reaction is apparently controlled more by steric effects than by electronic factors.† That is, the aryl group predominantly adds to the less substituted carbon, regardless of the

---

†Steric inhibition by the bulky phenyl group is likely to be more severe than that by the "PdOAc" group.

electronic nature of the substituents. Furthermore, the stereochemistry of addition is most likely *cis*, as is elimination of the "hydridopalladium salt."

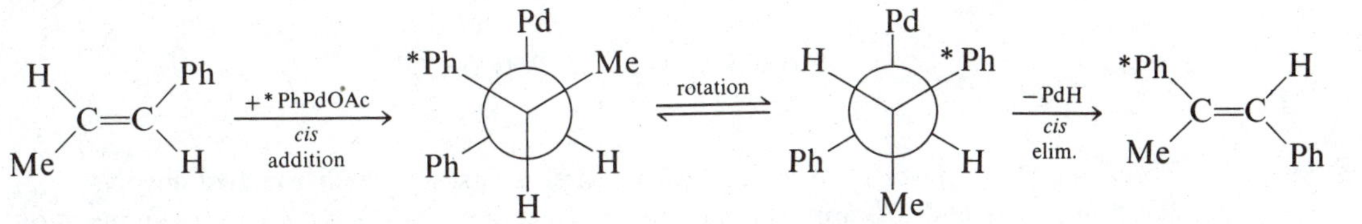

$$(17\text{-}38)$$

Yet another very important example of a 1,2-addition or insertion reaction is reaction 17-39. In this instance, the elements of Pt and H add across the double bond.

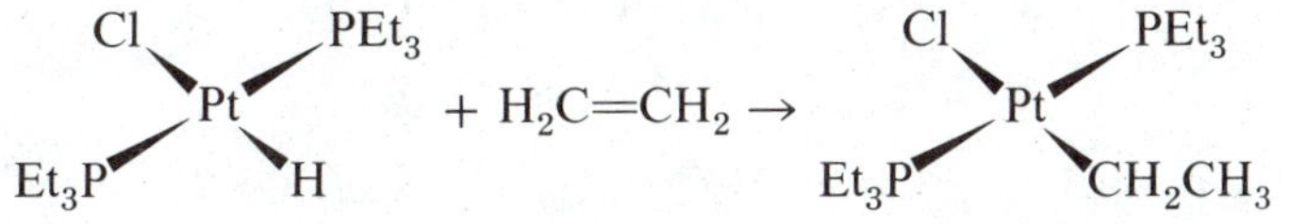

$$(17\text{-}39)$$

Such reactions are undoubtedly involved in metal-catalyzed olefin hydrogenations, and their mechanism is pursued in the next chapter. However, before dismissing the topic, we should mention *hydrozirconation*,[43] the addition of the elements of zirconium and hydrogen across double bonds, which has potential in organic synthesis.

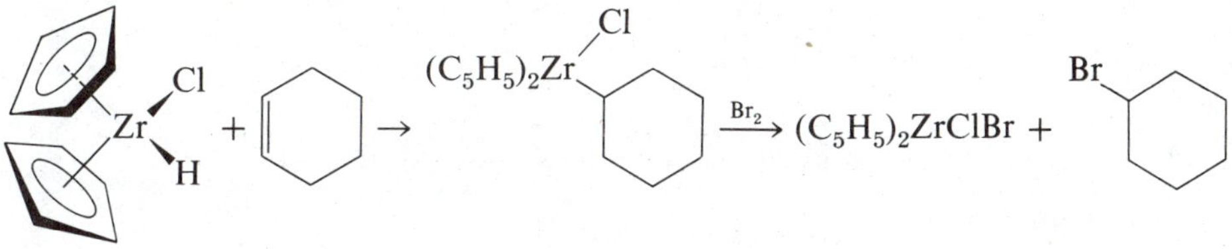

$$(17\text{-}40)$$

There are two distinct advantages of the hydrozirconation method when compared with hydroboration or hydroalumination: (i) unlike the boron and aluminum alkyls, dry air does not decompose the zirconium alkyl (but the latter can be hydrolyzed to give the appropriate alkane), and (ii) cleavage of the Zr—C bond with halogen or acyl halide to provide the organic derivative leaves the compound $(\eta^5\text{-}C_5H_5)_2ZrClX$, which can be recycled back to the starting metal hydride.†

1,1-Addition to CO: Carbonylation and Decarbonylation

The *oxo process* is one of the most important examples of homogeneous catalysis. In a system composed of a cobalt salt, CO, H_2, and an olefin, the main reaction that occurs is **hydroformylation**, the conversion of the olefin to an aldehyde.[44]

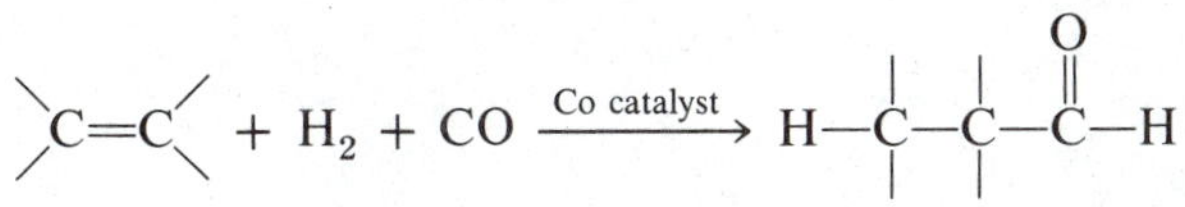

$$(17\text{-}41)$$

---

†Cp_2ZrClH can be produced from Cp_2ZrCl_2 (commercially available) by reduction of the latter with Vitride in THF. [Vitride = $NaAlH_2(OCH_2CH_2OCH_3)_2$, Eastman Organic Chemicals.] See Chapter 8 for a discussion of the replacement of X by H^-.

Under the reaction conditions (temperatures of 90° to 200°C and pressures of 100 to 400 atm), the cobalt is presumed to be in the form of $Co_2(CO)_8$ and $HCo(CO)_4$, these compounds being related by an equilibrium reaction involving hydrogen.

(17-42)
$$Co_2(CO)_8 + H_2 \rightleftharpoons 2HCo(CO)_4$$

The aldehyde is thought to be formed in three steps, of which the first involves 1,2-addition of the hydridocarbonyl to the olefin (reaction 17-43); the second is an **insertion of CO** into the cobalt-alkyl bond (**carbonylation**; reaction 17-44), and the third is a reaction of the acyl derivative with H_2 to give the aldehyde and regenerate catalyst (reaction 17-45).† It is the second of these steps in which we are now interested.

(17-43)
$$HCo(CO)_4 + H_2C{=}CH_2 \rightarrow H_3C{-}CH_2{-}Co(CO)_4$$

(17-44)
$$H_3C{-}CH_2{-}Co(CO)_4 + CO \rightarrow H_3C{-}CH_2{-}\overset{\overset{\text{O}}{\|}}{C}{-}Co(CO)_4$$

(17-45)
$$H_3C{-}CH_2{-}\overset{\overset{\text{O}}{\|}}{C}{-}Co(CO)_4 + H_2 \rightarrow H_3C{-}CH_2\overset{\overset{\text{O}}{\|}}{C}H + HCo(CO)_4$$

The insertion of CO into metal-carbon bonds has been extensively studied,[8,45] and one example is reaction 17-46 involving methylpentacarbonylmanganese (reversible when L = CO).

(17-46)
$$H_3C{-}Mn(CO)_5 + L \rightleftharpoons H_3C{-}\overset{\overset{\text{O}}{\|}}{C}{-}Mn(CO)_4L$$

(L = CO, amine, phosphine)

In general, a two-stage mechanism seems to operate (17-47), the reaction passing through an intermediate in which the acetyl group is formed by *migration of the R group* to one of the *cis* CO ligands.[45a]

(17-47)

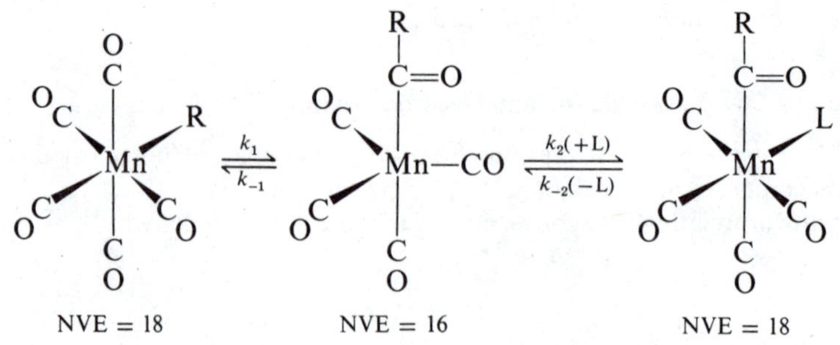

$$\text{NVE} = 18 \qquad \text{NVE} = 16 \qquad \text{NVE} = 18$$

---

†Reactions 17-43 to 17-45 are not meant to convey mechanistic details. This aspect of the overall process is considered on page 577.

In spite of the fact that such reactions apparently occur quite generally by R migration,

$$\overset{O}{\underset{\|}{}}$$

formation of M$-$C$-$R from R$-$M$-$CO is often referred to as a CO **insertion reaction**,[45] a terminology used in this chapter.

One of the most elusive aspects of the CO insertion reaction has been its stereo-chemistry. That is, does the CO insert in the R$-$Mn bond, or does the methyl group migrate to the carbonyl ligand? This question has been examined in many cases, by looking not only at the carbonylation reaction itself but also at the reverse, or decarbonylation, reaction (*i.e.*, the reverse of 17–47). The assumption has been that carbonylation and decarbonylation must follow the same reaction coordinate, and, if the alkyl group is the migrating group in the carbonylation process, it must also be the migrating group in the reverse.[45a] Such experiments as the decarbonylation of $CH_3COMn(CO)_4(^{13}CO)$ have unambiguously shown that it is the alkyl group which migrates.[46] For example, according to 17–48 and 17–49, the two possible pathways would give significantly different product distributions. In fact, *cis-* and *trans*-$CH_3Mn(CO)_4(^{13}CO)$ are obtained in a 2:1 ratio, thereby proving that methyl migration is the correct mechanism.

Carbonyl insertion

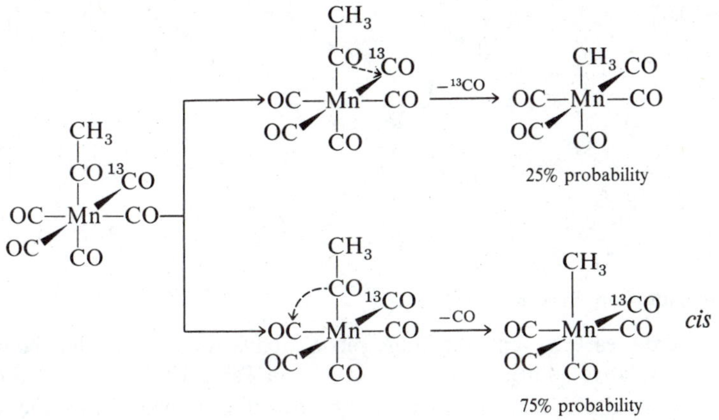

(17–48)

Methyl migration

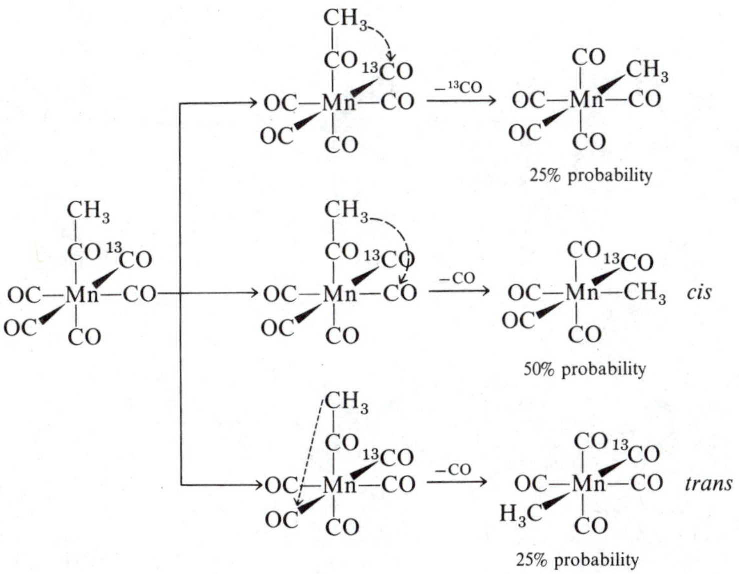

(17–49)

The migration of organic groups to produce acyl-metal complexes has also recently been applied to organic synthesis. Reaction of pentacarbonyliron with sodium in dioxane (to which benzophenone is added to produce $Ph_2CO^-Na^+$) gives an excellent yield of the white dioxane complex of $Na_2Fe(CO)_4$.[47]

(17-50)

$$Fe(CO)_5 + Na \xrightarrow{\text{dioxane, } 100°} Na_2Fe(CO)_4 \cdot 1.5 \text{ dioxane}$$

Because of its chemistry, this compound has been called a transition metal Grignard reagent. For example, it reacts with alkyl halides to give an alkyltetracarbonyliron complex, which may then be decomposed with acid or O_2/H_2O to give, respectively, the alkane or carboxylic acid. More important to the current discussion of addition reactions is the reaction with a Lewis base such as CO or a phosphine to give 10 *via* an alkyl migration. This acyl-metal complex can then be used to produce a variety of organic compounds.

(17-51)

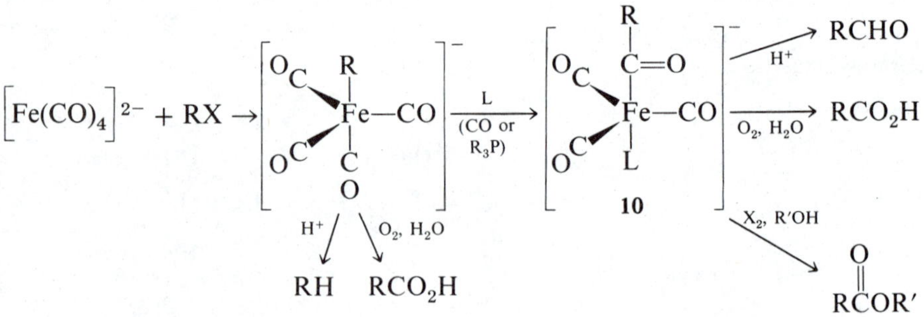

Oxidative Addition Reactions[48-52]

One of the earliest generalizations recognized in organometallic chemistry was the idea of an oxidative addition reaction. As seen in Table 17-1, such reactions involve an increase in formal oxidation state and in coordination number; hence the term oxidative addition. Some examples are given in reactions 17–52 and 17–53, in which all steps are labeled with their reaction types.

(17-52)

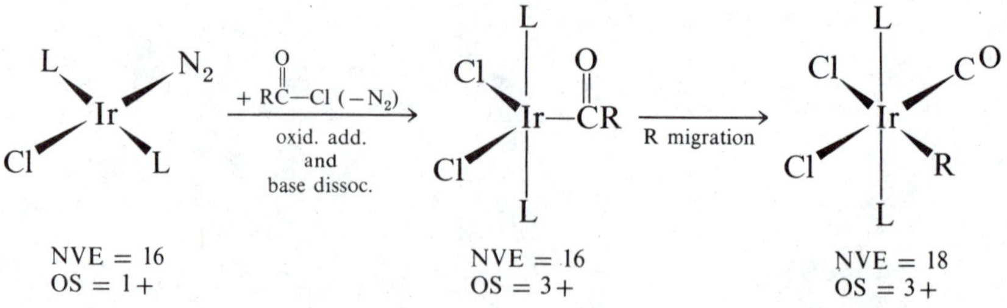

(17-53)

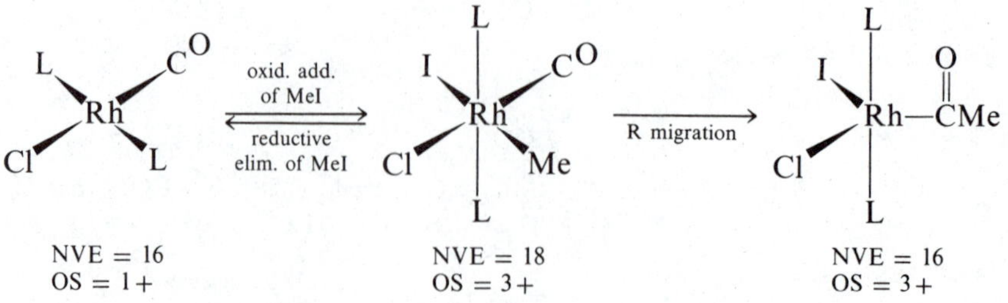

Yet another example of an oxidative addition is found in an industrial process for the conversion of methanol to acetic acid (reaction scheme 17-54).[52] The key step in this process is the oxidative addition of methyl iodide to Rh(I). You should notice that R migration (carbonyl insertion) is also involved in a later step, as is reductive elimination (the latter process is described in the next section of this chapter).

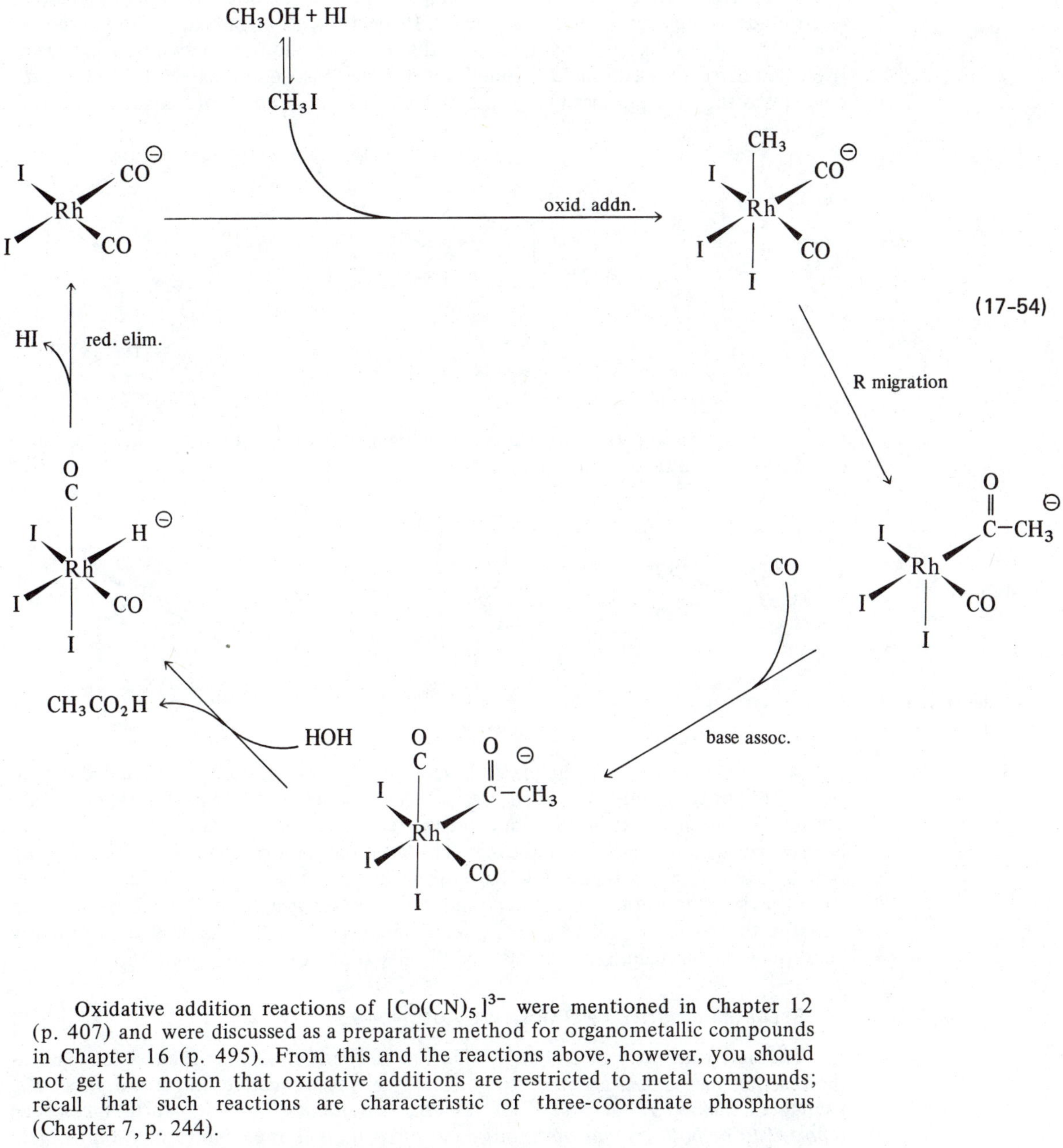

(17-54)

Oxidative addition reactions of $[Co(CN)_5]^{3-}$ were mentioned in Chapter 12 (p. 407) and were discussed as a preparative method for organometallic compounds in Chapter 16 (p. 495). From this and the reactions above, however, you should not get the notion that oxidative additions are restricted to metal compounds; recall that such reactions are characteristic of three-coordinate phosphorus (Chapter 7, p. 244).

Before examining in greater detail oxidative additions of d^8 and d^{10} transition metal organometallic compounds, we should make one additional point. That is, the addition reactions discussed in the previous sections involve addition of M—X to some unsaturated substrate (alternatively, the substrate inserts into the M—X bond); no metal oxidation occurs. In contrast, in oxidative addition reactions it is the metal complex that is unsaturated and can add various substrates (alternatively, the metal complex inserts into an E—E' bond).

GENERAL CONSIDERATIONS

Usually, d^8 complexes of Fe(0), Co(I), and Ni(II) and their congeners are four-coordinate, square planar, 16 electron complexes or five-coordinate, trigonal bipyramidal, 18 electron complexes (*cf.* Chapter 15). Both polar and non-polar species such as H_2, halogens, hydrogen halides, alkyl and acyl halides, organotin halides, mercuric halides, organosilicon halides and hydrides, and other similar compounds can add to these transition metal compounds, sometimes reversibly. In general, addition to square planar, 16 electron, d^8 complexes gives six-coordinate, 18 electron, d^6 complexes (17-53), whereas one of the original ligands must be lost in the oxidative addition of two new ligands to five-coordinate, 18 electron, d^8 complexes to give a six-coordinate, 18 electron, d^6 complex as the final product (17-55); in some cases the intermediate in such a reaction may be isolated.

(17-55)

$$\text{Os(CO)}_3\text{L}_2 \xrightarrow[\text{oxid. add.}]{I_2} \left[\text{Os(CO)}_3\text{L}_2\text{I} \right]^+ I^- \xrightarrow[\substack{\text{base} \\ \text{exch.}}]{-CO, +I^-} \text{Os(CO)}_2\text{L}_2\text{I}_2$$

NVE = 18	NVE = 18	NVE = 18
OS = 0	OS = 2+	OS = 2+

Frequently, d^{10} complexes are coordinatively unsaturated, being only three- or four-coordinate, so addition occurs readily.

(17-56)

$$(\eta^2\text{-}C_2H_4)\text{Pt(PPh}_3)_2 \xrightarrow[\text{oxid. add.}]{+MeI} (\eta^2\text{-}C_2H_4)\text{Pt(PPh}_3)_2(\text{CH}_3)(\text{I}) \xrightarrow[\text{base dissoc.}]{-C_2H_4} \text{Pt(PPh}_3)_2(\text{CH}_3)(\text{I})$$

NVE = 16	NVE = 18	NVE = 16
OS = 0	OS = 2+	OS = 2+

Additions to d^8 complexes have been most thoroughly studied using $IrCl(CO)(PPh_3)_2$, frequently called Vaska's compound† after Lauri Vaska, who first synthesized the material in 1961 and recognized its importance as a model compound for studies in homogeneous catalysis.[53] The reactions of this compound are generally first order in both reactants, and the entropy of activation ($\Delta S^{\ddagger}$) is usually negative. Although such activation parameters and the rate law suggest an A or I_a mechanism wherein the metal functions as a nucleophile, there is still considerable uncertainty concerning the mechanism of oxidative addition, a topic taken up on page 560.

STEREOCHEMISTRY OF OXIDATIVE ADDITIONS

The stereochemistry of oxidative additions has been closely scrutinized for clues to the reaction mechanism and the structure of the transition state. In general, *non-polar substances such as H_2 add in a cis fashion, whereas alkyl halide addition has been reported to be both cis and trans, but most often trans.* For example, addition of MeCl to $IrBr(CO)(PPh_2Me)_2$ gives a *trans* addition product (**11**) if the reaction is carried out in benzene. However, on refluxing in $MeOH$-C_6H_6, **11** is converted to **12**, a complex that would have resulted if the original oxidative addition had been *cis*. This result suggests

---

†Vaska's compound is synthesized by refluxing $IrCl_3$ with PPh_3 in an alcohol or ether; the CO apparently arises from the solvent.

that the stereochemistry of oxidative addition is the result of kinetic and not thermo-dynamic control.

(17-57)

11   **12**

Another aspect of the stereochemistry of oxidative additions is the stereo-chemistry of the carbon atom newly bound to the metal in alkyl halide reactions. That is, by noting the configuration about this carbon atom before and after addition, some insight may be gained into the nature of the transition state. One very thorough study of this aspect of oxidative additions involved the series of reactions outlined by 17-58.[52] S-(−)-α-phenethylbromide oxidatively adds to carbonyltris(triphenylphosphine) palladium to give, after a subsequent alkyl migra-tion (with retention of configuration), compound **13** with inversion of configura-tion. Notice that compound **13** can also be produced by reaction of the acyl halide of known configuration, **14**, with $Pd(PPh_3)_4$. This proves that configuration inversion occurred in the addition of Ph(Me)(Br)CH, since addition of **14** can occur only with retention.

(17-58)

S-(−)   **R-(+)**   **R-(+)**   **R-(−)**

13   **14**

INFLUENCE OF CENTRAL METAL, LIGANDS, AND ADDEND ON OXIDATIVE ADDITION

The one aspect of oxidative addition reactions that has been emphasized from the beginning is that the metal complex can be thought of as functioning as a nucleophile, and anything that affects the nucleophilicity (or basicity) of the metal therefore influences the course of the reactions. For this reason, the effect of the metal itself, the electronic and steric properties of the ligands, and the addend (the electrophile) have been extensively examined, and a reasonably consistent reactivity picture has begun to emerge.

Although numerous studies of the effects of various factors on oxidative additions could be cited, one specific example will serve well; this example also shows beautifully the twists and turns of real research problems.

Collman began a study of the reaction of Vaska's compound with acyl azides, expecting to observe an adduct, **15**. Instead, a molecular nitrogen adduct, **16**, was isolated![54] As this work was done shortly after the first molecular nitrogen complexes were discovered, the isolation of $IrCl(N_2)(PPh_3)_2$ was an especially important result and warranted further study. Therefore, the original reaction was repeated with carefully purified reagents and solvent (chloroform)—to the detriment of the reaction; **17** was formed instead. By observing the infrared spectrum of the reaction mixture during the course of the reaction, it was found that the ethyl alcohol that is present in commercial chloroform as an oxidation inhibitor traps the acyl isocyanate that is the product of the first stage of the reaction. If this is not trapped, the acyl isocyanate reacts further with the N_2 complex to produce **17**.

(17-59)

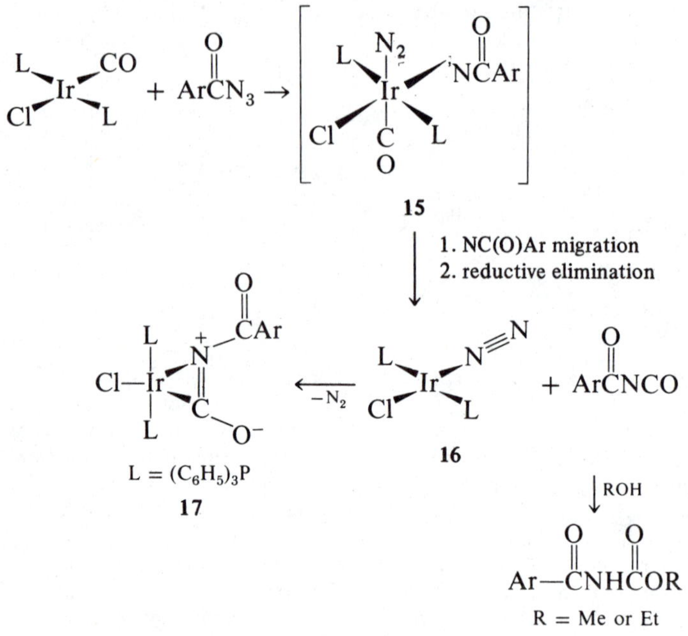

The importance of reaction 17–59 to the current discussion is the change in rate of formation of **15** with changes in metal, halogen, and azide. That is, the rate data show that the rate decreases as:

  i. the metal is changed from iridium to rhodium
  ii. the halogen is changed in the order $I > Br > Cl > N_3$
  iii. the phosphine is changed in the order $PhEt_2P > Ph_2MeP > Ph_3P$
  iv. the azide is changed in the order $PhCON_3 > PhN_3 > p\text{-}MeC_6H_4N_3$

It is readily apparent from these observed reactivity trends that the central metal functions as a base and the azide as an acid; any change in the metal complex that leads to an increase in electron density at the central metal leads to an increase in reactivity. These same general trends have been found, with a few exceptions, in numerous other studies of the kinetics of oxidative addition reactions as a function of ligand.

The reactivity trend suggested by item i is general and mirrors the periodic trend in metal complex basicity noted on p. 542. The following chart summarizes the periodic trend in ease of oxidation addition for d^8 metals.

MECHANISM OF OXIDATIVE ADDITION

The intimate mechanism of oxidative addition is a matter of considerable speculation. We can only reflect the current uncertainty in the literature,[52,55,55a] and emphasize that generalizations are difficult because the mechanism may be changed as a consequence of simple alterations in the ligand environment at the metal center.

Basically, three mechanistic pathways have been proposed: (i) bimolecular nucleophilic substitution (classical S_N2 mechanism with inversion at the carbon center in a reacting alkyl halide); (ii) concerted nucleophilic attack resulting in a three-center transition state with varying degrees of asymmetry and polarity (retention or even inversion of configuration); and (iii) free radical paths (leading to racemization). Prior to 1972 only the first two of these paths were considered. In general, oxidative additions involving *alkyl halides* and the Pt-group metals are first order in each reactant, the entropy of activation is usually quite negative, the activation parameters are dependent on solvent polarity, and the rate is dependent on the apparent nucleophilicity of the metal complex. Such observations suggest nucleophilic substitution and a polar transition state.

One example of a mechanism that meets the criteria noted just above, in addition to explaining configuration inversion and the lack of evidence for solvent-separated ions seen in reaction 17–58 and others like it, is outlined below.[52] You should recognize that other modes of attack and other transition state structures are possible.

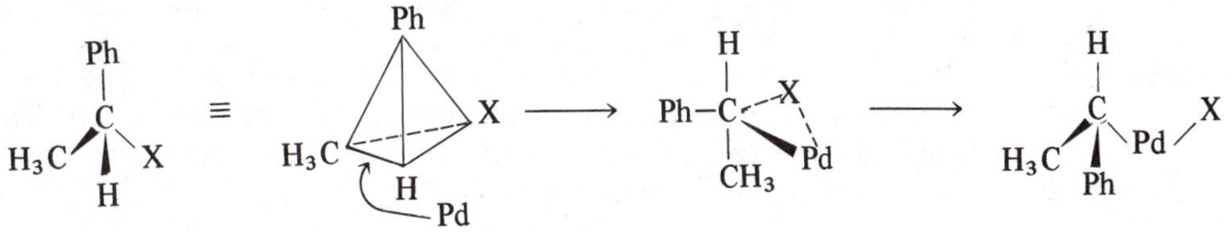

$$(17\text{-}60)$$

Pd = coordinatively unsaturated palladium compound

More recently, considerable experimental evidence has accumulated that suggests that radical mechanisms may compete with nucleophilic attack. The most unequivocal evidence thus far is reaction (17–61), wherein both radical products of initial one-electron transfer were observed.[56]†

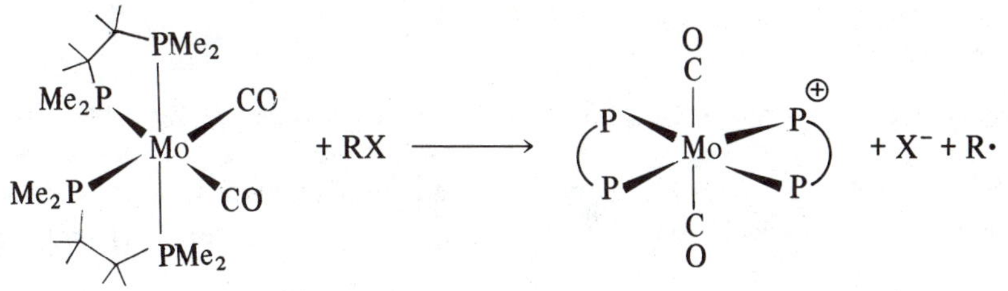

$$(17\text{-}61)$$

Elimination Reactions and the Stability of Metal–Carbon σ Bonds[57,58,58a]

Until recently, one of the least understood phenomena in organometallic chemistry was the wide range of stabilities observed for metal-carbon σ bonds. For example, whereas tetramethyltitanium decomposes at $-65°C$ in ether to give methane, quite stable metal-carbon σ bonds are observed in $Pt(PPh_3)_2Cl(CH_3)$ and other Pt(II) complexes, $Cr(CH_2CPhMe_2)_4$ (structure **3**, page 490), and vitamin B_{12} (structure **15**, page 507), among others. Many chemists had contended that transition metal–carbon σ bonds in particular were simply unstable in a thermodynamic sense, with kinetic stability being achieved by the presence of π bonding or bulky ligands. However, recent thermodynamic measurements have shown that transition metal–carbon σ bonds are often more stable

---

†The initial products of this reaction are actually thought to be $RX^{\cdot-}$ and $[cis\text{-}(Me_2PC_2H_4PMe_2)_2 Mo(CO)_2]^+$, but neither is observed. $RX^{\cdot-}$ very rapidly decays to $R^{\cdot}$ and X^-, and the *cis* metal-containing ion rapidly isomerizes to the *trans* form.

thermodynamically than their main group counterparts (*cf.* Figure 16-5). The wide range of stabilities of transition metal compounds having $M-C$ σ bonds must therefore be explicable, at least in part, in terms of relative kinetic stability. That is, unstable compounds must have low energy pathways available for their decomposition.

Two possible pathways for transition metal–carbon σ bond scission have been proposed: multi-step, radical mechanisms and single-step, non-radical mechanisms. These are discussed briefly at this point because such processes are surely important in metal-catalyzed processes such as the coupling of Grignard and organolithium reagents with alkyl halides (17–62), olefin hydrogenation (page 576), and the Wacker process (page 578).

(17–62)
$$RM + R'X \longrightarrow R-R' + MX$$

A radical pathway is apparently preferred by metals early in the transition series and especially by Hg(II), Hg-C bonds being notably weak in the thermodynamic sense. For example, isobutylneopentylmercury, when heated in CCl_4, produces isobutylene, neopentyl chloride, $CHCl_3$, and metallic mercury.[59] The mechanism for this reaction, which is known to be accelerated by catalytic amounts of free radicals, is a free radical chain process initiated by formation of $CCl_3\cdot$ (reaction 17–63; Np = neopentyl).

(17–63)
$$HC(Me)_2CH_2HgNp + CCl_3\cdot \longrightarrow \cdot C(Me)_2CH_2HgNp + HCCl_3$$

$$\cdot C(Me)_2CH_2HgNp \longrightarrow Me_2C=CH_2 + \cdot HgNp$$

$$\cdot HgNp \longrightarrow Hg^0 + Np\cdot$$

$$Np\cdot + CCl_4 \longrightarrow Np-Cl + CCl_3\cdot$$

Net Reaction

$$HC(Me)_2CH_2HgCH_2CMe_3 + CCl_4 \xrightarrow{100°C} Hg^0 + Me_2C=CH_2 + Cl-CH_2CMe_3 + CHCl_3$$

Two processes that appear to be more generally used by metals later in the transition series are **β-elimination** (17–64) (see also pages 244, 499, and 552) and **binuclear reductive elimination** (17–65).

(17–64) β-elimination†

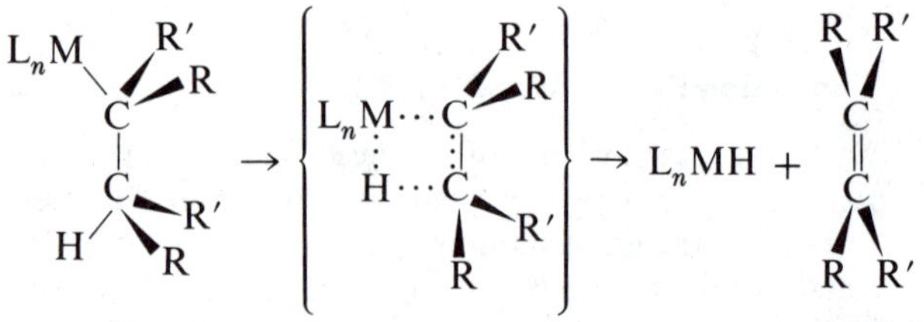

(17–65) **binuclear reductive elimination**[58a]
$$L_nM-H + R-ML_n \rightarrow R-H + 2M + 2nL$$

$$L_nM-R + R'-ML_n \rightarrow R-R' + 2M + 2nL$$

---

†This reaction is the reverse of the hydrometalation reaction.

An excellent example of these two paths is the thermal decomposition of n-butyl-(tri-n-butylphosphine)copper (17-66).[60]

$$2(H_3CCH_2CH_2CH_2)Cu(PBu_3) \xrightarrow[\text{ether}]{0°C}$$
$$H_3C-CH_2-CH=CH_2 + H_3CCH_2CH_2CH_3 + 2Cu + 2PBu_3 \qquad (17\text{-}66)$$

The proposed reaction sequence involves as a first step β-elimination of the olefin and a copper hydride (17-67a), followed by binuclear elimination in the second step (17-67b). This mechanism demands the formation of equal amounts of n-butene and n-butane, as is observed.

$$CH_3CH_2CH_2CH_2Cu(PBu_3) \rightarrow CH_3CH_2CH=CH_2 + HCu(PBu_3) \qquad (17\text{-}67a)$$
$$CH_3CH_2CH_2CH_2Cu(PBu_3) + HCu(PBu_3) \rightarrow CH_3CH_2CH_2CH_3 + 2Cu + 2PBu_3 \qquad (17\text{-}67b)$$

Further examples of β-elimination will be encountered in the chapter on catalysis.

Yet another process characteristic of later transition metals is **1,1-reductive elimination**.

1,1-reductive elimination
$$L_nM \begin{smallmatrix} R \\ \\ R' \end{smallmatrix} \rightarrow L_nM + R-R' \qquad (17\text{-}68)$$

One very well understood example of this process is the loss of alkane from Au(III) complexes of the type R_3AuL [e.g., $(CH_3)_2CD_3Au(PPh_3)$].[61] The first step in the elimination sequence is base dissociation to give a three-coordinate Au(III) complex.†

$$\begin{array}{c} CH_3 \\ | \\ CH_3-Au-PPh_3 \end{array} \rightleftharpoons \begin{array}{c} CH_3 \\ | \\ CH_3-Au \\ | \\ CD_3 \end{array} + PPh_3 \qquad (17\text{-}69)$$

According to the angular overlap model, a d^8, low spin Au(III) complex is expected to be T-shaped (Figure 10-1). However, a Y-shaped isomer (of C_{2v} symmetry) is calculated to be close in energy, and only minor bond angle changes can lead to its formation (see below). This distortion from an ideal trigonal shape seems to forecast an interaction between R groups as a hint of impending R_2 elimination. Reversion to a T-shaped isomer can give a different "geometric" isomer, and continuation of the process can lead ultimately to R group scrambling. Reductive elimination of CH_3-CH_3 or CH_3-CD_3 in our example can occur if the Y-shaped structures act as precursors for elimination; the particular alkane isolated depends on which two R groups are closest to one another in the Y structure.

†Note that the intermediate is a 14-electron species!

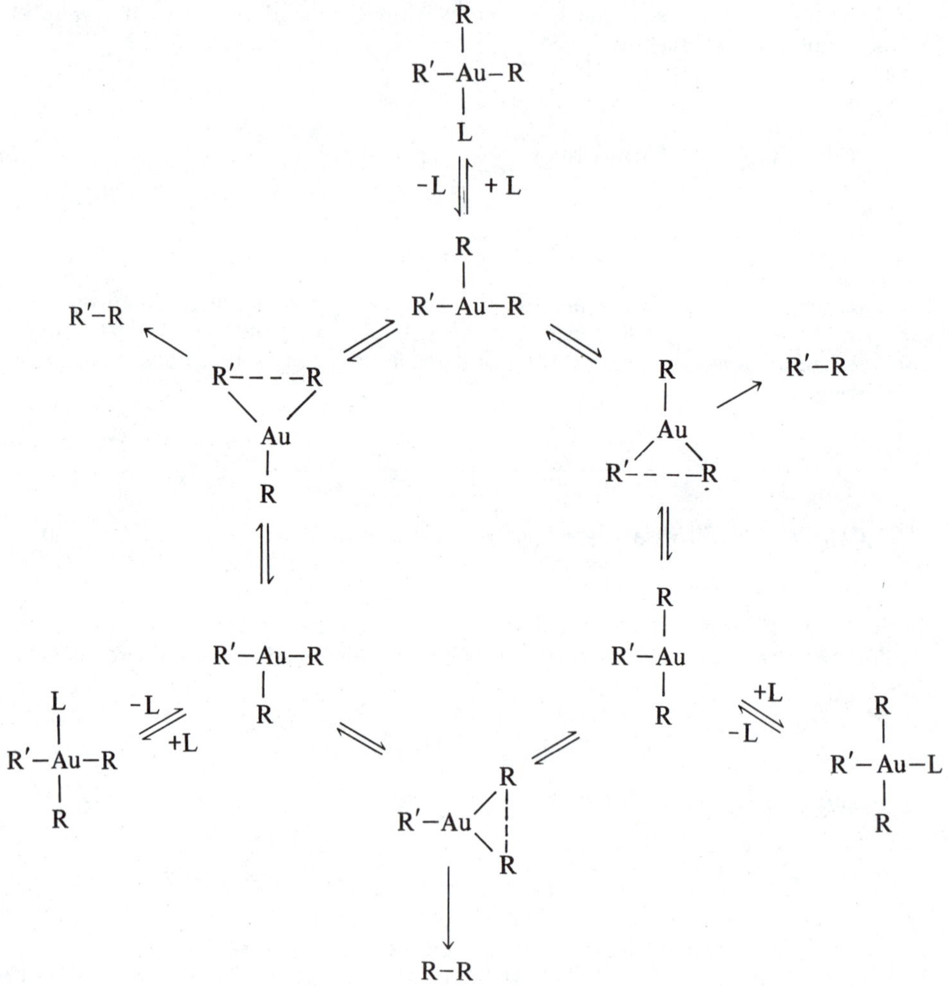

$$R$$
$$|$$
$$R'-Au-R$$
$$|$$
$$L$$

$$-L \big\Updownarrow +L$$

$$R$$
$$|$$
$$R'-Au-R$$

R'—R ←   R'----R   R   R'—R
              \ /   Au
              Au   R'----R
              | |
              R   R
                               ⇊
              ⇊
                            R
                            |
         R'—Au—R   R'—Au
            | |
            R   R

    L   -L   R   R
    | ⇄ | |
R'—Au—R   +L R'—Au—R   ⇄   +L   R'—Au—L
    | | -L |
    R   R   R

                R'—Au
                   |
                   R

                R—R

7. Using addition reactions, suggest syntheses for the following compounds:

a) 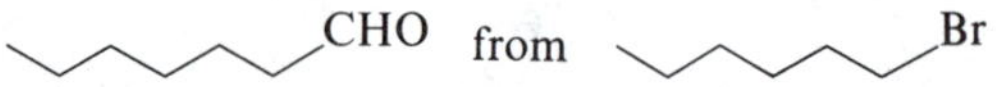 $\sim\sim\sim\sim$CHO   from $\sim\sim\sim\sim$Br

b) 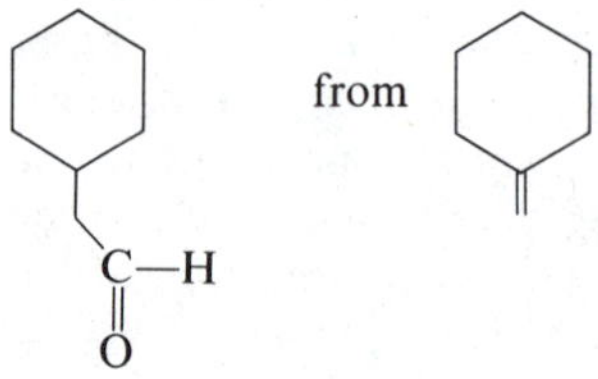 [cyclohexyl-CH$_2$-CHO structure]   from   [methylenecyclohexane structure]

c) 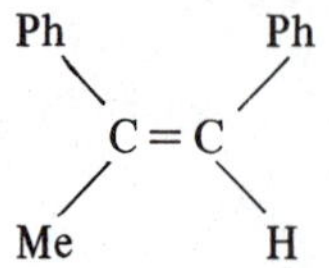 $\underset{Me}{\overset{Ph}{\diagdown}}C=C\underset{H}{\overset{Ph}{\diagup}}$ from   PhC≡CPh

d) 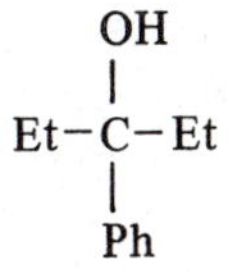
$$
\begin{array}{c}
\text{OH} \\
| \\
\text{Et}-\text{C}-\text{Et} \\
| \\
\text{Ph}
\end{array}
\quad \text{from Et}_2\text{C}=\text{O}
$$

8. If MeCl is added to $IrCl(CO)L_2$, six isomers are possible. Draw the structures of these isomers. [Hint: see p. 386.]

9. Explain why the reaction below proves that the CO lost on decarbonylation is not the acyl CO and that alkyl group (and not CO) migration has occurred.

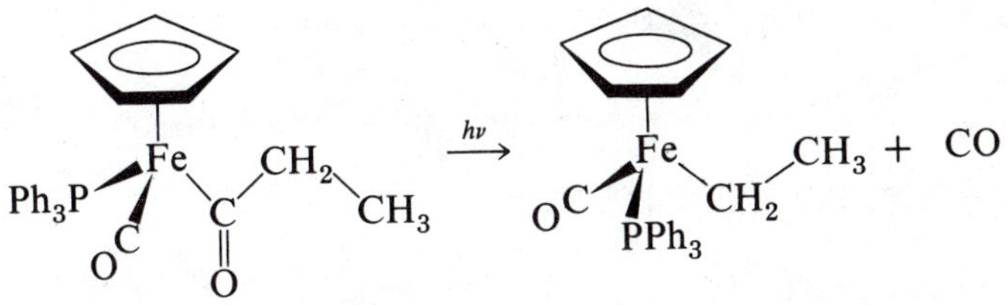

10. Account for the course of the following reaction.

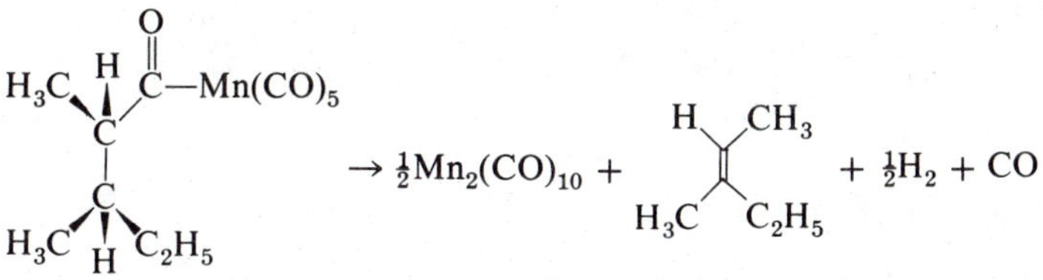

11. Consider the basic oxidative addition reaction shown below.

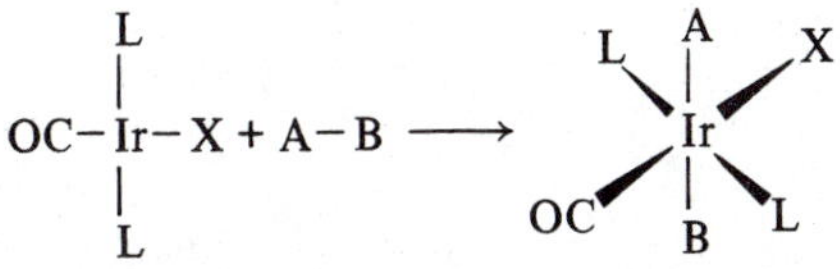

a) Why should the oxidative addition of H_2 generally be accelerated as X is changed from Cl to I (*i.e.*, the rate constant increases by about 10^2)?

b) Methyl iodide reacts faster when L is changed from $P(p\text{-}ClC_6H_4)_3$ to $P(p\text{-}CH_3OC_6H_4)_3$. Why?

12. Based on your knowledge of substitution reactions in square planar complexes from Chapter 13 and the discussion of insertion reactions in this chapter, propose a mechanistic pathway for the following reaction that is first order in both starting materials.

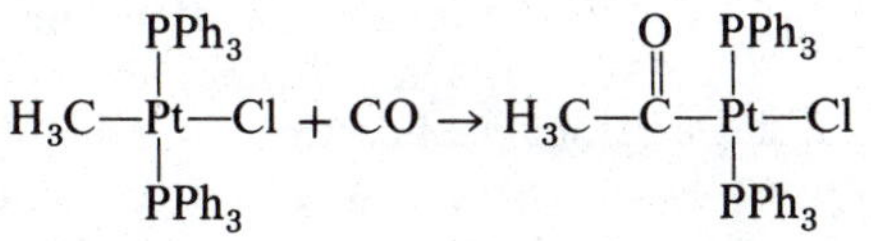

$$
\begin{array}{c}
\text{PPh}_3 \\
| \\
\text{H}_3\text{C}-\text{Pt}-\text{Cl} \\
| \\
\text{PPh}_3
\end{array}
+ \text{CO} \rightarrow
\begin{array}{c}
\quad\;\; \text{O}\;\; \text{PPh}_3 \\
\quad\;\; \| \quad | \\
\text{H}_3\text{C}-\text{C}-\text{Pt}-\text{Cl} \\
\quad\quad\quad | \\
\quad\quad\quad \text{PPh}_3
\end{array}
$$

[If you need some help, see P. E. Garrou and R. F. Heck, *J. Amer. Chem. Soc.*, **98**, 4115 (1976).]

13. Cl^- reacts with *trans*-$[Pt(PEt_3)_2(C_6H_5)_2]$ in the presence of acid to give *trans*-$[Pt(PEt_3)_2(C_6H_5)Cl]$ and benzene. No reaction occurs with LiCl in the absence of acid, and the rate decreases drastically on changing the solvent from methanol to anhydrous diethyl ether. The overall rate expression is

$$-\frac{d[PtL_2Ph_2]}{dt} = k[H^+][PtL_2Ph_2]$$

Suggest a mechanism for the reaction consistent with the information above. [If you need help, see U. Belluco, *et. al., Inorg. Chem.,* **6**, 718 (1967).]

14. In the reductive elimination of ethane from $(CH_3)_2CD_3AuPPh_3$ (reaction 17–69), what ratio of CH_3-CH_3 to CH_3-CD_3 should be observed?

REARRANGEMENT REACTIONS[62]

In this section we take up another reaction type that is prevalent in organometallic chemistry as well as in classical coordination chemistry and main group chemistry. That is, there are redistribution or exchange reactions, which are intermolecular processes, and fluxional isomerism or stereochemical non-rigidity, the intramolecular time-averaging of several possible molecular configurations. Exchange reactions in boron and silicon chemistry are discussed in Chapter 7, pp. 231 and 235, and non-rigidity of silicon and phosphorus compounds is presented in Chapter 7 and of classical coordination compounds in Chapter 14.

Redistribution Reactions[63-66]

There are at least three distinct types of organometallic redistribution reactions:
a) Exchange of a ligand between two molecules of the same type.

(17–70)
$$Me_3B \cdot NMe_3 + BMe_3^* \rightleftharpoons Me_3B + Me_3N \cdot BMe_3^*$$

b) Scrambling of a group within a molecule.

(17–71)

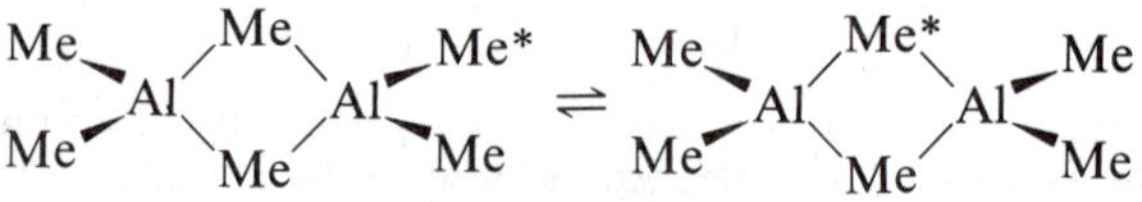

c) Interchange of groups or ligands between two molecules of the same or different types.

(17–72)
$$aMe_4Sn + bSnCl_4 \rightleftharpoons xMe_3SnCl + yMe_2SnCl_2 + zMeSnCl_3$$

One of the most thoroughly studied exchange reactions is the scrambling of the bridge and terminal methyl groups in dimeric trimethylaluminum and analogous organoaluminum compounds.[63-65,67] At room temperature the proton nmr spectrum of Me_6Al_2 in toluene consists of a single resonance line for all methyl groups, but cooling to $-40°$ and finally to $-65°$ causes the line to split into two resonance lines, one at higher field for the terminal methyl groups and one at lower field for the bridging groups (Figure 17-1). There are two possible mechanisms for the bridge-terminal exchange that occurs at room temperature.

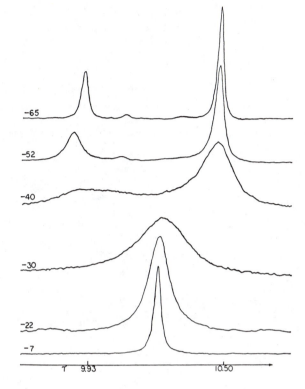

Figure 17-1. Proton nmr spectrum of Me_6Al_2 in toluene. [Reprinted with permission from K. C. Williams and T. L. Brown, *J. Amer. Chem. Soc.*, **88**, 5460 (1966). Copyright by the American Chemical Society.]

i) Rate determining dissociation of the dimer to monomers, followed by monomer recombination in a new orientation or combination of a new pair of monomers.

$$\text{(Me}_2\text{Al(Me)(Me)Al(Me*)(Me))} \rightleftharpoons Me_3Al + (Me*)Me_2Al \rightleftharpoons \text{(Me}_2\text{Al(Me*)(Me)Al(Me)(Me))}$$

ii) An intramolecular process involving the breaking of only one Al—Me—Al bridge followed by rotation about the surviving $Al–C_{bridge}$ bond and reformation of the doubly bridged dimer.

All of the available evidence points to mechanism (i).

Fluxional Isomerism or Stereochemical[62] Non-Rigidity

Another form of molecular rearrangement is the intramolecular time-averaging of two or more possible molecular configurations. This phenomenon has been called fluxional isomerism by Cotton[68] and stereochemical non-rigidity by Muetterties.[69]

We have already discussed this form of isomerism in the case of classical coordination complexes (Chapter 14) and for Si and P compounds (Chapter 7), and we now turn to this phenomenon as it manifests itself in organometallic chemistry.

There are several types of stereochemical non-rigidity observable in organometallic chemistry; among these are:

a) Rearrangement of σ-bonded polyenes.[70]

(17–73)

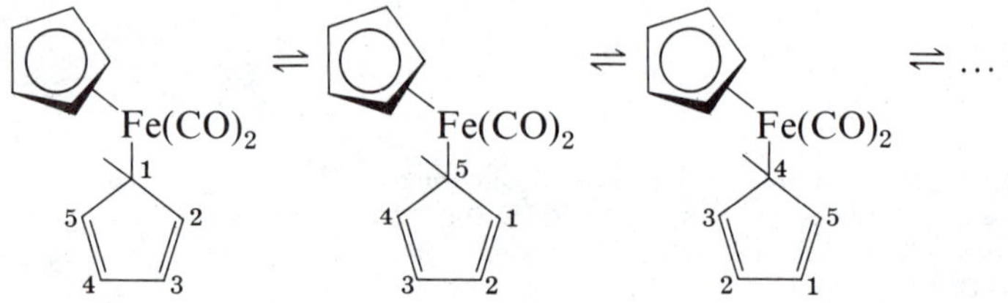

b) Rearrangement of acyclic π-bonded polyenes.[70]

(17–74)

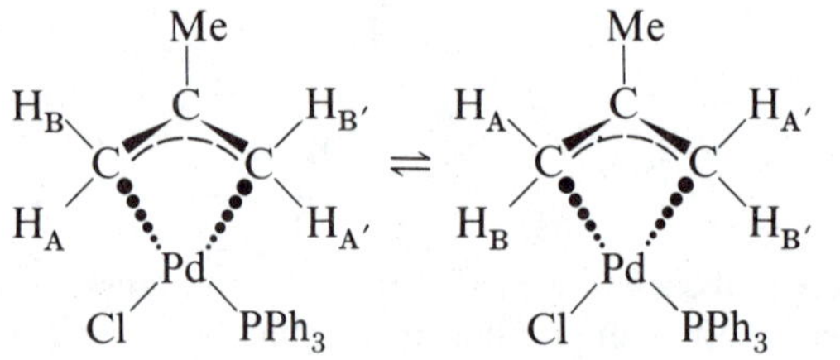

c) Rearrangement of metal carbonyls.[71]

(17–75)

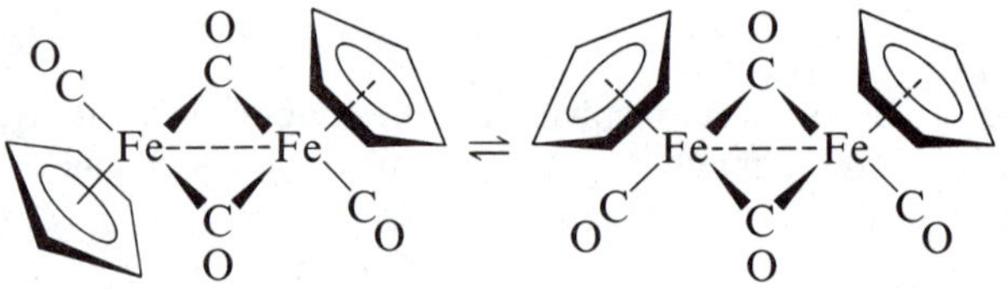

Only the second of these types will be considered here.

One of the more important forms of stereochemical non-rigidity is found in metal-η^3-allyl complexes (see Chapter 16, p. 521).[70] A stable complex of this type can exhibit a "static" proton nmr spectrum such as that illustrated in Figure 17-2. (A molecule corresponding to such a spectrum is often labeled AA'BB'X where, in the case of the η^3-allyl complex, X is the proton on carbon 2, A and A' are protons quite different from X and are those located *anti* to X, and B and B' are protons again quite different from X but similar to A/A' and located *syn* to X.) At higher temperatures or in the presence of a base, this spectrum collapses to a "dynamic" A_4X spectrum, the chief feature of which is the change of the doublets for the A and B protons to a single doublet, presumably owing to a rapid (on the nmr time scale) intramolecular rearrangement that causes the A/A' (*anti*) and B/B' (*syn*) protons to have the same chemical shift.

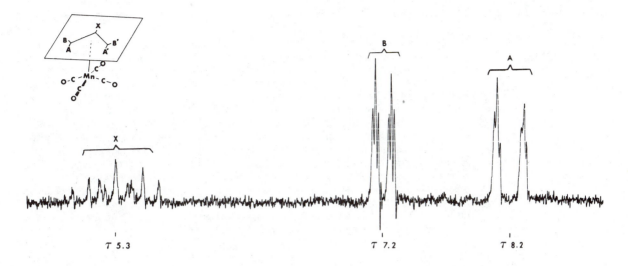

Figure 17-2. Proton nmr spectrum of (π^3-allyl)tetracarbonyl-manganese. [Reprinted with permission from F. A. Cotton, J. W. Faller, and A. Musco, *Inorg. Chem.,* **6**, 181 (1967). Copyright by the American Chemical Society.]

The mechanism of η^3-allyl rearrangement has been extensively examined, and a number of pathways for the scrambling of *syn* and *anti* protons have been proposed; two of the transition states or intermediates that can lead to scrambling in a typical complex are illustrated below for (η^3-allyl)MClL.

18a   18b

In the first of these (**18a**), the two terminal carbon atoms are momentarily bonded to the metal by normal σ bonds, and the middle carbon flips from a position above the plane of the complex to a position below the plane; in the process the distinction between *syn* and *anti* protons is lost. The same result is achieved if the allyl group becomes σ bonded to the metal through just one of its terminal carbons (**18b**); rotation about the M—C bond and re-formation of the η^3 form of bonding can lead to the observed terminal proton equilibration. Indeed, it is this latter mechanism that best accounts for terminal proton scrambling in cases where a static proton nmr spectrum for a η^3-allyl is observed to change to a dynamic spectrum. The importance of such a η^3 to η^1 structural change for allyl complexes (actually a base dissociation step to give a 16 e$^-$ intermediate) has been documented in the hydrogenation of benzene to cyclohexane, a process catalyzed by (η^3-C_3H_5)Co[P(OMe)$_3$]$_3$.[72]

STUDY QUESTIONS

15. One of the first examples of fluxional isomerism to be recognized was $(\eta^1\text{-}C_5H_5)(\eta^5\text{-}C_5H_5)Fe(CO)_2$ (reaction 17–73), a compound first synthesized in 1956. It was recognized immediately as unusual in that only two sharp lines were observed in the ^1H nmr spectrum at room temperature; one would, of course, expect the *pentahapto*cyclopentadienyl group to give a single line, and the *monohapto* group to give rise to three groups of lines. That the *monohapto* C_5H_5 ligand is seen only as a single sharp line at room temperature in this compound, and in many other similar systems isolated since, demands that there be available some low energy pathway for rapid rearrangement that renders all of the protons of $\eta^1\text{-}C_5H_5$ magnetically equivalent. Careful examination of the nmr spectra of $(\eta^1\text{-}C_5H_5)(\eta^5\text{-}C_5H_5)Fe(CO)_2$ and similar compounds has ruled out mechanisms whereby all protons simultaneously (through a dissociative mechanism, for example) or randomly become equivalent. The two most reasonable pathways for proton averaging that remain are either a 1,2- or a 1,3-metal shift; that is, movement of a metal atom around the C_5 ring from one carbon to the carbons α or β to it.

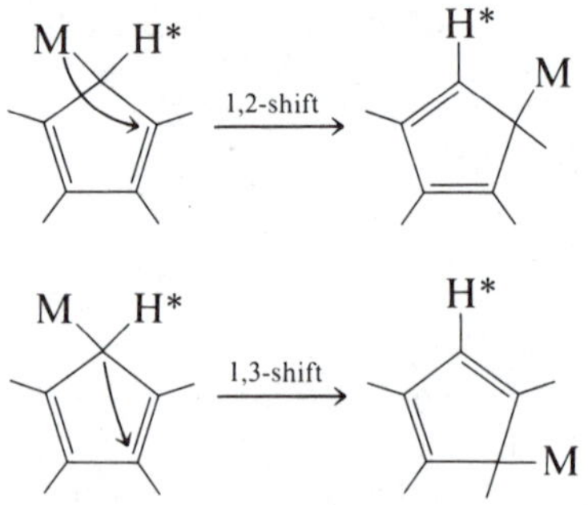

According to nmr convention, the protons of the ligand can be labeled as illustrated below.

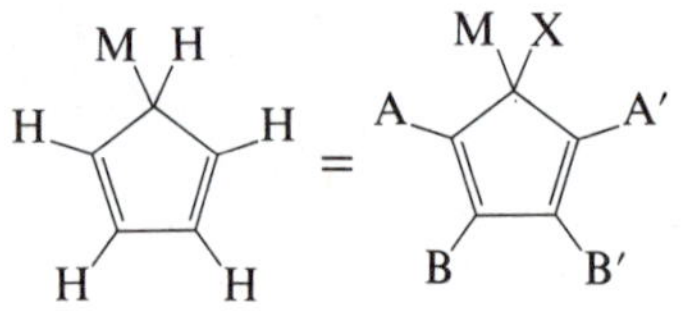

At low temperatures where the rate of metal shifting is slow, the three types of protons (A and A', B and B', and X) can be differentiated by an nmr spectrometer, and a spectrum consisting of basically three lines (one for each proton type) is seen. However, as the temperature is raised, the metal shifting becomes more rapid, the line for the X proton, and that for the A and A' protons, begins to collapse before that for the B and B' protons. That is, the X, A, and A' protons are more affected by metal shifting than are the B and B' protons. To account for this observation, and to decide between 1,2- and 1,3-shifts, show to what type of proton the X, A, A', B, and B' protons are transformed in the first 1,2- or 1,3-shift.

16. Stereochemical non-rigidity is observed in five-coordinate organometallic complexes just as it is in trigonal bipyramidal main-group compounds (*cf.* Chapter 7). For example, the following compound has one axial CO and two equatorial CO's.

$$P \smile P = Me_2P-C_2H_4-PMe_2$$

However, the room temperature ^{13}C nmr spectrum of the compound in the ^{13}CO region suggests that only one kind of CO is present because of a very rapid intramolecular averaging process. Apply the Berry pseudo-rotation process (*cf.* Chapter 7) to the compound above to show how the three CO ligands can move between axial and equatorial sites.

CHECKPOINTS

1. There are basic patterns of reactivity in organometallic chemistry, especially that of the transition metals: ligand association and dissociation; nucleophilic substitution of one ligand for another or electrophilic substitution of attached organic groups; 1,2- and 1,1-additions and eliminations; rearrangements and redistributions.

2. In transition metal chemistry, reactions generally proceed by elementary steps involving 16 or 18 valence electrons, although 14 electron species have been postulated in some cases.

3. In all instances of substitution at metal carbonyl centers in which there is a distribution of 18 electrons, the mechanism is dissociative (D or I_d).

4. Ferrocene has an extremely high propensity to undergo electrophilic substitution and thus has a very extensive derivative chemistry.

5. 1,2-Additions of metal-containing compounds to unsaturated substrates (*e.g.*, alkenes, alkynes, and carbonyls) are generally thought to proceed through a concerted mechanism involving a four-center transition state; addition is *cis*.

6. The addition of CO and H_2 to an olefin (hydroformylation) produces an aldehyde as the primary product. One step in the process is the 1,1-addition of a metal-alkyl to the carbon of CO. Mechanistically, the process is the migration of a metal-attached alkyl group to the carbon atom of a metal-attached CO.

7. Oxidative additions to unsaturated metal centers are important in many processes catalyzed by transition metals, especially those of the Group VIII metals (see the following chapter). Such reactions usually proceed by nucleophilic attack by the metal center, but there is strong evidence for free radical paths in some cases. Small changes in the ligand environment can produce significant mechanistic changes.

REFERENCES

[1] *Advances in Organometallic Chemistry*, F. G. A. Stone and R. West, eds., Academic Press, New York. Issued approximately annually.

[2] G. E. Coates, M. L. H. Green, and K. Wade, "Organometallic Compounds," Methuen, London. Vol. 1, "The Main Group Elements," G. E. Coates and K. Wade (1967); Vol. 2, "The Transition Elements," M. L. H. Green (1968).

[3] B. J. Wakefield, "The Chemistry of Organolithium Compounds," Pergamon Press, New York (1974).

[4] T. Mole and E. A. Jeffery, "Organoaluminum Compounds," Elsevier, New York (1972).

[5] U. Belluco, "Organometallic and Coordination Chemistry of Platinum," Academic Press, New York (1974); F. R. Hartley, "The Chemistry of Platinum and Palladium," John Wiley and Sons, New York (1973).

[6] J. Kochi, "Organometallic Mechanisms and Catalysis," Academic Press, New York (1978).

[7] J. P. Candlin, K. A. Taylor, and D. T. Thompson, "Reactions of Transition Metal Complexes," Elsevier, New York (1968).

[8] R. F. Heck, "Organotransition Metal Chemistry," Academic Press, New York (1974).

[9] R. C. Poller, "The Chemistry of Organotin Compounds," Academic Press, New York (1970).

[10] P. M. Maitlis, "The Organic Chemistry of Palladium," Vols. I and II, Academic Press, New York (1971).

[11] D. S. Matteson, "Organometallic Reaction Mechanisms of the Nontransition Elements," Academic Press, New York (1974).

[12] P. W. Jolly and G. Wilke, "The Organic Chemistry of Nickel," Academic Press, New York (1974).

[13] P. C. Wailes, R. S. P. Coutts, and H. Weigold, "Organometallic Chemistry of Titanium, Zirconium, and Hafnium," Academic Press, New York (1974).

[14] B. R. James, "Homogeneous Hydrogenation," John Wiley and Sons, New York (1973).

[15] M. Rosenblum, "Chemistry of the Iron Group Metallocenes," John Wiley and Sons, New York (1965).

[16] C. A. Tolman, *Chem. Soc. Rev.*, **1**, 337 (1972).

[17] To say that Lewis base dissociation and reductive elimination, insertion, and oxidative coupling are restricted to 18-electron molecules may be too dogmatic. There may be exceptions. See, for example, S. Komiya, P. A. Albright, R. Hoffmann, and J. K. Kochi, *J. Amer. Chem. Soc.*, **98**, 7255 (1976).

[18] J. Kotz and D. G. Pedrotty, *Organometal. Chem. Rev., A*, **4**, 479 (1969).

[19] D. F. Shriver, *Acc. Chem. Res.*, **3**, 321 (1970).

[20] R. B. King, *Adv. Organometal. Chem.*, **2**, 157 (1964); *Acc. Chem. Res.*, **3**, 417 (1970).

[21] D. F. Shriver and A. Alich, *Coord. Chem. Rev.*, **8**, 15 (1972).

[22] M. Brookhart, E. R. Davis, and D. L. Harris, *J. Amer. Chem. Soc.*, **94**, 7853 (1972).

[23] M. A. Haas, *Organometal, Chem. Rev., A*, **4**, 307 (1969).

[24] F. Basolo and R. G. Pearson, "Mechanisms of Inorganic Reactions," 2nd ed., John Wiley, New York (1967).

[25] R. J. Angelici, *Organometal. Chem. Rev. A*, **3**, 173 (1968).

[26] A. Z. Rubezhov and S. P. Gubin, *Adv. Organometal. Chem.*, **10**, 347 (1972).

[27] J. P. Day, F. Basolo, and R. G. Pearson, *J. Amer. Chem. Soc.*, **90**, 6927 (1968).

[28] G. R. Dobson, *Acc. Chem. Res.*, **9**, 300 (1976).

[29] G. R. Dobson, *Inorg. Chem.*, **13**, 1790 (1974).

[30] W. D. Covey and T. L. Brown, *Inorg. Chem.*, **12**, 2820 (1973).

[31] D. A. White, *Organometal. Chem. Rev. A*, **3**, 497 (1968).

[32] J. P. Collman, *Transition Metal Chemistry*, **2**, 1 (1966).

[33] D. E. Bublitz and K. L. Rinehart, Jr., *Organic Reactions*, **17**, 1 (1969).

[34] D. W. Slocum and C. R. Ernst, *Adv. Organometal. Chem.*, **10**, 106 (1972).

[35] D. W. Slocum and C. R. Ernst, *Organometal. Chem. Rev.*, **6**, 337 (1970).

[36] R. L. Sime and R. J. Sime, *J. Amer. Chem. Soc.*, **96**, 892 (1974).

[37] H. Zeiss, P. J. Wheatley, and H. J. S. Winkler, "Benzenoid-Metal Complexes," Ronald Press, New York (1966).

[38] M. F. Semmelhack, H. T. Hall, M. Yoshifuji, and G. Clark, *J. Amer. Chem. Soc.*, **97**, 1248 (1975).

[39] M. F. Lappert and B. Prokai, *Adv. Organometal. Chem.*, **5**, 225 (1967).

[39a] D. L. Thorn and R. Hoffmann, *J. Amer. Chem. Soc.*, **100**, 2079 (1978).

[40] K. Ziegler in "Organometallic Chemistry," H. H. Zeiss, ed., Reinhold Publishing Company, New York (1960), pp. 194 ff.; K. W. Egger and A. T. Cocks, *J. Amer. Chem. Soc.*, **28**, 193 (1971).

[41] J. J. Eisch and C. K. Hordis, *J. Amer. Chem. Soc.*, **93**, 2974, 4496 (1971).

[42] R. F. Heck, *J. Amer. Chem. Soc.*, **91**, 6707 (1969); *ibid.*, **93**, 6896 (1971).

[43] D. W. Hart and J. Schwartz, *J. Amer. Chem. Soc.*, **96**, 8115 (1974); J. Schwartz and J. A. Labinger, *Angew. Chem. Internat. Ed.*, **15**, 333 (1976).

[44] R. F. Heck, *Adv. Organometal. Chem.*, **4**, 243 (1966).

[45] A. Wojcicki, *Adv. Organometal. Chem.*, **11**, 87 (1973).

[45a] H. Berke and R. Hoffmann, *J. Amer. Chem. Soc.*, **100**, 7224 (1978).

[46] K. Noack and F. Calderazzo, *J. Organometal. Chem.*, **10**, 101 (1967).

[47] J. P. Collman, *Acc. Chem. Res.*, **8**, 342 (1975); J. P. Collman, R. G. Finke, J. N. Cawse, and J. I. Brauman, *J. Amer. Chem. Soc.*, **100**, 4766 (1978).

[48] J. P. Collman and W. R. Roper, *Adv. Organometal, Chem.*, **7**, 53 (1968).

[49] L. Vaska, *Acc. Chem. Res.*, **1**, 335 (1968).

[50] J. Halpern, *Acc. Chem. Res.*, **3**, 386 (1970).

[51] J. P. Collman, *Acc. Chem. Res.*, **1**, 136 (1968).

[52] J. K. Stille and K. S. Y. Lau, *Acc. Chem. Res.*, **10**, 434 (1977); D. Forster, *J.C.S. Dalton*, 1639 (1979).

[53] L. Vaska and J. W. DiLuzio, *J. Amer. Chem. Soc.*, **83**, 2784 (1961); *ibid.*, **84**, 679 (1962).

[54] J. P. Collman, M. Kubota, F. D. Vastine, J. Y. Sun, and J. W. Wang, *J. Amer. Chem. Soc.*, **90**, 5430 (1968).

[55] M. F. Lappert and P. W. Lednor, *Adv. Organometal. Chem.*, **14**, 345 (1976).

[55a] W. H. Thompson and C. T. Sears, Jr., *Inorg. Chem.*, **16**, 769 (1977); R. J. Mureinik, M. Weitzberg, and J. Blum, *ibid.*, **18**, 915 (1979).

[56] J. A. Connor and P. I. Riley, *Chem. Commun.*, 634 (1976); *J. C. S. Dalton*, 1318 (1979).

[57] P. J. Davidson, M. F. Lappert, and R. Pearce, *Chem. Rev.*, **76**, 219 (1976); R. R. Schrock and G. W. Parshall, *Chem. Rev.*, **76**, 243 (1976).

[58] P. S. Braterman and R. J. Cross, *J. C. S. Dalton*, 657 (1972); *Chem. Soc. Rev.*, **2**, 271 (1973).

[58a] J. R. Norton, *Acc. Chem. Res.*, **12**, 139 (1979).

[59] W. A. Nugent and J. K. Kochi, *J. Amer. Chem. Soc.*, **98**, 5405 (1976).

[60] G. M. Whitesides, E. R. Stedronsky, C. P. Casey, and J. San Filippo, Jr., *J. Amer. Chem. Soc.*, **92**, 1426 (1970).

[61] S. Komiya, T. A. Albright, R. Hoffmann, and J. K. Kochi, *J. Amer. Chem. Soc.*, **98**, 7255 (1976); see also *ibid.*, **99**, 8440 (1977) and references therein.

[62] Volume 16 (1977) of *Advances in Organometallic Chemistry* is completely devoted to the topic of molecular rearrangements.

[63] J. C. Lockhart, "Redistribution Reactions," Academic Press, New York, (1970).

[64] J. P. Oliver, *Adv. Organometal. Chem.*, **8**, 167 (1970); *ibid.*, **16**, 111 (1977).

[65] T. L. Brown, *Acc. Chem. Res.*, **1**, 23 (1968).

[66] K. Moedritzer, *Organometal. Chem. Rev.*, **1**, 179 (1966); *Adv. Organometal. Chem.*, **6**, 171 (1968).

[67] K. C. Williams and T. L. Brown, *J. Amer. Chem. Soc.*, **88**, 5460 (1966); D. S. Matteson, *Inorg. Chem.*, **10**, 1555 (1971).

[68] F. A. Cotton, *Acc. Chem. Res.*, **1**, 257 (1968).

[69] E. L. Muetterties, *Acc. Chem. Res.*, **3**, 266 (1970).

[70] J. W. Faller, *Adv. Organometal. Chem.*, **16**, 211 (1977).

[71] J. Evans, *Adv. Organometal. Chem.*, **16**, 319 (1977).

[72] E. L. Muetterties and F. J. Hirsekorn, *J. Amer. Chem. Soc.*, **96**, 4063 (1974).

18. CATALYSIS INVOLVING ORGANOMETALLIC COMPOUNDS

Much of the chemical industry is based on the catalytic combination of small molecules such as C_2H_4, CO, H_2, H_2O, and NH_3 to give larger molecules of economic importance, for example, ethylene glycol, acetaldehyde and acetic acid, and acrylonitrile. Homogeneous (and heterogeneous) catalysts activate the smaller molecules and promote their union. The intermediates are presumably organometallic compounds such as those described in Chapter 16, and the mechanistic steps involved are those discussed in Chapter 17. This chapter describes specific examples of catalyzed processes involving organometallic compounds.

Ethylene, acetylene, 1,4-butadiene, and other olefins are common products of commercial processes and are building blocks for the synthesis of many other compounds of commerce after hydrogenation, oxidation, rearrangement, or polymerization. However, severe conditions are often required to effect such reactions. For example, ethylene must be subjected to a series of vigorous addition-elimination reactions to produce commercially useful acetaldehyde. Such conditions may be circumvented by using a catalyst such as palladium chloride, as you shall see below.

$$\text{ethylene} + H_2O \xrightarrow[\substack{\textit{or}\text{ gas phase over phosphoric acid} \\ \text{absorbed on diatomaceous earth}}]{\text{liquid phase in the presence of } H_2SO_4} C_2H_5OH$$

$$C_2H_5OH \xrightarrow{260\text{–}290^\circ\text{C; copper catalyst}} CH_3CHO + H_2 \qquad (18\text{–}1)$$

The function of a catalyst—whether homogeneous or heterogeneous—is to increase the rate of a thermodynamically allowed reaction by lowering the activation energy barrier for the process. In addition, if several reaction paths are possible, a catalyst may increase product specificity by lowering the barrier for one path or by raising it for another. In the case of the hydrocarbons mentioned above, all form well-known complexes with a variety of low-valent transition metals, especially those late in the transition series. Coordination of the organic substrate can result in a lower activation energy in either of two ways: coordination of an olefin may make it more susceptible to nucleophilic attack (as in the $PdCl_2$-catalyzed oxidation of ethylene to acetaldehyde), or both reactants may be brought into proximity by being coordinated to adjacent sites on the catalytic metal (as in hydrogenations, polymerizations and cyclooligomerizations, and nitrogen fixation).

You have already seen some examples of catalyzed reactions involving organometallic compounds in Chapters 16 and 17: for example, the aluminum-catalyzed growth reaction (reaction 16–43), and the conversion of methanol to acetic acid with $[Rh(CO)_2I_2]^-$ (reaction 17–54). In addition, some of the best known catalytic processes involving organometallic compounds are:

a) **hydrogenation** of olefins in the presence of compounds of low-valent metals such as rhodium [*e.g.*, $RhCl(PPh_3)_3$, **Wilkinson's catalyst**].

b) **hydroformylation** of olefins using a cobalt or rhodium catalyst (**oxo process**).

c) **oxidation** of olefins to aldehydes and ketones (**Wacker process**).

d) **polymerization** of propylene using an organoaluminum-titanium catalyst (**Ziegler-Natta catalyst**) to give stereoregular polymers.

e) **cyclooligomerization** of acetylenes using nickel catalysts (**Reppe's** or **Wilke's catalysts**).

f) **olefin isomerization** using nickel catalysts.

The chemistry of some of these processes, and others such as the water-gas shift reaction and the Fischer-Tropsch process, are discussed below. You are referred to the literature for more detailed information.[1-5]

In your reading, notice that the elementary steps in the reactions described are those discussed in Chapter 17, and, as suggested by Tolman,[3] they generally involve species with 16 or 18 valence electrons.

OLEFIN HYDROGENATION[5,6]

A great many metals having d^5 to d^{10} configurations [Cu(II), Cu(I), Ag(I), Hg(II), Ru(II), Ru(III), Co(I), Co(II), Rh(I), Rh(III), Ir(I), Pd(II), and Pt(II)] are known to activate hydrogen, some of the best known of these being ruthenium(III) chloride (d^5), $[Co(CN)_5]^{3-}$ (d^7), and $RhCl(PPh_3)_3$ (d^8). These three compounds illustrate the three ways in which molecular hydrogen may be activated for hydrogenation:

a) oxidative addition

(18-2)
$$RhCl(PPh_3)_3 + H_2 \rightarrow Rh(H)_2Cl(PPh_3)_2 + PPh_3$$

b) heterolytic splitting

(18-3)
$$[Ru^{III}Cl_6]^{3-} + H_2 \rightarrow [Ru^{III}HCl_5]^{3-} + H^+ + Cl^-$$

c) homolytic splitting

(18-4)
$$2[Co^{II}(CN)_5]^{3-} + H_2 \rightarrow 2[Co^{III}H(CN)_5]^{3-}$$

The first of these reactions offers some insight into catalytic processes in general. While studying the oxidative addition reactions of a rhodium(I) compound, Wilkinson found that $RhCl(PPh_3)_3$ is an effective homogeneous hydrogenation catalyst.[7] [Figure 18-1 compares the rates of hydrogenation of 1-heptene using 10^{-3} M $RhCl(PPh_3)_3$ and a 10^{-3} M suspension of Adam's catalyst, a heterogeneous catalyst based on platinum oxide.] Although there is still considerable speculation, the pathway for the Rh(I)-catalyzed hydrogenation is thought to resemble that illustrated by the scheme below.[1,3] The first step of importance is formation of the oxidative addition product. This is followed by formation of an olefin complex and *cis* addition of Rh and H across the C=C bond. The process is completed by reductive elimination of the now-hydrogenated olefin and regeneration of the catalytic species.

(18-5a)

NVE = 16   NVE = 16
OS = 1+   OS = 3+

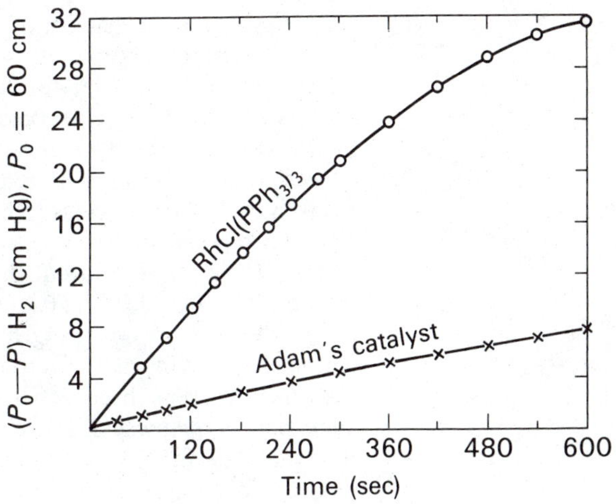

Figure 18-1. Comparison of the rates of homogeneous and heterogeneous hydrogenation. o = RhCl(PPh$_3$)$_3$ and x = Adam's catalyst. [From G. Wilkinson. *Proc. R. A. Welch Found. Conf. Chem. Res.,* **9**, 139 (1966).]

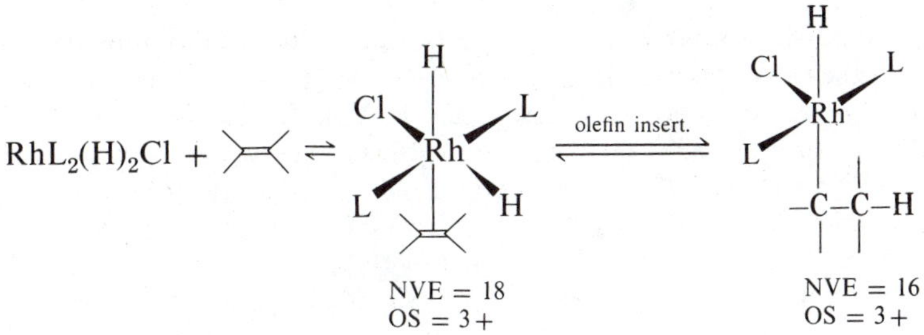

(18-5b)

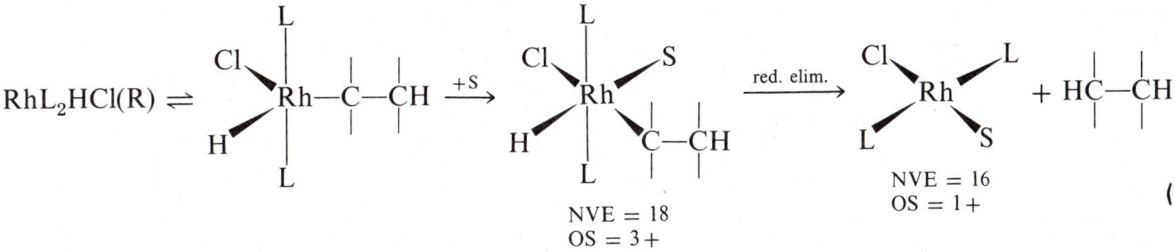

(18-5c)

HYDROFORMYLATION; THE OXO REACTION† [8-10]

In a system composed of cobalt salts, CO, H$_2$, and an olefin, the most important reaction that occurs is **hydroformylation**—the addition of the elements of formaldehyde to an olefin.

$$RHC{=}CH_2 + CO + H_2 \xrightarrow{Co_2(CO)_8} RH_2C{-}CH_2{-}\overset{\displaystyle O}{\overset{\|}{C}}{-}H$$

This process is the most important industrial synthesis using metal carbonyl catalysis. Hydroformylation annually produces about 8 to 10 billion pounds of aldehydes and their derivatives, largely alcohols; butyraldehyde from propylene is the largest single primary product, with about 6 billion pounds produced annually.

---

†A new rhodium-based oxo process was recently put into operation [Chemical and Engineering News, April 26, 1976]. If propene is used as a feedstock in the oxo process, for example, both normal and iso-butyraldehyde are produced. The advantage of the Rh-based process is that it produces a higher normal-to-iso ratio than the Co-based reaction.

Clues to the mechanism of the reaction are provided by the observations that the reaction is first order in olefin and approximately first order in the amount of cobalt present; the rate is faster for terminal olefins than for internal olefins; and excess CO inhibits the reaction. One possible scheme explaining these observations is shown in Figure 18-2. As we have previously mentioned (p. 554), the primary cobalt-containing species in the system is hydridotetracarbonylcobalt, $HCo(CO)_4$, a five-coordinate, 18 electron molecule. If one is to build a pathway involving only 16 and 18 electron species, the most rational first step is loss of CO to give a coordinatively unsaturated 16 electron species, $HCo(CO)_3$. Indeed, this initial loss of CO rationalizes the observation that the reaction is inhibited by high CO pressures. Addition of olefin then returns the molecule to an 18 electron complex. Insertion of olefin into the Co—H bond [step (3)] then presumably occurs *via* a four-center transition state. Step (5) is the now familiar alkyl migration reaction discussed for H_3C—$Mn(CO)_5$ in Chapter 17 (page 554). From this point on, the pathway is less clear. Oxidative addition of H_2 in step (6) may produce the six-coordinate (d^6), 18 electron species, which can then undergo reductive elimination to give the aldehyde and return the cycle to the catalytically active $HCo(CO)_3$.

THE WACKER PROCESS (SMIDT REACTION)[11],[12]

Acetaldehyde, vinyl acetate, and vinyl chloride are important industrial intermediates. Although they may be produced in a variety of ways, many of these methods suffer major disadvantages. For example, acetylene is considerably more expensive than ethylene, but addition of water to acetylene to give acetaldehyde was found earlier to be somewhat easier than production of acetaldehyde from ethylene. However, it was recently discovered that acetaldehyde may be produced by bubbling ethylene into an aqueous solution of a soluble palladium(II) salt in the presence of O_2 (18-6). Moreover, if the reaction is run in the presence of acetate, vinyl acetate is obtained instead (18-7),

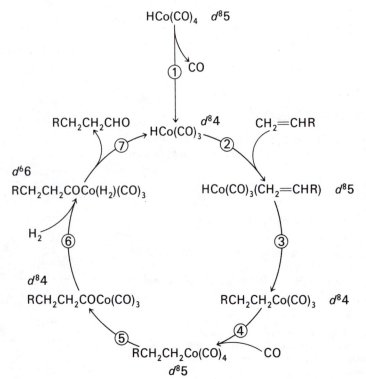

Figure 18-2. Suggested pathway for the hydroformylation of a terminal olefin ["OXO" reaction] by $HCo(CO)_4$. [From C. A. Tolman, *Chem. Soc. Revs.*, **1**, 337 (1972).] (The number of *d* electrons and the coordination number are given for each transition metal species.)

whereas vinyl chloride is among the products formed when ethylene is passed into a suspension of $PdCl_2$ in a non-aqueous solvent (18-8).

$$H_2C{=}CH_2 + \frac{1}{2}O_2 \xrightarrow{PdCl_4^{2-}/H_2O} H_3CCHO \qquad (18\text{-}6)$$

$$H_2C{=}CH_2 + PdCl_2 + OAc^- \rightarrow H_2C{=}CHOAc + Pd + H^+ + 2Cl^- \qquad (18\text{-}7)$$

$$H_2C{=}CH_2 + PdCl_2 + 2L \rightarrow H_2C{=}CHCl + [HPdClL_2] \qquad (18\text{-}8)$$

The oxidation of ethylene to acetaldehyde with platinum group metals has been known since 1894. When ethylene is bubbled into a solution of K_2PtCl_4, the first product is, of course, Zeise's salt, $K[PtCl_3(C_2H_4)]$, but heating this compound in water results ultimately in the formation of acetaldehyde and metallic Pt.

$$[PtCl_3(C_2H_4)]^- + H_2O \xrightarrow{heat} H_3C{-}\overset{\overset{\displaystyle O}{\displaystyle \|}}{C}{-}H + Pt(0) + 3Cl^- + 2H^+ \qquad (18\text{-}9)$$

In contrast, palladium chloride does not give an isolable intermediate analogous to Zeise's salt;† rather, the process moves rapidly to final products.

The palladium chloride-catalyzed production of acetaldehyde was first exploited commercially by Smidt at Wacker Chemie in Germany.[13] Basically, three reactions are involved in the overall process:

product formation

$$C_2H_4 + PdCl_4^{2-} + H_2O \rightarrow CH_3CHO + Pd(0) + 2HCl + 2Cl^- \qquad (18\text{-}10a)$$

catalyst regeneration

$$Pd(0) + 2CuCl_2 + 2Cl^- \rightarrow PdCl_4^{2-} + 2CuCl \qquad (18\text{-}10b)$$

co-catalyst regeneration

$$2CuCl + 2HCl + \frac{1}{2}O_2 \rightarrow 2CuCl_2 + H_2O \qquad (18\text{-}10c)$$

The mechanism of the Wacker process has attracted considerable attention, and the following sequence of reactions is proposed.

a) Formation of a metal-olefin π complex.

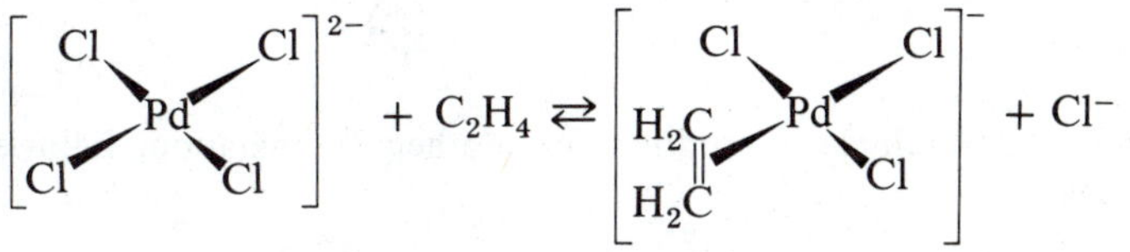

$$(18\text{-}11a)$$

---

†Note that the dimeric complex $[PdCl_2(C_2H_4)_2]_2$ may be isolated in non-aqueous solvents.

b) Substitution of another Cl^- by H_2O.†

(18-11b)

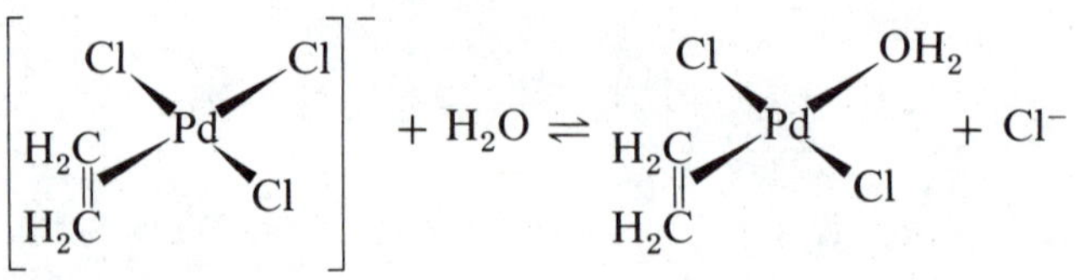

c) Attack on the coordinated olefin by water from the solution. This takes place with reversible formation of the σ-complex and gives a *trans* addition product as observed.[14]

(18-11c)

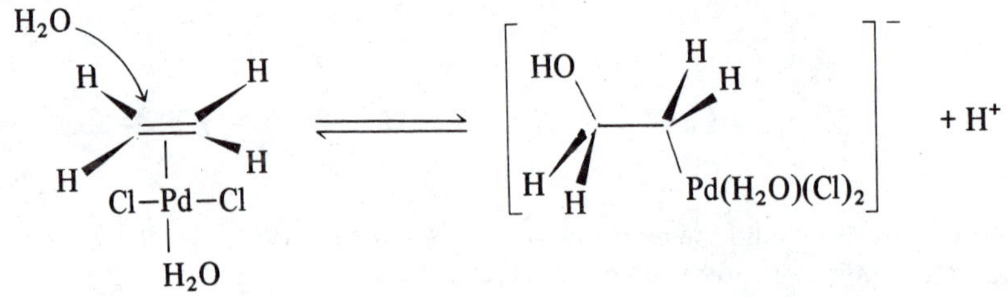

d) Rate determining loss of Cl^- from the σ-bound β-hydroxyethyl complex.

(18-11d)

$$[HOCH_2CH_2Pd(H_2O)(Cl)_2]^- \xrightarrow[+\,solvent]{} Cl^- + HOCH_2CH_2Pd(H_2O)Cl(S)$$

e) An important feature of palladium chemistry is the affinity of the metal for hydrogen.[15] Therefore, the next step in the Wacker process is thought to be β-elimination to produce a π-vinyl alcohol, which then reinserts into the Pd−H bond in the opposite sense.

(18-11e)

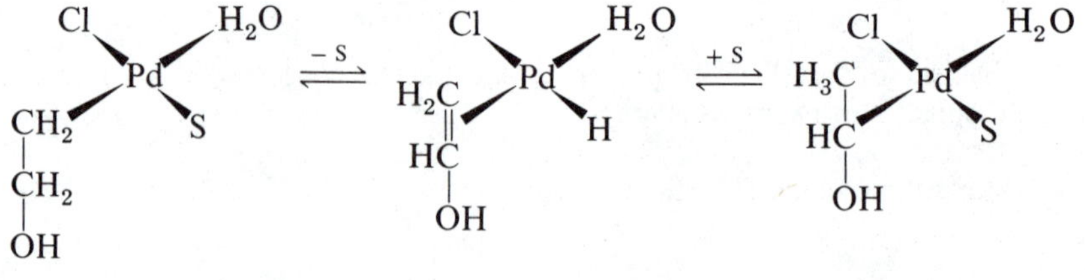

f) Finally, acetaldehyde is produced in another β-elimination, followed by a base dissociation reaction.

---

†This reaction is a good illustration of the *trans* effect discussed on pages 416–419. A coordinated olefin is known to have a greater *trans* labilizing effect than Cl^-, so it is the Cl^- opposite the ethylene that is replaced by water.

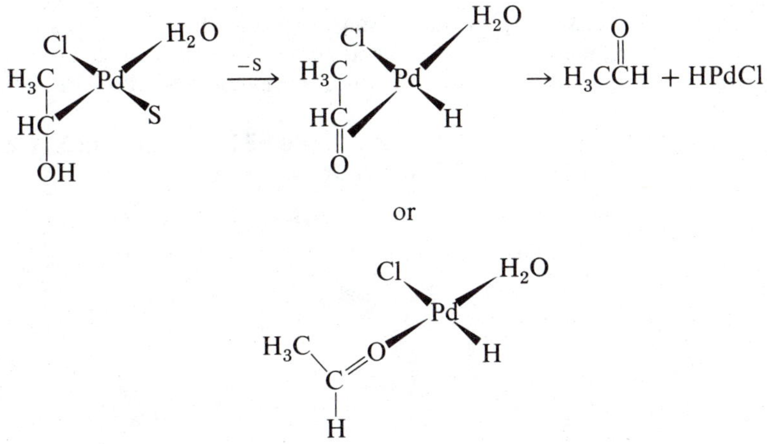

or

(18-11f)

One of the chief reasons for the manufacture of acetaldehyde is for conversion to the economically important product acetic acid; world-wide production of the acid was approximately 2,500,000 tons in 1977. Although some acid can be produced from the aldehyde product of the Wacker process, a Monsanto process based on the carbonylation of methanol with a rhodium carbonyl iodide catalyst (reaction 17-54) may soon be a major source, particularly as CH_3OH can be a major product of the synthesis gas obtained from coal gasification (page 588).

POLYMERIZATION[4,5,16,17]

The polymerization of olefins to give commercially useful fibers, resins, and plastics is a thermodynamically allowed process. However, the olefin must first be activated, or each step of the polymerization reaction must be activated. Compounds having metal-carbon bonds are frequently essential to this activation or catalytic process, and the reactions that occur are those we have been discussing in this chapter. Therefore, the discussion that follows outlines the important metal-olefin interactions leading to poly-merization and the function of such interactions in producing poly-olefins with unique properties.

There are basically two types of polymers—addition and condensation. The familiar material polystyrene is an addition polymer formed by the stepwise addition of each monomer unit to the growing polymer.

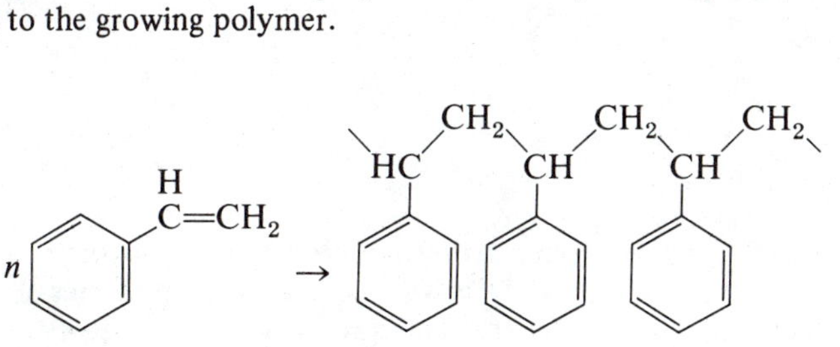

(18-12)

Nylon, however, is a condensation polymer; water is eliminated in the reaction of an amine with a carboxylic acid.† Addition polymers are the only type to be pursued in this chapter.

(18-13)

---

†Note the analogy here with the general "solvolysis" polymerization of non-metal compounds (discussed in the last half of Chapter 8).

Addition polymerizations are chain reactions for which three stages may be defined: initiation, propagation, and termination. The all-important initiation step may occur in any of several ways, some of which are associated intimately with the propagation step.

a) *Free radical initiation.* A free radical source is added to the system; the radical adds to one monomer, which forms the first unit of the growing polymer.

(18-14)

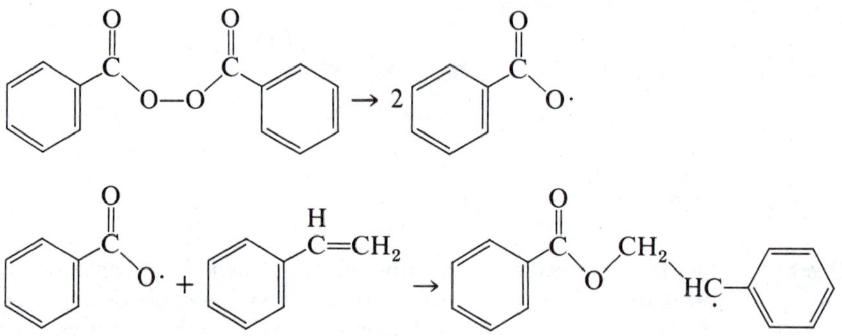

b) *Acid catalysis.* A very strong Lewis acid is added to a scrupulously dry non-aqueous solvent; the strong protonic acid that is formed protonates a monomer unit, thereby forming a carbonium ion that is the first unit of a polymer.

$$TiCl_4 + RH \; (\textit{e.g., ethylene}) \rightarrow H^+TiCl_4R^-$$

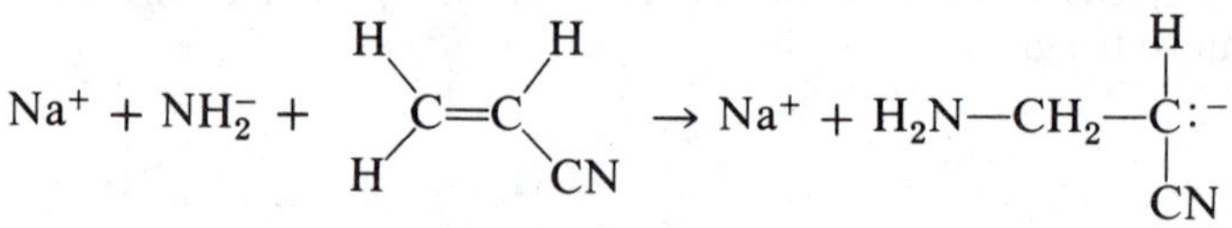

(18-15)

c) *Base catalysis.* The base first attacks a monomer unit, thereby forming a carbanion, which is again the first unit of the polymer.

(18-16)

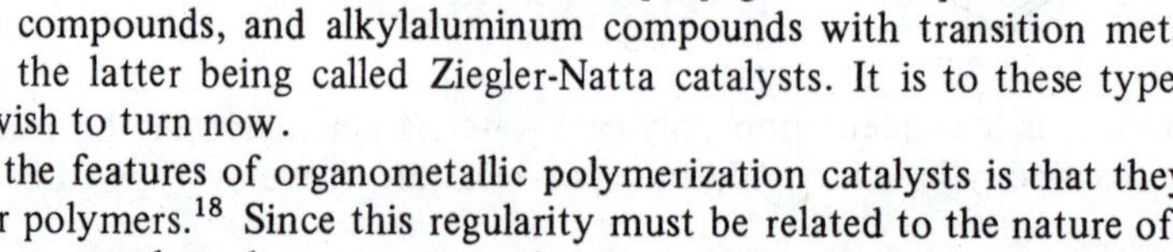

d) *Organometallic initiators and templates for chain propagation.* Examples of compounds that function as initiators and/or propagation templates are nickel salts, alkyllithium compounds, and alkylaluminum compounds with transition metal salts as co-catalysts, the latter being called Ziegler-Natta catalysts. It is to these types of compounds we wish to turn now.

One of the features of organometallic polymerization catalysts is that they produce stereoregular polymers.[18] Since this regularity must be related to the nature of the interaction between metal catalyst, growing polymer, and monomer, we must say something about polymer stereochemistry first. Stereoregular polymerization is not new. Nature has accomplished it beautifully for millions of years. Quartz, the polysaccharides starch and cellulose, and the polynucleotides DNA (deoxyribonucleic acid) and RNA (ribonucleic acid) are all examples of stereoregular polymerization. Olefins may also be polymerized with stereoregularity. Propylene, for example, may form two stereoregular polymers—either **isotactic** or **syndiotactic**—or a non-regular or **atactic** polymer (Figure 18-3). In atactic polypropylene the methyl groups have no regularity along the chain, whereas those of an isotactic polymer always lie on the same side of the chain. (Note that an isotactic polymer cannot lie "flat"; rather, it is forced into a helical arrangement so that the methyl groups do not interfere with one another.) Syndiotactic polymers are formed when the methyl groups alternate along the chain.

Stereoregularity often introduces some very desirable properties into a polymer. For example, stereoregular polypropylene is harder and tougher, and has a higher melting point than polyethylene, and it is stronger than nylon.

One of the products of the "growth reaction" in alkylaluminum chemistry is isoprene (Chapter 16, p. 500). This compound is of great commercial value, as it may be polymerized using organolithium initiators—or Ziegler-Natta catalysts of the type to be discussed momentarily—to produce *cis*-1,4-polyisoprene, a polymer identical to natural rubber.

trans-1,4-polyisoprene (gutta percha)

isoprene

cis-1,4-polyisoprene (natural rubber)

In 1954, Ziegler discovered that, on attempting to reproduce experiments on "growth reactions," ethylene gave only 1-butene, not the long chain hydrocarbon that was the usual product of such reactions.[19] Careful investigation revealed that a nickel salt had contaminated the autoclave used for the reaction, the nickel salt apparently catalyzing chain termination after formation of the C_4 chain. Ziegler guessed that if nickel catalyzed chain termination before substantial chain growth, perhaps another

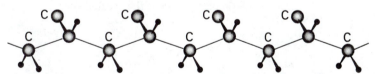

(A) Isotactic—all chain carbons have the same configuration

Figure 18-3. Possible configurations for linear polymers of propene.

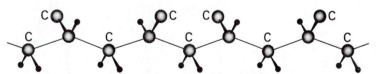

(B) Syndiotactic—regular alternation of configuration

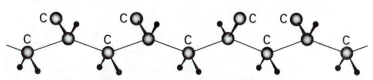

(C) Atactic—no regularity of configuration

metal would catalyze chain lengthening to form linear polyethylene with a usable molecular weight (ca. 30,000 to 50,000 instead of weights considerably less than 28,000 that are normally produced in the growth reaction). Indeed, Ziegler soon found that a system composed of ethylene, triethylaluminum, and *tris*(acetylacetonato)zirconium(III) gave linear polyethylene, and substitution of titanium chlorides for the zirconium compound gave stereoregular polypropylene from propylene.† Continuing work over the years since 1954 has shown that stereoregular polymerization of olefins may be accomplished readily by catalysts (both homo- and heterogeneous) formed from a Group I-III metal alkyl (usually Et_3Al or Et_2AlCl) and some transition metal compound (usually a halide of a metal early in the series; *e.g.*, $TiCl_3$ or VCl_4). In fact, Ziegler cites the following rule: "Take any organometallic, combine it with a compound of a transition metal, and you will have a good chance of finding a suitable catalyst for your special polymerization problem." For their work on metal-catalyzed, stereoregular polymerization, Karl Ziegler and Giulio Natta shared the Nobel Prize in Chemistry in 1963.

One hypothesis for the mechanism of Ziegler-Natta catalysis is outlined in the reactions that follow.[20,21] The first step is clearly the alkylation of the transition metal compound with the soluble Group I-III metal alkyl, thereby producing the catalytically active transition metal alkyl.

(18–17a)

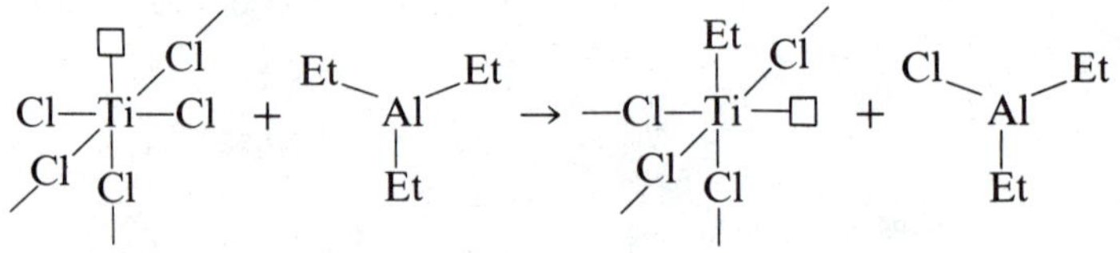

☐ = ligand vacancy

(When $TiCl_3$, VCl_4, or similar salts are used, a hetereogeneous catalyst is generated, since these compounds are not soluble in the reaction medium; therefore, polymerization occurs at a solid surface. However, homogeneous or soluble catalysts can be formed from Cp_2TiCl_2 or $Cp_2TiEtCl$.) If the alkylated metal halide has an empty coordination site available (or one occupied only by solvent), olefin can bind to the metal in the usual manner. In the next step, it is suggested that the alkyl group migrates to the olefin, thereby forming a new metal alkyl, that is, a growing polymer chain. A coordination site is therefore again vacated on the metal, and olefin binding may again occur. The polymerization process then continues by successive alkyl (= polymer) migration to the olefin and olefin occupation of the vacated site.

(18–17b)

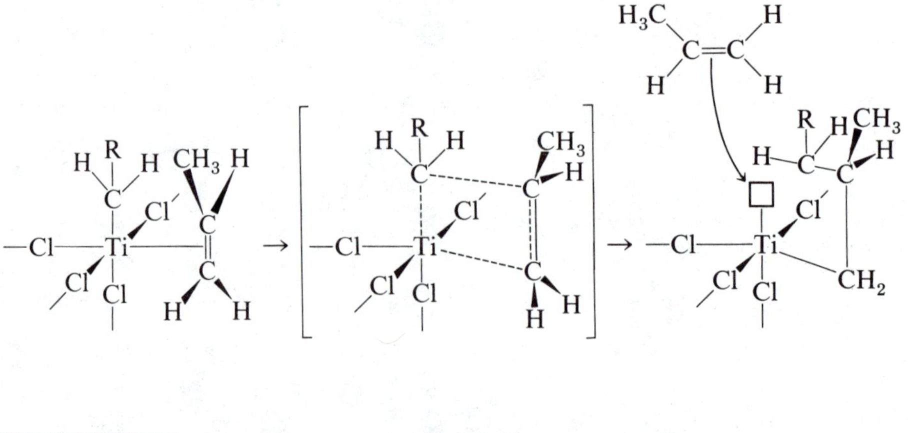

---

†Serendipitous results such as these are quite common in chemistry. It is often the lucky accident that, in the hands of a scientist of insight, imagination, and intuition, turns into an important and useful discovery.

The process may be interrupted by β-elimination of a metal hydride or by hydrolysis.

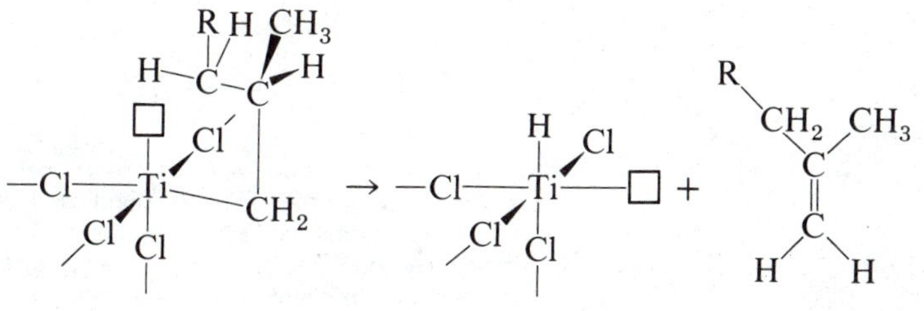

$$(18\text{-}17c)$$

Highly charged metal ions early in the transition series are of special significance in the Ziegler-Natta system for several reasons. α-TiCl$_3$, which produces an isotactic polymer, has a layer lattice structure (see Chapter 2, p. 62) and consists of a hexagonal close-packed array of Cl$^-$ ions, with Ti^{3+} ions occupying two-thirds of the octahedral holes in each alternate pair of layers [Figure 18-4(A)]. However, while the internal titanium ions are octahedrally coordinated, the surface ions are not. Therefore, the process of alkylation of a surface ion and coordination of an olefin to that ion (Figure 18-4(B)] brings the reactants into proximity for the ligand transfer reaction. More importantly, when an asymmetric olefin such as propene is used, the R group and five chloride ions of the hexagon allow the propene to approach the titanium ion in only one way, that is, with the methyl group protruding from the cavity [Figure 18-4(B)]. However, the propene position is even more restricted. The orientation in Figure 18-4(B) is considered more likely than any other orientation; for example if the olefin C=C bond is rotated by 90° about the Ti$\cdots\parallel$ axis, the CH$_2$ group of the olefin would protrude directly into the neighboring titanium. Therefore, addition of the alkyl group or growing polymer chain to the coordinated olefin will always occur with the same stereochemistry. In the next alkyl transfer step [Figure 18-4(C)], the olefin position is again restricted to the one shown, and alkyl transfer occurs with the same stereochemistry as in the previous step. The end result is the formation of an isotactic polymer.

A recently proposed alternative mechanism, which applies to both homo- and heterogeneous polymerization, merits serious consideration. Green and his co-workers assert that the "insertion of an olefin into a metal-carbon single bond [as in reaction scheme 18-17] may be the exception rather than the rule"; they claim there are no unambiguous examples of such a reaction occurring in isolated metal complexes.†[22] Thus, they have proposed that polymerization may proceed through the mechanism outlined below, a mechanism in which *every step* has a parallel in known organometallic chemistry.

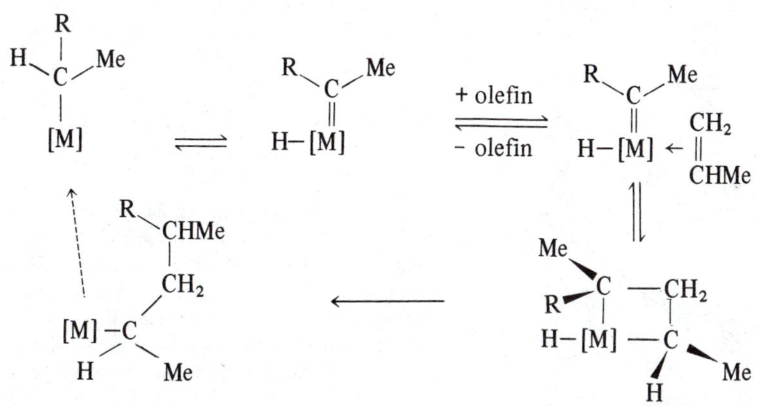

$$(18\text{-}18)$$

---

†But see reaction 17-27. This reaction is ostensibly the insertion of an olefin into a metal-carbon single bond, but the intimate mechanism is not known. [R. Cramer, *J. Amer. Chem. Soc.*, **87**, 4717 (1965).] It may be possible to write a mechanism such as that in 18-18 for 17-27.

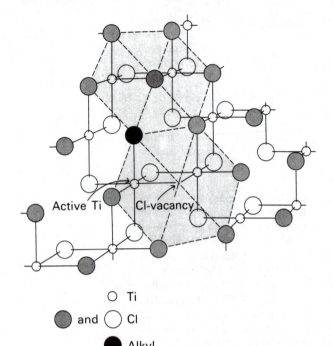

○ Ti

● and ○ Cl

● Alkyl

Figure 18–4. (A) One layer of the crystal structure of α-TiCl₃ showing the Ti-alkyl bond and the chlorine vacancy that forms the "active" center in the surface. The dotted lines show the hexagonal packing of the Cl⁻ ions. Note that the R group was introduced by reaction with AlR₃ in the initiation step.

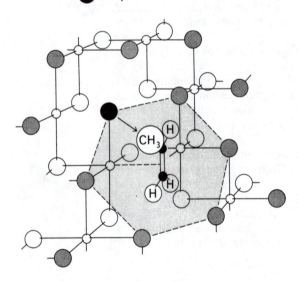

(B) A propene molecule has been inserted into the chlorine vacancy, acting as a π donor toward a titanium. An alkyl group or polymer chain can now migrate to the olefin. After migration, the sequence of groups is R → CH₃ → H clockwise as seen from the carbon attached to the metal.

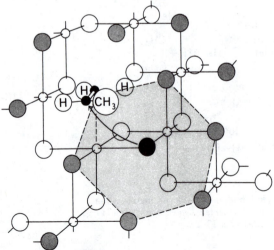

(C) Following step (B), another propene is inserted into the newly vacated metal coordination position. Migration of the growing polymer chain can occur, the sequence of groups again being R → CH₃ → H clockwise as seen from the carbon attached to the metal; thus, polymerization is stereoregular. Notice the alternation of olefin coordination between the two possible coordination sites on Ti.

The first step after alkylation of the metal catalyst is elimination of α-hydrogen to form a metal hydride with coordinated carbene. Binding of olefin is then followed by *cis* addition of the carbene-metal moiety across the olefinic double bond. (Such reversible additions are considered to be the pathway for olefin metathesis as described below.) The final result is a growing, stereospecific polymer.

As noted above, high valent metal ions of the early transition metals are of special significance in Ziegler-Natta catalysis, in that they can supply coordinatively unsaturated sites for heterogeneous polymerization. In addition, given the alternative mechanism above, the two *d* electrons in the early metals—and the consequent availability of empty *d* orbitals—means that the 1,2-hydrogen shift (α-elimination) can occur rapidly.

CATALYSIS BY METAL CLUSTERS[36]

One topic of considerable current interest is that of clusters formed by transition metals and organic ligands; examples include $Ru_3(CO)_{12}$ and $Ir_4(CO)_{12}$ below.[23] Not only are chemists interested in them because of the questions of structure and bonding that they pose, but it is recognized that they may be models for the surface or the chemisorbed surface state of metals involved in heterogeneous processes, the latter being still incompletely understood.[24]

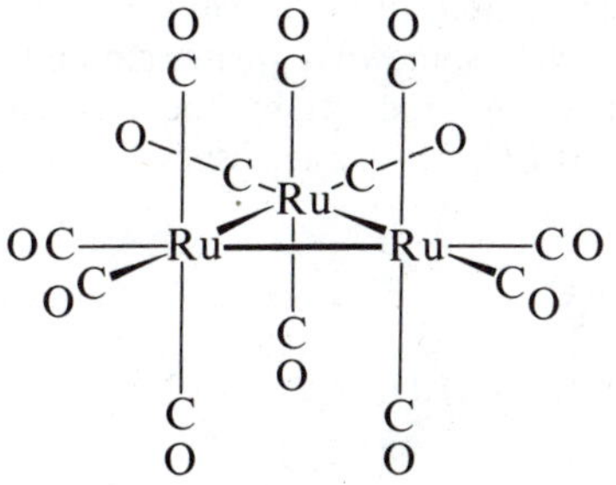

$Ru_3(CO)_{12}$

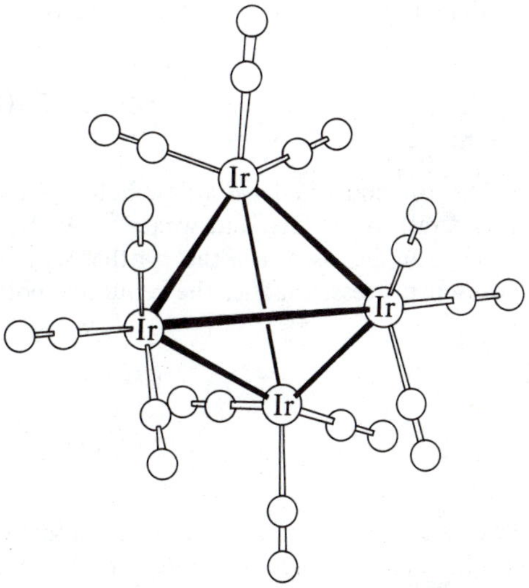

$Ir_4(CO)_{12}$

One of the latest studies involving the use of clusters as catalysts found that $M_3(CO)_{12}$ (M = Ru and Fe; see page 513 for the structure of the M = Fe cluster) and $Ru_3(CO)_9(PPh_3)_3$ all catalyze olefin isomerization upon visible light excitation.[25] Not

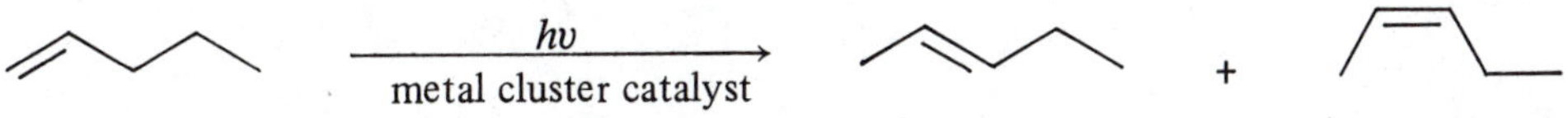

(18-19)

only was it found that quantum yields (the number of molecules isomerized per absorbed photon) were high, but the catalyst turnover numbers (the number of olefin molecules isomerized per metal atom originally present) were at least 10^3.

THE FISCHER-TROPSCH PROCESS[26,27]

The oil crises of the 1970's have focused attention on alternative energy sources, among them coal. Although there are many unknowns in coal technology,[28] one way to use the substance is to treat it with steam in the presence of oxygen at high temperatures, the so-called **coal gasification** reaction.

(18-20)
$$C(s) + H_2O(g) \longrightarrow CO(g) + H_2(g)$$

The product, synthesis gas, may be used for a variety of processes, particularly if the relative amount of H_2 is increased by using up the CO in the **water-gas shift** reaction.

(18-21)
$$CO(g) + H_2O(g) \longrightarrow H_2(g) + CO_2(g)$$

By carrying out the shift reaction numerous times, essentially pure H_2 may be generated for use in the Haber ammonia synthesis. A 2:1 ratio of H_2:CO can be used to synthesize methanol, which in turn may be used as a fuel, turned into acetic acid (see pages 488 and

(18-22)
$$CO(g) + 2 H_2(g) \longrightarrow CH_3OH$$

557),[37] or converted *via* a new heterogeneous catalyst invented by Mobil Oil Company into paraffins, olefins, and waxes.[26] An H_2/CO mixture with somewhat greater amounts of H_2 can be used for the synthesis of methane and other products by the **Fischer-Tropsch process**, that is, the reductive polymerization of CO. The basic Fisher-Tropsch

(18-23)
$$CO + 3 H_2 \longrightarrow CH_4 + H_2O$$

(18-24)
$$n\, CO + (2n + 1) H_2 \longrightarrow H(CH_2)_n H + n\, H_2O$$

process was patented in 1922; depending on the catalyst, pressure, and temperature, it can be oriented predominantly to liquid hydrocarbons, waxes, and alcohols. During World War II it was in fact used by Nazi Germany to supply most of its oil needs, and South Africa will soon have coal gasification/Fischer-Tropsch plants capable of providing almost half of its domestic fuel oil needs.

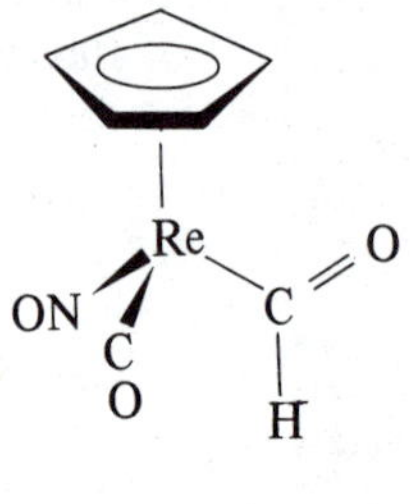

Organometallic chemists have recently become interested in two aspects of the reactions above: the water-gas shift and the conversion of CO and H_2 into heavier organic chemicals. For example, very recent work on the metal-formyl complex has suggested that metal-catalyzed reduction of CO to methanol could occur by the following sequence of steps:[29]

$$2\,MH + 2\,CO \rightleftharpoons 2\,MCHO$$

$$2\,MCHO \longrightarrow MCH_2OC(O)M$$

$$MCH_2OC(O)M + H_2O \rightleftharpoons MCH_2OH + MCO_2H$$

$$MCH_2OH + H_2 \longrightarrow MH + CH_3OH$$

$$MCO_2H \longrightarrow MH + CO_2$$

$$\overline{2\,CO + H_2O + H_2 \longrightarrow CH_3OH + CO_2}$$

That is, the initially formed metal-formyl complex undergoes a Cannizzaro-like condensation to give a dimeric metallo-ester. This ester reacts with water to produce a hydroxymethyl complex and metal-carboxy complex. The former gives CH_3OH (and returns the catalytic metal hydride) upon hydrogenolysis, and the carboxy complex decarboxylates to CO_2 and metal hydride (β-elimination; see reaction 18-11f).

The above scheme for methanol production involves two cooperating metal centers, and others have pursued the idea that reduction of CO may generally involve bimetallic complexation of CO, e.g., $M-C\equiv O-M'$.[26] Indeed, the statement has been made that "there [are no] reports of mononuclear complexes which catalyze the reductive polymerization of CO."† for example, $Ir_4(CO)_{12}$ (page 587) and $AlCl_3$, wherein a unit such as $[Ir-C\equiv O-Al]$ could form, has been found to give essentially complete conversion of CO to hydrocarbon in a matter of hours.[30]

$$CO + H_2 \xrightarrow[\substack{Ir_4(CO)_{12}\ \text{in molten} \\ AlCl_3 \cdot NaCl \text{ at } 180°C}]{} CH_4, C_2H_6, C_3H_8, C_4H_{10} \qquad (18\text{-}25)$$

Metal clusters have also been used to homogeneously catalyze the water-gas shift reaction (18-21). In particular, $Ru_3(CO)_{12}$ (page 587) in the presence of a large excess of KOH catalyzes the shift reaction at 100°C and 1 atmosphere of CO. However, the true nature of the catalytic species is not known.[31]

An economy based on fossil fuels other than petroleum is going to require a wide array of new technologies, not least among them new homogeneous catalysts for the manufacture of "high-value" chemicals such as acetic acid and ethylene glycol from starting materials such as CO and water.

OTHER METAL-CATALYZED PROCESSES

The metal-catalyzed reactions discussed in this chapter represent some important and well understood examples of the many such reactions known. So that you are aware of a few other metal-catalyzed processes, we present the following reactions that are of considerable current interest to academic and industrial chemists.

---

†See page 82 of ref. 6. However, we must note a recent paper claiming that $HCo(CO)_4$ converts CO and H_2 to methanol, perhaps by a free radical mechanism. J. W. Rathke and H. M. Feder, *J. Amer. Chem. Soc.*, **100**, 3623 (1978); J. S. Bradley, *J. Amer. Chem. Soc.*, **101**, 7419 (1979).

A. Cyclooligomerization of Olefins and Acetylenes

Cobalt-containing organometallic complexes can be used to transform alkynes into their cyclic trimers, that is, into benzene derivatives.[32] For example,

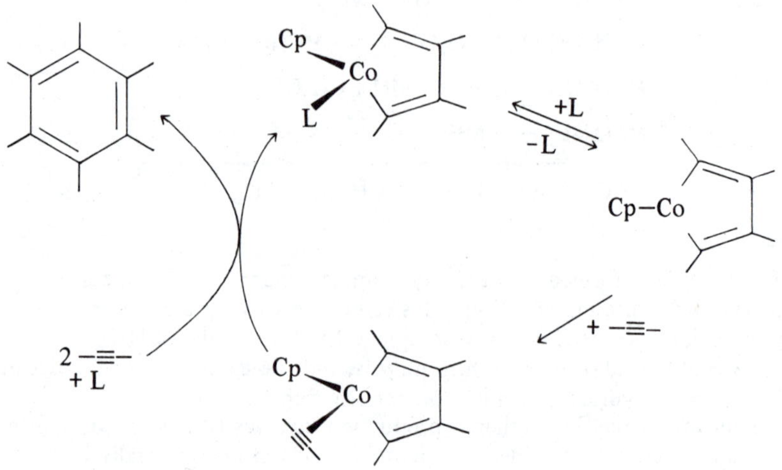

B. Olefin Metathesis[33]

A combination of $Re(CO)_5Cl$ and $EtAlCl_2$, for example, can catalyze the transformation of propylene into ethylene and 2-butene.[34] Considerable controversy has

$$2\,H_3C-CH=CH_2 \xrightarrow[\text{catalyst,95°C}]{} H_2C=CH_2 + H_3C-CH=CH-CH_3$$

arisen concerning the mechanism of such transformations,[35] but it would now appear that metal-carbene complexes (p. 514) are strongly implicated as intermediates. That is, a metal-carbene complex generated in any of several ways (*e.g.*, α-hydrogen elimination from a metal-alkyl as in reaction 18-18) interacts with an olefin to produce a metallo-cyclobutane; this latter compound may then convert to a new metal-carbene and olefin.

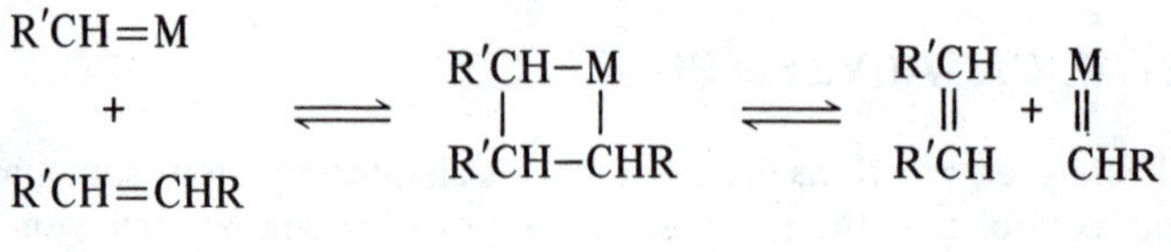

STUDY QUESTIONS

1. At about 160°C and with about 100 atm of CO, $Co_2(CO)_8$ in aqueous acetone will convert cyclohexene to cyclohexanecarboxylic acid in high yield. Assuming that $HCo(CO)_4$, formed from $Co_2(CO)_8$ and water, is the actual catalytic species, propose a mechanism for the olefin-to-acid conversion.

2. Show how two molecules of norbornadiene can dimerize in the presence of $Fe(CO)_5$. The steps involved in the reaction are ligand substitution, oxidative coupling, and reductive elimination.

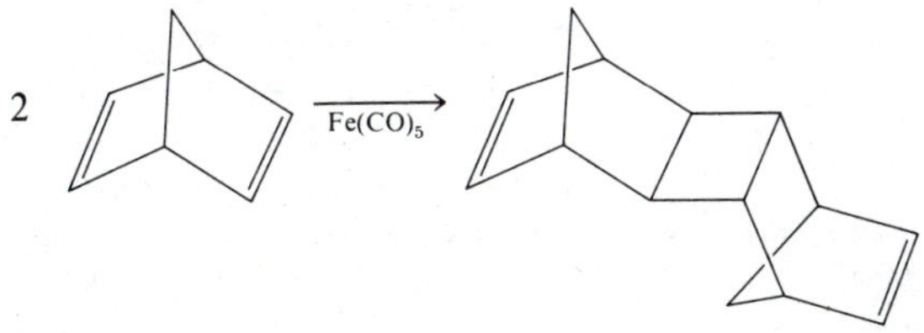

3. Show how 1-butene may be isomerized to 2-butene in the presence of catalytic amounts of $HCo(CO)_4$.

4. $Pd[P(OPh)_3]_4$ catalyzes the addition of HCN to olefins:

$$H_2C{=}CH_2 + HCN \rightarrow H_3C{-}CH_2{-}CN$$

Write out the various steps involved if they are base dissociation, oxidative addition, base association followed by insertion, and reductive elimination of product.

5. Cyclohexene is dehydrogenated to benzene in the presence of two molecules of $Pd(OAc)_2$ in acetic acid. The first stage of the process gives cyclohexadiene and is outlined below. Propose a mechanism for the conversion of the diene to benzene.

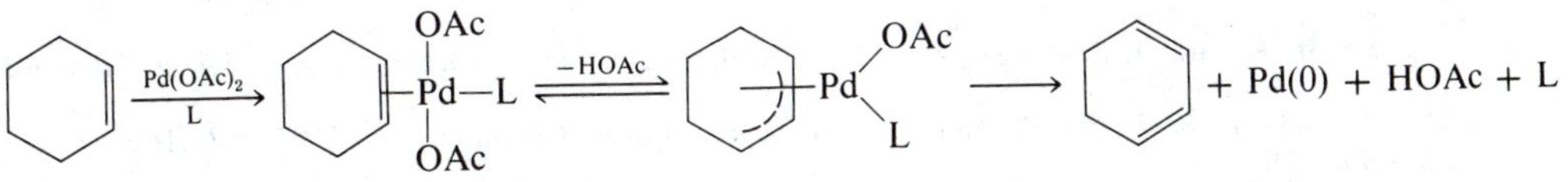

6. Write a mechanism for the homogeneous hydrogenation of olefins, using the Rh(I) compound below as the catalyst. As in reaction 18–5, the first step is base dissociation. However, the second step in this case is association of the rhodium complex with the olefin. With this beginning, complete the mechanism of hydrogenation.

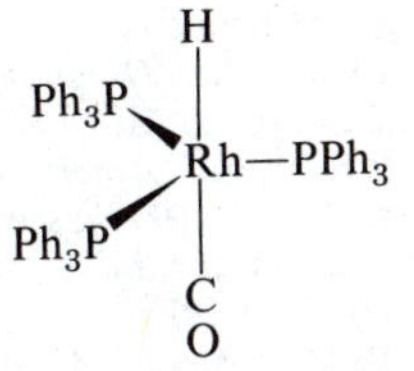

7. Ethylene and CO give the lactone illustrated below in the presence of dichlorobis-(triphenylphosphine)palladium. Assuming that the catalytically active species is $Pd(PPh_3)_2HCl$, write a mechanism for the production of the lactone.

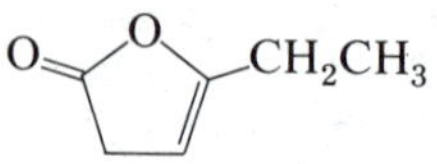

CHECKPOINTS

1. The metal catalyzed processes described in this chapter are seen to involve primarily metal-olefin complexes, because numerous reactions of industrial importance are based on the conversion of ethylene and other olefins to useful products such as aldehydes, alcohols, acids, and polymers.

2. Mechanisms involving homogeneous catalysts can be written with 16 and 18 electron species as intermediates.

3. Mechanistic steps common to many catalyzed processes are (i) oxidative addition-reductive elimination, (ii) olefin insertion (*i.e.*, addition of metal-containing fragments across C=C), (iii) metal-carbene formation, and (iv) alkyl migration.

REFERENCES

[1] R. F. Heck, "Organotransition Metal Chemistry," Academic Press, New York (1974).
[2] B. R. James, "Homogeneous Hydrogenation," Wiley, New York (1973).
[3] C. A. Tolman, *Chem. Soc. Rev.*, **1**, 337 (1972).
[4] M. L. Bender, "Mechanisms of Homogeneous Catalysis from Protons to Proteins," Wiley-Interscience, New York (1971). A lower level version of this book, but also an excellent introduction to catalysis, is: M. L. Bender and L. J. Brubacher, "Catalysis and Enzyme Action," McGraw–Hill, New York (1973).
[5] M. M. Taqui Khan and A. E. Martell, "Homogeneous Catalysis by Metal Complexes," Volumes I and II, Academic Press, New York (1974).
[6] B. R. James, *Adv. Organometal. Chem.*, **17**, 319 (1979).
[7] J. A. Osborne, F. H. Jardine, J. F. Young, and G. Wilkinson, *J. Chem. Soc., (A)*, 1711 (1966).
[8] A. J. Chalk and J. F. Harrod, *Adv. Organometal. Chem.*, **6**, 119 (1968).
[9] R. F. Heck, *Acc. Chem. Res.*, **2**, 10 (1969).
[10] R. L. Pruett, *Adv. Organometal. Chem.*, **17**, 1 (1979).
[11] P. M. Henry, *Acc. Chem. Res.*, **6**, 16 (1973); *Adv. Organometal. Chem.*, **13**, 363 (1975).
[12] F. R. Hartley, *J. Chem. Educ.*, **50**, 263 (1973).
[13] J. Smidt, R. Jira, J. Sedlmeier, R. Sieber, R. Turringer, and H. Kojer, *Angew. Chem. Inter. Ed. Engl.*, **1**, 80 (1962).
[14] J.-E. Bäckvall, B. Åkermark, and S. O. Ljunggren, *Chem. Commun.*, 262 (1977); *J. Amer. Chem. Soc.*, **101**, 2411 (1979).
[15] J. Tsuji, *Acc. Chem. Res.*, **2**, 144 (1969).
[16] M. L. Tobe, "Inorganic Reaction Mechanisms," Thomas Nelson and Sons Ltd., London (1972).
[17] J. C. W. Chien, "Coordination Polymerizations–A Memorial to Karl Ziegler," Academic Press, New York (1975).
[18] G. Natta, *Scientific American*, **205** (August), 33 (1961).
[19] K. Ziegler, *Adv. Organometal. Chem.*, **6**, 1 (1968).
[20] P. Cossee, *Trans. Faraday Soc.*, **58**, 1226 (1962); P. Cossee, *J. Catalysis*, **3**, 80 (1964); E. J. Arlman and P. Cossee, *J. Catalysis*, **3**, 99 (1964).
[21] G. Henrici-Olivé and S. Olivé, *Angew. Chem. Int. Ed. Engl.*, **6**, 790 (1967).
[22] K. J. Ivin, J. J. Rooney, C. D. Stewart, M. L. H. Green, and R. Mahtab, *J. C. S. Chem. Comm.*, 604 (1978).
[23] K. F. Purcell and J. C. Kotz, "Inorganic Chemistry," W. B. Saunders Company, Philadelphia (1977); P. Chini, G. Longoni, and V. G. Albano, *Adv. Organometal. Chem.*, **14**, 285 (1976).
[24] M. Moskovits, *Acc. Chem. Res.*, **12**, 229 (1979); H. F. Schaefer, III, *ibid.*, **10**, 287 (1977).
[25] J. L. Graff, R. D. Sanner, and M. S. Wrighton, *J. Amer. Chem. Soc.*, **101**, 273 (1979).
[26] C. Masters, *Adv. Organometal. Chem.*, **17**, 61 (1979); see also ref. 6.
[27] G. Henrici-Olivé and S. Olivé, *Angew. Chem. Int. Ed. Engl.*, **15**, 136 (1976).
[28] M. L. Gorbaty, *et al.*, *Science*, **206**, 1029 (1979).
[29] C. P. Casey, M. A. Andrews, and D. R. McAlister, *J. Amer. Chem. Soc.*, **101**, 3371 (1979).

[30] G. C. Demitras and E. L. Muetterties, *J. Amer. Chem. Soc.*, **99**, 2796 (1977).

[31] R. M. Laine, R. G. Rinker, and P. C. Ford, *J. Amer. Chem. Soc.*, **99**, 252 (1977).

[32] K. P. C. Volhardt, *Acc. Chem. Res.*, **10**, 1 (1977); D. R. McAlister, J. E. Bercaw, and R. G. Bergman, *J. Amer. Chem. Soc.*, **99**, 1666 (1977).

[33] T. J. Katz, *Adv. Organometal. Chem.*, **16**, 283 (1977); N. Calderon, J. P. Lawrence, and E. A. Ofstead, *ibid.*, **17**, 449 (1979).

[34] W. G. Greenlee and M. F. Farona, *Inorg. Chem.*, **15**, 2129 (1976).

[35] F. D. Mango, *J. Amer. Chem. Soc.*, **99**, 6117 (1977).

[36] A. K. Smith and J. M. Basset, *J. Molec. Catalysis*, **2**, 229 (1977).

[37] D. Forster, *J. C. S. Dalton*, 1639 (1979).

19. BIOCHEMICAL APPLICATIONS

One of the more rapidly developing areas of inorganic chemistry, indeed in all of chemistry, is that of its applications to living systems. The name "inorganic" applied to metal ions and polyphosphate anions, in particular, is certainly a misnomer. Such reagents are vital to the existence of living systems as sources of chemical potential to drive unfavorable organic reactions and as catalysts to ensure that the required reactions proceed with sufficient velocity. Even organometallic reactions have found a place in living systems, as have more traditional concepts of coordination chemistry in the areas of transport of O_2, electrons, and metal ions. In what follows you will find general overviews of selected in vivo processes. A key role for the inorganic chemist is to model as closely as possible the active prosthetic groups of metal-requiring in vivo enzymes. In this context, many of the key concepts of earlier chapters (mechanisms of substitution reactions at non-metal centers, electron transfer, ligand effects on metal ion coordination geometry and electron structure, organometallic reactions, and photochemical concepts) are intimately involved in attempts to understand the working pieces of the living system. The subjects chosen for this chapter reflect varying degrees of incomplete understanding by the interdisciplinary chemical community, and so constitute a fitting conclusion for a text about a subject with an exciting future.

THE CELL[1]

The reaction "flask" of living organisms is the cell (Figure 19-1). Each cell is characterized by an **outer membrane** whose function is to contain the highly organized chemical system and to monitor the influx of needed reagents. One function, for example, is to raise the concentration of K^+ within the cell while diminishing the concentration of Na^+. Mg^{2+} is another ion that is necessary for enzymatic action within the cell. Ca^{2+}, on the other hand, is excluded from the cell (but it is found in bones, teeth, and as an activator of extracellular enzymes).

Within the cell membrane are several organelles immersed in cellular fluid or **cytoplasm**. One of these organelles—the **mitochondrion**—is the focus of most of the chemistry discussed in this chapter. Mitochondria are the organelles within which occur the redox/electron-transfer processes so important in the combustion of glucose and the synthesis of adenosine triphosphate (ATP). Each mitochondrion consists of an outer membrane and a folded, inner membrane to which a complex (but necessarily highly organized) system of enzymes, their co-factors, and electron transfer agents is attached. In animals the mitochondria are the sole centers of energy generation in the cell; the cells of green plants contain, in addition, **chloroplasts**. These organelles contain chlorophyll, which makes possible the light-sensitized phosphorylation reactions of ATP regeneration. (In the dark the mitochondria of plant cells maintain this regeneration, though at a lower rate than in animals because the mitochondria are sparser in green plant cells.)

The lysosome and Golgi bodies shown in Figure 19-1 are involved in cell digestion and excretion, respectively, and will not be further mentioned. The endoplasmic reticulum defines an intracellular network of channels for transport of proteins synthesized by the **ribosomes** that stud the surfaces of the channels.

OXYGEN CARRIERS—HEMOGLOBIN AND MYOGLOBIN[2]

Probably the most familiar examples of bioinorganic chemistry are the O_2 carriers so necessary for getting oxygen to the cell (hemoglobin) and into the cell (myoglobin) for glucose oxidation. Before beginning a discussion of their modes of action, we need to

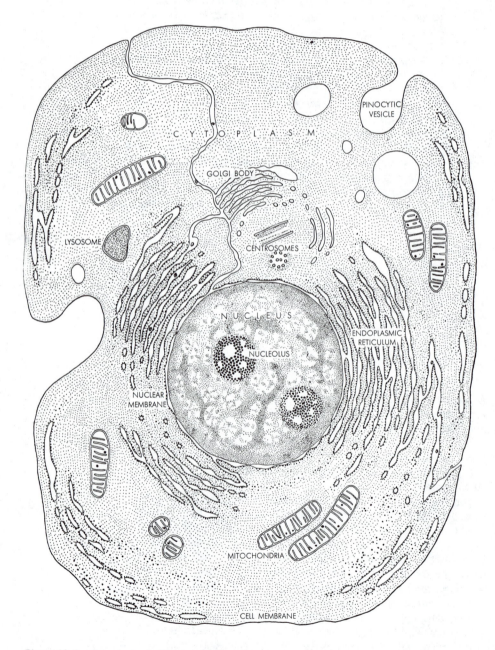

Figure 19-1.   A typical cell. [From J. Brachet, *Scientific American,* p. 55, September, 1961.]

characterize their composition. The iron complexes defining the heme group are those of **protoporphyrin IX (1)** (abbreviated PIX).† The porphyrins are a class of nitrogen macrocycles derived from **porphin (2)**, an unsubstituted tetra-pyrrole connected at the α-carbons by methylidyne (CH) bridges.§

PIX

1

porphin

2

Coordination of Fe^{2+} by the four nitrogens of PIX produces uncharged **heme** (heme carries a dinegative charge at pH 7 because the carboxyl groups are ionized). Since heme is irreversibly air oxidized in H_2O to the Fe^{3+} complex (called **hemin**), heme by itself cannot act as an O_2 carrier *in vivo*. Nature avoids this catastrophe by embedding heme in a protein crevice where it is surrounded by hydrophobic groups from amino acids. A schematic of heme buried in peptide is shown in Figure 19-2, where you see that the Fe^{2+} in hemoglobin finds itself close to a nitrogen of one imidazole group (the proximal histidine) and further from another (the distal histidine). The proximal histidine is coordinated to the Fe^{2+}, which actually lies ~80 pm above (toward the proximal histidine) the mean porphyrin plane (the five-coordinate Fe^{2+} finds itself in a square pyramidal environment of idealized C_{4v} symmetry).

The role of the peptide portion of hemoglobin in preventing the irreversible O_2 oxidation of Fe^{2+} is thought to be twofold: (i) the hydrocarbon environment of the heme pocket has a low dielectric constant and so acts as a nonpolar "solvent" that cannot support the ionic charge separation developed at the transition state of Fe^{2+} oxidation,[3] and (ii) the sterically protected environment of the heme prevents formation of an intermediate μ-oxo-heme dimer, which appears to be a mechanistically necessary intermediate in the irreversible oxidation.[4] (The μ-oxo dimer may well be preceded by a μ-peroxo dimer; the formation of $[(H_3N)_5CoO_2Co(NH_3)_5]^{4+}$ (Chapter 13, p. 441) is an analog from conventional coordination chemistry.)

Various studies support both concepts. The 1-(2-phenylethyl) imidazole heme diethylester embedded in an amorphous mixture of polystyrene shows reversible O_2 uptake.[5] Similarly, Fe(TPP) bound to an imidazole, itself attached to a silica gel support

---

†The dimethylester (PIXDME), with the carboxylic acid groups esterified, is a commonly encountered derivative.

§These methylidyne carbon positions are variously labeled the α, β, γ, δ and 5, 10, 15, 20 positions of porphin and porphyrins. A commonly used synthetic porphyrin is the 5,10,15,20-tetraphenyl derivative (TPP), desirable for *in vitro* studies because of its ease of synthesis and purification.

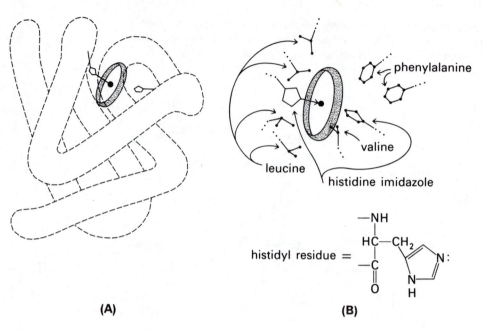

phenylalanine

valine

leucine

histidine imidazole

histidyl residue =

(A)   (B)

Figure 19-2. Hemoglobin schematic. (A) Heme in a peptide crevice; (B) a closer view to show the immediate hydrocarbon environment of Fe.

(as in **3**), also exhibits reversible O_2 uptake.[6] Recent work with the "picket fence" porphyrin **4** shows that it, too, provides a properly shielded environment for reversible O_2 coordination of Fe^{2+} without oxidation.[7]

3

4

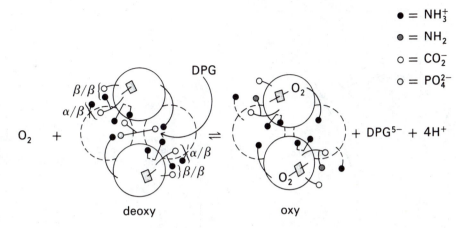

$\bullet = NH_3^+$

$\circleddash = NH_2$

$\circ = CO_2^-$

$\circ = PO_4^{2-}$

Figure 19-3. An idealization of the deoxyHb ⇌ oxyHb structure change, including the role of DPG^{5-} and H^+ ionization. Two of the four protons released on forming oxyHb arise from disruption of α/α salt bridges not shown. Each protein mass contains a heme unit (solid circle = β-chain, broken circle = α-chain).

Further indication of the importance of properties of the medium and thermal control of relative reaction rates comes from studies[8] showing that various Fe porphyrins may be caused to bind O_2 reversibly, without formation of the μ-oxo Fe^{3+} dimer, at low temperatures in non-aqueous solvents.

The protein structures in hemoglobin consist of a peptide† backbone with various side chains. These side chains define a "surface" for the polymer consisting of a variety of nonpolar (hydrocarbon), cationic (*e.g.*, $-NH_3^+$), and anionic (*e.g.*, $-CO_2^-$) groups. Hydrogen bonding between N—H and O=C groups of the peptide backbone units and interactions between nonpolar regions of the polymer surface determine the polymer structure (generally, of two kinds—a helix or a pleated sheet). Myoglobin (Mb) has a peptide surface not conducive to self-association, but hemoglobin (Hb) does. Consequently, Mb is monomeric whereas Hb is a tetrameric unit (of C_2 symmetry) consisting of two α- and two β-peptide chains (actually folded helices; Figure 19-2 depicts one of these units). Figure 19-3 depicts the tetramer segments as protein masses at the corners of a distorted tetrahedron.

X-ray studies of deoxyHb and oxyHb have revealed what appear to be highly significant changes in $-NH_3^+$ ⋯ $^-O_2C-$ interactions within and between peptide chains.[9] Specifically, there are eight of these salt bridges in deoxyHb, which are disrupted upon coordination of O_2 to Fe. Two disruptions occur *within* the β-chains, four *between* the α-units, and one *between* each α,β pair. Only the β/β and α/β bridges are depicted in Figure 19-3. Also shown is diphosphoglycerate (DPG), bridging the β helices in deoxyHb. To condense an otherwise complex description, the disruption of the DPG and $-NH_3^+$ ⋯ $^-O_2C-$ hydrogen bond interactions accompanies oxygenation of Hb, and the β chains move closer together along their common edge of the tetrahedron.

$$DPG^{5-} = \;\; ^{(2-)}O_3PO-\overset{\displaystyle O}{\overset{\|}{C}}-\overset{\displaystyle H}{\underset{\displaystyle O_{(-)}}{\overset{|}{C}}}-\overset{\displaystyle H}{\underset{\displaystyle H}{\overset{|}{C}}}-OPO_3^{(2-)}$$

†Peptides are polyamides made from α-amino acids, *i.e.*,

$$\left(-\overset{\displaystyle O}{\overset{\|}{C}}-\overset{|}{C_\alpha}-\overset{|}{\underset{\displaystyle H}{N}}-\right)_n.$$

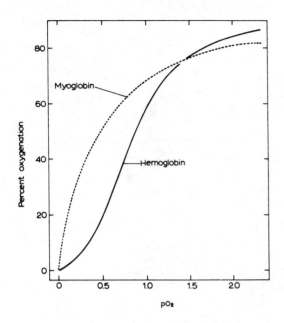

Figure 19-4. Oxygen saturation curves of myoglobin and hemoglobin. [From J. M. Rifkind, in "Inorganic Biochemistry," G. L. Eichhorn (Ed.), Vol. 2, Elsevier, New York (1973) p. 832.

It is presently believed that the peptide chains are constrained by these salt bridges (all eight of them) in deoxyHb. In this "tense" form the sixth coordination position about Fe^{2+} is directed "inward" and sterically blocked by a valine† group [Figure 19-2(B)]. The bridge disruption incurred on coordination of one or two heme units by O_2 is associated with a conformational change in which this sixth position is exposed and the inter- and intra-protein salt bridges are broken. The entire "tense" tetramer then relaxes so as to expose the sixth positions about the other heme iron atoms. This phenomenon is referred to as **cooperativity** between heme units. For example, myoglobin, a single-heme protein found within cells, takes up O_2 in a 1:1 ratio with Fe^{2+} according to a normal relationship between the degree of complexation and O_2 pressure. This is shown in Figure 19-4 as a plot of percentage oxymyoglobin as a function of O_2 pressure, p_{O_2}. By way of contrast, Figure 19-4 also depicts the same sort of plot for hemoglobin. Comparison of the curves should convince you that greater p_{O_2} is required to oxygenate Hb than Mb, or that hemoglobin is less efficient at O_2 uptake under low O_2 pressures, where myoglobin is very efficient. In muscle tissue, for example, where p_{O_2} is small, there is a thermodynamically favorable O_2 transfer from oxyhemoglobin to myoglobin to pass O_2 into the cell.

The oxygenation equilibrium for myoglobin is represented as

$$\text{Fe} + O_2 \rightleftharpoons \text{FeO}_2 \qquad \text{with } K = \frac{[\text{FeO}_2]}{[\text{Fe}]p_{O_2}}$$

Defining the fraction of O_2-bearing iron as $f = [\text{FeO}_2]/\{[\text{Fe}] + [\text{FeO}_2]\}$, then

$$K = f/\{(1 - f)p_{O_2}\}$$
$$f = Kp_{O_2}/(1 + Kp_{O_2})$$
$$\frac{1}{f} = 1 + \left(\frac{1}{K}\right)\frac{1}{p_{O_2}}$$

---

†Valine is a terminal group here, with a structure

$$\begin{array}{cc} \text{H} & \text{H} \\ | & | \\ -\text{N}-\text{C}-\text{CH(CH}_3)_2 \\ | & | \\ \oplus\text{H} & \text{CO}_2^{\ominus} \end{array}$$

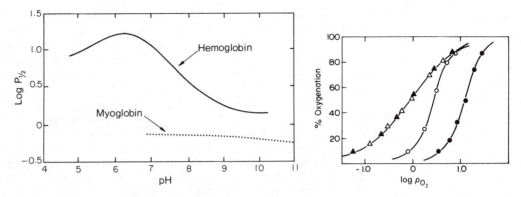

Figure 19-5. (A) The Bohr effect; (B) the effect of 2,3-diphospho-glyceric acid on Hb-O₂ uptake [the two leftmost curves are for the α and β chains of Hb alone, both with and without 2,3-DPG; the two rightmost curves are for hemoglobin without (○) and with (●) 2,3-DPG]. [From J. M. Rifkind, in "Inorganic Biochemistry," G. L. Eichhorn (Ed.), Elsevier, New York (1973), p. 832.]

The myoglobin curve in Figure 19–4 corresponds to this equation. The hemoglobin curve does not follow such an equation, but it does follow an empirically modified form with p_{O_2} replaced by $p_{O_2}^n$, where the exponent n is called the Hill constant. For hemoglobin, $n = 2.8$; this reflects the degree of cooperativity between heme units.[2]

The salt bridges between and within the peptide chains provide a means for control of O_2 transfer from oxyHb to Mb. First of all, the cooperativity effect is pH-dependent, as shown in Figure 19–5(A). This is called the **Bohr effect**, and it is known not to involve acid dissociation of ligands bound to Fe^{2+} in hemoglobin. Rather, $-NH_3^+$ sites involved in α/α and β/β bridges (see Figure 19–3 for the β/β sites) are more strongly acidic when they no longer form hydrogen bonds to the carboxylate groups. Consequently, basic conditions remove these protons and shift the deoxyHb ⇌ oxyHb equilibrium in favor of the oxy-form. The acid condition (lactic acid) in working muscle tissues therefore facilitates the release of O_2 from oxyHb. Secondly, deoxyHb has a greater affinity for phosphate ester anions (such as 2,3-diphosphoglycerate, DPG) and EDTA than does oxyHb. Both anions bear several negatively charged oxygen atoms and are large enough to span cationic sites (*e.g.*, $-NH_3^+$) between the β-protein chains. The effect of the hydrogen bonding of these anions to the cationic sites is to hold the β-chains apart, in their deoxy configuration, and facilitate O_2 loss. Again, in working tissues the DPG concentration is fairly high, a condition which shifts the equilibrium toward deoxy-Hb and favors the transfer of O_2 to myoglobin. Figure 19–5(B) confirms that these effects are peculiar to the tetrameric peptide chains of Hb, for in neither Mb nor the individual Hb peptide units is O_2 uptake affected by pH or DPG.

The link between O_2 coordination and salt bridge disruption is currently being studied. One hypothesis is centered on the characteristic feature that formation of oxyhemoglobin or oxymyoglobin causes the conversion of five-coordinate, high spin Fe^{2+} to six-coordinate, low spin Fe^{2+}, with Fe^{2+} being pulled into the porphyrin plane. It was originally thought that the proximal histidine, in following this Fe^{2+} displacement of ~80 pm, also shifted ~80 pm and so caused a conformational change throughout the peptide; this in turn was postulated to alter the interpeptide salt bridges (as noted above) and result in crevice opening.[10] More recent work[11] with Co^{2+} in place of Fe^{2+} (cobo-globin, CoHb) reveals, however, the same degree of cooperativity ($n \approx 2.5$) as found in hemoglobin, with less than one-half the histidine displacement. Consequently, the problem is not solved and the search continues.

Of more direct interest to the coordination chemist is the nature of the O_2-Fe^{2+} interaction and its stereochemistry. It would be desirable to study the Fe-O_2 interaction in oxyHb and various models, but definitive studies have proven difficult. As a result, studies of Co^{2+} analogs have been pursued, with interesting results. In both the Fe and Co systems the M—O—O group is angular[12] and a simple mo model[13] depicts a σ interaction between the metal d_{z^2} ao and one of the O_2 π^* mo's. A π interaction between d_{xz}, say, and the orthogonal O_2 π^* mo is also possible. (Another discussion of M—O_2 bonding can be found in Chapter 5, p. 136.)

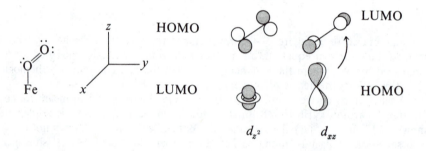

Viewing the oxy-complexes as adducts of O_2 with a square pyramidal complex, these M—O_2 orbital interactions result in the following orbital scheme:

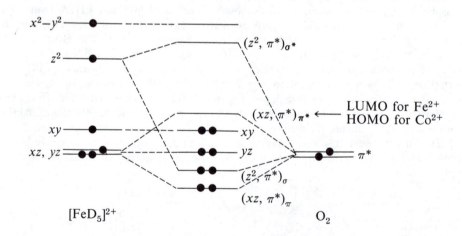

With Fe^{2+} as the metal, there are eight electrons to populate these mo's, giving a configuration...$(xy)^2$. Notice that both the $(z^2/\pi^*)_\sigma$ and $(xz/\pi^*)_\pi$ mo's are occupied. Since the σ and π bonding electron pairs are polarized in the sense $Fe^{\delta+}/O_2^{\delta-}$, there is a partial oxidation of Fe^{2+}, and the π interaction would seem to be an important feature of the M—O_2 bonding. With Co^{2+} you find nine electrons to occupy the mo's of the scheme, and this means that the $(xz/\pi^*)_{\pi*}$ mo is half-occupied.

The role of the other ligands bound to the metal has received some consideration. Again with cobalt, studies[11] with model five- and six-coordinate systems support the above concept, for they reveal an over-all correlation between O_2 uptake, ease of Co^{2+}

$$Co(L)B + O_2 \rightleftharpoons Co(L)B(O_2)$$

oxidation, and (roughly) the pK_a of the group axially coordinated to Co^{2+} (Figure 19-6).

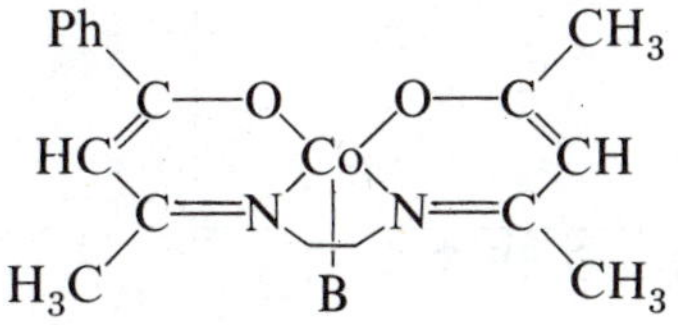

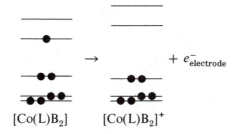

[see Figure 19-6(A)]   [see Figure 19-6(B)]

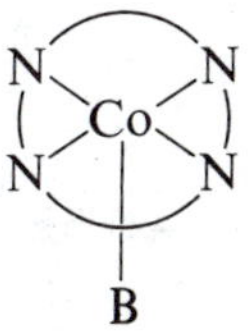

These correlations can be interpreted as follows. In Figure 19-6(A) is plotted the log of the equilibrium constant for oxygen uptake by [Co(L)B] against $E_{1/2}$ for reduction of $[Co(L)B_2]^+$. The more negative the $E_{1/2}$ for the reduction process, the more readily the opposite process, oxidation of $[Co(L)B_2]$, occurs. So it is seen that there is a correlation between log K_{O_2} and the ease of removal of the electron from the z^2 orbital. As depicted below, the greater the basicity of B, the higher the energy of the z^2 electron and the

$$[Co(L)B_2] \qquad \rightarrow \qquad [Co(L)B_2]^+ \quad + e^-_{electrode}$$

easier it is to remove. Since it is the "transfer" of the z^2 electron from [Co(L)B] to the O_2 π^* orbital that is important in the oxygenation of [Co(L)B], it would seem that there should be a correlation between ease of oxygenation and the basicity of the ligand (B) *trans* to the incoming O_2 ligand. This is also born out by Figure 19-6(B), although there are notable inconsistencies. For example, the order of decreasing K_{O_2} and $E_{1/2}$ of n-$BuNH_2$ > 1-methylimidazole > piperidine does not follow the order of amine basicities toward the proton (pK_a's: 10.6 > 7.3 < 11.3). Thus, 1-methylimidazole promotes O_2 coordination and $Co(L)B_2$ oxidation to an extent greater than anticipated from its proton basicity, and this is attributed to its π-donor nature. The ease of z^2 electron removal derives not only from direct σ interaction with the axial base but also from an indirect destabilization† of that electron by axial-base π-donor effects. The correlation of K_{O_2} with pK_a is generally less satisfactory, but is excellent for a series of related *para*-substituted pyridine ligands. Here again, increased axial ligand donor ability does tend to enhance O_2 uptake.

---

†This occurs through increased electron repulsion at the metal.

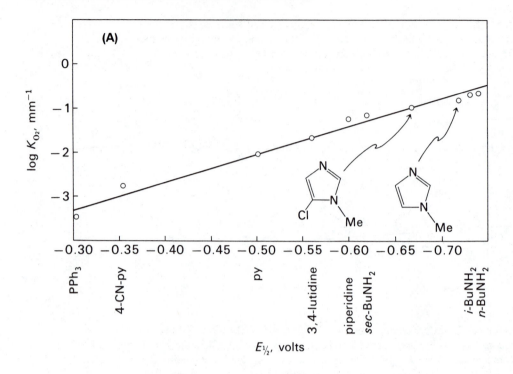

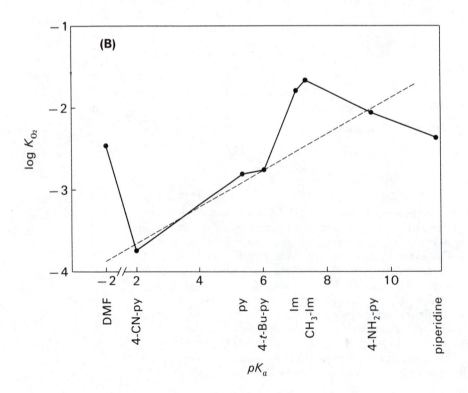

Figure 19-6. Correlations between K_{O_2} for Co(L)B (ordinate of both graphs) and (A) $E_{1/2}$ (polarographic half-wave reduction potentials) for Co(L)B$_2^+$ and (B) pK_a of ligand B in Co(L)B.

1. CO and CN$^-$ bind to heme with similar affinities. However, CO has greater affinity for hemoglobin and myoglobin than for heme, while CN$^-$ does *not* bind well to the heme group in the proteins [as an aside, CN$^-$ does bind to oxidized Fe(III)Hb, called methemoglobin]. How might solvation energies and the nature of the "heme pocket" in the proteins explain this marked reversal in CN$^-$ and CO ligation? In the context of your answer, does the heme pocket environment support CN$^-$ coordination of metHb? [Hint: see point (i), p. 597.]

2. Further regarding solvent effects on O$_2$ uptake, the K_{O_2} values for Co(L)B in toluene increase on changing the solvent to the more polar N,N-dimethylformamide. How does this support the idea that there is net Co → O$_2$ electron drift in the oxygenated complex?

3. With reference to the low spin, *five*-coordinate Co^{2+}(L)B complexes of Figure 19–6, is the metal-centered HOMO a σ or π type mo (axial bonding by B is weaker than equatorial by L, so that $e_\sigma^{ax} < e_\sigma^{eq}$)? How does the energy of this mo vary as the σ donor strength of B increases? If B is a good π donor, will the HOMO electron be easier or harder to remove? Should the HOMO energy also depend on the donor strengths of the equatorial ligands? If so, to a greater or lesser extent than on the axial ligands?

4. The fact that organically grown plants may contain high levels of nitrate/nitrite was recently realized in the case of an infant fed carrot juice from carrots grown in the Florida Everglades. The nitrite stored in the carrots oxidized most of the infant's hemoglobin to methemoglobin (Fe^{3+}Hb). This oxidation is dangerous because O$_2$ has little affinity for methemoglobin. What is there about the Fe−O$_2$ interaction that explains this reduced affinity with increased Fe charge?

COBALAMINS; VITAMIN B$_{12}$ COENZYME[14]

Cobalamin has a structure that is in some ways similar to that of myoglobin, but there are significant differences. The macrocycle (5) binding Co^{3+} in this case is like a porphyrin but is less regular. Specifically, one of the methylidyne carbons of porphyrin is missing and there is only one NH group at the center. This basic tetrapyrrole macrocycle is called a **corrin**.

5

Furthermore (see **6**), the carbon atoms at the periphery of the corrin ring are saturated and bear seven amide groups as substituents; three of these (a, b, d) are acetamide and three (c, e, f) are propionamide. The last amide (g) is an N-substituted propionamide. The amide substituent is a propyl group connected through the 2-carbon to an unusual ribonucleotide. The unusual feature of this nucleotide is that the base unit is 5,6-dimethylbenzimidazole. The cobalt corrin ring and its substituents down to and including the ribose phosphate are collectively called **cobamide**. Including the benzimidazole, the structure is a **cobalamin**. As in heme, the Co^{3+} ion finds itself coordinated to four "pyrrole" and one imidazole nitrogen atoms. In hemes, the sixth position normally is vacant or occupied by H_2O or O_2; in natural cobalamins, the sixth position is normally occupied by a carbon donor atom!

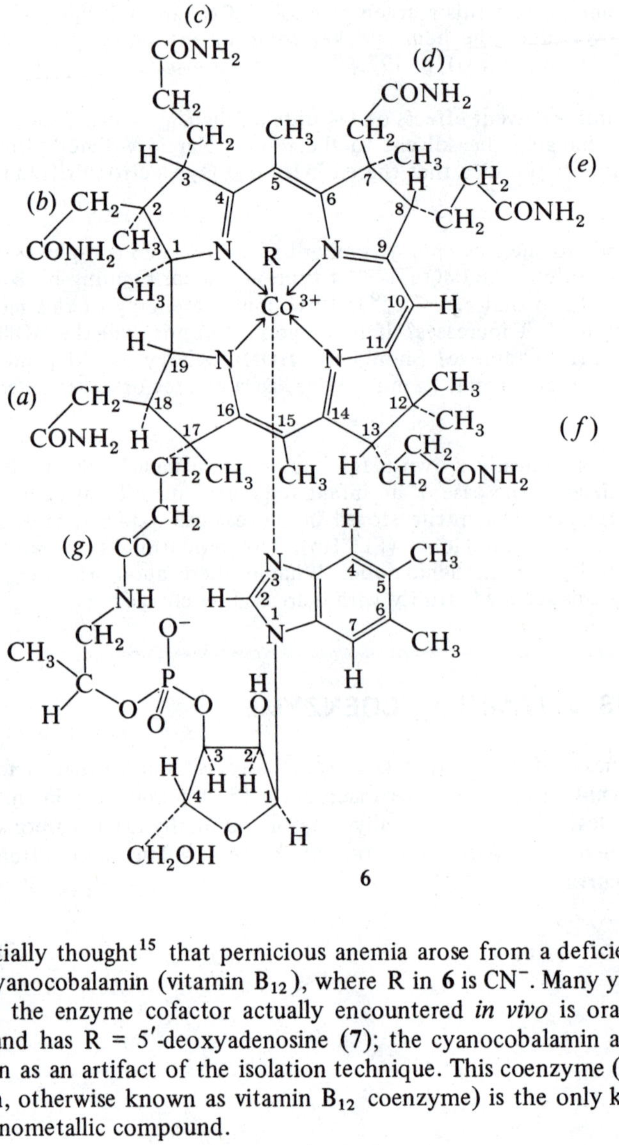

6

It was initially thought[15] that pernicious anemia arose from a deficiency of the red, diamagnetic cyanocobalamin (vitamin B_{12}), where R in **6** is CN^-. Many years later it was learned[16] that the enzyme cofactor actually encountered *in vivo* is orange-yellow and diamagnetic, and has R = 5′-deoxyadenosine (**7**); the cyanocobalamin arose during the earlier isolation as an artifact of the isolation technique. This coenzyme (5′-deoxyadenosine cobalamin, otherwise known as vitamin B_{12} coenzyme) is the only known naturally occurring organometallic compound.

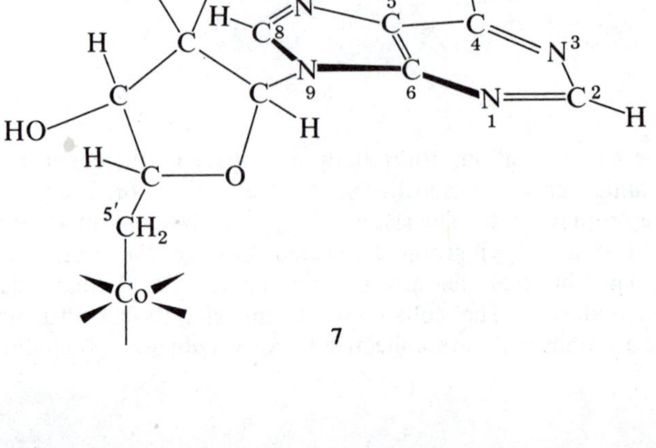

7

At present, the only known mammalian requirement for B_{12} coenzyme is as a cofactor for methylmalonyl coenzyme A mutase in the conversion of methylmalonyl coenzyme A into succinyl coenzyme A.

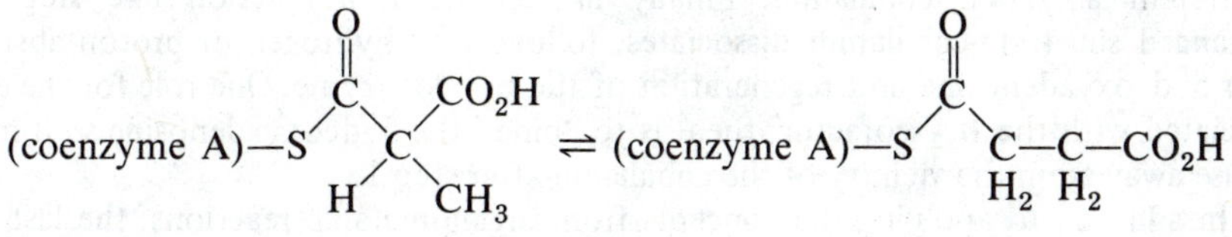

This is one of several reactions catalyzed by the B_{12} coenzyme in which there is, in effect, a swapping of H and X substituents on adjacent carbons:

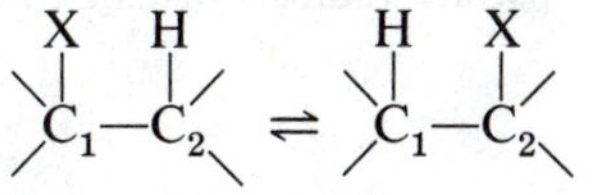

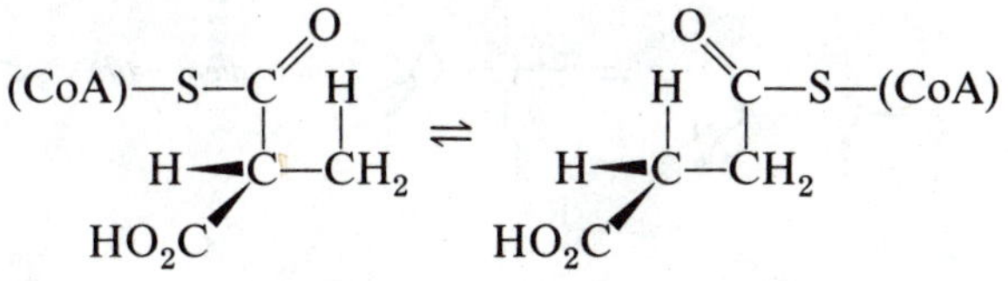

The mechanism of this unusual reaction is being vigorously pursued at present. A current proposal,[17] which draws on experience from metal catalysis of organic reactions (Chapters 16 to 18), is illustrated for the methylmalonyl mutase reaction below.

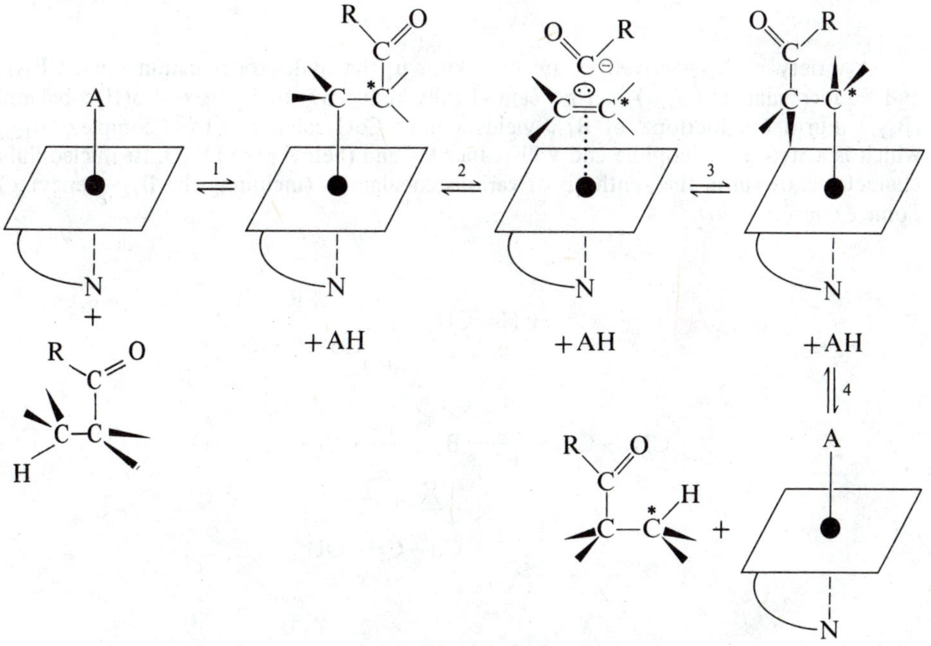

Step 1 in this scheme entails dissociation of the 5′-deoxyadenosyl group (as a radical or carbanion, it is not known for certain; recall, however, the known radical reactions of $Co(CN)_5^{3-}$ in Chapter 12) and consequent hydrogen or proton abstraction from the substrate, yielding transalkylated Co(III) and adenosine. Drawing from the well-known β-activation pathways in organometallic chemistry and the stability of olefin complexes, step 2 may involve dissociation of the β acyl group.† Step 3 accomplishes the migration of this group by its attack on the olefin complex at what was initially the α-carbon to yield a different alkylated cobalamin. Finally in step 4, in a reaction like step 1, the rearranged substrate-cobalamin dissociates, followed by hydrogen or proton abstraction from 5′-deoxyadenosine and regeneration of the B_{12} coenzyme. One role for the enzyme associated with the B_{12} cofactor, then, is to "bind" the 5′-deoxyadenosine so it may not diffuse away from the vicinity of the cobalamin after step 1.

In addition to applying the concepts from organometallic reactions, the last step of this mechanism is consistent with the fact that labeling of the 5′-deoxyadenosine carbon of the B_{12} cofactor with tritium (^3H) shows tritium to appear in the rearranged substrate.

This is not to say, however, that radical or carbanion pathways (*i.e.*, cleavage of the Co−C bond) are not possible for step 2, and it is not necessarily true that all substrates react by the same mechanism. In radical and carbanion pathways, β-group migration would occur in *uncoordinated* substrate radicals or carbanions after step 1 and in place of steps 2 and 3:

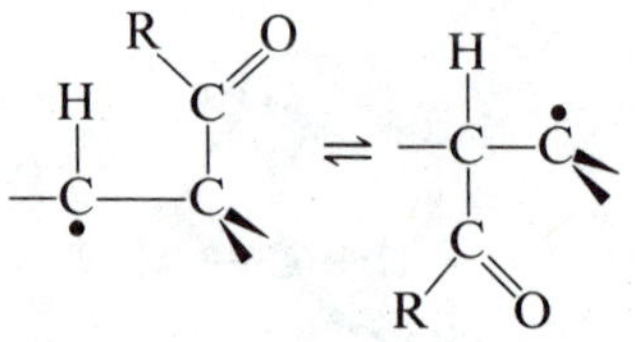

(radical path)

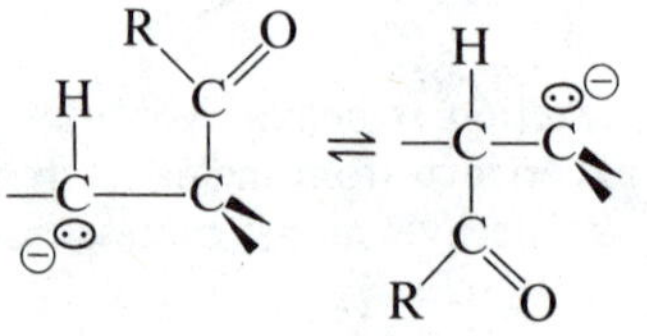

(carbanion path)

A variety of B_{12} derivatives are now known; the hydroxocobalamin (called B_{12b}) and aquocobalamin (B_{12a}) can be catalytically reduced § to hydroxo-Co(II)cobalamin (B_{12r}). Further reduction‡ of B_{12r} yields a green Co(I)cobalamin (d^8) complex (B_{12s}), which is a strong nucleophile and will reduce O_2 and (below pH 8) H_2O. Its nucleophilic character is useful in the synthesis of various cobalamins (including the B_{12} coenzyme). Some examples follow:

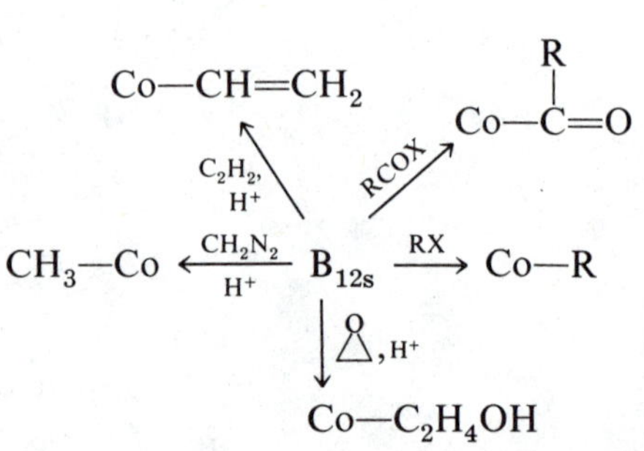

---

†Precedent for acyl migration comes from known 1,2 shifts in organic and organometallic chemistry (the alkyl migratory insertion reaction in metal carbonyls, p. 554).

§This reduction is mild and readily achieved by $SnCl_2$, Cr^{2+} (pH ~ 5), and H_2/PtO_2 for example.

‡Any one of Na, BH_4^-, Cr^{2+} (pH ~ 10), and electrochemical reduction (−1.4 v *vs.* S.C.E.) is sufficient for $B_{12r} \rightarrow B_{12s}$.

Since H^+ must be shown as an additional reagent in some of these reactions for material balance, it may well be that H^+ forms an adduct with B_{12s}, presenting another example (Chapters 5 and 17) of metal Lewis basicity,[18]

$$[Co^{III}-H]^+ \rightleftharpoons [Co^I] + H^+$$

Either species [five-coordinate Co(I) or six-coordinate Co(III)-H] could be effective in the reaction scheme above; the precedent for hydrometalation reactions was presented in Chapters 16 to 18, and the nucleophilicity of Co(II) in $Co(CN)_5^{3-}$ was established in Chapter 12. Knowing that B_{12r} and B_{12s} exist raises interesting questions, to which there are no firm answers as yet, concerning their appearance in *in vivo* reactions.

5. Devise a laboratory synthesis of methylcobalamin from aquocobalamin. A few years ago a great stir was caused by the realization that many bacteria contain Me-cobalamin; in natural waters these bacteria can convert otherwise relatively harmless (insoluble) Hg^{2+} salts into highly toxic (soluble) organomercury compounds, RHg^+. Describe a possible reaction mechanism for Me-cobalamin + Hg^{2+} that is consistent with the experimental rate law:

$$v = (3.7 \times 10^2 \, M^{-1}sec^{-1})[Me-Co][Hg^{2+}]$$

6. Refer to study question 9–28 and explain why exposure of B_{12} coenzyme to light produces 5′-8-cycloadenosine (the ribose 5′-carbon is bound to the 8-carbon of adenine). What cobalamin product is produced by this reaction? [In 7, the 5′ carbon is bound to Co.]

7. Describe the steps for propane-1,2-diol → propionaldehyde and for ethanolamine → acetaldehyde conversions by enzymes requiring B_{12} coenzyme.

8. What is the implication regarding a *trans* effect in cobalamins from the fact that the benzimidazole pK_a in Me-cobalamin and aquo-Co(II)cobalamin are the same (~ 2.5)? The pK_a of benzimidazole in B_{12a} (aquocobalamin) is -2.4.

ELECTRON TRANSFER AGENTS

Iron-Sulfur Proteins[19]

The non-heme iron proteins, as they are sometimes called, are known to participate in redox reactions in all forms of life tested to date; in particular, they play important roles in photosynthesis, nitrogen fixation, and mitochrondial respiration. More importantly, the Fe−S proteins exhibit reversible redox couples within a few hundred millivolts of the H^+/H_2 couple, in contrast to the aqueous Fe^{3+}/Fe^{2+} redox potential of 0.78 v.

The sulfur in iron-sulfur proteins is from cisteinyl units in the protein chain and from coordinated S^{2-} ions. Thus, iron-sulfur proteins are often referred to by the abbreviation

cysteinyl residue in proteins

n-Fe−S*, where n represents the number of Fe cations per protein and S* designates the presence of labile sulfur, usually also n in number. The labile sulfur is generally identified as the S^{2-} ions that are coordinated to Fe^{n+}; they are labile in the sense that (i) they generate H_2S on acidification of the protein and (ii) they are *readily* air-oxidized to elemental sulfur. Also, as far as is known, these proteins feature each Fe^{n+} in an approximately tetrahedral environment of sulfur atoms, at least one of which is a cysteinyl sulfur from the protein. Another potentially important structural feature of the iron-sulfur proteins is that n-Fe—S* proteins of the same n, but from different sources, have the same number of cysteinyl groups, in spite of the fact that the protein sequences are different. For example, the 2-Fe—S* proteins all have five cysteinyl residues, while the 8-Fe—S* proteins have eight such residues.

1-Fe—S AND 2-Fe—S

The so-called **rubredoxins** (1-Fe—S and 2-Fe—S proteins) lack labile sulfur, an example being the 1-Fe—S protein **8**. Here the protein chain is folded to create a pocket of four sulfur donor atoms to coordinate the Fe^{n+} (the chain termini are marked A and B).

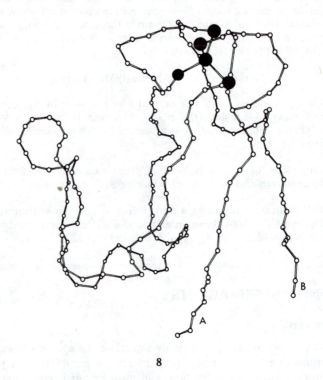

8

The rubredoxin **8** is a one-electron transfer agent, with both Fe^{2+} and Fe^{3+} having high spin d^n configurations. One of the interesting structural features of this complex is the C_1 symmetry of the S atoms about Fe^{n+}. As indicated by **9**, S_1 and S_4 appear radially equivalent while S_2 and S_3 are distinctly unique.[20] A "Jahn-Teller" distorted structure (Chapter 9, p. 326) is, of course, expected for T_d Fe^{2+} but not for T_d Fe^{3+}; the distortion in **8** is ligand-imposed, and some believe it to be critical for the low redox potential and fast electron transfer properties of rubredoxin (Chapter 12, p. 394). That is, the ligand-imposed coordination geometry imposes little "chemical activation" on the process $Fe^{2+} \rightleftharpoons Fe^{3+}$. Similarly, the distorted structure may render the Fe^{2+} HOMO neither strongly bonding nor anti-bonding, thereby facilitating (both kinetically and thermo-dynamically) the electron transfer.†

---

†Note the "open" 1-4 edge in **9**.

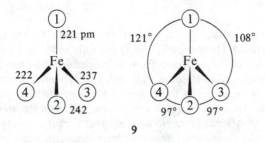

9

2-Fe–S*

These proteins are also one-electron transfer agents (in spite of having $2Fe^{n+}/$ molecule) and one of them is known to be involved in mitochondrial respiration. The oxidized protein bears two high spin Fe^{3+} ions in the form of a S^{2-} bridged dimer (the bridge sulfurs are labile); each ion is also coordinated by two terminal cysteinyl sulfurs (**10**).

10

The oxidized proteins are *diamagnetic* at very low temperatures, while the reduced form is paramagnetic to the extent of a single electron. How can two high spin Fe^{3+} ions yield a diamagnetic dimer? The diamagnetism indicates that the electrons of one ion have spin orientations opposite to those of the other, and so the electrons appear spin-paired. Normally this arises in covalent electron pair bond formation, but here the Fe^{3+} ions are not likely to be *directly* bonded by *five* electron pair bonds. Rather, they are *indirectly* coupled through the bridging sulfide ligands.† A consequence of spin correlation forces§, this phenomenon is well known and is called "superexchange" or "anti-ferromagnetic coupling" of electrons. It is entirely analogous to the coupling of nuclear spins (Chapter 3, p. 84). Upon one-electron reduction, the total number of d^* electrons in the protein is odd (= 11) and so the *reduced protein* behaves as a molecule with an unpaired electron. It is intriguing that the Fe^{n+} ions of the reduced protein are non-equivalent,[21] so the Fe_2S_6 unit cannot have an idealized D_{2h} symmetry; otherwise, the added electron would be found with equal probability at the two iron ions. The fact that the Fe^{3+} ions of the oxidized form are equivalent indicates that some structural non-equivalence is induced upon reduction and that the added electron is "localized" about one iron ion.‡

Synthetic models of the 2-Fe–S* proteins are beginning to appear[22] (*e.g.*, **11**). Their study over the next few years will hopefully augment and clarify the findings of biochemists regarding the natural proteins.

11

---

†This is not to say, however, that there is *no* direct overlap of Fe d electrons. In fact, the Fe ions in a structure like **10** are close enough (~278 pm) for such interaction to occur, superimposed upon reinforcing spin polarization of the sulfide-iron bond pairs.

§See p. 31.

‡At this stage of the community's understanding, there remains the possibility of a "slow" dynamical structure change to permit e^- transfer back and forth between Fe ions.

8-Fe—S*

Escalating in complexity, you encounter the important class of electron transfer agents called **ferredoxins**.† These proteins are found only in lower organisms and are most important constituents in the nitrogen fixation mechanism of plants. In such cases the ferredoxin is usually associated in some way with a molybdenum-containing protein (the Mo is also essential to the nitrogen fixation action). The ferredoxin protein contains two Fe_4(cysteine)$_4$S$_4^*$ clusters, each of which acts as a *one-electron* transfer site; the *complete protein*, therefore, can act as a *two-electron* agent. In what may represent a finding of great significance to the mechanistic action of ferredoxin, the non-cysteinyl sulfur atoms undergo exchange with radioactive S^{2-} and are labile in the usual sense of H_2S evolution on acidification. Recently, a good model of the Fe—S cluster has been synthesized.[23] The synthetic (**12**) and natural clusters (compare the 2-Fe—S* proteins) feature terminal cysteine and bridging sulfide groups with four Fe^{n+} and four S^{2-} at the eight corners of a D_{2d} "cube."

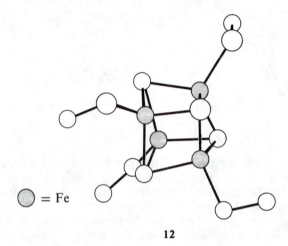

$\bigcirc$ = Fe

12

One way to view the Fe_4S_8 unit (so as to emphasize its relationship to the 2-Fe—S* dimer) is to look at any two opposite faces of the tetramer as 2-Fe—S* dimers now associated by substitution of one terminal cysteine at each iron by a sulfide of the other dimer. This view has the advantage of relating the "anti-ferromagnetic coupling" behaviors of 2-Fe—S* and 4-Fe—S* clusters. A view that two such facial sub-units are like two 2-Fe—S* units cannot be taken literally, however, because such a model would imply that each cube could act as a $2e^-$ redox unit—in clear contradiction with fact. Furthermore, the average Fe^{n+} oxidation states in the 2-Fe—S* protein are $Fe^{3+} \rightleftharpoons Fe^{2.5+}$, while the following changes have been found for $[Fe_4S_4(SR)_4]^z$:

$$[Fe_4S_4(SR)_4]^- \rightleftharpoons [Fe_4S_4(SR)_4]^{2-} \rightleftharpoons [Fe_4S_4(SR)_4]^{3-}$$

$$Fe^{2.75+} \underset{\text{like HIPIP}}{\rightleftharpoons} Fe^{2.5+} \underset{\text{like ferredoxin}}{\rightleftharpoons} Fe^{2.25+}$$

The second redox couple mimics the situation in ferredoxin; the first couple corresponds to the redox couple for another type of non-heme iron protein, the so-called "high potential iron protein" (HIPIP).† HIPIP is a 4-Fe—S* protein, of unknown function, containing a *single* $[Fe_4S_4(SR)_4]^z$ cluster. The fact that HIPIP and ferredoxin exhibit different redox couples reveals an important effect on the prosthetic group by the protein surrounding it. Whether this is a specific structural phenomenon or a general "medium" effect as in hemoglobin is unknown.

---

†Unfortunately for communication clarity, all Fe—S proteins are sometimes referred to as "ferredoxins," where "fer" = iron and "redoxin" = redox protein.

Another point of demarcation for 2-Fe—S* and the tetrameric cluster concerns the equivalence of Fe ions in the intermediate state. The D_{2d} symmetry of the cage (which is actually a smaller Fe_4 tetrahedron within a larger S_4 tetrahedron) insures the equivalency of the Fe atoms and S atoms.

9. Reasonably facile RS^- exchange with $[Fe_4S_4SR']_4]^{2-}$ has made possible the synthesis of various ferredoxin analogs. Is such exchange unexpected, given the oxidation state and spin condition of Fe^{n+} in such complexes? Does the concept of an $\{Fe_4S_4\}$ cage cluster make reasonable the possibility that its electron transfer reactions are of the "outer sphere" variety? Does the occurrence of RS^- exchange complicate this view?

STUDY QUESTION

Chlorophyll[24]

Plants *and* animals effect glucose oxidation by O_2. Plants, however, are unique in synthesizing glucose from CO_2 and H_2O by a photo-initiated reaction. Ferredoxins are involved in the electron transfer that is initiated by oxidation of photo-excited chlorophyll.

Structure **13** shows that chlorophyll (Chl) is a tetrapyrrole belonging to the porphyrin family, but with some significant modifications to the porphyrin ring complex (the various chlorophylls differ by the ring substituents at the 2, 3, and 10 positions).

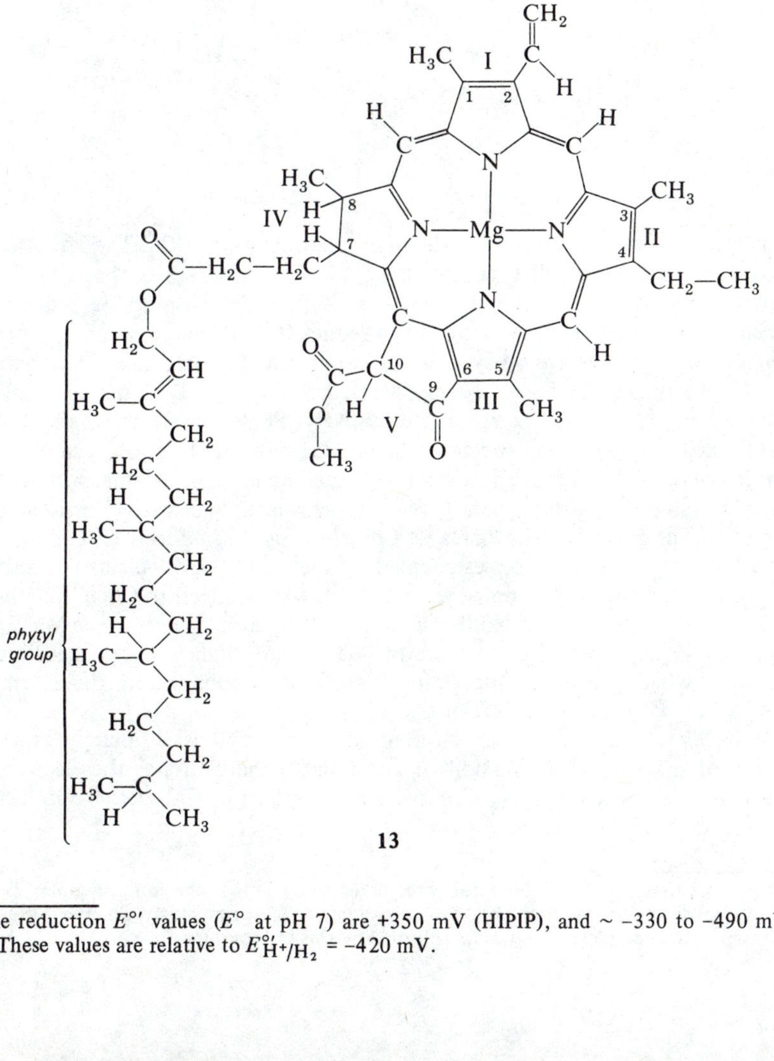

13

---

†The reduction $E^{\circ'}$ values (E° at pH 7) are +350 mV (HIPIP), and ~ −330 to −490 mV (ferredoxins). These values are relative to $E^{\circ'}_{H^+/H_2} = -420$ mV.

First of all, Chl is not a heme because the metal at the tetrapyrrole coordination site is Mg^{2+} which, incidentally, also is 30 to 50 pm above the non-planar, "ruffled" macrocycle plane), and Mg^{2+} is mysteriously unique among metal ions in conferring just the right physico-chemical properties necessary for the photocatalysis of glucose synthesis. The porphyrin modification that distinguishes Chl from other porphyrins is the connection of the methylidyne carbon between pyrrole rings III and IV with the 6-carbon in

ring III by the $\text{HC(CO}_2\text{Me)C}{=}\text{O}$ group. This group exhibits *keto-enol* equilibration strongly favoring the keto form, as shown in **13**. The long phytyl group serves to bind Chl to membranes in the cell.

Before delving into what little is known about the photo-induced electron transport mechanism, let us pause to view the overall process.[25] Basically, light quanta are utilized to photo-excite a chlorophyll *a* molecule to a strongly reducing excited state, from which electron transfer proceeds through an unknown species X (perhaps an Fe—S protein), to a ferredoxin (green plants appear to utilize a 2Fe—S ferredoxin of the type **10** or **11**, while bacteria use a 4Fe—S of the type **12**), to $NADP^{\oplus}$. $NADP^{\oplus}$ is an oxidizing agent with the following structure:

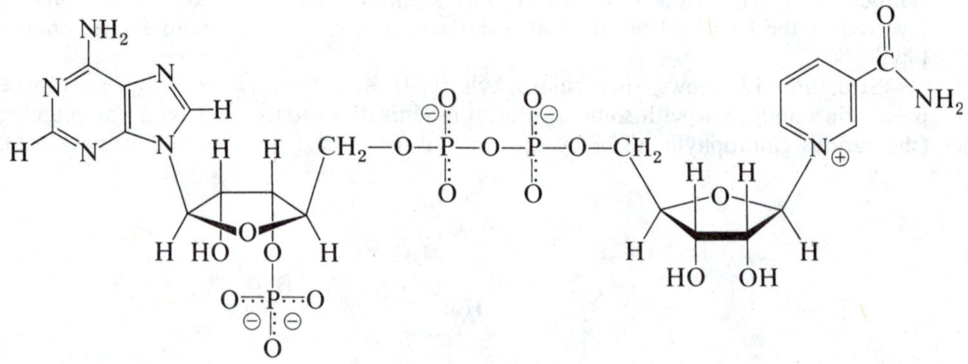

NADP is but one intermediate along the path to reduction of CO_2. The important consequence of this process (called **photosystem I** in Figure 19-7) is that Chl *a* has been oxidized to Chl *a*⁺. To sustain the action of Chl *a*, Chl *a*⁺ must be reduced. Water, unassisted, has too high a redox potential to reduce Chl a_1^+; thus, a second photo-induced electron transfer process (**photosystem II**) enters the picture. The Chl *a* site of this system absorbs quanta of higher energy than that of system I, so the two Chl sites are distinguished as Chl a_2 and Chl a_1 or, alternatively, P680 and P700, where 680 and 700 designate excitation photon wavelengths. Analogously with photosystem I, Chl a_2^* reduces some yet-to-be-identified reagent (Q) and the electron is transported through a system of a quinone (plastoquinone), two cytochromes, and a Cu^{2+} protein known as plastocyanin. The journey culminates at Chl a_1^+, which is reduced to Chl a_1. In short, there are two photo-redox processes coupled by a series of "dark" electron transfer steps. At this point, oxidized Chl a_2^+ remains, and it must be reduced to Chl a_2 for the process to continue. Chl a_2^+ is strongly oxidizing and, with the assistance of some Mn^{2+} protein,† is reduced by H_2O. This entire sequence of "dark" and "light" reactions is illustrated in Figure 19-7, where the reduction potentials of the various intermediates are roughly indicated by the E_0' scale to the left of the figure.

The highly conjugated porphyrin ring of chlorophyll is ultimately linked to the absorption of red light (transmission of green light) that initiates the electron transfer processes. It is known, however, that the concentration of Chl at the cell "active site"

---

†About all that is known about this prosthetic group is that the Mn^{2+} complex is labile, for attempts to isolate the protein with Mn^{2+} intact have failed. The principles of Chapter 13 regarding the lability of metal ion complexes seem to be followed by this bio-complex.

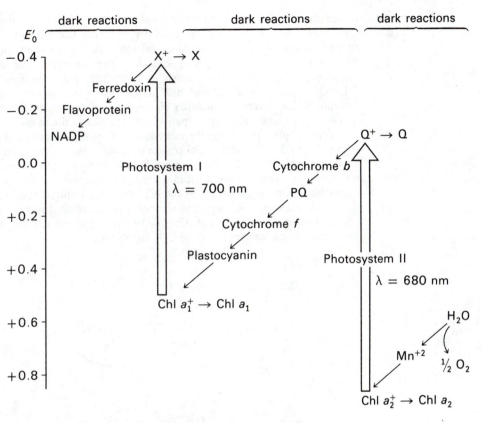

Figure 19-7. The "Z diagram" of the photosynthetic electron transport sequence. The photo processes are endergonic, while the dark reactions are exergonic. E_0' is the reduction potential of the oxidized form of each couple at pH = 7.

is actually too small (because of the required presence of the other proteins at that site) to capture photons efficiently. This explains the presence of carotenes in the cell; carotenes are highly conjugated hydrocarbon derivatives that undergo photo-excitation at higher frequencies (mid-visible region) than Chl. The photo-excitation energy of the carotenes is then passed on to the Chl oligomers (see later) about the "active site" and then on to the "active site" Chl itself.

One of the more interesting chemical/structural features of Chl is that it dimerizes in solvents that poorly coordinate Mg^{2+} (poor Lewis basicity) and that poorly solvate the porphyrin moiety. In such circumstances, the special keto group at carbon 9 of ring V coordinates the Mg^{2+} of a second Chl through the carbonyl oxygen. Structure 14 is a schematic of a possible dimer structure.

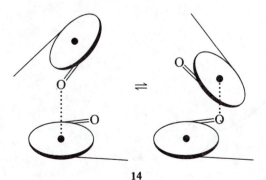

14

In aqueous solution these dimers add water to form $(Chl \cdot H_2O)_n$ oligomers. Because H_2O can act as both a Lewis base and acid, it may bridge Chl monomers as shown in **15**. The critical feature of such H_2O-bridged Chl molecules is that they (and not monomers or dimers) are the only Chl species known to become paramagnetic on photolysis. The origin of this paramagnetism seems to arise from photo-induced H atom transfer to $>C=O$ of Chl from a "sandwiched" H_2O. Electron transfer to the enzyme electron transport chain leaves an unpaired electron in the porphyrin ring system (actually, the odd electron appears to "hop" rapidly from one ring to the next).

Very simply, the current hypothesis for the photo-oxidation of Chl is as follows (see Figure 19–8). The "antenna" Chl consist of $(Chl)_n$ chains terminating in a $(Chl \cdot H_2O \cdot Chl)$ unit. The antenna Chl either harvest light quanta by direct photo-excitation or are sensitized by energy transfer from a carotene that was previously photo-excited. This excitation energy is passed on to the $(Chl \cdot H_2O \cdot Chl)$ unit at the "active site" containing the ferredoxins, cytochromes, and so forth. There, in photosystem I, the transfer noted above initiates the oxidation of $(Chl \cdot H_2O \cdot Chl)$ by quinones, which diffuse away with the electron to begin the conversion of CO_2 to glucose.

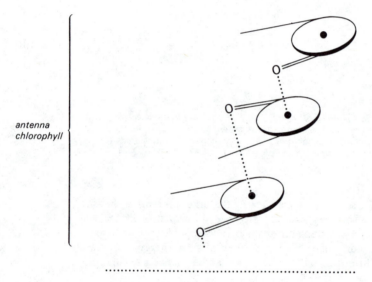

antenna chlorophyll

Figure 19–8. An idealization of photoelectron transfer in the cell.

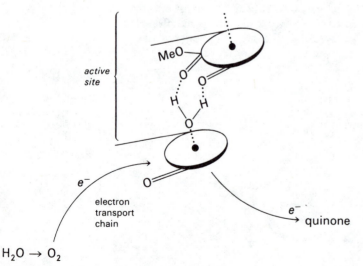

active site

e^-

electron transport chain

e^- → quinone

$H_2O \rightarrow O_2$

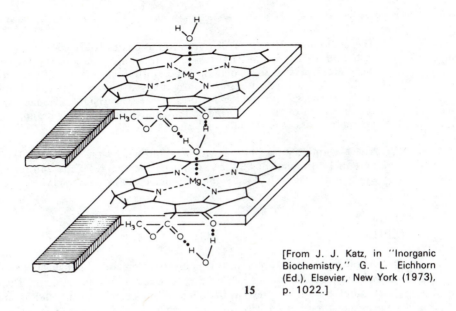

15

[From J. J. Katz, in "Inorganic Biochemistry," G. L. Eichhorn (Ed.), Elsevier, New York (1973), p. 1022.]

This concept of chlorophyll has been successfully applied to the formulation[26] of a "synthetic leaf." The construction of the leaf (Figure 19-9) is as follows: $(Chl \cdot H_2O)_n$ adduct is impregnated in a polymer membrane or deposited on a metal foil; this membrane or foil separates the two compartments of a photoelectric cell, of which one compartment contains an oxidizing agent (such as tetramethylphenylenediamine) and the other contains a reducing agent (sodium ascorbate). Each compartment is connected to an external circuit by electrodes. The cell responds to red light by generating a potential difference of 422 mV (the optimum obtained to date) and a current of 24 μamp.

Figure 19-9. The "synthetic leaf" photovoltaic cell under development at Argonne National Laboratory. [From J. J. Katz, et al., Chem. Eng. News, **54**, No. 7, 32 (1976).]

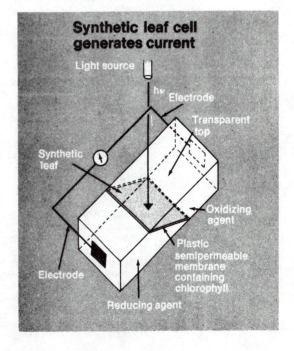

STUDY QUESTIONS

10. The addition of H_2O to hydrocarbon solutions of $(Chl)_2$ causes important changes in the infrared spectrum. The two keto stretching vibrations at 1655 cm^{-1} and 1700 cm^{-1} are depressed to a common band at 1638 cm^{-1}. The $C \dot{\dot{=}} O$ ester vibrations at ~1735 cm^{-1} are clearly resolved into bands at higher (1745 cm^{-1}) and lower (1727 cm^{-1}) energies. Identify the features of the $(Chl \cdot H_2O)_n$ oligomer that account for the directions of these i.r. band shifts. Similarly account for the appearance of O−H stretching vibrations at 3460 cm^{-1} and 3240 cm^{-1}.

11. The absorption of a photon with λ = 740 nm by the "synthetic leaf" generated a potential difference of 422 mV. Compare the energy of the 740 nm photon to that of the potential drop. What is the percentage energy conversion?

EPILOG

It should be clear to you at this point that a great deal remains to be discovered and unraveled about the role of "inorganic" processes in organic systems. There are a great many more areas of interest to the bioinorganic chemist than could be presented here, such as the storage and release of iron *in vivo*, the role of metals in amino acid/peptide synthesis, metal ion requirements in nerve tissues, and so on. Extension of the understanding of how living systems effect energy conversion is certainly one of the more pressing scientific issues for mankind today. A truly interdisciplinary effort will be required to push forward; and it is with an eye to the future, and the promise it holds, that we conclude this chapter and text.

REFERENCES

[1] An excellent, comprehensive treatment of this subject, and one to which we will frequently refer, is "Inorganic Biochemistry," G. L. Eichhorn, ed., Vols. 1 and 2, American Elsevier, New York (1973).

[1a] R. P. Hanzlik, "Inorganic Aspects of Biological and Organic Chemistry," Academic Press, New York, N.Y. (1976) gives further treatment of these topics at the level of this text.

[2] J. M. Rifkind in Vol. 2, p. 832 of ref. 1; a recent review of model systems is G. McLendon and A. E. Martell, *Coord. Chem. Revs.*, 19, 1 (1976).

[3] J. H. Wang, *et al.*, *J. Amer. Chem. Soc.*, 80, 1109 (1958).

[4] J. O. Alben, *et al.*, *Biochemistry*, 7, 624 (1968); I. A. Cohen and W. S. Caughey, *ibid.*, 7, 636 (1968).

[5] J. H. Wang, *J. Amer. Chem. Soc.*, 80, 3168 (1958).

[6] O. Leal, *et al.*, *J. Amer. Chem. Soc.*, 97, 5125 (1975).

[7] J. P. Collman, *et al.*, *J. Amer. Chem. Soc.*, 96, 6522 (1974).

[8] C. K. Chang and T. G. Traylar, *J. Amer. Chem. Soc.*, 95, 5810 (1973); see also ref. 11.

[9] R. E. Dickerson, *Ann. Rev. Biochm.*, 41, 807 (1972).

[10] M. F. Perutz and C. F. Ten Eyck, *Cold Spring Harbor Symp. Quant. Biol.*, 36, 315 (1971).

[11] F. Basolo, B. M. Hoffman, and J. A. Ibers, *Accts. Chem. Res.*, 8, 384 (1975) and references therein.

[12] L. D. Brown and K. N. Raymond, *Inorg. Chem.*, 14, 2595 (1975); see also ref. 11.

[13] A. Dedieu, *et al.*, *J. Amer. Chem. Soc.*, 98, 5789 (1976).

[14] H. A. O. Hill in Vol. 2, p. 1067 of ref. 1; J. M. Pratt, "Inorganic Chemistry of Vitamin B_{12}," Academic Press, New York (1972).

[15] E. L. Rickes, *et al.*, *Science*, 107, 396 (1948); E. L. Smith and L. F. J. Paker, *Biochem. J.*, 43, viii (1948); G. R. Minot and W. P. Murphy, *J. Amer. Med. Assoc.*, 87, 470 (1962).

[16] H. A. Barker, *et al.*, *Proc. Nat. Acad. Sci. U.S.A.*, 44, 1093 (1958).

[17] R. H. Abeles and D. Dolphin, *Accts. Chem. Res.*, 9, 114 (1976); R. B. Silverman and D. Dolphin, *J. Amer. Chem. Soc.*, 98, 4626 (1976).

[18] D. Dolphin, *et al.*, *Ann. N. Y. Acad. Sci.*, 112, 590 (1964); G. N. Schrauzer and R. J. Holland, *J. Amer. Chem. Soc.*, 93, 4060 (1971); see also ref. 17.

[19] W. H. Orme-Johnson, in Vol. 2, p. 710 of ref. 1.

[20] J. R. Herriott, *et al.*, *J. Mol. Biol.*, 50, 391 (1970).

[21] R. Dunham, *et al.*, *Biochim. Biophys. Acta*, 253, 134 (1971); M. Poe, *et al.*, *Proc. Nat. Acad. Sci. U.S.A.*, 68, 68 (1971).

[22] J. J. Mayerle, *et al.*, *J. Amer. Chem. Soc.*, 97, 1032 (1975); T. Herskovitz, *et al.*, *Inorg. Chem.*, 14, 1426 (1975).

[23] B. V. DePamphilis, *et al.*, *J. Amer. Chem. Soc.*, 96, 4159 (1974); L. Que, *et al.*, *ibid.*, 96, 4168 (1974).

[24] For a global overview of photosynthesis in relation to energy and material resources, the article by M. Calvin, *American Scientist,* **64**, 270 (1976) is recommended. For a technical analysis of data and model evolution, see J. J. Katz, in Vol. 2, p. 1022 of ref. 1.

[25] A very readable account of the critical experiments and their interpretation is given in Chapters 7 and 8 of D. W. Krogmann, "The Biochemistry of Green Plants," Prentice–Hall, Englewood Cliffs, N. J. (1973).

[26] J. J. Katz, *et al., Chem. Eng. News.,* **54**, No. 7, 32 (1976).

COMPOUND INDEX

If the point of interest in a compound is a metal atom, the compound is listed under the metal's heading. Compounds containing one atom of the metal per formula are listed first, followed by compounds with 2, 3, ..., n metal atoms. Within these groups, compounds are arranged by coordination number of the metal and, within coordination groups, alphabetically.

Compounds based on non-metals are grouped by number of substituents and, within those groups, alphabetically.

SUBJECT INDEX

Items are listed alphabetically letter by letter, disregarding spaces and prefixes.

The entries under "Functional groups" list references to atom-pairs defining key bonds in the compound of interest. Coupled with the formula index, these entries provide access to classes of compounds and their reactions.

COMMONLY USED ABBREVIATIONS

I. Organic Groups

Me methyl

Et ethyl

Pr propyl (whether *normal* or *iso* is specified by *n* or *i*)

Bu butyl (*n* = *normal*, *i* = *iso*, *t* = *tertiary*)

Ph phenyl

Ar aryl

Cp cyclopentadienyl (usually indicates a C_5H_5 anion bound to a metal *via* a π interaction involving all five carbon atoms)

II. Common Ligands

en ethylenediamine, $H_2N-C_2H_4-NH_2$

ox oxalate, $C_2O_4{}^{2-}$

EDTA ethylenediaminetetraacetate, $(^{-}OOCCH_2)_2NCH_2CH_2N(CH_2COO^{-})_2$

phen *o*-phenanthroline   bipy bipyridyl   acac acetylacetonate

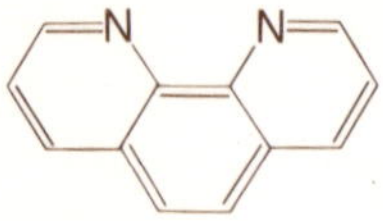

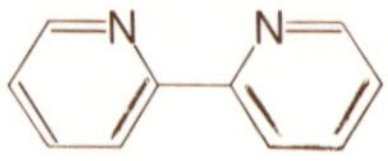

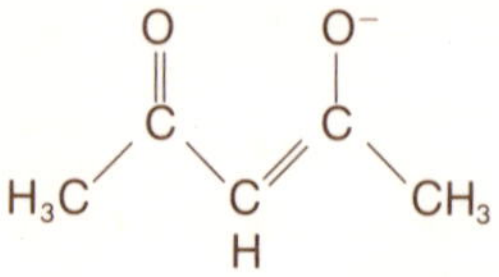

III. Theories, Concepts, Other

ao atomic orbital

mo molecular orbital

TASO terminal atom symmetry orbital(s) (Chapter 4)

HOMO highest occupied molecular orbital

LUMO lowest unoccupied molecular orbital

PMR principle of microscopic reversibility (Chapter 7, page 378)

AOM angular overlap model (a simplified molecular orbital approach applied to transition metal complexes in Chapter 9 and following)

SPE structure preference energy (Chapter 9)

LFSE ligand field stabilization energy

CONSTANTS

Avogadro's Number (L or N) $= 6.0225 \times 10^{23} \, \text{mol}^{-1}$

Boltzmann's Constant (k) $= 1.3805 \times 10^{-23} \, \text{J K}^{-1}$

Electron Rest Mass (m_e) $= 9.1091 \times 10^{-31} \, \text{kg}$

Faraday's Constant (F) $= 9.6487 \times 10^4 \, \text{C mol}^{-1}$

Planck's Constant (h) $= 6.6256 \times 10^{-34} \, \text{J s}$

Proton Rest Mass (m_H) $= 1.6725 \times 10^{-27} \, \text{kg}$

Light Velocity (vacuum) (c) $= 2.99795 \times 10^8 \, \text{m s}^{-1}$